Mathematical Connections

A Bridge to Algebra and Geometry

Francis J. Gardella

Patricia R. Fraze

Joanne E. Meldon

Marvin S. Weingarden

Cleo Campbell, CONTRIBUTING AUTHOR

TEACHER CONSULTANTS

Alma Cantu Aguirre

Cheryl Arevalo

Douglas Denson

Jackie G. Piver

Jocelyn Coleman Walton

Houghton Mifflin Company • BOSTON

Atlanta • Dallas • Geneva, Ill. • Palo Alto • Princeton • Toronto

AUTHORS

Francis J. Gardella, Supervisor of Mathematics and Computer Studies, East Brunswick Public Schools, East Brunswick, New Jersey

Patricia R. Fraze, former Mathematics Department Chairperson, Huron High School, Ann Arbor, Michigan

Joanne E. Meldon, Mathematics Teacher, Taylor Allderdice High School, Pittsburgh, Pennsylvania

Marvin S. Weingarden, Supervisor of Secondary Mathematics, Detroit Public Schools, Detroit, Michigan

Cleo Campbell, Coordinator of Mathematics, Anne Arundel County Public Schools, Annapolis, Maryland

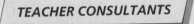

TEACHER CONSULTANTS

Alma Cantu Aguirre, Mathematics Department Chairperson, Thomas Jefferson High School, San Antonio, Texas

Cheryl Arevalo, Supervisor of Mathematics, Des Moines Public Schools, Des Moines, Iowa

Douglas Denson, Principal, Twin Bridges Elementary School, Twin Bridges, Montana

Jackie G. Piver, Mathematics Department Chairperson, Enloe High School, Raleigh, North Carolina

Jocelyn Coleman Walton, Supervisor of Mathematics, Plainfield High School, Plainfield, New Jersey

The authors wish to thank Robert H. Cornell, Mathematics Teacher, Milton Academy, Milton, Massachusetts, and James M. Sconyers, Mathematics Teacher, East Preston High School, Terra Alta, West Virginia, for their contributions to this textbook.

ISBN: 0-395-46150-2

EFGHIJ-D-998765432

Dear student,

You're invited . . .
Mathematical Connections is a bridge that will take you from where you are in your study of mathematics to algebra and geometry. Since topics in mathematics are connected, this course will also lead you to data analysis and probability. We have written this textbook because we want you to enjoy your continuing journey into mathematics.

to explore . . .
In this course you're expected to get actively involved in learning: explore, ask questions, discuss alternatives, and make connections between what's new and what's known. Exploration can be done by using mathematical models, by using a calculator or a computer, and by using your active, open mind.

together . . .
If you have difficulty, and you may at times, don't get discouraged. Your teacher wants you to succeed and will be working hard to help you. Your classmates are there with you, and you may be able to solve a problem more easily by working cooperatively in a group. We've also built help for you into the textbook. We encourage you to read the explanations in your book, study the *Examples*, answer the *Check Your Understanding* questions, and compare your answers with those at the back of your book.

for your future . . .
Don't close off any roads to the future at this stage of your journey. In tomorrow's world, the new situations you'll face as a citizen may call for decision-making skills you're building now, and your job may be one that you can't even imagine today. Through the study of mathematics, prepare yourself to cross any bridge along the way.

We wish you a successful, enjoyable journey.

Franus J Gardella

Cleo Campbell

Marvin Weingarden

Patricia R. Fraze

Joanne E. Meldon

CONTENTS

photo from page 85

v

photo from page 355

≣ **CHAPTER 6** ≣

INTRODUCTION TO GEOMETRY **234**

≣ **CHAPTER 7** ≣

NUMBER THEORY AND FRACTION CONCEPTS **288**

photo from page 331

=== CHAPTER 8 ===

RATIONAL NUMBERS 338

=== CHAPTER 9 ===

RATIO, PROPORTION, AND PERCENT 386

photo from page 278

photo from page 523

========= **CHAPTER 10** =========

CIRCLES AND POLYGONS 434

========= **CHAPTER 11** =========

STATISTICS AND CIRCLE GRAPHS 492

photo from page 698

photo from page 466

photo from page 679

TABLE OF SYMBOLS

		Page			Page
$\cdot$	$\times$ (times)	12	$\parallel$	is parallel to	249
$\dfrac{x}{y}$	$\div$ (division)	12	$\leq$	is less than or equal to	325
$>$	is greater than	18	$\geq$	is greater than or equal to	325
$<$	is less than	18	$\dfrac{b}{a}$	reciprocal of $\dfrac{a}{b}$	340
$=$	equals, is equal to	18	$a:b$	ratio of a to b	388
$(\ )$	parentheses—a grouping symbol	22	$\%$	percent	404
$-$	negative	94	π	pi, a number approximately equal to 3.14 and $\dfrac{22}{7}$.	441
$+$	positive	94	$\cong$	is congruent to	462
$-n$	opposite of n	94	$\triangle ABC$	triangle ABC	463
$|n|$	absolute value of n	95	$\sim$	is similar to	466
(x, y)	ordered pair of numbers	124	$\sqrt{}$	positive square root	475
$\neq$	is not equal to	142	$P(E)$	probability of event E	541
$\approx$	is approximately equal to	213	$n!$	n factorial	559
$\overleftrightarrow{AB}$	line AB	236	$_nP_r$	number of permutations of n items taken r at a time	560
$\overline{AB}$	line segment AB	237	$_nC_r$	number of combinations of n items taken r at a time	560
$\overrightarrow{AB}$	ray AB	237	$f(x)$	f of x, the value of f at x	613
$\angle A$	angle A	237	$\sin A$	sine of angle A	716
$^\circ$	degree(s)	240	$\cos A$	cosine of angle A	716
$m\angle A$	measure of angle A	240	$\tan A$	tangent of angle A	716
$\perp$	is perpendicular to	249			

TABLE OF MEASURES

Time

60 seconds (s) = 1 minute (min)
60 minutes = 1 hour (h)
24 hours = 1 day
7 days = 1 week
4 weeks (approx.) = 1 month

365 days ⎫
52 weeks (approx.) ⎬ = 1 year
12 months ⎭
10 years = 1 decade
100 years = 1 century

Metric	United States Customary

Length

10 millimeters (mm) = 1 centimeter (cm)

100 cm ⎫
1000 mm ⎬ = 1 meter (m)

1000 m = 1 kilometer (km)

Length

12 inches (in.) = 1 foot (ft)

36 in. ⎫
3 ft ⎬ = 1 yard (yd)

5280 ft ⎫
1760 yd ⎬ = 1 mile (mi)

Area

100 square millimeters = 1 square centimeter
(mm^2)   (cm^2)
10,000 cm^2 = 1 square meter (m^2)
10,000 m^2 = 1 hectare (ha)

Area

144 square inches (in.^{2}) = 1 square foot (ft^2)
9 ft^2 = 1 square yard (yd^2)

43,560 ft^2 ⎫
4840 yd^2 ⎬ = 1 acre (A)

Volume

1000 cubic millimeters = 1 cubic centimeter
(mm^3)   (cm^3)
1,000,000 cm^3 = 1 cubic meter (m^3)

Volume

1728 cubic inches (in.^{3}) = 1 cubic foot (ft^3)
27 ft^3 = 1 cubic yard (yd^3)

Liquid Capacity

1000 milliliters (mL) = 1 liter (L)
1000 L = 1 kiloliter (kL)

Liquid Capacity

8 fluid ounces (fl oz) = 1 cup (c)
2 c = 1 pint (pt)
2 pt = 1 quart (qt)
4 qt = 1 gallon (gal)

Mass

1000 milligrams (mg) = 1 gram (g)
1000 g = 1 kilogram (kg)
1000 kg = 1 metric ton (t)

Weight

16 ounces (oz) = 1 pound (lb)
2000 lb = 1 ton (t)

Temperature–Degrees Celsius (°C)

0°C = freezing point of water
37°C = normal body temperature
100°C = boiling point of water

Temperature–Degrees Fahrenheit (°F)

32°F = freezing point of water
98.6°F = normal body temperature
212°F = boiling point of water

TECHNOLOGY IN OUR DAILY LIVES

Today's World. Over the past 20 years, technology has dramatically changed the way that we live and work. From hand-held calculators to microwave ovens and VCRs to large-scale computers that process financial data or help plan space explorations, calculators and computers have become an important part of our lives.

Computers are well known for their ability to perform numerical calculations quickly. An even more important contribution of computers may be their ability to transmit information (words, pictures, sounds) electronically. By means of computers, information can be shared almost instantly with people throughout the world.

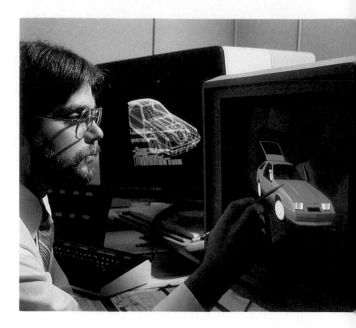

USING CALCULATORS

Calculators for Everyday Use. Since calculators are useful in a wide variety of different situations, specialized calculators have been designed to meet particular home, school, or job needs. For example, a calculator that does arithmetic operations is useful for everyday needs such as checking an itemized bill, balancing a checkbook, or preparing a tax form.

Tomorrow's World. In the 21st century, even more jobs than today will involve the use of calculators or computers. Becoming familiar with these technological tools will help you prepare for your future work. It will also help you become a well-informed citizen. In order to make effective decisions about the public and private uses of technology, every citizen needs a general understanding of what calculators and computers can—and cannot—do.

Scientific Calculators. A calculator that can perform more advanced mathematical operations, such as those studied in advanced high school or college courses, is particularly useful to scientists, engineers, and others who use mathematics in their daily work.

A calculator that can perform financial and statistical operations is especially useful to bankers, financial analysts, and statisticians.

Fraction Calculators. Calculators have also been designed especially for school use. For example, a calculator that performs fraction operations has been developed to help students understand work with fractions.

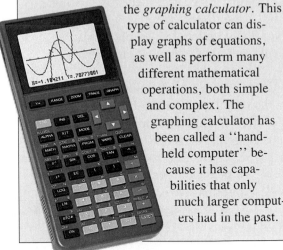

Graphing Calculators. An exciting recent advance in calculator technology is the *graphing calculator*. This type of calculator can display graphs of equations, as well as perform many different mathematical operations, both simple and complex. The graphing calculator has been called a "hand-held computer" because it has capabilities that only much larger computers had in the past.

USING COMPUTERS

Computer Software. Programmers can write special programs to solve particular problems. However, many programs used in schools and businesses are written in a general form, with instructions for inserting formulas or values. Such programs may be supplied on disks to be transferred to the computer's memory. Here are some general categories:

Function Graphing and Automatic Drawing. Function graphing software allows the user to quickly plot graphs of functions by entering an equation or a formula.

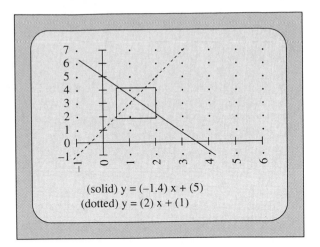

(solid) $y = (-1.4)x + (5)$
(dotted) $y = (2)x + (1)$

Most function graphing programs contain options for changing the scale, or "zooming," to better investigate small portions of the graph, such as the region where the graph crosses an axis. Automatic drawing software can be used to draw and measure geometric figures.

Word Processing. A word processing program accepts typed text and makes corrections as requested, adjusting the lines to fit. The text may be divided into paragraphs and pages, and words may be made italic or boldface. The methods of page make-up can produce pages ready to be printed directly. This process is called "desktop publishing."

Manipulating databases. With database software, collections of data, such as membership lists with addresses, telephone numbers, and other information, can be sorted and analyzed in various ways. For instance, mailings can be sorted by zip codes.

Spreadsheets. With an electronic spreadsheet like the one pictured on the next page, a user can enter a number, a formula, or a label (word) into each compartment, or *cell*. Such a program can be used to produce financial statements, solve equations, or determine function values. An especially useful feature is that by

Spreadsheet program provides a network of cells.

	A	B	C	D	E	F	G
1							
2							
3							

using formulas, the entire layout can be recomputed automatically whenever a given value is changed. The effects of such changes can be determined immediately.

Simulations. Simulation programs can be constructed to take certain assumptions and compute what the results would be. These results

can then be used to predict what will happen in the real-world situation. Simulation techniques are currently used in a wide range of fields, including financial forecasting, astronaut training, consumer-buying patterns, and building and city planning. Automobile manufacturers often use wind-tunnel simulations to investigate the effects of different design modifications. Simulations are particularly useful when the situation under consideration is difficult, impossible, costly, or very dangerous to observe directly.

USING TECHNOLOGY IN THIS COURSE

Tools for Learning. Calculators and computers can help you learn mathematics. They can help you explore mathematical concepts as well as do calculations. Like a typewriter, they are *tools*. You must decide when and how to use them. Just as a typewriter cannot tell you what to write or whether what you have written is good, a calculator cannot tell you which mathematical operation to perform or whether the answer makes sense. You must choose the mathematical operations and interpret the result.

An important part of learning to use technological tools is deciding when it is efficient to use them and when it is better to use some other method, such as mental math, paper and pencil, or estimation. Since this decision-making process is so important, there is a lesson right in Chapter 1 (see page 28) on choosing the most efficient method of computation.

Computers. Applications of computers are included throughout the book—in the exercises called "Computer Application" and in the sections called "Focus on Computers."

Here are some page references:

BASIC Programs	6, 298, 485, 567, 688
CAD Software	631
Database Software	220, 349
Geometric Drawing Software	270, 470
Graphing Software	
Function Graphing	596, 598
Geometric Graphing	131
Statistical Graphing	212, 512, 521
Logo Programs	265
Spreadsheets	55, 70, 174, 411

Calculators. Calculators are featured in many places in this course—in the lessons, in the exercises, and in special explorations. The following chart lists some topics and the associated calculator keys that you will learn about in this course.

Topics	Keys	Explained, Pages
Using Memory	M+ M− MR MC	xviii
Basic Operations	+ − × ÷ =	7–8, 11–12
Exponents	y^x x^2	31, 453
Order of Operations	( )	37
Repeated Operations	K	63
Scientific Notation	EE EXP	81
Operations with Integers	+/−	108
Finding Statistics	Σx n $\bar{x}$	231
Fraction Operations	$a^b/_c$	341
Finding Reciprocals	1/X	345
Finding Percents	%	409
Geometric Measurement	π	441
Finding Square Roots	$\sqrt{x}$	476
Finding Permutations	x!	561
Graphing Functions	Y= X|T GRAPH (−) ∧	586, 617
Trigonometric Functions	SIN COS TAN INV	719, 723

Getting Started with Your Calculator

Calculators can help you to do complicated computations easily. In order to use a calculator effectively, you should become familiar with its basic keys. Some particularly useful keys are those that control the memory.

M+ adds the number displayed to the number in the memory.

M− subtracts the number displayed from the number in the memory.

MR or **MRC** recalls the number stored in the memory.

MC clears the memory by setting it to zero.

Activity I   *Using the Memory Keys*

1 Turn on your calculator and be sure that the display and memory are clear.

2 Store the number 101 in the memory using the following key sequence:

3 Clear the display, then multiply the number in the memory by 53 using the following key sequence:

Record the result on a piece of paper.

4 Repeat Step 3 using other two-digit numbers. What pattern do you notice in the results?

5 Predict what might happen when you multiply a three-digit number by the number 1001. Test your prediction using your calculator.

6 What number should you store in memory to obtain a similar pattern when you multiply a four-digit number? Use a calculator to test your prediction.

Activity II   *Looking for Patterns*

1 Turn on your calculator and be sure that the display and memory are clear.

2 Store the number 11 in the memory.

3 Clear the display. Then divide 1 by the number in the memory using the following key sequence.

Record the result on a piece of paper.

4 Repeat Step 3 using the digits from 2 through 9. What pattern do you notice in the results?

Activity III   *Problem Solving*

Using the key sequence shown below, you will obtain the displayed result.

1 Copy the diagram shown below on your paper. Using each of the digits from 5 through 9, the plus key, and the minus key, fill in the boxes to obtain the displayed result. Keep a record of your attempts.

☐ ☐ ☐ ☐ ☐ ☐ ☐ ☐ = | 20.|

2 Repeat Step 1 using the diagram shown below.

☐ ☐ ☐ ☐ ☐ ☐ ☐ ☐ ☐ ☐ = | 5.|

3 Repeat Step 2, but this time obtain the number 9.

4 Using each of the digits from 5 through 9, the plus key, and minus key, what is the greatest possible result you can obtain using the diagram shown below?

☐ ☐ ☐ ☐ ☐ ☐ ☐ ☐ = | ?|

Activity IV   *Problem Solving*

1 Copy the diagram at the right on your paper.

2 Using each of the digits from 5 through 9, fill in the boxes to obtain the greatest possible product. Do the computations on your calculator. Make a table to record your attempts.

3 Repeat Step 2, but this time obtain the least possible product.

$$\begin{array}{ccc} \square & \square & \square \\ \text{X} & \square & \square \end{array}$$

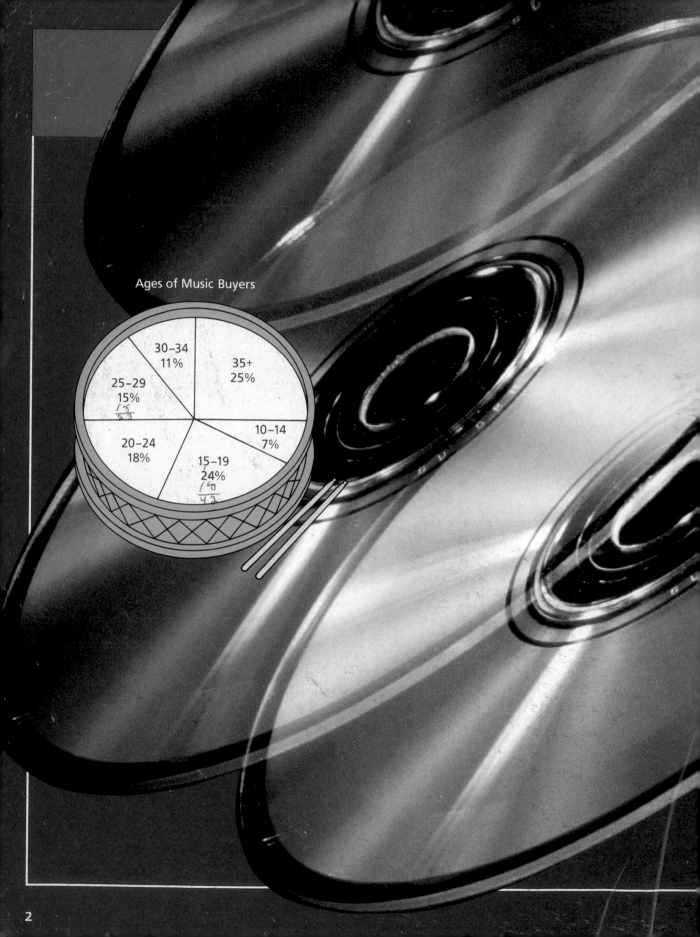

Ages of Music Buyers

30–34
11%

35+
25%

25–29
15%

10–14
7%

20–24
18%

15–19
24%

Connecting Arithmetic and Algebra 1

In order for music artists to receive the Gold Music Award, they must sell either five hundred thousand albums or one million singles. To receive the Platinum Music Award, artists must sell either one million albums or two million singles.

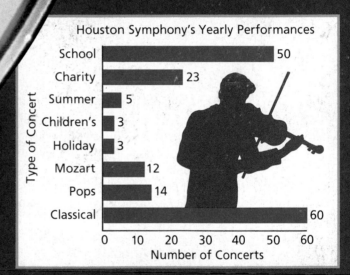

Houston Symphony's Yearly Performances

Type of Concert	Number of Concerts
School	50
Charity	23
Summer	5
Children's	3
Holiday	3
Mozart	12
Pops	14
Classical	60

Number of Concerts: 0 10 20 30 40 50 60

Shipments of Audio Recordings (Millions)

	1973	1978	1983	1988
LP's	280.0	341.3	209.6	72.4
CD's	0	0	0.8	149.7
Cassettes	15.0	61.3	236.8	450.1
8-Tracks	91.0	133.6	6.0	0

Variables and Variable Expressions

Objective: To evaluate variable expressions involving whole numbers.

Terms to Know
- *variable*
- *variable expression*
- *evaluate*
- *value of a variable*

Jan earns $5 per hour working at The Music Shop. The amount of money she earns changes, or *varies,* with the amount of time she works. The table below shows how much money Jan earns for working 1, 2, 3, 4, and 5 hours.

Hours Worked	Money Earned ($)
1	5 × 1 = 5
2	5 × 2 = 10
3	5 × 3 = 15
4	5 × 4 = 20
5	5 × 5 = 25

QUESTION How much money does Jan earn for working 11 hours?

One way to find the answer is to continue the chart until you reach 11 for the number of hours worked. Another way is to use the pattern 5 × 1, 5 × 2, 5 × 3, 5 × 4, ... (The three dots mean *and so on.*) If you do, you can say that the amount of money earned is 5 × *n,* where *n* represents the number of hours worked. The letter *n* is called a *variable.*

A symbol that represents a number is called a **variable.** An expression that contains a variable is called a **variable expression.** Variable expressions involving multiplication are usually written without the × sign.

$$5 \times n \text{ is usually written as } 5n.$$

When you **evaluate** a variable expression, you substitute a number for the variable. This number is called the **value** of the variable.

Example 1

Solution

Evaluate the expression 5*n* when *n* = 11.

5*n* = 5 × 11   ⟵ Substitute 11 for *n.*
  = 55

Check Your Understanding

1. Describe how Example 1 would be different if Jan works 20 hours.
2. Describe how the variable expression in Example 1 would be different if Jan earned $6 per hour.

ANSWER Jan earns $55 for working 11 hours.

Some variable expressions contain more than one variable. To evaluate these expressions you substitute the value given for each variable.

Example 2	Evaluate the expression $y + 9 + z$ when $y = 2$ and $z = 8$.
Solution	$y + 9 + z = 2 + 9 + 8$ ⟵ Substitute 2 for y
	$\qquad\quad = 11 + 8$ and 8 for z.
	$\qquad\quad = 19$

Guided Practice

COMMUNICATION «*Reading*

« **1.** Replace each __?__ with the correct word.

A __?__ is a symbol that represents a number. An expression that contains a __?__ is called a variable expression. Any number that you substitute for a variable is called a __?__ of the variable.

« **2.** Explain the meaning of the word *variable* in the following sentence. *The weather report today calls for variable cloudiness.*

Assume that $a = 3$, $b = 9$, and $c = 5$. Replace each __?__ with a number that makes the statement true.

3. $7a = 7 \times \underline{} = \underline{}$ **4.** $c + 13 = \underline{} + 13 = \underline{}$

5. $c - a = \underline{} - \underline{} = \underline{}$ **6.** $b \div a = \underline{} \div \underline{} = \underline{}$

Evaluate each expression when $x = 6$, $y = 8$, and $z = 4$.

7. $8x$	**8.** $11 + z$	**9.** $16 \div z$
10. $y - 7$	**11.** $y + 2 + z$	**12.** $y - x$
13. $2xz$	**14.** $y \div z$	**15.** $x + 4 + x$
16. $5zx$	**17.** $z + z + 1$	**18.** $x + z + y$

Exercises

Evaluate each expression when $a = 3$, $b = 12$, and $c = 4$.

1. $10a$	**2.** $12c$	**3.** $b - 2$
4. $27 - c$	**5.** $8 + c$	**6.** $a + 13$
7. $c \div 2$	**8.** $b \div 6$	**9.** $5ba$
10. $3ac$	**11.** $a + 5 + b$	**12.** $c + 2 + a$
13. $b \div a$	**14.** $b \div c$	**15.** $b - c$
16. $c - a$	**17.** $b + c + a$	**18.** acb

19. Rosita was born on her brother's sixth birthday. Write a variable expression that represents how many years old Rosita is when her brother is *b* years old.

20. A ream of paper contains five hundred sheets. Write a variable expression that represents the number of reams in *s* sheets of paper.

PATTERNS

Write a variable expression for each pattern. Use *n* as the variable.

21. $7 \times 1, 7 \times 2, 7 \times 3, 7 \times 4, \ldots$

22. $4 + 1, 4 + 2, 4 + 3, 4 + 4, \ldots$

23. $1 - 1, 2 - 1, 3 - 1, 4 - 1, \ldots$

24. $9 - 1, 9 - 2, 9 - 3, 9 - 4, \ldots$

25. $60 \div 1, 60 \div 2, 60 \div 3, 60 \div 4, \ldots$

26. $1 \div 5, 2 \div 5, 3 \div 5, 4 \div 5, \ldots$

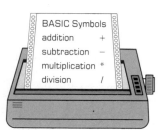

COMPUTER APPLICATION

BASIC Symbols
addition   +
subtraction   −
multiplication *
division   /

In a computer, the term *variable* indicates a location in the memory. When you write a BASIC computer program, you use capital letters for variables. To perform operations on variables, you create BASIC expressions using the operation symbols shown at the left.

Evaluate each BASIC expression when A = 2, B = 6, and C = 12.

27. A + 5   **28.** 3 * C   **29.** A + B + C   **30.** C/A

Write a BASIC expression for each variable expression.

31. $6m$ **32.** $p \div q$ **33.** $j + 12 + k$ **34.** rst

LOGICAL REASONING

Find a value of *r* and a value of *s* that satisfy the given conditions.

35. $rs = 96$ and $r - 4 = s$ **36.** $rs = 64$ and $r \div s = 16$

37. $r \div 2 = s$ and $r + s = 45$ **38.** $r + s = 17$ and $rs = 60$

SPIRAL REVIEW

39. Write the word form of 9.0002.   *(Toolbox Skill 5)*

40. Estimate the product: 3185×79 *(Toolbox Skill 1)*

41. Evaluate $4cd$ when $c = 5$ and $d = 9$.   *(Lesson 1-1)*

42. Is 430 divisible by 2? by 3? by 4? by 5? by 10?   *(Toolbox Skill 16)*

43. Find the product: $\frac{1}{2} \times \frac{4}{5}$ *(Toolbox Skill 20)*

44. Round 69.553 to the nearest tenth.   *(Toolbox Skill 6)*

Addition and Subtraction Expressions

Objective: To evaluate variable expressions involving addition and subtraction of whole numbers and decimals.

APPLICATION

John Carter works in a shoe store. He earns $287 per week in base pay plus a bonus, or commission, on his total sales.

QUESTION Last week John's commission was $116.49. What was his total pay for the week?

You can find John's total pay for any week by evaluating the addition expression $287 + n$, where n represents his commission for the week.

Example 1

Solution

Evaluate $287 + n$ when $n = 116.49$.

First substitute 116.49 for n. $287 + n = 287 + 116.49$

Next estimate the sum by adjusting the sum of the front-end digits.

$$\begin{array}{r} 287.00 \\ +116.49 \end{array} \rightarrow \text{about } 100$$

$300 \; + \; 100 \longrightarrow \textbf{about } 400$

Then add. Remember to line up the decimal points.

$$\begin{array}{r} \overset{1\,1}{287.00} \\ +116.49 \\ \hline 403.49 \end{array}$$

The answer 403.49 is close to the estimate of 400, so 403.49 is reasonable.

Check Your Understanding

1. In Example 1, explain why you estimate before adding.
2. In Example 1, explain how you get the estimate *about 100*.
3. Why is 287 written as 287.00 in Example 1?

You might decide to use a calculator to find the sum in Example 1. This is the key sequence you would use.

| 403.49 |
| %on | off |

287. + 116.49 =

ANSWER John's total pay last week was $403.49.

To evaluate subtraction expressions, you use a similar procedure.

Example 2

Evaluate $x - y$ when $x = 67.8$ and $y = 48.63$.

Solution

First substitute 67.8 for x and 48.63 for y.

$$x - y = 67.8 - 48.63$$

Next estimate the difference by rounding each number to the place of the leading digit.

$$
\begin{array}{rcr}
67.8 & \longrightarrow & 70 \\
-48.63 & \longrightarrow & -50 \\
\hline
& & \text{about } 20
\end{array}
$$

Then subtract. Remember to line up the decimal points.

$$
\begin{array}{r}
{\scriptstyle 5\ 17\ 7\ 10} \\
\cancel{6}\cancel{7}.\cancel{8}\cancel{0} \\
-\ 4\ 8.6\ 3 \\
\hline
1\ 9.1\ 7
\end{array}
$$

The answer 19.17 is close to the estimate of 20, so 19.17 is reasonable.

If you decide to use a calculator to find the difference in Example 2, you would use this key sequence.

$\boxed{67.8}\ \boxed{-}\ \boxed{48.63}\ \boxed{=}$

Guided Practice

COMMUNICATION « *Reading*

« **1.** Replace each __?__ with the correct word.
When you add, the result is called the __?__.
When you subtract, the result is called the __?__.

« **2.** What is the objective of this lesson? Name two ways that it is different from the objective of Lesson 1-1.

COMMUNICATION « *Writing*

Describe how to obtain a reasonable estimate of each answer.

« **3.** $42.1 + 37.7$ « **4.** $714.6 + 81$

« **5.** $5.14 - 2.81$ « **6.** $6.9037 - 0.751$

Evaluate each expression when $x = 26.4$, $y = 163.5$, and $z = 39$.

7. $z + 88$ **8.** $103 - z$ **9.** $y + 92$

10. $83 - x$ **11.** $x - 7.6$ **12.** $x + y$

Evaluate each expression when $a = 94$, $b = 21.4$, $c = 12.86$, and $d = 106.4$.

1. $17.7 + b$
2. $c + 9.37$
3. $d + 97.81$
4. $c + 43.9$
5. $102.5 - b$
6. $a - 14.8$
7. $114.5 - c$
8. $d - 44.6$
9. $a + d$
10. $c + b$
11. $d - b$
12. $a - c$

13. Bill Mihalko spent $149.95 for a CD player, $227 for an amplifier, and $199.49 for a pair of speakers. How much did he spend for this audio equipment?

14. DATA, *pages 2–3*   How many more classical concerts than school concerts does the Houston Symphony perform yearly?

PROBLEM SOLVING/APPLICATION

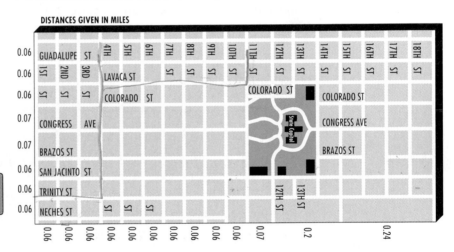

DISTANCES GIVEN IN MILES

Use the map of a part of Austin, Texas, above.

15. What is the length of the part of Trinity Street shown on the map?

16. How much longer is the part of San Jacinto Street shown on the map than the part of Neches Street?

17. You go from the corner of 2nd Street and Brazos Street to the corner of 2nd Street and Guadalupe Street. How far have you gone?

18. You go from the corner of Trinity Street and 4th Street to the corner of Trinity Street and 11th Street to the corner of 11th Street and Lavaca Street. How far have you gone?

19. ESTIMATION   Estimate the length of 12th Street from Guadalupe Street to Colorado Street.

COMMUNICATION «*Writing*

Write a variable expression for each phrase.

«**20.** the sum of 7.9 and a number n

«**21.** 43.61 added to a number p

«**22.** a number x minus 270.5

«**23.** a number z subtracted from 8.7

Write a phrase for each variable expression.

«**24.** $x + 6.2$ «**25.** $15.8 + z$

«**26.** $31.7 - a$ «**27.** $b - 16.4$

THINKING SKILLS

28. Determine a value of p and a value of q for which $p + q = p - q$ is a true statement.

29. Compare your answer to Exercise 28 to your classmates' answers. Make a generalization about the statement $p + q = p - q$.

SPIRAL REVIEW

30. Find the quotient: $5 \div \frac{1}{5}$ *(Toolbox Skill 21)*

31. Estimate the sum: $6.9 + 7.2 + 6.7 + 6.9 + 7.4$ *(Toolbox Skill 4)*

32. Evaluate $a + b$ when $a = 6.3$ and $b = 8.9$.   *(Lesson 1-2)*

33. Find the difference: $\frac{7}{8} - \frac{1}{2}$ *(Toolbox Skill 17)*

34. Evaluate $x \div y$ when $x = 48$ and $y = 4$.   *(Lesson 1-1)*

35. Find the sum: $69{,}483 + 35{,}670$ *(Toolbox Skill 7)*

36. Find the product: 4.93×1000 *(Toolbox Skill 15)*

Historical Note

In the past, mathematicians have used many different forms of variables. Records show that before 1600 B.C., Egyptians used a word meaning *heap* in much the same way that a variable is used today. The French mathematician François Viète, who is pictured at the left, used letters as variables. Viète lived from 1540 to 1603.

Research

Both the ancient Greeks and Hindus used symbols for unknown quantities. Find out what these symbols were and how they originated.

Multiplication and Division Expressions

Objective: To evaluate variable expressions involving multiplication and division of whole numbers and decimals.

DATA ANALYSIS

The pictograph at the right shows the value of home sales in Leeville.

QUESTION What was the value of home sales in 1990?

You can find the value of home sales in a given year by evaluating the multiplication expression 2.5*n*, where *n* represents the number of symbols on the graph for that year.

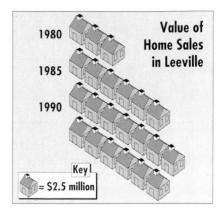

Value of Home Sales in Leeville

1980
1985
1990

Key
= $2.5 million

Example 1

Solution

Evaluate 2.5*n* when *n* = 13.

First substitute 13 for *n*.

$$2.5n = 2.5 \times 13$$

Next estimate the product by rounding each factor to the place of the leading digit.

$$2.5 \longrightarrow 3$$
$$\times 13 \longrightarrow \times 10$$
$$\text{about } 30$$

Then multiply. Remember to include the correct number of decimal places in the product.

$$\begin{array}{r} 2.5 \\ \times 13 \\ \hline 75 \\ 250 \\ \hline 32.5 \end{array}$$

The answer 32.5 is close to the estimate of 30, so 32.5 is reasonable.

 **Check Your Understanding**

1. In Example 1, how do you know the correct number of decimal places to show in the product?

You might decide to use a calculator to find the product in Example 1. This is the key sequence you would use.

| 13. | × | 2.5 | = |

ANSWER The value of home sales in 1990 was $32.5 million.

In algebra, it is important to realize that you can show both multiplication and division in a number of different ways. For example, each of these symbols represents multiplication.

times sign	raised dot	parentheses
7×5	$7 \cdot 5$	$(7)5$ or $7(5)$ or $(7)(5)$

Similarly, these expressions all represent division.

division sign	division house	fraction bar
$42 \div 7$	$7\overline{)42}$	$\dfrac{42}{7}$

Example 2
Solution

Evaluate $\dfrac{y}{x}$ when $y = 17.4$ and $x = 6$.

First substitute 17.4 for y and 6 for x.

$$\frac{y}{x} = \frac{17.4}{6}$$

Next estimate the quotient using compatible numbers.

$$6\overline{)17.4} \longrightarrow \overset{\text{about 3}}{6\overline{)18}}$$

Then use the division process: divide, multiply, subtract, bring down the next digit.

$$
\begin{array}{r}
2.9 \\
6\overline{)17.4} \\
\underline{12} \\
5\,4 \\
\underline{5\,4} \\
0
\end{array}
$$

The answer 2.9 is close to the estimate of 3, so 2.9 is reasonable.

✔ Check Your Understanding

2. In Example 2, how do you know where to place the decimal point in the quotient?

If you decide to use a calculator to find the quotient in Example 2, you would use this key sequence.

$$\boxed{17.4} \; \boxed{\div} \; \boxed{6.} \; \boxed{=}$$

Guided Practice

COMMUNICATION «*Reading*

«1. Choose all the words that are associated with *multiplication*.
 a. factor **b.** sum **c.** quotient **d.** product

«2. Choose all the words that are associated with *division*.
 a. addend **b.** divisor **c.** quotient **d.** difference

COMMUNICATION « *Writing*

Write each expression in two other ways.

«**3.** $27 \cdot 15$ 　　«**4.** $884 \div 22$ 　　«**5.** $6.5\overline{)124.2}$ 　　«**6.** $(12.4)(26.9)$

Describe how to obtain a reasonable estimate of each answer.

«**7.** $(3.7)81$ 　　«**8.** $4.3(5.1)$ 　　«**9.** $2.2\overline{)12.32}$ 　　«**10.** $14.73 \div 2.8$

Evaluate each expression when $a = 18$, $b = 3.2$, **and** $c = 0.8$.

11. $17a$ 　　**12.** $22b$ 　　**13.** $385.2 \div a$ 　　**14.** $27.2 \div c$

15. ab 　　**16.** bc 　　**17.** $\dfrac{b}{c}$ 　　**18.** $\dfrac{a}{c}$

Exercises

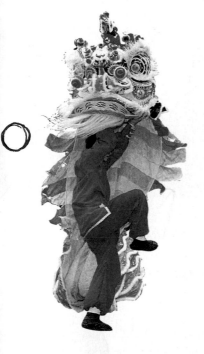

Evaluate each expression when $w = 63$, $x = 1.6$, $y = 62.72$, **and** $z = 18.27$.

1. $87x$ 　　**2.** $12.4w$ 　　**3.** $27y$ 　　**4.** $3.4z$

5. $30.87 \div w$ 　　**6.** $z \div 30$ 　　**7.** $\dfrac{y}{32}$ 　　**8.** $\dfrac{35.28}{x}$

9. wx 　　**10.** xy 　　**11.** $z \div w$ 　　**12.** $y \div x$

13. Sandy Haig drives 37.4 mi round trip to work and back home. Sandy works five days each week. How many miles does she drive to work and back home each week?

14. Martin Chun purchased nine tickets to a dance festival. He paid a total of $139.50 for the tickets. How much did each ticket cost?

▦ CALCULATOR

Estimate to place the decimal point in each calculator answer.

15. $(5.92)(7.15)$ 　　| 42328. |

16. $92 - 79.65$ 　　| 1235. |

17. $48.678 \div 915$ 　　| 000532. |

18. $(32.5)(8.64)$ 　　| 2808. |

19. $862.37 + 1158.93$ 　　| 20213. |

20. $21.973 \div 0.73$ 　　| 3010. |

Estimate to tell whether each calculator answer is *reasonable* **or** *unreasonable*. **If an answer is unreasonable, find the correct answer.**

21. $(3.32)(21.6)$ 　　| 71712. |

22. $17.4 - 9.38$ 　　| 8.02 |

23. $59.78 \div 98$ 　　| 6.1 |

24. $1121 \div 19$ 　　| 590. |

25. $0.35 + 1.9$ 　　| 2.25 |

26. $(52.4)(2.3)$ 　　| 120.52 |

Write a variable expression for each phrase.

« **27.** 15.4 times a number *z*

« **28.** the product of a number *x* and 1.01

« **29.** 986.4 divided by a number *n*

« **30.** a number *y* divided by 2.4

Write a phrase for each variable expression.

« **31.** 12.3*a*   « **32.** 9.7*b*   « **33.** $x \div 5.4$ « **34.** $26.8 \div z$

GROUP ACTIVITY

Suppose that numbers are assigned to the letters of the alphabet as follows:

A = 1   B = 2   C = 3   D = 4   E = 5

and so on to Z = 26.

Using this code, you find the value of a name by multiplying the values of its digits. For example:

$$ED = 5 \cdot 4 = 20$$

Find the numerical value of each name.

35. BOB   **36.** AMY

37. SUE   **38.** JUAN

39. Find the value of your name. Compare it with the values of the names of others in your group. Are there two names with the same value?

40. Find each of the following.
   a. a man's name that has three letters and a value of 140
   b. a woman's name that has three letters and a value of 140
   c. a person's name that has four letters and a value of 700
   d. a person's name that has four letters and a value of 1400

41. Find a value other than 140, 700, and 1400 that can represent two different names. Give two names that have this value.

SPIRAL REVIEW

42. Find the product: $35 \times \frac{3}{5}$ *(Toolbox Skill 20)*

43. Estimate the quotient: $2674 \div 9$ *(Toolbox Skill 3)*

44. Evaluate 16*n* when *n* = 5.4.   *(Lesson 1-3)*

45. Give the place of the underlined digit: 13,207.9586   *(Toolbox Skill 5)*

46. Find the difference: $6\frac{3}{4} - 4\frac{1}{2}$ *(Toolbox Skill 19)*

1-4

Reading for Understanding

Objective: To read problems for understanding.

To solve a problem, you must first understand it. This means that you may need to read the problem several times to determine what information is given, what you must find, and whether any facts are not needed.

Problem

Use this paragraph for parts (a)–(d).

Aaron had $234 in his savings account on December 31. During the next two months, he made deposits of $53, $65, and $40. He also made withdrawals of $25 and $37. The account earned $1.95 in interest during the same period.

a. What is the paragraph about?

b. How much interest did the account earn during the two months?

c. Does the paragraph tell how much money Aaron spent during the two months?

d. Identify any facts that are not needed to find the total amount of money added to Aaron's account during the two months.

Solution

a. The paragraph is about the money in Aaron's savings account.

b. The account earned $1.95 interest during the two months.

c. No. Aaron withdrew $25 and $37, but the paragraph does not tell how much he spent.

d. The following facts are not needed.
  the amount of money in his account [$234]
  the withdrawals [$25 and $37]

Look Back What if Aaron had deposited $17 instead of withdrawing $25? How would the amount in his savings account be changed?

Guided Practice

Use this paragraph for Exercises 1–3. Choose the letter of the correct answer.

When Keisha filled the gas tank of her car on June 5, the odometer showed 7251.3 mi. She bought 11.7 gal of gasoline. On June 18, Keisha filled the gas tank with 14.2 gal of gasoline and the odometer showed 7588.7 mi.

1. How many gallons of gasoline did Keisha buy on June 5?
   a. 11.7 gal   b. 14.2 gal   c. (14.2 − 11.7) gal

2. How many miles did the car's odometer show on June 18?
   a. 7251.3 mi   b. 7588.7 mi   c. (7251.3 + 7588.7) mi

3. Identify any facts that are not needed to find the number of miles traveled from June 5 to June 18.
   a. the amounts of gasoline bought on June 5 and on June 18
   b. the number of miles shown on the odometer on June 5
   c. the number of miles shown on the odometer on June 18

Use this paragraph for Exercises 4–7.

On August 23, Brett bought two pairs of jeans for $24.99 each, a shirt for $15, and a sweater for $34.49. Brett paid with five $20 bills.

4. What is the paragraph about?

5. How many items of clothing did Brett buy?

6. Identify any facts not needed to find the total cost of the clothes.

7. Describe how you would find the amount of change Brett received.

Problem Solving Situations

Use this paragraph for Exercises 1–4.

Cathy earns $6.50 per hour working part-time as a cashier. She works one hour more on Tuesday than on Thursday. On Wednesday she works 4 h. On Thursday she works half as long as on Wednesday.

1. What is the paragraph about?

2. How many hours does Cathy work on Wednesday?

3. Identify any facts that are not needed to find the number of hours Cathy works during the week.

4. Describe how you would find the number of hours Cathy works on Tuesday.

Use this paragraph for Exercises 5–8.

Paul is saving money to buy a video game system. The game system costs $105 and comes with one free game cartridge. Game cartridges cost $35.49 each. Paul saved $34 in May, $45 in June, $61 in July, and $33 in August.

5. What is the paragraph about?

6. How much did Paul save in July?

7. Does the paragraph tell how many game cartridges Paul plans to buy?

8. Identify any facts that are not needed to find the month in which Paul will be able to buy the game system.

Use this paragraph for Exercises 9–12.

Carol Jigargian bought a stereo system for $600. The tax on the stereo system was $30. She made a down payment of $100 and agreed to pay the remainder in 10 equal payments.

9. What is the paragraph about?

10. How much tax did Carol Jigargian pay?

11. Identify any facts that are not needed to find the total cost of the stereo system.

12. Describe how you would find the amount of each payment.

SPIRAL REVIEW

13. Find the difference: $16.53 - 0.5319$ *(Toolbox Skill 12)*

14. Evaluate $a + b$ when $a = 7.65$ and $b = 12.4$.   *(Lesson 1-2)*

15. Find the product: 13.87×1000 *(Toolbox Skill 15)*

16. Identify any facts that are not needed to solve this problem: Last baseball season Todd's batting average was 0.350 with 21 hits. This season his batting average was 0.331 with 19 hits. How many hits did Todd have during the two baseball seasons?   *(Lesson 1-4)*

Self-Test 1

Evaluate each expression when $a = 2$, $b = 5$, and $c = 8$.

1. $16a$ 2. $65 \div b$ 3. $a + 6 + a$ 4. $4ac$ 1-1

Evaluate each expression when $w = 13.67$, $x = 42$, $y = 75.4$, and $z = 9.04$.

5. $97 + y$ 6. $w + z$ 7. $w - 8.9$ 8. $y - z$ 1-2

9. $52w$ 10. xz 11. $x \div 0.6$ 12. $268.8 \div x$ 1-3

Use this paragraph for Exercises 13–16.

Jill rented 48 movies and 15 video games last year. Each movie rental cost $2.99 and each video game rental cost $1.50.

13. What is the paragraph about?   1-4

14. How many video games did Jill rent last year?

15. Identify any facts that are not needed to find the amount of money Jill spent on movie rentals last year.

16. Describe how you would find the total amount Jill spent on video game rentals last year.

1-5

The Comparison Property

Objective: To use the comparison property to compare and to order numbers.

 **EXPLORATION**

1 Al's height is 5 ft 2 in. and Bob's height is 4 ft 11 in. What is the relationship between their heights?
  a. Al is taller than Bob.
  b. Al is shorter than Bob.
  c. Al is the same height as Bob.

2 Is there a different way to express the relationship between their heights? Explain.

3 List all the possible relationships between their *weights*.

4 How many possible relationships are there between their *ages*?

When you compare any two measurements, such as heights, weights, or ages, there are only three possible relationships between them. The **comparison property** of numbers summarizes these relationships.

Comparison Property

For any two numbers a and b, exactly one of the following is true.

In Words	**In Symbols**
a is greater than b	$a > b$
a is less than b	$a < b$
a is equal to b	$a = b$

The symbols $>$ and $<$ are called **inequality symbols**.

Example 1

Replace each __?__ with $>$, $<$, or $=$.
a. 15,824 __?__ 15,794   **b.** 3.059 __?__ 3.51

Solution

Compare the numbers place-by-place from left to right. Find the first place in which the digits are different.

a. 1 5 , 8 2 4
     1 5 , 7 9 4
     $8 > 7$
     So $15,824 > 15,794$.

b. 3 . 0 5 9
     3 . 5 1
     $0 < 5$
     So $3.059 < 3.51$.

✓ **Check Your Understanding**

1. In Example 1(a), why is there a box around the digits 8 and 7?
2. Describe two ways that Example 1(b) is different from Example 1(a).

Usually there are two ways to express an order relationship.

$$a > b \text{ has the same meaning as } b < a.$$

You can use what you know about comparing two numbers when you need to put three or more numbers in numerical order.

Example 2	Write 0.47, 0.4, 0.247 in order from least to greatest.
Solution	

0 . $\boxed{4}$ 7   2 < 4, so   0 . 4 $\boxed{7}$ 7 > 0, so
0 . $\boxed{4}$ 0.247 is the   0 . 4 $\boxed{0}$ 0.47 is the
0 . $\boxed{2}$ 4 7 *least* number.   *greatest* number.

From the least to greatest, the numbers are: 0.247; 0.4; 0.47
You write this with two inequality symbols as $0.247 < 0.4 < 0.47$.

✓ Check Your Understanding

3. In the Solution of Example 2, why is 0.4 written as 0.40?
4. Write the numbers in Example 2 in order from *greatest* to *least*. Use two inequality symbols.

There are two ways to read a statement that contains two inequality symbols. For example, you can read $0.247 < 0.4 < 0.47$ as follows.

0.247 is less than 0.4, and 0.4 is less than 0.47,

or

0.4 is between 0.247 and 0.47.

Guided Practice

COMMUNICATION « *Writing*

Write each sentence in symbols.

« **1.** Ninety-five is greater than seventeen.

« **2.** Six and fifty-nine hundredths is less than eight and one tenth.

« **3.** Twelve and forty hundredths is equal to twelve and four tenths.

« **4.** Five is greater than four and nine thousandths, and four and nine thousandths is greater than four.

Write each statement in words.

« **5.** $5001 < 5100$ « **6.** $0.6 = 0.600$

« **7.** $9.03 > 9.007$ « **8.** $550 < 581 < 600$

Express each relationship in another way.

 9. Sue is older than Tom.   **10.** $8.9 < 10.5$

11. $14.0 = 14$ **12.** $0 < 0.861 < 1$

Replace each __?__ with >, <, or =.

13. 80,492 __?__ 80,942   **14.** 14.69 __?__ 14.690   **15.** 16.40 __?__ 16.04

Write in order from least to greatest. Use two inequality symbols.

16. 3674; 372; 3476   **17.** 5.01; 0.51; 0.015   **18.** 2.09; 0.209; 2

« **19.** COMMUNICATION « *Discussion*   Describe a way to remember which inequality symbol represents *is less than* and which represents *is greater than*.

Exercises

Replace each __?__ with >, <, or =.

1. 11,388 __?__ 11,614   **2.** 27,459 __?__ 26,721   **3.** 6.40 __?__ 6.4

4. 5 __?__ 0.54   **5.** 18.073 __?__ 13.562   **6.** 45.09 __?__ 45.090

Write in order from least to greatest. Use two inequality symbols.

7. 0.26; 0.2; 0.238   **8.** 0.57; 0.5; 0.519

9. 14.36; 12.03; 14   **10.** 36; 31.86; 36.93

11. 25.60; 25.08; 25.04   **12.** 69.43; 69.48; 69.45

13. In a recent year, the population of Canada was 26.1 million and the population of Mexico was 84 million. Which country had the greater population?

14. The best three times in a 100-m dash were as follows: Julie → 13.24 s; Anne → 12.92 s; Sonya → 13.41 s. Which woman finished first? second? third?

Let $w = 6$, $x = 12$, $y = 6.3$, and $z = 0.85$. Replace each __?__ with >, <, or =.

15. w __?__ 21   **16.** 16 __?__ x **17.** 6.21 __?__ y

18. z __?__ 0.805   **19.** z __?__ 8.5   **20.** 6.30 __?__ y

21. w __?__ y **22.** z __?__ y **23.** $x - 6$ __?__ w

24. $z + 5$ __?__ w **25.** $2y$ __?__ x **26.** $\frac{w}{6}$ __?__ z

NUMBER SENSE

Without multiplying, replace each __?__ with >, <, or =.

27. 57 __?__ 57(0.3)   **28.** 4.8 __?__ (4.8)(0.91)

29. 5.42 __?__ (5.42)1   **30.** 33 __?__ (33)(1.08)

31. 2.05 __?__ (2.05)(1.899) **32.** 84 __?__ 84(13.004)

CAREER/APPLICATION

A *statistician* collects, organizes, and interprets numerical facts, called *data*. Often the statistician must compare data in order to identify trends and analyze the characteristics of a population.

Use the data on the clipboard at the right.

33. Which region had the least population in 1960? in 1980?

34. Which region had the least change in population from 1960 to 1980?

35. Find the total population of the United States in 1960 and in 1980. About how many more people were there in 1980 than in 1960?

36. **RESEARCH** Find the region with the greatest change in population from 1960 to 1980. Find two probable reasons for this change.

Population of the United States
(Millions)

	1960	1980
Northeast	44.7	49.1
Midwest	51.6	58.9
South	55.0	75.4
West	28.1	43.2

LOGICAL REASONING

Assume that *x* represents any of the numbers 1, 2, 3, 4, 5,... .
Tell whether each statement is *always*, *sometimes*, or *never* true.

37. $x + 1 > x$

38. $x + 3 > x + 1$

39. $x + 1 = 2x$

40. $3x = 6$

41. $x < 2x$

42. $x \cdot x > 2x$

43. $1 > x \div x$

44. $x \div 1 < x \cdot 1$

SPIRAL REVIEW

45. Evaluate the expression $2yz$ when $y = 3$ and $z = 1.2$.   *(Lesson 1-3)*

46. Find the quotient: $389{,}760 \div 96$ *(Toolbox Skill 10)*

47. Estimate the sum: $657.2 + 194 + 34.91$ *(Toolbox Skill 2)*

48. Round 9.975 to the nearest hundredth.   *(Toolbox Skill 6)*

49. Find the sum: $\frac{6}{7} + \frac{4}{7}$ *(Toolbox Skill 17)*

50. Replace the __?__ with >, <, or =: 1.72 __?__ 1.072 *(Lesson 1-5)*

51. Find the product: 462×709 *(Toolbox Skill 9)*

52. **DATA,** *pages 2–3* How many pops concerts does the Houston Symphony perform each year?   *(Toolbox Skill 24)*

1-6

Properties of Addition

Objective: To recognize the properties of addition and to use them to find sums mentally.

Terms to Know

* *commutative property*
* *associative property*
* *additive identity*
* *identity property of addition*

CONNECTION

Diego Sanchez travels 16 mi from his home to his office. After work he travels the same 16 mi from his office to his home. Reversing the order in which he *commutes* does not change the distance that he commutes. This idea is similar to the **commutative property** of addition.

Commutative Property of Addition

Changing the order of the terms does not change the sum.

In Arithmetic
$36 + 10 = 10 + 36$

In Algebra
$a + b = b + a$

You can sit between two friends and say the same thing first to one and then to the other. The result is the same no matter which friend you speak to first. This idea of *associating* first with one friend and then with the other is similar to the **associative property** of addition.

Associative Property of Addition

Changing the grouping of the terms does not change the sum.

In Arithmetic
$(36 + 10) + 5 = 36 + (10 + 5)$

In Algebra
$(a + b) + c = a + (b + c)$

Parentheses show you how to group the numbers in an expression. Do the work within the parentheses first.

Example 1

Replace each __?__ with the number that makes the statement true.
a. $29 + 46 = 46 +$ __?__
b. $20 + (18 + 7) = ($__?__$+ 18) + 7$

Solution

a. Use the commutative property of addition.
$29 + 46 = 46 + 29$

b. Use the associative property of addition.
$20 + (18 + 7) = (20 + 18) + 7$

The number 0 has a special addition property. When 0 is added to any number, the sum is *identical* to the original number. For this reason, the number 0 is called the **additive identity.**

> ### Identity Property of Addition
> The sum of any number and zero is the original number.
>
> **In Arithmetic**   **In Algebra**
> $14 + 0 = 14$ $a + 0 = a$

You can use the properties of addition to help you find sums mentally.

Example 2

Use the properties to find each sum mentally.

a. $24 + 0$ b. $76 + 29 + 14$

Solution

a. $24 + 0 = 24$

b. $76 + 29 + 14 = (76 + 14) + 29$ ⟵ Group numbers whose sum is
$ = 90 + 29$ 10, 20, 30, and so on.
$ = 119$

✔ **Check Your Understanding**

1. Which property was used to find the sum in Example 2(a)?
2. In Example 2(b), explain why it is helpful to group 76 and 14.

Guided Practice

COMMUNICATION « *Reading*

Refer to the text on pages 22–23.

«**1.** State the identity property of addition in words.

«**2.** Give an everyday word that is related to *commutative*.

Name the property shown by each statement.

3. $17 + 22 = 22 + 17$ 4. $89 + 0 = 89$

5. $(8 + 4) + 5 = 8 + (4 + 5)$ 6. $n + 0 = n$

7. $b + c = c + b$ 8. $(x + y) + z = x + (y + z)$

Replace each ? with the number that makes the statement true.

9. $17 + 56 = \underline{\ ?\ } + 17$ 10. $\underline{\ ?\ } + 43 = 43 + 29$

11. $\underline{\ ?\ } + (2.3 + 9) = (31 + 2.3) + 9$

12. $(42 + 15) + \underline{\ ?\ } = 42 + (15 + 0.29)$

Use the properties to find each sum mentally.

13. $36 + 48 + 14$ 14. $0 + 147$ 15. $5.9 + 3.7 + 3.1$

16. $104 + 47 + 53$ 17. $15 + 67 + 55$ 18. $0 + 2.8 + 11.2$

Replace each __?__ with the number that makes the statement true.

1. $37 + 15 = 15 + \underline{\quad?\quad}$

2. $62 + 47 = \underline{\quad?\quad} + 62$

3. $(7 + 5) + 2 = 7 + (\underline{\quad?\quad} + 2)$

4. $0 + 3.2 = 3.2 + \underline{\quad?\quad}$

5. $\underline{\quad?\quad} + 8.3 = 8.3 + 29$

6. $(6 + \underline{\quad?\quad}) + 7 = 6 + (8 + 7)$

Use the properties to find each sum mentally.

7. $29 + 14 + 71$

8. $25 + 31 + 35$

9. $4.4 + 1.9 + 3.6$

10. $4.8 + 2.7 + 4.2$

11. $0 + 155$

12. $23.4 + 0$

13. $41 + 15 + 9 + 50$

14. $23 + 12 + 7 + 68$

15. $5.5 + 2.4 + 3.6 + 7.5$

16. $9.1 + 7.4 + 2.9 + 4.6$

Use the properties to evaluate each expression when $a = 1.3$, $b = 4.7$, and $c = 26$.

17. $c + 49 + 64$

18. $1.7 + 2.2 + a$

19. $0 + c + 54$

20. $b + 0 + 3.3$

21. $a + 6.2 + b$

22. $c + a + b$

23. $11.2 + c + 8.8 + 14$

24. $c + 0 + 24 + 12$

25. $a + 3.1 + 6.9 + b$

26. **WRITING ABOUT MATHEMATICS** Suppose that your friend is having difficulty understanding the commutative and associative properties of addition. Write a paragraph that describes for your friend at least two ways in which the properties are different.

CONNECTING MATHEMATICS AND LANGUAGE ARTS

Once you understand the meaning of a mathematical term like *commutative* or *associative*, you may be able to figure out the meaning of other, unfamiliar words that share the same root word.

Choose the correct word for each definition. Then use a dictionary to check your answer.

27. to reverse the direction of an electric current
    a. comminute   **b.** commutate   **c.** concatenate   **d.** communicate

28. capable of being joined in a relationship
    a. assessable   **b.** accessible   **c.** associable   **d.** asocial

SPIRAL REVIEW

29. Find the quotient: $66.69 \div 5.4$ *(Toolbox Skill 14)*

30. Find the sum mentally: $54 + 32 + 46$ *(Lesson 1-6)*

Properties of Multiplication

Objective: To recognize the properties of multiplication and to use them to find products mentally.

EXPLORATION

Replace each __?__ with the number or word that makes the statement true.

1. The products $15 \times 30 =$ __?__ and $30 \times 15 =$ __?__ suggest that multiplication is __?__.

2. The products $(15 \times 30) \times 2 =$ __?__ and $15 \times (30 \times 2) =$ __?__ suggest that multiplication is __?__.

Multiplication, like addition, has commutative and associative properties.

Terms to Know
- *commutative property*
- *associative property*
- *multiplicative identity*
- *identity property of multiplication*
- *multiplication property of zero*

Commutative Property of Multiplication

Changing the order of the factors does not change the product.

In Arithmetic
$15 \times 30 = 30 \times 15$

In Algebra
$ab = ba$

Associative Property of Multiplication

Changing the grouping of the factors does not change the product.

In Arithmetic
$(15 \times 30) \times 2 = 15 \times (30 \times 2)$

In Algebra
$(ab)c = a(bc)$

Example 1

Replace each __?__ with the number that makes the statement true.

a. $18 \cdot 31 = 31 \cdot$ __?__

b. $10 \cdot (28 \cdot 5) = ($ __?__ $\cdot 28) \cdot 5$

Solution

a. Use the commutative property of multiplication.
$18 \cdot 31 = 31 \cdot 18$

b. Use the associative property of multiplication.
$10 \cdot (28 \cdot 5) = (10 \cdot 28) \cdot 5$

The number 1 has a special multiplication property. When any number is multiplied by 1, the product is *identical* to the original number. For this reason, the number 1 is called the **multiplicative identity.**

Identity Property of Multiplication

The product of any number and 1 is the original number.

In Arithmetic
$14 \times 1 = 14$

In Algebra
$a \cdot 1 = a$

The number 0 also has a special multiplication property.

> **Multiplication Property of Zero**
>
> The product of any number and zero is zero.
>
> | **In Arithmetic** | **In Algebra** |
> |---|---|
> | $14 \times 0 = 0$ | $a \cdot 0 = 0$ |

You can use the properties of multiplication to find products mentally.

Example 2

MENTAL MATH

Use the properties to find each product mentally.

a. $3(0)(8)(4)$ **b.** $15 \cdot 1 \cdot 3$ **c.** $25 \cdot 13 \cdot 4$

Solution

a. $3(0)(8)(4) = 0$

b. $15 \cdot 1 \cdot 3 = 15 \cdot 3 = 45$

c. $25 \cdot 13 \cdot 4 = (25 \cdot 4) \cdot 13$ ⟵ Group numbers whose
             $= 100 \cdot 13$ product is easy to find.
             $= 1300$

✔ **Check Your Understanding**

1. Which property was used to find the product in Example 2(a)? Example 2(b)?

2. In Example 2(c), explain why you find the product of 25 and 4 first.

Can you divide a number, such as 8, by zero? If so, you could find a number n such that $8 \div 0 = n$. But this means that $n \cdot 0 = 8$. However, the multiplication property of zero states that $n \cdot 0 = 0$. So you cannot divide by zero, and you say that the quotient $8 \div 0$ is *undefined*.

Guided Practice

COMMUNICATION «*Reading*

Refer to the text on pages 25–26.

«**1.** What is the main idea of the lesson?

«**2.** List four major points that support the main idea of the lesson.

Name the property shown by each statement.

3. $12 \cdot 46 = 46 \cdot 12$ **4.** $(7.3)(1) = 7.3$

5. $29(0) = 0$ **6.** $(6 \cdot 9) \cdot 3 = 6 \cdot (9 \cdot 3)$

Replace each __?__ with the number that makes the statement true.

7. $54 \cdot 63 = \underline{\,?\,} \cdot 54$ **8.** $(\underline{\,?\,})(8.2) = (8.2)(2.6)$

9. $\underline{\,?\,} \cdot (18 \cdot 7) = (71 \cdot 18) \cdot 7$ **10.** $(23 \cdot 5) \cdot 32 = 23 \cdot (\underline{\,?\,} \cdot 32)$

Use the properties to find each product mentally.

11. $(2)(4.2)(1)$ **12.** $2 \cdot 78 \cdot 5$ **13.** $5(1.2)(20)$ **14.** $25(11)(4)(0)$

Exercises

Replace each __?__ with the number that makes the statement true.

1. $26 \cdot 38 = 38 \cdot \underline{}$ **2.** $(27)(1.2) = (\underline{})(27)$

3. $(9 \cdot 7) \cdot 3 = 9 \cdot (\underline{} \cdot 3)$ **4.** $6 \cdot (8 \cdot 9) = (6 \cdot 8) \cdot \underline{}$

5. $(\underline{})(52) = (52)(2.1)$ **6.** $(5 \cdot \underline{}) \cdot 4 = 5 \cdot (6 \cdot 4)$

Use the properties to find each product mentally.

7. $2 \cdot 32 \cdot 5$ **8.** $2(8)(50)$ **9.** $1 \cdot 89$ **10.** $56(1)$

11. $24(0)(5)$ **12.** $0 \cdot 22 \cdot 5$ **13.** $(24)(1)(5)$ **14.** $33 \cdot 2 \cdot 1$

15. $(0.6)(1.1)(5)$ **16.** $(0.5)(3.7)(2)$ **17.** $(4.4)(0)(5)$ **18.** $(1)(0.5)(8)$

Use the properties to evaluate each expression when $x = 5$, $y = 12$, and $z = 3$.

19. $(20)(43)(x)$ **20.** $(25)(z)(4)$ **21.** $x \cdot 7 \cdot y$ **22.** $x \cdot z \cdot 1$

23. $y \cdot 1 \cdot z$ **24.** $z \cdot 0 \cdot y$ **25.** $x \cdot y \cdot 0$ **26.** xzy

27. $(20)(y)(0.2)$ **28.** $(0.5)(x)(14)$ **29.** $0.5zy$ **30.** $0.2zx$

THINKING SKILLS

In Exercises 31 and 33, do the work in parentheses first.

31. Find $49 - (27 - 14)$. Then find $(49 - 27) - 14$.

32. Examine the results of Exercise 31. Is subtraction associative? Create another example to support your answer.

33. Find $24 \div (6 \div 2)$. Then find $(24 \div 6) \div 2$.

34. Examine the results of Exercise 33. Is division associative? Create another example to support your answer.

SPIRAL REVIEW

35. Evaluate $a \div b$ when $a = 10$ and $b = 2$. *(Lesson 1-1)*

36. Complete: 6 ft 4 in. $= \underline{}$ in. *(Toolbox Skill 22)*

37. Find the product mentally: $12(5)(0)(8)$ *(Lesson 1-7)*

38. Find the sum: $3\frac{2}{3} + 4\frac{5}{6}$ *(Toolbox Skill 18)*

39. Evaluate $b - a$ when $a = 4.6$ and $b = 11.2$. *(Lesson 1-2)*

40. Is 6240 divisible by 2? by 3? by 4? by 5? by 8? by 9? by 10? *(Toolbox Skill 16)*

1-8

Choosing the Most Efficient Method of Computation

Objective: To decide whether it would be most efficient to use mental math, paper and pencil, or a calculator to solve a given problem.

Before doing a computation, you should *inspect* the problem and decide whether to use mental math, paper and pencil, or a calculator.

- *Mental math* may be most efficient when you see sums of ten or products of ten, when you do not need to rename, or when you can *add on* or *count back* easily.

- *Paper and pencil* may be most efficient when the computation seems simple or involves numbers with few digits.

- A *calculator* may be most efficient when the computation involves many numbers. You also might decide to use a calculator when accuracy is very important.

Example

Tell whether it is most efficient to find each answer using *mental math*, *paper and pencil*, or a *calculator*. Then find each answer.

a. $147 - 98$ **b.** $23.47 + 10.82 + 16.09$

Solution

a. If you know how to *add on*, use mental math.
    $98 + 2 = 100; \quad 100 + 47 = 147$
    Because $2 + 47 = 49$, $147 - 98 = 49$.

b. There are three terms and each term has four digits. A calculator would be the most efficient method.

$\boxed{23.47}$ $\boxed{+}$ $\boxed{10.82}$ $\boxed{+}$ $\boxed{16.09}$ $\boxed{=}$ $\boxed{50.38}$

☑ Check Your Understanding

Suppose you do not know how to *add on*. What is another method to use in part (a) of the Example? Explain.

Guided Practice

COMMUNICATION «*Discussion*

Explain why it is easy to find each answer using mental math.

«**1.** $(2000)(30)$ «**2.** $236 - 99$ «**3.** $67 + 23$ «**4.** $540 \div 60$

«**5.** $4 + 73.6$ «**6.** $0.9 \cdot 100$ «**7.** $1.5 \div 0.3$ «**8.** $4.3 - 1.2$

In each set, choose the exercise that might be done more efficiently using paper and pencil instead of a calculator. Give a reason for your choice.

9. $948 + 1003$ **10.** $724.3 - 6.531$ **11.** $852 \cdot 791$ **12.** $8.208 \div 3$
    $948 + 9765$ $7.243 - 6.531$ $(852)(3)$ $8.208 \div 342$

Tell whether it is most efficient to find each answer using *mental math,* *paper and pencil,* or a *calculator.* Then find each answer using the method that you chose.

13. $49.5 \div 3$ **14.** $100 \cdot 2 \cdot 3$ **15.** $5414 - 678$ **16.** $2.5 + 7$

Exercises

Tell whether it is most efficient to find each answer using *mental math,* *paper and pencil,* or a *calculator.* Then find each answer using the method that you chose.

1. $13 \cdot 6$ **2.** $165 - 45$ **3.** $380 + 135$

4. $84 \div 6$ **5.** $783 \cdot 35$ **6.** $1.2 \div 0.4$

7. $521(0.9)$ **8.** $10.93 - 2.981$ **9.** $6.6 \div 2.75$

10. $143.9 + 2.3$ **11.** $(0.4)(0.7)$ **12.** $45.7 \div 100$

Tell whether it is most efficient to solve each problem using *mental math, paper and pencil,* or a *calculator.* Then solve each problem using the method that you chose.

13. Greg spent $2.70 on 3 packages of seeds. How much did each package of seeds cost?

14. Sharla has $12.50. She wants to buy as many subway tokens as possible. Tokens cost 75¢ each. How many can she buy?

15. Alan bought items priced at $.89, $2.49, $5.45, and $1.49 at the grocery store. Estimate the total cost of the items.

16. Mai purchased seven flower arrangements for a party. Each flower arrangement cost $19.85. About how much was the total cost?

SPIRAL REVIEW

17. Find the product: 0.04×0.05 *(Toolbox Skill 13)*

18. Is it most efficient to find the product $(4.82)(5.64)$ using *mental math, paper and pencil,* or a *calculator?* *(Lesson 1-8)*

19. Replace the ___?___ with $>$, $<$, or $=$: 2.04 ___?___ 2.040 *(Lesson 1-5)*

20. Find the quotient: $8.48 \div 100$ *(Toolbox Skill 15)*

Exponents

Objective: To compute with exponents and to evaluate expressions involving exponents.

EXPLORATION

1 Enter each key sequence on a calculator. Record the results.

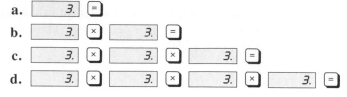

2 Suppose that n represents the number of times you entered 3 as a factor. Find the value of n that makes each statement true.

a. $3^n = 3$ **b.** $3^n = 9$ **c.** $3^n = 27$ **d.** $3^n = 81$

3 What number is represented by 3^{10}?

4 Find the value of x that makes this statement true: $3^x = 1{,}594{,}323$

You can write a multiplication expression in which all the factors are the same in a shortened form called **exponential form.**

$$\underbrace{3 \cdot 3 \cdot 3 \cdot 3 \cdot 3 \cdot 3}_{6 \text{ factors}} = 3^{\underset{\uparrow}{6}} \longleftarrow \text{exponent}$$
$$\text{base}$$

The number 3^6, or 729, is called a **power** of 3. The **exponent** 6 shows that the **base** 3 is used as a factor six times.

> You read 3^6 as *three to the sixth power* or *the sixth power of three.*
> You read 3^2 as *three to the second power* or *three squared.*
> You read 3^3 as *three to the third power* or *three cubed.*

Any number to the first power is equal to that number, as in $3^1 = 3$. The number 1 to any power equals 1, as in $1^7 = 1$.

Terms to Know
- *exponential form*
- *power*
- *exponent*
- *base*

Example 1

Solution

✓ **Check Your Understanding**

Find each answer: **a.** 6^3 **b.** 1^4 **c.** 10^1

a. $6^3 = 6 \cdot 6 \cdot 6 = 216$ **b.** $1^4 = 1$ **c.** $10^1 = 10$

1. Describe how 6^3 is different from $6 \cdot 3$.
2. Describe how 6^3 is different from 3^6.
3. In Example 1(b), explain why $1^4 = 1$.

You can use exponents with variables as well as with numbers. For instance, you can write $x \cdot x \cdot x \cdot x \cdot x$ as x^5. You read x^5 as *a number x to the fifth power* or *the fifth power of a number x.*

You may need to evaluate variable expressions that contain exponents.

Example 2

Evaluate each expression when $n = 4$.

a. $3n^2$ **b.** $(3n)^2$

Solution

a. $3n^2 = 3 \cdot n \cdot n = 3 \cdot 4 \cdot 4 = 48$

b. $(3n)^2 = (3n)(3n) = (3 \cdot 4)(3 \cdot 4) = (12)(12) = 144$

Some calculators have a $\boxed{y^x}$ key. If you have this key on your calculator, you may wish to use it when you work with exponents. For example, you can use the following key sequence to find 8^6.

$$\boxed{8.} \quad \boxed{y^x} \quad \boxed{6.} \quad \boxed{=}$$

Guided Practice

COMMUNICATION « *Reading*

Each word is followed by four correct meanings. Choose the meaning that is used in this lesson.

«**1.** base
    a. the lowest point
    b. a supporting layer
    c. a number that is raised to a power
    d. a corner of the infield in softball

«**2.** power
    a. strength
    b. authority
    c. electricity
    d. exponent

COMMUNICATION « *Writing*

Write an expression for each phrase.

«**3.** three to the fifth power

«**4.** a number x to the sixth power

«**5.** a number x cubed multiplied by six

«**6.** six times a number x, cubed

Write a phrase for each expression.

«**7.** 8^3 «**8.** a^4 «**9.** $3x^5$ «**10.** $(3x)^5$

Give the exponential form of each expression.

11. $(7)(7)(7)(7)(7)$ **12.** $x \cdot x \cdot x \cdot x$

13. $5 \cdot d \cdot d \cdot d$ **14.** $4y \cdot 4y \cdot 4y$

Give the multiplication that each expression represents.

15. 6^7 **16.** c^5 **17.** $5x^4$ **18.** $(2c)^3$

Find each answer.

19. 4^3 **20.** 1^5 **21.** 2^5 **22.** 10^4 **23.** 11^2 **24.** 2^8

Evaluate each expression when $n = 3$.

25. n^5 **26.** $5n^2$ **27.** $3n^3$ **28.** $(8n)^1$ **29.** $(4n)^2$ **30.** $(3n)^4$

Exercises

Find each answer.

1. 7^3 **2.** 1^{14} **3.** 9^1 **4.** 12^2 **5.** 10^3 **6.** 6^4

Evaluate each expression when $n = 2$.

7. n^4 **8.** n^6 **9.** n^7 **10.** n^9

11. $6n^3$ **12.** $10n^5$ **13.** $4n^4$ **14.** $7n^2$

15. $(2n)^2$ **16.** $(9n)^2$ **17.** $(4n)^3$ **18.** $(5n)^3$

19. The memory of a computer is measured in a unit called a byte. The letter K represents 2^{10} bytes. Write this number without an exponent. Then write the number of bytes represented by 40K of memory.

20. DATA, *pages 666–667* Calculate 11^1, 11^2, 11^3, and 11^4. Where can these powers of 11 be found in Pascal's triangle?

Evaluate each expression for the given value of the variable.

21. 5^x, when $x = 2$ **22.** 3^v, when $v = 5$

23. 4^m, when $m = 4$ **24.** 9^w, when $w = 3$

25. 12^x, when $x = 1$ **26.** 1^c, when $c = 12$

Find each answer.

27. $(0.9)^2$ **28.** $(0.7)^2$ **29.** $(0.5)^3$ **30.** $(0.6)^3$

31. $(0.2)^3$ **32.** $(0.3)^3$ **33.** $(0.1)^3$ **34.** $(0.1)^4$

CALCULATOR

Match each expression with the correct calculator key sequence.

35. $(1.2)^5$ **A.** [0.5] [y^x] [12.] [=]

36. 12^5 **B.** [5.] [y^x] [12.] [=]

37. 5^{12} **C.** [12.] [y^x] [5.] [=]

38. $(0.5)^{12}$ **D.** [1.2] [y^x] [5.] [=]

CONNECTING ALGEBRA AND GEOMETRY

Why are the exponents 2 and 3 read as *squared* and *cubed*? The reason is their connection to geometry. Evaluating x^2 gives you the area of a square with side of length x, and evaluating x^3 gives you the volume of a cube with edge of length x. The figures below illustrate the geometric meaning of 4^2 and 4^3.

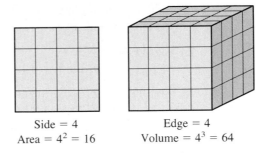

Side = 4
Area = 4^2 = 16

Edge = 4
Volume = 4^3 = 64

39. Find the area of a square with side of length 9.

40. Find the volume of a cube with edge of length 7.

41. The area of a square is 121. What is the length of a side?

42. The volume of a cube is 125. What is the length of an edge?

THINKING SKILLS

Replace each __?__ with >, <, or =.

43. 5^3 __?__ 3^5 **44.** 4^3 __?__ 3^4 **45.** 1^8 __?__ 8^1 **46.** 2^7 __?__ 7^2

Assume that a and b represent any of the numbers 2, 3, 4, 5, ... , and $a > b$. Determine a value of a and a value of b that make each statement true.

47. $a^b = b^a$ **48.** $a^b > b^a$

SPIRAL REVIEW

49. Tell what this paragraph is about: Emily Ling earns $9 per hour at Technology Industries. Each week she works five days for a total of 42 h. *(Lesson 1-4)*

50. Find the sum mentally: $118 + 0$ *(Lesson 1-6)*

51. Find the quotient: $\frac{7}{8} \div \frac{3}{4}$ *(Toolbox Skill 21)*

52. Evaluate $4x^3$ when $x = 5$. *(Lesson 1-9)*

Challenge

Describe how to find this sum using only mental math.

$$20 + 21 + 22 + 23 + 24 + 25 + 26 + 27 + 28 + 29 + 30$$

Modeling Powers of Two

Objective: To use paper folding to model powers of two.

Materials

- letter-size paper
- scissors

You can use a sheet of paper to show how quickly powers of numbers increase. To do this, you simply keep folding the paper again and again.

Activity I *Collecting Data*

1. Fold a sheet of paper in half. Unfold the paper. Into how many parts has the paper been separated?

2. Refold the paper. Now fold it in half again. Unfold the paper. Into how many parts has the paper been separated this time?

3. Continue the process of folding and unfolding the paper. After each new fold, open the paper and count the number of parts into which it has been separated. To record your results, copy and complete the table below.

Number of Folds	1	2	3	4	5
Number of Parts	2	?	?	?	?

Activity II *Problem Solving*

1. Look at the data in the table from Activity I. Describe the numbers that appear in the row labeled "Number of Parts."

2. At this point, you may not be able to fold the paper anymore. However, you can use the data in the table from Activity I to predict the number of parts that the paper would have if you could fold it six times. How many parts would the paper have?

3. How many parts would the paper have if you could fold it seven times? eight times? nine times? ten times?

4. What do the numbers you found in the previous step represent mathematically?

Origami, the art or process of folding paper into shapes, originated in Japan.

5 How many folds would you have to make for the paper to have 4096 parts?

6 After how many folds would the number of parts be more than one million?

Activity III *Visual Thinking*

1 Each piece of paper has been folded twice. The dark lines indicate cuts that have been made through the folded paper. Sketch how each piece of paper would look when unfolded.

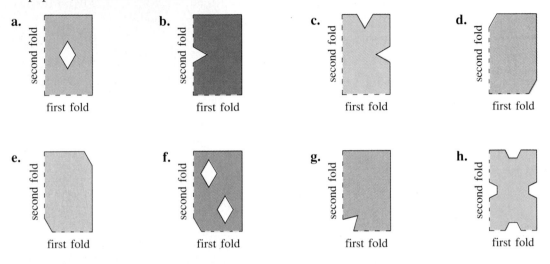

2 Fold a letter-size piece of paper twice. Then cut the folded paper so that you can create each design when it is unfolded.

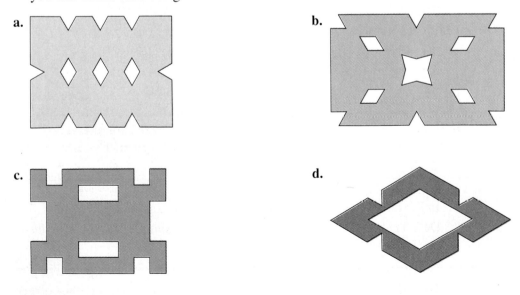

1-10

Order of Operations

Objective: To use the order of operations to compute and to evaluate expressions.

Terms to Know
• *order of operations*

APPLICATION

Jerry and Sondra entered this exercise on different calculators.

$$30 + 24 \div 6 \cdot 2$$

The result on Jerry's calculator was 18, while the result on Sondra's calculator was 38.

QUESTION Which calculator displayed the correct answer?

In mathematics, you must perform operations in an agreed-upon order to make sure that an expression has only one answer.

Order of Operations

1. First do all work inside any parentheses.
2. Then find each power.
3. Then do all multiplications and divisions in order from left to right.
4. Then do all additions and subtractions in order from left to right.

Example 1

Find each answer.

a. $30 + 24 \div 6 \cdot 2$ **b.** $7 \cdot 3^2 - (5 + 6)$

Solution

a. $30 + 24 \div 6 \cdot 2 = 30 + 4 \cdot 2$ ←— Divide.
$\qquad\qquad\qquad\quad = 30 + 8$ ←— Multiply.
$\qquad\qquad\qquad\quad = 38$ ←— Add.

b. $7 \cdot 3^2 - (5 + 6) = 7 \cdot 3^2 - (11)$ ←— Work inside parentheses first.
$\qquad\qquad\qquad\quad = 7 \cdot 9 - 11$ ←— Find the power ($3^2 = 9$).
$\qquad\qquad\qquad\quad = 63 - 11$ ←— Multiply.
$\qquad\qquad\qquad\quad = 52$ ←— Subtract.

☑ Check Your Understanding

1. In Example 1(a), why do you first divide and then multiply?
2. Describe how you would find the answer to Example 1(a) if the expression were $(30 + 24) \div 6 \cdot 2$.

ANSWER Example 1(a) shows that Sondra's calculator displayed the correct answer.

The fraction bar acts like parentheses in the order of operations.

Example 2

Find the answer: $\dfrac{24 + 12}{13 - 4}$

Solution

$\dfrac{24 + 12}{13 - 4} = \dfrac{36}{9}$ ← Work above the fraction bar, then below the fraction bar.

$= 4$ ← Divide.

✔️ **Check Your Understanding**

3. Write the expression in Example 2 with parentheses instead of with a fraction bar.

You can use a calculator to find an answer when order of operations is involved, but the key sequence that you use depends on the type of calculator you have. For instance, if your calculator has parenthesis keys, you can find $18 \div (7 - 4)$ using this key sequence.

| 18. | $\div$ | (| 7. | $-$ | 4. |) | $=$ |

If your calculator does not have parenthesis keys, you will need to use the memory keys.

| 7. | $-$ | 4. | $=$ | M+ | 18. | $\div$ | MR | $=$ |

You must also use the order of operations when you evaluate variable expressions.

Example 3

Evaluate each expression when $a = 5$ and $b = 4$.

a. $(8 + b^2) \div 3 + a$ **b.** $\dfrac{43 + a}{10 - b}$

Solution

a. $(8 + b^2) \div 3 + a = (8 + 4^2) \div 3 + 5$
$= (8 + 16) \div 3 + 5$
$= 24 \div 3 + 5$
$= 8 + 5$
$= 13$

b. $\dfrac{43 + a}{10 - b} = \dfrac{43 + 5}{10 - 4}$

$= \dfrac{48}{6}$

$= 8$

✔️ **Check Your Understanding**

4. Describe how you would evaluate the expression in Example 3(a) if it were $(8 + b)^2 \div 3 + a$.

5. Describe how you would evaluate the expression in Example 3(a) if it were $(8 + b)^2 \div (3 + a)$.

Guided Practice

« 1. **COMMUNICATION** « *Reading*   Without referring to the book, write the correct order of operations on a sheet of paper. Then check your answer against the text on pages 36–37.

« 2. **COMMUNICATION** « *Writing*   Write an expression in which you can work from left to right to find the answer. Write an expression in which you *cannot* work from left to right to find the answer.

List the operations in the order you would perform them.

3. $34 - 10 + 4 \cdot 2$

4. $(53 - 8) \div 5$

5. $4 \cdot 5^2 + 7$

6. $\dfrac{16 + 14}{10 - 5}$

Find each answer.

7. $7 + 45 \div 9$

8. $4^3 + (15 - 7)$

9. $8 + 32 \div 4 \cdot 5$

10. $\dfrac{10 \cdot 4}{14 - 9}$

Evaluate each expression when $a = 2$, $b = 6$, and $c = 12$.

11. $3b + 7$

12. $(2a + c) \div 2$

13. $3(b - a^2) + 16$

14. $\dfrac{36 + c}{b - a}$

Exercises

Find each answer.

1. $9^2 + 3 \cdot 5$

2. $4^3 - 30 \div 2$

3. $9 + 45 \div 9 \cdot 8$

4. $7^2 - 14 + 5 \cdot 2$

5. $8^2 \div (8 - 4)$

6. $11 + 3(19 + 2)$

7. $2^2 + 6(3 + 10)$

8. $14 + (3^3 - 7)$

9. $\dfrac{18 + 10}{7 - 3}$

10. $\dfrac{51 - 9}{11 - 4}$

11. $\dfrac{9 \cdot 5}{3 + 6}$

12. $\dfrac{54 + 18}{2 \cdot 4}$

Evaluate each expression when $a = 2$, $b = 5$, and $c = 4$.

13. $9 + 7a$

14. $8c - 13$

15. $(14 - a) \cdot c$

16. $b + c^2 \div a + 4$

17. $81 - (b + 9) + a^2$

18. $65 \div b - a + 6$

19. $\dfrac{b + 19}{a + 4}$

20. $\dfrac{15c}{1 + b}$

21. $b^2 - a^3$

22. June runs 37 mi each week, twelve miles of which she runs on the weekend. She runs the same number of miles each weekday. How many miles does June run each weekday?

23. **DATA,** *pages 2–3*   The Rolling Pebbles had five gold albums and two platinum albums last year. What is the least number of albums they could have sold last year?

Find each answer. Work inside the parentheses first. Then work inside the square brackets.

24. $[(12 - 4) \cdot 2 + 11] \div 3$ **25.** $[(6 + 8) \cdot 3 - 12] + 6$

26. $110 - [20 + (4^2 - 9)]$ **27.** $45 + [6^2 - (12 + 6)]$

28. $34 + [(16 + 12) \div 4] - 16$ **29.** $48 - [36 \div (4 + 5)] + 11$

THINKING SKILLS

Classify each statement as *True* or *False*. Then make each false statement true by inserting parentheses where necessary.

30. $4 \cdot 5 + 6 = 44$ **31.** $24 - 4 \cdot 2 = 40$

32. $24 \div 3 + 5 \cdot 2 = 6$ **33.** $6 \cdot 12 - 5 \cdot 8 = 32$

34. $4 \cdot 6 + 3 \cdot 3 = 33$ **35.** $3 \cdot 4 - 2 \cdot 3 = 18$

36. $4 + 4^2 \div 2 = 32$ **37.** $12 - 2^2 \div 4 = 2$

SPIRAL REVIEW

38. Find the answer: 5^5 *(Lesson 1-9)*

39. Find the difference: $40{,}007 - 6218$ *(Toolbox Skill 8)*

40. Find the sum: $8.635 + 27.9$ *(Toolbox Skill 11)*

41. Evaluate $5u - 18 \div b$ when $u = 9$ and $b = 6$.   *(Lesson 1-10)*

Self-Test 2

Replace each ___?___ with >, <, or =.

1. $2874 \underline{?} 2957$ **2.** $7.02 \underline{?} 7.020$ **3.** $4.19 \underline{?} 4.0187$ 1-5

Use the properties of addition and multiplication to find each answer mentally.

4. $81 + 42 + 19$ **5.** $2.3 + 3.9 + 4.7$ **6.** $106 + 0$ 1-6

7. $(0.5)(1.7)(20)$ **8.** $15 \cdot 1 \cdot 10$ **9.** $7(3)(0)(4)$ 1-7

Tell whether it is most efficient to find each answer using *mental math*, *paper and pencil*, or a *calculator*. Then find each answer using the method that you chose.

10. $3.487 + 69.88$ **11.** $84 - 18$ **12.** $(59.1)(10)$ 1-8

Evaluate each expression when $a = 5$, $b = 2$, and $c = 7$.

13. a^4 **14.** $3b^6$ **15.** $(2c)^2$ 1-9

16. $17 - 8 + b^2$ **17.** $48 \div (16 - 4) \cdot b$ **18.** $\dfrac{28 - c}{a + b}$ 1-10

Chapter Review

variable (p. 4)
variable expression (p. 4)
evaluate (p. 4)
value of a variable (p. 4)
comparison property (p. 18)
inequality symbols (p. 18)
commutative property of addition
   (p. 22)
associative property of addition
   (p. 22)
additive identity (p. 22)
identity property of addition (p. 23)

commutative property of
   multiplication (p. 25)
associative property of multiplication
   (p. 25)
multiplicative identity (p. 25)
identity property of multiplication
   (p. 25)
multiplication property of zero
   (p. 26)
exponential form (p. 30)
power (p. 30)
exponent (p. 30)
base (p. 30)
order of operations (p. 36)

**Choose the correct term from the list above to complete each
sentence.**

1. The expression $x + 9$ is an example of a(n) __?__.

2. The __?__ states that changing the order of the terms does not
change the sum.

3. The number 1 is called the __?__.

4. A symbol that represents a number is called a(n) __?__.

5. In the expression 6^4, the 6 is called the __?__.

6. $4 \cdot 0 = 0$ is an example of the __?__.

Evaluate each expression when $a = 3$ and $b = 8$. *(Lesson 1-1)*

7. $7a$ **8.** $4b$ **9.** $39 - b$ **10.** $25 + a$

11. $27 \div a$ **12.** $b \div 2$ **13.** $b + b$ **14.** ab

Evaluate each expression when $x = 5.6$, $y = 85$, $m = 1.32$, and $n = 24$.
(Lessons 1-2, 1-3)

15. $31.9 + x$ **16.** xm **17.** $m \div 0.3$ **18.** $x - m$

19. $2.3n$ **20.** $y - 4.7$ **21.** $x + y$ **22.** $m \div n$

Replace each __?__ with >, <, or =. *(Lesson 1-5)*

23. $14{,}259$ __?__ $14{,}312$ **24.** 4.86 __?__ 4.860 **25.** 0.12 __?__ 0.012

Use this paragraph for Exercises 26–29. *(Lesson 1-4)*

Yvonne earns $7 per hour typing term papers. Last week, she typed 6 h on Monday, 5 h on Tuesday, and 7 h on Wednesday. On Thursday, Yvonne typed the same number of hours as on Monday.

26. What is the paragraph about?

27. How many hours did Yvonne type on Thursday?

28. Identify any facts that are not needed to find the total number of hours Yvonne typed.

29. Describe how you would find the total number of hours that Yvonne typed last week.

Write in order from least to greatest. Use two inequality symbols.
(Lesson 1-5)

30. 12,456; 5642; 12,375   **31.** 0.62; 0.078; 0.102   **32.** 6.10; 6; 0.611

Replace each __?__ with the number that makes the statement true.
(Lessons 1-6, 1-7)

33. $25 + 0.72 = \underline{} + 25$ **34.** $\underline{} \cdot 31 = 31 \cdot 44$ **35.** $\underline{} \cdot (5 \cdot 14) = (12 \cdot 5) \cdot 14$

Use the properties of addition and multiplication to find each answer mentally. *(Lessons 1-6, 1-7)*

36. $37 + 49 + 63$ **37.** $6(3)(0)(8)$ **38.** $(4)(0.7)(2.5)$ **39.** $2.9 + 0 + 4.1$

Tell whether it is most efficient to find each answer using *mental math*, *paper and pencil*, or a *calculator*. Then find each answer using the method that you chose. *(Lesson 1-8)*

40. $56 + 87$ **41.** $98.25 - 3.667$ **42.** $(79)(6.32)$ **43.** $1200 \div 6$

Find each answer. *(Lessons 1-9, 1-10)*

44. 3^5 **45.** 2^7 **46.** 1^8 **47.** 4^5 **48.** 10^4 **49.** 8^3

50. $7^2 - 8 \cdot 3 + 6$ **51.** $\dfrac{3 \cdot 14}{10 - 4}$ **52.** $3^2 + 5(2 + 9)$

Evaluate each expression when $a = 3$, $b = 5$, and $c = 6$.
(Lessons 1-9, 1-10)

53. b^2 **54.** a^5 **55.** $4c^3$ **56.** $2a^5$ **57.** $(4b)^2$ **58.** $(2a)^3$

59. $25 - 3c$ **60.** $64 - (c + 9) + a^2$ **61.** $\dfrac{b + a}{c - 2}$

Evaluate each expression when $x = 45$ and $y = 9$.

1. $8y$ **2.** $23 + x$ **3.** $57 - x$ **4.** $x \div y$ 1-1

Evaluate each expression when $a = 1.6$, $b = 1.44$, $c = 48$, and $d = 8.1$.

5. $c + 12.7$ **6.** $a + b$ **7.** $c - 26.4$ **8.** $d - a$ 1-2

9. $7b$ **10.** ad **11.** $d \div 3$ **12.** $b \div a$ 1-3

Use this paragraph for Exercises 13–16.

Dale bought two cassette tapes for $9.99 each and a compact disc for $13.95. An album costs $8.95. Dale gave the clerk two $20 bills.

13. What is the paragraph about?   1-4

14. How much did Dale pay for the compact disc?

15. Identify any facts that are not needed to find the total amount of Dale's purchase.

16. Describe how you would find the amount of change Dale received.

Replace each __?__ with >, <, or =.

17. 2324 __?__ 2243   **18.** 3.16 __?__ 3.106   **19.** 2.50 __?__ 2.5   1-5

Write in order from least to greatest. Use two inequality symbols.

20. 7634; 779; 7073   **21.** 8.65; 0.0865; 0.865

22. Name four numbers that are between 3.1 and 3.15.

23. How many numbers in all are there between 3.1 and 3.15?

Use the properties of addition and multiplication to find each answer mentally.

24. $76 + 38 + 24$ **25.** $3.5 + 5.2 + 4.5$ **26.** $0 + 12.8$ 1-6

27. $8(12)(1)$ **28.** $(50)(9)(0.2)$ **29.** $15 \cdot 8 \cdot 0$ 1-7

Tell whether it is most efficient to find each answer using *mental math*, *paper and pencil*, or a *calculator*. Then find each answer using the method that you chose.

30. $800 + 755$ **31.** $8(17)$ **32.** $13.58 \div 1.4$ 1-8

Evaluate each expression when $x = 2$, $y = 5$, and $z = 9$.

33. y^3 **34.** x^5 **35.** $2z^2$ **36.** $5x^4$ **37.** $(3y)^2$ 1-9

38. $45 - y^2$ **39.** $(33 - z) \div x \cdot 6$ **40.** $\dfrac{y + 39}{x + z}$ 1-10

Other Operations

Objective: To create and to use new operations.

You have used the operations of addition, subtraction, multiplication, and division. By combining these four basic operations, it is possible to create other operations. For instance, the example below shows an operation that might be called "diamond."

Example

For all numbers a and b, $a \blacklozenge b = ab - b$. Find each of the following.
a. $7 \blacklozenge 3$ **b.** $3 \blacklozenge 7$

Solution

a. $7 \blacklozenge 3 = (7)(3) - 3$
$= 21 - 3$
$= 18$

b. $3 \blacklozenge 7 = (3)(7) - 7$
$= 21 - 7$
$= 14$

Exercises

For all numbers a and b, $a \heartsuit b = a^2 + b^2$. Find each of the following.

1. $2 \heartsuit 3$ **2.** $3 \heartsuit 2$ **3.** $1 \heartsuit 6$ **4.** $6 \heartsuit 1$

5. Does $a \heartsuit b = b \heartsuit a$? Explain.

6. Does $a \heartsuit (b \heartsuit c) = (a \heartsuit b) \heartsuit c$? Give an example.

For all numbers x and y, $x \blacksquare y = x + y^2$. Find each of the following.

7. $5 \blacksquare 4$ **8.** $4 \blacksquare 5$ **9.** $6 \blacksquare 8$ **10.** $8 \blacksquare 6$

11. Is $\blacksquare$ commutative? Give an example to support your answer.

12. Is $\blacksquare$ associative? Give an example to support your answer.

13. Write a variable expression for the operation $\star$. Use a and b as the variables.

$$1 \star 2 = 5$$
$$2 \star 3 = 8$$
$$7.1 \star 5 = 17.1$$
$$12 \star 10 = 32$$

14. **GROUP ACTIVITY** Make up an operation, $\triangle$. List the results when you perform your operation on several pairs of numbers. Trade results with a partner and determine the operation your partner created. Write your partner's operation in the form "For all numbers a and b, $a \triangle b = \ldots$"

Cumulative Review

Standardized Testing Practice

Choose the letter of the correct answer.

1. What information is not needed to solve this problem?

   Joe earns $9 per hour and works 8 h per day. He works 40 h per week. How much does Joe earn per week?

   A. earns $9 per hour
   B. works 8 h per day
   C. works 40 h per week
   D. all the information is needed

2. Evaluate $45.97 + x$ when $x = 32.5$.

   A. 13.47   B. 49.22
   C. 4922   D. 78.47

3. Evaluate $54.4 - a$ when $a = 17.9$.

   A. 72.3
   B. 46.5
   C. 36.5
   D. 37.5

4. Tran, Kay, Jarreau, and Lea live 0.61 mi, 0.061 mi, 0.601 mi, and 0.16 mi from school, respectively. Who lives closest to school?

   A. Tran   B. Kay
   C. Jarreau   D. Lea

5. Evaluate $247.04 \div a$ when $a = 6.4$.

   A. 38.6   B. 253.44
   C. 240.64   D. 386

6. Which of the following do you know is true for all numbers x, y, and z?

   I. $(xy)z = x(yz)$
   II. $(xyz)^2 = xyz^2$
   III. $(x \div y) \cdot z = x \div y \cdot z$

   A. I only
   B. II only
   C. I and II
   D. I and III

7. Evaluate $16m$ when $m = 4.3$.

   A. 688   B. 6.88
   C. 0.688   D. 68.8

8. List 0.847, 0.0847, 8.47, 0.1847 in order from least to greatest.

   A. 8.47, 0.847, 0.1847, 0.0847
   B. 0.1847, 0.0847, 0.847, 8.47
   C. 0.0847, 0.1847, 0.847, 8.47
   D. 8.47, 0.1847, 0.0847, 0.847

9. $1 + 0 = 1$ is an example of which property?

   A. commutative property of addition
   B. associative property of addition
   C. identity property of addition
   D. identity property of multiplication

10. Which is a statement of the commutative property of addition?

   A. $ab = ba$ B. $a + b = b + a$
   C. $a = b$ D. $a(bc) = (ab)c$

11. Use the properties of addition to find $68 + 0 + 42$ mentally.
  A. 100
  B. 120
  C. 0
  D. 110

12. $7(0) = 0$ is an example of which property?
  A. comparison property
  B. identity property of addition
  C. multiplication property of zero
  D. identity property of multiplication

13. Which is a statement of the associative property of multiplication?
  A. $ab = ba$ B. $a + b = b + a$
  C. $a > b$ D. $a(bc) = (ab)c$

14. Use the properties of multiplication to find $4(1)(6)(25)$ mentally.
  A. 1600
  B. 0
  C. 600
  D. 2400

15. Write in exponential form.
  $4 \cdot 4 \cdot 4 \cdot 4 \cdot 4$
  A. 5^4 B. 4^5
  C. 20   D. 1024

16. The volume of a cube is 8^3 cubic feet. How many cubic feet is that?
  A. 24
  B. 83
  C. 6561
  D. 512

17. Evaluate a^4 when $a = 6$.
  A. 24
  B. 4096
  C. 1296
  D. 10

18. Choose the best estimate.
  $716.54 - 388.16$
  A. about 300
  B. about 400
  C. about 500
  D. about 1100

19. Find the answer.
  $24 \div 8 + 4 \times 3^2$
  A. 63
  B. 147
  C. 39
  D. 441

20. Evaluate $a^4 - (33 + a)$ when $a = 3$.
  A. 45
  B. 51
  C. 118
  D. 70

Notable Firsts in Space

1959	Space vehicle lands on the moon.
1961	Manned flight orbits Earth.
1965	Closeup pictures of Mars.
1965	Humans "walk in space."
1968	Space vehicle orbits the sun.
1968	Live TV transmissions from orbit.
1969	Space vehicle lands on Venus.
1969	Manned flight lands on the moon.
1973	Closeup pictures of Jupiter.
1974	Closeup pictures of Mercury.
1979	Closeup pictures of Saturn.
1986	Closeup pictures of Uranus.
1989	Closeup pictures of Neptune.

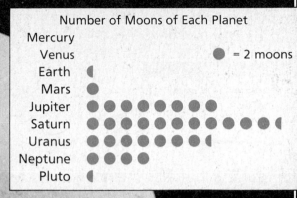

Number of Moons of Each Planet

● = 2 moons

Planet	Average Distance from the Sun (miles)
Mercury	3.6×10^7
Venus	6.727×10^7
Earth	9.3×10^7
Mars	1.4171×10^8
Jupiter	4.8388×10^8
Saturn	8.8714×10^8
Uranus	1.78398×10^9
Neptune	2.79546×10^9
Pluto	3.67527×10^9

In 1973 NASA launched its first space station *Skylab.* Some of the experiments conducted aboard *Skylab* were designed by high school students.

NASA is working on a design for a permanent space station to be called *Freedom.* This station will contain about two to three times as much usable space as *Skylab* contained. *Freedom* will be 153 m long and will house four to eight astronauts. The station is expected to orbit about 483 km above Earth.

2-1 Simplifying Multiplication Expressions

Objective: To simplify multiplication expressions using the product of powers rule and the properties of multiplication.

Terms to Know

- *product of powers rule*
- *simplify*

EXPLORATION

1. Using a calculator, replace each __?__ with the number that makes the statement true.

a.	b.	c.
$4^3 = \underline{?}$	$5^2 = \underline{?}$	$2^3 = \underline{?}$
$4^2 = \underline{?}$	$5^4 = \underline{?}$	$2^6 = \underline{?}$
$4^3 \cdot 4^2 = \underline{?}$	$5^2 \cdot 5^4 = \underline{?}$	$2^3 \cdot 2^6 = \underline{?}$
$4^5 = \underline{?}$	$5^6 = \underline{?}$	$2^9 = \underline{?}$
$4^3 \cdot 4^2 = 4^?$	$5^2 \cdot 5^4 = 5^?$	$2^3 \cdot 2^6 = 2^?$

2. Use your results from Step 1. Complete this statement: To multiply powers that have the same base, you __?__ the exponents.
3. Use your result from Step 2. Find the value of n that makes each statement true. Use a calculator to check your answers.
   a. $6^3 \cdot 6^7 = 6^n$ **b.** $3^5 \cdot 3^9 = 3^n$ **c.** $7^5 \cdot 7^4 = 7^n$

You can use exponents to count factors when finding a product of powers.

$$7^5 \cdot 7^4 = \underbrace{(7 \cdot 7 \cdot 7 \cdot 7 \cdot 7)}_{\text{5 factors}} \cdot \underbrace{(7 \cdot 7 \cdot 7 \cdot 7)}_{\text{4 factors}} = \underset{\text{9 factors}}{7^9}$$

Generalization: *Product of powers rule*

To multiply powers having the same base, add the exponents.

$$a^m \cdot a^n = a^{m+n}$$

You can use the product of powers rule to *simplify* an expression. You **simplify** an expression by performing as many of the indicated operations as possible.

Example 1 **Simplify:** **a.** $x^3 \cdot x^5$ **b.** $w^6 \cdot w$ **c.** $c^5 \cdot c^2 \cdot c^4$

Solution

a. $x^3 \cdot x^5 = x^{3+5}$
$= x^8$

b. $w^6 \cdot w = w^6 \cdot w^1$
$= w^{6+1}$
$= w^7$

c. $c^5 \cdot c^2 \cdot c^4 = c^{5+2+4}$
$= c^{11}$

Check Your Understanding

In Example 1(b), why is w rewritten as w^1?

To simplify some expressions, you might need to use the product of powers rule together with the commutative and associative properties.

Example 2 | **Simplify:** | **a.** $6a^2 \cdot 4a^3$ | **b.** $(5n)(7n^2)$
Solution | **a.** $6a^2 \cdot 4a^3 = (6 \cdot 4)(a^2 \cdot a^3)$ | | **b.** $(5n)(7n^2) = (5 \cdot 7)(n \cdot n^2)$
| $= (24)(a^{2+3})$ | | $= (5 \cdot 7)(n^1 \cdot n^2)$
| $= 24a^5$ | | $= (35)(n^{1+2})$
| | | $= 35n^3$

You can also simplify multiplication expressions that involve more than one variable.

Example 3 | **Simplify:** | **a.** $(7x)(2y)$ | **b.** $5y \cdot 3x \cdot 2y$
Solution | **a.** $(7x)(2y) = (7 \cdot 2)(x \cdot y)$ | | **b.** $5y \cdot 3x \cdot 2y = (5 \cdot 3 \cdot 2)(y \cdot x \cdot y)$
| $= 14xy$ | | $= (5 \cdot 3 \cdot 2)(x \cdot y \cdot y)$
| | | $= 30xy^2$

Guided Practice

COMMUNICATION «*Reading*

Refer to the text on pages 48–49.

«**1.** Describe the product of powers rule in words.

«**2.** What does it mean to *simplify* an expression?

Choose the letter of the correct answer.

3. $q^6 \cdot q^2$ **a.** q^{12} **b.** q^8 **c.** q^4 **d.** q^3

4. $t \cdot t^4$ **a.** $2t^4$ **b.** $2t^5$ **c.** t^4 **d.** t^5

5. $(5c^4)(3c^6)$ **a.** $8c^{10}$ **b.** $15c^{10}$ **c.** $8c^{24}$ **d.** $15c^{24}$

6. $9w \cdot 4z$ **a.** $36wz$ **b.** $13wz$ **c.** $94wz$ **d.** $36w^2z^2$

Simplify.

7. $y^7 \cdot y^3$ **8.** $x^5 \cdot x$ **9.** $w^1 \cdot w^5 \cdot w^1$

10. $d \cdot d^3 \cdot d^6$ **11.** $3k^4 \cdot 4k^2$ **12.** $(2c^3)(10c^4)$

13. $(6z^3)(6z)$ **14.** $5p \cdot 9p^7$ **15.** $(5p)(4q)$

16. $3a \cdot 9b$ **17.** $4w \cdot 2z \cdot 6w$ **18.** $(7a)(5b)(4b)$

Exercises

Simplify.

1. $c^6 \cdot c^4$
2. $b^8 \cdot b^3$
3. $n^2 \cdot n$

4. $y \cdot y^8$
5. $x^3 \cdot x^7 \cdot x^3$
6. $a^5 \cdot a^8 \cdot a^6$

7. $4d^2 \cdot 3d^3$
8. $5k^6 \cdot 3k^2$
9. $(6b^2)(5b)$

10. $(8c)(4c^3)$
11. $(2c)(14d)$
12. $(12q)(4r)$

13. $(7w)(4w)(2y)$
14. $5b \cdot 3a \cdot 2b$
15. $(5w)(6x)(8x)$

16. Find the product when $2a$ is multiplied by $3a^5$.

17. Find the product of a number z squared and the same number z cubed.

Replace each ___?___ with the expression that makes the statement true.

18. $x^5 \cdot \underline{\ ?\ } = x^{13}$
19. $c^7 \cdot \underline{\ ?\ } = c^{11}$
20. $\underline{\ ?\ } \cdot n^{12} = n^{19}$

21. $(5c)(\underline{\ ?\ }) = 15c^2$
22. $8p \cdot \underline{\ ?\ } = 48p^2$
23. $(\underline{\ ?\ })(4a) = 32ab^2$

Find the value of n that makes each statement true.

24. $z^3 \cdot z^n = z^{12}$
25. $x^n \cdot x^2 = x^{10}$
26. $d^2 \cdot d^4 \cdot d^n = d^{18}$

27. $y^5 \cdot y^n \cdot y^3 = y^{10}$
28. $q^n \cdot q = q^4$
29. $t \cdot t^n = t^7$

30. $a^n \cdot a^n = a^{10}$
31. $k^n \cdot k^n \cdot k^n = k^6$
32. $y^n \cdot y \cdot y^3 \cdot y^n = y^{18}$

LOGICAL REASONING

Exercises 33–36 refer to these statements:
$$\textbf{I.}\ \ 2^m \cdot 2^0 = 2^{m+0} = 2^m \qquad \textbf{II.}\ \ 2^m \cdot 1 = 2^m$$

33. Which property or rule is illustrated by Statement I?

34. Which property or rule is illustrated by Statement II?

35. Explain why this statement must be true: $2^m \cdot 2^0 = 2^m \cdot 1$

36. What number do you think is represented by 2^0? Give a convincing argument to support your answer.

SPIRAL REVIEW

37. Tell what this paragraph is about: Amy Gold bought six tickets to a basketball game and five tickets to a hockey game. The basketball tickets cost $67.50 and the hockey tickets cost $53.75.   *(Lesson 1-4)*

38. Find the answer: $4^3 - (12 + 9) \div 7$ *(Lesson 1-10)*

39. Simplify: $(5c^2)(6c)$ *(Lesson 2-1)*

40. DATA, *pages 46–47* How many moons does Jupiter have? *(Toolbox Skill 23)*

Modeling Expressions Using Tiles

Objective: To use algebra tiles to model expressions.

Materials

■ algebra tiles

You can use algebra tiles to visualize expressions. For example, let the tile ▬ represent the variable n and the tile ▢ represent the number 1. You can represent the expression $3n + 2$ as shown in the diagram below. Group the tiles by circling them.

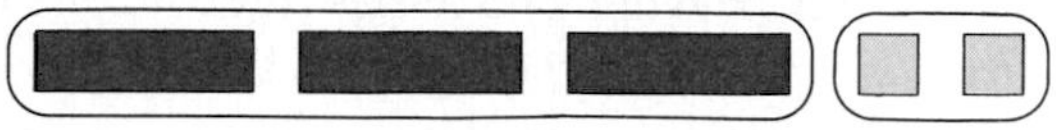

Activity I *Expressions Involving Similar Tiles*

1 Write the expression represented by each diagram.

a.

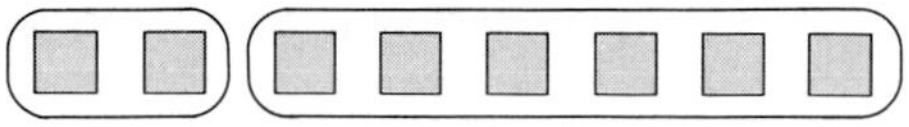

b.

2 Show how to use tiles to represent each expression.

a. $4 + 2$ b. $5 + 3$ c. $1 + 6$

d. $3n + 3n$ e. $4n + n$ f. $2n + 7n$

3 Show two different ways to separate each set of tiles into two groups. Write the expression represented by each grouping.

a.

b.

Activity II *Expressions Involving Unlike Tiles*

1 Write the expression represented by each diagram.

a.

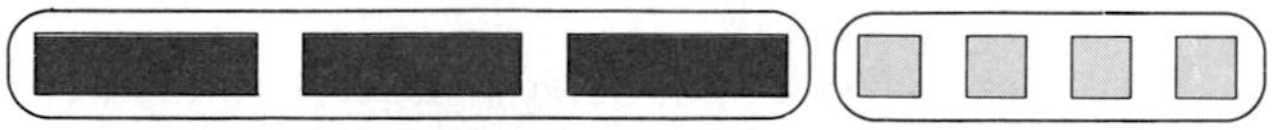

b.

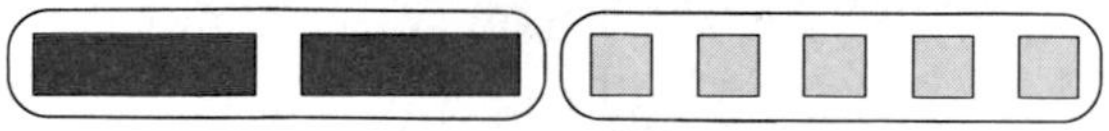

2 Show how to use tiles to represent each expression.

a. $n + 4$ b. $6 + n$ c. $5n + 2$

d. $4n + 1$ e. $3 + 2n$ f. $5 + 6n$

3 Show two different ways to separate this set of tiles into two groups. Write the expression represented by each grouping.

a.

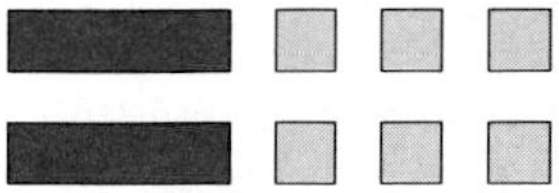

b.

Introduction to Algebra **51**

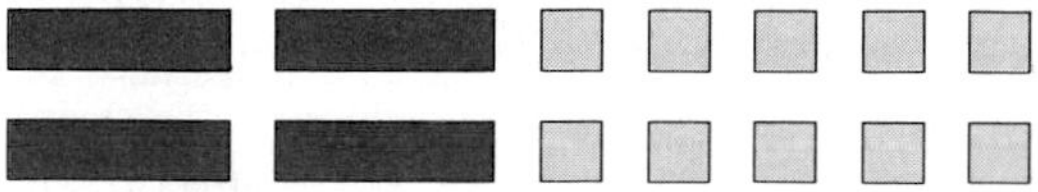

The Distributive Property

2-2

Objective: To use the distributive property to compute mentally and to simplify expressions.

A parking garage has three levels. On each level, there are 10 parking spaces for the disabled and 80 other parking spaces. To determine the total number of parking spaces in the garage, you can perform the calculations in two different ways.

$$3(80 + 10) = 3(90) = 270$$
$$3(80) + 3(10) = 240 + 30 = 270$$

The fact that you get the same answer either way is an application of the *distributive property*.

The **distributive property** allows you to multiply each term inside a set of parentheses by a factor outside the parentheses. You say that multiplication is *distributive* over addition and over subtraction.

The Distributive Property

In Arithmetic

$$3(80 + 10) = 3(80) + 3(10)$$
$$3(80 - 10) = 3(80) - 3(10)$$

In Algebra

$$a(b + c) = ab + ac$$
$$a(b - c) = ab - ac$$

The distributive property can help you to perform some calculations mentally.

Example 1

Use the distributive property to find each answer mentally.

a. $8 \cdot 36 - 8 \cdot 16$ **b.** $7(108)$

Solution

a. $8 \cdot 36 - 8 \cdot 16 = 8(36 - 16)$
$$= 8(20)$$
$$= 160$$

b. $7(108) = 7(100 + 8)$
$$= 7(100) + 7(8)$$
$$= 700 + 56$$
$$= 756$$

✒️ **Check Your Understanding**

1. In Example 1(b), why is 108 written as $100 + 8$ and not as $110 - 2$?
2. Suppose Example 1(b) had been $7(98)$. Describe a method to find this product mentally.

You can use the distributive property to simplify variable expressions.

Example 2

Solution

Simplify $3(n + 2)$.

Let 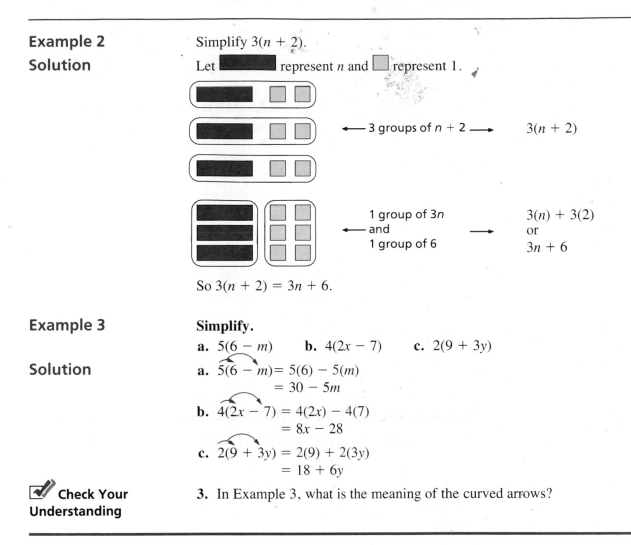 represent n and ☐ represent 1.

⟵ 3 groups of $n + 2$ ⟶ $3(n + 2)$

1 group of $3n$
and   ⟶ $3(n) + 3(2)$
1 group of 6   or
          $3n + 6$

So $3(n + 2) = 3n + 6$.

Example 3

Simplify.

a. $5(6 - m)$ **b.** $4(2x - 7)$ **c.** $2(9 + 3y)$

Solution

a. $5(6 - m) = 5(6) - 5(m)$
     $= 30 - 5m$

b. $4(2x - 7) = 4(2x) - 4(7)$
     $= 8x - 28$

c. $2(9 + 3y) = 2(9) + 2(3y)$
     $= 18 + 6y$

Check Your Understanding

3. In Example 3, what is the meaning of the curved arrows?

Guided Practice

COMMUNICATION « *Reading*

« **1.** Replace each __?__ with the correct word or phrase.
   The __?__ allows you to multiply each term inside a set of parentheses
   by a __?__ outside the parentheses. The distributive property involves
   two operations. The operations are multiplication and either __?__ or
   subtraction.

« **2.** Explain the meaning of the word *distributed* in the following sentence.
   The teacher distributed the tests to the students.

« **3.** Let 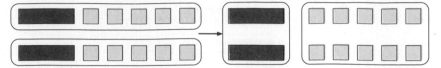 represent n and ☐ represent 1. Write the statement that is represented by this diagram.

« **4.** Draw a diagram similar to the one in Exercise 3 to represent this statement: $3(2n + 1) = 6n + 3$.

Replace each __?__ with the number that makes the statement true.

5. $8(43) + 8(7) = 8(\underline{})$

6. $7(199) = 7(\underline{}) - 7(1)$

7. $4(n + 6) = (\underline{})n + 4(6)$

8. $2(3c - 12) = 2(3c) - \underline{}(12)$

Use the distributive property to find each answer mentally.

9. $6(62) - 6(12)$

10. $9 \cdot 13 + 9 \cdot 7$

11. $4(307)$

Simplify.

12. $6(x + 4)$

13. $7(3 - n)$

14. $2(5 - 12a)$

Exercises

Use the distributive property to find each answer mentally.

1. $7(68) + 7(12)$

2. $8(31) + 8(19)$

3. $15(8) - 15(6)$

4. $6(13) - 6(7)$

5. $4(109)$

6. $7(206)$

7. $9(197)$

8. $5(296)$

9. $8(398)$

Simplify.

10. $5(n + 12)$

11. $9(10 + a)$

12. $6(8 - y)$

13. $5(x - 9)$

14. $7(7 + 2c)$

15. $8(6m + 9)$

16. $4(4w - 6)$

17. $3(9 - 4a)$

18. $5(3c + 7)$

19. Peter bought two pairs of pants for \$29.99 each and two shirts for \$19.99 each. How much more did he spend on pants than on shirts?

20. Last week Lauren ran 5 mi three times. This week she ran 5 mi four times. Did she run more than 40 mi during the two weeks? Explain.

Replace each __?__ with the number that makes the statement true.

21. $4(2n - \underline{}) = 8n - 16$

22. $2(5t + \underline{}) = 10t + 2$

23. $\underline{}(2n + 1) = 6n + 3$

24. $\underline{}(6z - 5) = 12z - 10$

COMPUTER APPLICATION

Like the distributive property, the spreadsheet below finds an amount in two ways. It calculates weekly earnings both by finding the product of the total hours and the hourly rate and by finding the sum of the daily earnings.

	A	B	C	
1		Hours	Earnings	
2	Monday	6	$54	
3	Tuesday	5	$45	
4	Wednesday	8	$72	
5	Thursday	7	$63	
6	Friday	5	$45	
7				
8	Total Hours	31		
9	Hourly Rate	$9		
10				
11	Weekly Earnings	$279	$279	

Use an electronic spreadsheet or draw a spreadsheet similar to the one shown to find each person's weekly earnings in two ways.

25. Leotie earns $8/h. She worked 7 h on Monday, 4 h on Tuesday, 8 h on Wednesday, 5 h on Thursday, and 6 h on Friday.

26. Ellen earns $7/h. She worked 6 h on Monday, 3 h on Tuesday, 6 h on Wednesday, 8 h on Thursday, and 7 h on Friday.

SPIRAL REVIEW

27. Evaluate ab when $a = 7$ and $b = 9$.   *(Lesson 1-1)*

28. Find the sum: $1.63 + 0.96$ *(Toolbox Skill 11)*

29. Simplify: $5(3x + 5)$ *(Lesson 2-2)*

30. Is 177 divisible by 2? by 3? by 4? by 5? by 9?   *(Toolbox Skill 16)*

31. Evaluate $4n$ when $n = 2.9$.   *(Lesson 1-3)*

32. Estimate the quotient: $46,287 \div 59$ *(Toolbox Skill 3)*

Combining Like Terms

Objective: To simplify expressions by combining like terms.

Terms to Know

- *like terms*
- *unlike terms*
- *combining like terms*

CONNECTION

Barry simplified the process of locating items in his music collection by combining like items. He put CDs in one box, cassettes in another box, and record albums in a third box.

In mathematics, sometimes you can simplify an expression by *combining like terms*. The expression $2n + 5m + 4n$ contains three *terms*: $2n$, $5m$, and $4n$. The terms $2n$ and $4n$ have identical variable parts, so they are called **like terms.** The terms $2n$ and $5m$ have different variable parts, so they are called **unlike terms.**

The process of adding or subtracting like terms is often called **combining like terms.** Unlike terms cannot be combined.

Example 1

Simplify: **a.** $2n + 4n$ **b.** $4n - n$ **c.** $3n + 4 + 2n$

Solution

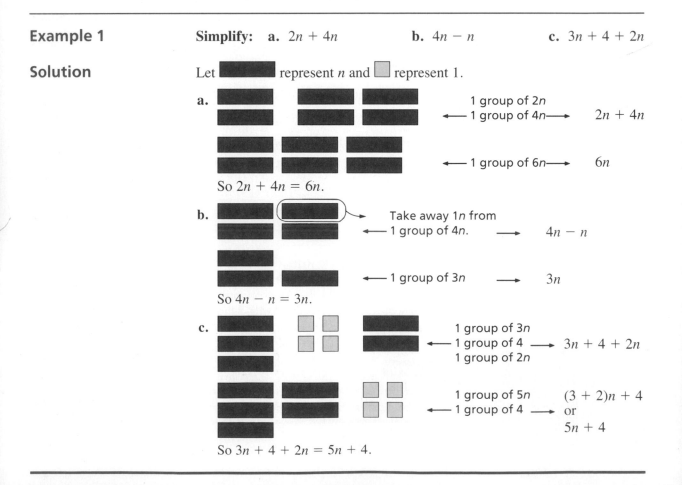

Let ▬ represent n and ☐ represent 1.

a.

1 group of 2n
⟵ 1 group of 4n ⟶ $2n + 4n$

⟵ 1 group of 6n ⟶ $6n$

So $2n + 4n = 6n$.

b.

Take away 1n from
⟵ 1 group of 4n.   ⟶ $4n - n$

⟵ 1 group of 3n   ⟶ $3n$

So $4n - n = 3n$.

c.

1 group of 3n
⟵ 1 group of 4   ⟶ $3n + 4 + 2n$
1 group of 2n

1 group of 5n $(3 + 2)n + 4$
⟵ 1 group of 4   ⟶   or
$5n + 4$

So $3n + 4 + 2n = 5n + 4$.

When you combine like terms, you are applying the distributive property.

Example 2	**Simplify:** **a.** $9x + 7x$	**b.** $11c + c - 8b$

Solution

a. $9x + 7x = (9 + 7)x$
$\qquad\qquad = 16x$

b. $11c + c - 8b = 11c + 1c - 8b$ ⟵ Recall that $c = 1 \cdot c$.
$\qquad\qquad\qquad = (11 + 1)c - 8b$
$\qquad\qquad\qquad = 12c - 8b$

✔️ **Check Your Understanding**

1. In Example 2(b), why do you combine only $11c$ and c?
2. What property is used in Example 2(b) to write c as $1c$?

Guided Practice

«**1.** COMMUNICATION «*Reading* Replace each _?_ with the correct word.

In an expression, terms that can be combined are called _?_ terms. Terms that cannot be combined are called _?_ terms.

Tell whether the following are *like terms* or *unlike terms*.

  2. $4a; 6a$ **3.** $5d; 8b$ **4.** $6xy; 6x$ **5.** $2w; 2$ **6.** $54r, r$

Simplify.

  7. $7a + 4a$ **8.** $c + 3c$ **9.** $12x - 4x$
  10. $10y - y$ **11.** $3a + 9 + 11a$ **12.** $5b + 8b - 7$
  13. $c + 4n + 6n$ **14.** $8d - d - 5m$ **15.** $6z + z + 9z$

Exercises

Simplify.

  1. $6x + 8x$ **2.** $7a + 10a$ **3.** $5m + m$
  4. $c + 21c$ **5.** $12w - w$ **6.** $14p - 5p$
  7. $8x - 3x$ **8.** $9y - y$ **9.** $2z + 7 + 6z$
  10. $6n + 9n + 4$ **11.** $6k + 3k - 6$ **12.** $5x + 2x - 5$
  13. $4w + 4r - 3r$ **14.** $8y - 7y + 4$ **15.** $2a + 4b + 5b$
  16. $3x + 7x + 9y$ **17.** $5n + 12n + n$ **18.** $m + 4m + 6m$

Simplify, if possible. If not possible, write *cannot be simplified*.

19. $4a + 5b + 6a + 8b$

20. $b + 8c + 2x + 5z$

21. $7c + 6w - 2w + 4c$

22. $4y + 3y + 12z - 6z$

23. $a + 9m - n + 4x$

24. $7n - 2n + 4 + 3n$

25. $d - 4n + 3w - 7x$

26. $2h + 8k + 10h + 9k$

27. $5a + 6b + 3x + 7b$

28. $4a + 8b + 9c + z$

29. $12c - 3c + 8 + 5c$

30. $2x + 7w - 5w - 6z$

31. WRITING ABOUT MATHEMATICS   In Lessons 2-1, 2-2, and 2-3 you learned four methods for simplifying expressions. Write a brief report that summarizes these methods. Be sure to include an example for each method. (*Hint:* Refer to the objective of each lesson.)

Simplify.

32. $2(x + 7) + 8x$

33. $7y + 9(4 + 9y)$

34. $6(2a + 1) + 3(2 + 3a)$

35. $3(6 + b) + 5(b + 1)$

SPIRAL REVIEW

36. Amy drove 255 mi in four days and used 10 gal of gasoline. Identify any facts that are not needed to find miles per gallon.   (*Lesson 1-4*)

37. Evaluate x^4 when $x = 5$.   (*Lesson 1-9*)

38. Simplify: $4c + 7c - 5d$ (*Lesson 2-3*)

39. Find the sum: $\frac{3}{4} + \frac{7}{8}$ (*Toolbox Skill 17*)

40. Find the product: $(0.37)(1000)$ (*Toolbox Skill 15*)

Self-Test 1

Simplify.

1. $x^4 \cdot x^3$

2. $c \cdot c^7$

3. $a^3 \cdot a^8 \cdot a^5$ 2-1

4. $3w^2 \cdot 9w^5$

5. $(6x)(15z)$

6. $(4a)(7b)(3a)$

Use the distributive property to find each answer mentally.

7. $6 \cdot 12 + 6 \cdot 8$

8. $5(107)$

9. $8(196)$ 2-2

Simplify.

10. $2(x + 9)$

11. $8(7 - 5x)$

12. $2(11 + 7c)$

13. $3n + 4n$

14. $14x - x$

15. $7x - 4x - 5y$ 2-3

16. $4a + 2b + 9b$

17. $6z + 1 + 2z$

18. $w + 3w + 8w$

2-4

Using a Four-Step Plan

Objective: To use a four-step plan to solve problems.

In Lesson 1-4 you learned how to read a problem for understanding. Understanding a problem is the first step to solving the problem. Below are all four steps you should carry out when solving a problem.

Four-Step Plan

UNDERSTAND	**Read and understand the problem.** Know what is given and what you have to find.
PLAN	**Make a plan.** Choose a problem solving strategy.
WORK	**Carry out the plan.** Use the strategy and do any necessary calculations.
ANSWER	**Check any calculations and answer the problem.** Interpret the answer, if necessary.

One strategy you might decide to use in the second step—making a plan—is *choosing the correct operation*. Sometimes it takes more than one operation to solve a problem. In that case, you must decide not only which operations are needed, but also in which order to use them.

Problem

Two weeks ago Michael worked 35 h. Last week he worked 31 h. Michael earns an hourly wage of $7.15. How much did Michael earn during these two weeks?

Solution

UNDERSTAND The problem is about Michael's earnings.
Facts: worked 35 h
worked 31 h
earns $7.15 per hour
Find: the total amount Michael earned during two weeks

PLAN First add to determine the number of hours Michael worked. Then multiply the total number of hours by the hourly wage.

WORK $35 + 31 = 66$
$7.15(66) = 471.9$

ANSWER Check the calculations: $7.15(35 + 31) = 7.15(66) = 471.9$
Michael earned $471.90 during the two weeks.

Look Back An alternative method for solving the problem is to first find the amount earned each week and then add. Solve the problem again using this alternative method.

Guided Practice

Use this paragraph for Exercises 1–3.

In ice hockey a team earns two points for a win, one point for a tie, and no points for a loss. The Hornets won 36, tied 12, and lost 16 games.

«**1.** How many points does a team earn for a win?

«**2.** Identify any facts that are not needed to find the number of games the Hornets played.

«**3.** Describe how you would find the number of points the Hornets earned.

Choose the letter of the key sequence needed to solve each problem.

4. Keisha bought a dress for $36.98 and two blouses for $20 each. What was the total cost of her purchases?

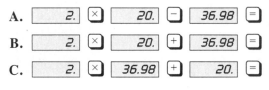

A. ⬚ 2. ⊠ ⬚ 20. ⊖ ⬚ 36.98 🟰

B. ⬚ 2. ⊠ ⬚ 20. ⊕ ⬚ 36.98 🟰

C. ⬚ 2. ⊠ ⬚ 36.98 ⊕ ⬚ 20. 🟰

5. Kay bought a sweater for $36.98. She gave the clerk two $20 bills. How much change did she receive?

Solve.

6. Larry bought six pairs of socks for $3.99 per pair and three shirts for $19.95 each. What was the total cost of the socks?

7. ABC Company bought 145 new office chairs. The office manager put 26 new chairs in each of four small conference rooms. How many new chairs remained for the manager to put in the large conference room?

Problem Solving Situations

Solve.

1. Ron has a 30-ft sailboat. He sailed 7 mi from Kingsport to Elk Island. After sailing back to Kingsport, Ron then sailed 12 mi to Cape West. How many miles did Ron sail in all?

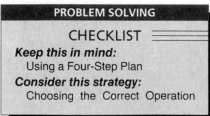

PROBLEM SOLVING
CHECKLIST
Keep this in mind:
Using a Four-Step Plan
Consider this strategy:
Choosing the Correct Operation

2. Gail's bus trip from home to work is 3 mi and usually takes 20 min. Bus tokens cost $.75. Gail bought twelve tokens. How much did she spend?

3. Jamie bought two loaves of bread for $1.29 each and three heads of lettuce for $.95 each. What was the total cost?

4. King High School has budgeted $4350 for new desks. Each desk costs $115 including tax. How many desks can the school buy?

5. The student council bought 400 sweatshirts, 650 T-shirts, and 1100 notebooks to sell during the school year. At the end of the year the council had 96 sweatshirts, 139 T-shirts, and 227 notebooks left. How many items did the student council sell during the school year?

6. Jill Panov bought a television that cost $299 and a VCR that cost $349.98. The sales tax was $38.94. She made a down payment of $140 and agreed to pay the rest in eight equal payments. What was the amount of each payment?

DATA ANALYSIS

Use the diagram at the right.

7. How many more one-room schools are there in Pennsylvania than in South Dakota?

8. About how many times the number of one-room schools in Montana are there in Nebraska?

9. The total number of one-room schools in the United States is 1279. How many one-room schools are not in one of the four states shown?

10. There are 84 one-room schools in Ohio. Is the number of one-room schools in Nebraska *greater than* or *less than* the total number in Ohio, South Dakota, and Montana? by how many?

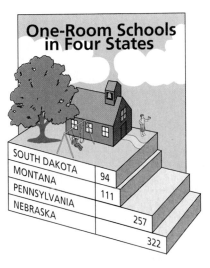

One-Room Schools in Four States

SOUTH DAKOTA	
MONTANA	94
PENNSYLVANIA	111
NEBRASKA	257
	322

WRITING WORD PROBLEMS

Using the information given, write a word problem for each exercise. Then solve the problem.

11. $600 - 180 = 420; \ 420 \div 12 = 35$

12. $(10.75)42 = 451.50; \ (451.50)52 = 23{,}478$

SPIRAL REVIEW

13. Simplify: $4(3a - 8)$ *(Lesson 2-2)*

14. Carlos bought six greeting cards and six stamps. Each card cost $1.25 and each stamp cost $.25. How much did he spend?   *(Lesson 2-4)*

15. Find the sum mentally: $73 + 0 + 27 + 19$ *(Lesson 1-6)*

16. DATA, *pages 2–3* How many summer concerts does the Houston symphony perform each year?   *(Toolbox Skill 24)*

2-5 Patterns

Objective: To recognize and to extend patterns.

Many patterns occur naturally in the real world. In 1202, Leonardo Fibonacci wrote about a pattern from nature that is called the *Fibonacci sequence*. These are the first ten numbers of that pattern.

$$1, 1, 2, 3, 5, 8, 13, 21, 34, 55$$

Beginning with the third number, 2, each number in the pattern is the sum of the two numbers immediately preceding it. You can find this pattern in the spirals of the seeds on most sunflowers and in the spirals of the scales on many pineapples.

Recognizing a pattern is a useful way to solve some problems in mathematics. Many mathematical patterns involve addition, subtraction, multiplication, and division.

Example 1

Find the next three numbers in each pattern.

a. 3, 6, 12, 24, _?_, _?_, _?_   **b.** 5, 7, 10, 14, _?_, _?_, _?_

Solution

a. Each number is 2 times the preceding number. The pattern is *multiply by 2*.

3　　6　　12　　24　　48　　96　　192
　＼×2↗＼×2↗＼×2↗＼×2↗＼×2↗＼×2↗

The next three numbers are 48, 96, and 192.

b. The second number is 2 more than the first number. The third number is 3 more than the second number. The fourth number is 4 more than the third number. The pattern is *add 2, add 3, add 4, and so on*.

5　　7　　10　　14　　19　　25　　32
　＼+2↗＼+3↗＼+4↗＼+5↗＼+6↗＼+7↗

The next three numbers are 19, 25, and 32.

✓ Check Your Understanding

1. Suppose you extend the pattern in Example 1(a). What would be the next two numbers after 192?

2. Suppose you extend the pattern in Example 1(b). What would you add to 32 to get the next number? What would be the next number?

Example 2

Find the next three expressions in each pattern.

a. $x, x + 3, x + 6, \underline{\ ?\ }, \underline{\ ?\ }, \underline{\ ?\ }$

b. $12a + 5, 10a + 5, 8a + 5, \underline{\ ?\ }, \underline{\ ?\ }, \underline{\ ?\ }$

Solution

a. Each expression is 3 more than the preceding expression. The pattern is *add 3*.

$$x \quad\quad x + 3 \quad\quad x + 6 \quad\quad x + 9 \quad\quad x + 12 \quad\quad x + 15$$
$$_{+3}\quad_{+3}\quad_{+3}\quad_{+3}\quad_{+3}$$

The next three expressions are $x + 9$, $x + 12$, and $x + 15$.

b. Each expression is $2a$ less than the preceding expression. The pattern is *subtract 2a*.

$$12a + 5 \quad\quad 10a + 5 \quad\quad 8a + 5 \quad\quad 6a + 5 \quad\quad 4a + 5 \quad\quad 2a + 5$$
$$_{-2a}\quad_{-2a}\quad_{-2a}\quad_{-2a}\quad_{-2a}$$

The next three expressions are $6a + 5$, $4a + 5$, and $2a + 5$.

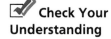 **Check Your Understanding**

3. Suppose the pattern in Example 2(b) continued. What would be the next expression after $2a + 5$?

You can use a calculator to work with patterns that involve a repeated operation. For instance, you can use a calculator to find the pattern 6, 9, 12, 15, 18. If your calculator has a *constant key*, $\boxed{K}$, you can use this key sequence.

$$\boxed{\ 3.\ } \ \boxed{+} \ \boxed{K} \ \boxed{\ 6.\ } \ \boxed{=} \ \boxed{=} \ \boxed{=} \ \boxed{=}$$

If your calculator does not have a constant key, you might be able to use the equals key.

$$\boxed{\ 6.\ } \ \boxed{+} \ \boxed{\ 3.\ } \ \boxed{=} \ \boxed{=} \ \boxed{=} \ \boxed{=}$$

To find out how the calculator you are using handles constants, consult the user's manual.

Guided Practice

COMMUNICATION « *Writing*

Write the first five numbers or expressions in each pattern.

«**1.** 8; *add 6* «**2.** 2; *multiply by 2* «**3.** 128; *subtract 15*

«**4.** 400; *divide by 2* «**5.** 8x; *multiply by 5* «**6.** $x - 2$; *add 4x*

Describe each pattern in words.

«**7.** 240, 120, 60, 30, 15 «**8.** 5, 8, 12, 17, 23

«**9.** $c, 3c, 9c, 27c$ «**10.** $a + 22, a + 18, a + 14, a + 10$

Find the next three numbers or expressions in each pattern.

11. 7, 11, 15, 19, _?_, _?_, _?_

12. 128, 64, 32, 16, _?_, _?_, _?_

13. 2, 2, 4, 12, _?_, _?_, _?_

14. $3x$, $6x$, $9x$, _?_, _?_, _?_

15. $2x + 11$, $2x + 9$, $2x + 7$, _?_, _?_, _?_

16. $7c + 1$, $6c + 1$, $5c + 1$, _?_, _?_, _?_

Exercises

Find the next three numbers or expressions in each pattern.

1. 12, 22, 32, 42, _?_, _?_, _?_

2. 88, 85, 82, 79, _?_, _?_, _?_

3. 1, 5, 25, 125, _?_, _?_, _?_

4. 729, 243, 81, 27, _?_, _?_, _?_

5. 6, 8, 11, 15, _?_, _?_, _?_

6. 18, 23, 29, 36, _?_, _?_, _?_

7. 23, 22, 20, 17, _?_, _?_, _?_

8. 1, 2, 6, 24, _?_, _?_, _?_

9. $7m$, $14m$, $28m$, _?_, _?_, _?_

10. $2y$, $8y$, $32y$, _?_, _?_, _?_

11. a, $a + 5$, $a + 10$, _?_, _?_, _?_

12. $n + 12$, $n + 15$, $n + 18$, _?_, _?_, _?_

13. $n + 3$, $2n + 3$, $3n + 3$, _?_, _?_, _?_

14. $13x - 2$, $11x - 2$, $9x - 2$, _?_, _?_, _?_

15. John needs $550 to buy a television. He had $150 in May, $200 in June, $250 in July, and $300 in August. If this pattern continues, in what month will he be able to buy the television?

16. Suppose that you told a secret to three friends. On Monday, each friend told the secret to three other people. On Tuesday, each of these people told the secret to three other people, and so on. How many people are told the secret on Saturday?

17. Find the next two numbers in the Fibonacci sequence on page 62.

18. Find the next two numbers in the pattern $\frac{1}{2}, \frac{1}{4}, \frac{1}{8}, \frac{1}{16}, \frac{1}{32}$.

Some number patterns are associated with geometric figures. For instance, the figure below represents the first four *triangular numbers*. This number pattern has been used since ancient times.

19. Draw the figures that represent the fifth and sixth triangular numbers.

20. Without drawing the figures, find the seventh and eighth triangular numbers.

21. Describe the pattern of the triangular numbers in words.

22. The figure at the right represents the first four *square numbers*. Use the pattern to make a list of the first ten square numbers.

GROUP ACTIVITY

Give two different ways to describe each pattern.

23. 1, 4, 9, 16, 25, 36, 49, . . .

24. 2, 4, 8, 16, 32, 64, 128, . . .

Find two different ways to continue each pattern.

25. 2, 4, 8, 10, $\underline{\ ?\ }$, $\underline{\ ?\ }$, $\underline{\ ?\ }$

26. 1, 3, 9, 11, $\underline{\ ?\ }$, $\underline{\ ?\ }$, $\underline{\ ?\ }$

27. Make up a number pattern, then trade patterns with a partner.
    a. Describe your partner's pattern in words.
    b. Find the next three numbers in the pattern.
    c. Determine if there is more than one way to continue the pattern.

SPIRAL REVIEW

28. Evaluate $2a + 3b$ when $a = 6$ and $b = 4$.   *(Lesson 1-10)*

29. Find the next three numbers in the pattern:
    1, 6, 36, 216, $\underline{\ ?\ }$, $\underline{\ ?\ }$, $\underline{\ ?\ }$ *(Lesson 2-5)*

30. Find the product: $(6.7)(3.2)$ *(Toolbox Skill 13)*

31. Find the quotient: $4\frac{3}{8} \div 1\frac{2}{5}$ *(Toolbox Skill 21)*

Challenge

Find the next three items in each pattern.

a. J, F, M, A, M, J, J, $\underline{\ ?\ }$, $\underline{\ ?\ }$, $\underline{\ ?\ }$

b. O, T, T, F, F, S, S, $\underline{\ ?\ }$, $\underline{\ ?\ }$, $\underline{\ ?\ }$

c. K, ⅄, Ʞ, $\underline{\ ?\ }$, $\underline{\ ?\ }$, $\underline{\ ?\ }$

d. ⥘,, ⥙, ⥚, ⥛, $\underline{\ ?\ }$, $\underline{\ ?\ }$, $\underline{\ ?\ }$

Functions

Objective: To complete function tables and to find function rules using tables.

EXPLORATION

1 Emily Scott bought a bouquet of flowers for $4. Can you tell how much change she received from the florist? Why or why not?

2 How much change would Emily receive if she gave the florist a $5 bill? a $10 bill? a $20 bill?

3 Complete this statement: The amount of change that Emily receives depends on __?__.

There are many times when one quantity depends on another. For instance, the amount of change that you receive when you make a purchase depends on the amount of money that you give the salesperson. In mathematics, it is said that the amount of change is a *function* of the amount of money you give the salesperson. A **function** is a relationship that pairs each number in a given set of numbers with *exactly one* number in a second set of numbers.

Often you can describe a function by a **function rule.** For instance, suppose your purchase costs $4. If you give the salesperson x dollars, the amount of change you receive is $(x - 4)$ dollars. You can use the variable expression $x - 4$ to create a function rule.

x	$x - 4$
5	1
10	6
20	?
50	?
100	?

You say: *x is paired with x − 4*
You use **arrow notation** to write: $x \rightarrow x - 4$

You can also use the variable expression to make a **function table,** like the one at the right.

Example 1

Solution

Complete the function table shown above.

The function rule is $x \rightarrow x - 4$.
Substitute 20, 50, and 100 for x in the expression $x - 4$.

$$x - 4 = 20 - 4 = 16$$
$$x - 4 = 50 - 4 = 46$$
$$x - 4 = 100 - 4 = 96$$

x	$x - 4$
5	1
10	6
20	16
50	46
100	96

✓ Check Your Understanding

1. In Example 1, what number does the function pair with 50?

2. What number would be paired with 500 using the function rule in Example 1?

You can sometimes use a function table to find a pattern that can help you write the function rule.

Example 2

Write a function rule for the function table shown at the right.

x	?
1	8
2	16
5	40
6	48
10	80

Solution

Notice the pattern: Each number in the second column is eight times the number in the first column.

$8(1) = 8 \qquad 8(2) = 16 \qquad 8(5) = 40$
$\qquad 8(6) = 48 \qquad 8(10) = 80$

The function rule is $x \rightarrow 8x$.

Guided Practice

Is the first quantity a function of the second quantity? Write *Yes* or *No*.

1. the cost of a long-distance phone call; the time of day you place the call

2. the cost of a long-distance phone call; the color of your phone

3. the cost of mailing a package; the temperature

4. the cost of mailing a package; the weight of the package

COMMUNICATION «*Writing*

Write each function rule using arrow notation.

«**5.** x is paired with $2x + 15$

«**6.** x is paired with $3x - 1$

Write each function rule in words.

«**7.** $x \rightarrow x + 7$ «**8.** $x \rightarrow 7x$ «**9.** $x \rightarrow \dfrac{x}{3}$ «**10.** $x \rightarrow 6x - 9$

Choose the letter of the correct function rule for each function table.

A. $x \rightarrow x + 2$ **B.** $x \rightarrow 4x - 4$ **C.** $x \rightarrow 4x$
D. $x \rightarrow x + 4$ **E.** $x \rightarrow x^2$ **F.** $x \rightarrow 2x$

11.

x	?
2	4
3	6
4	8
5	10
6	12

12.

x	?
2	4
4	6
6	8
8	10
10	12

13.

x	?
2	4
3	8
4	12
5	16
6	20

14. Complete the function table.

x	$x - 6$
10	4
12	6
14	?
16	?
18	?

15. Use the table to find the function rule.

x	?
7	1
14	2
21	3
28	4
35	5

Exercises

Complete each function table.

1.

x	$10x$
1	10
2	20
4	?
6	?
8	?

2.

x	$\frac{x}{6}$
6	1
12	2
18	?
24	?
30	?

3.

x	$2x + 9$
2	13
4	17
6	?
8	?
10	?

Use each function table to find the function rule.

4.

x	?
5	8
6	9
7	10
8	11
9	12

5.

x	?
1	5
2	10
3	15
4	20
5	25

6.

x	?
8	1
10	3
12	5
14	7
16	9

7. A person's weekly pay often depends on the number of hours the person works that week. Daniel earns $8.25 per hour. Use the function rule $t \rightarrow 8.25t$ to complete the function table at the right. Find Daniel's weekly pay for weeks that he works 7, 9, 12, 13, and 16 hours.

t	$8.25t$
7	?
9	?
12	?
13	?
16	?

8. Suppose Daniel has received a raise and now earns $9.75 per hour. Make a function table to find his weekly pay for weeks that he works 5, 8, 12, 15, and 20 hours.

The number of times a cricket chirps per minute is thought to be a function of the air temperature. If c is the air temperature in degrees Celsius (°C), then the function rule for the number of chirps per minute is $c \rightarrow 7c - 30$.

9. The air temperature is 20°C. How many times per minute does a cricket chirp?

10. The air temperature is 5°C. How many times does a cricket chirp in 1 h?

11. How many more times per minute does a cricket chirp at 28°C than at 25°C?

12. Make a function table for a cricket's chirps per minute when the air temperature is 8°C, 10°C, 12°C, 14°C, 16°C, and 18°C.

PROBLEM SOLVING/APPLICATION

The Center Town Library charges $.15 for each day a book is overdue.

13. What does it mean for a library book to be *overdue*?

14. Write the function rule that describes the relationship between the number of days a book is overdue and the amount charged by the library. Represent the number of days by d and write the function rule using arrow notation.

15. Use the function rule from Exercise 14 to find the charge for a book that is three days overdue.

16. A book that was due on Monday is returned on Friday. Use the function rule from Exercise 14 to find the overdue charge.

17. RESEARCH Find out the amount that your school library or public library charges for overdue books. If possible, write a function rule to represent the overdue charge.

SPIRAL REVIEW

18. Complete the function table at the right. *(Lesson 2-6)*

19. Find the quotient: $146.94 \div 6.2$ *(Toolbox Skill 14)*

20. Find the answer: 13^2 *(Lesson 1-9)*

21. Simplify: $8x - 3x + 5y$ *(Lesson 2-3)*

x	$6x - 5$
1	1
3	13
5	?
7	?
9	?

Using Spreadsheet Software

Objective: To use a spreadsheet to analyze patterns in function tables.

Spreadsheet software is ideal for making function tables and for analyzing patterns among functions. The diagram below shows a portion of a spreadsheet displaying function tables for the function rules $x \rightarrow 2x + 5$ and $x \rightarrow 3x + 5$.

	A	B	C
1	x	2x + 5	3x + 5
2	1	7	8
3	2	9	11
4	3	11	14
5	4	13	17
6	5	15	20

Exercises

Use computer software to create a spreadsheet similar to the one above for the function rules $x \rightarrow 2x + 5$ and $x \rightarrow 3x + 5$. Extend your spreadsheet to include values for x from 1 to 10.

1. Using computer notation, write the formula in cell B4.

2. Using computer notation, write the formula in cell C10.

3. Describe in words the pattern of the numbers in cells B2 through B11.

4. What is the relationship between the pattern of numbers in cells B2 through B11 and the expression in cell B1?

5. Describe in words the pattern of the numbers in cells C2 through C11.

6. What is the relationship between the pattern of numbers in cells C2 through C11 and the expression in cell C1?

7. Use your results from Exercises 3–6. Predict what will be the pattern in a function table for the function rule $x \rightarrow 4x + 5$. Create a function table in your spreadsheet to verify your prediction.

Use computer software to create function tables on a spreadsheet for the function rules $x \rightarrow 5x + 1$ and $x \rightarrow 5x + 2$. Use values for x from 1 to 10.

8. Describe the patterns in the function tables for the two functions. How are the two function tables alike? How are they different?

9. Predict how the function table for the function rule $x \rightarrow 5x + 3$ will compare with the given functions. Create a function table in the spreadsheet to verify your prediction.

2-7

Checking the Answer

Objective: To check the answer to a problem.

When you find an answer to a problem, it is important to check that your answer is correct, or at least reasonable. There are several ways to do this.

- Check your calculations. Compare your work to the wording of the problem to make sure you performed the operations correctly.
- Solve the problem again using an alternative method. Compare the new answer to your original answer to see if they match.
- Solve the problem again using estimation. Compare the estimate to your answer to see if your answer is reasonable.

Problem

Kelly earns $9.75 per hour. Last week he worked 8 h on Monday, 7 h on Tuesday, 5 h on Friday, and 9 h each on Wednesday and Thursday. Kelly said he earned $282.75 for the week. Is this correct? Explain.

Solution

You can use estimation to check Kelly's answer. Round the hourly wage to the leading digit. Then multiply by the total number of hours.

Hourly wage: $9.75 → $10

Number of hours: $8 + 7 + 5 + 9 + 9 = 38$

Multiply:
$$\begin{array}{r} 38 \\ \times 10 \\ \hline 380 \end{array} \rightarrow \text{about } \$380$$

Because $282.75 is not close to $380, the answer is not correct.

The hours worked on Thursday were not included in the total number of hours for the week.

Look Back Check the answer by using an alternate method to solve the problem. First multiply the number of hours for each day by $9.75. Then add these amounts to find the total pay for the week.

Guided Practice

COMMUNICATION «*Discussion*

«**1.** Why is it important to check your answer?

«**2.** List two ways of checking an answer.

«**3.** What types of errors might someone make when solving a problem?

«**4.** Why should you answer a problem with a complete sentence?

5. Bob bought two pairs of jeans for $24.90 each and one sweatshirt for $32.95. Bob said that he spent $115.70. This is not correct. What error did Bob make?
   a. He multiplied $32.95 by 2.
   b. He forgot to multiply $24.90 by 2.
   c. He added before multiplying.

6. Lisa had $243.50 in her checking account on June 4. During June she deposited $80 and wrote checks for $42.10 and $27.80. Lisa said her new balance at the end of June was $93.60. Is this correct? Explain.

Problem Solving Situations

1. Daniel spent $23.40 at the supermarket, $6.52 at the drug store, and rented three video tapes for $2.50 each. Daniel said he spent $37.42. Is this correct? Explain.

2. Pearl scored seven two-point baskets in a basketball game. In the next game, she scored five two-point baskets and two one-point baskets. Pearl said she scored 24 points. Is this correct? Explain.

Solve. Check your answer.

3. Chris exchanged two pairs of shoes that cost $36.85 each for one pair of boots that cost $65 and one pair of sneakers that cost $29. How much more money did Chris need to pay?

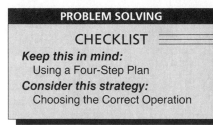

PROBLEM SOLVING

CHECKLIST

Keep this in mind:
Using a Four-Step Plan
Consider this strategy:
Choosing the Correct Operation

4. Ninety-one students volunteered to clean parks and visit nursing homes. The students were evenly assigned to five parks and two nursing homes. How many students were assigned to each place?

5. On March 2 Sue spent $11.60 for gasoline. She paid $13.25 for gasoline on March 10 and $87.43 for a tune-up on March 13. On March 21 Sue spent $9.75 for gasoline. She paid $10.50 for gasoline on March 29. How much did Sue spend for gasoline during March?

6. Paul earns $12.25 for each new cable TV customer he recruits, plus a bonus of $8.50 for each subscriber to the arts channel. Last week Paul recruited 30 new customers, 12 of whom subscribed to the arts channel. How much did Paul earn last week?

7. A math test contained two parts and was worth 100 points. Each item on Part A was worth eight points and each item on Part B was worth six points. Rosa got four items correct on Part A and eight items correct on Part B. What was Rosa's score on the test?

8. Trang wants to budget money for telephone service. His telephone bills for the last six months were $21.50, $18.78, $17.52, $22.15, $23.60, and $19.20. He assumes these amounts are typical. About how much should Trang expect to pay for telephone service for a year?

WRITING WORD PROBLEMS

Using the information given, write a word problem for each exercise. Then solve the problem and check your answer.

9. 250 mi; 5 days 10. a cassette costs $9.95; a compact disc costs $14.50

SPIRAL REVIEW

11. Replace the ? with >, <, or =: 0.309 ? 0.39 *(Lesson 1-5)*

12. Estimate the difference: 76.81 − 39.5 *(Toolbox Skill 1)*

13. Jamal bought nine packs of baseball cards. Each pack cost $.79. How much did Jamal spend? Check your answer. *(Lesson 2-7)*

14. Simplify: $(8a)(3a)$ *(Lesson 2-1)*

Self-Test 2

1. The price of a sofa is $389.99. It can be bought on an installment plan for a $50 down payment and 18 payments of $25. How much more does the sofa cost on the installment plan? **2-4**

Find the next three numbers in each pattern.

2. 5, 8, 11, 14, ? , ? , ? 3. 1, 4, 16, 64, ? , ? , ? **2-5**

4. Complete the function table.

x	$2x + 6$
1	8
2	10
5	?
9	?
11	?

5. Find the function rule. **2-6**

x	?
3	21
5	35
7	49
9	63
10	70

6. Marvin earns $7.50 per hour. If he works more than 40 h per week, he earns $10.75 per hour for each extra hour. Last week Marvin worked 45 h. He says he earned $337.50. Is this correct? Explain. **2-7**

Finding an Estimate or an Exact Answer

Objective: To decide whether an estimate or an exact answer is needed to solve a problem.

Some real-life problems require only an *estimate* for a solution. Others require an *exact answer*. Before attempting to solve a problem, you should decide whether an estimate or an exact answer is needed.

Example

Decide whether an *estimate* or an *exact answer* is needed for parts (a) and (b). Then solve.

Best Clothes is having a sale on sweaters. The sale price is $17.95 each. Ellen wants to buy three sweaters.

a. How much money should Ellen take to the store to make sure that she has enough to pay for the sweaters?

b. What will be the total cost of the three sweaters excluding tax?

Solution

a. Ellen needs to *estimate* the amount of money to bring. To be sure that she has enough money, Ellen should *round up*.
cost of one sweater = $17.95 → about $20
$20 · 3 = $60
Ellen should bring about $60.

b. Finding the total cost of the sweaters requires an *exact answer*.
$17.95 · 3 = $53.85
The total cost of the sweaters excluding tax is $53.85.

✓ Check Your Understanding

In part(a) of the Example, why did Ellen round $17.95 up to $20 instead of up to $18?

Guided Practice

Decide whether an *estimate* or an *exact answer* is needed in each situation.

1. the number of hours a trip will take

2. the amount an employee is paid

3. the width of a new window shade

4. the number of books in a library

Decide whether an *estimate* or an *exact answer* is needed. Then solve.

5. Ana Rivera earns $14.25 per hour as a researcher. She works 40 h per week. How much does she earn each week?

« 6. COMMUNICATION *« Discussion*   Describe a recent situation in which you needed only an estimate to make a decision. Explain why you did not need an exact answer.

Decide whether an *estimate* or an *exact answer* is needed. Then solve.

1. Eli wants to plant lettuce seeds in a garden. Packets of seeds cost $.85 each. How much money should Eli take to the store to buy six packets?

2. Helen pumped 11.5 gal of gasoline into her car's gas tank. Gasoline costs $1.10 per gallon. How much did Helen pay for the gasoline?

3. On Monday, Sky Airlines flew 120 flights. Of these, 111 arrived on time. How many flights did not arrive on time?

4. It is recommended that restaurant customers leave a tip of $.15 for each dollar spent. Cy's bill is $19.97. How much tip should he leave?

5. The coach of a baseball team has $400 with which to buy equipment. Including tax, a shirt costs $8.95, a cap costs $4.95, and a bat costs $12.75. Can the coach buy a shirt and a cap for each of the 15 players?

6. Use the information in Exercise 5. Can the coach buy a shirt, a cap, and a bat for each of the 15 players?

GROUP ACTIVITY

Your class decides to remodel your classroom by painting the walls and retiling the floor. Paint costs $12.50 per gallon, and each gallon covers 400 ft². Floor tile costs $1.27 per square foot.

7. Find the length, width, and height of the classroom in feet. Use these values to find the areas of the floor and walls. Remember to subtract the areas of the doors and windows.

8. How many square feet of tile will be needed? What will this cost?

9. How many gallons of paint will be needed? What will this cost?

10. Give examples of when your group used estimates and when your group used exact answers.

SPIRAL REVIEW

11. Find the product mentally: $(25)(1)(6)(4)$ *(Lesson 1-7)*

12. Find the sum: $5\frac{2}{3} + 4\frac{3}{4}$ *(Toolbox Skill 18)*

2-9 The Metric System of Measurement

Objective: To recognize and to use metric units of measure.

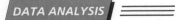

Sam Foster has a recipe that calls for 0.25 liter of milk. However, Sam's measuring cup is marked in milliliters.

QUESTION How many milliliters is 0.25 liter?

To find the answer, Sam used the table below, which he found in a reference book.

Metric System of Measurement

Prefix and Meaning		Length	Liquid Capacity	Mass
kilo-	1000	kilometer (km)	kiloliter (kL)	kilogram (kg)
hecto-	100	hectometer (hm)	hectoliter (hL)	hectogram (hg)
deka-	10	dekameter (dam)	dekaliter (daL)	dekagram (dag)
	1	meter (m)	liter (L)	gram (g)
deci-	0.1	decimeter (dm)	deciliter (dL)	decigram (dg)
centi-	0.01	centimeter (cm)	centiliter (cL)	centigram (cg)
milli-	0.001	millimeter (mm)	milliliter (mL)	milligram (mg)

Note: The most frequently used metric units are in red.

The table shows that the three basic units of measure in the metric system are the **meter** (m) for length, the **liter** (L) for liquid capacity, and the **gram** (g) for mass. Units beginning with *kilo-* are the largest units in the table and units beginning with *milli-* are the smallest.

The table lists enough data so that you can change one unit of measure to another. Each unit in the table is 10 times as large as the unit immediately below it. For example, 1 cm is equal to 10 mm. Therefore, to change from a *larger* metric unit to a *smaller* metric unit, you multiply by 10, 100, 1000, and so on.

Example 1

Write 0.25 L in mL.

Solution

Liters are larger than milliliters. Multiply by 1000.
0.25(1000) = 250   ◄── You can use mental math.
0.25 L = 250 mL

✔ Check Your Understanding

1. Why do you multiply by 1000 in Example 1?

ANSWER 0.25 liter is equal to 250 milliliters.

In order to change from a *smaller* metric unit to a *larger* metric unit, you multiply by 0.1, 0.01, 0.001, and so on.

Example 2	**a.** Write 48 mm in cm. **b.** Write 37.5 g in kg.
Solution	**a.** Multiply by 0.1. **b.** Multiply by 0.001.

Example 2

a. Write 48 mm in cm.   **b.** Write 37.5 g in kg.

Solution

a. Multiply by 0.1.
$48(0.1) = 4.8$
48 mm = 4.8 cm

b. Multiply by 0.001.
$37.5(0.001) = 0.0375$
37.5 g = 0.0375 kg

✔️ **Check Your Understanding**

2. In Example 2(b), how many places do you move the decimal point?

Guided Practice

COMMUNICATION « *Reading*

Choose the correct phrase to complete each definition.

«**1.** A milligram is:
   a. one thousand grams
   b. one million grams
   c. one thousandth of a gram
   d. one millionth of a gram

«**2.** A kilometer is:
   a. one hundred meters
   b. one thousand meters
   c. one hundredth of a meter
   d. one thousandth of a meter

Multiply.

3. $(2.43)(10)$ **4.** $(31.5)(100)$ **5.** $(6.83)(1000)$

6. $(4.4)(1000)$ **7.** $(13.7)(0.1)$ **8.** $(11.8)(0.01)$

9. $(12.5)(0.01)$ **10.** $(6.19)(0.001)$ **11.** $(58.6)(0.001)$

Write each measure in the unit indicated.

12. 350 cm; m   **13.** 6.2 L; mL   **14.** 0.75 kg; g

15. 12.5 mm; cm   **16.** 735 mL; L   **17.** 65 m; km

Exercises

Write each measure in the unit indicated.

1. 30 mm; cm   **2.** 4150 mL; L   **3.** 450 m; km

4. 2000 mg; g   **5.** 25 g; mg   **6.** 400 cm; m

7. 615 mm; m   **8.** 1.2 m; mm   **9.** 2345 mL; L

10. 25 m; cm   **11.** 3.4 L; mL   **12.** 45.2 g; kg

13. 0.74 m; cm   **14.** 0.07 L; mL   **15.** 0.098 km; m

16. 21.6 cm; mm   **17.** 0.88 km; m   **18.** 5.6 g; mg

19. The mass of a hummingbird is about 0.002 kg. About how many grams is the mass of a hummingbird?

20. **DATA** *pages 46–47* The proposed space station *Freedom* will orbit Earth at a height of how many meters?

Replace each __?__ with >, <, or =.

21. 540 mL __?__ 0.54 L

22. 0.87 km __?__ 87 m

23. 0.25 g __?__ 2500 mg

24. 34 cm __?__ 0.034 m

25. 907 g __?__ 0.0907 kg

26. 125 mm __?__ 12.5 cm

MENTAL MATH

You can use mental math to multiply numbers by 0.1, 0.01, 0.001, and so on. You simply move the decimal point of the number being multiplied to the left the same number of decimal places as there are in the number by which you are multiplying. For instance, 0.01 has two decimal places, so $(231.4)(0.01) = 2.314$.

Find each product mentally.

27. $(43.6)(0.1)$

28. $(764.4)(0.01)$

29. $(57.8)(0.01)$

30. $(891.3)(0.001)$

31. $(8.09)(0.1)$

32. $(0.56)(0.1)$

33. $(1.36)(0.01)$

34. $(3.08)(0.001)$

35. $(24.5)(0.001)$

NUMBER SENSE

It is sometimes necessary to estimate measures. You can make more reasonable estimates using familiar measures as references. For instance, a dime is about one millimeter thick, a fingernail is about one centimeter wide, and the length of twelve city blocks is about one kilometer.

Select the most reasonable measure for each item.

36. length of a soccer field
    a. 100 cm   **b.** 100 m   **c.** 100 km

37. height of a person
    a. 175 mm   **b.** 175 cm   **c.** 175 m

38. width of a computer screen
    a. 23 cm   **b.** 23 m   **c.** 23 km

39. distance from New York to London
    a. 5567 mm   **b.** 5567 cm   **c.** 5567 km

40. The normal mass of an adult male is about 70 kg. Use this fact to estimate the normal mass of a five-year-old child.

41. The capacity of a bucket is about 8 L. Use this fact to estimate the capacity of a drinking glass.

Automotive technicians maintain and repair cars. They must be familiar with metric units because many cars are manufactured using parts with metric measures.

42. The measure of the head of the oil drain plug on a car is 2.2 cm. Write this measure in millimeters.

43. A radiator holds 17.1 L of coolant. Write this measure in milliliters.

44. The capacity of the oil crankcase on a car is 4.2 L. Motor oil comes in containers that hold 946 mL. At least how many containers of motor oil does the technician need to fill the crankcase?

45. The tread on an old tire is 185 mm wide. The tread on a new tire is 20.5 cm wide. How many centimeters wider is the tread on the new tire than on the old tire?

RESEARCH

Find the meaning of the metric unit used in each exercise.

46. In June 1989, an asteroid passed within 16 gigameters of Earth.

47. Scientists can make a scratch in aluminum 1 micrometer wide.

48. A transistor can open and close a pathway of electrons in 0.01 nanosecond.

49. Ultraviolet light has a wavelength of 300 nanometers.

50. Laser pulses last about 1 picosecond.

51. Soil surrounding the roots of one giant redwood tree can hold 0.5 megaliter of water.

SPIRAL REVIEW

52. Complete the function table at the right. *(Lesson 2-6)*

53. Write 15 lb in oz. *(Toolbox Skill 22)*

54. Round to the nearest hundredth: 12.8728 *(Toolbox Skill 6)*

55. Simplify: $5(2x - 4)$ *(Lesson 2-2)*

56. Write 187.4 mm in cm. *(Lesson 2-9)*

x	$x + 5$
2	7
4	9
6	?
8	?
10	?

Scientific Notation

Objective: To write numbers in scientific notation.

EXPLORATION

Look at the following pattern.

$$9.3 \times 10^1 = 9.3 \times 10 = 93$$
$$9.3 \times 10^2 = 9.3 \times 100 = 930$$
$$9.3 \times 10^3 = 9.3 \times 1000 = 9300$$
$$9.3 \times 10^4 = 9.3 \times 10,000 = 93,000$$

1 Replace each __?__ with the number that makes the statement true.
  a. $9.3 \times 10^5 = 9.3 \times 100,000 = $ __?__
  b. $9.3 \times 10^6 = 9.3 \times $ __?__ $ = 9,300,000$
  c. $9.3 \times 10^? = 9.3 \times 10,000,000 = 93,000,000$

2 Write the number represented by 9.3×10^8. What is the relationship between the exponent of 10 and the number of places the decimal point in 9.3 is moved to the right?

Greater numbers can be difficult to read and to write. Scientists and other people who use these numbers often write them in *scientific notation*. A number is written in **scientific notation** when it is written as a number that is at least one but less than ten multiplied by a power of ten.

$$6 \times 10^5$$
at least 1, but less than 10 ⟵⟶ 2.3×10^8 ⟷ a power of 10
$$1.52 \times 10^9$$

Example 1

Write each number in scientific notation.
a. 13,000   **b.** 24,500,000

Solution

Move the decimal point to get a number that is at least 1, but less than 10.
a. $13,000 = 1.3 \times 10^4$

    4 places
b. $24,500,000 = 2.45 \times 10^7$

    7 places

Check Your Understanding

1. In Example 1(a), why do you multiply by 10^4?

A number such as 24,500,000 is said to be in **decimal notation.** Numbers written in scientific notation can be rewritten in decimal notation.

Example 2	Write each number in decimal notation.

a. 7.16×10^5 **b.** 5.2×10^6

Solution	Move the decimal point to the right.

a. $7.16 \times 10^5 = 716{,}000$ **b.** $5.2 \times 10^6 = 5{,}200{,}000$

5 places   6 places

 **Check Your Understanding**

2. In Example 2(b), why do you move the decimal point 6 places?

3. Describe how Example 2(a) would be different if the number were 7.16×10^2.

Some calculators display greater numbers in scientific notation. For instance, a calculator will display 480,000,000,000 as | *4.8 11* |.

Many calculators have an (EE) or (EXP) key. You use this key to enter numbers in scientific notation. For instance, to enter 4.8×10^{11} you use this key sequence.

| *4.8* | (EE) | *11* |

Guided Practice

COMMUNICATION « *Reading*

Refer to the text on pages 80–81.

« **1.** Describe the form of a number written in scientific notation.

« **2.** List at least two reasons for writing numbers in scientific notation.

Is each number written in scientific notation? Write *Yes* or *No*.

3. 5.8×10^9 **4.** 1.07×5^{10} **5.** 0.64×10^7 **6.** 12.7×10^6

Replace each __?__ with the number that makes the statement true.

7. $560 = 5.6 \times 10^?$ **8.** $7000 = 7 \times 10^?$

9. $67{,}500 = 6.75 \times 10^?$ **10.** __?__ $\times 10^4 = 24{,}000$

11. __?__ $\times 10^3 = 2758$ **12.** __?__ $\times 10^5 = 900{,}000$

Write each number in scientific notation.

13. $34{,}000$ **14.** $150{,}000$ **15.** $600{,}000$ **16.** $1{,}420{,}000$

Write each number in decimal notation.

17. 4.2×10^3 **18.** 1.65×10^5 **19.** 2.173×10^8 **20.** 8×10^9

Exercises

Write each number in scientific notation.

1. 1,200,000   **2.** 6000   **3.** 5700   **4.** 254,000

5. 45,200   **6.** 95,000,000   **7.** 851,400,000   **8.** 3,680,000

Write each number in decimal notation.

9. 3.8×10^3 **10.** 7×10^5 **11.** 1.52×10^6 **12.** 7.5×10^4

13. 5×10^8 **14.** 6.15×10^9 **15.** 3.425×10^5 **16.** 1.06×10^7

17. Scientists use solar telescopes to study the sun. The diameter of the sun is approximately 1,392,000 km. Write this number in scientific notation.

18. The distance that light travels in one year is approximately 9.46×10^{15} m. Write this number in decimal notation.

DATA, *pages 46–47*

Write in decimal notation the average distance from each planet to the sun.

19. Mercury   **20.** Venus   **21.** Saturn   **22.** Pluto

 CALCULATOR

For each calculator display, write the number first in scientific notation and then in decimal notation.

23. | 3.5 11 | **24.** | 7.9 09 |

25. | 4.853 07 | **26.** | 1.963 10 |

Choose the key sequence you would use to enter each number on a calculator.

27. 1.2×10^{12} **A.** | 1.4 | [EE] | 14 |

28. 1.4×10^{14} **B.** | 12 | [EE] | 1.2 |

     C. | 14 | [EE] | 1.4 |

     D. | 1.2 | [EE] | 12 |

NUMBER SENSE

Without writing in decimal notation, replace each __?__ with >, <, or =.

29. 4×10^5 __?__ 4×10^2 **30.** 6×10^6 __?__ 6×10^8

31. 1.7×10^5 __?__ 3.2×10^5 **32.** 8×10^3 __?__ 7×10^3

33. 0×10^7 __?__ 2×10^4 **34.** 0×10^4 __?__ 0×10^6

Kitt Peak National
Observatory, Arizona

Replace each ? with the number that makes the statement true.

35. $(2 \times 10^2)(4 \times 10^3) = (2 \times \underline{\ ?\ })(10^2 \times 10^3) = 8 \times 10^?$

36. $(3 \times 10^2)(5 \times 10^4) = (3 \times 5)(10^? \times 10^4)$
$$= 15 \times 10^6 = 1.5 \times 10^7$$

37. Compare Exercises 35 and 36. How are they alike? How are they different?

38. Develop a method for multiplying numbers that are written in scientific notation. Use your method to find each product.
a. $(3 \times 10^4)(2 \times 10^7)$
b. $(6 \times 10^2)(8 \times 10^5)$

39. Find the product: $2\frac{1}{3} \times 3\frac{1}{4}$ *(Toolbox Skill 20)*

40. Evaluate $3.65 + m$ when $m = 4.8$.   *(Lesson 1-2)*

41. Write 42,540,000 in scientific notation.   *(Lesson 2-10)*

42. Is it most efficient to find the quotient $1400 \div 20$ by using *paper and pencil, mental math*, or a *calculator*?   *(Lesson 1-8)*

43. Simplify: $5a - 2a + 7b$ *(Lesson 2-3)*

44. Lou works five days per week, eight hours per day. She earns $11.50 per hour. How much does Lou earn per week?   *(Lesson 2-4)*

Self-Test 3

Decide whether an *estimate* or an *exact answer* is needed. Then solve.

1. Grapes cost $1.49 per pound. Sergei needs to buy 12 lb of grapes for a fruit platter. He has $20. Does he have enough money to buy 12 lb of grapes?   2-8

Write each measure in the unit indicated.

2. 480 mm; cm   **3.** 3.2 L; mL   **4.** 0.5 kg; g   2-9

Write each number in scientific notation.

5. 44,000   **6.** 9,870,000   **7.** 21,300,000   2-10

Write each number in decimal notation.

8. 5.6×10^3 **9.** 3.78×10^5 **10.** 6.43×10^7

Converting Customary Units to Metric Units

Objective: To convert customary units to metric units.

Because most other countries use the metric system, many United States companies convert the customary measurements of their products into metric units before exporting them. You can do this type of conversion using a chart like the one below.

When You Know	Multiply By	To Find
inches	2.54	centimeters
feet	0.3	meters
yards	0.91	meters
miles	1.61	kilometers
ounces	28.35	grams
pounds	0.454	kilograms
fluid ounces	29.573	milliliters
pints	0.473	liters
quarts	0.946	liters
gallons	3.785	liters

Example

Anderson Architects designs buildings worldwide. They recently designed a house that might be built in both the United States and Canada. The United States plans show that the length of the house is 57 ft and the width of the house is 33 ft. What are the length and width in meters that should be shown on the Canadian plans?

Solution

To convert from feet to meters, multiply by 0.3.

$$57(0.3) = 17.1$$
$$33(0.3) = 9.9$$

Estimate to check the reasonableness of these answers.

$$57(0.3) \rightarrow 60(0.3) = 18$$
$$33(0.3) \rightarrow 30(0.3) = 9$$

The answers 17.1 and 9.9 are close to the estimates of 18 and 9, so 17.1 and 9.9 are reasonable.

On the Canadian plans, the length should be 17.1 m and the width should be 9.9 m.

Exercises

Write each measure in the unit indicated.

1. 6 pt; L **2.** 27 mi; km **3.** 62.5 in.; cm **4.** 9.4 qt; L

Estimate each measure in the unit indicated.

5. 7 fl oz; mL   **6.** 5 ft; m   **7.** 98 lb; kg   **8.** 21 oz; g

9. Which metric measure is approximately equal to one yard?

10. Which metric measure is approximately equal to one quart?

11. Which customary measure is approximately equal to four liters?

12. Which customary measure is approximately equal to two and one half centimeters?

PROBLEM SOLVING/APPLICATION

13. According to the owner's manual, a car has a capacity for 4.3 qt of oil and 6.9 pt of transmission fluid. The car is to be exported to Mexico. What capacities in liters should be listed in the owner's manual?

14. A set of plans for a house indicates that its width is 28 ft and its length is 42 ft. The length of the driveway is 19 yd. What are these dimensions in meters?

15. The plans for a new medical building show that the dimensions of the lobby will be 16 ft 9 in. by 14 ft 3 in. What dimensions should the plans show in meters?

16. The label on a jar of fruit juice indicates that the capacity is 2 qt 10 fl oz. What is the capacity of the jar in milliliters?

17. An adjustable wrench is advertised for use with bolts that measure from 0.25 in. through 1.75 in. The wrench is being exported to South America. How should the sizes be advertised in millimeters?

18. The label on a package of modeling clay indicates that the weight is 12 lb 6 oz. What is the mass of the package in kilograms?

19. RESEARCH When was the metric system created? Name two advantages of using the metric system.

Chapter Review

product of powers rule (p. 48)
simplify (p. 48)
distributive property (p. 52)
like terms (p. 56)
unlike terms (p. 56)
combining like terms (p. 56)
function (p. 66)
function rule (p. 66)

arrow notation (p. 66)
function table (p. 66)
meter (p. 76)
gram (p. 76)
liter (p. 76)
scientific notation (p. 80)
decimal notation (p. 80)

Choose the correct term from the list above to complete each sentence.

1. A relationship that pairs each number in a given set of numbers with exactly one number in a second set of numbers is called a(n) __?__.

2. The __?__ allows you to multiply each term inside parentheses by a factor outside the parentheses.

3. The expressions $7x$ and $4z$ are called __?__.

4. The basic unit of mass in the metric system is the __?__.

5. The number 3.6×10^5 is written in __?__.

6. $x \to 2x + 1$ is written in __?__.

Simplify.   *(Lessons 2-1, 2-2, 2-3)*

7. $a^5 \cdot a$

8. $c^4 \cdot c^2 \cdot c^7$

9. $6(4x + 5)$

10. $2(12 + x)$

11. $8n + 7m + 16n$

12. $11a + 6a$

13. $9x + 12 + x$

14. $(5a)(11b)$

15. $b^5 \cdot b \cdot b^6$

16. $x^7 \cdot x^9$

17. $9(6 - z)$

18. $8(3c - 9)$

19. $3x^2 \cdot 5x^6$

20. $17c - 8c$

21. $(8w)(3x)(2x)$

22. $15 + 13z - z$

23. $18d - 11d - 5c$

24. $6c \cdot 7c^3$

Use the distributive property to find each answer mentally.   *(Lesson 2-2)*

25. $6(23) + 6(7)$

26. $4 \cdot 48 - 4 \cdot 8$

27. $8(104)$

28. $3(94)$

Chapter 2

Solve. *(Lesson 2-4)*

29. Laurie bought two sets of screwdrivers at $19.49 each, three boxes of screws at $4.98 each, and a drill for $39.95. What was the total cost of Laurie's purchase?

Find the next three numbers or expressions in each pattern.
(Lesson 2-5)

30. 3, 10, 17, 24, __?__, __?__, __?__

31. 2, 6, 18, 54, __?__, __?__, __?__

32. $a + 6$, $a + 7$, $a + 8$, __?__, __?__, __?__

33. $38m$, $35m$, $32m$, __?__, __?__, __?__

Complete each function table. *(Lesson 2-6)*

34.

x	$x - 8$
12	4
16	8
20	?
24	?
28	?

35.

x	$9x$
2	18
4	36
6	?
8	?
10	?

Find each function rule. *(Lesson 2-6)*

36.

x	?
3	1
6	2
9	3
12	4
15	5

37.

x	?
5	10
10	15
15	20
20	25
25	30

Tell whether the final statement is correct. Explain. *(Lesson 2-7)*

38. The mortgage loan, property tax, and maintenance fee on Myeesha's condominium total $1135 per month. The maintenance fee is $110 per month and the property tax is $75 per month. Myeesha said that her mortgage loan is $950 per month.

Decide whether an *estimate* or an *exact answer* is needed. Then solve.
(Lesson 2-8)

39. Tim charges $1.50 per page to type term papers. How much will he charge to type a 33-page paper?

Write each measure in the unit indicated. *(Lesson 2-9)*

40. 1700 g; kg
41. 750 mL; L
42. 0.76 km; m
43. 1.9 m; cm

Write each number in scientific notation. *(Lesson 2-10)*

44. 350,000
45. 12,700
46. 6,550,000
47. 48,000,000

Write each number in decimal notation. *(Lesson 2-10)*

48. 1.3×10^5
49. 9.7×10^7
50. 2.64×10^4
51. 5.88×10^8

Simplify.

1. $a^7 \cdot a^6$ 2. $6x \cdot 7x^3$ 3. $(5m)(8n)$ 4. $(4b)(7c)(3b)$ **2-1**

5. $5(x + 11)$ 6. $3(9 - a)$ 7. $6(7 + 4c)$ 8. $9(5z - 8)$ **2-2**

9. $6m + 8m$ 10. $12x - 5x$ 11. $c + 14 + 8c$ 12. $6n - 4n + 3m$ **2-3**

13. Write an addition expression that contains two unlike terms.

14. Explain why $5x$ and $5x^2$ are unlike terms.

15. Michi bought two hardcover books for $17.98 each and four paper-   **2-4**
    back books for $3.95 each. How much more did the hardcover books
    cost than the paperback books?

Find the next three numbers or expressions in each pattern.

16. 1, 3, 9, 27, __?__, __?__, __?__   17. $x - 1, 3x - 1, 5x - 1$, __?__, __?__,   **2-5**
    __?__

18. Complete the function table.   19. Find the function rule.   **2-6**

x	$x - 3$
3	0
6	3
9	?
12	?
15	?

x	?
5	12
6	13
7	14
8	15
9	16

20. Levi lives 12 mi from work. He works five days each week. When   **2-7**
    asked how many miles he commutes each week, Levi said he com-
    mutes 60 mi. Is this answer correct? Explain.

Decide whether an *estimate* or *exact answer* is needed. Then solve.

21. Ground beef sells for $2.09 per pound. Sue needs to buy 36 lb for a   **2-8**
    picnic. How much money should she take to the store?

Write each measure in the unit indicated.

22. 655 mL; L   23. 8.3 kg; g   24. 56.4 cm; mm   **2-9**

25. Write 2,400,000 in scientific   26. Write 453,000 in scientific   **2-10**
    notation.   notation.

27. Write 4.7×10^5 in decimal   28. Write 1.62×10^7 in decimal
    notation.   notation.

Other Rules for Exponents

Objective: To use the rules of exponents.

1 Replace each __?__ with the number that makes the statement true.
 a. $(x^5)^3 = x^5 \cdot x^5 \cdot x^5 = x^{5 + 5 + 5} = x^?$
 b. $(a^3)^4 = a^3 \cdot a^3 \cdot a^3 \cdot a^3 = a^{3 + 3 + 3 + 3} = a^?$

2 Use your results from Step 1. Complete this statement:
 To find a power of a power, you __?__ the exponents.

3 Use your result from Step 2. Find the value of n that makes each statement true. Use a calculator to check your answer.
 a. $(5^2)^3 = 5^n$ **b.** $(3^2)^7 = 3^n$ **c.** $(2^4)^3 = 2^n$

When you simplify expressions involving exponents, there are a number of rules that may make your work easier. For instance, in Lesson 2-1 you learned the *product of powers rule*.

$$a^m \cdot a^n = a^{m + n}$$

Another helpful rule for exponents is the *power of a power rule*.

Power of a Power Rule

To find the power of a power, multiply the exponents.
$$(a^m)^n = a^{mn}$$

Example $(c^2)^7 = c^{2 \cdot 7} = c^{14}$

Exercises

Simplify.

 1. $(b^2)^5$ **2.** $(x^5)^3$ **3.** $(c^7)^4$ **4.** $(m^3)^9$ **5.** $(x^6)^6$ **6.** $(z^4)^8$

THINKING SKILLS

Find each answer.

 7. a. $(3 \cdot 4)^2$ **b.** $3^2 \cdot 4^2$ **8. a.** $(2 \cdot 3)^4$ **b.** $2^4 \cdot 3^4$

 9. Examine the results of Exercises 7 and 8. Create a rule for simplifying $(ab)^m$. Call it the *power of a product rule*.

 10. Simplify each expression by combining the rules for exponents.
  a. $(xy)^5$ **b.** $(2k)^3$ **c.** $(a^2b)^3$ **d.** $(m^3n^2)^4$ **e.** $(3c^3d^5)^2$

Cumulative Review

Standardized Testing Practice

Choose the letter of the correct answer.

1. **How can you use the distributive property to find 8(105) mentally?**
   A. $8(10 + 5) = 8 \times 15 = 120$
   B. $8(100) + 8(5) = 800 + 40 = 840$
   C. $5(105) + 3(105) = 525 + 315 = 840$
   D. $8(100) - 8(5) = 800 - 40 = 760$

2. **Find the next number in the pattern:**
   $3, 6, 10, 15, \underline{\ ?\ }$
   A. 30   B. 18
   C. 21   D. 20

3. **Evaluate $412.5 + n$ when $n = 86$.**
   A. 498.5
   B. 326.5
   C. 421.1
   D. 403.9

4. **Write 76,500 in scientific notation.**
   A. 7.65×10^4
   B. 76.5×10^3
   C. 7.65×10^3
   D. 765×10^2

5. **Which expression has a value of 216 when $n = 3$?**
   A. $2n^3$ B. $(2n)^3$
   C. $3n^2$ D. $(3n)^2$

6. **Complete:** The exact weight of a package rather than the estimated weight is needed to __?__.
   A. store the package on a shelf
   B. carry the package on a bike rack
   C. mail the package
   D. all of the above

7. **Evaluate $a + b + 2$ when $a = 4$ and $b = 8$.**
   A. 64   B. 12
   C. 34   D. 14

8. **Simplify:** x^2y^3
   A. $(xy)^5$
   B. xy^5
   C. $(xy)^6$
   D. already simplified

9. **Which property enables you to find $(4 + 1)(0)$ mentally?**
   A. commutative property of addition
   B. multiplication property of zero
   C. identity property of addition
   D. identity property of multiplication

10. **Choose the correct relationship:**
    $x = 6(2b + 3); y = 2(9 + 6b)$
    A. $x < y$ B. $x = y$
    C. $x > y$ D. cannot determine

11. During the last 3 days, Ruth drove 120 mi, 380 mi, and 250 mi. Gas costs $1.10 per gallon. Her car used 30 gal of gas. Which of the following cannot be determined?

 A. number of mi/gal car averages
 B. number of miles driven
 C. capacity of gas tank
 D. total cost of gas used

12. Which number is greatest?

 A. 0.2346   B. 0.3246
 C. 0.3264   D. 0.3624

13. Jorge bought 3 lb of apples at $.89/lb and 2 lb of grapes at $2.49/lb. Find the total cost.

 A. $7.65   B. $9.25
 C. $3.38   D. $8.45

14. Evaluate the expression $5 + 3(x - y^2)$ when $x = 10$ and $y = 2$.

 A. 197
 B. 48
 C. 512
 D. 23

15. Write 6.45 kg in g.

 A. 645 g
 B. 64.5 g
 C. 6450 g
 D. 64,500 g

16. Find the function rule.

 A. $x \rightarrow x + 1$
 B. $x \rightarrow 3x - 3$
 C. $x \rightarrow 2x - 1$
 D. $x \rightarrow 4x + 5$

x	?
2	3
3	5
4	7
5	9
6	11

17. Simplify: $3x + 5y + 4x$

 A. $12xy$ B. $7x + 5y$
 C. $12(x + y)$ D. $5x + 7y$

18. Evaluate the quotient $7.5 \div b$ when $b = 1.5$.

 A. 6   B. 5
 C. 9   D. 0.2

19. A bill for two $38-sweaters and one $24-shirt came to $62. Find the error.

 A. $38 was not multiplied by 2.
 B. $24 was multiplied by $38.
 C. $24 was multiplied by 2.
 D. No error was made.

20. Write 43.5 mm in cm.

 A. 435 cm
 B. 0.435 cm
 C. 4.35 cm
 D. 4350 cm

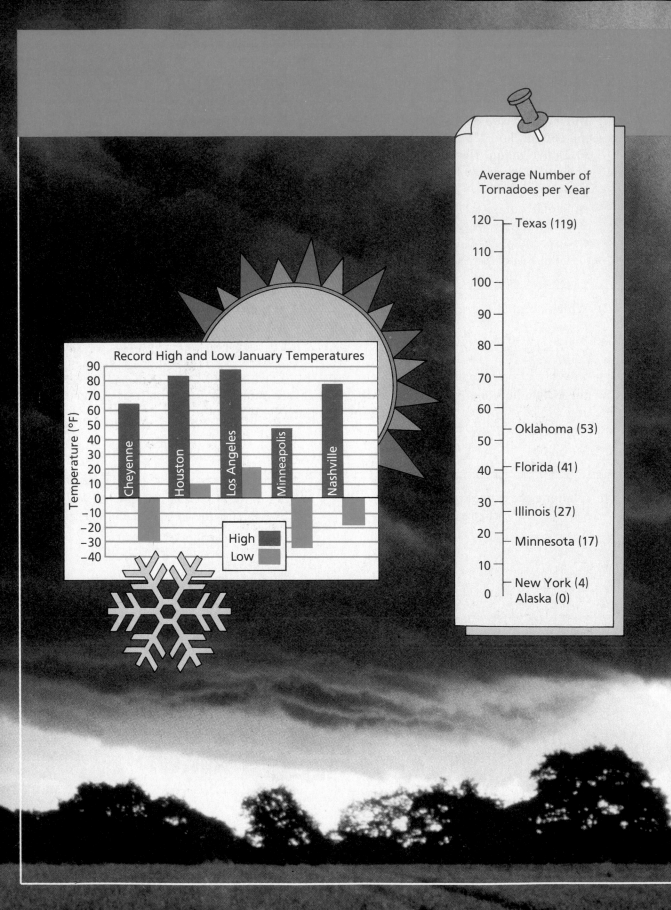

Average Number of Tornadoes per Year

120 — Texas (119)
110 —
100 —
90 —
80 —
70 —
60 —
50 — Oklahoma (53)
40 — Florida (41)
30 — Illinois (27)
20 — Minnesota (17)
10 —
0 — New York (4)
    Alaska (0)

Record High and Low January Temperatures

Temperature (°F)

Cheyenne
Houston
Los Angeles
Minneapolis
Nashville

High
Low

Normal Low January Temperatures (°F)

50 40   30 20   10   0 −5   −10   −5 0   10

60

70   60   50   40

30

40
50
60
70

40   50   40
60
70

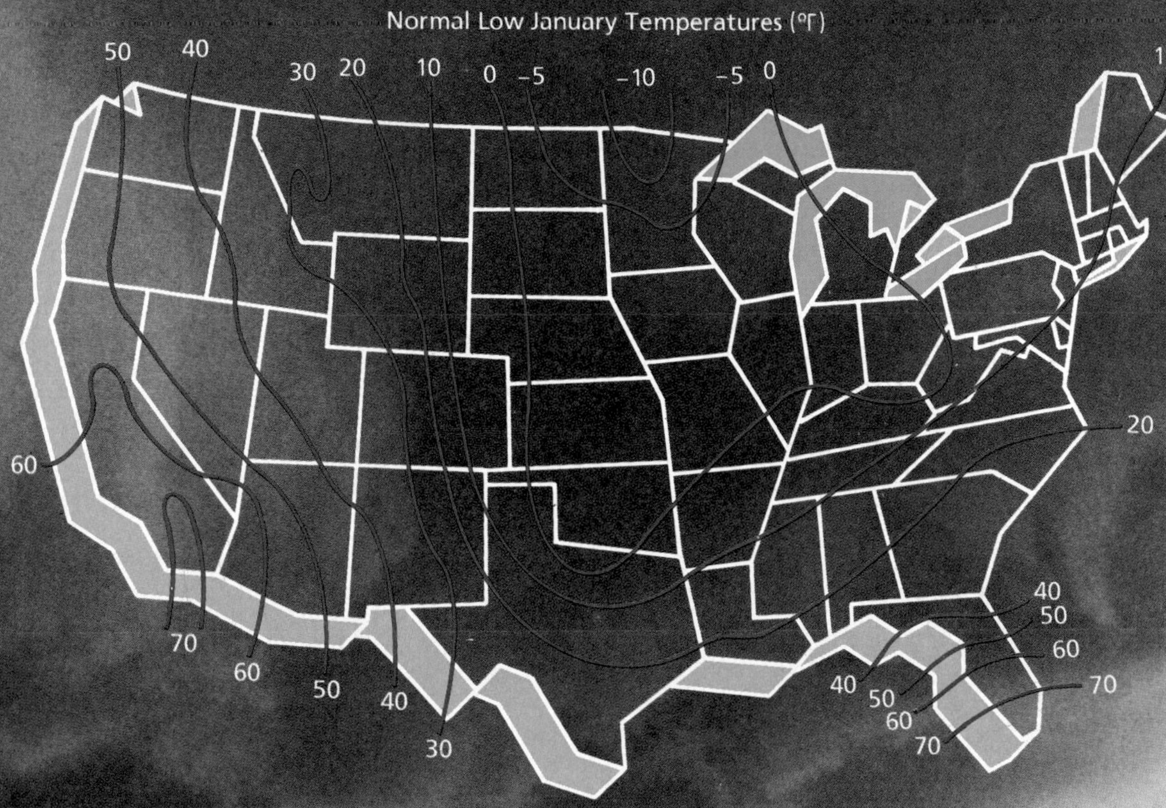

Did You Know?

A tropical storm has a wind speed that is greater than 39 mi/h. Storms in the Atlantic and eastern Pacific oceans are identified by a person's first name. These names have an international flavor because storms are tracked by many nations.

When the wind speed of a tropical storm reaches 74 mi/h, the storm is upgraded. An upgraded storm is called a *hurricane* in the Atlantic and eastern Pacific oceans, a *typhoon* in the western Pacific Ocean, and a *cyclone* in the Indian Ocean.

Integers on a Number Line

Objective: To recognize and compare integers and to find opposites and absolute values.

Terms to Know
- *integers*
- *positive integers*
- *negative integers*
- *opposites*
- *absolute value*

DATA ANALYSIS

The diagram at the right shows information about the highest and lowest points on Earth. The positive sign on $^+29{,}028$ tells you that the top of Mount Everest is 29,028 ft *above* sea level. The negative sign on $^-38{,}635$ tells you that the bottom of the Marianas Trench is 38,635 ft *below* sea level. Sea level is represented by 0. The numbers $^+29{,}028$, $^-38{,}635$, and 0 are examples of *integers*.

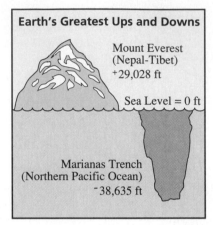

Earth's Greatest Ups and Downs

Mount Everest (Nepal-Tibet) $^+29{,}028$ ft

Sea Level = 0 ft

Marianas Trench (Northern Pacific Ocean) $^-38{,}635$ ft

An **integer** is any number in the following set.

$$\{\ldots,\ ^-4,\ ^-3,\ ^-2,\ ^-1,\ 0,\ ^+1,\ ^+2,\ ^+3,\ ^+4,\ \ldots\}$$ ◄── The braces { } mean *the set that contains.*

Integers greater than zero are called **positive integers.** Integers less than zero are called **negative integers.** Zero is neither positive nor negative. To make notation simpler, you generally write positive integers without the positive sign.

Another way to show the integers is to locate them as points on a number line. On a horizontal number line, *positive integers* are to the right of zero and *negative integers* are to the left.

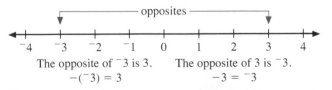

negative   positive

$^-4$ $^-3$ $^-2$ $^-1$ 0 1 2 3 4

Numbers that are the same distance from zero, but on opposite sides of zero, are called **opposites.** To indicate the opposite of a number n, you write $-n$. You read $-n$ as "the opposite of n."

opposites

$^-4$ $^-3$ $^-2$ $^-1$ 0 1 2 3 4

The opposite of $^-3$ is 3.   The opposite of 3 is $^-3$.
$-(^-3) = 3$ $-3 = {}^-3$

On the number line above, you see that the symbols -3 and $^-3$ represent the same number, negative three. To make notation simpler, this textbook will use the *lowered* sign to indicate a negative number. From this point on, you will see negative three written as -3.

The distance that a number is from zero on a number line is the **absolute value** of the number. You use the symbol | | to indicate absolute value. You read $|n|$ as "the absolute value of n."

Example 1

Find each absolute value.

a. $|3|$ **b.** $|-4|$

Solution

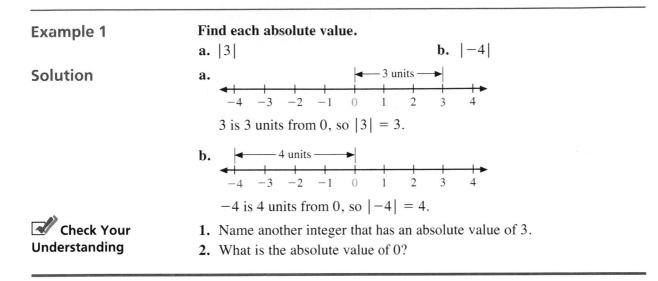

a.

3 is 3 units from 0, so $|3| = 3$.

b.

-4 is 4 units from 0, so $|-4| = 4$.

✓ Check Your Understanding

1. Name another integer that has an absolute value of 3.
2. What is the absolute value of 0?

When you compare numbers, you may want to picture them on a number line. On a horizontal number line, numbers increase in order from left to right.

Example 2

Replace each __?__ with >, <, or =.

a. $1 \underline{\ ?\ } -3$ **b.** $-4 \underline{\ ?\ } -2$

Solution

a.

1 is *to the right of* -3, so $1 > -3$.

b.

-4 is *to the left of* -2, so $-4 < -2$.

Guided Practice

COMMUNICATION « *Reading*

Replace each __?__ with the correct phrase.

«**1.** The __?__ of a number is its distance from zero on a number line.

«**2.** The expression $-b$ is read as __?__.

Write an integer that represents each situation.

«**3.** At night, temperatures on Mars can reach 130°F below zero.

«**4.** Jacksonville, Florida, is located at sea level.

«**5.** Linda deposited $25 in her savings account.

«**6.** The Bay City Bengals football team lost a total of 80 yd in one game.

Describe a situation that can be represented by each integer.

«**7.** 44   «**8.** −6   «**9.** −15   «**10.** 0

Write the opposite of each integer.

11. 5   **12.** −9   **13.** −11   **14.** 0

For each pair of integers, tell which integer is farther to the right on a horizontal number line.

15. −6, 2   **16.** 0, −1   **17.** 10, −10   **18.** −11, −8

Find each absolute value.

19. $|-8|$ **20.** $|33|$ **21.** $|0|$ **22.** $|-23|$

Replace each __?__ with >, <, or =.

23. 0 __?__ 6   **24.** 5 __?__ −2   **25.** −10 __?__ −12   **26.** −13 __?__ −9

Exercises

Find each absolute value.

1. $|-5|$ **2.** $|7|$ **3.** $|13|$

4. $|-6|$ **5.** $|-1|$ **6.** $|10|$

Replace each __?__ with >, <, or =.

7. 8 __?__ −9   **8.** −4 __?__ 4

9. ⁻18 __?__ −18   **10.** 12 __?__ 0

11. −11 __?__ −7   **12.** −1 __?__ −8

13. 0 __?__ −6   **14.** −5 __?__ 6

15. Two different depths in the Carlsbad Caverns are 754 ft below sea level and 900 ft below sea level. Which depth is closer to sea level?

16. DATA, *pages 92–93* How many states have regions with normal low January temperatures less than or equal to 0°F?

Write in order from least to greatest. Use two inequality symbols.

17. $4, -3, 9$ **18.** $1, 0, -1$ **19.** $-10, -8, -6$ **20.** $2, -2, 0$

21. WRITING ABOUT MATHEMATICS In some textbooks, *absolute value* is defined as follows.

If a is positive, $|a| = a$.
If a is negative, $|a| = -a$.
If a is zero, $|a| = 0$.

Write a paragraph that compares this definition to the definition given in this lesson.

DATA ANALYSIS

Use the graph at the right.

22. How much profit does the Value Company show?

23. How much loss does the Do Right Company show?

24. How much loss does the Hearty Company show?

25. Which company shows a greater loss?

26. List the companies in order from the company with the greatest profit to the company with the greatest loss.

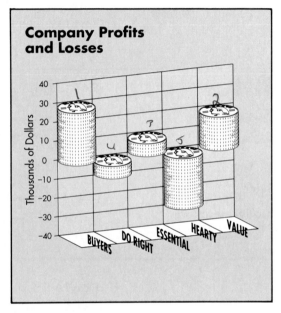

LOGICAL REASONING

Assume that *n* represents a negative integer and *p* represents a positive integer. Tell whether each statement is *always, sometimes,* or *never* true.

27. $n > p$ **28.** $p > n$

29. $0 > p$ **30.** $-n < 0$

31. $|p| = p$ **32.** $|n| < 0$

33. $|n| = |p|$ **34.** $|n| < |p|$

SPIRAL REVIEW

35. Find the sum: $1529 + 674$ *(Toolbox Skill 7)*

36. Evaluate $8qr$ when $q = 5$ and $r = 4$.   *(Lesson 1-1)*

37. Find the absolute value: $|-2|$ *(Lesson 3-1)*

38. Evaluate $51.3 \div p$ when $p = 3$.   *(Lesson 1-2)*

3-2

Adding Integers with the Same Sign

Objective: To add integers with the same sign.

APPLICATION

In a football game, Todd lost three yards on one play and four yards on the next play.

QUESTION How many yards did Todd lose in all on the two plays?

To find the total, think of the loss of three yards as -3 and the loss of four yards as -4. You can then use arrows along a number line to add these two integers. Arrows pointing to the left represent negative numbers. Arrows pointing to the right represent positive numbers.

Example 1

Solution

Find the sum $-3 + (-4)$.

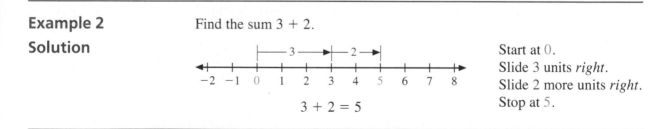

$$-3 + (-4) = -7$$

Start at 0.
Slide 3 units *left*.
Slide 4 more units *left*.
Stop at -7.

✓ Check Your Understanding

1. In Example 1, why do the arrows point to the left?
2. How would Example 1 be different if you were asked to find the sum $-4 + (-3)$? How would it be the same?

ANSWER Todd lost a total of seven yards.

You can also use the number line as a model for the more familiar situation of adding *positive* integers.

Example 2

Solution

Find the sum $3 + 2$.

Start at 0.
Slide 3 units *right*.
Slide 2 more units *right*.
Stop at 5.

$$3 + 2 = 5$$

In each example above, notice that the integer representing the sum is related to the total distance from 0 represented by the arrows. Because an integer's distance from zero on a number line is its absolute value, you can use absolute value to add integers with the same sign.

Example 3
Solution

Find each sum: **a.** $-10 + (-14)$ **b.** $18 + 9$

a. $|-10| = 10$ and $|-14| = 14$
$10 + 14 = 24$, so
$-10 + (-14) = -24$

b. $|18| = 18$ and $|9| = 9$, so
$18 + 9 = 27$

Guided Practice

« **1.** COMMUNICATION «*Reading* Replace each ___?___ with the correct word.
The sign of the sum of two negative integers is ___?___.
The sign of the sum of two positive integers is ___?___.

COMMUNICATION «*Writing*

Write the addition represented on each number line.

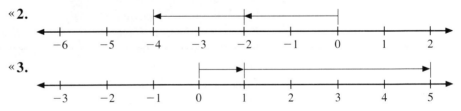

« **2.**

« **3.**

Represent each addition on a number line.

« **4.** $3 + 3$ « **5.** $-4 + (-4)$ « **6.** $-1 + (-2)$ « **7.** $6 + 2$

Write an addition that represents each situation.

« **8.** Yesterday the temperature rose 6°F and then rose 7°F.

« **9.** Lia withdrew $5 from her account one day and $5 the next day.

Describe a situation that can be represented by each addition.

« **10.** $-15 + (-45)$ « **11.** $4 + 4$ « **12.** $27 + 60$ « **13.** $-12 + (-10)$

Find each sum.

14. $-12 + (-23)$	**15.** $-7 + (-15)$	**16.** $42 + 9$
17. $16 + 31$	**18.** $39 + 86$	**19.** $-55 + (-34)$
20. $-97 + (-27)$	**21.** $48 + 53$	**22.** $-44 + (-19)$

Exercises

Find each sum.

1. $22 + 7$

2. $-14 + (-13)$

3. $-5 + (-25)$

4. $8 + 34$

5. $-53 + (-10)$

6. $-72 + (-8)$

7. $62 + 6$

8. $42 + 52$

9. $-29 + (-16)$

10. $68 + 49$

11. $75 + 37$

12. $-32 + (-18)$

13. Rich has $13 in his savings account. He deposits $25. How much money is in his savings account after the deposit?

14. The football team lost 11 yd, 11 yd, and 12 yd on three plays. How many yards did the team lose in all on the plays?

15. During Jenna's experiment, the temperature of water dropped three times. She recorded losses of 6°F, 12°F, and 15°F. What was the total loss?

16. Shari earned 29 points on Part I of a test, 31 points on Part II, and 29 points on Part III. What was the total number of points Shari earned on Parts II and III?

Evaluate each expression when $p = 8$, $q = -9$, $r = -5$, and $s = 10$.

17. $p + s$

18. $r + q$

19. $q + (-3)$

20. $16 + s$

21. $-2 + r + (-15)$

22. $12 + 24 + p$

23. $q + (-2) + r + (-1)$

Simplify.

24. $-3z + (-5z)$

25. $6y + 9y$

26. $-5b + 7c + (-2b)$

27. $-8m + (-2m)$

28. $7 + 14x + 29$

29. $-3r + (-7r) + 4s$

30. $4d + 2c + 11c$

31. $8m + 3n + 5n$

32. $-9 + (-2y) + (-8)$

SPIRAL REVIEW

33. Estimate the quotient: $6445 \div 83$ *(Toolbox Skill 3)*

34. Find the sum: $-5 + (-5)$ *(Lesson 3-2)*

35. Simplify: $3(4y - 1)$ *(Lesson 2-2)*

3-3 Adding Integers with Different Signs

Objective: To add integers with different signs.

Terms to Know
- addition property of opposites
- additive inverse

Carl enters an elevator at a garage level that is two floors below the lobby level. He leaves the elevator after going up five floors.

QUESTION At which floor does Carl leave the elevator?

To find the answer, you can represent the garage level as -2 and the number of floors Carl goes up as $+5$. Then add the integers. Sometimes it is helpful to use a number line to add integers that have different signs.

Example 1

Solution

Find each sum.

a. $-2 + 5$ **b.** $1 + (-3)$

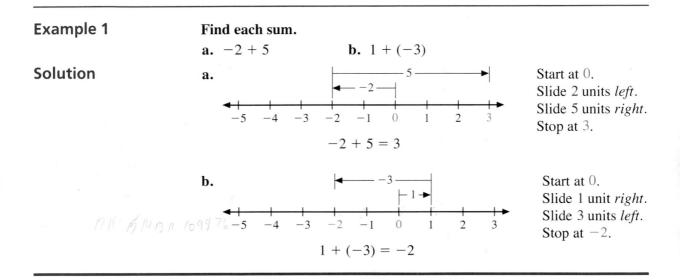

a.

$-2 + 5 = 3$

Start at 0.
Slide 2 units *left*.
Slide 5 units *right*.
Stop at 3.

b.

$1 + (-3) = -2$

Start at 0.
Slide 1 unit *right*.
Slide 3 units *left*.
Stop at -2.

ANSWER Part (a) shows that Carl leaves the elevator at the third floor.

In Example 1, notice that the *difference* between the distances the arrows represent is the absolute value of the sum of the two integers. You can use the difference of absolute values to add integers with different signs.

Generalization: *Adding Integers with Different Signs*

To add two integers with different signs, first find their absolute values. Then subtract the lesser absolute value from the greater absolute value. Give the result the sign of the integer with the greater absolute value.

Example 2

Solution

Find each sum: **a.** $10 + (-16)$ **b.** $-7 + 12$

a. $|10| = 10$ and $|-16| = 16$
Subtract: $16 - 10 = 6$
The negative integer has the
greater absolute value, so the
sum is negative.
$10 + (-16) = -6$

b. $|-7| = 7$ and $|12| = 12$
Subtract: $12 - 7 = 5$
The positive integer has the
greater absolute value, so the
sum is positive.
$-7 + 12 = 5$

✔️ **Check Your**
Understanding

1. Describe how Example 1(a) would be different if you found the sum
$-10 + 16$.

2. Describe how to find the sum $3 + (-3)$.

In the case of adding opposites, the sum will always be zero. This fact is
so useful in algebra that it is identified as a *property* of opposites.

> **Addition Property of Opposites**
>
> The sum of a number and its opposite is zero.
> $$a + (-a) = 0 \text{ and } -a + a = 0$$

Because the sum of a number and its opposite is zero, the opposite of a
number is sometimes called the **additive inverse** of the number.

Guided Practice

COMMUNICATION « *Reading*

Refer to the text on pages 98–99 and 101–102.

« **1.** Compare the generalizations for adding integers with the same sign
and adding integers with different signs. How are the generalizations
alike? How are they different?

« **2.** Describe the addition property of opposites in words.

COMMUNICATION « *Writing*

Write the addition represented on each number line.

« **3.**

« **4.**

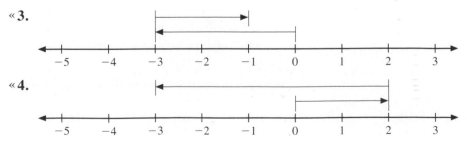

Represent each addition on a number line.

«**5.** $-1 + 4$ «**6.** $-4 + 6$ «**7.** $5 + (-1)$ «**8.** $4 + (-7)$

Tell whether each sum is *positive*, *negative*, or *zero*.

9. $88 + (-67)$ **10.** $-22 + 19$ **11.** $25 + (-26)$ **12.** $14 + (-14)$

Find each sum.

13. $-13 + 2$ **14.** $4 + (-9)$ **15.** $8 + (-8)$

16. $-4 + 4$ **17.** $3 + (-7) + 9$ **18.** $-2 + 6 + (-10)$

Exercises

Find each sum.

1. $-12 + 12$ **2.** $19 + (-19)$ **3.** $5 + (-26)$

4. $-1 + 27$ **5.** $9 + (-9)$ **6.** $-6 + 6$

7. $-2 + 7$ **8.** $-25 + 18$ **9.** $-8 + 18$

10. $26 + (-15)$ **11.** $-13 + 6$ **12.** $-28 + 24$

13. $14 + (-40)$ **14.** $41 + (-45)$ **15.** $20 + (-6) + (-7)$

16. $-8 + 23 + (-14)$ **17.** $-8 + (-13) + 7$ **18.** $-25 + 11 + 5$

19. Sue deposited \$45 in her savings account and then withdrew \$53. After the withdrawal, did she have more or less money than when she started? How much more or less?

20. Pete lost three yards on one play and then gained eight yards on the next play. Find the total number of yards gained or lost.

Evaluate each expression when $a = 4$, $b = -6$, and $c = -3$.

21. $-1 + a + c$ **22.** $a + c + b$ **23.** $b + a + b$

24. $-7 + a + c$ **25.** $b + 10 + a$ **26.** $c + b + 9$

MENTAL MATH

To add integers mentally, it is helpful to look for opposites. You can also group positive and negative integers.

Find each sum mentally.

27. $-3 + 5 + (-8) + (-6) + 9 + 3$

28. $-10 + (-7) + 12 + (-12) + 8 + 8$

29. $-2 + (-11) + 5 + 11 + (-7) + 16$

30. $6 + 10 + (-4) + (-10) + (-2)$

PROBLEM SOLVING/APPLICATION

The time for maneuvers during a space shuttle launch is given relative to liftoff, which is called *T*. The seconds before liftoff are assigned negative numbers, the seconds after liftoff are assigned positive numbers, and liftoff itself is zero. For instance, when you hear a mission controller refer to *T minus 45*, the time is 45 s before liftoff.

31. What is the meaning of *T plus 50*?

32. Give the expression for one half minute before liftoff.

33. Do you think that the times for these maneuvers are *estimates* or *exact times*? Explain.

34. A maneuver begins at *T minus 40* and requires 25 s for completion. When should this maneuver be completed?

35. A maneuver begins at *T plus 15*. Preparation begins one minute earlier. At what time must the preparation begin?

36. **RESEARCH** Find out how the expression *T minus ten and counting* is different from the expression *T minus ten and holding*.

GROUP ACTIVITY

Suppose that ■ represents one integer in each exercise. Replace each ■ with the integer that makes the statement true.

37. $6 + ■ = -2$ **38.** $-4 + ■ = 7$ **39.** $■ + 10 = 0$

40. $5 + ■ + ■ = -3$ **41.** $-8 + ■ + ■ = -12$

42. $■ + (-1,000,000) = 1$ **43.** $■ + 1,000,000,000 = -1$

Suppose that ■ and ▲ represent different integers.

44. List four pairs of integers for which $■ + ▲ = -1$ is true. Compare your list with other lists in the group. Are all lists the same?

45. For how many different pairs of integers is $■ + ▲ = -1$ a true statement? Give a convincing argument to support your answer.

SPIRAL REVIEW

46. Find the sum mentally: $27 + 21 + 43$ *(Lesson 1-6)*

47. Continue the pattern: 1, 5, 9, 13, __?__, __?__, __?__   *(Lesson 2-5)*

48. Write 2,700,000 in scientific notation.   *(Lesson 2-10)*

49. Find the sum: $-44 + 16$ *(Lesson 3-3)*

Modeling Integers Using Integer Chips

Objective: To use integer chips to model integers.

Materials

▪ integer chips

A ⊞ integer chip represents positive 1. A ⊟ integer chip represents negative 1. Because $1 + (-1) = 0$, the pair of integer chips ⊞⊟ represents 0.

Activity I *Representing Integers*

1 Name the integer represented in each diagram.

a. ⊞ ⊞ ⊞ ⊞ ⊞ **b.** ⊟ ⊟ ⊟ **c.** ⊟ **d.** ⊞ ⊞

2 Show how to use integer chips to represent each integer.
  a. 4 **b.** −2 **c.** −5 **d.** 3

3 Show how to represent the number 0 using the number of chips indicated.
  a. 2 **b.** 4 **c.** 6 **d.** 8

Activity II *Combining Chips*

The combination of ⊞ and ⊟ chips in the diagram below represents −2.

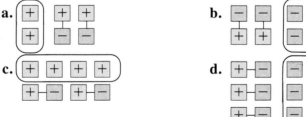

1 Name the integer represented by each diagram.

a. **b.**

c. **d.**

2 Show how to combine as many integer chips as possible in each diagram. What integer does the diagram represent?

a. ⊞ ⊞ ⊞ / ⊟ ⊟ ⊟

b. ⊞ ⊞ ⊞ ⊞ / ⊟ ⊟ ⊟

c. ⊟ ⊟ ⊟ / ⊞ ⊟

d. ⊞ ⊞ ⊞ ⊞ / ⊞ ⊞ ⊞ ⊞ / ⊞ ⊟ ⊟ ⊟

Activity III   *Using Pairs of Zero*

1 Use the diagram below. Find the minimum number of ⊞ or ⊟ chips you would have to add to the diagram to obtain each integer.

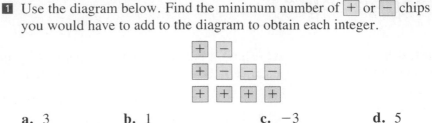

a. 3   **b.** 1   **c.** −3   **d.** 5

2 Use the diagram from Step 1. Explain how you could add two integer chips without changing the integer that is represented.

3 Show how to represent the integer −3 using the number of chips indicated.

a. 3   **b.** 5   **c.** 7   **d.** 9

4 Is it possible to represent the integer −3 using an even number of chips? Explain.

Activity IV   *Adding Integers*

To add integers using integer chips, bring all the chips together. Combine any pairs of ⊞ and ⊟ chips. The remaining integer chips represent the sum.

5 negative chips   8 positive chips   After combining 5 pairs of 0, you have 3 positive chips.

−5   +   8   =   3

1 Write the addition that is represented by each diagram. Combine the chips to find the sum.

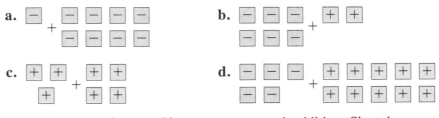

2 Show how to use integer chips to represent each addition. Show how you would combine the chips to find the sum. What is the sum?

a. 4 + 6   **b.** −2 + (−7)   **c.** −8 + 4   **d.** 6 + (−5)

3 Suppose you are trying to find a sum using integer chips. Explain how you can tell what the sign of the sum will be.

Subtracting Integers

Objective: To find the difference of two integers.

APPLICATION

An atom is made up of tiny particles: electrons, protons, and neutrons. An electron has a charge of -1, a proton has a charge of $+1$, and a neutron has no charge. During chemical reactions, only electrons move to other atoms. If an atom loses or gains electrons, its charge changes.

QUESTION During a chemical reaction, an atom with a charge of -3 loses one electron. What is the charge of the atom after the reaction?

To find the charge, represent the electron by -1. Then subtract -1 from the atom's charge of -3.

Example 1

Solution

Find the difference $-3 - (-1)$.

One way to subtract integers is to use integer chips.

Start with 3 negative chips.	Take away 1 negative chip.	2 negative chips remain.
-3	$-(-1)$ =	-2

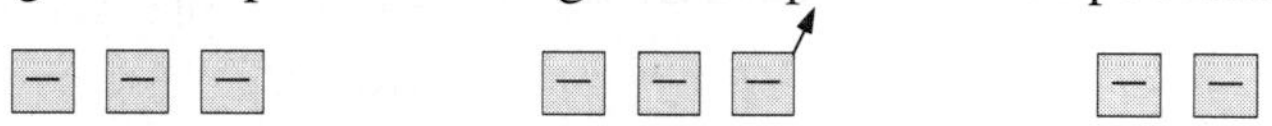

ANSWER The atom has a charge of -2 after the reaction.

Example 2

Solution

Find the difference $4 - 6$.

Start with 4 positive chips.	In order to take away 6 positive chips, add 2 sets of 0.	2 negative chips remain.
4	-6 =	-2

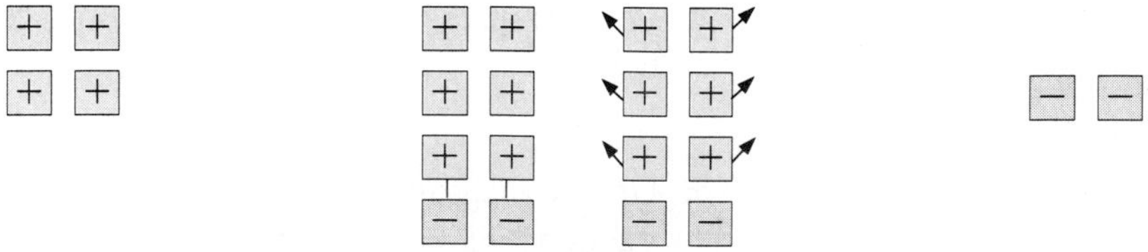

✔ **Check Your Understanding**

1. In Example 2, why is the answer negative?

When you find the sum $-3 + 1$, the answer is -2. In Example 1 you found that the difference $-3 - (-1)$ also is -2. You can see that subtracting -1 is the same as adding 1.

> **Generalization:** *Subtracting Integers*
>
> Subtracting an integer is the same as adding its opposite.
> $$a - b = a + (-b)$$

Example 3

Solution

Find each difference: **a.** $-3 - 4$ **b.** $-7 - (-8)$

a. $-3 - 4 = -3 + (-4)$
       $= -7$

b. $-7 - (-8) = -7 + 8$
       $= 1$

✓ **Check Your Understanding**

2. In Example 3(b), why is $-7 - (-8)$ written as $-7 + 8$?

Some calculators have a change-sign key, $\boxed{+/-}$. Pressing this key changes the sign of the number that is displayed. You use this key to enter negative numbers. For example, you can use this key sequence to add -49 and -78.

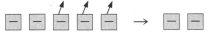

Guided Practice

«**1.** COMMUNICATION «*Reading* Replace the __?__ with the correct word.

To subtract an integer, add its __?__.

COMMUNICATION «*Writing*

«**2.** Let $\boxed{-}$ represent -1. Write the subtraction represented by the diagram below.

«**3.** Draw a diagram similar to the one in Exercise 2 to illustrate this subtraction: $-3 - (-2) = -1$

Replace each __?__ with the integer that makes the statement true.

4. $6 - (-8) = 6 + $ __?__

5. $-12 - 14 = -12 + $ __?__

Find each difference.

6. $12 - (-3)$ **7.** $-4 - 14$ **8.** $32 - 40$

9. $-11 - (-15)$ **10.** $-24 - (-16)$ **11.** $19 - 7$

Exercises

Find each difference.

1. $-5 - 4$

2. $-21 - 3$

3. $-12 - (-18)$

4. $-3 - (-3)$

5. $10 - (-15)$

6. $7 - (-14)$

7. $9 - 9$

8. $16 - 5$

9. $8 - 11$

10. $6 - 35$

11. $-13 - (-9)$

12. $-28 - (-8)$

13. $-18 - (-18)$

14. $-24 - (-32)$

15. $-15 - 37$

16. $38 - 47$

17. $11 - 7$

18. $23 - 23$

19. The record low temperature in Denver for March is $-11°F$. The normal low temperature for March is $25°F$. How much greater is the normal than the record temperature?

20. The peak of Mount Everest is 8848 m above sea level. The shore of the Dead Sea is 400 m below sea level. How much higher is the peak of Mount Everest than the shore of the Dead Sea?

Evaluate each expression when $c = -3$ and $d = 9$.

21. $c - (-6)$

22. $-4 - d$

23. $7 - d - 14$

24. $5 - 19 - c$

25. $c - d$

26. $d - c$

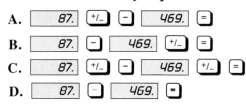 **CALCULATOR**

Match each subtraction with the correct calculator key sequence.

27. $87 - (-469)$

28. $87 - 469$

29. $-87 - (-469)$

30. $-87 - 469$

A. [87.] [+/-] [-] [469.] [=]

B. [87.] [-] [469.] [+/-] [=]

C. [87.] [+/-] [-] [469.] [+/-] [=]

D. [87.] [-] [469.] [=]

PATTERNS

Find the next three integers in each pattern.

31. $16, 11, 6, 1, \underline{\quad?\quad}, \underline{\quad?\quad}, \underline{\quad?\quad}$

32. $2, 0, -3, -7, \underline{\quad?\quad}, \underline{\quad?\quad}, \underline{\quad?\quad}$

33. $-23, -19, -15, -11, \underline{\quad?\quad}, \underline{\quad?\quad}, \underline{\quad?\quad}$

34. $-1, -3, -5, -7, \underline{\quad?\quad}, \underline{\quad?\quad}, \underline{\quad?\quad}$

Meteorologists analyze weather data and forecast the weather. They report the *wind-chill factor* on windy days. The wind-chill factor is the temperature that results from the air temperature combined with the wind speed. Meteorologists use a chart like the one below to find the wind-chill factor.

Wind Speed in Miles per Hour	Air Temperature (°F)			
	20	10	0	−10
10	3	−9	−22	−34
20	−10	−24	−39	−53
30	−18	−33	−49	−64

35. The air temperature is 10°F and the wind speed is 30 mi/h. What is the wind-chill factor?

36. The air temperature is −10°F. How much colder does it feel when the wind speed is 10 mi/h than when there is no wind?

37. The air temperature is 0°F. The wind speed increases from 10 mi/h to 20 mi/h. By how much does the temperature seem to drop?

38. ESTIMATION The wind speed is 10 mi/h and the air temperature is 15°F. Estimate the wind-chill factor.

SPIRAL REVIEW

39. DATA, *pages 92–93* What is the record low January temperature for Houston? *(Toolbox Skill 24)*

40. Find the difference: $5 - 9$ *(Lesson 3-4)*

41. Write 107 cm in mm. *(Lesson 2-9)*

42. Find the quotient: $\frac{9}{10} \div \frac{2}{5}$ *(Toolbox Skill 21)*

43. Estimate the sum: $7.9 + 8.4 + 8.1 + 7.8$ *(Toolbox Skill 4)*

44. Evaluate $a + 10.9$ when $a = 17.5$. *(Lesson 1-2)*

Multiplying and Dividing Integers

Objective: To find products and quotients of integers.

DATA ANALYSIS

Companies that make a profit are said to be operating "in the black." A company that loses money is operating "in the red." The pictograph at the right uses black symbols for profits and red symbols for losses. For 1988, you can calculate the loss in millions of dollars by finding $2(-3)$.

Markets, Inc.			
1988	$ $		
1989	$		
1990		$ $	
1991		$ $ $	

$ = -\$3$ million $ = \$3$ million

Example 1

Find the product $2(-3)$.

Solution

You can use integer chips to multiply integers.

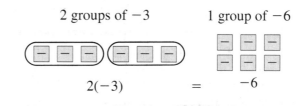

2 groups of -3 1 group of -6

$2(-3)$ $=$ -6

Example 1 shows that, when you multiply a negative integer by a positive integer, the product is *negative*. What is the result when you multiply a positive integer by a negative integer? From Example 1 and the commutative property of multiplication, you know these two facts.

$$2(-3) = -6$$
$$2(-3) = (-3)2$$

You can conclude that $(-3)2$ must also equal -6. So, when you multiply a positive integer by a negative integer, the product is *negative*.

What is the result when you multiply two negative integers? Using Example 1 as a starting point, you can generate the pattern at the right. Notice that, when -3 is multiplied by a negative integer, the product is *positive*.

$$2(-3) = -6$$
$$1(-3) = -3$$
$$0(-3) = 0$$
$$(-1)(-3) = 3$$
$$(-2)(-3) = 6$$

Generalization: *Multiplying Integers*

The product of two integers with the same sign is positive.
The product of two integers with different signs is negative.

Example 2

Find each product: **a.** $(-11)(4)$ **b.** $(-7)(-10)$ **c.** $(-5)(4)(-2)$

Solution

a. $(-11)(4)$ **b.** $(-7)(-10)$ **c.** $(-5)(4)(-2)$
$\quad = -44$ $\qquad\qquad\quad = 70$ $\qquad\qquad\qquad = (-20)(-2)$
$\qquad\qquad\qquad\qquad\qquad\qquad\qquad\qquad\qquad\qquad = 40$

☑ **Check Your Understanding**

1. In Example 2(c), why is the product positive?

You can use the relationship between multiplication and division to divide integers.

$$3(-7) = -21 \qquad \rightarrow \qquad -21 \div (-7) = 3$$
$$(-3)7 = -21 \qquad \rightarrow \qquad -21 \div 7 = -3$$
$$(-3)(-7) = 21 \qquad \rightarrow \qquad 21 \div (-7) = -3$$

> **Generalization:** *Dividing Integers*
>
> The quotient of two integers with the same sign is positive.
> The quotient of two integers with different signs is negative.

Keep in mind that division by zero is *undefined*.

Example 3

Find each quotient: **a.** $-15 \div 5$ **b.** $\dfrac{-48}{-12}$

Solution

a. $-15 \div 5 = -3$ **b.** $\dfrac{-48}{-12} = 4$

☑ **Check Your Understanding**

2. In Example 3(a), how do you know the quotient is negative?

Guided Practice

«**1.** COMMUNICATION «*Reading* Replace each __?__ with the correct word.

The product of two integers is positive if both factors are __?__ or both factors are __?__. The quotient of two integers is __?__ if the two integers have different signs.

Tell whether each answer is *positive*, *negative*, *zero*, or *undefined*.

2. $(-7)14$ **3.** $(-19)(0)(-3)$ **4.** $8 \div (-2)$

5. $\dfrac{0}{-5}$ **6.** $-6 \div 0$ **7.** $\dfrac{-24}{-6}$

Find each answer.

8. $2(-5)$

9. $(-8)(9)$

10. $4(-32)(-3)$

11. $0(-2)(-5)$

12. $-63 \div (-9)$

13. $-36 \div 6$

14. $\frac{39}{-13}$

15. $\frac{-84}{-4}$

16. $(-7)(-11)(-2)$

Exercises

Find each answer.

1. $(-4)(-1)$

2. $2(-30)$

3. $-77 \div 7$

4. $-42 \div (-6)$

5. $(6)(0)(-15)$

6. $-70 \div (-5)$

7. $32 \div (-8)$

8. $-10(12)$

9. $(-7)(-22)$

10. $144 \div (-3)$

11. $-250 \div 25$

12. $3(3)(-7)$

13. $(-4)(5)(-5)$

14. $(-6)(3)(0)$

15. $(-7)(-6)(-2)$

16. $\frac{75}{-5}$

17. $\frac{-48}{3}$

18. $\frac{-169}{-13}$

19. Tina Landon noticed an unusual temperature change of $-20°$F in just 4 h. Assume that the temperature changed at a steady rate. What was the change in temperature each hour?

20. Maria Savino's watch lost 2 min every day. How many minutes did the watch lose over a period of three days?

Use the order of operations to find each answer.

21. $-18 \div 3 + 5(-6)$

22. $16 \div 4 + 2(-8)$

23. $-3(1 - 8) + 2^3$

24. $\frac{2}{8 - 10}$

25. $\frac{-39 + 3}{-4}$

Evaluate each expression when $x = -12$.

26. $2x + 1$

27. $3x - 5$

28. $8 - 5x$

29. $\frac{x}{-2} + 5$

30. $\frac{x}{-3} - 6$

31. $9 - \frac{x}{4}$

Simplify.

32. $2(5d - 2)$

33. $4(-3c + 6)$

34. $-2(-3a + 5)$

35. $-7(-2y - 8)$

36. $6(-7 + x)$

37. $-5(4 - 4b)$

 CALCULATOR

Match each calculator key sequence with the correct result. Assume that the calculator follows the order of operations.

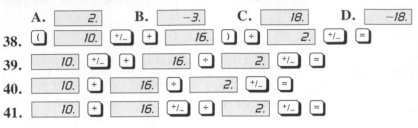

A. `2.` B. `-3.` C. `18.` D. `-18.`

38. `(` `10.` `+/-` `+` `16.` `)` `÷` `2.` `+/-` `=`

39. `10.` `+/-` `+` `16.` `÷` `2.` `+/-` `=`

40. `10.` `+` `16.` `÷` `2.` `+/-` `=`

41. `10.` `+` `16.` `+/-` `÷` `2.` `+/-` `=`

CONNECTING MATHEMATICS AND SCIENCE

Ions have electrical charges. If the sum of their charges is zero, ions combine to form compounds. For example, calcium chloride has one calcium ion and two chloride ions, because $1(+2) + 2(-1) = 0$.

Ion	Charge
sodium	+1
calcium	+2
aluminum	+3
chloride	−1
oxide	−2

Find the number and types of ions in each compound.

42. calcium oxide

43. aluminum chloride

44. A compound contains three oxide ions. How many aluminum ions does it contain? What is the compound's name?

45. RESEARCH   Find the common name and use for each compound.
a. sodium chloride   **b.** sodium carbonate   **c.** sodium bicarbonate

SPIRAL REVIEW

46. Find the sum: $4\frac{1}{9} + 6\frac{1}{6}$ *(Toolbox Skill 18)*

47. Find the quotient: $-60 \div 5$ *(Lesson 3-5)*

Self-Test 1

Replace each __?__ with >, <, or =.

1. 5 __?__ -6	**2.** -9 __?__ -4	**3.** -3 __?__ -3	**3-1**

Find each answer.

4. $-2 + (-2)$	**5.** $-5 + (-17)$	**6.** $-4 + (-14)$	**3-2**
7. $-4 + 9$	**8.** $7 + (-13)$	**9.** $-16 + 16$	**3-3**
10. $-26 - 5$	**11.** $14 - 28$	**12.** $-1 - (-7)$	**3-4**
13. $8(-4)$	**14.** $(-7)(5)(-2)$	**15.** $36 \div (-3)$	**3-5**

3-6

Evaluating Expressions Involving Integers

Objective: To evaluate expressions involving integers.

CONNECTION

In previous chapters you learned how to evaluate expressions. In this chapter you learned how to perform the four basic operations with integers. You can combine these skills to evaluate expressions in which the values of the variables are integers. These expressions might also involve exponents, absolute value, and opposites.

Example 1

Solution

Evaluate each expression when $y = -4$:   **a.** y^2 **b.** $2y^3 + 18$

a. $y^2 = (-4)^2$
$\qquad = 16$ ⟵ $(-4)^2 = (-4)(-4) = 16$

b. $2y^3 + 18 = 2(-4)^3 + 18$
$\qquad\qquad = 2(-64) + 18$
$\qquad\qquad = -128 + 18$
$\qquad\qquad = -110$

☑ **Check Your Understanding**

1. In Example 1(b), why does $(-4)^3$ equal -64?

Absolute value signs have the same priority as parentheses in the order of operations. When evaluating expressions involving absolute value, you evaluate any expression within absolute value signs first.

Example 2

Evaluate each expression when $c = -9$ and $d = 4$.
a. $|c + d|$ **b.** $|c| + |d|$

Solution

a. $|c + d| = |-9 + 4|$
$\qquad\quad = |-5|$
$\qquad\quad = 5$

b. $|c| + |d| = |-9| + |4|$
$\qquad\qquad = 9 + 4$
$\qquad\qquad = 13$

☑ **Check Your Understanding**

2. In Example 2(b), why do you add 9 and 4 instead of -9 and 4?

You will sometimes find it necessary to use the *multiplication property of −1* when evaluating expressions involving integers.

> **Multiplication Property of −1**
>
> The product of any number and −1 is the opposite of the number.
>
> $$-1n = -n \quad \text{and} \quad -n = -1n$$

Example 3

Evaluate each expression when $p = -3$ and $q = 5$.

a. $-q + 9$ **b.** $-pq$

Solution

a. $\begin{aligned} -q + 9 &= -1q + 9 \\ &= (-1)(5) + 9 \\ &= -5 + 9 \\ &= 4 \end{aligned}$ **b.** $\begin{aligned} -pq &= -1pq \\ &= (-1)(-3)(5) \\ &= 3(5) \\ &= 15 \end{aligned}$

✔️ **Check Your Understanding**

3. In Example 3(b), why can $-pq$ be written as $-1pq$?

Guided Practice

COMMUNICATION « *Reading*

Refer to the text on pages 115–116.

«**1.** State the multiplication property of −1 in words.

«**2.** Where does absolute value fit in the order of operations?

Find each answer.

3. $(-7)^2$ **4.** $(-1)^5$ **5.** $|-16 + 23|$ **6.** $|-5| - |-4|$

Evaluate each expression when $a = -5$, $b = -10$, and $c = 7$.

7. b^3 **8.** $4a^2$ **9.** $b^2 + (-6)$

10. $|a - b|$ **11.** $|a| - |b|$ **12.** $-10 - |c|$

13. $-ac$ **14.** $-b - 18$ **15.** $-ab + 12$

Exercises

Evaluate each expression when $m = -3$, $n = 8$, and $s = -6$.

1. s^4 **2.** m^5 **3.** $5m^3$

4. $-10s^2$ **5.** $s^3 + 50$ **6.** $m^4 - 2$

7. $-9 + 4m^2$ **8.** $30 - 4s^2$ **9.** $m^2 - n$

10. $2s^2 + n$ **11.** $|m - n|$ **12.** $|m + s|$

Evaluate each expression when $m = -3$, $n = 8$, and $s = -6$.

13. $|n| + |s|$

14. $|m| + |n|$

15. $|m + 3| - 6$

16. $7 - |n - 8|$

17. $|m| - |n|$

18. $|s| - |n|$

19. $-mn$

20. $-ns$

21. $-s + (-7)$

22. $-n - 11$

23. $-ns - 6$

24. $-mn + 14$

25. Find the sum of 10 squared and -4 cubed.

26. Find the difference when -11 is subtracted from -3 squared.

27. Find the difference when the absolute value of 24 is subtracted from the opposite of -9.

28. Find the sum when the absolute value of -8 is added to the opposite of 5.

LOGICAL REASONING

Choose the correct relationship between the expressions in columns I and II. Assume that each variable can represent any integer except zero.

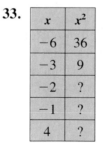

	I	II
29.	pq	$-pq$
30.	$(3m)^2$	$(-3m)^2$
31.	$|c|$	$-|c|$
32.	$-5a^2$	$6a^2$

A. I > II

B. I < II

C. I = II

D. cannot determine

FUNCTIONS

Complete each function table.

33.

x	x^2
-6	36
-3	9
-2	?
-1	?
4	?

34.

x	$-x + 2$
-5	7
-4	6
-1	?
0	?
2	?

Use each table to find the function rule.

35.

x	?
-6	24
-5	20
-4	16
1	-4
3	-12

36.

x	?
-4	-10
-1	-7
0	-6
3	-3
8	2

Evaluate each expression when $x = -1$.

37. x^2 **38.** x^3 **39.** x^{10} **40.** x^{15} **41.** x^{47} **42.** x^{100}

43. Assume that n is a positive integer. Explain how you know whether the value of $(-1)^n$ is 1 or -1.

44. Assume that n is a positive integer and that a is any negative integer. Explain how you know whether the value of a^n is positive or negative.

Without computing, tell whether each answer is *positive* or *negative*.

45. $(-2)^{14}$ **46.** $(-3)^9$ **47.** $(-68)^3$

48. $(-122)^2$ **49.** $(-7)^{24}$ **50.** $(-8)^{15}$

51. Use the line graph at the right. What was the population of the United States in 1950? *(Toolbox Skill 25)*

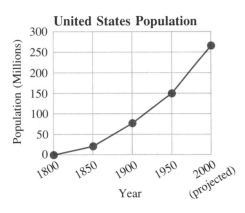

United States Population

52. Estimate the sum: $547 + 298 + 369$ *(Toolbox Skill 2)*

53. Evaluate $-2x + 6$ when $x = -5$.   *(Lesson 3-6)*

54. Find the difference: $-34 - 14$ *(Lesson 3-4)*

55. Julia rode her bicycle in a charity bike-a-thon. She collected $.10/mile, $.75/mile, and $1/mile from each of three friends. Two others paid $.25/mile each. Her brother and sister paid $.50/mile each. Julia rode 25 mi. She said that she raised $71.25 for charity. Is this correct? Explain.   *(Lesson 2-7)*

56. Replace the __?__ with >, <, or =: -9 __?__ 4 *(Lesson 3-1)*

57. Multiply mentally: $10(6)(0)(5)$ *(Lesson 1-7)*

58. Find the difference: $4\frac{2}{5} - 2\frac{3}{10}$ *(Toolbox Skill 19)*

Challenge

There are many ways to write an expression equal to -2. For example, $-6 \div 2 + 5 \cdot 1 - 4 = -2$. Find two other expressions that equal -2. Use only the integers from -9 to 9, do not repeat any integer, and use all four operations exactly once.

3-7

Strategy:
Making a Table

Objective: To solve problems by making a table.

Some problems ask you to find all possibilities in a given situation. To help you solve this type of problem, make a table to organize the information. You list the possibilities without regard to order. For example, a listing of two quarters and one dime is the same as one dime and two quarters.

Problem

Benito Alomar is a sales clerk in a shoe store. The cash register in the store contains only quarters, dimes, and nickels. In how many different ways could he make 35¢ in change?

Solution

UNDERSTAND The problem is about a sales clerk making change.
Facts: only quarters, dimes, and nickels
        35¢ change needed
Find: number of ways to make the change

PLAN You can make a table that lists each type of coin. To find the total value of the coins, multiply the number of quarters by 25, the number of dimes by 10, and the number of nickels by 5. Then add.

WORK

Number of Quarters	1	1	0	0	0	0
Number of Dimes	1	0	3	2	1	0
Number of Nickels	0	2	1	3	5	7
Total Value of Coins	35	35	35	35	35	35

ANSWER He could make 35¢ in change in 6 different ways.

Look Back What if Benito Alomar used four coins? How many different amounts could he make using only quarters, dimes, and nickels?

Guided Practice

Use the problem below for Exercises 1–4.

Last weekend Pets, Inc. cared for 12 animals, all of which were either dogs or birds. The workers at Pets, Inc. counted the number of animal legs. How many different totals are possible?

« **1.** How many animals were at Pets, Inc. last weekend?

« **2.** What was the greatest number of dogs possible at Pets, Inc. last weekend?

« **3.** What was the least number of birds possible at Pets, Inc. last weekend?

   4. Solve the problem by making a table.

   5. Solve by making a table: Basketball shots are worth one, two, or three points. In a basketball game Mei Lin scored 7 points. In how many different ways could she have scored the 7 points?

« **6.** COMMUNICATION « *Discussion*   There are several ways to organize information in a table. Discuss these ways. Which seems most efficient to you? Explain.

Problem Solving Situations

Solve by making a table.

   1. Lewanda Warner has to pay $27 for food at the grocery store. In how many different ways can she pay the $27 with bills of $10, $5, and $1?

   2. Three darts are thrown at the target at the right. All three darts hit the target. How many different point totals are possible?

   3. Carpenter Joe Gibbs makes three-legged tables and four-legged tables. Both kinds of tables have the same type of leg. Last month Joe made 14 tables and then to-taled the number of legs he made. How many different totals are possible?

   4. Stanley Weldman scored 11 points in a basketball game. Basketball shots are worth one, two, or three points. In how many different ways could Stanley have scored the 11 points?

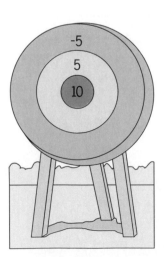

Solve using any problem solving strategy.

5. Four darts are thrown at the target shown on page 120. All four darts hit the target. How many different point totals are possible?

> **PROBLEM SOLVING**
> **CHECKLIST**
> ***Keep this in mind:***
>   Using a Four-Step Plan
> ***Consider these strategies:***
>   Choosing the Correct Operation
>   Making a Table

6. Daniel Kuo bought two gallons of milk for $1.99 each, a dozen eggs for $1.19, and a loaf of bread for $1.09. He gave the cashier a $10 bill. How much change did Daniel receive?

7. Angela Marini wants to buy 8 tickets to a concert. Tickets cost $20 each in advance and $24 each on the day of the concert. There is a $2 discount for sales of 10 tickets or more. How much will Angela save if she buys the tickets in advance?

8. Juniors at Bentley High School held a car wash and earned $100. They charged $6 per truck and $4 per car. In how many different ways could the juniors have earned the $100?

9. In ice hockey, a team earns 2 points for a win, 1 point for a tie game, and no points for a loss. The Ice Pirates earned 12 points and lost 3 games in January. In how many different ways could the team have earned the 12 points?

10. A restaurant contains two-person tables and four-person booths. There are 15 tables and 28 booths in the restaurant. How many people can dine at the restaurant at one time?

WRITING WORD PROBLEMS

For each exercise, use one piece of information at the right. Write a word problem that you would solve by making a table. Then solve the problem.

11. Use quarters, dimes, and nickels in the problem.

12. Use dimes, nickels, and pennies in the problem.

SPIRAL REVIEW

13. Find the sum: $-8 + (-4)$ *(Lesson 3-2)*

14. Bill Blanchette has to pay $43. In how many different ways could he pay the $43 with bills worth $20, $10, and $1?   *(Lesson 3-8)*

15. Evaluate $x - |y|$ when $x = -3$ and $y = -8$.   *(Lesson 3-6)*

16. Find the product: $(11.5)(3.2)$ *(Toolbox Skill 13)*

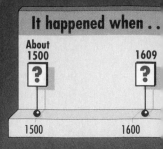

MIXED REVIEW
Operations with Integers

On September 4, 1781, forty-four pioneers ended a seven-month journey from Sonora, Mexico, to the site of a new town in Alta California. The group—made up of Africans, Native Americans, Spaniards, and people of mixed ancestry—had come to California to found what would become a famous American city.

To find out the name of the new settlement, complete the code boxes below. Begin by solving Exercise 1. The answer to Exercise 1 is −9, and the letter associated with Exercise 1 is *U*. Write *U* in the box above −9 as shown below. Continue in this way with Exercises 2–24.

1. $-4 + (-5)$ (U)
2. $12 + 7$ (A)
3. $35 + (-97)$ (L)
4. $-28 - (-5)$ (D)
5. $8(-4)$ (E)
6. $(-7)(-9)$ (G)
7. $\frac{55}{5}$ (E)
8. $\frac{-96}{12}$ (E)
9. $728 \div (-14)$ (O)
10. $-69 \div (-3)$ (B)
11. $5(-7)$ (L)
12. $(-8)(10)$ (P)
13. $-17 - 17$ (O)
14. $-31 - (-31)$ (A)
15. $(-15)(6)$ (T)
16. $(-5)(-7)$ (S)
17. $-11 + 19$ (A)
18. $14 + (-27)$ (L)
19. $\frac{-64}{16}$ (O)
20. $\frac{-48}{-8}$ (A)
21. $46 - 83$ (D)
22. $33 - (-21)$ (U)
23. $57 + (-5)$ (A)
24. $(-9)(3)$ (L)

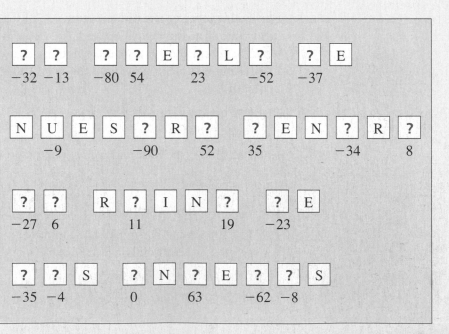

?	?		?	?	E	?	L	?		?	E
−32	−13		−80	54		23		−52		−37	

N	U	E	S	?	R	?		?	E	N	?	R	?
	−9					−90		52		35		−34	8

?	?		R	?	I	N	?		?	E
−27	6			11			19		−23	

?	?	S		?	N	?	E	?	?	S
−35	−4			0		63		−62	−8	

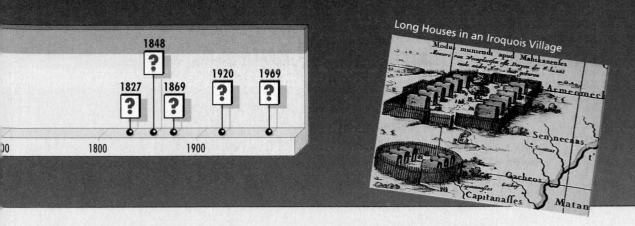

1848 ?

1827 ? 1869 ? 1920 ? 1969 ?

1800 1900

Long Houses in an Iroquois Village

The timeline on this page gives seven important dates in American history. Each date is associated with one of the events listed next to Exercises 25–31. To match each date with the correct event, first solve all seven exercises. Then put your answers in order from least to greatest. The order of your answers will correspond to the order of the dates on the timeline, from 1500 to 1969. For example, the event associated with the smallest answer took place first.

25. $-20 + (-4) + 38$

Women win voting rights with the passage of the 19th Amendment to the Constitution.

27. $5(-3)(4)$

The Cayuga, Mohawk, Oneida, Onondaga, and Seneca nations form the Iroquois Confederacy, whose system of representative government later inspires the drafters of the United States Constitution.

29. $\dfrac{-36}{-9}$

Men and women meet at Seneca Falls, New York, to draw up a declaration of rights for women.

26. $-5 + 17 + (-3)$

Workers, most of them Chinese and European immigrants, complete the first transcontinental railroad.

28. $(-2)(0)(-5)$

The United States' first newspaper run by African-Americans, *Freedom's Journal*, is founded in New York.

30. $19 + (-28) + 37$

American pioneers in space first walk on the moon.

31. $-8 + (-15) + (-3)$

Hispanic settlers from Mexico establish Santa Fe as the capital of the colony of New Mexico.

Supporters of Women's Voting Rights

Oldest House, Santa Fe, New Mexico

First Transcontinental Railroad

3-8 The Coordinate Plane

Objective: To find coordinates and graph points on a coordinate plane.

The grid at the right is called a **coordinate plane.** A coordinate plane is formed by two number lines called **axes.** The horizontal number line is the **x-axis,** and the vertical number line is the **y-axis.** The point where the axes meet is called the **origin.** The axes separate the coordinate plane into four sections called **quadrants.**

You can assign an **ordered pair** of numbers to any point on the plane. The first number in an ordered pair is the **x-coordinate.** The second number is the **y-coordinate.** The origin has coordinates (0, 0).

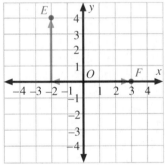

Coordinate Grid

Example 1

Use the coordinate plane at the right. Write the coordinates of each point.

a. *E* b. *F*

Solution

a. Start at the origin. Point *E* is 2 units left (negative) and 4 units up (positive). The coordinates are $(-2, 4)$.

b. Start at the origin. Point *F* is 3 units right (positive) and 0 units up or down. The coordinates are (3, 0).

☑ Check Your Understanding

1. In Example 1(a), why is the *x*-coordinate negative?
2. In Example 1(b), why is the *y*-coordinate zero?

When you **graph a point** $A(x, y)$ on a coordinate plane, you show the point that is assigned to the ordered pair (x, y).

Generalization: *Graphing a point A(x, y)*

1. Start at the origin.
2. Move *x* units horizontally along the *x*-axis.
3. Then move *y* units vertically.
4. Draw the point and label it *A*.

Example 2

Graph each point on a coordinate plane.

   a. $A(4, 1)$ **b.** $B(1, 4)$

   c. $C(-3, 0)$ **d.** $D(0, -4)$

Solution

In each case, start at the origin.

   a. Move 4 units to the right and then 1 unit up.

   b. Move 1 unit to the right and then 4 units up.

   c. Move 3 units to the left and then 0 units up or down.

   d. Move 0 units to the left or right and then 4 units down.

✓ **Check Your Understanding**

3. In Example 2(d), how do you know point D is on the y-axis?

4. Describe how the graph of point A would be different if the coordinates were $(4, -1)$.

Guided Practice

COMMUNICATION «*Reading*

Explain the meaning of the underlined word in each sentence.

«**1.** The members of the team agreed to <u>coordinate</u> their efforts.

«**2.** The <u>origin</u> of the Mississippi River is Lake Itasca, Minnesota.

«**3.** The prefix *quadr-*, as in *quadrant*, means "four."

   a. What is the meaning of the familiar term *quadruplet*?

   b. What is the meaning of the mathematical term *quadrant*?

   c. How can knowing the meaning of *quadruplet* help you remember the meaning of *quadrant*?

To graph each point, tell how many units and in what direction from the origin you would move for the underlined coordinate.

   4. $Q(\underline{2}, 3)$ **5.** $R(-8, \underline{8})$

   6. $S(\underline{-5}, -9)$ **7.** $T(4, \underline{-2})$

   8. $U(\underline{0}, 6)$ **9.** $V(-3, \underline{0})$

Describe how you would move from the origin to graph each point.

   10. $A(-7, 2)$ **11.** $B(5, -3)$

   12. $C(0, 2)$ **13.** $D(-4, 0)$

   14. $E(-2, -4)$ **15.** $F(3, 5)$

Use the coordinate plane at the right.
Write the coordinates of each point.

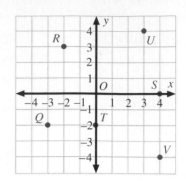

16. Q **17.** R

18. S **19.** T

20. U **21.** V

Graph each point on a coordinate plane.

22. $J(3, 5)$ **23.** $K(3, -2)$

24. $L(0, -1)$ **25.** $M(2, 0)$

Exercises

Use the coordinate plane at the right.
Write the coordinates of each point.

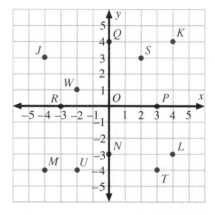

1. J **2.** K

3. L **4.** M

5. N **6.** P

7. R **8.** Q

9. S **10.** T

11. U **12.** W

Graph each point on a coordinate plane.

13. $A(5, 1)$ **14.** $B(3, 3)$ **15.** $C(6, -2)$ **16.** $O(0, 0)$

17. $E(-3, -3)$ **18.** $F(-2, 4)$ **19.** $G(-6, 1)$ **20.** $H(-4, -1)$

Tell in which quadrant or on which axis each point lies.

21. $P(1, 9)$ **22.** $Q(5, -14)$ **23.** $R(-12, 6)$ **24.** $S(-8, -3)$

25. $A(0, -3)$ **26.** $B(-12, 0)$ **27.** $C(1, 0)$ **28.** $D(0, 9)$

THINKING SKILLS

**Develop a rule that describes what is true of all points in the given part
of the coordinate plane.**

29. Quadrant I   **30.** Quadrant II

31. Quadrant III   **32.** Quadrant IV

33. Quadrants II and III   **34.** Quadrants III and IV

35. the x-axis   **36.** the y-axis

CONNECTING ALGEBRA AND GEOMETRY

In *coordinate geometry*, you represent geometric figures on a coordinate plane. For example, at the right you see a square on a coordinate plane. The points $A(-2, -2)$, $B(-2, 2)$, $C(2, 2)$, and $D(2, -2)$ are called the *vertices* of the square.

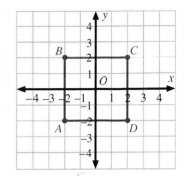

37. What is the shape of the figure that has vertices $Q(-3, 5)$, $R(-3, 1)$, $S(4, 1)$, and $T(4, 4)$?

38. Figure *RSTV* is a square. Its vertices are $R(-1, 4)$, $S(-1, 1)$, $T(2, 1)$ and a point V. What are the coordinates of V?

39. Triangle *ABC* has vertices $A(4, -3)$, $B(2, 1)$, and $C(0, -1)$. Add 2 to the *x*-coordinate and 3 to the *y*-coordinate of all the vertices. Call the new points D, E, and F. Compare the shape and location of triangle *DEF* to the shape and location of triangle *ABC*.

40. Rectangle *JKLM* has vertices $J(2, 2)$, $K(2, 5)$, $L(8, 5)$, and $M(8, 2)$. Multiply the *x*-coordinate and *y*-coordinate of all the vertices by -1. Call the new points N, P, Q, and R. Compare the shape and location of rectangle *NPQR* to the shape and location of rectangle *JKLM*.

SPIRAL REVIEW

41. Simplify: $4a - 7b + 3a$ *(Lesson 2-3)*

42. Find the product: $(-16)(7)$ *(Lesson 3-5)*

43. Graph the point $A(-5, 4)$ on a coordinate plane.   *(Lesson 3-8)*

44. Do you need an *estimate* or an *exact answer* to find the amount of money you should bring to school to buy lunch?   *(Lesson 2-8)*

Historical Note

In the ancient world, the Egyptians and the Romans used coordinates in surveying. The idea of using a coordinate plane to graph points assigned to ordered pairs came much later. René Descartes, a French mathematician, developed it in a book published in 1637. This idea revolutionized mathematics because it tied together geometry and algebra.

Research

Find out why it was important to the Egyptians to learn how to use coordinates to survey the land.

Graphing Functions

Objective: To graph FUNC-TIONS on a coordinate plane.

 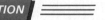

EXPLORATION

1 Make a list of the ordered pairs that are graphed at the right.

2 Complete this statement: In this list, each *x*-coordinate is paired with exactly one *y*-coordinate, so the relationship is a ___?___.

3 Use the ordered pairs in your list to make a function table.

4 Write the function rule.

In Lesson 2-6, a function was defined as a relationship that pairs each number in a given set of numbers with exactly one number in a second set of numbers. You learned to represent this pairing of numbers using either a *function table* or a *function rule*. For instance, the function shown in the table at the right can be represented by the rule $x \rightarrow x - 2$.

x	$x - 2$
-3	-5
-2	-4
0	-2
1	-1
5	3

Because it is a pairing of numbers, a function can also be represented by a set of ordered pairs. Therefore, you can draw the **graph of the function** by graphing the points that correspond to all the ordered pairs.

Example 1
Solution

Graph the function shown in the table above.

• Use the table to get a set of ordered pairs.

• Graph the ordered pairs on a coordinate plane.

$(-3, -5)$
$(-2, -4)$
$(0, -2)$
$(1, -1)$
$(5, 3)$

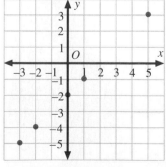

☑ Check Your Understanding

1. In Example 1, how do you determine the *x*-coordinates of the ordered pairs? the *y*-coordinates?

Example 2	Graph the function $x \rightarrow x^2 - 1$ when $x = -2, -1, 0, 2$, and 3.

Solution

- Make a function table.
- Use the table to get a set of ordered pairs.
- Graph the ordered pairs on a coordinate plane.

x	$x^2 - 1$
-2	$(-2)^2 - 1 = 3$
-1	$(-1)^2 - 1 = 0$
0	$0^2 - 1 = -1$
2	$2^2 - 1 = 3$
3	$3^2 - 1 = 8$

$\longrightarrow$

$(-2, 3)$
$(-1, 0)$
$(0, -1)$
$(2, 3)$
$(3, 8)$

✔ Check Your Understanding

2. In Example 2, how do you determine which numbers to enter in the first column of the table?

Guided Practice

COMMUNICATION «*Reading*

Refer to the text on pages 128–129.

«**1.** What is the main idea of the lesson?

«**2.** What is the main difference between Example 1 and Example 2?

«**3.** Name three different ways to represent a function.

Match each graph with the correct function rule.

4.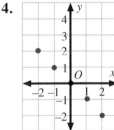

5.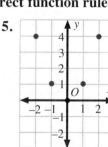

A. $x \rightarrow x$

B. $x \rightarrow -x$

C. $x \rightarrow |x|$

D. $x \rightarrow x^2$

6.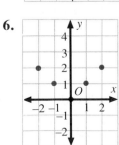

7.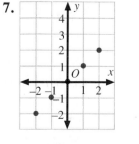

Graph each function.

8.

x	−2x
−2	4
−1	2
0	0
3	−6
4	−8

9.

x	|x| + 1
−5	6
−2	3
0	1
1	2
3	4

10. $x \to 5 - x$, when $x = -4, -2, 0, 3$, and 5

11. $x \to -x - 3$, when $x = -2, -1, 0, 3$, and 6

Exercises

Graph each function.

1.

x	x + 3
−4	−1
−3	0
0	3
1	4
2	5

2.

x	x ÷ (−2)
−6	3
−4	2
0	0
2	−1
4	−2

3. $x \to \dfrac{6}{x}$, when $x = -3, -2, 1, 2$, and 6

4. $x \to -5 + x$, when $x = -3, -1, 0, 2$, and 5

5.

x	2x − 3
−2	−7
−1	−5
0	−3
2	1
3	3

6.

x	x² + 2
−2	6
−1	3
0	2
2	6
3	11

7. $x \to -x + 2$, when $x = -5, -3, 0, 2$, and 4

8. $x \to |x - 2|$, when $x = -3, -1, 0, 4$, and 5

9. Graph the function that satisfies these conditions: The values of x are −4, −2, −1, 2, and 5, and the function rule is *x is paired with* $|x|$.

10. Graph the function that pairs the number x with $x - 4$. The values of x are −5, −1, 0, 2, and 4.

Using −2, −1, 0, 1, and 2 as values for x, graph five ordered pairs that satisfy each condition. Use the ordered pairs to make a function table. Then give the rule for the function.

11. The y-coordinate is twice the x-coordinate.

12. The y-coordinate is six more than the x-coordinate.

13. The y-coordinate is the opposite of the x-coordinate.

14. The y-coordinate is the third power of the x-coordinate.

COMPUTER APPLICATION

A type of computer software called *geometric graphing software* allows you to use a computer to graph points on a coordinate plane. For Exercises 15 and 16, use this type of software if it is available, or use graph paper.

15. Graph the function $x \rightarrow x + 5$ when $x = -4, 0$, and 2. Draw a line through the points that you graphed.
   a. Name three other points that appear to lie on this line.
   b. Does the same function rule represent the relationship between the coordinates of these three points?

16. Graph the function $x \rightarrow x - 3$ when $x = -1, 0$, and 6. Draw a line through the points that you graphed. On the same coordinate plane, graph the function $x \rightarrow -x + 1$ when $x = -4, 0$, and 3. Draw a line through these three points.
   a. Name the point where the two lines meet.
   b. Which function rule represents the relationship between the coordinates of this point?

SPIRAL REVIEW

17. Graph the function shown in the table at the right. *(Lesson 3-9)*

18. Find the sum: $13.6 + 9.4 + 7.13$ *(Toolbox Skill 11)*

19. Ed bought three shirts for $13.50 each and a pair of pants for $27.50. What was the total cost? *(Lesson 2-4)*

20. Find the difference: $-9 - (-12)$ *(Lesson 3-4)*

x	$3x - 1$
-2	-7
-1	-4
0	-1
1	2
2	5

Self-Test 2

Evaluate each expression when $a = -5$ and $b = 2$.

1. $a^2 + 3b$ 2. $|a| - b$ 3. $-ab$ 3-6

4. Solve by making a table: Pilar has quarters and dimes. She wants to buy a sandwich that costs $2.95. In how many different ways could Pilar pay for the sandwich?   3-7

Graph each point on a coordinate plane.

5. $D(-2, 3)$ 6. $E(0, -4)$ 3-8

7. Graph the function $x \rightarrow |x + 2|$ when $x = -3, -2, 0, 1$, and 4.   3-9

Map Reading

Objective: To read and
interpret a map.

Like a coordinate plane, most maps are divided into square sections by
horizontal and vertical lines. A location on a map is identified by a set of
coordinates that consists of a letter and a number. Unlike coordinates in a
coordinate plane, these map coordinates identify a square, not a single
point. A *map index* like the one shown below lists the coordinates for each
city on a map.

Below is a map of a section of northern New Jersey and adjoining states.

New Jersey Index			
Morristown	E4	New Brunswick	E2
Princeton	D1	Paterson	F5
Ramsey	F5		

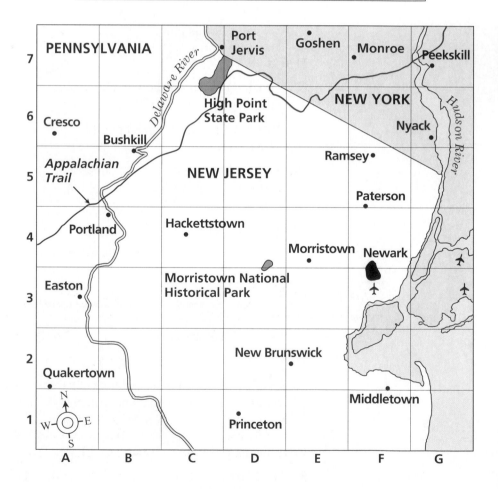

Example	Write the coordinates of Middletown, New Jersey.
Solution	Find the square that contains Middletown. Move *down* the column to find the letter: F. Move *left* on the row to find the number: 2. Combine the letter and number to give the coordinates: F2.

Exercises

Use the map shown on page 132.

1. Write the coordinates of Bushkill, Pennsylvania.

2. Write the coordinates of Hackettstown, New Jersey.

3. Name two New Jersey cities in Column F.

4. Name two New York cities in Row 7.

5. In which squares is Newark, New Jersey located?

6. In which squares do Pennsylvania and New Jersey share a border?

7. In which squares is Morristown National Historical Park located?

8. Name the squares in Pennsylvania shown on the map that include the Appalachian Trail.

9. Which rivers are shown on the map?

10. How many airports are indicated on the map? In which squares are they located?

11. Which town is farther west, Morristown or Newark?

12. Which town is farther north, Paterson or Ramsey?

13. Make a map index for the five Pennsylvania towns shown.

14. Make a map index for the five New York towns shown.

15. GROUP ACTIVITY   Make a map of your school and the area around your school. Label points of interest. Include a map index of the points of interest.

16. WRITING ABOUT MATHEMATICS   Write a paragraph that describes three similarities and three differences between a map and a coordinate plane.

Chapter Review

integers (p. 94)
positive integers (p. 94)
negative integers (p. 94)
opposites (p. 94)
absolute value (p. 95)
addition property of opposites
   (p. 102)
additive inverse (p. 102)
multiplication property of −1
   (p. 116)
coordinate plane (p. 124)

axes (p. 124)
origin (p. 124)
x-axis (p. 124)
y-axis (p. 124)
quadrant (p. 124)
ordered pair (p. 124)
x-coordinate (p. 124)
y-coordinate (p. 124)
graph a point (p. 124)
graph of a function
   (p. 128)

Choose the correct term from the list above to complete each sentence.

1. On a number line, numbers that are the same distance from zero but on opposite sides of zero are called __?__.

2. On a number line, the distance of a number from zero is the __?__ of the number.

3. The __?__ are the numbers in the set {. . . , −3, −2, −1, 0, 1, 2, 3, . . .}.

4. The __?__ states that the sum of a number and its opposite is zero.

5. The __?__ is the point in a coordinate plane where the axes intersect.

6. In an ordered pair, the first number is called the __?__.

Find each absolute value.   *(Lesson 3-1)*

7. $|-11|$ **8.** $|5|$ **9.** $|0|$ **10.** $|-2|$ **11.** $|4|$ **12.** $|-3|$

Replace each __?__ with >, <, or =.   *(Lesson 3-1)*

13. $-12 \underline{} 3$ **14.** $-8 \underline{} -8$ **15.** $-2 \underline{} 20$ **16.** $-5 \underline{} 0$

Solve by making a table.   *(Lesson 3-7)*

17. Corey Watson, a florist, sells daisies for 50¢ each, carnations for 75¢ each, and roses for $1 each.

   a. In how many different ways can he make a bouquet that sells for $3?

   b. A customer wants to buy four flowers. How many different price totals are possible?

Chapter 3

Find each answer. *(Lessons 3-2, 3-3, 3-4, 3-5)*

18. $-2 + (-3)$ **19.** $14 + 8$ **20.** $48 - (-28)$ **21.** $-16 - (-4)$

22. $4(-5)$ **23.** $(-6)(-8)$ **24.** $\frac{88}{8}$ **25.** $\frac{-84}{12}$

26. $1001 \div (-13)$ **27.** $-72 \div (-6)$ **28.** $3(-6)(-6)$ **29.** $(-4)(-1)(-4)$

30. $-23 - 23$ **31.** $-24 - (-24)$ **32.** $(-9)(5)$ **33.** $(-2)(7)(0)$

34. $-5 + 16$ **35.** $12 + (-32)$ **36.** $\frac{-32}{4}$ **37.** $\frac{-63}{-9}$

38. $26 - 49$ **39.** $38 - 74$ **40.** $26 + (-26)$ **41.** $-45 + 3$

Evaluate each expression when $a = 6$, $b = -2$, and $c = -7$.
(Lesson 3-6)

42. b^2 **43.** a^3 **44.** $3b^3$ **45.** $-5a^2$

46. $c^2 - 9$ **47.** $b^4 + 12$ **48.** $|a| + |c|$ **49.** $|b| - |c|$

50. $|c| - b$ **51.** $a + |b|$ **52.** $|a + b|$ **53.** $|c - 2|$

Use the coordinate plane at the right.
Write the coordinates of
each point. *(Lesson 3-8)*

54. Q **55.** R

56. S **57.** T

58. U **59.** V

Graph each point on a coordinate
plane. *(Lesson 3-8)*

60. $D(5, -3)$ **61.** $E(-2, 0)$

62. $F(1, 3)$ **63.** $G(-5, 4)$

Graph each function. *(Lesson 3-9)*

64.

x	$x - 3$
-3	-6
-2	-5
0	-3
3	0
5	2

65.

x	$3x + 1$
-3	-8
-1	-2
0	1
2	7
3	10

66. $x \rightarrow -2 + x$,
  when $x = -3, -1, 0, 3,$ and 5

67. $x \rightarrow x^2 + 1$,
  when $x = -2, -1, 0, 1,$ and 2

Chapter Test

Find each absolute value.

1. $|-2|$ 2. $|9|$ 3. $|21|$ 4. $|-14|$ 3-1

5. Which expression is greater: x or $-x$? Explain.

6. For what values of n is $|n| = -n$?

Find each answer.

7. $-3 + (-5)$ 8. $4 + 26$ 9. $18 + 32$ 10. $-7 + (-36)$ 3-2

11. $-18 + 72$ 12. $49 + (-49)$ 13. $12 + (-30)$ 14. $-7 + 35$ 3-3

15. $-6 - 25$ 16. $14 - (-53)$ 17. $29 - 30$ 18. $-12 - (-6)$ 3-4

19. $(-7)(-11)$ 20. $5(-2)(9)$ 21. $-64 \div 8$ 22. $\dfrac{-6}{-2}$ 3-5

Evaluate each expression when $v = 3$, $w = -2$, and $y = -4$.

23. $-vy$ 24. $y^3 - 10$ 25. $|w| - |v|$ 26. $|w + y|$ 3-6

Solve by making a table.

27. Dwayne has only nickels, dimes, and quarters. The *Weekly Times*   3-7
    costs 50¢. In how many different ways could he pay the 50¢ to buy a
    copy of the *Weekly Times*?

Use the coordinate plane at the right. Write the coordinates of each point.

28. A 29. B 30. C 31. D 3-8

Graph each point on a coordinate plane.

32. $P(1, -1)$ 33. $Q(0, 3)$ 34. $R(-3, 5)$

Graph each function.

35.

x	$x + 4$
-4	0
-2	2
0	4
1	5
3	7

36.

x	$3x - 1$
-2	-7
-1	-4
0	-1
1	2
2	5

37. $x \rightarrow -x - 1$ 3-9
    when $x = -4, -2, -1, 0$, and 1

38. $x \rightarrow |x + 3|$
    when $x = -6, -4, 0, 2$, and 3

Closure

Objective: To determine whether a set of numbers is closed under a given operation.

The numbers in the set {0, 1, 2, 3, . . .} are called **whole numbers.** You already know that when you add any two whole numbers, the sum is also a whole number.

$$1 + 2 = 3$$
$$35 + 35 = 70$$
$$146 + 0 = 146$$

For this reason, the set of whole numbers is said to have *closure* under addition.

A set of numbers has **closure** under a given operation when performing the operation on any numbers in the set results in a number that is also in the set. You describe a set of numbers as being either *closed* or *not closed* under a given operation.

Terms to Know
- *whole numbers*
- *closure*

Example

Tell whether the set of whole numbers is *closed* or *not closed* under the operation of subtraction.

Solution

Look for a whole-number difference that is *not* a whole number.

$$8 - 5 = 3 \qquad 25 - 25 = 0 \qquad 7 - 12 = -5$$

Because $7 - 12 = -5$, and -5 is not a whole number, the whole numbers are *not closed* under subtraction.

There are, of course, many other examples of whole-number differences that are negative integers. However, to show that the set of whole numbers is not closed under subtraction, you need to find only one of these differences.

Exercises

Tell whether the given set of numbers is *closed* or *not closed* under each operation.
   a. **addition**
   b. **subtraction**
   c. **multiplication**
   d. **division**

1. the set of whole numbers   **2.** the set of integers

3. the set of positive integers   **4.** the set of negative integers

5. {0, 1}   **6.** {1, 3, 5, . . .}

7. {2, 4, 6, . . .}   **8.** {. . ., −5, −3, −1}

Cumulative Review

Standardized Testing Practice

Choose the letter of the correct answer.

1. **Write 53 m in km.**
   A. 5300 km   B. 53,000 km
   C. 0.53 km   D. 0.053 km

2. **Find the coordinates of point D.**
   A. $(-2, 1)$
   B. $(2, -1)$
   C. $(-1, 2)$
   D. $(1, -2)$

3. **In 24 h the temperature went from $-12°C$ to $10°C$. Find the change in temperature.**
   A. $22°C$
   B. $-22°C$
   C. $2°C$
   D. $-2°C$

4. **Tyson bought two ties at $12.99 each. The total tax was $1.30. Tyson got a discount of $2.60. Find the total amount he paid.**
   A. $16.89   B. $24.68
   C. $29.88   D. $27.28

5. **Find the answer mentally.**
   $7(28) + 7(52)$
   A. 1480   B. 560
   C. 700   D. 490

6. **Simplify:** $3(x + 5y) + 2x + y$
   A. $5x + 16y$ B. $5x + 6y$
   C. $9x + 18y$ D. $21xy$

7. **Express the relationship $6 < 8 < 10$ in another way.**
   A. $6 > 8 > 10$
   B. $6 < 10 < 8$
   C. $10 > 8 > 6$
   D. $10 > 6 > 8$

8. **Amy has $2.05 in dimes and quarters. It is *not* possible for Amy to have which grouping of coins?**
   A. 7 quarters, 3 dimes
   B. 5 quarters, 8 dimes
   C. 3 quarters, 13 dimes
   D. 6 quarters, 5 dimes

9. **Find the function rule.**
   A. $x \rightarrow x - 5$
   B. $x \rightarrow 3x - 13$
   C. $x \rightarrow 3x - 1$
   D. $x \rightarrow -x + 5$

x	?
-2	-7
-1	-4
0	-1
3	8
5	14

10. **Evaluate $b + a$ when $a = 43.87$ and $b = 116.5$.**
    A. 55.52   B. 555.2
    C. 160.37   D. 72.63

11. **Find the sum:** $-42 + 18$
    A. -60
    B. -24
    C. 60
    D. 24

12. **Find the quotient:** $\frac{-32 + 8}{-4}$
    A. 10
    B. -10
    C. 6
    D. -6

13. **Evaluate** $3 - 2c^2$ **when** $c = -5.$
    A. -47
    B. -97
    C. 53
    D. -22

14. **Find the answer:** $\frac{64 - 48}{8 - 6}$
    A. 0
    B. -2
    C. 24
    D. 8

15. **Write in order from least to greatest:**
    $|12|, |0|, |-15|, |-1|$
    A. $|0|, |-1|, |12|, |-15|$
    B. $|-15|, |-1|, |0|, |12|$
    C. $|-15|, |12|, |-1|, |0|$
    D. $|12|, |0|, |-15|, |-1|$

16. **Simplify:** $-4(-2t + 3)$
    A. $-8t + 12$
    B. $8t - 12$
    C. $8t + 12$
    D. $-8t - 12$

17. **Evaluate** $xy + z$ **when** $x = 13$, $y = 4$, **and** $z = 2.$
    A. 19
    B. 78
    C. 104
    D. 54

18. **Which word or phrase, used twice, correctly completes "The product of any number and __?__ is __?__" ?**
    A. one   B. zero
    C. itself   D. its inverse

19. **Find the sum:** $-32 + (-18)$
    A. -14
    B. 14
    C. -50
    D. 50

20. **Complete:** The cost of three tickets at $3 each and two tickets at $9.50 each is __?__.
    A. $28   B. $12.50
    C. $38   D. $34.50

Maximum Speeds of Animals (mi/h)

70 Cheetah
35 Rabbit
30 Grizzly Bear
25 Elephant
9 Chicken

United States
Metroliner
Baltimore to
Wilmington

Great Britain
High Speed Train
Swindon to Reading

Japan
Yamabiko
Morioka to
Sendai

France
TGV
Paris to Mâcon

Average Rates of Trains

97.8 mi/h   108.3 mi/h   127.6 mi/h   135.4 mi/h

Equations 4

Frequently Used Rates		
		Formula
	Distance	Distance = rate x time
	Price	Price = rate x weight
	Earnings	Earnings = rate x time

Did You Know? Light travels at a rate of 186,282 mi/s. Electric impulses are slower and travel at a rate of 126,263 mi/s. Sound is much slower. At 32°F and sea level, sound travels at a rate of only 1086 ft/s.

Equations

Objective: To find solutions of equations by substitution and mental math.

Terms to Know
- *equation*
- *solution of an equation*

CONNECTION

People who work with precious metals such as gold and silver use a balance scale like the one at the right to weigh the metal. A balance scale shows a relationship between two quantities.

In mathematics, an **equation** is a statement that two numbers or two expressions are equal. You can represent an equation by a balance scale, with each pan holding one of the two *sides* of the equation.

Some equations, such as $x + 1 = 6$, contain a variable. A value of the variable that makes an equation true is called a **solution of the equation.**

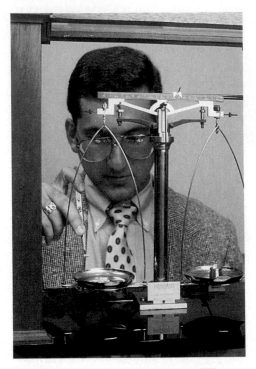

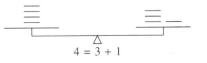

$$4 = 3 + 1 \qquad\qquad x + 1 = 6$$

Example 1

Is the given number the solution of the equation? Write *Yes* or *No*.

  a. $x + 1 = 6$; 5   **b.** $15 = 5k$; 2

Solution

a. Substitute 5 for x in the equation.

$$x + 1 = 6$$
$$5 + 1 \stackrel{?}{=} 6$$
$$6 = 6 \checkmark$$

Yes, 5 is the solution of $x + 1 = 6$.

b. Substitute 2 for k in the equation.

$$15 = 5k$$
$$15 \stackrel{?}{=} 5 \cdot 2 \qquad \text{The symbol} \neq \text{means}$$
$$15 \neq 10 \longleftarrow \text{is not equal to.}$$

No, 2 is not the solution of $15 = 5k$.

✓ **Check Your Understanding**

1. Why is the scale in Example 1(b) not in balance?

Sometimes you can use mental math to find a solution. State the equation as a question and see if you can find the answer easily.

Example 2

Solution

MENTAL MATH

Use mental math to find the solution of $2n - 6 = 14$.

First, think about the value of $2n$:
  What number minus six equals fourteen?
  $20 - 6 = 14$, so $2n = 20$.

Then think about the value of n:
  Two times *what number* equals 20?
  $2 \cdot 10 = 20$, so $n = 10$.

The solution of $2n - 6 = 14$ is 10.

✔️ **Check Your Understanding**

2. What questions would you ask if the equation in Example 2 were $4n - 6 = 14$?

3. Describe how Example 2 would be different if the equation were $2n + 6 = 14$.

Guided Practice

COMMUNICATION « *Reading*

Does each of the following fit the definition of an equation? Write *Yes* or *No*. If *No*, explain.

« **1.** $3 + 4 = 7$　　　　　　« **2.** $a + 12$

« **3.** $x + 19 - 17$　　　　　« **4.** $18 = 3b - 6$

COMMUNICATION « *Writing*

Write the equation represented by each balance scale.

« **5.**　　　　　　　　　　　　« **6.**

Draw a balance scale similar to those in Exercises 5 and 6 to represent each equation.

« **7.** $5 = 4 + 1$　　　« **8.** $x + 6 = 9$　　　« **9.** $7 = 3n + 1$

Is the given number a solution of the equation? Write *Yes* or *No*.

10. $n - 8 = 16; 8$　　　**11.** $-2 = x + 5; -7$　　　**12.** $6y = 20; -2$

Use mental math to find each solution.

13. $5 + r = 37$　　　**14.** $-18 = z - 5$　　　**15.** $\frac{c}{5} = 10$

16. $4m + 10 = -30$　　　**17.** $11 = 2t - 1$　　　**18.** $-1 = \frac{w}{3} - 1$

Exercises

Is the given number a solution of the equation? Write *Yes* or *No*.

1. $x + 6 = 9$; 3

2. $18 = 3n$; 15

3. $-20 = \frac{c}{-4}$; -5

4. $b - 3 = -4$; -1

5. $-1 = 1 + d$; 0

6. $12w = 12$; 1

7. $\frac{k}{3} = -6$; -18

8. $5 = m - 9$; 4

9. $-7y = 28$; -4

Use mental math to find each solution.

10. $p + 8 = 9$

11. $z - 4 = -1$

12. $-12 = 6v$

13. $5 = \frac{h}{-9}$

14. $5 + n = 5$

15. $-3 = 1 + \frac{a}{2}$

16. $-2 = \frac{w}{3} - 1$

17. $3r - 11 = 4$

18. $\frac{n}{2} + 2 = 0$

19. Is -4 the solution of $x + 3 = 7$? Write *Yes* or *No*.

20. Which of the integers -1, 0, and 1 is the solution of $3x + 3 = 0$?

Choose the letter of the solution of each equation.

    A. -1 **B.** 0   **C.** 1   **D.** no solution

21. $a - 8 = -8$

22. $0 = d + 1$

23. $-5p = 5$

24. $0 = 7q$

25. $\frac{4}{k} = 4$

26. $\frac{0}{r} = 9$

27. $-3y = -3$

28. $-6p + 6 = 6$

29. $-2n - 2 = 0$

LOGICAL REASONING

Tell whether each statement is *True* or *False*, or if you *cannot determine*.

30. $5 + 7 = 12$

31. $54 = 7 \cdot 8$

32. $15 - 8 \neq 7$

33. $\frac{20}{-5} \neq 4$

34. $3m + 2 = 5$

35. $3(2) - 7 = -1$

36. $(-5)(-7) \neq -12$

37. $-5n \neq 15$

38. $\frac{-24}{3} + 2 \neq -6$

THINKING SKILLS

Create two equations that have the same solution as the given equation.

39. $8 + m = 7$

40. $-35 = 7p$

41. $12 = \frac{c}{2}$

42. $a - 8 = -3$

SPIRAL REVIEW

43. Justin worked five days and received $23, $44, $36, $50, and $27. Justin said he earned $130. Is this correct? Explain.　*(Lesson 2-7)*

44. Find the difference: $\frac{7}{9} - \frac{1}{6}$　*(Toolbox Skill 17)*

45. Find the sum: $-8 + 11$　*(Lesson 3-3)*

46. Use mental math to find the solution: $5z - 2 = 23$　*(Lesson 4-1)*

4-2

Strategy:
Guess and Check

Objective: To solve problems by guessing and checking.

One method for solving a problem is to use the guess-and-check strategy. Guess an answer and check it against the information in the problem to see if it is correct. If your first guess is not correct, use the results to make a second guess. Keep guessing and checking until you find the correct answer.

Problem

Denzel spent $17.53 on pet food for his cats and dogs. The cat food cost $.89 per can and the dog food cost $1.19 per can. How many cans of each type of pet food did he buy?

Solution

UNDERSTAND The problem is about buying cans of pet food.

Facts: cat food cans at $.89 each
        dog food cans at $1.19 each
        total value of $17.53

Find: the number of cans of each type of pet food bought

PLAN First guess the number of cans of pet food, say 8 cat food cans and 8 dog food cans. Check the total value. Continue guessing and checking until you find the correct value of $17.53.

WORK Make a table to organize the work.

	Cat	Dog	Total Value Spent	
First Guess	8	8	$.89(8) + $1.19(8) = $16.64	too low
Second Guess	9	9	$.89(9) + $1.19(9) = $18.72	too high
Third Guess	9	8	$.89(9) + $1.19(8) = $17.53	correct

ANSWER Denzel bought 9 cans of cat food and 8 cans of dog food.

Look Back What if Denzel spent $20.80 on pet food? How many cans of each type of pet food did he buy?

You can also use the guess-and-check strategy to find the solution of an equation. Check your guesses by substitution.

Suppose you guess that 12 is ⟶ the solution.

$$3x - 14 = 13$$
$$3(12) - 14 \stackrel{?}{=} 13$$
$$36 - 14 \stackrel{?}{=} 13$$
$$22 \neq 13, \text{ so 12 is incorrect.}$$

Because 22 is greater than 13, the guess of 12 is too high. The next guess should be less than 12. Continue guessing and checking until you find the solution, 9.

Guided Practice

Use the problem below for Exercises 1–4.

Sarah charges $7 per truck and $4 per car to wash her neighbors' vehicles. She earned $41 last weekend. How many vehicles of each type did she wash?

« **1.** What is the paragraph about?

« **2.** How much did Sarah earn last weekend?

3. Suppose you guess that Sarah washed 4 cars and 4 trucks. Did she earn *less than* or *more than* $41?

4. Solve the problem using the guess-and-check strategy.

Solve using the guess-and-check strategy.

5. $14v - 8 = 90$

6. $13 + 2m = 57$

7. Rene Valmont collected fees from entrants in a race. He collected $5.50 each from runners who registered before the day of the race and $8 each from those who registered on the day of the race. Rene collected $177. How many runners registered before the day of the race and how many on the day of the race? (There is more than one answer.)

Problem Solving Situations

Solve using the guess-and-check strategy.

1. $5n - 18 = 102$

2. $23 + \frac{x}{8} = 48$

3. $94 = \frac{b}{12} - 16$

4. $87 = 15 + 6y$

5. Pedro Sonez bought sweaters at $26 each and shirts at $16 each. He paid a total of $242. How many of each did Pedro buy?

6. In a two-day bike race, Rachel Davis traveled 30 mi. She biked 6 mi farther on the first day than she did on the second day. How many miles did Rachel travel each day?

7. Linda Chang has fifteen quarters and dimes in all. Their total worth is $2.55. How many of each type of coin does she have?

8. A bunch of tulips costs $4 and a bunch of daffodils costs $2.50. Paula Wilcox bought some tulips and daffodils for prom decorations. She spent $47. How many bunches of each type of flower did Paula buy? (There is more than one answer.)

Solve using any problem solving strategy.

9. Rose Perrini has only quarters and dimes. In how many different ways can she pay a $1.35 tip?

10. A child's bank contains 39 nickels and pennies in all. Their total worth is $.87. How many of each type of coin are in the bank?

PROBLEM SOLVING
CHECKLIST

Keep this in mind:
  Using a Four-Step Plan
Consider these strategies:
  Choosing the Correct Operation
  Making a Table
  Guess and Check

11. Michael Olin bought several baseball cards for $10 each. When the price increased to $17 per card, he sold all of the cards except one. His profit was $46. How many baseball cards did Michael buy?

12. Bradley Ellis went on a trip to visit some friends in another state. She traveled 350 mi on each of the first two days and 400 mi on the third day. How far did she travel?

13. The top three students in a creative writing contest shared a $10,000 college scholarship award. The second-place student received $4000 and the third-place student received $1500. How much did the first-place student receive?

WRITING WORD PROBLEMS

Using the information given, write a word problem that you could solve by using the guess-and-check strategy. Then solve the problem.

14. apples at $.50 each and bananas at $.75 each

15. sweaters at $17 each and shirts at $11 each

SPIRAL REVIEW

16. Find the difference: $11.06 - 7.98$ *(Toolbox Skill 12)*

17. DATA, *pages 2–3* How many charity concerts does the Houston Symphony perform each year?   *(Toolbox Skill 24)*

18. Find the answer: $-7 + 5 \times 6 \div (-2)$ *(Lesson 3-6)*

19. Sanjee bought compact discs at $14 each and cassette tapes at $9 each. He paid a total of $83. How many of each did Sanjee buy? *(Lesson 4-2)*

Using Addition or Subtraction

Objective: To solve equations using addition or subtraction.

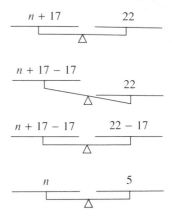

EXPLORATION

1. What operation could you use to undo the addition on the left pan of this balance scale?

2. What happens to the balance scale if you subtract 17 from the expression on the left pan?

3. What happens to the balance scale if you subtract 17 from both the left and the right pans at the same time?

4. What is in each pan after you simplify the expressions in Step 3?

To **solve** an equation, you find all values of the variable that make the equation true. Recall that each of these values is called a *solution*. When you solve an equation, you get the variable alone on one side of the equals sign. To do this, you need to undo any operations on the variable.

Example 1
Solution

Solve $6 + k = 31$. Check the solution.

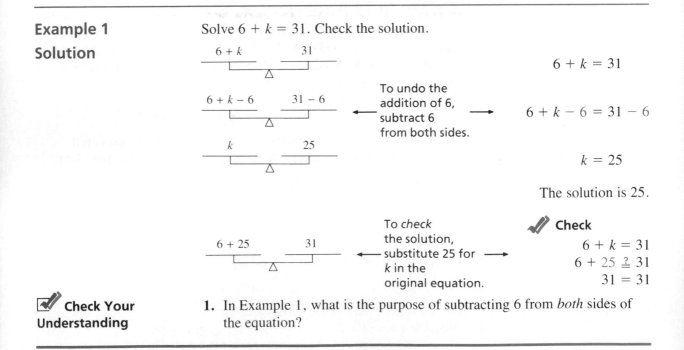

To undo the addition of 6, subtract 6 from both sides.

$$6 + k = 31$$

$$6 + k - 6 = 31 - 6$$

$$k = 25$$

The solution is 25.

To *check* the solution, substitute 25 for k in the original equation.

✔️ **Check**
$$6 + k = 31$$
$$6 + 25 \stackrel{?}{=} 31$$
$$31 = 31$$

✔️ **Check Your Understanding**

1. In Example 1, what is the purpose of subtracting 6 from *both* sides of the equation?

Just as you used subtraction to undo addition, you can use addition to undo subtraction.

Example 2	Solve $-29 = s - 15$. Check the solution.
Solution	

$$-29 = s - 15$$
$$-29 + 15 = s - 15 + 15 \longleftarrow$$
$$-14 = s$$

To undo the subtraction of 15, add 15 to both sides.

The solution is -14.

✔ **Check**

$$-29 = s - 15$$
$$-29 \stackrel{?}{=} -14 - 15$$
$$-29 = -29$$

✔ **Check Your Understanding**

2. How would Example 2 be different if the right side of the equation were $s + 15$?

Because addition and subtraction undo each other, they are called **inverse operations.** Recognizing inverse operations can help you to solve equations.

> **Generalization:** *Solving Equations Using Addition or Subtraction*
>
> If a number has been *added* to the variable, subtract that number from both sides of the equation.
>
> If a number has been *subtracted* from the variable, add that number to both sides of the equation.

You can use a calculator to check the solution of an equation. For instance, to check if -508 is the solution of $n + 163 = -345$, substitute -508 for n and use this key sequence.

$$\boxed{508.} \;\; \boxed{+/-} \;\; \boxed{+} \;\; \boxed{163.} \;\; \boxed{=}$$

Guided Practice

«**1.** COMMUNICATION «*Reading* Replace each __?__ with the correct word.

To solve an equation, you undo any operation on the variable by using a(n) __?__ operation. To undo the addition of 8, for example, you would __?__ 8 from both sides of the equation.

Explain how you would solve each equation.

2. $x - 16 = 18$

3. $y + 6 = 27$

4. $-41 = 14 + w$

5. $-62 = r - 37$

Solve. Check each solution.

6. $w + 8 = 6$ **7.** $16 = a - 17$ **8.** $4 + n = 9$

9. $12 + t = -6$ **10.** $-11 = z - 5$ **11.** $m - 8 = -8$

Exercises

Solve. Check each solution.

1. $a + 2 = 11$ **2.** $5 + b = -19$ **3.** $-2 = 7 + c$

4. $45 = d + 8$ **5.** $1 = r - 3$ **6.** $w - 15 = -40$

7. $h - 10 = -21$ **8.** $-10 = c - 20$ **9.** $a - 9 = 9$

10. $-9 = a - 9$ **11.** $a + 9 = 9$ **12.** $-9 = a + 9$

13. The sum of 16 and a number k is 5. Find k.

14. When nine is subtracted from a number c, the result is twelve. What is the value of c?

CALCULATOR

Solve. Then use a calculator to check the solution.

15. $c + 463 = 859$ **16.** $n - 862 = 98$

17. $-1026 = 554 + d$ **18.** $-523 = 146 + w$

19. $562 = 999 + t$ **20.** $h - 772 = 3480$

THINKING SKILLS

Complete.

21. If $|x| = 5$, then $x = \underline{\ ?\ }$ or $x = \underline{\ ?\ }$.

22. If $|x + 2| = 7$, then $x + 2 = \underline{\ ?\ }$ or $x + 2 = \underline{\ ?\ }$.

Analyze your answers to Exercises 21 and 22. Then find the solution(s) of each of the following equations. If there is no solution, write *no solution*.

23. $|x| = 3$ **24.** $|m| = 0$ **25.** $|d - 2| = 0$

26. $|c + 4| = 6$ **27.** $|h| = -5$ **28.** $|y - 4| = -14$

29. $|p| + 5 = 25$ **30.** $|n| - 8 = 0$ **31.** $|z| + 7 = 0$

SPIRAL REVIEW

32. Find the next three numbers: 4, 8, 12, $\underline{\ ?\ }$, $\underline{\ ?\ }$, $\underline{\ ?\ }$ *(Lesson 2-5)*

33. Solve: $m - 19 = 24$ *(Lesson 4-3)*

34. Simplify: $(4x)(3y)(2x)$ *(Lesson 2-1)*

35. Find the product: $(-9)(13)$ *(Lesson 3-5)*

4-4 Using Multiplication or Division

Objective: To solve equations using multiplication or division.

 **APPLICATION**

A pair of running shoes weighs 860 g. To find the weight of each shoe you can solve an equation.

two times	weight of each shoe	is	weight of the pair
↓ ↓	↓	↓	↓
2 ·	s	=	860

The equation $2s = 860$ involves multiplication. Just like addition and subtraction, multiplication and division are inverse operations. You can use division to undo multiplication and use multiplication to undo division.

Example 1

Solve $2s = 860$. Check the solution.

Solution

To undo the multiplication by 2, divide both sides by 2.

$$2s = 860$$
$$\frac{2s}{2} = \frac{860}{2}$$
$$s = 430$$

The solution is 430.

Substitute 430 for s in the original equation.

✔ **Check**
$$2s = 860$$
$$2 \cdot 430 \stackrel{?}{=} 860$$
$$860 = 860$$

Example 2

Solve $18 = \frac{n}{-4}$. Check the solution.

Solution

$$18 = \frac{n}{-4}$$
$$18(-4) = \frac{n}{-4}(-4)$$
$$-72 = n$$

The solution is −72.

To undo the division by −4, multiply both sides by −4.

✔ **Check**
$$18 = \frac{n}{-4}$$
$$18 \stackrel{?}{=} \frac{-72}{-4}$$
$$18 = 18$$

As you have seen in Examples 1 and 2, you can use inverse operations to help you decide whether to multiply or divide to solve an equation.

> **Generalization:** *Solving Equations Using Multiplication or Division*
>
> If a variable has been *multiplied* by a nonzero number, divide both sides by that number.
>
> If a variable has been *divided* by a number, multiply both sides by that number.

Before multiplying or dividing to solve an equation, you sometimes might need to use the multiplication property of -1.

Example 3

Solution

Solve $-n = 9$. Check the solution.

$$-n = 9$$
$$-1n = 9 \quad \longleftarrow \text{ Use the multiplication property of } -1.$$
$$\frac{-1n}{-1} = \frac{9}{-1}$$
$$n = -9$$

The solution is -9.

✔ **Check**
$$-n = 9$$
$$-1n = 9$$
$$(-1)(-9) \stackrel{?}{=} 9$$
$$9 = 9$$

✔ **Check Your Understanding**

1. Why is $-n$ rewritten as $-1n$ in Example 3?
2. Describe how Example 3 would be different if the equation were $-3n = -9$.

Guided Practice

«**1.** COMMUNICATION «*Discussion*   Many everyday actions, like tying and untying a shoelace, can be thought of as inverse operations. List other everyday actions that can be thought of as inverse operations.

Tell whether you would *multiply* or *divide* to solve each equation.

2. $-8b = -64$ **3.** $\frac{y}{7} = 14$ **4.** $19 = \frac{t}{-5}$ **5.** $-81 = 9d$

Replace each __?__ with the number that makes the statement true.

6. $3h = 45$

$\frac{3h}{3} = \frac{45}{?}$

$h = \underline{\ ?\ }$

7. $-54 = 2v$

$\frac{-54}{?} = \frac{2v}{?}$

$\underline{\ ?\ } = v$

8. $\frac{z}{-3} = 32$

$\frac{z}{-3}(\underline{\ ?\ }) = 32(\underline{\ ?\ })$

$z = \underline{\ ?\ }$

Use this paragraph for Exercises 9 and 10.

The term *nonzero* is very important in mathematics. All numbers except zero are nonzero numbers. When you use division in a computation or when you use division to solve an equation, you can divide only by non-zero numbers. Division by zero is said to be *undefined*.

«**9.** What is a nonzero number?

«**10.** Name two instances in mathematics in which you may use only non-zero numbers.

Solve. Check each solution.

11. $12n = 24$ **12.** $-15 = -5c$ **13.** $11 = -y$ **14.** $-a = 15$

15. $\frac{m}{-7} = -4$ **16.** $-100 = \frac{d}{10}$ **17.** $12 = \frac{r}{-8}$ **18.** $\frac{n}{2} = 25$

Exercises

Solve. Check each solution.

1. $2b = 30$ **2.** $-6c = 108$ **3.** $-64 = 8d$

4. $-55 = -5v$ **5.** $-q = 0$ **6.** $-6 = -k$

7. $-14 = \frac{x}{4}$ **8.** $\frac{y}{5} = -20$ **9.** $\frac{a}{-3} = 63$

10. $\frac{b}{-7} = 7$ **11.** $-31 = \frac{r}{-2}$ **12.** $3 = \frac{n}{12}$

13. $-7t = -105$ **14.** $-10u = 120$ **15.** $17 = -w$

16. $-19 = -h$ **17.** $-9y = 9$ **18.** $1 = \frac{x}{-5}$

19. If $2r = 0$, what is the value of r?

20. The quotient $-63 \div m$ is equal to -9. Find the value of m.

▦ CALCULATOR

The solution of each equation is -12. Match each equation with the calculator key sequence needed to check the solution.

21. $\frac{72}{n} = 6$

22. $-72 = \frac{864}{n}$

23. $864 = -72n$

24. $864n = -10{,}368$

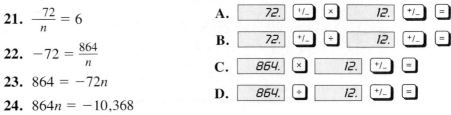

A. [72.] [¹/₋] [×] [12.] [⁺/₋] [=]

B. [72.] [⁺/₋] [÷] [12.] [⁺/₋] [=]

C. [864.] [×] [12.] [⁺/₋] [=]

D. [864.] [÷] [12.] [⁺/₋] [=]

ESTIMATION

To estimate the solution of an equation, first decide which inverse operation to use. Then use an appropriate estimation method for that operation.

Choose the letter of the best estimate for the solution of each equation.

25. $346 = -49m$ **a.** -5 **b.** -7 **c.** -70 **d.** -50

26. $\frac{n}{67} = 205$ **a.** 3 **b.** 30 **c.** 1400 **d.** $14{,}000$

27. $x - 402 = 234$ **a.** 200 **b.** $80{,}000$ **c.** -2 **d.** 600

28. $2070 = 486 + d$ **a.** 1500 **b.** 2500 **c.** -1500 **d.** 4

Estimate the solution of each equation.

29. $\frac{b}{88} = 17$

30. $314 = 21p$

31. $1597 = z + 206$

32. $n - 482 = -695$

SPIRAL REVIEW

33. **DATA,** *pages 46–47*   How many moons does Saturn have? *(Toolbox Skill 23)*

34. Solve: $-8x = 72$ *(Lesson 4-4)*

35. Graph points $A(2, 4)$, $B(-3, 2)$, and $C(-1, -3)$ on a coordinate plane.   *(Lesson 3-8)*

36. Estimate the difference: $903 - 719$ *(Toolbox Skill 1)*

37. Estimate the sum: $3.8 + 4.1 + 4.4 + 3.7$ *(Toolbox Skill 4)*

Self-Test 1

Use mental math to find each solution.

1. $r + 4 = -3$ 2. $\frac{h}{-3} = 3$ 3. $13 = 3w - 5$ 4-1

Solve by using the guess-and-check strategy.

4. Lucia bought birthday cards at \$1.50 each and thank-you notes   4-2
   at \$1.10 each. She paid a total of \$10.80. How many of each
   did she buy?

Solve. Check each solution.

5. $c + 12 = 9$ 6. $m - 8 = -10$ 7. $15 = 13 + w$ 4-3

8. $-3z = 90$ 9. $-4 = \frac{a}{6}$ 10. $-n = 7$ 4-4

Modeling Equations Using Tiles

Objective: To use algebra tiles to model equations.

Materials

■ algebra tiles

Just as you can use algebra tiles to model variable expressions, you also can use tiles to model equations. Remember to use the tile ▬ to represent the variable n and the tile ☐ to represent the number 1. The equation $n + 2 = 7$ then can be represented as follows.

Activity I *Modeling Equations with Tiles*

1 Write the equation represented by each diagram.

a.

b.

2 Show how to use tiles to represent each equation.

    a. $6 = n$ **b.** $n + 1 = 4$ **c.** $3n = 9$ **d.** $12 = 2n$

Activity II *Solving Equations Involving Addition*

1 Describe the situation that you think is being represented in each diagram.

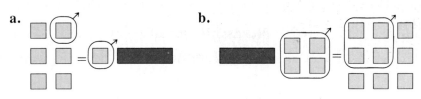

a.   **b.**

2 Show how to use tiles to solve each equation. What is the solution?

    a. $n + 5 = 7$ **b.** $10 = n + 2$ **c.** $3 + n = 11$

Activity III *Solving Equations Involving Multiplication*

1 Describe the situation that you think is being represented in each diagram.

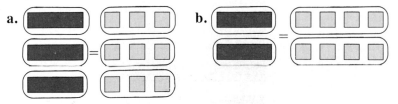

a.   **b.**

2 Show how to use tiles to solve each equation. What is the solution?

    a. $2n = 10$ **b.** $8 = 4n$ **c.** $6n = 6$

Two-Step Equations

Objective: To solve equations using two steps.

1. Suppose that ▬▬▬ represents n and ☐ represents 1. What equation is represented by the diagram at the right?

2. Can you get one ▬▬▬ alone in just one step by removing the same types of tiles from each side of the diagram? Explain.

3. Can you get one ▬▬▬ alone in just one step by separating the tiles on each side of the diagram into identical groups? Explain.

When you solve an equation, your goal is to get the variable alone on one side of the equals sign. If an equation involves two operations, then you need to use two steps to achieve that goal.

Example 1

Solve $2n + 1 = 7$.

Solution

Let ▬▬▬ represent n and ☐ represent 1.

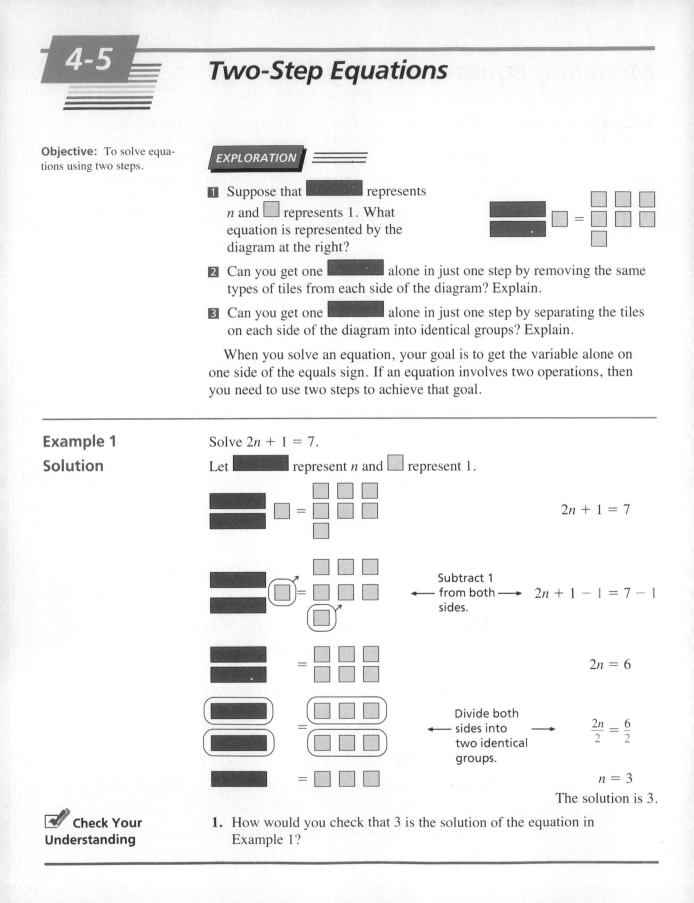

$2n + 1 = 7$

Subtract 1 from both sides. ⟶ $2n + 1 - 1 = 7 - 1$

$2n = 6$

Divide both sides into two identical groups. ⟶ $\dfrac{2n}{2} = \dfrac{6}{2}$

$n = 3$

The solution is 3.

Check Your Understanding

1. How would you check that 3 is the solution of the equation in Example 1?

Example 2

Solution

Solve $-20 = \frac{t}{3} - 4$. Check the solution.

$$-20 = \frac{t}{3} - 4$$

$$-20 + 4 = \frac{t}{3} - 4 + 4 \quad \longleftarrow \begin{array}{l}\text{Add 4 to} \\ \text{both sides.}\end{array}$$

$$-16 = \frac{t}{3}$$

$$-16 \cdot 3 = \frac{t}{3} \cdot 3 \quad \longleftarrow \begin{array}{l}\text{Multiply both} \\ \text{sides by 3.}\end{array}$$

$$-48 = t$$

The solution is -48.

✔ **Check**

$$-20 = \frac{t}{3} - 4$$

$$-20 \stackrel{?}{=} \frac{-48}{3} - 4$$

$$-20 \stackrel{?}{=} -16 - 4$$

$$-20 = -20$$

✔ **Check Your Understanding**

2. What two operations are involved in the equation in Example 2?

3. In Example 2, what two operations were used to solve the equation?

To solve equations involving two steps, such as the equations in Examples 1 and 2, you need to use *two* inverse operations.

> **Generalization:** *Solving Two-Step Equations*
>
> First undo the addition or subtraction, using the inverse operation.
> Then undo the multiplication or division, using the inverse operation.

 You can use a calculator to solve an equation. Use inverse operations as you normally would. To solve $11m - 44 = -121$, use this key sequence.

Guided Practice

COMMUNICATION « *Writing*

« **1.** Let ⬛ represent n and ▢ represent 1. Write the equation that is represented by the diagram below.

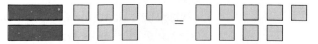

« **2.** Draw a diagram similar to the diagram in Exercise 1 to represent this equation: $8 = 3n + 5$

Choose the letter of the equation that has the same solution as the given equation.

3. $3x - 12 = 18$ **a.** $x - 12 = 6$ **b.** $3x = 6$ **c.** $3x = 30$

Choose the letter of the equation that has the same solution as the given equation.

4. $\frac{w}{5} + 10 = 20$ **a.** $\frac{w}{5} = 10$ **b.** $w + 10 = 100$ **c.** $\frac{w}{5} = 2$

Replace each _?_ with the number that makes the statement true.

5.
$$6n + 3 = -39$$
$$6n + 3 - \underline{\,?\,} = -39 - \underline{\,?\,}$$
$$6n = -42$$
$$\frac{6n}{?} = \frac{-42}{?}$$
$$n = \underline{\,?\,}$$

6.
$$25 = \frac{h}{2} - 9$$
$$25 + \underline{\,?\,} = \frac{h}{2} - 9 + \underline{\,?\,}$$
$$34 = \frac{h}{2}$$
$$34 \cdot \underline{\,?\,} = \frac{h}{2} \cdot \underline{\,?\,}$$
$$\underline{\,?\,} = h$$

Solve. Check each solution.

7. $4c - 2 = 10$

8. $-28 = 12 + 2z$

9. $15 = \frac{t}{4} - 1$

10. $4 + \frac{x}{-2} = 1$

Exercises

Solve. Check each solution.

1. $6n + 4 = 28$ 2. $3t + 1 = -8$ 3. $7w - 5 = -19$

4. $8b - 5 = 35$ 5. $37 = -5c + 2$ 6. $-21 = -9b + 6$

7. $28 = -3x - 2$ 8. $35 = 9m - 10$ 9. $\frac{r}{5} - 1 = 7$

10. $\frac{n}{-6} - 4 = 6$ 11. $7 = \frac{a}{3} - 6$ 12. $\frac{w}{-3} + 8 = 14$

13. $\frac{t}{6} - 6 = -32$ 14. $-53 = \frac{k}{4} - 53$ 15. $86 = \frac{u}{-2} + 86$

16. The sum $6 + 5n$ is equal to 16. Find the value of n.

17. If $3x - 11 = 4$, what is the value of x?

📇 CALCULATOR

Write the calculator key sequence you would use to solve each equation. Solve the equation.

18. $16c - 71 = 153$ 19. $187 = \frac{k}{8} + 135$

20. $\frac{x}{13} - 32 = 58$ 21. $-42z + 161 = 1505$

22. $1480 + 7w = 2040$ 23. $\frac{b}{-15} + 112 = -88$

24. Describe in your own words the process of solving an equation.

25. Tell how solving an equation like $x + 6 = 10$ is different from solving an equation like $2x + 3 = 15$.

CONNECTING MATHEMATICS AND PHYSICAL SCIENCE

A chemical reaction has a natural balance: there are the same number of atoms of each element after the reaction as there were before. For this reason, *chemical equations* are used to represent reactions. For example, the diagram below shows how hydrogen and oxygen react to form water.

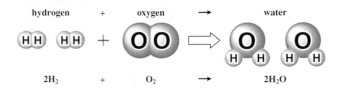

hydrogen + oxygen → water

$$2H_2 + O_2 \rightarrow 2H_2O$$

26. What does it mean to say that a chemical equation *balances*?

27. If there are two oxygen atoms before a reaction, how many must there be after the reaction? Explain.

28. Use the diagram below to complete the chemical equation.

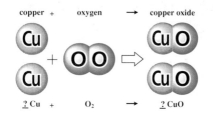

copper + oxygen → copper oxide

$$\underline{2}\ Cu + O_2 \rightarrow \underline{2}\ CuO$$

29. The elements sodium and oxygen combine to form sodium oxide. Complete the chemical equation for this reaction.

$$\underline{\quad?\quad} Na + O_2 = \underline{\quad?\quad} Na_2O$$

SPIRAL REVIEW

30. Find the sum: $6\frac{5}{8} + 4\frac{2}{3}$ *(Toolbox Skill 18)*

31. Write 8.32×10^9 in decimal notation. *(Lesson 2-10)*

32. Solve: $\frac{x}{4} - 15 = -12$ *(Lesson 4-5)*

33. Find the product mentally: $(9)(6)(0)(5)(2)$ *(Lesson 1-7)*

MIXED REVIEW
Solving Equations

Jan Matzeliger
(1852-1889)
United States

Marie Curie
(1867-1934)
Poland/France

Katherine Johnsc
(b. 1918)
United States

Chandrasekhara
Venkata Raman
(1888-1970)
India

Men and women from around the world have made important scientific and technical contributions. To find out what Chandrasekhara Venkata Raman and Marie Curie have in common, complete the code boxes below. Begin by solving Exercise 1. The answer to Exercise 1 is −3, and the letter associated with Exercise 1 is N. Write N in the box above −3 as shown below. Continue in this way with Exercises 2–24.

1. $6n = -18$ (N)

2. $-a = 0$ (E)

3. $14 = -3x + 2$ (E)

4. $\dfrac{b}{4} = 7$ (N)

5. $5 = t + 7$ (Y)

6. $13 + x = 16$ (E)

7. $-15 + 9d = 21$ (R)

8. $8 = \dfrac{m}{6} + 2$ (P)

9. $-y - 18 = -13$ (I)

10. $12t = 108$ (W)

11. $11 = \dfrac{a}{-4}$ (T)

12. $-21 = \dfrac{-n}{2}$ (O)

13. $10c - 18 = 42$ (H)

14. $54 = -6g$ (N)

15. $\dfrac{d}{7} = -9$ (B)

16. $-8 = 16 - 3w$ (Z)

17. $x + 12 = 38$ (A)

18. $\dfrac{n}{5} - 2 = 0$ (R)

19. $32 = b - 18$ (I)

20. $-a - 19 = 27$ (E)

21. $3w - 8 = 13$ (S)

22. $\dfrac{m}{-8} - 2 = 0$ (L)

23. $-7c = -84$ (E)

24. $4 + g = 56$ (R)

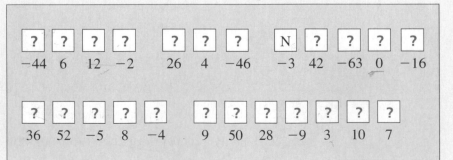

?	?	?	?		?	?	?		N	?	?	?	?
−44	6	12	−2		26	4	−46		−3	42	−63	0	−16

?	?	?	?	?		?	?	?	?	?	?	?
36	52	−5	8	−4		9	50	28	−9	3	10	7

Juan de la Cierva
(1895-1936)
Spain

Chien-Shiung Wu
(b. 1912)
United States

To find out more about the contributions of the scientists and inventors on these pages, solve the equations and match your solutions to the answers at the bottom of the page.

25. $\frac{x}{3} + 1 = 5$

Marie Curie

26. $2n - 6 = 8$

Jan Matzeliger

27. $-6 = a + 4$

Chandrasekhara Venkata Raman

28. $7t = -21$

Chien-Shiung Wu

29. $55 = -11g$

Juan de la Cierva

30. $7 + 6n = 61$

Katherine Johnson

---

a. 7

Inventor who patented a machine that made possible the mass production of shoes.

b. 9

Aerospace engineer who helped design systems used in tracking space missions.

c. -10

Physicist who made new observations about the properties of light frequencies.

d. -5

Inventor who designed the autogiro, a cross between a helicopter and an airplane.

e. -3

Physicist whose innovative experiment disproved the law that like nuclear particles always act alike.

f. 12

Chemist who discovered the elements polonium and radium.

Writing Variable Expressions

Objective: To write variable expressions for word phrases.

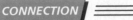

CONNECTION

When you visit another country, you may need to ask directions in a language other than your own. You can translate what you want to say by using words and phrases from the other language that have similar meanings.

Mathematics has a language of its own, the language of symbols. In many problem solving situations you need to translate from a word phrase to a variable expression.

Example 1

Write a variable expression that represents the phrase *eight increased by five times a number n.*

Solution

Increased by suggests addition. *Times* suggests multiplication.

$$8 + 5n$$

✓ **Check Your Understanding**

1. Does the expression $5n + 8$ also represent the phrase in Example 1? Explain.

When writing a variable expression that represents a word phrase, you first choose a variable to represent the unknown number.

Example 2

Write a variable expression that represents the phrase *$35 less than twice Mary's salary.*

Solution

Choose a variable to represent the unknown number.

Let s = Mary's salary.
Then $2s$ = twice Mary's salary.
So $2s - 35$ = $35 less than twice Mary's salary.

If Mary's salary is s dollars, then $(2s - 35)$ dollars represents $35 less than twice that salary.

✓ **Check Your Understanding**

2. Does the expression $35 - 2s$ also represent the phrase in Example 2? Explain.

COMMUNICATION « *Reading*

Choose all the terms that are associated with each operation.

«**1.** addition

«**2.** subtraction

«**3.** multiplication

«**4.** division

A. times

C. combined

E. difference

G. sum

I. product

B. quotient

D. fewer than

F. increased by

H. shared equally

J. decreased by

Replace each __?__ with the correct number or variable expression.

5. $r -$ __?__ represents *one less than a number r*.

6. __?__ n represents *twice a number n*.

7. __?__ $+$ __?__ represents *six more than three times a number m*.

8. __?__ $+$ __?__ represents *the sum of seven and a number w divided by two*.

Write a variable expression that represents each phrase. If necessary, choose a variable to represent the unknown number.

9. a number v divided by 30

10. $4 less than last paycheck

11. six more than twice as many hits

12. the sum of four times a number r and two

Write a variable expression that represents each phrase. If necessary, choose a variable to represent the unknown number.

1. five more than a number x

2. seven times a number y

3. twice as old as Walter

4. six divided by a number n

5. the total of three times a number c and eight

6. a number x decreased by three

7. four less than six times a number d

8. seven more sunny days than in May

9. twelve fewer apples on the tree than yesterday

10. three inches shorter than Francesca

11. students separated into six equal teams

12. double the number of points Derek scored

13. seventeen points more than Tina scored

14. grapes shared equally by four people

15. Make a chart with four columns. Write the name of one of the four arithmetic operations at the top of each column. Under each operation, list all the words and phrases you associate with the operation.

16. Is the expression *four less than a number x* different from *four is less than a number x*? Give a convincing argument to support your answer.

PROBLEM SOLVING/APPLICATION

Use this information for Exercises 17–20.

At the weekday rate, the cost of a long-distance telephone call from Centerville to Newtown is 26¢ for the first minute and 12¢ for each additional minute.

17. Let *x* represent the number of additional minutes. Write a variable expression that represents the cost in cents of a telephone call made at the weekday rate.

18. How would the variable expression that you wrote for Exercise 17 be different if it represented the cost in dollars rather than the cost in cents?

19. Suppose you make a call that lasts *x* additional minutes at the weekday rate, then extends *y* additional minutes into the evening period. At the evening rate, the cost of each additional minute is 7¢. Write a variable expression that represents the cost in cents of your call.

20. FUNCTIONS   Make a function table that shows the cost in cents of a long-distance telephone call at the weekday rate for each whole number of minutes from 1 min through 10 min. What is the function rule?

SPIRAL REVIEW

21. Complete the function table at the right. *(Lesson 2-6)*

22. Estimate the quotient: $38)\overline{2749}$ *(Toolbox Skill 3)*

23. Write a variable expression that represents this phrase: the sum of seven and three times a number *x*   *(Lesson 4-6)*

24. Evaluate $x \div 12$ when $x = 1.44$.   *(Lesson 1-3)*

x	4x − 1
2	7
4	15
6	?
8	?
10	?

4-7

Writing Equations

Objective: To write equations for sentences.

CONNECTION

In the English language, you can form simple sentences by using a verb to join phrases. For instance, you join the phrases "My art class" and "in Room 112" by a linking verb such as "is" to form a complete thought: "My art class is in Room 112." Similarly, in mathematics you form an equation by using an equals sign to join mathematical expressions.

Example 1

Write an equation that represents each sentence.

a. Nine more than a number x is 12.

b. Twenty-four is a number t divided by 3.

Solution

a. $x + 9 = 12$ **b.** $24 = \frac{t}{3}$

☑ **Check Your Understanding**

1. In Example 1, what verb is translated into an equals sign?

In problem solving situations, you may need to translate a sentence into an equation. To do this, first find the verb that you can represent by an equals sign. Then translate the remaining phrases into two mathematical expressions and join them by the equals sign.

Example 2

Write an equation that represents the relationship in the following sentence.

A financial software package costs $115, which is $25 more than the cost of a game software package.

Solution

Choose a variable to represent the unknown number.

The cost of the financial package is $115.
Let g = cost of the game package.
Then $g + 25$ = cost of the financial package.
So $g + 25 = 115$.

The equation $g + 25 = 115$ represents the relationship between the costs of the financial package and the game package.

☑ **Check Your Understanding**

2. In Example 2, why would you decide to let the variable represent the cost of the game package rather than the financial package?

Guided Practice

COMMUNICATION « *Reading*

« **1.** What is the objective of this lesson? How is it different from the objective of Lesson 4-6?

« **2.** In this lesson, how is Example 1 different from Example 2?

Use this information for Exercises 3–5.

Karen has $46, which is twice as much money as Sal has.

3. Let s = number of dollars Sal has. Then Karen has __?__ · s dollars.

4. The actual amount of money Karen has is __?__.

5. An equation that represents this situation is __?__ = __?__.

Write an equation that represents each sentence.

6. Three times a number x is 18.

7. A number t more than 9 is 17.

8. Sixteen is a number m divided by 3.

9. A number z decreased by 3 is 39.

Write an equation that represents the relationship in each sentence.

10. Eve sold 45 tickets, which is three times as many tickets as Ben sold.

11. At the elementary school there are 22 teachers, which is nine fewer teachers than at the high school.

Exercises

Write an equation that represents each sentence.

1. Four more than a number m is 5.

2. Sixty-three is 7 times a number n.

3. A number a divided by 6 is 12.

4. Thirty-four is a number t increased by 7.

5. The product of 15 and a number k is 105.

6. Two subtracted from a number b is 9.

Write an equation that represents the relationship in each sentence.

7. In his locker Alexander has seven books, which is the number of books Joanne has in her locker divided by two.

8. At South High School there are 534 girls, which is 89 more than the number of boys.

9. The low temperature on Monday was 10°F, which is 15°F less than the low temperature on Sunday.

10. Last week Manuel earned $297, which is twice the amount that Tony earned.

Each of the situations in Exercises 9–14 can be represented by one of the equations below. Choose the letter of the correct equation.

A. $x + 8 = 24$ **B.** $x - 8 = 24$ **C.** $8x = 24$ **D.** $\frac{x}{8} = 24$

11. Rich scored 8 more points than Teresa. Rich scored 24 points.

12. Mario is 8 times as old as his sister. Mario is 24 years old.

13. Paula has 24 albums, which is the number of cassettes Richard has divided by 8.

14. Christine worked a total of 24 h during the last two weeks. The first week she worked 8 h.

15. The total rainfall in Warwick last year was 8 times the amount of rain that fell during March. Warwick had 24 cm of rainfall last year.

16. Raoul's score of 24 on today's quiz is 8 points lower than his score on last week's quiz.

COMMUNICATION «*Writing*

Write a sentence that could be represented by each equation.

«**17.** $m - 8 = 16$ «**18.** $y + 9 = 32$

«**19.** $4w = 48$ «**20.** $\frac{u}{9} = 14$

SPIRAL REVIEW

21. Is it most efficient to find the quotient $1800 \div 90$ using *mental math, paper and pencil,* or a *calculator*? *(Lesson 1-8)*

22. Write an equation to represent this sentence: The product of 5 and a number x is 42. *(Lesson 4-7)*

23. Graph the function $x \rightarrow 2x$ when $x = -3, -1, 0, 1,$ and 3. *(Lesson 3-9)*

24. Find the product: $2\frac{6}{7} \times 3\frac{3}{5}$ *(Toolbox Skill 20)*

25. Is 4 a solution of $6x = -24$? Write *Yes* or *No*. *(Lesson 4-1)*

Challenge

Use four of these weights to balance the scale.

     11   19   47   56   92

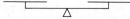

Can you balance the scale using all five weights?

4-8

Strategy:
Using Equations

Objective: To solve problems by using equations.

Some problems describe a relationship between two or more numbers. To solve this type of problem, choose a variable to represent one of the unknown numbers in the problem. Use that variable to write expressions for the other unknown numbers. Then use the facts of the problem to write an equation. You solve the problem by solving this equation and finding the unknown numbers.

Problem

A parking garage charges $3 for the first hour and $2 for each additional hour. On a recent day, a motorist paid $17 to park a car in the garage. How many hours was the car parked in the garage?

Solution

UNDERSTAND The problem is about the cost of parking a car in a parking garage.
Facts: $3 for the first hour
$2 for each additional hour
total of $17 paid
Find: the number of hours the car was parked

PLAN Choose a variable and decide what the variable will represent. Use the variable to write expressions and then an equation for the problem. Solve the equation to answer the question.

WORK Let h = the number of additional hours the car was parked. Then $2h$ = the cost of parking for the additional hours.

The cost for the first hour plus the cost for the additional hours is 17.

$$3 \qquad + \qquad 2h \qquad = 17$$

$$3 + 2h = 17$$
$$3 + 2h - 3 = 17 - 3$$
$$2h = 14$$
$$\frac{2h}{2} = \frac{14}{2}$$
$$h = 7$$

✓ **Check**

$$3 + 2h = 17$$
$$3 + 2(7) \stackrel{?}{=} 17$$
$$3 + 14 \stackrel{?}{=} 17$$
$$17 = 17$$

ANSWER Because $h = 7$, the car was parked for 7 additional hours. So, including the first hour, the car was parked for a total of 8 h.

Look Back An alternative method for solving the problem is to use the guess and check strategy. Solve the problem again using this alternative method.

Guided Practice

Use the problem below for Exercises 1–7.

A construction site has carpenters and electricians. The number of carpenters is eight fewer than twice the number of electricians. If there are twelve carpenters on the job site, how many electricians are there?

« **1.** What do you need to find out?

« **2.** Use the variable n. What will you let n represent?

« **3.** Write a variable expression that represents the number of carpenters.

« **4.** How many carpenters are there?

« **5.** Use your answers in Exercises 3 and 4 to write an equation.

  6. Solve the equation. Check the solution.

  7. Use the solution of the equation to write an answer to the problem.

Solve using an equation.

  8. The number of drummers in a band is two more than three times the number of tuba players. If there are eight drummers in the band, how many tuba players are there?

  9. A number is divided by two. Then five is added to the result. If the answer is -2, what was the original number?

 10. Hector bought a computer system for $989. He made a $125 down payment and paid the remaining amount in twelve equal payments. What was the amount of each payment?

Problem Solving Situations

Solve using an equation.

  1. At her new job Mary earns $100 more than twice the salary she earned as a college intern. If she now earns $2100, how much did she earn as an intern?

  2. The greater of two numbers is nine less than four times the other number. If the greater number is 71, find the lesser number.

  3. Fran's second bowling score was 72 points less than twice her first score. If her second score was 156, what was her first score?

  4. On his tenth birthday, Tom's monthly allowance was doubled. When he turned twelve, his allowance increased by $10 to $34 per month. What was Tom's allowance before his tenth birthday?

Solve using any problem solving strategy.

5. Scott Bradley bought five shirts at $14 each and six pairs of socks at $3.50 each. What was the total cost of his purchase?

6. Nineteen more than twice a number is 17. What is the number?

7. Jamie Karp has dimes and quarters in her wallet. In how many different ways can she pay the $1.25 fare for the bus?

8. The volleyball team is holding a month-long fund drive to raise $1350 for a trip to New York. During the first week the team raised $275. The second week the team raised $490. How much does the team still need to raise?

9. Niabi Thomas bought a used car for $4800. The tax on the car was $216. Niabi made a down payment of $1200 and agreed to pay the remainder in 24 equal payments. What was the amount of each payment?

10. Alfonso Gomez went on a cross-country bus trip. On the first day he traveled 250 mi. He traveled 320 mi on each of the next two days. During the fourth day he traveled 280 mi. Alfonso says that by the end of the fourth day he had traveled a total of 850 mi. Is this correct? Explain.

11. Samantha Marshall bought some paperback books for $5 each and some magazines for $3 each. She paid a total of $32. How many of each did she buy?

DATA ANALYSIS

Use the table at the right. Solve using any problem solving strategy.

12. How many more milligrams of calcium are in a 240-mL glass of milk than in one egg?

13. How many milligrams of calcium are in three tomatoes?

Food	Calcium (mg)
1 egg	27
1 orange	54
1 tomato	24
1 glass (240 mL) whole milk	288

14. The Recommended Daily Dietary Allowance (RDA) of calcium for Melania's age group is 800 mg. Melania has already had two oranges and one 240-mL glass of whole milk today. How many more glasses of milk must she have to meet her RDA for calcium?

15. RESEARCH What is the RDA of calcium for your age group?

Write a word problem that you could solve by using each equation. Then solve the problem.

16. $16 = x + 5$ **17.** $3x - 1 = 26$

18. Complete: $64 \text{ oz} = \underline{\quad ? \quad} \text{ lb}$ *(Toolbox Skill 22)*

19. Is 426 divisible by 2? by 3? by 4? by 5? by 8? by 9? by 10? *(Toolbox Skill 16)*

20. Find the difference: $4\frac{7}{12} - 2\frac{5}{6}$ *(Toolbox Skill 19)*

21. The sum of three times a number and seven is 55. Find the number. *(Lesson 4-8)*

Self-Test 2

Solve. Check each solution.

1. $6m + 3 = -15$ **2.** $\frac{n}{-4} + 8 = 2$ **3.** $89 = 10q - 11$ **4-5**

Write a variable expression that represents each phrase. If necessary, choose a variable to represent the unknown number.

4. a number z divided by fourteen **4-6**

5. three times as many ripe tomatoes as yesterday

Write an equation that represents the relationship in each sentence.

6. Seven more than a number n is 35. **4-7**

7. Sandra has twenty-two video cassettes, which is nine fewer than Thomas has.

Solve using an equation.

8. Lynn's second test score was 60 points less than twice her first score. If her second score was 86, what was Lynn's first test score? **4-8**

9. Pat Burns is planning to compete in a 300-km bicycle race. He plans to travel 90 km on the first day and then to travel the same distance on each of the next three days. How far does he plan to travel on the fourth day?

More Equations

Objective: To solve equations by simplifying expressions involving combining like terms and the distributive property.

 DATA ANALYSIS

In the pictograph at the right, the key is missing.

QUESTION Suppose you know that 560 students graduated from Conway and Wheaton together. What is the key?

Let the variable x represent the number in the key. Since there are four symbols next to Conway, the number of graduates from that town can be represented by the expression $4x$. Similarly, the number of graduates from Wheaton can be represented by the expression $3x$. The expression $4x + 3x$ represents the total number of graduates from Conway and Wheaton. To find the key you must solve the equation $4x + 3x = 560$.

Graduates from Area Towns

Conway

Hampton

Stoneham

Wheaton

= students

Example 1

Solution

Solve $4x + 3x = 560$. Check the solution.

$$4x + 3x = 560$$
$$7x = 560 \qquad \longleftarrow \text{Combine like terms.}$$
$$\frac{7x}{7} = \frac{560}{7}$$
$$x = 80$$

The solution is 80.

✓ **Check**

$$4x + 3x = 560$$
$$4(80) + 3(80) \stackrel{?}{=} 560$$
$$320 + 240 \stackrel{?}{=} 560$$
$$560 = 560$$

✓ **Check Your Understanding**

1. What are the like terms that were combined in Example 1?

ANSWER The key is 🎓 = 80 students.

You should simplify the sides of an equation before solving. This may involve combining like terms, as in Example 1. It may also involve using the distributive property.

Example 2

Solution

Solve $2(4x + 3) = 46$. Check the solution.

$$2(4x + 3) = 46$$
$$8x + 6 = 46 \qquad \leftarrow \text{Use the distributive property.}$$
$$8x + 6 - 6 = 46 - 6$$
$$8x = 40$$
$$\frac{8x}{8} = \frac{40}{8}$$
$$x = 5$$

The solution is 5.

Check

$$2(4x + 3) = 46$$
$$2(4[5] + 3) \stackrel{?}{=} 46$$
$$2(20 + 3) \stackrel{?}{=} 46$$
$$2(23) \stackrel{?}{=} 46$$
$$46 = 46$$

☑ **Check Your Understanding**

2. Why was the distributive property used in Example 2?

Guided Practice

COMMUNICATION « *Reading*

Refer to the text on pages 172–173.

« **1.** What is the main idea of the lesson?

« **2.** List two major points that support the main idea of the lesson.

Match each equation with the equation that has the same solution.

3. $3(t + 2) = 30$ **A.** $t = 30$

4. $2(t + 3) = 30$ **B.** $3t + 6 = 30$

5. $3t + 2t = 30$ **C.** $5t = 30$

6. $3t - 2t = 30$ **D.** $2t + 6 = 30$

Solve. Check each solution.

7. $8n + 5n = 39$ **8.** $2p - 6p = 40$

9. $5(3x + 4) = -10$ **10.** $6 = 2(7t - 4)$

11. $3a - 6 + 4a = 22$ **12.** $9b + 7 - b = 71$

Exercises

Solve. Check each solution.

1. $7n + 4n = 132$ **2.** $12a + 3a = 60$

3. $2(3v + 4) = -40$ **4.** $5(2x + 7) = 45$

5. $-5c + 9c = -20$ **6.** $14g - 17g = -21$

Solve. Check each solution.

7. $3(2q - 7) = 3$ **8.** $6(7k - 10) = 24$

9. $-12 = 3(2x - 10)$ **10.** $36 = 4(z + 11)$

11. $4y + 7 + 8y = 43$ **12.** $36 = 6b - 6 + b$

13. The sum of twice a number and four times the number is 54. Find the number.

14. A number subtracted from five times the number is -28. Find the number.

15. Twice the difference of 7 subtracted from a number is -70. Find the number.

16. Three times the sum of a number and 4 is 108. Find the number.

Solve. Check each solution.

17. $5(v + 1) + 5v = 35$ **18.** $4(g + 2) + 8g = 56$

19. $12 = 2(a - 8) + 5a$ **20.** $24 = 7(n - 3) + 8n$

21. $3(4x + 1) - 2x = 23$ **22.** $6(2k + 5) - 3k = 66$

💻 COMPUTER APPLICATION

Using an electronic spreadsheet, you can solve equations that you have not learned how to solve using paper and pencil. You use the spreadsheet to find the value of the variable that makes the expressions on either side of the equation equal.

	A	B	C
1	x	2(4x − 7)	5(x + 2)
2	1	−6	15
3	2	2	20
4	3	10	25
5	4	18	30

Using the equation $2(4x - 7) = 5(x + 2)$, create an electronic spreadsheet or draw a spreadsheet like the one shown above.

23. For which values of x is the expression in cell B1 less than the expression in cell C1?

24. For which values of x is the expression in cell B1 greater than the expression in cell C1?

25. Find the solution of the equation.

26. What would be the solution of the equation if the expression on the right side were $5(x - 4)$?

27. On his first day of a new job, Steve Rosadini worked for 7 h. On the next three days he worked for 8.5 h, 7 h, and 2 h. His hourly wage is $9.25. *(Lesson 1-4)*
   a. What is the paragraph about?
   b. How long did Steve work on the first day?
   c. Identify any facts that are not needed to find the total number of hours that Steve worked for the four days.
   d. Describe how you would find the total amount that Steve earned for the four days.

28. Use the bar graph below. Find the height of the waterfall named Tugela. *(Toolbox Skill 24)*

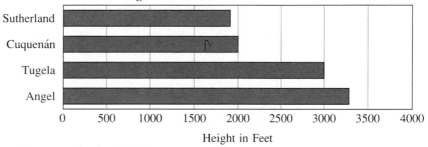

The Four Highest Waterfalls

Height in Feet

29. Estimate the sum: $156 + 203 + 242$ *(Toolbox Skill 2)*

30. Solve: $2(4x + 5) = 66$ *(Lesson 4-9)*

31. PATTERNS Find the next three numbers in the pattern: 18, 15, 12, 9, __?__, __?__, __?__ *(Lesson 2-5)*

32. Ed has only $10, $5, and $1 bills. In how many different ways could he pay $26? *(Lesson 3-7)*

Historical Note

Stars that change in brightness are called *variable stars*. Cecilia Payne-Gaposchkin (1900–1979), an astronomer from the United States, perfected the technique of studying variable stars photographically. Her findings provided important clues to the structure of our galaxy, the Milky Way.

RESEARCH

Recent astronomical studies indicate that there may be a black hole at the center of the Milky Way. Find out what a black hole is.

Formulas

Objective: To work with formulas.

A **formula** is an equation that states a relationship between two or more quantities. The quantities are usually represented by variables. The variable used is often the first letter of the word it represents.

For instance, the *distance formula* represents a relationship between distance (D), rate (r), and time (t).

$$\text{distance} = \text{rate} \times \text{time}$$
$$D = rt$$

Example 1

Solution

Use the formula $D = rt$. Find the distance when $r = 55$ mi/h and $t = 3$ h.

Substitute 55 for r and 3 for t. Solve.

$$D = rt$$
$$D = 55(3)$$
$$D = 165$$

The distance is 165 mi.

Sometimes you might need to use inverse operations to find an unknown value in a formula.

Example 2

Solution

Use the formula $D = rt$. Find the time when $D = 240$ mi and $r = 40$ mi/h.

Substitute 240 for D and 40 for r. Solve.

$$D = rt$$
$$240 = 40t$$
$$\frac{240}{40} = \frac{40t}{40}$$
$$6 = t$$

The time is 6 h.

✔️ **Check Your Understanding**

1. In Example 2, how do you get the equation $240 = 40t$?
2. How would Example 2 be different if a value for t, not r, were given?

Guided Practice

COMMUNICATION «*Reading*

Explain the meaning of the word *formula* in each sentence.

«**1.** Melissa was fed baby formula every four hours.

«**2.** Andrew's formula for success on tests included reviewing his notes.

Choose the letter of the equation that reflects the given information.

3. $D = rt$; $r = 20$ mi/h and $D = 600$ mi

    a. $20 = 600t$ **b.** $600 = \dfrac{t}{20}$

    c. $600 = 20t$ **d.** $600 = 20 + t$

4. $P = 2l + 2w$; $w = 3$ m and $P = 24$ m

    a. $3 = 2l + 48$ **b.** $24 = 2l + 6$

    c. $24 = 6 + 2w$ **d.** $3 = 48 + 2w$

Use the formula $C = p - d$, where C represents cost, p represents price, and d represents discount. Replace each ? with the correct value.

5. $p = \$50$, $d = \$5$, $C = \$\underline{\ ?\ }$ **6.** $p = \$240$, $d = \$60$, $C = \$\underline{\ ?\ }$

7. $C = \$604$, $d = \$151$, $p = \$\underline{\ ?\ }$ **8.** $C = \$76$, $d = \$4$, $p = \$\underline{\ ?\ }$

Exercises

The amount of force (F) applied to move an object a certain distance (d) is defined as work (W). When a force is applied in the direction that the object moves, the relationship between these three variables is $W = Fd$.

Use the formula $W = Fd$. Replace each ? with the correct value.

	Work (W)	Force (F)	Distance (d)
1.	_?_ ft-lb	40 lb	55 ft
2.	520 ft-lb	13 lb	_?_ ft
3.	2400 ft-lb	_?_ lb	60 ft

4. An elevator does 62,400 ft-lb of work lifting a load 120 ft. How heavy is the load?

The speed that an airplane travels in the air depends upon the speed of the wind. The relationship between ground speed (g), air speed (a), and head wind speed (h) is given by the formula $g = a - h$.

Use the formula $g = a - h$. Replace each ? with the correct value.

	Ground Speed (g)	Air Speed (a)	Head Wind Speed (h)
5.	_?_ km/h	620 km/h	10 km/h
6.	575 mi/h	_?_ mi/h	25 mi/h

7. An airplane travels into a 15 mi/h head wind at a ground speed of 460 mi/h. What is the airplane's air speed?

8. An airplane traveling into a 35 mi/h head wind has an air speed of 325 mi/h. What is the airplane's ground speed?

CONNECTING ALGEBRA AND GEOMETRY

Geometry is a branch of mathematics that involves many formulas. These include the formulas for perimeter (P) and area (A). In order to do problems in geometry, you will often have to work with formulas.

Use the given formula. Replace each __?__ with the correct value.

9.

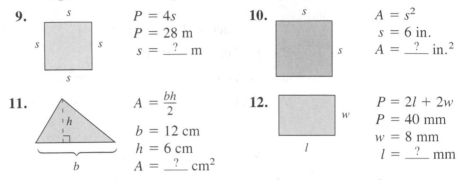

$P = 4s$
$P = 28$ m
$s = $ __?__ m

10.

$A = s^2$
$s = 6$ in.
$A = $ __?__ in.^{2}

11.

$A = \dfrac{bh}{2}$
$b = 12$ cm
$h = 6$ cm
$A = $ __?__ cm^2

12.

$P = 2l + 2w$
$P = 40$ mm
$w = 8$ mm
$l = $ __?__ mm

CAREER/APPLICATION

A *nurse* or *health practitioner* might determine blood pressure. This is a measure of the pressure of the blood within the arteries. *Systolic* pressure is the highest level in the pressure cycle; *diastolic* pressure is the lowest level.

13. The normal systolic blood pressure (P) for a person of a given age (a) is 110 more than the quotient when that person's age is divided by two. Write a formula for systolic blood pressure.

14. Use the formula in Exercise 13. What would you expect the normal systolic blood pressure to be for an 18-year-old person?

15. Use the formula in Exercise 13. If a person has a normal systolic blood pressure of 150, how old might you expect that person to be?

16. **RESEARCH** Blood pressure is often reported as a relationship between the systolic and diastolic pressures. Find out how this relationship is expressed.

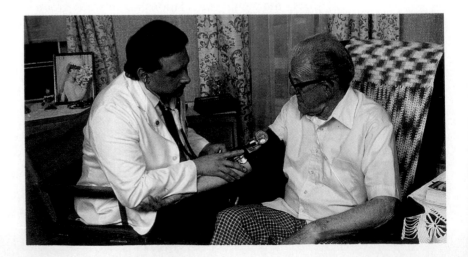

DATA, pages 140–141

Use the formula $D = rt$.

17. How far will the *Metroliner* normally travel in 3 h?

18. How long will it take an impulse of electricity to travel 631,315 mi?

RESEARCH

19. Find out the formula representing Ohm's law.

20. Find out the Einstein equation.

21. Find out what the variables represent in the formula for Ohm's law.

22. Find out what the variables represent in the Einstein equation.

SPIRAL REVIEW

23. Find the quotient: $31.08 \div 3.7$ *(Toolbox Skill 14)*

24. Use the formula $W = Fd$. Let $W = 70$ ft-lb and $F = 14$ lb. Find d.
(Lesson 4-10)

25. Evaluate $|x + y|$ when $x = 7$ and $y = -9$.   *(Lesson 3-6)*

26. Complete: 1.23 km $=$ __?__ m   *(Lesson 2-9)*

27. Simplify: $9a + 5b - b$ *(Lesson 2-3)*

28. Solve: $\frac{x}{9} = 14$ *(Lesson 4-4)*

29. Evaluate n^5 when $n = 4$.   *(Lesson 1-9)*

30. Solve: $x + 9 = -6$ *(Lesson 4-3)*

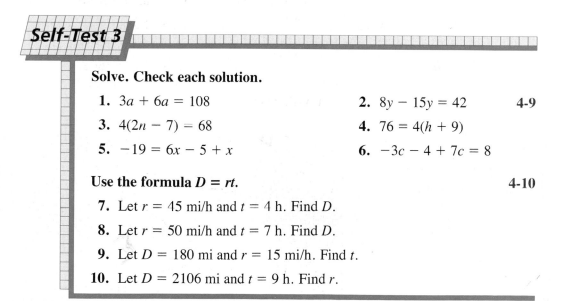

Self-Test 3

Solve. Check each solution.

1. $3a + 6a = 108$ **2.** $8y - 15y = 42$ **4-9**

3. $4(2n - 7) = 68$ **4.** $76 = 4(h + 9)$

5. $-19 = 6x - 5 + x$ **6.** $-3c - 4 + 7c = 8$

Use the formula $D = rt$.   **4-10**

7. Let $r = 45$ mi/h and $t = 4$ h. Find D.

8. Let $r = 50$ mi/h and $t = 7$ h. Find D.

9. Let $D = 180$ mi and $r = 15$ mi/h. Find t.

10. Let $D = 2106$ mi and $t = 9$ h. Find r.

Chapter Review

equation (p. 142)
solution of an equation (p. 142)
solve (p. 148)

inverse operations (p. 149)
formula (p. 176)

Choose the correct term from the list above to complete each sentence.

1. Multiplication and division are __?__ because each can be used to undo the other in solving an equation.

2. A(n) __?__ is a statement that two numbers or two expressions are equal.

3. To __?__ an equation, you find all values of the variable that make the equation true.

4. A(n) __?__ is an equation that states a relationship between two or more quantities that are usually represented by variables.

5. A value of the variable that makes an equation true is a(n) __?__.

Is the given number a solution of the equation? Write *Yes* or *No*. *(Lesson 4-1)*

6. $x - 3 = 7$; 10

7. $\frac{n}{8} = -2$; 6

8. $9 = t + 7$; 16

9. $36 = -4a$; -9

Use mental math to find each solution. *(Lesson 4-1)*

10. $11 = r + 6$

11. $8m - 4 = -60$

12. $13 = \frac{b}{5} + 3$

13. $x - 2 = -2$

Solve using the guess-and-check strategy. *(Lesson 4-2)*

14. The prices at Salons Now are $11 for a haircut and $34 for a perm. On Tuesday Salons Now collected $279 from customers who were given haircuts or perms. How many of each were given on Tuesday?

15. Malik bought some pens for $1.25 each. He also bought some notebooks for $2.75 each. He spent $14.75. How many of each item did he buy?

Solve. Check each solution. *(Lessons 4-3, 4-4, and 4-5)*

16. $r + 12 = 15$

17. $b - 4 = -6$

18. $8 = 13 + y$

19. $14 = a - 9$

20. $-q = 16$

21. $\frac{v}{7} = -2$

22. $15 = \frac{m}{3}$

23. $11x = 132$

24. $1 + 4n = 9$

25. $5 = \frac{w}{3} + 2$

26. $-13 = 5 + 2t$

27. $\frac{a}{-8} - 5 = 6$

Write a variable expression that represents each phrase. If necessary, choose a variable to represent the unknown number. *(Lesson 4-6)*

28. 6 less than 8 times a number q

29. a number h divided by 19

30. 7 times as many books as Ann read last year

31. 4 more than twice the number of tickets sold yesterday

Write an equation that represents the relationship in each sentence. *(Lesson 4-7)*

32. A number z decreased by 15 is 33.

33. The school debating team has 8 members, which is 5 fewer than the number of students on the school mathematics team.

34. Jack has 48 trophies, which is three times as many as Mark has.

Solve using an equation. *(Lesson 4-8)*

35. Beth MacCurdy paid a total of $355 for two identical wool coats, including tax of $17. What was the price of each coat?

36. Andrew Morrison takes a group of students on a fishing trip each year. This year they caught a record 27 fish, which is one less than four times as many fish as the group caught last year. How many fish were caught last year?

Solve. Check each solution. *(Lesson 4-9)*

37. $-2x - 7x = 18$

38. $6m + 4 + 5m = -29$

39. $4(v - 10) = 24$

40. $110 = 5(1 + 3k)$

Use the formula $C = np$, where C is the total cost, n is the number of items purchased, and p is the price per item. *(Lesson 4-10)*

41. Let $C = \$182$ and $p = \$13$. Find n.

42. Let $n = 24$ and $p = \$1.99$. Find C.

43. Lucas bought 12 posters at $10.50 per poster. How much did he spend?

44. Robin spent $60 for 8 tickets. What was the price of each ticket?

Chapter Test

Is the given number a solution of the equation? Write *Yes* or *No*.

1. $-6x = 18$; -3 **2.** $9 = u - 4$; 5 **3.** $\frac{a}{7} = -7$; 49 **4-1**

Solve using the guess-and-check strategy.

4. Florina has eighteen dimes and nickels in all. Their total worth is   **4-2**
$1.40. How many of each type of coin does she have?

Solve. Check each solution.

5. $g + 15 = 7$ **6.** $9 = b - 17$ **7.** $28 + n = -10$ **4-3**

8. $-13 = \frac{v}{-5}$ **9.** $18c = 180$ **10.** $-r = 20$ **4-4**

11. $3z + 5 = -4$ **12.** $15 + \frac{m}{8} = 20$ **13.** $-72 = 6p - 30$ **4-5**

14. Explain the meaning of *inverse operations*. Give two examples.

Write a variable expression that represents each phrase. If necessary, choose a variable to represent the unknown number.

15. eighteen more than four times a number j **4-6**

16. ten times the number of boxes Jo sold

Write an equation that represents the relationship in each sentence.

17. A number n divided by 8 is 90.   **4-7**

18. Salim has 15 model cars in his collection, which is 8 fewer model cars than Franco has.

19. Explain the difference between an equation and an expression.

Solve using an equation.

20. Ali's mathematics textbook contains 690 pages. This is 248 pages less   **4-8**
than twice the number of pages in Ali's science textbook. How many
pages are in the science textbook?

Solve. Check each solution.

21. $-12x + 3 + 5x = 38$ **22.** $6(8 + 3q) = 66$ **23.** $-10 = 5(7m - 16)$ **4-9**

Use the formula $g = a - h$.

24. Let $a = 720$ km/h and $h = 15$ km/h. Find g.   **4-10**

25. Let $g = 650$ mi/h and $h = 25$ mi/h. Find a.

Equations Having Variables on Both Sides

Objective: To solve equations having variables on both sides.

Some equations have the variable on both sides. To solve, you need to get the variable alone on one side.

Example

Solution

Solve $5a - 9 = 2a - 3$. Check the solution.

Subtract a variable term from both sides.

$$5a - 9 = 2a - 3$$
$$5a - 9 - 2a = 2a - 3 - 2a$$
$$3a - 9 = -3$$
$$3a - 9 + 9 = -3 + 9$$
$$3a = 6$$
$$\frac{3a}{3} = \frac{6}{3}$$
$$a = 2$$

The solution is 2.

✔️ **Check**

$$5a - 9 = 2a - 3$$
$$5(2) - 9 \stackrel{?}{=} 2(2) - 3$$
$$10 - 9 \stackrel{?}{=} 4 - 3$$
$$1 = 1$$

Exercises

Solve. Check the solution.

1. $7n + 10 = 3n + 2$

2. $5h - 7 = 2h + 2$

3. $4v - 9 = 2v + 1$

4. $9x - 4 = 6x + 23$

5. $3t + 12 = 2t - 14$

6. $2y + 2 = y - 8$

7. $5a = 2a + 18$

8. $8u = 6u - 20$

9. $-5 + 12v = 11v - 7$

10. $-9 + 16g = 5g + 13$

11. $1 + 9h = 4h + 11$

12. $8 + 32t = 31t + 17$

13. $-5p = 2p + 14$

14. $-7g = 2g + 36$

15. $5a - 7 = -3a + 17$

16. $9b + 7 = -6b + 52$

17. $2v + 7 = 4v - 19$

18. $8x + 17 = 9x - 8$

GROUP ACTIVITY

19. The first step in solving the Example above involved subtracting a variable term from both sides of the equation. What other first step could have been taken to solve the equation?

20. Describe four different ways to begin solving the equation $3x + 5 = 8x - 10$.

Standardized Testing Practice

Choose the letter of the correct answer.

1. **Which equation could you use to solve the problem?**
   The Tigers' score was 35 points less than twice that of the Colonials. The Tigers scored 74 points. How many points did the Colonials score?
   A. $2c - 35 = 74$
   B. $2c + 35 = 74$
   C. $2c - 74 = 35$
   D. $74 - 2c = 35$

2. **Solve:** $16 = \frac{t}{4} + 4$
   A. $t = 80$ B. $t = 60$
   C. $t = 192$ D. $t = 48$

3. **Which number is to the left of -6 on a number line?**
   A. -8 B. 0
   C. $|-6|$ D. 7

4. **Write the phrase *7 less than 4 times a number* as a variable expression.**
   A. $7 - 4n$
   B. $7 < 4n$
   C. $4n - 7$
   D. $4n < 7$

5. **Evaluate $24 - a$ when $a = -6$.**
   A. 18 B. 30
   C. -18 D. -4

6. **Eldon bought 3 apples at 45¢ each, 2 pears at 38¢ each, and a juice drink for 65¢. He gave the cashier $5. Which question can be answered?**
   A. How much did he spend?
   B. How much change did he get?
   C. How many items did he buy?
   D. all of the above

7. **Solve:** $-13n = 52$
   A. $n = -4$ B. $n = 4$
   C. $n = -676$ D. $n = 676$

8. **Evaluate the quotient $41.6 \div z$ when $z = 8$.**
   A. 52 B. 5.2
   C. 332.8 D. 33.28

9. **How would you move the decimal point to change 47.5 mm to m?**
   A. 2 places to the right
   B. 2 places to the left
   C. 3 places to the right
   D. 3 places to the left

10. **Evaluate $(2b)^3$ when $b = 4$.**
    A. 512 B. 128
    C. 32 D. 216

11. In four weeks, Emma worked 35 h, 42 h, 22 h, and 18 h. Choose the best estimate of her total hours.

A. about 80 h
B. about 100 h
C. about 120 h
D. about 140 h

16. Complete: If $x + a = b$, then $x = $ ___?___ .

A. $b - a$
B. $a - b$
C. ab
D. $\dfrac{b}{a}$

12. Solve: $2(y + 6) = -24$

A. $y = -15$　　B. $y = -18$
C. $y = 18$　　D. $y = -6$

17. Solve: $\dfrac{x}{-16} = 8$

A. $x = 128$　　B. $x = -128$
C. $x = 2$　　D. $x = -2$

13. Find the function rule.

A. $x \rightarrow 5x - 4$
B. $x \rightarrow 3x + 6$
C. $x \rightarrow 5x + 1$
D. $x \rightarrow 4x + 2$

x	?
1	6
2	10
3	14
5	22
6	26

18. A rectangle has perimeter 28 cm and width 3 cm. Use the formula $P = 2l + 2w$ to find the length of the rectangle.

A. 25 cm　　B. 22 cm
C. 17 cm　　D. 11 cm

14. Find the sum: $-12 + (-18)$

A. -6　　B. 6
C. -30　　D. 30

19. Solve: $15 - 4y = 3$

A. $y = 3$　　B. $y = -3$
C. $y = 4.5$　　D. $y = -4.5$

15. Of which equation(s) is -5 a solution?

I. $x - 1 = 4$　　II. $12 - x = 17$
III. $3x = -15$

A. I and II only
B. I and III only
C. II and III only
D. I, II, and III

20. Which statement shows the commutative property of addition?

A. $a + (b + c) = (a + b) + c$
B. $a(b + c) = ab + ac$
C. $a + (b + c) = a + (c + b)$
D. $a(b + c) = (b + c)a$

Sites of Summer Olympic Games

Year	Location	Year	Location
1900	Paris	1952	Helsinki
1904	St. Louis	1956	Melbourne
1906	Athens	1960	Rome
1908	London	1964	Tokyo
1912	Stockholm	1968	Mexico City
1920	Antwerp	1972	Munich
1924	Paris	1976	Montreal
1928	Amsterdam	1980	Moscow
1932	Los Angeles	1984	Los Angeles
1936	Berlin	1988	Seoul
1948	London	1992	Barcelona

Did You Know?

The average speed of runners in a 200 m race is greater than the average speed in a 100 m race. In both races runners spend time accelerating, but in a 200 m race runners spend more time running at top speed, so the average speed increases. The record for a 100 m race is an average speed of 22.5 mi/h and the record for a 200 m race is 22.7 mi/h.

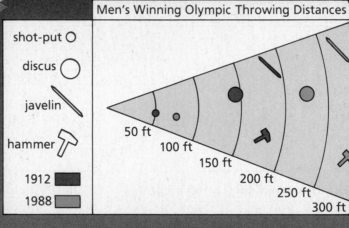

Men's Winning Olympic Throwing Distances

shot-put
discus
javelin
hammer
1912
1988

50 ft
100 ft
150 ft
200 ft
250 ft
300 ft

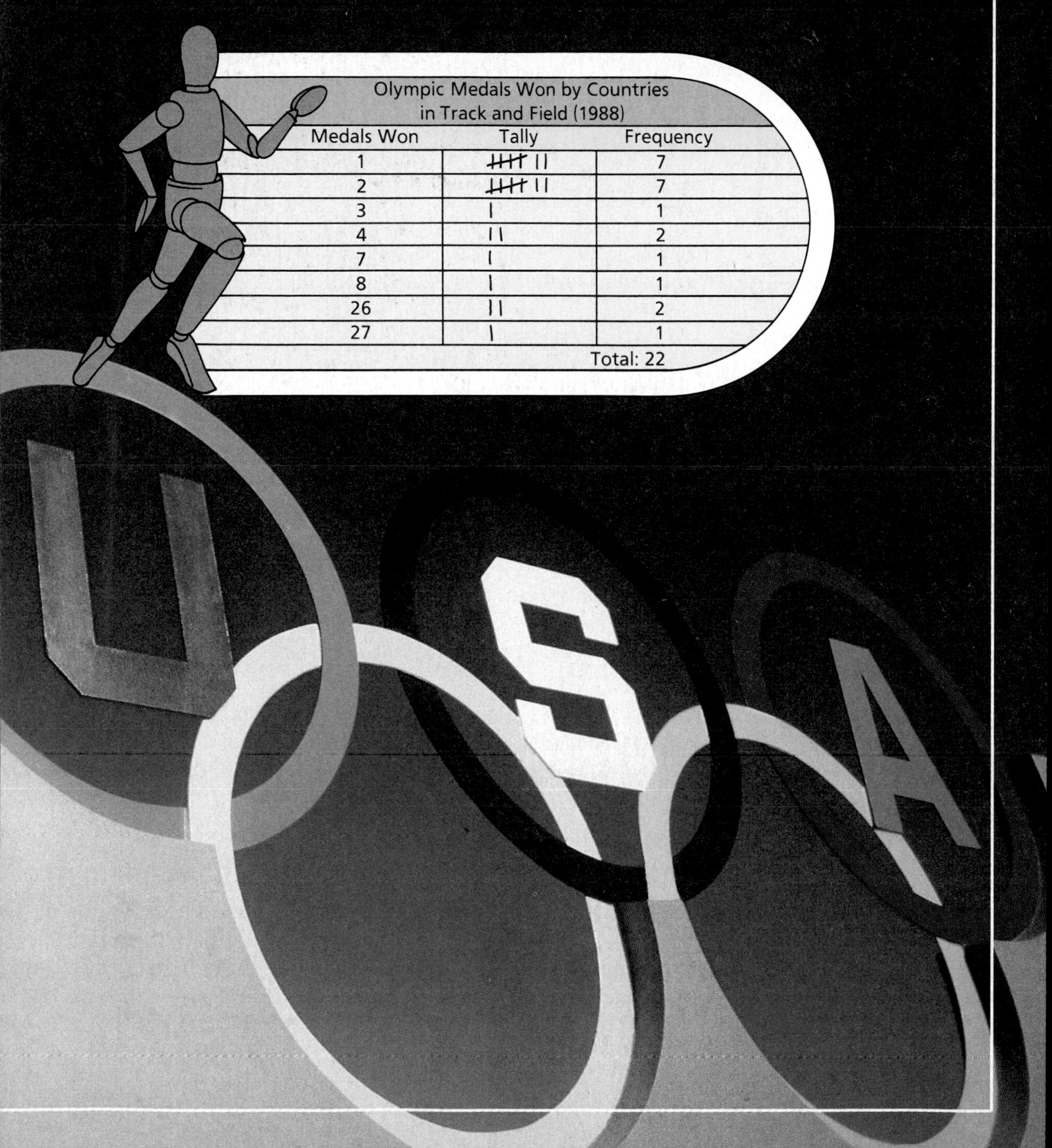

Olympic Medals Won by Countries in Track and Field (1988)		
Medals Won	Tally	Frequency
1	⊞⊞ ‖	7
2	⊞⊞ ‖	7
3	‖	1
4	‖	2
7	‖	1
8	‖	1
26	‖	2
27	‖	1
		Total: 22

5-1 Pictographs

Objective: To interpret and draw pictographs.

A **graph** is a picture that displays numerical facts, called **data.** One type of graph that you often see in newspapers and magazines is a **pictograph.** In this type of graph, a symbol is used to represent a given number of items. A **key** on the graph tells you how many items the symbol represents. It is important to note that pictographs often show only *approximations* of the data.

Example 1

Use the pictograph above.

a. About how many lime cones were sold?

b. About how many more chocolate cones were sold than orange cones?

Solution

a. Each symbol represents 150 cones.
There are $2\frac{1}{2}$, or 2.5, symbols on the lime cone.
Multiply.
$2.5(150) = 375$
About 375 lime cones were sold.

b. Compare the numbers of symbols.
The chocolate cone has 4 more symbols than the orange cone.
Multiply.
$4(150) = 600$
About 600 more chocolate cones were sold than orange cones.

✓ Check Your Understanding

1. Describe how Example 1(a) would be different if each symbol represented 500 cones.

2. Describe a different way to find the answer to Example 1(b).

In order to draw a pictograph from a set of data like the one below, you must first choose a symbol and find an appropriate amount for that symbol to represent.

Farms in the United States (Millions)

Year	1900	1920	1940	1960	1980
Farms	5.7	6.5	6.1	4.0	2.4

Example 2

Draw a pictograph to display the data in the table at the bottom of page 188.

Solution

- Choose a symbol to represent 1 million farms. Half a symbol represents 0.5 million farms.

- Round each number to the nearest half-million to find the number of symbols for each year.

  5.7 ⟶ 5.5 symbols
  6.5 ⟶ 6.5 symbols
  6.1 ⟶ 6 symbols
  4.0 ⟶ 4 symbols
  2.4 ⟶ 2.5 symbols

- Draw the graph. Include a title and the key.

Farms in the United States

1900 🚜🚜🚜🚜🚜🚜
1920 🚜🚜🚜🚜🚜🚜🚜
1940 🚜🚜🚜🚜🚜🚜
1960 🚜🚜🚜🚜
1980 🚜🚜🚜

🚜 = 1 million farms

Fresh
Farm

✓ Check Your Understanding

3. In Example 2, explain why 5.7 is rounded to 5.5 instead of 6.

Guided Practice

«1. COMMUNICATION «*Reading*   Replace each __?__ with the correct word.   A graph in which a symbol is used to represent a certain number of items is called a __?__. In this type of graph, the __?__ tells you how many items each symbol represents.

Use the table below. Assume that an hourglass is the symbol used in a pictograph displaying the data.

2. How many years might an hourglass represent?

3. How many years would half of the hourglass represent?

4. To what number would you round the life spans?

5. Draw a pictograph to display the data.

Use the pictograph on page 188.

6. About how many banana cones were sold?

7. About how many more strawberry cones were sold than lime cones?

Maximum Life Spans of Certain Animals

Animal	horse	sea lion	goat	fox	chipmunk
Years	46	28	18	14	8

Use the pictograph at the right. About how many passengers were there?

1. in 1992   2. in 1984

3. About how many more passengers were there in 1988 than in 1980?

4. In which years were there fewer than 2,000,000 passengers?

5. Name the year that there were about twice as many passengers as in 1984.

6. How many symbols represent 5,000,000 passengers?

Passengers on Calypso Cruise Lines

1992
1988
1984
1980

🛞 = 500,000 passengers

Draw a pictograph to display the data.

7. **Airport Limousine Company Earnings per Quarter**

Quarter	first	second	third	fourth
Earnings	$80,477	$99,812	$86,040	$51,252

8. **Number of Women in State Legislatures**

Year	1978	1980	1982	1984	1986	1988
Women	767	871	911	996	1101	1175

MENTAL MATH

Use the pictograph below. Find each answer mentally.

9. About how many rolls of film were developed altogether?

10. About how many more rolls were developed into textured prints than into glossy prints?

Foto Finish received a set amount of money for each roll developed into prints. About how much was received altogether?

11. small glossy prints: $12 per roll   12. large glossy prints: $16 per roll

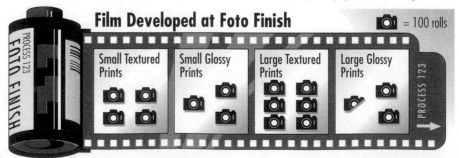

Film Developed at Foto Finish   📷 = 100 rolls

Small Textured Prints   Small Glossy Prints   Large Textured Prints   Large Glossy Prints

The pictograph at the right was drawn using the data in this table.

Airport Traffic
(Passengers Arriving and Departing)

Airport Traffic (Passengers Arriving and Departing)

City	Passengers
Atlanta	47,649,470
Dallas/Fort Worth	41,875,444
Denver	32,355,000
Boston	23,283,047
Pittsburgh	17,457,801

= 20 million passengers

Use the table and the pictograph above.

13. Explain why the key that was used (= 20 million passengers) does not represent the data accurately.

14. Which of the following keys do you think most people would agree is the most appropriate for this set of data? Give a convincing argument to support your choice.

   a. = 2 million passengers

   b. = 10 million passengers

   c. = 30 million passengers

   d. = 40 million passengers

15. Use the key that you chose in Exercise 14. Draw a more appropriate pictograph for the data in the table above.

16. WRITING ABOUT MATHEMATICS   Use the pictograph that you drew for Exercise 15. Write five questions that involve interpreting your pictograph.

SPIRAL REVIEW

17. Solve $y - 8 = -2$. Check the solution.   *(Lesson 4-3)*

18. Evaluate $62.66 \div k$ when $k = 2.6$.   *(Lesson 1-3)*

19. Find the answer: $2 \cdot 4^3 + 8$ *(Lesson 1-10)*

20. Use the pictograph on page 190 entitled *Film Developed at Foto Finish*. About how many more rolls of film were developed into large prints than into small prints?   *(Lesson 5-1)*

21. Find the difference: $-2 - (-6)$ *(Lesson 3-4)*

22. Estimate the product: $379 \cdot 42$ *(Toolbox Skill 1)*

23. Simplify: $8(5 - 3n)$ *(Lesson 2-2)*

Interpreting Bar Graphs

Objective: To interpret single and double bar graphs.

Terms to Know

- *bar graph*
- *scale*
- *legend*

The results of a survey about music preferences were published in the Alltown High School newspaper in the form of a *bar graph*. This made it easier for students to compare preferences.

A **bar graph** has two axes. One axis is labeled with a numerical **scale.** The other is labeled with the categories. When reading a bar graph, you might find it is often necessary to estimate where the bars end.

Music Preferred at Alltown High School

Example 1

Use the bar graph above.

a. Estimate the number of students who prefer folk music.

b. About how many more students prefer jazz than prefer hard rock?

Solution

a. Locate the bar for *folk* music.
It ends about halfway between 300 and 400, near 350.
About 350 students prefer folk music.

b. Compare the heights of the bars for *jazz* and *hard rock*.
The jazz bar is about 3 intervals higher than the hard rock bar.
Multiply: 3(100) = 300
About 300 more students prefer jazz than prefer hard rock.

✓ **Check Your Understanding**

Describe a different way to find the answer to Example 1(b).

There are many kinds of bar graphs. The graph at the right is a *double bar graph.* In this type of bar graph, two bars appear in each category. A **legend** is included on the graph to identify the two types of bars.

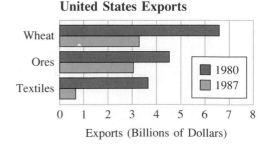

United States Exports

Example 2

Use the double bar graph on page 192. About how much greater was the value of ores exported in 1980 than in 1987?

Solution

Locate the bars for ores. The 1980 bar is *red*. It ends at about 4.5. The 1987 bar is *blue*. It ends at about 3.
Subtract: 4.5 billion − 3 billion = 1.5 billion

The value of ores exported was about $1.5 billion greater in 1980 than in 1987.

Guided Practice

« **1.** COMMUNICATION « *Reading* Choose the definition of the word *scale* that is used in this lesson.

scale (skāl) *noun* **1.** A dry, thin flake or crust. **2.** A series of marks placed at fixed distances, used for measuring. **3.** The relationship between the actual size of something and the size of a model or drawing that represents it. **4.** A series of musical tones that goes up or down in pitch. **5.** An instrument for weighing.

Use the double bar graph below.

2. What does each interval on the scale represent?

3. What do the red bars represent?

4. Which country is expected to have the greatest population in 2100?

5. Which country is expected to have the least increase in population?

6. Tell whether the bar for Brazil in 1988 ends closer to 0.1, 0.15, or 0.2.

7. Tell whether the bar for Mexico in 2100 ends closer to 0.1, 0.15, or 0.2.

8. Estimate the population of the United States in 1988.

9. Which country is expected to have a population of about 0.2 billion in 2100?

10. Estimate the projected increase in population for Brazil.

11. For the year 2100, about how much greater is the projected population for the United States than for Mexico?

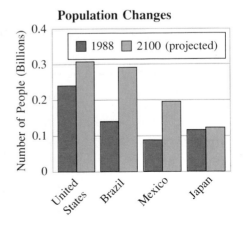

Exercises

Use the bar graph at the right. Estimate the number of balls sold for each sport.

1. soccer
2. football

3. basketball
4. volleyball

5. Name all the sports for which fewer than 2 million balls were sold.

6. About how many times as many soccer balls were sold as volleyballs?

7. About how many more basketballs were sold than bowling balls?

8. About how many more footballs were sold than volleyballs?

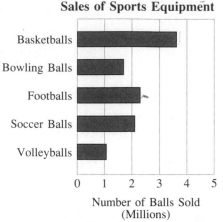

Sales of Sports Equipment

Use the double bar graph at the right.

9. In 1975, travelers from which region spent about $0.6 billion in the United States?

10. In 1985, travelers from which region spent about $0.6 billion in the United States?

11. About how much more did travelers from Japan spend in the United States in 1985 than in 1975?

12. Travelers from which region increased their spending in the United States the most between 1975 and 1985?

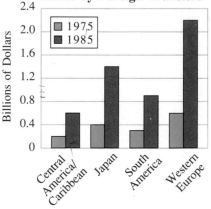

Money Spent in the United States by Foreign Travelers

DATA, *pages 92–93*

13. Estimate the record low temperature in Minneapolis.

14. About how much warmer is the record high temperature in Cheyenne than the record low temperature?

15. About how much warmer is the record high temperature in Los Angeles than the record high temperature in Minneapolis?

16. About how much colder is the record low temperature in Nashville than the record low temperature in Houston?

Newspapers and magazines often put the numerical fact at the end of each bar on a bar graph and omit the scale. This type of graph is called an *annotated* bar graph.

Use the annotated bar graph at the right.

17. How many more calculators were shipped in 1987 than in 1975?

18. In which four-year interval did the number of calculators shipped increase by less than 3 million?

19. Estimate the income from the sales of all the calculators shipped in 1975.

20. Estimate the income from the sales of all the calculators shipped in 1987.

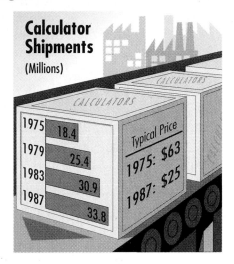

Data given in pairs, such as wins and losses, are sometimes graphed on a *sliding* bar graph. On this type of graph there are two scales. One scale extends to the right of zero and the other extends to the left.

Use the sliding bar graph at the right.

21. About how many games did the Cincinnati team lose?

22. About how many more games did the San Francisco team win than the Atlanta team?

23. Make a double bar graph that displays the same set of data as is shown in this sliding bar graph.

24. **WRITING ABOUT MATHEMATICS** Write a paragraph that compares the graph you drew for Exercise 23 to the graph shown at the right.

SPIRAL REVIEW

25. Simplify: $6u + 5v + 11u$ *(Lesson 2-3)*

26. Use the bar graph at the top of page 194. About how many more basketballs were sold than soccer balls?   *(Lesson 5-2)*

Interpreting Line Graphs

Objective: To interpret single and double line graphs.

A **line graph** shows an *amount* and a *direction* of change in data over a period of time. In a line graph the data are represented by points. These points are connected by line segments.

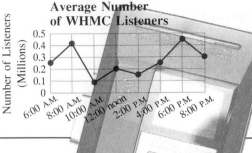

Average Number of WHMC Listeners

Example 1

Use the line graph above.

a. About how many people listen to WHMC at 5:00 P.M.?

b. Estimate the decrease in the number of listeners from 8:00 A.M. to 10:00 A.M.

Solution

a. Locate 5:00 P.M. halfway between 4:00 P.M. and 6:00 P.M. Move up to the red line, then left to the vertical axis.
The number for 5:00 P.M. is between 0.3 million and 0.4 million, at about 0.35 million.
About 0.35 million people listen at 5:00 P.M.

b. Estimate the number of listeners at each of the two hours.

$$8:00 \text{ A.M.} \longrightarrow \text{about } 0.4 \text{ million}$$
$$10:00 \text{ A.M.} \longrightarrow \text{about } 0.1 \text{ million}$$

Subtract: $0.4 - 0.1 = 0.3$
The number of listeners decreases by about 0.3 million from 8:00 A.M. to 10:00 A.M.

Terms to Know

- *line graph*
- *increasing trend*
- *decreasing trend*

If a series of segments on a line graph slopes upward over a given interval, there is an **increasing trend** in the data over that interval. If a series of segments slopes downward, there is a **decreasing trend** over that interval.

Double line graphs such as the one shown at the right are useful for comparing trends in two sets of data. This particular double line graph has a gap in the scale, marked with a jagged line (⌇).

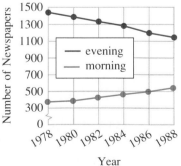

Daily Newspapers in the United States

— evening
— morning

Year

Example 2	Use the double line graph at the bottom of page 196.
	a. Estimate the number of morning newspapers in 1982.
	b. Describe the overall trend in the number of evening newspapers.
Solution	**a.** Locate the point for 1982 on the *blue* line.
	The number for 1982 is between 300 and 500, at about 400.
	There were about 400 morning newspapers in 1982.
	b. Each *red* segment slopes downward. Overall, there was a decreasing trend in the number of evening newspapers.
✓ **Check Your Understanding**	Describe how Example 2(a) would be different if you were asked to find the number of evening newspapers in 1982.

Guided Practice

«**1.** COMMUNICATION «*Reading*   Explain the meaning of the word *trends* in the following sentence.

Reading popular magazines can help a person keep up with the latest trends.

Use the line graph below. Tell whether there was an *increase* or a *decrease* in the data over each interval.

2. 1960–1965   **3.** 1980–1985

4. What does each interval on the horizontal axis represent?

5. What does each interval on the vertical axis represent?

6. Which interval along the vertical axis contains the point for 1965?

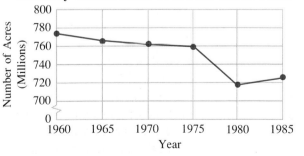

Federally Owned Land within the United States

Use the double line graph at the bottom of page 196.

7. In which given year were there about 1100 evening newspapers?

8. About how many more evening newspapers than morning newspapers were there in 1986?

9. Describe the overall trend in the number of morning newspapers.

Exercises

Use the line graph at the right. Estimate the number of members in each year.

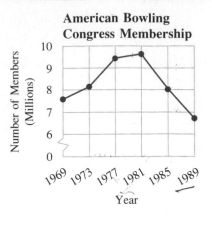

American Bowling Congress Membership

1. 1985
2. 1969
3. 1979
4. 1975

5. In which given years were there about 9.5 million members in the American Bowling Congress?

6. In which given year were there about 7 million members in the American Bowling Congress?

7. Estimate the decrease in membership from 1981 to 1989.

8. During which four-year interval did membership decrease by about one million people?

9. During which four-year interval did the greatest increase in membership occur?

10. Describe the trend in membership from 1969 to 1981.

Use the double line graph at the right. Estimate the number of public school teachers in each year at each level.

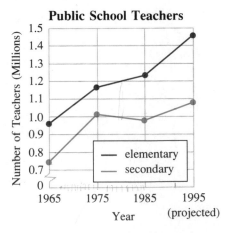

Public School Teachers

11. 1975
12. 1995
13. 1970
14. 1990

15. About how many more elementary than secondary public school teachers were there in 1965?

16. About how many more elementary than secondary public school teachers were there in 1985?

17. Estimate the total number of public school teachers in 1980.

18. Estimate the total number of public school teachers in 1970.

19. Estimate the increase in the number of secondary public school teachers from 1965 to 1995.

20. Describe the overall trend in the number of elementary public school teachers.

Line graphs are often shown without labels on the axes. To interpret these graphs, you must find the values of the intervals.

Use the graph at the right.

21. What does each interval on the horizontal axis represent?

22. What does each interval on the vertical axis represent?

23. In which year were there about 4 million fewer trucks from 3 to 5 years old than there were trucks 12 years and over?

24. Estimate the year in which there were as many trucks from 3 to 5 years old as there were trucks 12 years and over.

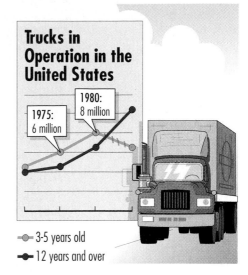

Trucks in Operation in the United States

1975: 6 million
1980: 8 million

— 3-5 years old
— 12 years and over

CONNECTING DATA ANALYSIS AND ALGEBRA

Each table and graph in this chapter represents a relationship between two sets of data. Each item in the first set of data is paired with exactly one item in the second set, so each relationship is a function.

Use the line graph below to complete each ordered pair.

25. (1910, _?_)   26. (_?_, 10 million)   27. (1955, _?_)

28. (1925, _?_)   29. (1940, _?_)   30. (_?_, 20 million)

31. Use the ordered pairs from Exercises 25–30 to make a function table for the data.

32. Use the function table that you made for Exercise 31 to complete this statement: The number of sheep on farms in the United States is a function of _?_.

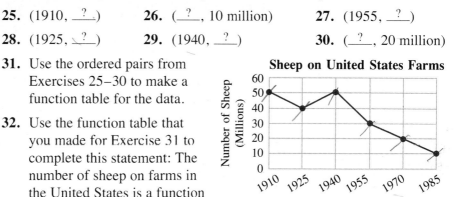

Sheep on United States Farms

Number of Sheep (Millions)

Year

SPIRAL REVIEW

33. Simplify: $(4x)(60x)$ *(Lesson 2-1)*

34. Find the product: $8(-6)(-3)$ *(Lesson 3-5)*

35. Estimate the sum: $0.59 + 0.61 + 0.627 + 0.584$ *(Toolbox Skill 4)*

36. Use the line graph on page 197. In which five-year interval did the greatest decrease occur?   *(Lesson 5-3)*

Drawing Bar and Line Graphs

Objective: To draw a bar or line graph for a given set of data.

While doing a research project for her history class, Lynne Greyson discovered that some counties in the United States have a president's name. She first recorded the information in a table like the one at the right. Lynne then decided to display these data in a bar graph so that she could compare values more easily.

County Name	Number of Counties
Washington	32
Adams	12
Jefferson	25
Madison	20
Monroe	17

Example 1

Solution

Draw a bar graph to display the data in the table above.

• Draw the axes. Position the names of the presidents on the vertical axis. This will make the names easier to read.

• Choose a scale. The greatest number is 32, so draw and label a scale from 0 to 35, using intervals of five.

• Draw a bar for each entry in the table. Use the scale to decide where to end the bars.

• Label the horizontal axis and title the graph.

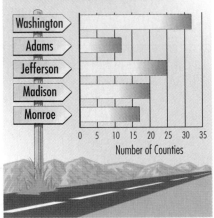

☑ Check Your Understanding

Why is it better to use intervals of five rather than intervals of ten on the graph in Example 1?

As part of his science project, John Nodlem wanted to compare his growth to that of the average male. To make it easier to compare the data and analyze trends, he drew a line graph.

Growth Record for John Nodlem

Age (years)	birth	4	8	12	16
Recorded Height (cm)	55	85	108	147	182
Height of Average Male (cm)	51	99	126	147	169

Example 2

Solution

Draw a line graph to display the data at the bottom of page 200.

- Draw the axes, then write the ages at evenly spaced intervals on the horizontal axis.

- Choose a scale. Notice that the heights range from 51 cm to 182 cm. Start with a gap in the scale. Then, using intervals of twenty-five, number the scale from 50 to 200.

- Place a point on the graph for John's height at each age. Connect the points from left to right with solid line segments. Repeat this process for the average heights, using dashes or a different color.

- Include a legend on the graph. Label both axes and title the graph.

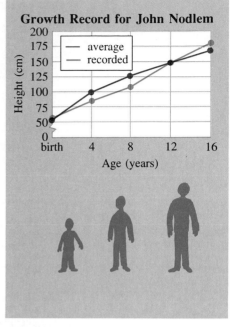

Guided Practice

« **1.** COMMUNICATION « *Reading*   Refer to the text on pages 200–201. Using your own words, develop two outlines for your notebook: *How to Draw a Bar Graph* and *How to Draw a Line Graph*.

Consider drawing a line graph to display the data below.

In-State Tuition and Fees at Public Two-Year Colleges

Year	1979	1982	1985	1988
Cost	$327	$434	$584	$690

2. Describe the scale you would use.

3. What labels would you place on the axes?

4. Draw a line graph to display the data.

5. Draw a bar graph to display the data below.

Appliance Shipments (Millions)

Year	1975	1979	1983	1987
Electric Ranges	2.1	3.0	2.8	3.6
Microwave Ovens	0.8	2.8	6.0	12.7

Exercises

Draw a bar graph to display the data.

1. **Adults Participating in Leisure-Time Activities (Millions)**

Activity	bicycling	softball	swimming	volleyball
Adults	60	39	75	34

2. **Tourism in the United States and Overseas (Millions)**

Year	1981	1983	1985	1987
Tourists in the U.S.	9.1	7.9	7.5	10.4
U.S. Tourists Overseas	8.0	9.6	12.3	13.7

Draw a line graph to display the data.

3. **Average Payment Period, Finance Company Loans on New Cars**

Year	1984	1985	1986	1987	1988	1989
Number of Months	48.3	53.5	50.0	53.5	56.2	54.2

4. **Recreational Boats in the United States (Millions)**

Year	1975	1979	1983	1987
Outboard Motor Boats	5.7	6.7	7.2	7.8
Canoes and Rowboats	2.4	2.7	3.4	4.1

5. DATA, *pages 92–93* Draw a bar graph to display the data in the diagram entitled *Average Number of Tornadoes per Year*.

6. DATA, *pages 2–3* Draw a bar graph to display the data in the table entitled *Shipments of Audio Recordings*.

CONNECTING MATHEMATICS AND SOCIAL STUDIES

7. Draw a line graph to display the data in the table at the right.

Use your graph from Exercise 7.

8. Do the retail and farm prices of oranges always change by the same amount? Explain.

9. Describe the overall relationship between the retail and farm prices of oranges.

10. THINKING SKILLS Compare the graph to the table. Explain why an economist might find the graph more useful than the table.

California Orange Prices
Cents per Pound

Year	Retail Price	Farm Price
1979	44	13
1981	39	8
1983	39	5
1985	53	12
1987	55	10
1989	56	11

11. Solve $15 = \frac{c}{11} + 17$. Check the solution.   *(Lesson 4-5)*

12. Write 4,900,000 in scientific notation.   *(Lesson 2-10)*

13. Estimate the quotient: $31.69 \div 4.2$ *(Toolbox Skill 3)*

14. Draw a bar graph to display the data.   *(Lesson 5-4)*

Egg Production in Five States (Billions)

State	Alabama	Arkansas	Georgia	Indiana	North Carolina
Eggs	2.6	3.8	4.3	5.6	3.4

Self-Test 1

1. Draw a pictograph to display the data.   5-1

   Average Annual Snowfall in Four Locations (Inches)

Location	Blue Canyon	Flagstaff	Marquette	Stampede Pass
Snowfall	242.0	96.9	122.0	430.8

2. Use the double bar graph below. Estimate the number of mathematics degrees awarded in 1985.

3. Use the double bar graph below. About how many fewer com-   5-2
   puter science degrees were awarded in 1975 than in 1985?

4. Use the line graph below. Estimate the number of state and   5-3
   local government employees in 1970.

5. Use the line graph below. Describe the overall trend in the
   number of state and local employees.

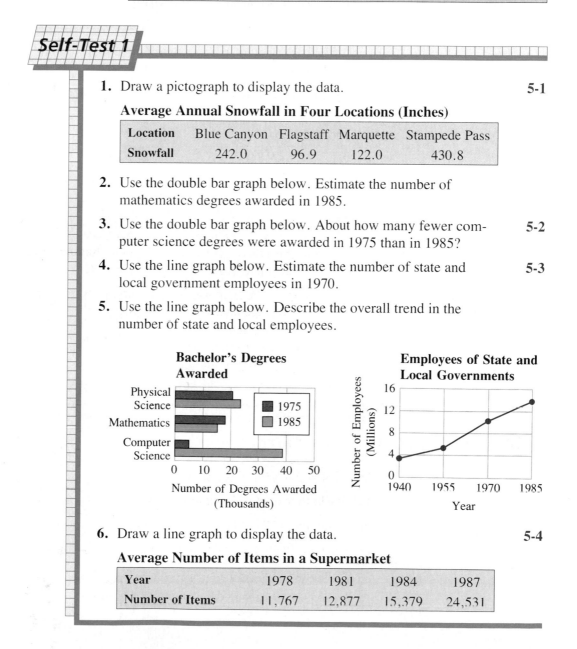

6. Draw a line graph to display the data.   5-4

   Average Number of Items in a Supermarket

Year	1978	1981	1984	1987
Number of Items	11,767	12,877	15,379	24,531

5-5

Choosing the Appropriate Type of Graph

Objective: To decide whether it would be more appropriate to draw a bar graph or a line graph to display a set of data.

Before displaying a set of data, you should first inspect it to decide whether to draw a bar graph or a line graph.

• A *bar graph* may be most appropriate when the data fall into distinct categories and you want to compare the totals.

• A *line graph* may be most appropriate when you want to emphasize trends in data that change continuously over time.

Example

Decide whether it would be more appropriate to draw a *bar graph* or a *line graph* to display the data. Then draw the graph.

National League Home Runs

Year	1985	1986	1987	1988	1989
Home Runs	1424	1523	1824	1279	1365

Solution

Bar graph. Yearly totals are distinct amounts. They are not carried over from one year to the next. In other words, yearly totals do not change continuously over time. For this reason, a line graph is not appropriate.

☑ Check Your Understanding

Suppose the data in the Example represented the average attendance per game in a given stadium. Would it be appropriate to draw a line graph to display the data? Explain.

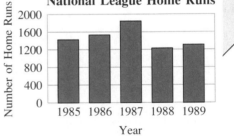

Guided Practice

COMMUNICATION «*Discussion*

Decide whether each type of data changes continuously. Explain.

« **1.** the number of graduates from a certain high school

« **2.** the enrollment at a certain high school

« **3.** the amount of money in a savings account

« **4.** the amount of money deposited to a savings account

5. Decide whether it would be more appropriate to draw a *bar graph* or a *line graph* to display the data. Then draw the graph.

Sales in Foreign Countries of Goods from the United States (Billions of Dollars)

Product	aircraft	auto parts	computers	autos	gold
Billions	$20.3	$13.2	$11.6	$9.1	$5.2

Exercises

Decide whether it would be more appropriate to draw a *bar graph* or a *line graph* to display the data. Then draw the graph.

1. Money Households Spend Annually on Reading Materials

Metropolitan Area	Houston	Boston	Chicago	Anchorage
Dollars	164	202	178	246

2. Women and the Labor Force (Millions)

Year	1965	1970	1975	1980	1985
Women in the Labor Force	26	32	37	45	51
Women out of the Labor Force	41	41	43	43	43

3. Theater Box Office Receipts (Millions)

Year	1980	1982	1984	1986
Broadway Receipts	$143	$221	$227	$191
Road Show Receipts	$181	$250	$206	$236

4. THINKING SKILLS Explain why someone might decide to draw a pictograph rather than a bar graph to display the data.

Pieces of Mail Handled Annually in Major Cities (Millions)

City	Atlanta	Denver	Louisville	Tampa
Pieces of Mail	2340	1660	948	1470

SPIRAL REVIEW

5. Simplify: $5(3n + 4)$ *(Lesson 2-2)*

6. Write as a variable expression: 5 less than twice the number a *(Lesson 4-6)*

7. Decide whether it would be more appropriate to draw a *bar graph* or a *line graph* to display data about the population of the United States over a period of years. Explain.   *(Lesson 5-5)*

8. Estimate the sum: $213 + 548 + 194 + 362$ *(Toolbox Skill 2)*

5-6

Too Much or Not Enough Information

Objective: To solve problems involving too much or not enough information.

Sometimes you need to use data from a graph to solve a problem. Graphs might contain more information than you need. They also might not contain enough information to solve a problem. When trying to solve a problem, you should always read graphs carefully.

Problem

Use the diagram below. Which state in the United States has the greatest number of daily newspapers?

Number of Daily Newspapers and Their Circulation

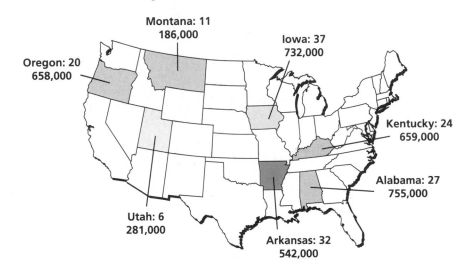

Montana: 11
186,000

Iowa: 37
732,000

Oregon: 20
658,000

Kentucky: 24
659,000

Alabama: 27
755,000

Utah: 6
281,000

Arkansas: 32
542,000

Solution

UNDERSTAND The problem is about daily newspapers.
Facts: the number of daily newspapers in seven states
the circulation of daily newspapers in seven states
Find: the state in the United States with the greatest number of daily newspapers

PLAN Compare the number of daily newspapers in every state in the United States.

WORK Not enough information. There is information for only seven states.

Look Back What if you are asked to find how much greater daily newspaper circulation is in Iowa than in Montana? What information about those states should you ignore?

Guided Practice

Tell whether each problem contains *too much* or *not enough* information. Identify any extra or missing information.

«**1.** Carla wants to buy five notebooks. How much will she spend?

«**2.** Michael mows lawns for six clients. He earns $72 each week. In how many weeks will Michael earn $360?

Use the table at the right. Solve, if possible. If there is not enough information, tell what facts are needed.

3. How many types of jets are listed?

4. What facts are given about the jets?

5. Does the diagram show the actual number of flights made by each type of jet?

6. How many DC-10's are there?

7. How many more 727's are there than 737's?

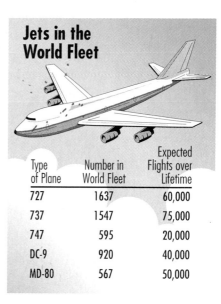

Jets in the World Fleet

Type of Plane	Number in World Fleet	Expected Flights over Lifetime
727	1637	60,000
737	1547	75,000
747	595	20,000
DC-9	920	40,000
MD-80	567	50,000

Problem Solving Situations

Use the double bar graph below. Solve, if possible. If there is not enough information, tell what facts are needed.

1. How many more grocery stores in Midtown had a bakery in 1990 than in 1980?

2. How many more grocery stores in Midtown had a video department in 1990 than in 1985?

3. How many more books were sold in Midtown grocery stores in 1990 than in 1980?

4. By how much more did the number of deli departments increase than the number of catering departments in Midtown grocery stores from 1980 to 1990?

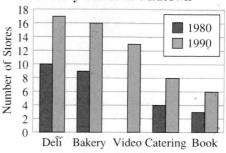

Specialty Departments at Grocery Stores in Midtown

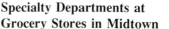

Solve using any problem solving strategy.

PROBLEM SOLVING
CHECKLIST
Keep these in mind:
  Using a Four-Step Plan
  Too Much or Not Enough
   Information
Consider these strategies:
  Choosing the Correct Operation
  Making a Table
  Guess and Check
  Using Equations

5. Sarah has only dimes and quarters. In how many different ways could she make $1.45?

6. Jeremy paid $10.80 for a phone call. The rates were $2.40 for the first minute, and $.60 for each additional minute. For how many additional minutes was Jeremy charged?

7. A soccer team sold 496 raffle tickets and collected $372.50. Expenses totaled $75.98. How much money did the team make after expenses?

8. A clothing manufacturer distributed 300 suits to several department stores. The manufacturer sent each store the same number of suits. How many suits did each store receive?

WRITING WORD PROBLEMS

For each exercise, write two word problems. One problem should contain not enough information. Then solve the problems, if possible.

9. Use the bar graph at the top of page 192.

10. Use the line graph at the top of page 196.

SPIRAL REVIEW

11. Find the difference: $5\frac{5}{6} - 3\frac{5}{12}$ *(Toolbox Skill 19)*

12. Replace the __?__ with >, <, or =: -3 __?__ -7 *(Lesson 3-1)*

13. Use the diagram at the top of page 207. How many more DC-9's are there than MD-80's in the world fleet?   *(Lesson 5-6)*

14. Is -3 a solution of the equation $x - 5 = -8$?   *(Lesson 4-1)*

Historical Note

Working as a nurse in British hospitals, Florence Nightingale (1820–1910) was shocked by the crude treatment patients received. To convince people of the need for reform, she developed new types of statistical graphs that highlighted the poor conditions. Her efforts led to substantial improvements in hospital health care.

Research

In a recent health or biology textbook, find two graphs that illustrate data related to health or fitness.

Misleading Graphs

Objective: To recognize how bar graphs and line graphs can be misleading.

1. Use *Graph A*. Estimate the attendance at the rock concert in Franklin, in Midway, and in Sunville.

2. Repeat Step 1, but use *Graph B* instead.

3. Describe how *Graph A* and *Graph B* are different.

Bar graphs illustrate data in a way that makes it easy to compare values. This is one of the advantages of displaying data in bar graphs rather than in tables. A bar graph is misleading, however, when it creates a false visual impression.

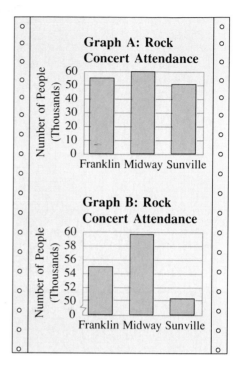

Example 1

Use *Graph B* above.

a. Explain why the graph is visually misleading.

b. Explain why someone who wants to attract performers to Midway might use this graph.

Solution

a. Because there is a gap in the scale, the graph shows only a part of each bar. This creates the false impression that there were great differences in attendance in the three cities.

b. The graph gives a visual impression that attendance in Midway was not only much greater than attendance in either Franklin or Sunville, but was also greater than the combined attendance in the two cities.

✓ Check Your Understanding

In *Graph B,* describe the visual impression of the relationship between the attendance in Midway and the attendance in Sunville.

Line graphs are useful for analyzing changes in data. However, a line graph is misleading when it makes these changes appear more dramatic than they really are.

Example 2

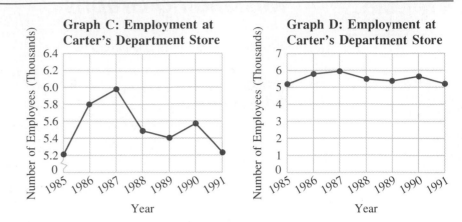

a. Contrast the visual impressions given by the graphs above.

b. Which graph might the department store present to new employees to show them that the store provides job security? Explain.

Solution

a. *Graph C* makes it appear that the employment level changed dramatically between 1985 and 1991. *Graph D* gives the opposite impression.

b. *Graph D*. It makes the employment level appear fairly stable.

Guided Practice

COMMUNICATION « *Reading*

Refer to the text on pages 209–210.

« **1.** State the main idea of the lesson.

« **2.** List two major points that support the main idea.

Use the bar graph at the right.

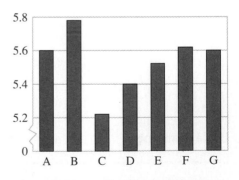

3. Do not use the scale. About how many times as high as bar C does bar F appear to be?

4. Use the scale. About how much greater is the value represented by bar F than by bar C?

5. Use the graphs on page 209. Explain why someone who wants to attract performers to Franklin would use *Graph A* rather than *Graph B*.

6. Use the graphs in Example 2 above. Which graph might employees use to show that they lack job security? Explain.

Exercises

Use the bar graph at the right.

1. Describe the visual impression of the relationship between production levels at Plants C and D.

2. Explain why Plant B might use this graph.

3. Explain how the graph might be redrawn to give a better impression of Plant A's performance.

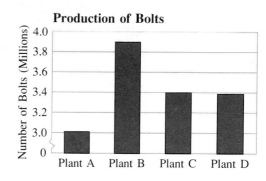

Production of Bolts

Use the line graphs below.

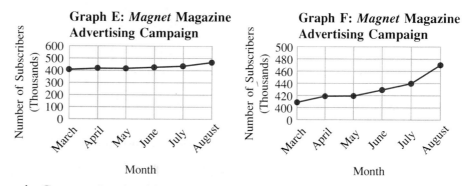

Graph E: *Magnet* Magazine Advertising Campaign

Graph F: *Magnet* Magazine Advertising Campaign

4. Contrast the visual impressions given by the two graphs.

5. Which graph might the magazine owner use to show displeasure at the results of the advertising campaign? Explain.

6. Which graph might the advertising agency use to attract new clients? Explain.

THINKING SKILLS

Use the bar graph at the right. Classify each statement as *accurate* or *misleading*.

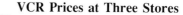
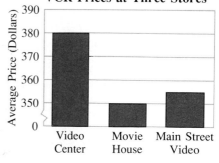

VCR Prices at Three Stores

7. VCRs cost four times as much at Video Center as they do at Movie House.

8. Prices at the stores vary widely.

9. Video Center charges about $30 more than Movie House charges for a VCR.

COMPUTER APPLICATION

Use graphing software or graph paper. Refer to *Graph F* on page 211.

10. Copy the graph as it is shown. (If you are using a computer, you may not be able to show the gap in the scale.)

11. Redraw the graph, using a scale from 350 to 550 with intervals of 50. Keep the dimensions of the graph the same as in Exercise 10.

12. Repeat Exercise 11, this time using a scale from 200 to 600 with intervals of 100.

13. Use the graphs you drew for Exercises 10–12 to explain how changing the scale on a line graph affects its visual impact.

SPIRAL REVIEW

14. Write 36 mL in L.   *(Lesson 2-9)*

15. Find the sum: $2\frac{3}{10} + 7\frac{3}{5}$ *(Toolbox Skill 18)*

16. Solve $-20 = 8q + 6q + 8$. Check the solution.   *(Lesson 4-9)*

17. Use the bar graph on page 211 entitled *Production of Bolts*. Explain why the manager of Plant C might not want to use this graph. *(Lesson 5-7)*

Self-Test 2

1. Decide whether it would be more appropriate to draw a *bar graph* or a *line graph* to display the data. Then draw the graph.   **5-5**

   Median Prices of New One-Family Homes in the Northeast

Year	1976	1978	1980	1982	1984	1986
Price	$47,300	$58,100	$69,500	$78,200	$88,600	$125,000

Solve, if possible. If there is not enough information, tell what facts are needed.

2. Sabrina had $987.63 in her checking account. She made two deposits of $472.38 each. How much did Sabrina deposit altogether?   **5-6**

3. Miguel drove 379 mi to visit a friend in Chicago. How many miles did the odometer show when he finished his trip?

4. Use the bar graph on page 211 entitled *Production of Bolts*. Describe the visual impression of the relationship between the levels of production at Plants A and B.   **5-7**

Mean, Median, Mode, and Range

Objective: To find the mean, median, mode, and range of a set of data.

The branch of mathematics that deals with collecting, organizing, and analyzing data is called **statistics.** Statisticians use graphs and a variety of *statistical measures* to describe a set of data.

The **mean,** or *average*, of a set of data is the sum of the data items divided by the number of items.

Terms to Know

- *statistics*
- *mean*
- *median*
- *mode*
- *range*

The **median** of a set of data is the middle item when the data are listed in numerical order. If there is an even number of items, the median is the average of the two middle items.

The **mode** of a set of data is the item that appears most often. There can be more than one mode. There can also be no mode, if each item appears only once.

The **range** of a set of data is the difference between the greatest and least values of the data.

Example

Find the mean, median, mode(s), and range of the data.

Marie's Bowling Scores: 149, 183, 149, 193, 147, 193

Solution

- Find the sum of Marie's scores.
  $149 + 183 + 149 + 193 + 147 + 193 = 1014$

  $\text{mean} = \dfrac{1014}{6} = 169$ ◄—— Divide by the number of scores.

- List the scores in numerical order.
  147, 149, 149, 183, 193, 193

  $\text{median} = \dfrac{149 + 183}{2} = 166$ ◄—— Find the average of the two middle scores.

- modes = 149 and 193 ◄—— Both 149 and 193 appear twice.

- range = $193 - 147 = 46$ ◄—— Subtract the least score from the greatest score.

☑ Check Your Understanding

1. In the Example, what does the 6 appearing in the first denominator represent?
2. In the Example, why is it necessary to average 149 and 183 to find the median?

When you divide to find the mean, you may get an answer with many decimal places. When this happens, you should round the answer to the nearest tenth. Indicate that your answer is rounded by using the approximation symbol $\approx$.

A calculator can be used to find the mean of a set of data. You would use this key sequence to find the mean of 234, 175, and 488.

234 [+] 175 [+] 488 [=] [÷] 3 [=]

Guided Practice

COMMUNICATION « *Reading*

Identify the statistical measure most closely related to each term.

«**1.** average   «**2.** difference   «**3.** middle   «**4.** most often

Find the average of the two numbers.

5. 8 and 14   **6.** −15 and 21   **7.** 85 and 88   **8.** 0.9 and 1.7

List the data in order, then find the middle number(s).

9. 3.2, 2.7, 2.3, 3.8, 0.3, 3.0

10. 304, 403, −3040, 3004, 4003, −340, 3400

Find the mean, median, mode(s), and range of the data.

11. **Typical Number of Prime Time TV Viewers Each Day (Millions):**
101.1, 94.5, 93.9, 97.0, 89.4, 87.1, 105.8

12. **Points Scored by Steve during a Week of Basketball Practice:**
2, 7, 7, 10, 12

13. **Deposits to and Withdrawals from a Savings Account:**
$35, $80, −$25, $95, −$50, $60, −$105, −$50, $80, $20

Exercises

Find the mean, median, mode(s), and range of the data.

1. **Luggage Space in Large Cars (Cubic Feet):**
20, 16, 22, 16, 15, 20, 21, 16

2. **Jay's Charitable Contributions in a Recent Year:**
$35, $50, $25, $55, $30, $35, $25, $70, $50, $30, $25, $50

3. **Profits and Losses from School Plays:**
−$79, $24, −$118, $37, $349, $48

4. **Base Prices for Two-Door Hatchbacks:**
$5500, $8300, $6300, $6400, $9900, $7000, $5850, $4350, $6700

Find the mean, median, mode(s), and range of the entire set of data in each diagram.

5. Typical February Temperatures in New England

6. Typical February Temperatures in the Mountain States

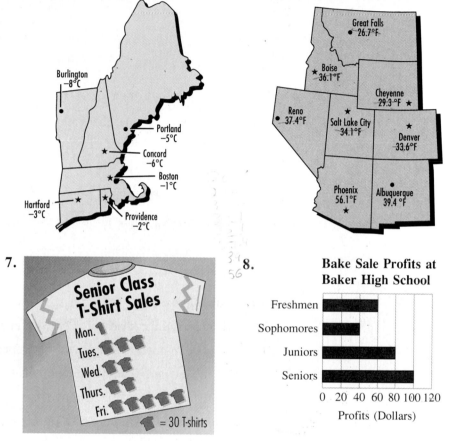

Burlington
-8°C

Portland
-5°C

Concord
-6°C

Boston
-1°C

Hartford
-3°C

Providence
-2°C

Great Falls
26.7°F

Boise
36.1°F

Cheyenne
29.3°F

Reno
37.4°F

Salt Lake City
34.1°F

Denver
33.6°F

Phoenix
56.1°F

Albuquerque
39.4°F

7.

Senior Class
T-Shirt Sales

Mon.
Tues.
Wed.
Thurs.
Fri.

= 30 T-shirts

8. Bake Sale Profits at Baker High School

Freshmen
Sophomores
Juniors
Seniors

0 20 40 60 80 100 120
Profits (Dollars)

9. Jack's scores on the first four days of a golf tournament were 72, 76, 71, and 74. What score must he receive on the last day of the tournament in order to have a mean score of 73?

10. Palladin spent a total of $100 for five shirts. Later he bought another shirt. He spent an average of $18.78 per shirt for the six shirts. What did Palladin pay for the sixth shirt?

11. The youngest person in an audience of 400 people is 26 years old. The range of ages in the audience is 38 years. Find the age of the oldest person in the audience.

12. In May the average of the agents' commissions at Country Realty was $2680.35. Find the agents' total commissions for May.

 **CALCULATOR**

Tell if each calculator mean is *reasonable* or *unreasonable*. If a mean is unreasonable, use your calculator to find the correct mean.

13. 5, 7, 9, 8, 5, 6, 5 $\boxed{4.}$

14. 116, 84, 123, 97, 65 $\boxed{150.}$

15. 12.4, 15.7, 11.9, 14.6 $\boxed{13.65}$

16. 1.25, 3.6, 8.12, 6.9 $\boxed{0.78}$

PROBLEM SOLVING/APPLICATION

In gymnastics competitions, there are usually four judges and a head judge. They rate performances on a scale of 0 to 10. The highest and lowest scores from the four judges are dropped. The final score is the mean of the two remaining scores. The head judge's score is used only if there is some question about the fairness of the scoring.

	Head Judge	Judge 1	Judge 2	Judge 3	Judge 4
Sachi	8.8	8.4	8.9	8.5	9.0
Janice	9.1	9.3	7.5	7.4	9.2

Use the table above.

17. What score did the head judge give to Sachi?

18. Which two of Sachi's scores will be dropped?

19. Using the method described above, what will be Sachi's final score?

20. Using the method described above, what will be Janice's final score?

21. Explain how the method of scoring explained above compares with finding the median of the four judges' scores.

22. THINKING SKILLS   If you were the head judge, how would you rule on the scores that the other four judges gave to Janice? Explain.

SPIRAL REVIEW

23. Is 8460 divisible by 2? by 3? by 4? by 5? by 8? by 9? by 10? *(Toolbox Skill 16)*

24. Find the mean, median, mode(s), and range: 4, −4, −18, 26 *(Lesson 5-8)*

25. Evaluate $2n^3$ when $n = 5$.   *(Lesson 1-9)*

Challenge

Find a set of five numbers with a mean of 24, a median of 15, a mode of 10, and a range of 41.

Modeling the Mean

Objective: To build a concrete model to represent the
mean of a set of numbers.

Mark off a strip of cardboard as shown in the diagram at the left below.
Punch holes one inch apart and one-eighth inch from the edge. Using
string and tape, suspend the strip of cardboard over the edge of a desk.

Materials

- strip of cardboard
  1" × 12"
- ruler
- hole punch
- string
- tape
- 7 paper clips

Activity I *Exploring the Mean*

1 Hang paper clips in the holes for
the given numbers. To reach a
balance, hang in the appropriate
holes as many additional paper
clips as there are blanks. Record
the results. You may not use the
hole in the center.

  a. 71, 74, 81, __?__
  b. 71, 74, 81, __?__ , __?__
  c. 71, 74, 80, __?__
  d. 71, 74, 80, __?__ , __?__
  e. 71, 74, 80, __?__ , __?__ , __?__

2 Find the mean of the data in each part of Step 1. How do the means
compare with the number at the center of the strip of cardboard?

3 Would any of the means in Step 2 change if you hung a paper clip
in the hole in the center of the strip of cardboard? Explain.

Activity II *Exploring the Median and the Mode*

1 Find the median and the mode of the data:
71, 73, 74, 76, 78, 78, 79

2 Hang a paper clip in each hole corresponding to a number in
Step 1. If a number appears more than once, use a paper clip for
each time it appears. Do the data balance around the median?

3 Do you think the data balance around the mode found in Step 1?

4 Under what condition(s) do you think a set of data would balance
around the mean, the median, and the mode? Use the strip of
cardboard to produce such a set of data.

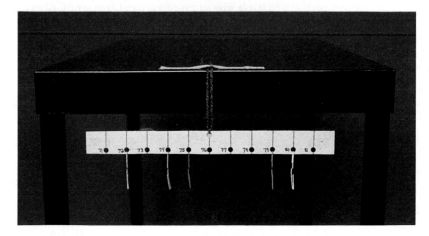

Choosing the Appropriate Statistic

Objective: To decide whether a given measure of central tendency is appropriate for a set of data.

The mean, median, and mode are sometimes called **measures of central tendency** because they describe how data are *centered*. Each of these measures describes a set of data in a slightly different way.

- The *mean* is the most familiar statistical measure. It is an appropriate measure of central tendency when the data are reasonably centered around it. However, the mean can be distorted by an *extreme* value, one much greater or less than the other values.

Terms to Know
- *measure of central tendency*

- When there is an extreme value that distorts the mean, the *median* is an appropriate measure of central tendency.

- When data cannot be averaged or listed in numerical order, the *mode* is an appropriate measure of central tendency.

Example

Ace Mechanics has seven employees. Their salaries are $18,000, $19,000, $19,000, $22,000, $24,000, $30,000, and $120,000. Decide whether the mean describes these data well. Explain.

Solution

No. The mean is $36,000 and is greater than six of the seven salaries. It is distorted by the extreme value of $120,000.

☑ Check Your Understanding

In the Example, how do you find that the mean is $36,000?

Guided Practice

COMMUNICATION «*Discussion*

Which measure of central tendency do you think is generally used in each situation? Explain.

«**1.** reporting yearly incomes for company employees

«**2.** summarizing scores on a math test

«**3.** determining the size of shoe in greatest demand

4. Manuel reports that people prefer Brand X more often than any other brand. Which measure of central tendency is he using?

5. Sharon says that half her employees earn more than $30,000 per year and half earn less. Which measure of central tendency is she using?

6. The breeds of the six dogs in a certain neighborhood were retriever, poodle, collie, retriever, bulldog, and beagle. Decide which measure of central tendency is appropriate for these data. Explain.

Exercises

1. The heights of the five starting players on a basketball team are 75 in., 74 in., 73 in., 70 in., and 68 in. Decide whether the mean describes these data well. Explain.

2. The monthly car payments for Ted Nelson's last eight customers were $266, $285, $285, $285, $285, $315, $325, and $344. Decide whether the mode describes these data well. Explain.

3. At the end of the winter, the seven sweaters left in stock at Angie's Outlet Store were brown, orange, orange, green, brown, orange, and orange. Decide which measure of central tendency is appropriate for these data. Explain.

4. The profits at five schools that sold Gems & Jewels products were $318.22, $440.79, $607.16, $1090.38, and $4790.15. Decide whether the mean or the median is a better measure of central tendency for these data. Explain.

5. The sizes of the thirteen families living on Wood Road are 2, 3, 3, 3, 4, 4, 5, 5, 6, 6, 7, 8, and 9. Decide whether the median or the mode is a better measure of central tendency for these data. Explain.

6. The favorite vegetables of the six members of the Walker family are peas, carrots, beans, peas, carrots, and carrots. Decide which measure of central tendency is appropriate for these data. Explain.

7. The numbers of hours worked by the seven nurses at Dale General Hospital in one week were 20, 20, 20, 29, 48, 52, and 56. Decide which measure of central tendency might encourage a nurse who wants to work at least 35 h per week to apply for a job at Dale General. Explain.

SPIRAL REVIEW

8. Solve $13 = \frac{b}{12}$. Check the answer.   *(Lesson 4-4)*

9. Graph the points on a coordinate plane: $A(3, 0)$, $B(2, -4)$, $C(-3, -2)$
   (Lesson 3-8)

10. A student's scores on five tests were 84, 78, 89, 94, and 74. Decide whether the mean describes these data well. Explain.   *(Lesson 5-9)*

11. **PATTERNS** Find the next three numbers in the pattern:
    5, 8, 12, 17, __?__, __?__, __?__   *(Lesson 2-5)*

Using Database Software

Objective: To use database software to analyze a set of data.

People who have to analyze great amounts of data will often use a **database management system.** They can then more easily sort through tables of data like the one below. In some cases, database software can also be used to calculate statistical measures.

Average Annual Household Expenditures (Dollars)

Metropolitan Area	Region	Food	Housing	Transportation
Atlanta	South	3472	9462	5578
Buffalo	Northeast	3956	6087	4048
Cleveland	Midwest	4012	7462	6084
Dallas/Fort Worth	South	4006	8953	6384
Honolulu	West	5126	9386	6023
Milwaukee	Midwest	3982	8488	4343
New York City	Northeast	4484	10191	4677
Philadelphia	Northeast	4262	8942	4700
Portland	West	3939	7961	4231
San Francisco	West	4422	11174	5600
St. Louis	Midwest	3757	7584	4590
Washington, D.C.	South	4388	12069	6751

Exercises

Use database software to create a database containing the information in the table above.

1. What names did you give the fields in the database?

2. How many records did you enter in the database?

3. Find all the records whose region is *South*. Which metropolitan areas do they represent?

4. Find all the records with food expenditures less than $4000. Which metropolitan areas do they represent?

5. Sort the records in decreasing order of transportation expenditures. Which record appears first?

6. Sort the records alphabetically both by region and by metropolitan area. Which record appears first?

7. Find the mean of the food expenditures for all the metropolitan areas.

8. Describe how the database software can help you find the median of the housing expenditures for all the metropolitan areas.

9. Which metropolitan area do you think has the greatest total costs for food, housing, and transportation? Explain how you made your decision.

5-10

Data Collection and Frequency Tables

Objective: To calculate statistical measures for data in a frequency table.

APPLICATION

Mrs. Fischer's first period gym class was doing a unit on health. The students were asked to measure their pulse rates. They organized the data in a **frequency table** like the one shown below.

Terms to Know
• *frequency table*

Pulse Rates (beats/min)

Rate	Tally	Frequency
69	ⅢⅠ	6
70	ⅢⅠ	6
71	Ⅰ	1
72	ⅢⅢ	10
73	ⅢⅠ	6
Total:		29

Example

Find the mean, median, mode(s), and range of the data in the frequency table above.

Solution

• Multiply each *rate* by its *frequency*:

$$69 \cdot 6 = 414$$
$$70 \cdot 6 = 420$$
$$71 \cdot 1 = 71$$
$$72 \cdot 10 = 720$$
$$73 \cdot 6 = \underline{438}$$

Add: 2063

⟵ In finding the mean, a calculator may be helpful.

Divide by the total of the frequencies:

$$2063 \div 29 \approx 71.1$$

The mean is about 71.1 beats/min.

• There are twenty-nine items. The median is the middle item, so look for the fifteenth item.

The median is 72 beats/min.

• Look for the rate with the most tally marks.

The mode is 72 beats/min.

• Subtract: $73 - 69 = 4$

The range is 4 beats/min.

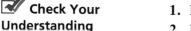 **Check Your Understanding**

1. In the Example, why was the number 29 used to find the mean?
2. In the Example, how do you determine that 72 is the fifteenth item?

Guided Practice

«1. COMMUNICATION «*Writing*   Make a frequency table for the data below.
Hours Worked: 5, 6, 6, 5, 7, 8, 6, 5, 7, 5, 5, 6, 6, 8, 5, 6, 8, 6

Use the frequency table at the right.

2. What is the total of the frequencies?

3. What was the total price for all of the soup?

4. Find the mean, median, mode(s), and range of the data.

Soup Prices	Tally	Frequency
$.69	IIII	4
$.79	I	1
$.89	II	2
$.99	III	3
	Total:	10

Exercises

Find the mean, median, mode(s), and range of the data.

1.

Miles per Gallon	Tally	Frequency
30	HHT	5
31	III	3
32	III	3
33	I	1
	Total:	12

2.

Bat Lengths (Inches)	Tally	Frequency
30	III	3
31	III	3
32	HHT II	7
34	I	1
	Total:	14

CAREER/APPLICATION

A *market researcher* collects data from consumers. These data are used to help people who provide goods and services decide how best to satisfy the needs of consumers.

Number of Movies Seen by Students in a Month

4	2	4	4	3	5
5	3	4	12	4	5
4	5	3	4	2	4
3	4	4	3	3	15

Use the data at the right.

3. Make a frequency table for the data.

4. Use the frequency table that you made for Exercise 3. Find the mean, median, mode(s), and range of the data.

5. Decide whether the mean describes the data well. Explain.

6. WRITING ABOUT MATHEMATICS   Ask ten of your classmates how many movies they have seen in the past month. Write a paragraph comparing your data to the data given in the table.

7. Use an almanac to find the minimum age requirement for obtaining a moped license in each of the fifty states. Make a frequency table for the data.

8. Use the frequency table that you made for Exercise 7. Find the mean, median, mode(s), and range of the data.

SPIRAL REVIEW

9. Use the formula $g = a - h$. Let $g = 756$ km/h and $h = 14$ km/h. Find a.   *(Lesson 4-10)*

10. Find the quotient: $4\frac{1}{2} \div \frac{3}{4}$ *(Toolbox Skill 21)*

11. The number of laps that Olga swam on Friday was four more than twice the number of laps she swam on Monday. If Olga swam 76 laps on Friday, how many laps did she swim on Monday?   *(Lesson 4-8)*

12. DATA, *pages 186–187*   Find the mean, median, mode(s), and range of the data in the frequency table about Olympic medals.   *(Lesson 5-10)*

Self-Test 3

1. The amounts withdrawn in one hour from an automatic teller machine were $20, $10, $50, $70, $30, $50, $100, and $20. Find the mean, median, mode(s), and range of these data.   **5-8**

2. The ages of the twelve students in a dance class were 14, 13, 15, 14, 15, 13, 16, 15, 16, 15, 15, and 13. Decide whether the mean describes these data well. Explain.   **5-9**

3. The five residents in a dormitory suite were from New York, Maine, New York, Iowa, and New York. Decide which measure of central tendency is appropriate for these data. Explain.

4. Find the mean, median, mode(s), and range of the data.   **5-10**

Hours of Sleep	Tally	Frequency
5					3
6	⊦⊦⊦		6
7	⊦⊦⊦ ||||	9
8	⊦⊦⊦ ⊦⊦⊦ ||	12
	Total:	30

Using Graphs in Business

Objective: To analyze profits and losses using a double line graph.

Veronica studied cosmetology at South Vocational High School and graduated as a licensed beautician. She opened her own manicure salon at a nearby mall. She kept careful records of the amount of money she collected, the amount of money she spent, and the number of manicures. After two months she drew the double line graph shown at the right to display her progress.

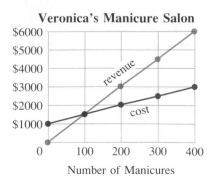

Veronica's Manicure Salon

The amount of money that a company collects is called **revenue.** The amount of money that the company spends is called **cost.** When revenue is greater than cost, the company makes a *profit*.

$$\text{profit} = \text{revenue} - \text{cost}$$

When cost is greater than revenue, the company experiences a *loss*.

$$\text{loss} = \text{cost} - \text{revenue}$$

When the company's cost and revenue are equal, the company is said to *break even*.

Example

Use the double line graph above.

  a. What was the revenue from 400 manicures?

  b. During which interval did Veronica lose money?

  c. How many manicures had Veronica done when she broke even?

  d. Estimate the profit from 300 manicures.

Solution

  a. Locate the point for 400 on the revenue line, which is *blue*. The revenue from 400 manicures was $6000.

  b. Veronica lost money during the interval where the line representing cost is *above* the line representing revenue. She lost money during the interval from 0 up to 100.

  c. Veronica broke even at the point where the two lines meet. She had done 100 manicures when she broke even.

  d. The revenue from 300 manicures was about $4500. The cost for 300 manicures was about $2500.
Subtract: $4500 - 2500 = 2000$
The profit from 300 manicures was about $2000.

Exercises

Jack opened a photo shop specializing in passport photos. He monitored his business for three months. Then he displayed his progress by drawing the double line graph at the right.

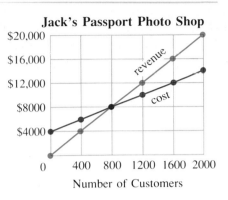

Jack's Passport Photo Shop

Use the double line graph at the right.

1. Estimate the cost for 2000 customers.

2. Estimate the revenue from 1800 customers.

3. During which interval did Jack lose money?

4. During which interval did Jack make a profit?

5. Estimate Jack's loss from photographing 400 people.

6. Estimate Jack's profit from photographing 1600 people.

7. How many people had Jack photographed when he broke even?

8. Estimate Jack's revenue at the point when he broke even.

9. Estimate Jack's profit per customer for 1200 customers.

10. Estimate Jack's profit per customer for 2000 customers.

11. Jack's cost for 2800 customers was $18,000. His revenue was $28,000. What was his profit?

12. Jack's cost for 2400 customers was $16,000. His profit was $8,000. What was his revenue?

GROUP ACTIVITY

13. Make a list of four things that might be included in the initial cost of opening a passport photo shop.

14. Make a list of four things that might be included in the initial cost of starting any business. Compare this list to the list you made for Exercise 13.

Chapter Review

Terms to Know

graph (p. 188)
data (p. 188)
pictograph (p. 188)
key (p. 188)
bar graph (p. 192)
scale (p. 192)
legend (p. 192)
line graph (p. 196)
increasing trend (p. 196)

decreasing trend (p. 196)
statistics (p. 213)
mean (p. 213)
median (p. 213)
mode (p. 213)
range (p. 213)
measure of central tendency
  (p. 218)
frequency table (p. 221)

Choose the correct term from the list above to complete each sentence.

1. The ___?___ of a set of data is the item that appears most often.

2. A(n) ___?___ is used to show data that change continuously over time.

3. A branch of mathematics that deals with collecting, organizing, and analyzing data is called ___?___.

4. A group of numerical facts is often referred to as ___?___.

5. In a double bar graph, the ___?___ is included to help you distinguish between the two types of bars.

6. The ___?___ on a line graph allows you to estimate the values represented by the points on the line.

Use the pictograph at the right. Solve, if possible. If there is not enough information, tell what facts are needed. *(Lessons 5-1, 5-6)*

7. About how many bushels of soybeans were produced in 1980?

8. About how many more bushels of soybeans were produced in 1985 than in 1970?

9. In 1980, about how much more corn was produced than soybeans?

10. Estimate the income from the sale of all the soybeans produced in 1975.

Soybean Production on United States Farms	Average Price per Bushel
1970	$2.85
1975	$4.92
1980	$7.57
1985	$5.05

= 400 million bushels

11. Draw a pictograph to display the data. *(Lesson 5-1)*

Parks in Four Major Cities

City	Fort Worth	Minneapolis	New Orleans	Omaha
Parks	136	153	250	99

Use the bar graph at the right. *(Lesson 5-2)*

12. About how much money was spent for advertising on television?

13. About how much more money was spent for advertising in newspapers than for advertising in magazines?

14. For which method of advertising was the amount of money spent about $20 billion?

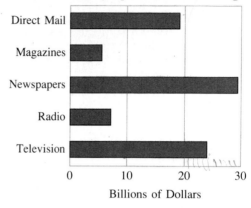

Money Spent for Advertising

Use *Graph A* below. *(Lesson 5-3)*

15. During which five-year interval did the greatest increase in the average number of visits occur?

16. Estimate the average number of visits in 1975.

17. In which given year was the average number of visits about 6?

Use *Graph A* and *Graph B* below. *(Lesson 5-7)*

18. Contrast the visual impressions given by the two graphs.

19. Which graph might someone use to show that the average number of visits to physicians did not change very much between 1970 and 1985? ·

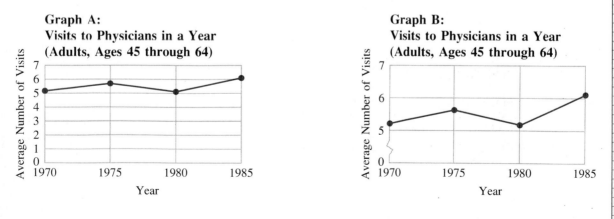

Graph A:
Visits to Physicians in a Year
(Adults, Ages 45 through 64)

Graph B:
Visits to Physicians in a Year
(Adults, Ages 45 through 64)

Decide whether it would be more appropriate to draw a *bar graph* or a *line graph* to display the data. Then draw the graph.
(Lessons 5-4, 5-5)

20. Attendance at Professional Basketball Playoffs (Millions)

Year	1975	1979	1983	1987
Attendance	0.7	0.9	0.6	1.1

21. Class I Locomotives in Service (Thousands)

Year	1975	1979	1983	1987
Locomotives	28	28	26	20

Solve, if possible. If there is not enough information, tell what facts are needed.   *(Lesson 5-6)*

22. The price per pound for apples is $.99. How much will a bag of 12 apples cost?

23. The 16 members of a swim team swam a total of 1000 laps to raise money for charity. They raised $250. How much money did they raise per lap?

Find the mean, median, mode(s), and range of the data.
(Lessons 5-8, 5-10)

24.

Sweater Prices
xxxxx
xxxx
xxxx
xxx $25.49
$28.99
$34.60
$47.99
$79.48

25.

Number of Siblings	Tally	Frequency
0	︲︲︲︲ ︲	6
1	︲︲︲︲	5
2	︲︲︲︲ ︲︲︲	8
3	︲	1
	Total:	20

26. Typical February temperatures in the West North-Central States are −10°C, −7°C, −8°C, −2°C, −4°C, 1°C, and 2°C. Find the mean, median, mode(s), and range of these data.   *(Lesson 5-8)*

27. The seven faculty members in an art department specialized in photography, painting, sculpting, printmaking, painting, painting, and graphic design. Decide which measure of central tendency is appropriate for these data. Explain.   *(Lesson 5-9)*

Chapter Test

1. Draw a pictograph to display the data. 5-1

Hospital Beds

State	Alabama	Florida	Georgia	Mississippi
Beds	24,900	61,500	33,700	16,900

2. Use the double bar graph below. Estimate the expected increase in the number of jobs in Los Angeles from 1985 to 2010. 5-2

3. Use the double bar graph below. In 1985, about how many more jobs were there in New York than in Seattle?

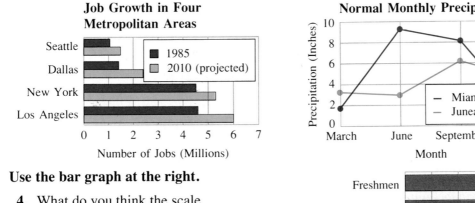

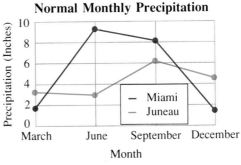

Use the bar graph at the right.

4. What do you think the scale could represent?

5. Use your answer to Exercise 4. What do you think is an appropriate title for the graph?

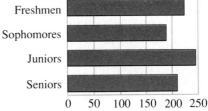

Exercises 6 and 7 refer to the double line graph above.

6. In which given month does Juneau normally get about six inches of precipitation? 5-3

7. In June, about how much greater is precipitation in Miami than in Juneau?

8. Draw a bar graph to display the data. 5-4

Maximum Depths of the Great Lakes

Lake	Erie	Huron	Michigan	Ontario	Superior
Depth in Feet	210	750	923	802	1333

Chapter 5

Decide whether it would be more appropriate to draw a *bar graph* or a *line graph* to display the data. Then draw the graph.

9. **Median Age of United States Citizens**

5-5

Year	1920	1940	1960	1980
Median Age	25.3	29.0	29.5	30.0

10. **Money Spent for Radio and Television Repair (Billions)**

Year	1970	1975	1980	1985
Dollars	1.4	2.2	2.6	3.1

Use the bar graph at the right. Solve, if possible. If there is not enough information, tell what facts are needed.

Points Scored in a Recent NFL Season

11. About how many more points were scored in the second quarter than in the first quarter?

5-6

12. About how many points were scored in overtime?

13. Describe the visual impression of the relationship between the number of points scored in the first and fourth quarters.

5-7

14. The cruising speeds in miles per hour of three-engine jets used by domestic airlines are 622, 600, 622, 620, 615, 580, 608, 615, and 615. Find the mean, median, mode(s), and range of these data.

5-8

15. The hourly wages for Mrs. Doyle's six employees are $7.50, $10.50, $11.00, $7.50, $12.50, and $29.00. Decide whether the mean describes these data well. Explain.

5-9

16. The favorite colors of the seven people who worked in an office were red, blue, purple, blue, red, blue, and blue. Decide which measure of central tendency is appropriate for these data. Explain.

5-10

17. Find the mean, median, mode(s), and range of the data in the frequency table at the right.

Class Sizes	Tally	Frequency
22	卌 II	7
24	卌 I	6
25	卌 III	8
27	II	2
	Total:	23

Using Variables in Statistics

Objective: To recognize and use statistical symbols associated with the mean.

Statistical measures are used so frequently that mathematicians have developed special symbols for them. Certain scientific calculators even have a *statistical mode* and show statistical symbols directly on the key pad. Three of these symbols are associated with the mean.

n represents the *number* of items in a set of data.

Σx read "the sum of all x," represents the *sum* of the data.

$\bar{x}$ read "x bar," represents the *mean* of the data.

Example

Solution

Find n, Σx, and $\bar{x}$ for the data: 30, 28, 36, 45, 48, 38

- n = the number of items = 6

- $\Sigma x = 30 + 28 + 36 + 45 + 48 + 38 = 225$

- $\bar{x} = \dfrac{\Sigma x}{n} = \dfrac{225}{6} = 37.5$

Exercises

Find n, Σx, and $\bar{x}$ for each set of data. Use a calculator with a statistical mode if you have one. (Refer to the manual that accompanies the calculator to determine the correct key sequence.)

1. **New England's Electoral College Votes:**

   8, 4, 13, 4, 4, 3

2. **Ages of Presidents from Ohio at Inauguration:**

   46, 54, 49, 55, 54, 51, 55

3. **Areas of the Great Lakes (Thousands of Square Miles):**

   31.8, 23.0, 22.4, 9.9, 7.5

4. **Park and Recreation Area Visitors in Pacific States (Millions):**

   46.7, 37.2, 72.9, 5.3, 20.2

Cumulative Review

Standardized Testing Practice

Choose the letter of the correct answer.

1. $a + (b + c) = (a + b) + c$ is an example of which property?
   A. commutative property of addition
   B. associative property of addition
   C. distributive property
   D. identity property of addition

2. Write 3.45×10^4 in decimal notation.
   A. 138   B. 3,450,000
   C. 34,500   D. 0.000345

3. Chernak's Chicken Pies sold 5500 pies in March, 5000 in April, and 3000 in May. If the symbol on a pictograph of these data represents 1000 pies, how many symbols will there be for March?
   A. 5   B. 3
   C. 11   D. 5.5

4. Write the coordinates of the point 3 units to the left of the y-axis and 4 units up from the x-axis.
   A. $(-3, -4)$ B. $(-3, 4)$
   C. $(3, -4)$ D. $(3, 4)$

5. Find the answer.
   $4^2 \cdot 3 - (5 - 2)$
   A. 0   B. 45
   C. 21   D. 41

6. Choose the most appropriate graph to display a patient's temperature over a period of twelve hours.
   A. bar graph
   B. pictograph
   C. line graph
   D. double bar graph

7. Continue the pattern.
   2, 9, 16, 23, ___?___, ___?___, ___?___
   A. 24, 26, 29   B. 28, 33, 38
   C. 30, 37, 44   D. 32, 42, 53

8. An auto magazine listed the wheelbases of 4 cars. They were the Archer (261.5 cm), the Bella (248.72 cm), the Cara (238 cm), and the Dante (230.8 cm). Which car has the longest wheelbase?
   A. Archer   B. Bella
   C. Cara   D. Dante

9. Choose the correct relationship.
   $x = (-12)(3)(-5)(0)(2)$
   $y = (-7)(20)(-3)(2)$
   A. $x > y$ B. $x < y$
   C. $x = y$ D. can't determine

10. Write the phrase as a variable expression: 7 less than $5n$
   A. $7 - 5n$ B. $5n - 7$
   C. $7 < 5n$ D. $5n < 7$

11. **Hours Spent Studying**

Who studied for about 5 h?

A. Al   B. Jo
C. Ty   D. Li

12. Simplify: $x^4 \cdot x^3$

A. x^{12} B. x^{24}
C. x^7 D. $2x^7$

13. Find the mean of the data.
10, 27, 10, 15

A. 17   B. 12.5
C. 10   D. 15.5

14. Solve: $y - 9 = -12$

A. $y = 3$ B. $y = -3$
C. $y = -21$ D. $y = 21$

15. The temperature increased from $-5°F$ to $12°F$ in 4 h. Find the change in temperature.

A. $7°F$ B. $-7°F$
C. $-17°F$ D. $17°F$

16. Cherise bought a tennis racket and some tennis balls. What information is needed to find the total amount she spent?

A. the price of the tennis racket
B. the price of each tennis ball
C. the number of tennis balls
D. all of the above

17. Solve: $24 = -3t$

A. $t = -8$ B. $t = 8$
C. $t = -72$ D. $t = 72$

18. Decide which statistical measure best describes the data.
62, 86, 88, 94, 94

A. the mean   B. the median
C. the range   D. the mode

19. Solve: $21 = 3m + 6$

A. $m = 9$ B. $m = 1$
C. $m = 45$ D. $m = 5$

20. Write an equation for the situation.
Ann has 32 tapes. She has 6 fewer tapes than Ted. How many tapes does Ted have?

A. $t + 6 = 32$
B. $6t = 32$
C. $t - 6 = 32$
D. $t + 32 = 6$

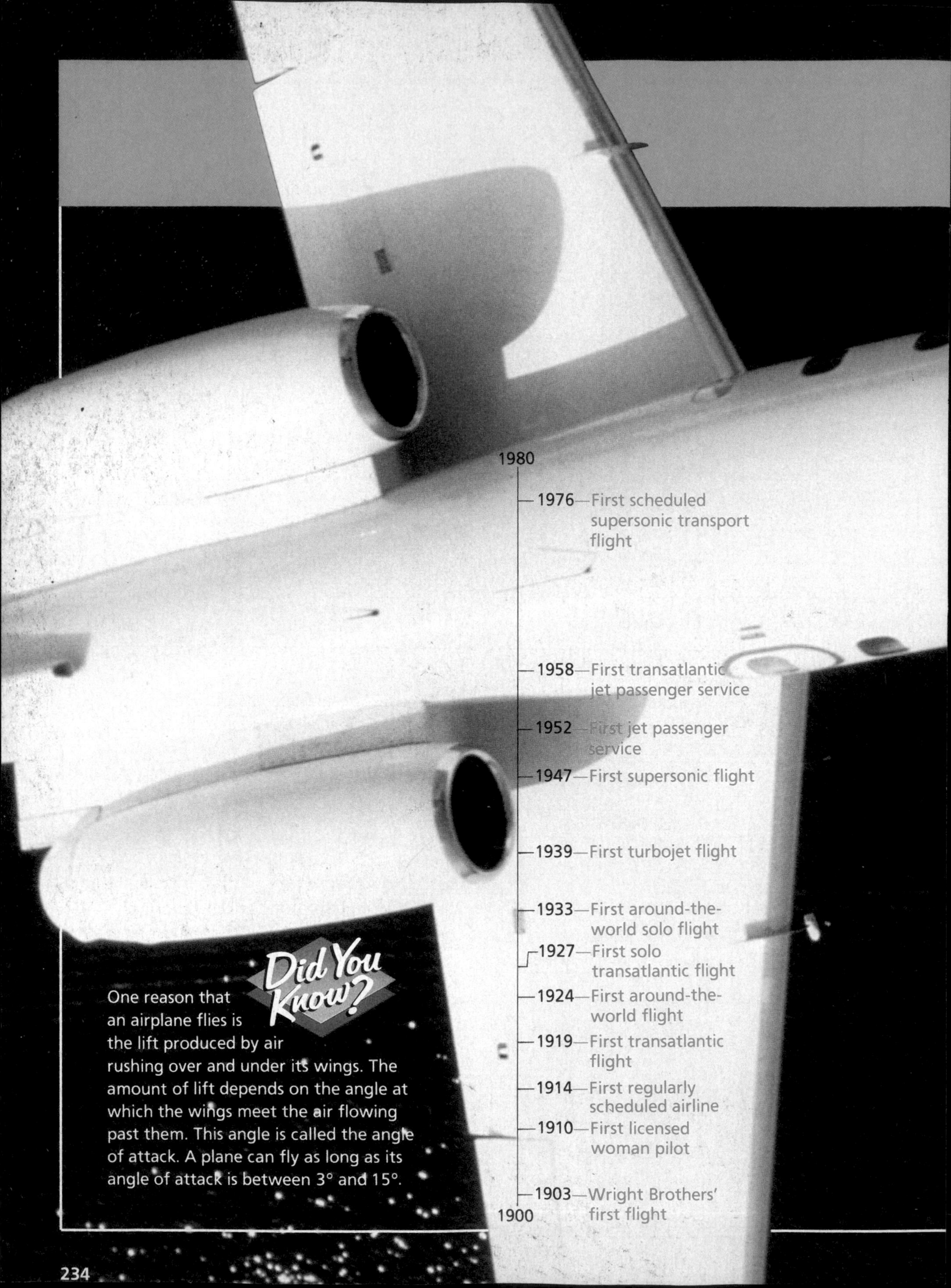

1980

— 1976—First scheduled
supersonic transport
flight

— 1958—First transatlantic
jet passenger service

— 1952—First jet passenger
service

— 1947—First supersonic flight

— 1939—First turbojet flight

— 1933—First around-the-
world solo flight

— 1927—First solo
transatlantic flight

— 1924—First around-the-
world flight

— 1919—First transatlantic
flight

— 1914—First regularly
scheduled airline

— 1910—First licensed
woman pilot

— 1903—Wright Brothers'
1900   first flight

Did You Know?

One reason that
an airplane flies is
the lift produced by air
rushing over and under its wings. The
amount of lift depends on the angle at
which the wings meet the air flowing
past them. This angle is called the angle
of attack. A plane can fly as long as its
angle of attack is between 3° and 15°.

Introduction to Geometry 6

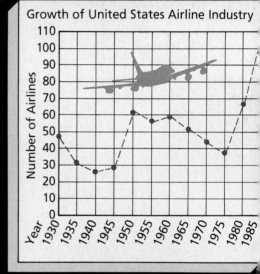

Growth of United States Airline Industry

Twelve Busiest United States Airports

Airport	Location	Passengers (Millions)
O'Hare	Chicago	56.3
Hartsfield	Atlanta	47.6
Los Angeles	Los Angeles	44.9
Dallas / Fort Worth	Dallas / Fort Worth	41.9
Stapleton	Denver	32.4
Kennedy	New York	30.2
San Francisco	San Francisco	29.8
LaGuardia	New York	24.2
Miami	Miami	24.0
Newark	Newark	23.5
Logan	Boston	23.3
Honolulu	Honolulu	20.4

6-1 Points, Lines, and Planes

Objective: To identify the basic geometric figures.

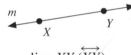

CONNECTION

When you write a report, an essay, or even a letter to a friend, your work is based on ideas that you developed over many years. For instance, the words you write are created using the alphabet that you learned as a child, and sentences and paragraphs flow from the rules of grammar and composition that you learned in school. You will find that geometry is similar because it involves organizing several basic ideas into useful and interesting mathematical structures.

The geometry that you will study in this course has its foundation in three undefined terms: *point, line,* and *plane*.

Terms to Know
- *point*
- *line*
- *plane*
- *collinear*
- *intersect*
- *line segment*
- *endpoint*
- *ray*
- *angle*
- *vertex of an angle*
- *side of an angle*

A **point** is an exact location in space. A point has no size, but you use a dot to represent a point. You name a point by a capital letter.

points A and B

A **line** is a straight arrangement of points that extends forever in opposite directions. You can name a line using any two points on the line. Another way to name a line is to use a single lowercase letter.

line XY ($\overleftrightarrow{XY}$), line YX ($\overleftrightarrow{YX}$) or line m

A **plane** is a flat surface that extends forever. A plane has no edges, but you use a four-sided figure to represent a plane. You can name a plane using a capital letter.

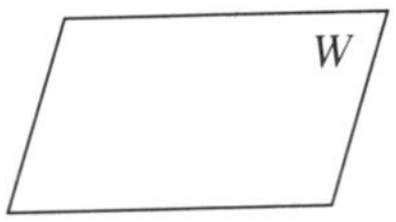

plane W

Points that lie on the same line are called **collinear** points. In the figure below, points P, Q, R, S, and T are collinear.

P   Q   R   S   T

Two lines that meet at one point are said to **intersect** in that point.

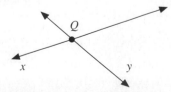

When two distinct planes intersect, their intersection is a line.

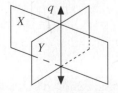

A **line segment** is a part of a line that consists of two **endpoints** and all the points between. You name a line segment using its endpoints.

line segment MN ($\overline{MN}$)
line segment NM ($\overline{NM}$)

A **ray** is a part of a line that has one endpoint and extends forever in one direction. You name a ray by writing the endpoint first, then writing one other point on the ray.

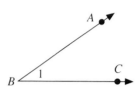

ray YZ ($\overrightarrow{YZ}$)

When two rays share a common endpoint, the figure that is formed is an **angle.** The endpoint is called the **vertex** of the angle, and the rays are called the **sides.** In the angle at the right, point B is the vertex. The sides are $\overrightarrow{BA}$ and $\overrightarrow{BC}$.

The symbol for an angle is $\angle$. You name an angle using letters or numbers, usually depending on what is most appropriate in the given situation. For the angle above, any of these is an appropriate name.

$$\angle ABC \qquad \angle CBA \qquad \angle B \qquad \angle 1$$

To name an angle using three letters, the vertex letter must be in the center. It would *not* be correct to refer to the angle above as $\angle BAC$.

Example

Write the name of each figure.

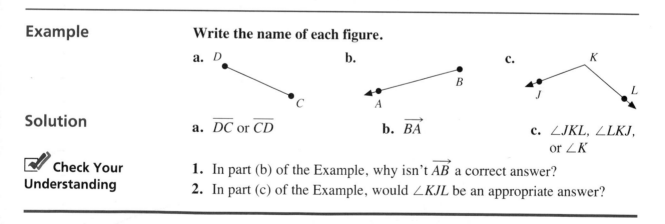

a. D, C

b. B, A

c. K, J, L

Solution

a. $\overline{DC}$ or $\overline{CD}$

b. $\overrightarrow{BA}$

c. $\angle JKL$, $\angle LKJ$, or $\angle K$

☑ **Check Your Understanding**

1. In part (b) of the Example, why isn't $\overrightarrow{AB}$ a correct answer?

2. In part (c) of the Example, would $\angle KJL$ be an appropriate answer?

Guided Practice

COMMUNICATION «*Reading*

Refer to the text on pages 236–237.

«**1.** Which three undefined terms form the foundation of geometry?

«**2.** How are a ray and a line segment alike? How are they different?

Choose the letters of all the correct names for each figure.

3.

a. line n
b. $\overleftrightarrow{PQ}$
c. $\overleftrightarrow{QP}$
d. line Q

4.

a. $\overleftrightarrow{GH}$
b. $\overrightarrow{GH}$
c. $\overrightarrow{HG}$
d. $\overleftrightarrow{HG}$

5.

a. $\angle ONM$
b. $\angle NOM$
c. $\angle O$
d. $\angle NMO$

6.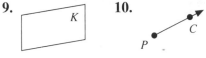

a. point T
b. plane T
c. angle T
d. line T

Write the name of each figure.

7.

8.

9.

10.

COMMUNICATION « *Discussion*

« **11.** Make a list of all the line segments that you can name in the figure below.

« **12.** Explain why $\angle M$ is not an appropriate name for the figure below.

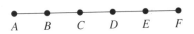

Exercises

Write the name of each figure.

1.

2.

3.

4.

5. Explain why the symbols $\overrightarrow{AG}$ and $\overrightarrow{GA}$ do not represent the same geometric figure.

6. Give four different names for the angle shown at the right.

Make a sketch of each figure.

7. $\angle RXM$　　**8.** $\overline{RS}$　　**9.** $\overrightarrow{ZN}$　　**10.** plane A

11. lines j and k, which intersect at point D

12. $\angle CBA$ and $\angle ABD$, which share $\overrightarrow{BA}$ as a common side

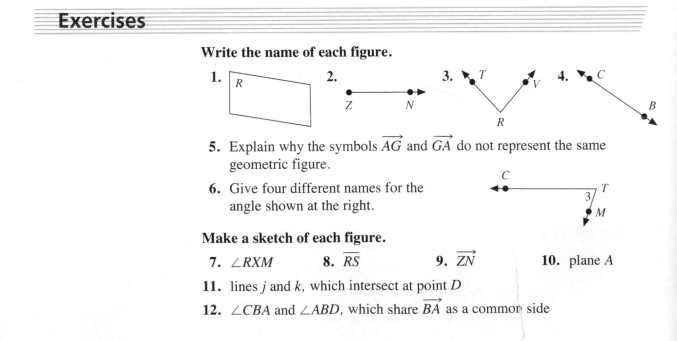

Use the figure at the right. Name another point that lies in the same plane as the given points.

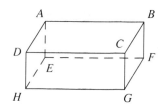

13. points A, C, D

14. points D, C, H

15. points A, B, F

16. points D, H, E

Use the figure at the right above. Name three line segments that intersect at the given point.

17. point C **18.** point A **19.** point F **20.** point E

MENTAL MATH

First answer each question by picturing the geometric figure in your mind. Then check if your mental picture was accurate by making a sketch of the figure on paper.

21. In angle AMT, which point is the vertex?

22. What are the sides of angle SYQ?

23. Points A, B, and C are noncollinear. What three different line segments have these points as their endpoints?

24. What figure is formed by the intersection of ray GH and ray HG?

SPIRAL REVIEW

25. Find the sum: $\frac{7}{8} + \frac{9}{16}$ *(Toolbox Skill 17)*

26. Write the name of the figure shown at the right.   *(Lesson 6-1)*

27. Sue sold tickets for a choral performance. Adult tickets cost $5.75 each and student tickets cost $2 each. Sue collected $72. How many of each type of ticket did she sell?   *(Lesson 4-2)*

28. Find the answer: $-2 + 8 \div (-2) - 10$ *(Lesson 3-6)*

Historical Note

Archie Alexander, one of the first African-American technicians to be internationally famous, was an outstanding structural engineer. After graduating from the University of Iowa in 1912, he studied bridge design at the University of London. He then became a specialist in the design and construction of concrete and steel bridge spans. Geometry played a very important role in his work.

Research

Find out two ways that geometry is used in the design and construction of bridges.

Measuring and Drawing Angles

Objective: To measure and draw angles using a protractor.

APPLICATION

Terms to Know
- *degree*
- *protractor*

The architect who plans a building usually presents the plan in a blueprint. A blueprint shows not only the sizes of pieces such as walls and built-in cabinets, but also their positions in relation to each other. To show positions accurately, the architect indicates the size of the angle formed where these pieces meet.

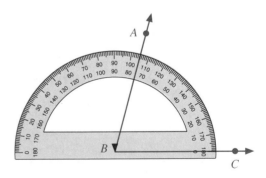

The unit that is commonly used to measure the size of an angle is the **degree.** The number of degrees in an angle's measure indicates the amount of openness between the sides of the angle. To measure an angle, you use the geometric tool called a **protractor.**

Example 1

Use a protractor to measure ∠ABC.

Solution

Put the center mark of the protractor on the vertex of the angle, which is B. Place the 0° mark on one side of the angle, $\overrightarrow{BC}$. Then read the number where the other side, $\overrightarrow{BA}$, crosses the scale. The measure of ∠ABC is 75°.

Check Your Understanding

1. In Example 1, why do you read the measure of the angle from the bottom scale of numbers rather than the top scale?

To indicate the degree measure of an angle, you use a small letter *m*.

In Words	In Symbols
The measure of angle *ABC* is seventy-five degrees.	$m\angle ABC = 75°$

Example 2	Use a protractor to draw ∠RST with measure 160°.
Solution	Draw a ray, $\overrightarrow{ST}$, to represent one side of the angle. Put the center of the protractor on the endpoint, point S, and line up the 0° mark along the ray. Make a mark at 160° and remove the protractor. Draw a ray, $\overrightarrow{SR}$, through the mark.

✅ **Check Your Understanding**

2. How would Example 2 be different if $\overrightarrow{ST}$ were drawn in the opposite direction?

Guided Practice

COMMUNICATION « *Reading*

Explain the meaning of the word *degree* in each sentence.

« **1.** Jerrold can compute mentally with a high degree of accuracy.

« **2.** The temperature increased ten degrees in just one hour.

« **3.** The pilot made a two-degree correction in the airplane's flight path.

« **4.** Lawanda earned a master's degree in English literature.

Refer to the sentences in Exercises 1–4.

« **5.** What do all these meanings of the word *degree* have in common?

« **6.** In which sentence is the meaning of *degree* closest to the meaning used in this lesson?

Use a protractor to measure each angle.

7.

8.

Use a protractor to draw an angle of the given measure.

9. 35°   **10.** 155°   **11.** 90°   **12.** 67°

Use a protractor to measure each angle.

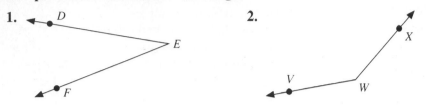

1.

2.

Use a protractor to draw an angle of the given measure.

3. 115° **4.** 20° **5.** 88° **6.** 164°

7. Draw angle *JKL,* which has a measure of sixty degrees.

8. The measure of the angle formed by joining rays *XY* and *XZ* is 172°. Draw the angle.

ESTIMATION

Match each angle with the best estimate of its measure. Then use a protractor to check your answer.

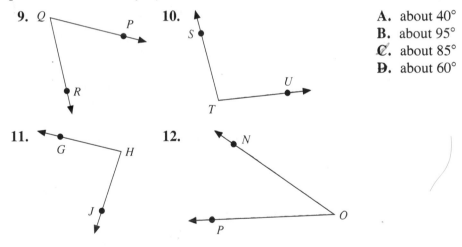

9.

10.

11.

12.

A. about 40°
B. about 95°
C. about 85°
D. about 60°

SPIRAL REVIEW

13. Solve: $x - 12 = -3$ *(Lesson 4-3)*

14. Estimate the sum: $47 + 51 + 44 + 58$ *(Toolbox Skill 4)*

15. Draw an angle that measures 104°. *(Lesson 6-2)*

16. The weekly salaries at Office Rite Company are $315, $340, $295, $305, $850, $800, and $345. Decide whether the mean describes the data well. Explain. *(Lesson 5-9)*

Constructing and Bisecting Line Segments and Angles

Objective: To use a
straightedge and a compass
to construct and bisect line
segments and angles.

Materials

▧ straightedge
▧ compass

In this exploration, you will *construct* geometric figures using only
two tools, a *straightedge* and a *compass*. You may use a ruler as a
straightedge, but you may not use the marks on the ruler for measuring. You will use a compass to draw a circle, or part of a circle called
an *arc*.

Activity I   *Constructing a Line Segment the Same Length as a Given
Line Segment*

1 Use a straightedge to draw any segment $\overline{AB}$.
   a. Use a straightedge to draw a line. Call it ℓ.
   b. Choose any point on ℓ and label it C.
   c. Open your compass so that the sharp tip is
      on A and the writing tip is on B. Keeping
      the same compass opening, place the sharp
      tip on C and draw an arc crossing line ℓ.
      Label the point of intersection D. $\overline{CD}$ is the
      same length as $\overline{AB}$.

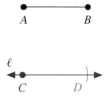

2 **a.** Draw any line segment $\overline{EF}$. Then use a straightedge and a compass to construct $\overline{MN}$ so that the length of $\overline{MN}$ is twice the length of $\overline{EF}$.
   b. Describe how you would construct $\overline{MN}$ so that the length of $\overline{MN}$ is three times the length of $\overline{EF}$.

Activity II   *Constructing an Angle with the Same Measure as a
Given Angle*

1 Draw any $\angle B$.
   a. To copy $\angle B$, begin by using a straightedge
      to construct any ray with endpoint E.
   b. Using any compass opening, place the sharp
      tip of your compass on point B and draw an
      arc intersecting the sides of $\angle B$. Label the
      points of intersection A and C.
   c. Using the same compass opening, place the
      sharp tip of your compass on E and draw an
      arc. Label the point of intersection F.
   d. Open your compass so that the sharp tip is on C and the writing
      tip is on A. Using the same compass opening, place the point
      of your compass on F and draw an arc intersecting the first arc.
      Label the point of intersection D. Draw $\overrightarrow{ED}$. $m\angle E$ equals $m\angle B$.

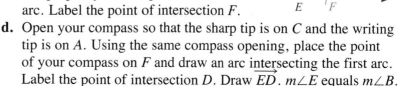

2 Draw any obtuse angle. Use a straightedge and a compass to
construct an angle with the same measure.

You can also use a straightedge and a compass to *bisect* a geometric figure, that is, to divide it into two parts of equal measure.

Activity III   *Bisecting a Line Segment*

1 Draw any line segment $\overline{AB}$.

  a. Choose a point C on $\overline{AB}$, so that the length of $\overline{AC}$ clearly is more than half the length of $\overline{AB}$. Set the compass so that the sharp tip is on A and the writing tip is on C. Draw an arc.

  b. Using the same compass opening, draw an arc with the sharp tip of the compass on B. This arc should intersect the first arc at two points. Label them D and E. Draw $\overleftrightarrow{DE}$. M is halfway between A and B on $\overline{AB}$, so the length of $\overline{AM}$ equals the length of $\overline{MB}$.

  c. Which line segment bisects $\overline{AB}$: $\overline{AC}$, $\overline{CB}$, or $\overline{DE}$?

2 Trace the segment $\overline{FG}$ below onto your paper. Then use a straightedge and a compass to divide the segment into four segments of equal length.

$$F \qquad\qquad\qquad\qquad\qquad G$$

Activity IV   *Bisecting an Angle*

1 Draw any $\angle B$.

  a. Using any compass opening, place the sharp tip of your compass on point B and draw an arc intersecting the sides of $\angle B$. Label the points of intersection A and C.

  b. Then place the sharp tip of the compass on A and draw an arc. Using the same compass opening, place the sharp tip of your compass on C and draw an arc that intersects the arc you just drew. Label the point of intersection X. Draw $\overrightarrow{BX}$. $m\angle ABX$ equals $m\angle XBC$.

2 How would the measures of $\angle ABX$ and $\angle XBC$ compare if $m\angle ABC$ increased? if $m\angle ABC$ decreased?

3 Trace each angle onto your paper. Bisect each angle.

  a.

  A

  b.

  B

Types of Angles

Objective: To find measures of acute, right, obtuse, straight, complementary, supplementary, vertical, and adjacent angles.

Terms to Know

- *acute angle*
- *right angle*
- *obtuse angle*
- *straight angle*
- *complementary*
- *supplementary*
- *adjacent*
- *vertical*

EXPLORATION

1 In medicine, a pain that is very sharp is called *acute* pain. What do you think an *acute angle* looks like?

2 In sailing, you *right* a capsized boat when you set it upright again. What do you think a *right angle* looks like?

3 In botany, a leaf with a blunt or rounded tip is called an *obtuse* leaf. What do you think an *obtuse angle* looks like?

4 In art, a line that is not curved or bent is called a *straight* line. What do you think a *straight angle* looks like?

Angles are often classified by their measures. The table below gives a summary to help you identify and compare the basic types.

Angle	Name	Measure
	acute angle	greater than 0°, less than 90°
	right angle	equal to 90° (*Note:* A small square indicates a right angle.)
	obtuse angle	greater than 90°, less than 180°
	straight angle	equal to 180°

Names are also given to special pairs of angles. For instance, some pairs are classified by the sum of their measures.

Two angles are **complementary** when the sum of their measures is 90°.

Two angles are **supplementary** when the sum of their measures is 180°.

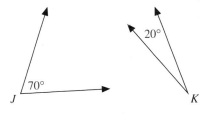

$m\angle J + m\angle K = 70° + 20°$
$= 90°$
$\angle J$ and $\angle K$ are complementary.

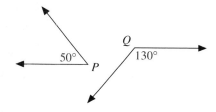

$m\angle P + m\angle Q = 50° + 130°$
$= 180°$
$\angle P$ and $\angle Q$ are supplementary.

Example 1	**a.** Find the measure of an angle complementary to a 23° angle.
	b. Find the measure of an angle supplementary to a 170° angle.
Solution	**a.** Subtract the given measure from 90°.
	$90° - 23° = 67°$
	b. Subtract the given measure from 180°.
	$180° - 170° = 10°$

Check Your Understanding

1. How would part (a) of Example 1 be different if you were asked to find the measure of a *supplement* of an angle that measures 23°?
2. In part (b) of Example 1, would it be possible to find a *complement* of an angle that measures 170°? Explain.

Two angles that share a common side, but do not overlap each other, are called **adjacent** angles. In the figure below at the left, $\angle ABX$ and $\angle XBC$ are adjacent, but $\angle ABC$ and $\angle XBC$ are not adjacent. Two special situations occur when complementary and supplementary angles are adjacent.

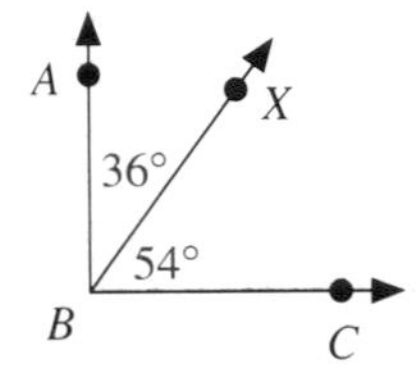

$m\angle ABX + m\angle CBX = 90°$,
so $\angle ABC$ is a right angle.

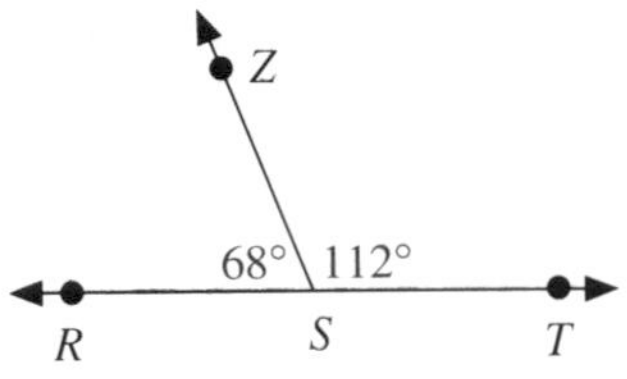

$m\angle RSZ + m\angle TSZ = 180°$,
so $\angle RST$ is a straight angle.

When two lines intersect, the angles that are *not* adjacent to each other are called **vertical** angles. Vertical angles are always equal in measure. In the figure at the right, $\angle 1$ and $\angle 2$ are vertical angles.

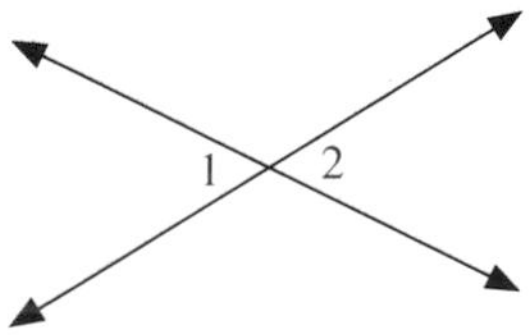

Example 2	**Use the figure at the right.**
	Find the measure of each angle.
	a. $\angle 2$ **b.** $\angle 3$
Solution	**a.** $\angle 2$ and $\angle 4$ are vertical angles,
	so $m\angle 2 = m\angle 4 = 60°$.
	b. $\angle 3$ and $\angle 4$ are supplementary angles,
	so $m\angle 3 = 180° - m\angle 4 = 180° - 60° = 120°$.

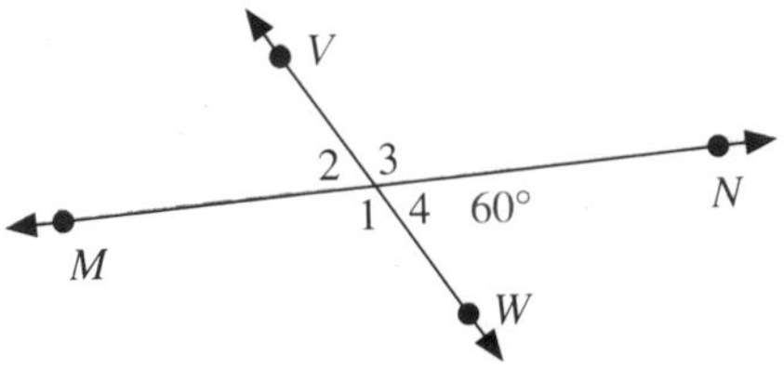

Check Your Understanding

3. Name the two pairs of vertical angles in the figure for Example 2.
4. In the figure for Example 2, name three pairs of supplementary angles other than $\angle 3$ and $\angle 4$.

Guided Practice

COMMUNICATION « *Reading*

«**1.** Explain the meaning of *supplement* in the following sentence.
Elena works part time to earn money to supplement her college scholarship.

«**2.** How is the meaning of the word *complementary* different from the meaning of *complimentary*?

Find each difference.

3. 90 − 74 **4.** 90 − 15 **5.** 90 − 53

6. 180 − 166 **7.** 180 − 82 **8.** 180 − 59

Tell whether two angles with the given measures are *complementary*, *supplementary*, or *neither*.

9. 40°, 50° **10.** 125°, 55° **11.** 12°, 88° **12.** 90°, 90°

Find the measure of a complement of an angle of the given measure.

13. 15° **14.** 79° **15.** 60° **16.** 33°

Find the measure of a supplement of an angle of the given measure.

17. 41° **18.** 145° **19.** 88° **20.** 27°

Use the figure at the right. Give the measure of each angle.

21. ∠5 **22.** ∠6

23. ∠7 **24.** ∠8

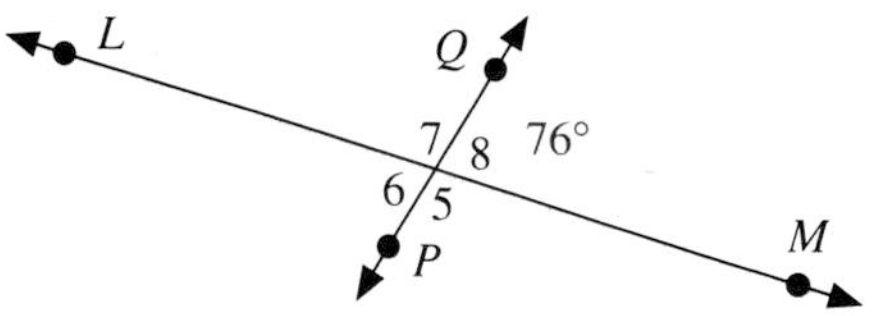

Exercises

Find the measure of a complement of an angle of the given measure.

1. 30° **2.** 5° **3.** 29° **4.** 87°

Find the measure of a supplement of an angle of the given measure.

5. 10° **6.** 65° **7.** 102° **8.** 90°

Use the figure at the right. Give the measure of each angle.

9. ∠9 **10.** ∠10

11. ∠11 **12.** ∠12

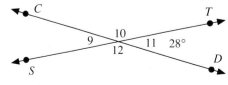

13. Angle *XYZ* has the same measure as its complement. Find $m\angle XYZ$.

14. **DATA**, *pages 234–235* Determine whether an airplane's angle of attack is an *acute*, *right*, or *obtuse* angle.

LOGICAL REASONING

Replace each __?__ with *always*, *sometimes*, or *never* to make a true statement.

15. A supplement of an obtuse angle is __?__ an acute angle.

16. A complement of an acute angle is __?__ an obtuse angle.

17. The measure of an angle is __?__ equal to the measure of its supplement.

18. Two vertical angles that are supplementary are __?__ right angles.

CONNECTING GEOMETRY AND ALGEBRA

Sometimes the measures of angles in a figure are given only as variable expressions. If you are given enough information, though, you can determine what the exact measures must be by using an equation.

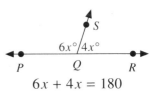
$$6x + 4x = 180$$

Exercises 19–22 refer to the figure at the right.

19. Why was the number 180 used as the right side of the equation?

20. Why are $6x$ and $4x$ added on the left side of the equation?

21. What value of x is the solution of the equation?

22. What are the measures of $\angle PQS$ and $\angle RQS$?

In each figure, find $m\angle ABC$.

23.   24.

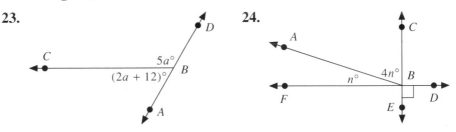

SPIRAL REVIEW

25. Write an equation that represents the sentence: The sum of twice a number *n* and eleven is negative thirty-one.   *(Lesson 4-7)*

26. Find the measure of an angle that is supplementary to a 56° angle.   *(Lesson 6-3)*

27. Find the sum: $-17 + (-14)$ *(Lesson 3-2)*

Perpendicular and Parallel Lines

Objective: To identify perpendicular and parallel lines and calculate the measures of the angles formed when a transversal intersects parallel lines.

APPLICATION

Many cities were planned so that the streets lie in an orderly grid pattern. At the point where streets intersect, they form right angles and are said to be *perpendicular*. Streets that do not intersect are *parallel*.

Terms to Know

- *perpendicular*
- *parallel*
- *transversal*
- *alternate interior angles*
- *corresponding angles*

New Orleans, circa 1765

Mississippi River

Two lines that intersect to form right angles are **perpendicular** lines. In the figure at the right, you see that ∠*AMY* is a right angle, and so you know that $\overleftrightarrow{AB}$ and $\overleftrightarrow{XY}$ are perpendicular. The symbol for *is perpendicular to* is ⊥.

In Words	In Symbols
Line AB is perpendicular to line XY.	$\overleftrightarrow{AB} \perp \overleftrightarrow{XY}$

Two lines in the same plane that do not intersect are **parallel** lines. In the figure at the right, $\overleftrightarrow{PQ}$ and $\overleftrightarrow{RS}$ will always remain the same distance apart, and so they are parallel. The symbol for *is parallel to* is ∥.

In Words	In Symbols
Line PQ is parallel to line RS.	$\overleftrightarrow{PQ} \parallel \overleftrightarrow{RS}$

The words *parallel* and *perpendicular* also can be used to refer to line segments and rays.

Example 1	Use the figure at the right. Tell whether each statement is *True* or *False*.

a. $\overleftrightarrow{AB} \parallel \overleftrightarrow{CD}$ **b.** $\overleftrightarrow{MN} \perp \overleftrightarrow{CD}$

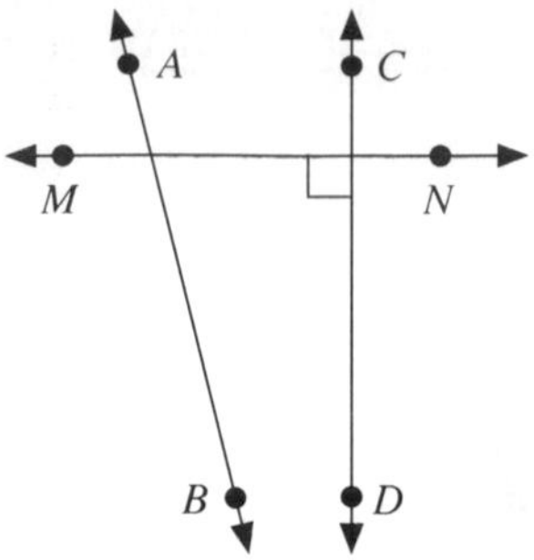

Solution

a. False. If $\overleftrightarrow{AB}$ and $\overleftrightarrow{CD}$ were extended downward, they would intersect.

b. True. A right angle is formed at the point where $\overleftrightarrow{MN}$ intersects $\overleftrightarrow{CD}$.

✓ **Check Your Understanding**

1. In Example 1, how many right angles are formed at the point in the figure where $\overleftrightarrow{MN}$ intersects $\overleftrightarrow{CD}$? Explain.

A **transversal** is a line that intersects two or more lines in the same plane at different points. In the figure at the right, line ℓ is a transversal that intersects $\overleftrightarrow{JK}$ and $\overleftrightarrow{MN}$.

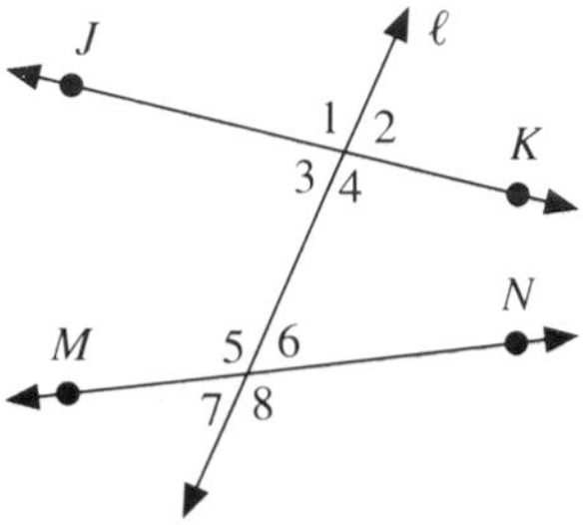

Alternate interior angles are *interior* to the two lines, but on *alternate* sides of the transversal. In the figure, these are the pairs of alternate interior angles.

$$\angle 3 \text{ and } \angle 6 \qquad \angle 4 \text{ and } \angle 5$$

Corresponding angles are in the same position with respect to the two lines and the transversal. In the figure above, these are the pairs of corresponding angles.

$$\angle 1 \text{ and } \angle 5 \quad \angle 2 \text{ and } \angle 6 \quad \angle 3 \text{ and } \angle 7 \quad \angle 4 \text{ and } \angle 8$$

There are special properties when the lines intersected by a transversal are parallel.

The angles in a pair of alternate interior angles are equal in measure.
The angles in a pair of corresponding angles are equal in measure.

Example 2

In the figure at the right, $\overleftrightarrow{EF} \parallel \overleftrightarrow{GH}$. Find the measure of each angle.

a. $\angle 6$ **b.** $\angle 8$

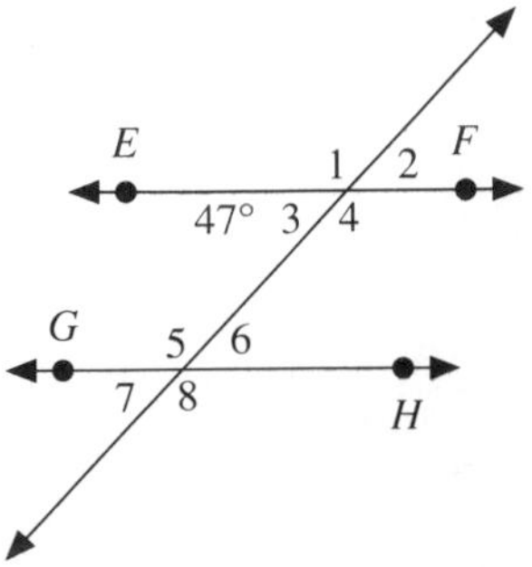

Solution

a. $\angle 3$ and $\angle 6$ are alternate interior angles, so $m\angle 6 = m\angle 3 = 47°$.

b. $\angle 8$ and $\angle 6$ are supplementary:
$$m\angle 8 = 180° - m\angle 6$$
$$= 180° - 47° = 133°$$

✓ **Check Your Understanding**

2. In Example 2, which angle in the figure forms a pair of corresponding angles with $\angle 3$? What is the measure of this angle?

COMMUNICATION « *Writing*

Write each sentence in symbols.

« **1.** Line *ST* is perpendicular to line *YZ*.

« **2.** Line segment *PQ* is parallel to line segment *MN*.

Write each statement in words.

« **3.** $\overleftrightarrow{AB} \parallel \overleftrightarrow{ST}$ « **4.** $\overleftrightarrow{FG} \perp \overleftrightarrow{UV}$ « **5.** $\overline{XY} \perp \overline{ZW}$

« **6.** Draw a diagram that represents the statement $\overrightarrow{MN} \perp \overrightarrow{GH}$.

Use the figure at the right. Tell whether the given angles are a pair of corresponding angles or alternate interior angles.

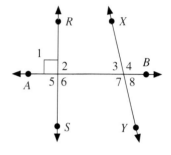

7. $\angle 1$ and $\angle 3$ **8.** $\angle 2$ and $\angle 7$

9. $\angle 6$ and $\angle 3$ **10.** $\angle 4$ and $\angle 2$

Use the figure above at the right. Tell whether each statement is *True* or *False*.

11. $\overleftrightarrow{RS} \parallel \overleftrightarrow{XY}$ **12.** $\overleftrightarrow{AB} \perp \overleftrightarrow{RS}$

In the figure at the right, $\overleftrightarrow{EF} \parallel \overleftrightarrow{GH}$ and $m\angle 9 = 75°$. Find the measure of each angle.

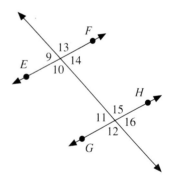

13. $\angle 10$ **14.** $\angle 11$ **15.** $\angle 12$

16. $\angle 13$ **17.** $\angle 14$ **18.** $\angle 15$

« **19.** COMMUNICATION « *Discussion*
Why do you think so many cities are laid out as a grid of parallel and perpendicular streets?

Exercises

Use the figure at the right. Tell whether each statement is *True* or *False*.

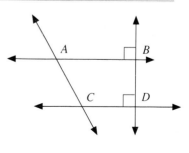

1. $\overleftrightarrow{AB} \perp \overleftrightarrow{CD}$ **2.** $\overleftrightarrow{AB} \perp \overleftrightarrow{DB}$

3. $\overleftrightarrow{CD} \parallel \overleftrightarrow{AB}$ **4.** $\overleftrightarrow{AC} \parallel \overleftrightarrow{BD}$

5. $\overleftrightarrow{AC} \perp \overleftrightarrow{CD}$ **6.** $\overleftrightarrow{AC} \parallel \overleftrightarrow{CD}$

7. $\overleftrightarrow{AB} \parallel \overleftrightarrow{AC}$ **8.** $\overleftrightarrow{CD} \perp \overleftrightarrow{BD}$

In the figure at the right,
$\overleftrightarrow{JK} \parallel \overleftrightarrow{QR}$ and $m\angle 7 = 143°$.
Find the measure of each angle.

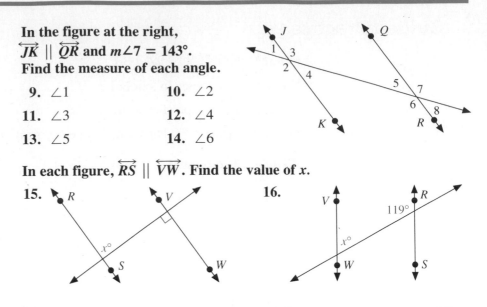

9. $\angle 1$ 10. $\angle 2$

11. $\angle 3$ 12. $\angle 4$

13. $\angle 5$ 14. $\angle 6$

In each figure, $\overleftrightarrow{RS} \parallel \overrightarrow{VW}$. Find the value of x.

15.   16.

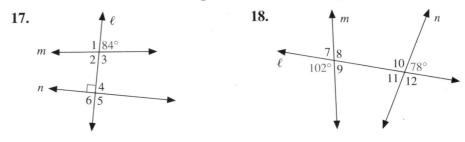

In each figure, first find the measures of all the numbered angles. Then tell whether lines m and n are parallel or intersecting lines.

17.   18.

THINKING SKILLS

Lines that do not lie in the same plane are called **skew** lines.

19. Compare the definition of skew lines, given above, to the definition of parallel lines given on page 249. How are skew lines and parallel lines different? How do you think they are alike?

20. The figure at the right represents a box, similar to an ordinary packing carton. Classify each pair of line segments as *parallel*, *perpendicular*, or *skew*.

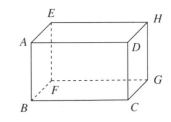

    a. $\overline{AB}$ and $\overline{DC}$ **b.** $\overline{EF}$ and $\overline{DH}$
    c. $\overline{CG}$ and $\overline{FG}$ **d.** $\overline{AB}$ and $\overline{GH}$

21. Is it possible for two skew lines to be perpendicular? Give a convincing argument to support your answer.

22. Determine whether the following statement is true or false.
Any two lines in space are related in exactly one of these ways: the lines either intersect, they are parallel, or they are skew.

23. Estimate the quotient: $3495 \div 38$
(Toolbox Skill 3)

24. In the figure at the right,
$\overleftrightarrow{AB} \parallel \overleftrightarrow{CD}$. Find the measure of
$\angle 8$. *(Lesson 6-4)*

25. Evaluate $-4a^2b$ when $a = 5$ and
$b = -2$. *(Lesson 3-6)*

26. Complete: $148 \text{ mm} = \underline{\ ?\ } \text{ m}$
(Lesson 2-9)

27. Draw a bar graph to display the
data in the table at the right.
(Lesson 5-4)

28. This week the Panthers scored
37 points. This is 5 points less
than twice the number of
points they scored last week.
How many points did the
Panthers score last week?
(Lesson 4-8)

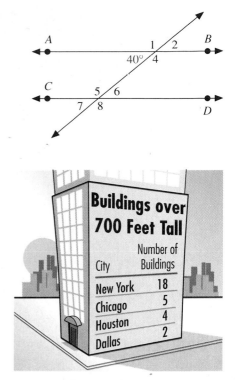

Buildings over 700 Feet Tall

City	Number of Buildings
New York	18
Chicago	5
Houston	4
Dallas	2

Self-Test 1

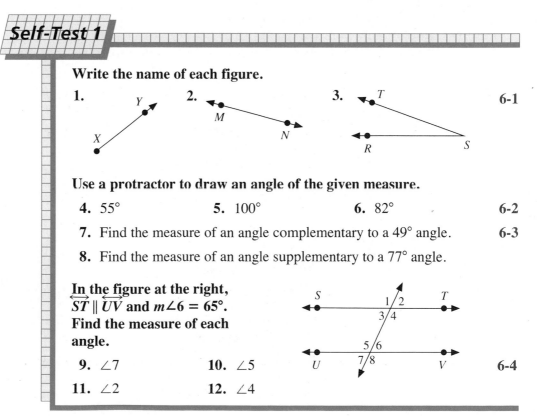

Write the name of each figure.

1.

2.

3.

6-1

Use a protractor to draw an angle of the given measure.

4. $55°$ **5.** $100°$ **6.** $82°$ 6-2

7. Find the measure of an angle complementary to a $49°$ angle.   6-3

8. Find the measure of an angle supplementary to a $77°$ angle.

In the figure at the right,
$\overleftrightarrow{ST} \parallel \overleftrightarrow{UV}$ **and** $m\angle 6 = 65°$.
Find the measure of each
angle.

9. $\angle 7$ **10.** $\angle 5$

11. $\angle 2$ **12.** $\angle 4$

6-4

Constructing Perpendicular and Parallel Lines

Objective: To construct a line perpendicular to a given line and to construct a line parallel to a given line.

Graphic designers, architects, and engineers often draw perpendicular and parallel lines in their work. The construction of these lines can be done by computer-aided design programs. You can also construct perpendicular and parallel lines with a straightedge and a compass.

Materials

■ straightedge
■ compass

Activity I   *Constructing a Line Perpendicular to a Given Line through a Point on the Line*

1 Draw a line ℓ.

   a. Choose any point P on the line. Put the sharp tip of your compass on P and draw arcs that intersect the line in two points. Label the points of intersection A and B.
   b. Does $AP = BP$?
   c. Widen the opening of your compass. With the sharp tip on A, draw an arc.
   d. Without changing the compass setting, place the sharp tip on B and draw an arc that intersects the arc you just drew. Label this point of intersection Q. Draw $\overleftrightarrow{PQ}$. $\overleftrightarrow{PQ} \perp$ line ℓ.

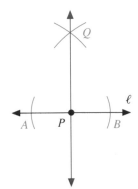

Activity II   *Constructing a Line Perpendicular to a Given Line through a Point not on the Line*

1 Draw a line ℓ.

   a. Draw a point N, not on ℓ. With the sharp tip of your compass on N, draw arcs that intersect line ℓ in two points. Label the points of intersection S and T.
   b. Widen the opening of your compass. With the sharp tip on S, draw an arc not on the line.
   c. Without changing the compass setting, put the sharp tip of your compass on T and draw an arc that intersects the arc you just drew. Label the point of intersection R. Draw $\overleftrightarrow{NR}$. $\overleftrightarrow{NR} \perp$ line ℓ.

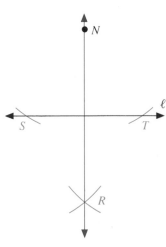

Activity III *Constructing a Line through a Given Point and Parallel to a Given Line*

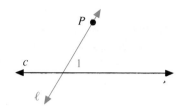

1. Draw a line *c*.
   a. Through any point *P* not on line *c*, draw any line ℓ that intersects line *c*. Label the angle formed ∠1 as shown.
   b. Construct angle ∠2, corresponding to ∠1, at *P*. Extend the side of ∠2 through *P* and label that line *d*.

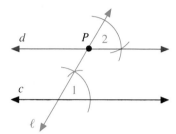

2. Explain why line *d* is parallel to line *c*.

3. Draw a point *Q* on line ℓ, far below line *c*. Construct line *e* that is parallel to line *c*. Is line *e* parallel to line *d*? Explain.

Activity IV *More Constructions*

1. Draw a line ℓ. Construct a line *m* that is perpendicular to line ℓ through a point *A* that is not on ℓ.

2. Construct a line *q* that is perpendicular to line ℓ through a point *B* that is not on ℓ or *m*. Is line *m* parallel to line *q*? Explain.

3. Make a general statement about two lines that are perpendicular to the same line.

4. Construct a line *r* parallel to line *l* through point *B*. Is line *r* perpendicular to line *q*? line *m*?

5. If a line is perpendicular to one of two parallel lines, is it also perpendicular to the other parallel line? Explain.

6. Make a generalization about the kind of angles formed by the intersection of parallel lines with lines perpendicular to them.

7. Trace triangle *ABC*. Construct a line parallel to $\overline{AB}$ at *C*. Construct a line parallel to $\overline{BC}$ at *A*. Construct a line parallel to $\overline{AC}$ at *B*.
   a. What geometric shape do these new lines form?
   b. What do you notice about the relationship between the sizes of the new shape and triangle *ABC*?

8. Trace triangle *DEF*. Construct lines parallel to $\overline{DE}$ at *F*, parallel to $\overline{EF}$ at *D*, and parallel to $\overline{DF}$ at *E*. Make two observations about the largest new shape you have constructed.

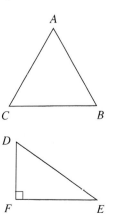

Polygons

Objective: To identify types of polygons and find the measures of their angles.

Terms to Know

- *polygon*
- *side of a polygon*
- *vertex of a polygon*
- *diagonal*
- *regular polygon*

EXPLORATION

1 Using a pencil and a straightedge, draw a triangle on a piece of paper. Then use scissors to cut out the triangle.

2 Tear off the three angles of the triangle. Position the angles side-by-side on a flat surface as shown at the right.

3 Compare your results from Step 2 with those of other students in your class. What do all the triangles seem to have in common?

4 Make a guess about triangles by completing this statement: The sum of the measures of the angles of a triangle is __?__. Your guess is called a *conjecture*.

5 Now test your conjecture: Draw a different triangle on a piece of paper and measure each of the three angles with a protractor. What is the sum of the measures? Was your conjecture correct?

A **polygon** is a closed figure formed by joining three or more line segments in a plane at their endpoints, with each line segment joining exactly two others. Each line segment is called a **side of the polygon.** Each point where two sides meet is a **vertex of the polygon.** (The plural of *vertex* is *vertices*.) A **diagonal** is a line segment that joins two nonconsecutive vertices.

These figures are all examples of polygons.

These figures are *not* polygons.

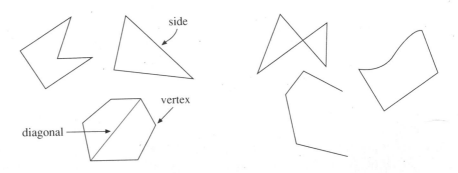

Polygons are identified by the numbers of their sides, using names given to them by the early Greek mathematicians. The table on the next page lists the names for some polygons. In each name, the prefix is underlined.

Name of Polygon	Meaning of Prefix	Number of Sides
triangle	three	3
quadrilateral	four	4
pentagon	five	5
hexagon	six	6
heptagon	seven	7
octagon	eight	8
nonagon	nine	9
decagon	ten	10

A polygon in which all sides have the same length and all angles have the same measure is called a **regular polygon.** In the figure at the right, the red marks show that all sides are equal in length and all angles are equal in measure, so the figure is a regular pentagon.

To name a polygon, you list its vertices in order. Two of the many names for the figure at the right are pentagon *PQRST* and pentagon *RSTPQ*. It should *not* be called pentagon *PRQST*.

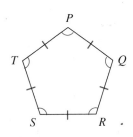

Example 1

Identify each polygon. Then list all its diagonals.

a.

b.

Solution

a.

Polygon *ABCD* is a quadrilateral.
Its diagonals are
$\overline{AC}$ and $\overline{BD}$.

b.

Polygon *RSTUVW* is a regular hexagon.
Its diagonals are
$\overline{RT}$, $\overline{RU}$, $\overline{RV}$, $\overline{SU}$, $\overline{SV}$, $\overline{SW}$, $\overline{TV}$, $\overline{TW}$, and $\overline{UW}$.

The simplest of all polygons is the triangle. The word *triangle* means "three angles." The Exploration on page 256 leads to the following conclusion about the three angles in a triangle.

The sum of the measures of the angles of a triangle is 180°.

You can use this fact to find the sum of the measures of the angles of any polygon. Simply use the diagonals from one vertex to separate the polygon into triangles.

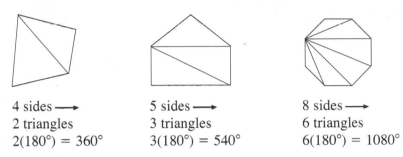

4 sides ⟶
2 triangles
2(180°) = 360°

5 sides ⟶
3 triangles
3(180°) = 540°

8 sides ⟶
6 triangles
6(180°) = 1080°

In each case, the number of triangles is two fewer than the number of sides of the polygon. This leads to the following conclusion.

Generalization: *Finding the Sum of the Angles of a Polygon*

To find the sum of the measures of the angles of a polygon with n sides, subtract two from n and multiply the result by 180°.

$$\text{sum of angles of a polygon} = (n - 2)(180°)$$

Example 2

Find the unknown angle measure in the polygon at the right.

Solution

First find the sum of the measures of all the angles. The polygon has 6 sides.

$$(n - 2)(180°) = (6 - 2)(180°)$$
$$= 4(180°) = 720°$$

Add the known measures.

$$110° + 107° + 140° + 120° + 132° = 609°$$

Subtract this sum from 720°.

$$720° - 609° = 111°$$

The unknown angle measure is 111°.

✓ **Check Your Understanding**

1. In Example 2, what number was substituted for n in the expression $(n - 2)(180°)$? Why?

2. What is represented by 609° in Example 2?

Guided Practice

Match the name of each polygon with the everyday word that has the same prefix. Then give the meaning of the everyday word.

« **1.** triangle   **A.** quadruple

« **2.** quadrilateral   **B.** octopus

« **3.** octagon   **C.** decade

« **4.** decagon   **D.** tricycle

Evaluate the expression $(n - 2)(180)$ for each value of the variable.

5. when $n = 7$ **6.** when $n = 9$ **7.** when $n = 12$

Find the sum of the measures of the angles of each type of polygon.

8. pentagon   **9.** decagon   **10.** octagon

Identify each polygon. Then list all its diagonals.

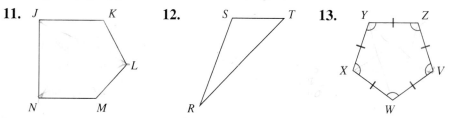

Find each unknown angle measure.

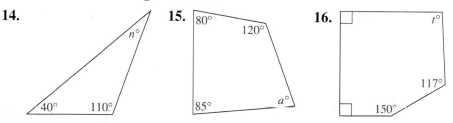

Exercises

Identify each polygon. Then list all its diagonals.

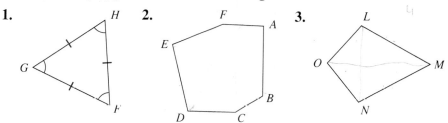

Find each unknown angle measure.

4.

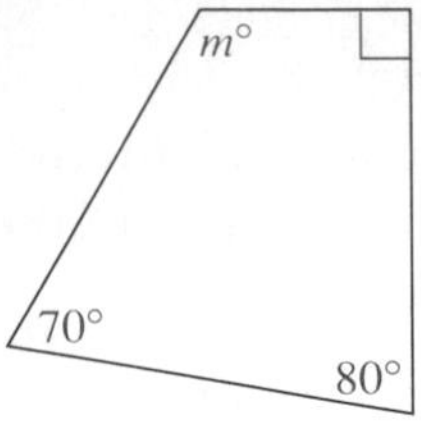

5.

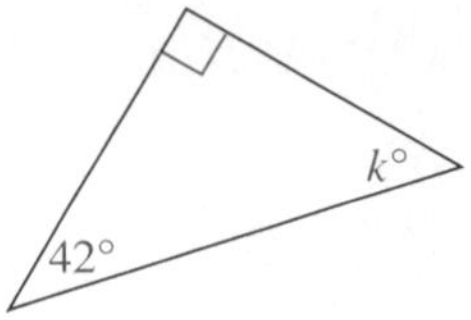

6.

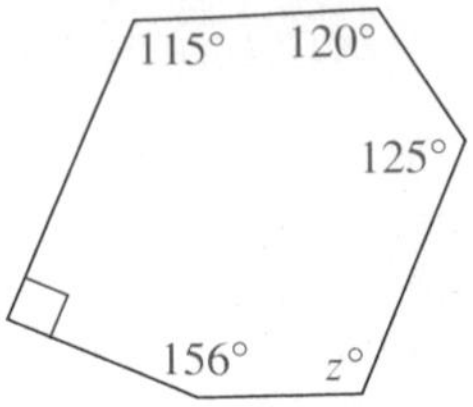

7.

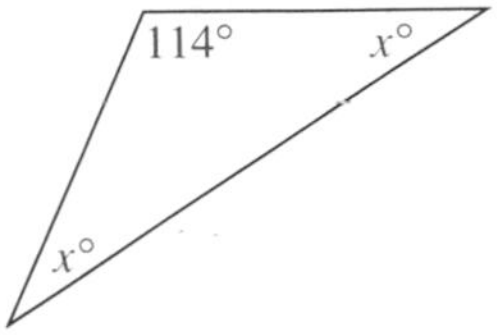

8.

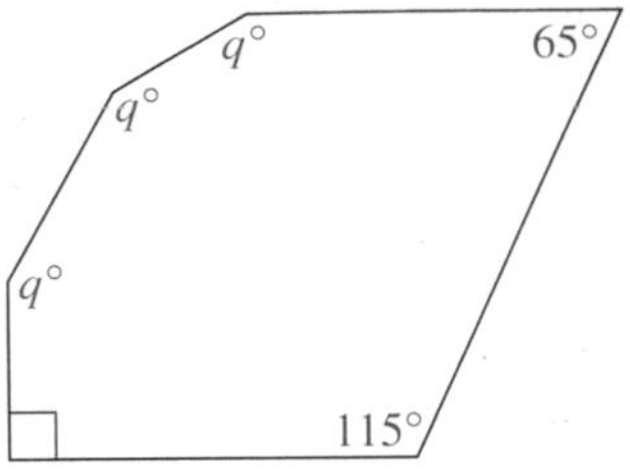

CONNECTING MATHEMATICS AND LANGUAGE ARTS

The prefixes *quadr-* and *quadri-*, as in quadrilateral, mean "four." Remembering this fact might help you figure out the meanings of other words with the same prefix.

Match each word with its correct meaning.

  9. quadraphonic   **A.** one of four major areas of the coordinate plane

10. quadrennial   **B.** having four channels that reproduce sound

11. quadrant   **C.** a four-footed animal

12. quadruped   **D.** occurring every four years

THINKING SKILLS

13. Replace each __?__ with the number that makes the sentence true.
A regular decagon has __?__ angles. The sum of the measures of the angles is __?__ degrees. To find the measure of just one angle, find the quotient __?__ ÷ __?__. The measure of each angle is __?__ degrees.

14. Generalize the results of Exercise 13. Develop a formula to find the measure of one angle of a regular polygon that has n sides.

SPIRAL REVIEW

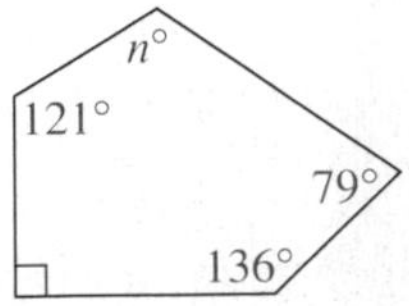

15. Solve: $3x - 8 = 25$ *(Lesson 4-5)*

16. Find the unknown angle measure in the polygon at the right.   *(Lesson 6-5)*

17. Find the difference: $7\frac{3}{8} - 5\frac{3}{4}$ *(Toolbox Skill 19)*

18. Find the mean, median, mode(s), and range for the data:
27, 31, 26, 31, 24, 27, 32, 27, 31, 32, 26   *(Lesson 5-8)*

6-6

Triangles

Objective: To classify triangles by their sides and angles and to use the triangle inequality.

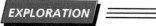

Terms to Know

- *scalene triangle*
- *isosceles triangle*
- *equilateral triangle*
- *Triangle Inequality*
- *acute triangle*
- *right triangle*
- *obtuse triangle*

1 Cut three narrow strips of paper so that their lengths are 4 in., 5 in., and 8 in. Then lay the papers on a flat surface and try to form a triangle. Is it possible?

2 Now repeat Step 1, this time using papers of lengths 3 in., 5 in., and 8 in. Is it possible to make a triangle?

3 Without actually cutting them out, predict whether strips of paper of the given measures would form a triangle.

    a. 2 in., 3 in., 5 in.   **b.** 2 in., 4 in., 5 in.
    c. 3 in., 4 in., 8 in.   **d.** 4 in., 5 in., 7 in.

4 Compare your answers to Step 3 with those of other students in your class. Are your answers the same? If they are not the same, can you give a convincing argument to support your answers?

5 Complete this statement: In a triangle, the sum of the lengths of any two sides must be __?__ the length of the third side.

Triangles are often classified by the measures of their sides.

scalene triangle	**isosceles triangle**	**equilateral triangle**
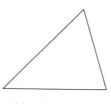	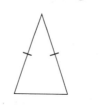	
No sides have the same length.	At least two sides have the same length.	All three sides have the same length.

An equilateral triangle is also *equiangular*. This means that all the angles have the same measure. Because all sides have the same length and all angles have the same measure, the term *equilateral triangle* is really a special name for a *regular* triangle.

For three line segments to be the sides of a triangle, there must be a specific relationship among their lengths. This relationship among the sides of a triangle is important in so many applications of geometry that it is given the special name of *The Triangle Inequality*.

The Triangle Inequality

In any triangle, the sum of the lengths of any two sides is greater than the length of the third side.

Example 1

Tell whether line segments of the given lengths *can* or *cannot* be the sides of a triangle. If they can, tell whether the triangle would be *scalene*, *isosceles*, or *equilateral*.

a. 8 ft, 7 ft, 9 ft   **b.** 9 m, 3 m, 4 m

Solution

Compare each sum of two lengths to the third length.

a. 8 + 7 _?_ 9   8 + 9 _?_ 7   7 + 9 _?_ 8
            15 > 9   17 > 7   16 > 8

Each sum of two lengths is greater than the third, so the line segments *can* be the sides of a triangle. No lengths are the same, so the triangle is *scalene*.

b. 9 + 3 _?_ 4   9 + 4 _?_ 3   3 + 4 _?_ 9
            12 > 4   13 > 3   7 < 9

The sum 3 + 4 is less than the third length, 9, so the line segments *cannot* be the sides of a triangle.

7 ft, 9 ft, 8 ft

4 m, 3 m, 9 m

✔ Check Your Understanding

1. In part (b) of Example 1, describe a way to form a triangle by changing just one of the given lengths.

Another way to classify triangles is by the measures of their angles.

acute triangle	**right triangle**	**obtuse triangle**
All angles are acute angles.	One angle is a right angle.	One angle is an obtuse angle.

Example 2

The measures of two angles of a triangle are 28° and 40°. Tell whether the triangle is *acute*, *right*, or *obtuse*.

Solution

Find the measure of the third angle.

Add the known measures: 28° + 40° = 68°

Subtract this sum from 180°: 180° − 68° = 112°

The third angle measure is 112°, so the triangle is *obtuse*.

28°, 40°, 112°

✔ Check Your Understanding

2. What does 180° represent in the solution of Example 2?
3. In Example 2, how do you know that the triangle is obtuse?

You can find the measure of the third angle of a triangle using the memory keys on a calculator. Here is a key sequence that you might use for Example 2.

Guided Practice

COMMUNICATION « *Writing*

Classify each triangle first by its sides, then by its angles.

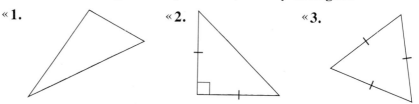

« **1.**   « **2.**   « **3.**

Use a pencil and straightedge to sketch an example of a triangle of the given classifications.

« **4.** scalene, right   « **5.** isosceles, acute   « **6.** equilateral, acute

Tell whether line segments of the given lengths *can* or *cannot* be the sides of a triangle. If they can, tell whether the triangle would be *scalene*, *isosceles*, or *equilateral*.

7. 12 ft, 3 ft, 11 ft

8. 4 m, 1.3 m, 2.7 m

The measures of two angles of a triangle are given. Tell whether the triangle is *acute*, *right*, or *obtuse*.

9. 48°, 16°

10. 61°, 29°

Exercises

Tell whether line segments of the given lengths *can* or *cannot* be the sides of a triangle. If they can, tell whether the triangle would be *scalene*, *isosceles*, or *equilateral*.

1. 3 km, 1 km, 4 km

2. 1 ft, 1 ft, 1 ft

3. 5 in., 10 in., 13 in.

4. 24 yd, 13 yd, 5 yd

5. 8 m, 3.1 m, 3.1 m

6. 6 cm, 4.5 cm, 4.5 cm

7. 7.1 cm, 6.4 cm, 7.1 cm

8. 9.81 m, 16 m, 6.19 m

9. 1.5 mi, 1.5 mi, 1.5 mi

10. 6 mm, 9 mm, 4 mm

The measures of two angles of a triangle are given. Tell whether the triangle is *acute*, *right*, or *obtuse*.

11. 27°, 141° **12.** 90°, 52° **13.** 69°, 53°

14. 24°, 26° **15.** 67°, 23° **16.** 39°, 66°

17. 50°, 50° **18.** 34°, 56° **19.** 7°, 39°

20. Is it possible for a triangle to have two right angles? Explain.

21. Can an isosceles triangle have sides that measure 8 cm, 4 cm, and 6 cm? Explain.

Use the figure at the right. Name all the triangles of the given classification.

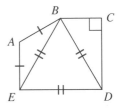

22. acute **23.** obtuse

24. scalene **25.** isosceles

26. right **27.** equilateral

In the figure at the right, $\overleftrightarrow{AB} \parallel \overleftrightarrow{XY}$. Find the measure of each angle.

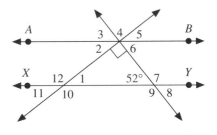

28. ∠1 **29.** ∠2

30. ∠3 **31.** ∠4

32. ∠5 **33.** ∠6

34. ∠7 **35.** ∠8

36. ∠9 **37.** ∠10

38. ∠11 **39.** ∠12

LOGICAL REASONING

Tell whether each statement is *True* or *False*. If the statement is false, give a reason that supports your answer.

40. An obtuse triangle can have a right angle.

41. An equilateral triangle is also an isosceles triangle.

42. A right triangle can be a scalene triangle.

43. An obtuse triangle can have more than one obtuse angle.

44. An equilateral triangle can be a scalene triangle.

45. An isosceles triangle can be a right triangle.

46. A scalene triangle can never be an isosceles triangle.

47. An acute triangle can never be an equilateral triangle.

COMPUTER APPLICATION

The program at the right is written in the Logo computer language. It will instruct a computer to draw an equilateral triangle like the one shown.

FD 40 RT 120
FD 40 RT 120
FD 40 RT 120

In this program, the command FD instructs the Logo "turtle" to go forward a specified distance. The command RT turns the turtle to the right a specified number of degrees. Notice that the angles used in this program must be the *supplements* of the angles of the triangle.

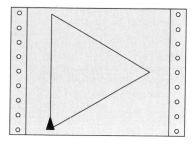

48. Write the Logo command that instructs the turtle to move forward a distance of 65.

49. Write the Logo command that creates a 45° angle for the triangle in the program above.

Use what you know about the properties of triangles. Explain why each of these Logo programs will *not* output a triangle.

50.
FD 20 RT 100
FD 50 RT 100
FD 20 RT 160

51.
FD 62 RT 115
FD 62 RT 115
FD 62 RT 115

Sketch the triangle that you think will be output when each program is run. Then, if you have access to a computer and the Logo language, use them to run the program and check your answer.

52.
FD 40 RT 90
FD 30 RT 127
FD 50 RT 143

53.
FD 40 RT 103
FD 90 RT 154
FD 90 RT 103

SPIRAL REVIEW

54. Use the formula $P = 2l + 2w$. Let $l = 11$ cm and $P = 34$ cm. Find w. *(Lesson 4-10)*

55. The measures of two angles of a triangle are 73° and 17°. Tell whether the triangle is *acute, right,* or *obtuse.* *(Lesson 6-6)*

56. Find the quotient: $\frac{6}{7} \div \frac{2}{3}$ *(Toolbox Skill 21)*

57. Find the difference: $-13 - (-24)$ *(Lesson 3-4)*

Challenge

How many times each day are the hands of a clock perpendicular?

Quadrilaterals

Objective: To identify special quadrilaterals and apply their properties.

DATA ANALYSIS

Many mathematics texts contain a reference chart similar to this one.

Terms to Know

- *parallelogram*
- *trapezoid*
- *rectangle*
- *rhombus*
- *square*

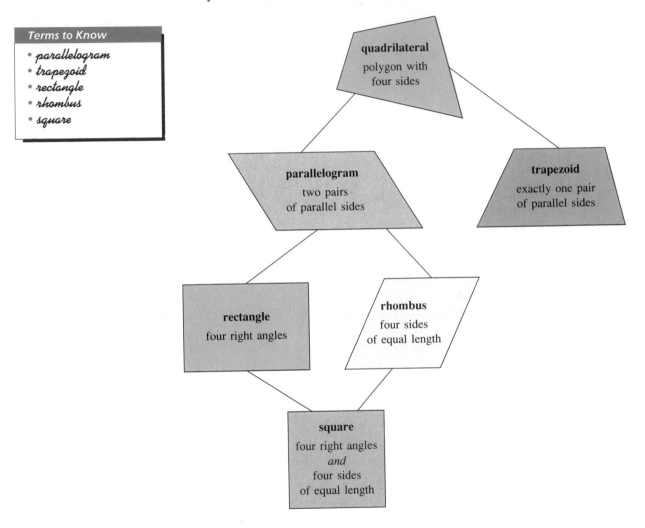

The chart contains data about quadrilaterals. Here is how you read the definition of a *rectangle* from the chart.

A rectangle is a quadrilateral with four right angles.

You can also use the chart to determine relationships among the special types of quadrilaterals. Here is an example.

A rectangle is a type of parallelogram.

A chart like the one on the previous page can contain only a limited amount of data. Many quadrilaterals have other properties that are not listed. Here are two important properties of the parallelogram.

The opposite sides of a parallelogram have the same length.
The opposite angles of a parallelogram have the same measure.

Example

In the figure at the right, ABCD is a parallelogram.
a. Find the length of $\overline{AB}$.
b. Find $m\angle 1$.

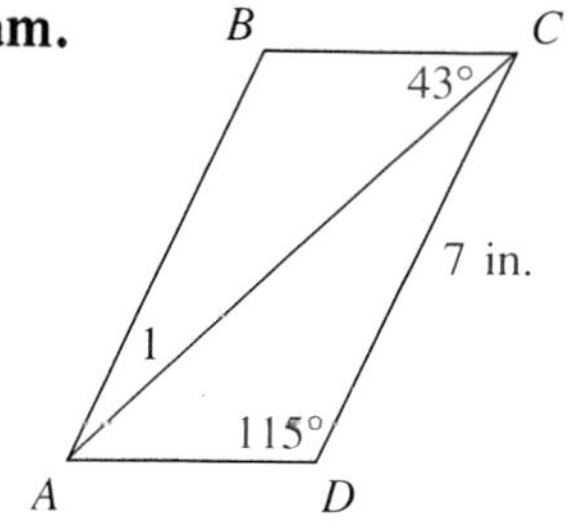

Solution

a. Opposite sides of a parallelogram have the same length, so $\overline{AB}$ is equal in length to $\overline{CD}$. The length of $\overline{AB}$ is 7 in.

b. Opposite angles of a parallelogram have the same measure, so $m\angle B = m\angle D = 115°$.
  In triangle ABC, the sum of the measures of the angles must be 180°. The measures of $\angle B$ and $\angle ACB$ are known, so subtract the sum of their measures from 180°.
    $m\angle B + m\angle ACB = 115° + 43° = 158°$
    $m\angle 1 = 180° - 158° = 22°$

Check Your Understanding

1. In part (a) of the Example, is it possible to find $m\angle DAC$? Explain.
2. Explain why 180° was used in part (b) of the Example.

Guided Practice

COMMUNICATION « *Reading*

Refer to the chart on page 266.

«1. What is the definition of a *square*?

«2. What is the difference between a parallelogram and a trapezoid?

«3. Is a rectangle a type of rhombus?

Refer to the terms defined in the chart on page 266. List *all* the names that apply to each figure.

4.   5.   6.

In the figure at the right, *RSTU* is a trapezoid.

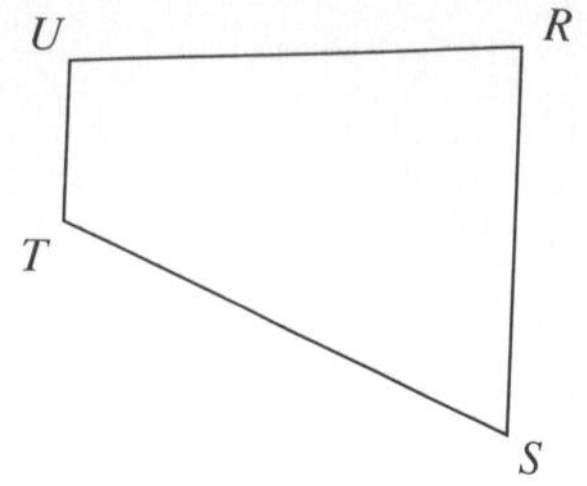

7. Name all pairs of parallel sides.

8. Name the side that is opposite $\overline{RU}$.

9. Name all pairs of opposite angles.

10. Are opposite angles equal in measure?

In the figure at the right, *JKLM* is a rhombus. Find each measure.

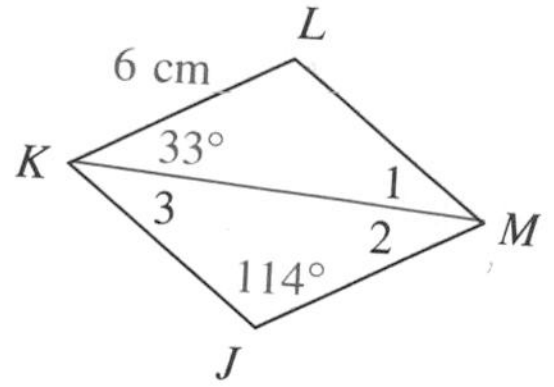

11. the length of $\overline{JM}$ 12. $m\angle L$

13. the length of $\overline{LM}$ 14. $m\angle 1$

15. $m\angle 2$ 16. $m\angle 3$

Exercises

In the figure at the right, *XYZW* is a rectangle. Find each measure.

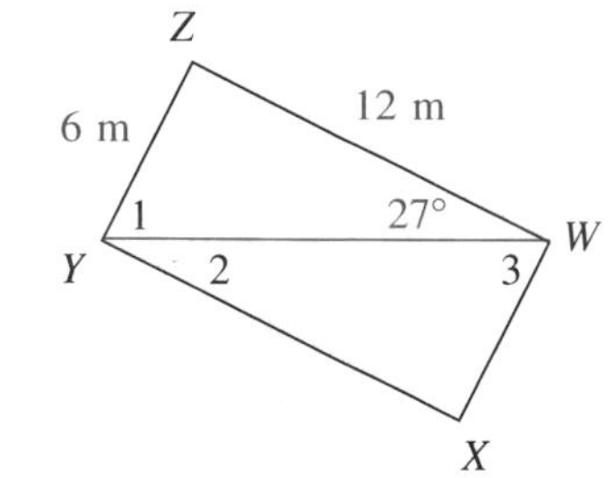

1. the length of $\overline{WX}$ 2. $m\angle Z$

3. the length of $\overline{XY}$ 4. $m\angle 1$

5. $m\angle 2$ 6. $m\angle 3$

In the figure at the right, *ACDF* is a parallelogram and *ABDE* is a square. Find each measure.

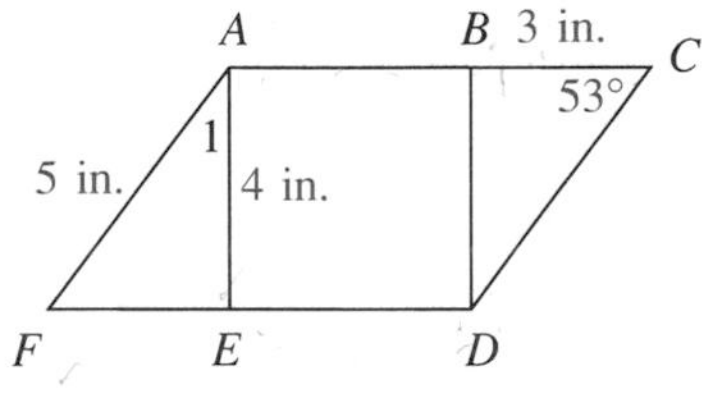

7. the length of $\overline{BD}$ 8. $m\angle BAE$

9. the length of $\overline{DE}$ 10. $m\angle F$

11. the length of $\overline{CD}$ 12. $m\angle 1$

13. the length of $\overline{DF}$ 14. $m\angle FDC$

15. In parallelogram *GHJK*, $m\angle J = 29°$. Find $m\angle G$, $m\angle H$, and $m\angle K$.

16. In rectangle *PQRS*, name the side equal in length to $\overline{QR}$.

17. *MNOP* is a trapezoid. Find the measures of all the numbered angles.

18. *QRST* is a parallelogram. Find the measures of all the numbered angles.

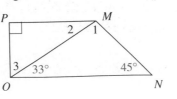

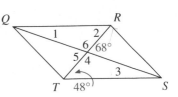

Rectangle
Definition: a quadrilateral with four right angles

Parallel sides: none   one pair   (two pairs)
Opposite sides equal: (yes)   no
Opposite angles equal: (yes)   no
Four right angles: (yes)   no
Four equal sides: yes   (no)

The index card shown above catalogs the major characteristics of a rectangle. For Exercises 19–21, you can use index cards similar to the one shown, or you might have access to computer database software.

19. Begin a quadrilateral database by copying the card above. Then create a card similar to it for each of the following figures: quadrilateral, parallelogram, trapezoid, rhombus, square.

20. Sort through the database that you created in Exercise 19. List all the figures that have the given characteristic.
    a. two pairs of parallel sides   **b.** four right angles
    c. opposite sides equal   **d.** opposite angles equal

21. In a quadrilateral, two angles that are not opposite are *consecutive angles*. Add the following entry to each card in your file.

Consecutive angles supplementary: none   two pairs   four pairs

Working with others in your group, determine how to complete this entry for each figure in your database.

SPIRAL REVIEW

22. Estimate the difference: $887 - 519$
(Toolbox Skill 1)

23. Solve: $\frac{x}{3} = -17$ *(Lesson 4-4)*

24. In the figure at the right, *CDEF* is a rhombus. Find the length of $\overline{EF}$ and $m\angle 1$.
(Lesson 6-7)

25. *True or False:* $\overline{XY}$ is the symbol for line *XY*.   *(Lesson 6-1)*

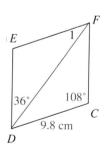

Using Geometric Drawing Software

Objective: To use geometric drawing software to test conjectures.

With **geometric drawing software,** you can use a computer to make accurate drawings of plane geometric figures. You also can instruct the computer to measure line segments and angles in the figures. Because the computer performs these tasks so quickly, you can use this software to test conjectures through experimentation.

Exercises

In Exercises 1–4, use geometric drawing software to help you test this conjecture: *The diagonals of a rectangle are equal in measure.*

Experiment Number	$\overline{AC}$	$\overline{BD}$
1	?	?
2	?	?
3	?	?
4	?	?

1. In your notebook, make a table like the one shown at the right.

2. Draw a rectangle *ABCD* on the computer screen. Find the measure of diagonals $\overline{AC}$ and $\overline{BD}$. Record the results in your table as Experiment 1.

3. Repeat Exercise 2 for three different rectangles. Record these results in your table as Experiments 2, 3, and 4.

4. Arrive at a conclusion: Is the conjecture *true* or *false*?

In Exercises 5 and 6, use geometric drawing software to test the conjecture. Then tell whether the conjecture is *true* or *false*.

5. The diagonals of a parallelogram are equal in measure.

6. The diagonals of a rhombus are perpendicular.

7. To **bisect** a line segment means to separate it into two line segments that are equal in measure. Using the following statement, make a conjecture by replacing the __?__ with the name of one of the special quadrilaterals.

    The diagonals of a __?__ bisect each other.

    Now test your conjecture. Was it true or false?

8. In an **isosceles trapezoid,** the nonparallel sides are equal in measure. Make a conjecture about the diagonals of an isosceles trapezoid. Now test your conjecture. Was it true or false?

Strategy:
Identifying a Pattern

Objective: To solve problems by identifying a pattern.

A useful strategy for solving many problems in mathematics is to look for a pattern in the given information. This strategy may be especially helpful when the problem is one that involves geometric figures.

Problem

The figures below show all the diagonals of a triangle, quadrilateral, pentagon, and hexagon. How many diagonals does a decagon have?

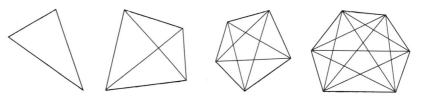

Solution

UNDERSTAND The problem is about the diagonals of polygons.
Facts: all diagonals are shown for 3, 4, 5, and 6 sides
Find: the number of diagonals when there are 10 sides

PLAN Count the diagonals in each of the given figures. Make a table that pairs the number of diagonals with the number of sides for each figure. Then look for a pattern among the numbers in the table.

WORK

Number of Sides	3	4	5	6
Number of Diagonals	0	2	5	9

$+2 \quad +3 \quad +4$

The pattern of the number of diagonals is *add 2, add 3, add 4,* and so on. Using this pattern, extend the table until the number of sides is 10.

Number of Sides	3	4	5	6	7	8	9	10
Number of Diagonals	0	2	5	9	14	20	27	35

$+2 \quad +3 \quad +4 \quad +5 \quad +6 \quad +7 \quad +8$

ANSWER A decagon has 35 diagonals.

Look Back Often you can *generalize* a pattern by using a variable expression. For instance, to find the number of diagonals of a polygon with n sides, you can use the expression $\frac{n(n-3)}{2}$. Show how to use this expression to find how many diagonals a decagon has.

Guided Practice

Use this problem for Exercises 1–6.

The figures below show a pattern of acute angles formed by 2, 3, 4, and 5 rays with a common endpoint. In this pattern, how many acute angles are formed by 8 rays with a common endpoint?

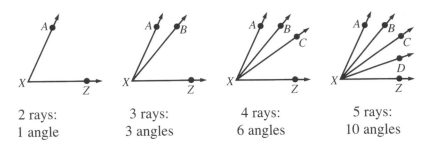

2 rays: 3 rays: 4 rays: 5 rays:
1 angle   3 angles   6 angles   10 angles

COMMUNICATION «*Reading*

«**1.** What is the problem about?

«**2.** What is the question that has to be answered?

3. Complete the following statement: In the figure that shows 3 rays with a common endpoint, the 3 acute angles are ∠*AXB*, ∠*BXZ*, and ∠__?__.

4. In the figure that shows 4 rays with a common endpoint, name the 6 acute angles.

5. Copy and complete this table.

Number of Rays	?	3	4	5	?	?	8
Number of Acute Angles	1	3	?	10	?	?	?

6. What is the solution of the problem?

The figures below represent the first four *oblong numbers*.

7. What is the eighth oblong number?

«**8.** **COMMUNICATION** «*Discussion*   What variable expression can be used to represent the *n*th oblong number?

Problem Solving Situations

Solve by identifying a pattern.

1. The figures below show the first four figures in a pattern of squares. How many squares are in the tenth figure in this pattern?

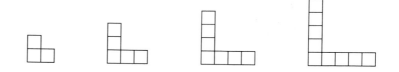

2. The figures below show the number of line segments that can be drawn between 2 points and between 3, 4, and 5 noncollinear points. How many line segments can be drawn between 9 noncollinear points?

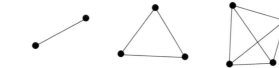

Solve using any problem solving strategy.

3. When a number is decreased by six, the result is −27. What is the number?

4. Jennifer Ng bought a bicycle helmet that cost $31.29. She gave the clerk two $20 bills. How much change did Jennifer receive?

PROBLEM SOLVING

CHECKLIST

Keep these in mind:
Using a Four-Step Plan
Too Much or Not Enough
Information

Consider these strategies:
Choosing the Correct Operation
Making a Table
Guess and Check
Using Equations
Identifying a Pattern

5. Darnel Jefferson has three $1 bills, two $5 bills, and one $20 bill. How many different amounts of money can Darnel make using these bills?

6. The figures below show all the diagonals that can be drawn from one vertex of a quadrilateral, pentagon, hexagon, and heptagon. How many diagonals can be drawn from one vertex of a decagon?

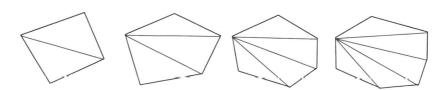

Often you can use a geometric problem to model a real-life situation. For example, Exercises 7–10 refer to the following problem.

Eight people meet at a party. Each person shakes hands with each of the others exactly once. How many handshakes are exchanged?

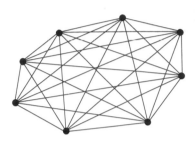

7. A geometric model for this problem is shown at the right. What do the black dots and the red line segments represent?

8. Which geometry problem in this lesson seems most closely related to the handshake problem?

9. Describe how to solve the handshake problem by identifying a pattern. What is the solution?

10. Describe how to solve the handshake problem using a strategy other than identifying a pattern.

WRITING WORD PROBLEMS

Write a problem that you could solve by identifying each pattern. Then solve the problem.

11.

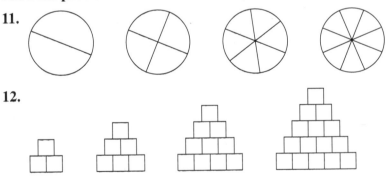

12.

SPIRAL REVIEW

13. Estimate the sum: $238 + 177 + 315 + 150$ *(Toolbox Skill 2)*

14. Solve: $3(2x - 5) = 15$ *(Lesson 4-9)*

15. Draw an angle that measures 82°.   *(Lesson 6-2)*

16. How many small triangles ($\triangle$) are in the ninth figure in this pattern of triangles?   *(Lesson 6-8)*

6-9

Symmetry

Objective: To recognize and find lines of symmetry in plane figures.

Terms to Know
- line symmetry
- line of symmetry

EXPLORATION

1 Fold a rectangular sheet of paper that is not a square three different times. Each time, make the fold as indicated by the dashed lines in the figures below. In each case, does one half of the folded paper fit exactly over the other half?

a.   b.   c.

2 Suppose that your sheet of paper were shaped like a square. How would the results of Step 1 be different?

3 Suppose you had a sheet of paper shaped like the parallelogram in the figure at the right. Do you think you could fold this paper so that one half fits exactly over the other?

4 Suppose your paper were shaped like a circle. How many ways could you fold it so that one half fits exactly over the other?

The figure at the right shows a butterfly with its wings in two different positions. When the butterfly flaps its wings, it is as if the butterfly "folds" itself along $\overleftrightarrow{AB}$. One half of the butterfly seems to fit exactly over the other half. Mathematically, the butterfly's shape has **line symmetry**, and $\overleftrightarrow{AB}$ is called a **line of symmetry**.

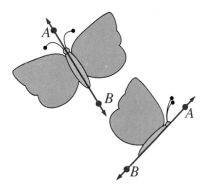

Example 1

Is $\overleftrightarrow{AB}$ a line of symmetry? Write *Yes* or *No*.

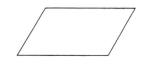

Solution

✓ **Check Your Understanding**

a. Yes.   b. No.   c. Yes.

1. In part (b) of Example 1, why isn't $\overleftrightarrow{AB}$ a line of symmetry?
2. In part (c) of Example 1, is $\overleftrightarrow{AB}$ the only line of symmetry?

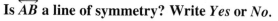

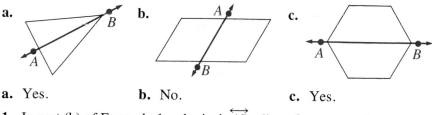

Some geometric figures have many lines of symmetry, while others have no lines of symmetry.

Example 2   **Draw all the lines of symmetry in each figure.**

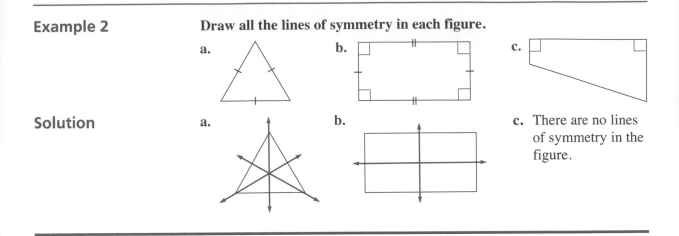

a.

b.

c.

Solution   a.

b.

c. There are no lines of symmetry in the figure.

Guided Practice

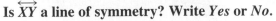

COMMUNICATION «*Reading*

«**1.** What is the main idea of this lesson?

«**2.** What is the main difference between Example 1 and Example 2?
(*Hint:* It may help to refer to the objective of the lesson.)

Does each pictured item have line symmetry? Explain.

3.   4.

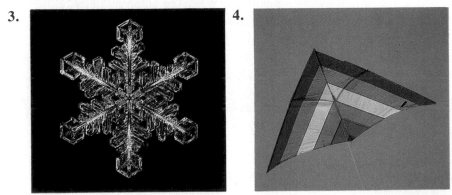

Is $\overleftrightarrow{XY}$ a line of symmetry? Write *Yes* or *No*.

5.   6.   7.

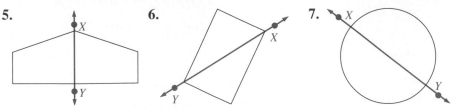

Trace each figure onto a piece of paper. Then draw all the lines of symmetry. If there are no lines of symmetry, write *None*.

8. **9.** **10.**

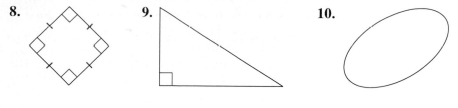

Exercises

Is $\overleftrightarrow{MN}$ a line of symmetry? Write *Yes* or *No*.

1. **2.** **3.**

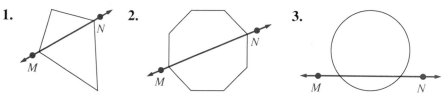

Trace each figure onto a piece of paper. Then draw all the lines of symmetry. If there are no lines of symmetry, write *None*.

4. **5.** **6.**

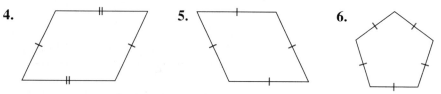

7. Identify the type of quadrilateral that has four lines of symmetry.

8. What type of triangle has exactly one line of symmetry?

Trace each figure onto a piece of paper. Then complete the figure so that $\overleftrightarrow{PQ}$ is a line of symmetry.

9. **10.**

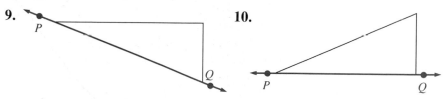

11. PATTERNS Complete this table for regular polygons.

Number of Sides	3	4	5	6	7	8
Number of Lines of Symmetry	?	?	?	?	?	?

12. FUNCTIONS Write a function rule to represent the relationship between the number of sides of a regular polygon and the number of lines of symmetry. Use n to represent the number of sides. (*Hint:* You might want to use the results from Exercise 11.)

CAREER/APPLICATION

A *craftsperson* is skilled in work that is done by hand rather than by machine. Practicing a craft often requires a knowledge of the line symmetry of traditional designs. Pottery, woodworking, sewing, and quilting are examples of the many crafts that maintain cultural heritages.

Each of the following is a sketch of the basic *template* for a quilt design. How many lines of symmetry does each template have?

13.   **14.**   **15.**

16. RESEARCH   Find a Native American design that has line symmetry. Trace the design and draw in all the lines of symmetry.

THINKING SKILLS

A figure has **rotational symmetry** if it fits exactly over its original position after being rotated less than a complete turn. A figure does not have to have line symmetry in order to have rotational symmetry. For instance, the figure at the right illustrates the rotational symmetry of a parallelogram. Given a half-turn, with point *A* as the center of the turn, the entire parallelogram *WXYZ* fits exactly over its original position.

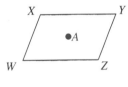

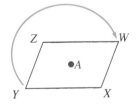

Determine whether each figure has *line symmetry*, *rotational symmetry*, or *both*.

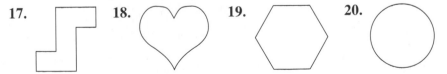

17.   **18.**   **19.**   **20.**

21. WRITING ABOUT MATHEMATICS   Suppose that you are a reporter for a high school mathematics newsletter. Write a feature article entitled *Symmetry and the Special Quadrilaterals*.

22. Find the product: $(-21)(8)$ *(Lesson 3-5)*

23. In the figure at the right, is $\overleftrightarrow{RS}$ a line of
symmetry?   *(Lesson 6-9)*

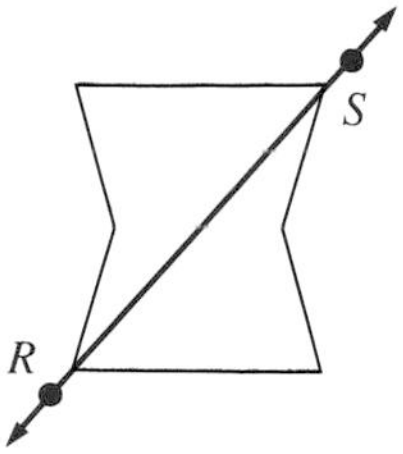

24. Decide whether it would be more appropriate
to draw a *bar graph* or a *line graph* to display
the data. Then draw the graph.   *(Lesson 5-5)*

Evening TV Viewers

Time	6:00 P.M.	8:00 P.M.	10:00 P.M.	12:00 A.M.
Viewers	45,000	65,000	50,000	10,000

Self-Test 2

Identify each polygon. Then list all its diagonals.

1.   **2.**   **6-5**

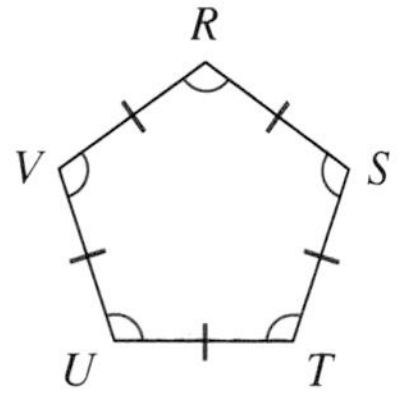

3. Tell whether line segments of lengths 9 ft, 7 ft, and 2 ft *can* or   **6-6**
cannot be the sides of a triangle. If they can, tell whether the
triangle would be *scalene*, *isosceles*, or *equilateral*.

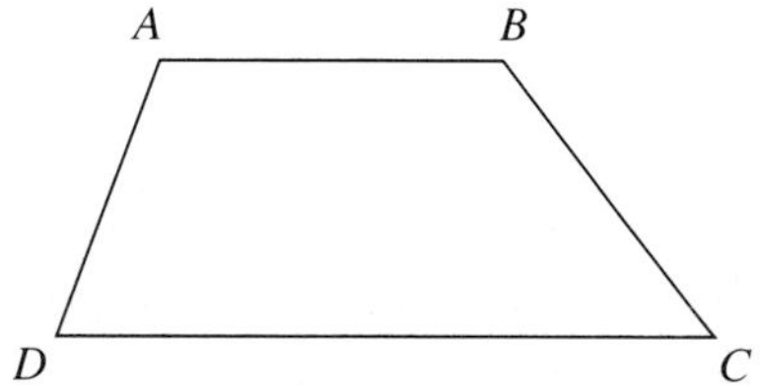

4. The measures of two angles of a triangle are 57° and 44°. Tell
whether the triangle is *acute*, *right*, or *obtuse*.

In the figure at the right,
ABCD **is a rectangle.**

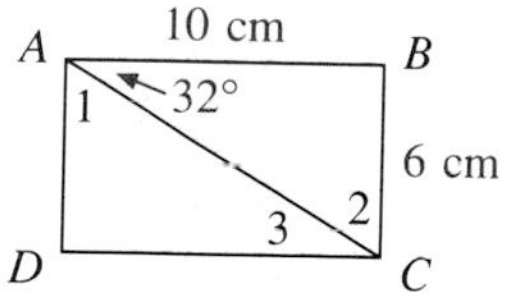

5. Find the length of $\overline{CD}$.   **6-7**

6. Find $m\angle 2$.

7. How many line segments are there if there are 8 points?   **6-8**

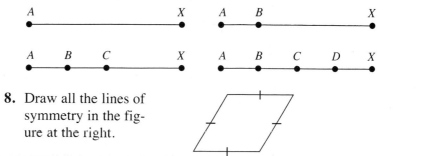

8. Draw all the lines of   **6-9**
symmetry in the fig-
ure at the right.

point (p. 236)
line (p. 236)
plane (p. 236)
collinear (p. 236)
intersect (p. 236)
line segment (p. 237)
endpoint (p. 237)
ray (p. 237)
angle (p. 237)
vertex of an angle (p. 237)
side of an angle (p. 237)
degree (p. 240)
protractor (p. 240)
acute angle (p. 245)
right angle (p. 245)
obtuse angle (p. 245)
straight angle (p. 245)
complementary (p. 245)
supplementary (p. 245)
adjacent (p. 246)
vertical (p. 246)
perpendicular (p. 249)

parallel (p. 249)
transversal (p. 250)
alternate interior angles (p. 250)
corresponding angles (p. 250)
polygon (p. 256)
side of a polygon (p. 256)
vertex of a polygon (p. 256)
diagonal (p. 256)
regular polygon (p. 257)
scalene triangle (p. 261)
isosceles triangle (p. 261)
equilateral triangle (p. 261)
The Triangle Inequality (p. 261)
acute triangle (p. 262)
right triangle (p. 262)
obtuse triangle (p. 262)
parallelogram (p. 266)
trapezoid (p. 266)
rectangle (p. 266)
rhombus (p. 266)
square (p. 266)
line symmetry (p. 275)
line of symmetry (p. 275)

Choose the correct term from the list above to complete each sentence.

1. A part of a line with two endpoints is called a(n) __?__.

2. Two lines in the same plane that do not intersect are __?__.

3. A(n) __?__ has a measure of 180°.

4. A rectangle with four sides that are the same length is a(n) __?__.

5. A quadrilateral with exactly one set of parallel sides is a(n) __?__.

6. A(n) __?__ has three sides of the same length.

7. Two angles whose measures have a sum of 90° are called __?__.

8. A(n) __?__ is an instrument used to measure angles.

9. A(n) __?__ has a measure greater than 90°, but less than 180°.

10. A triangle all of whose angles are less than 90° is a(n) __?__.

Write the name of each figure.   *(Lesson 6-1)*

11. G • —————— • H

12. Q ∠ P, R

13. (parallelogram) J

Use a protractor to measure each angle. *(Lesson 6-2)*

14.

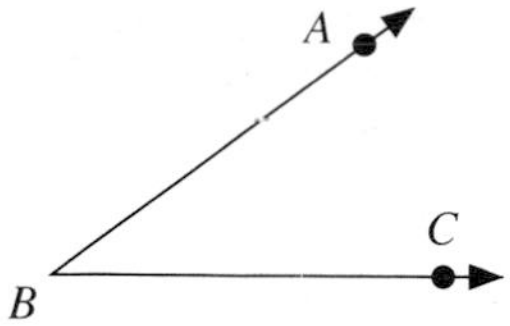

15.

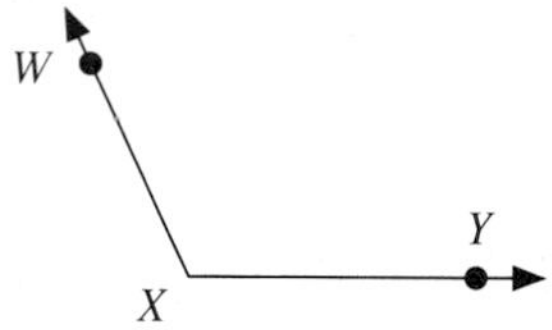

Use a protractor to draw an angle of the given measure. *(Lesson 6-2)*

16. 25°   **17.** 95°   **18.** 165°   **19.** 62°

Find the measure of a complement of an angle of the given measure.
(Lesson 6-3)

20. 24°   **21.** 45°   **22.** 73°   **23.** 88°

Find the measure of a supplement of an angle of the given measure.
(Lesson 6-3)

24. 30°   **25.** 75°   **26.** 85°   **27.** 122°

**In the figure at the right, $\overleftrightarrow{AB} \parallel \overleftrightarrow{CD}$. Find the
measure of each angle.** *(Lessons 6-3 and 6-4)*

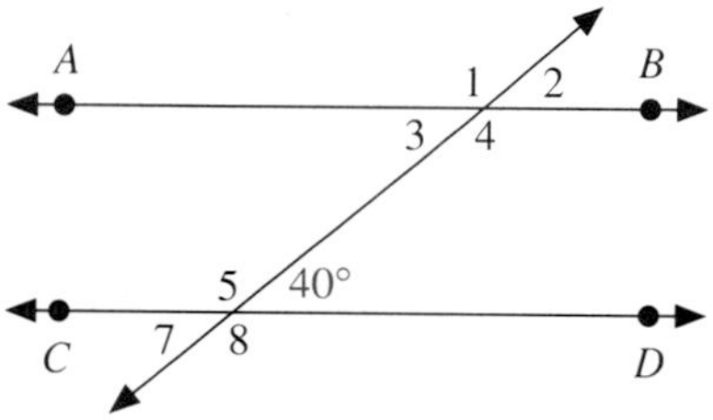

28. $\angle 2$ **29.** $\angle 5$

30. $\angle 4$ **31.** $\angle 1$

32. $\angle 7$ **33.** $\angle 8$

**Use the figure at the right. Tell whether each
statement is *True* or *False*.** *(Lesson 6-4)*

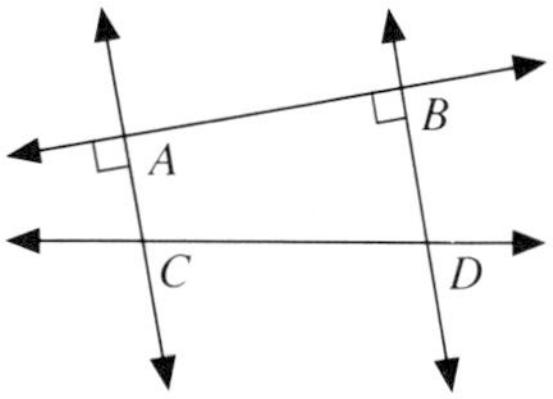

34. $\overleftrightarrow{AB} \parallel \overleftrightarrow{CD}$ **35.** $\overleftrightarrow{AC} \parallel \overleftrightarrow{BD}$

36. $\overleftrightarrow{BD} \perp \overleftrightarrow{AB}$ **37.** $\overleftrightarrow{CD} \perp \overleftrightarrow{AB}$

38. $\overleftrightarrow{AC} \perp \overleftrightarrow{AB}$ **39.** $\overleftrightarrow{AC} \parallel \overleftrightarrow{AB}$

Identify each polygon. Then list all its diagonals. *(Lesson 6-5)*

40.

41.

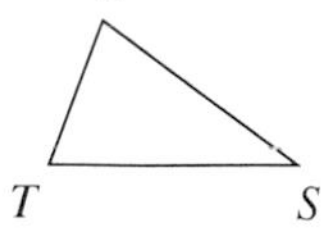

42.

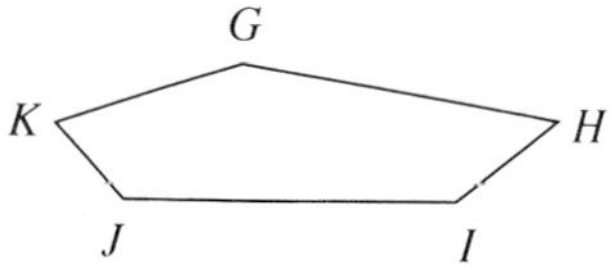

Find each unknown angle measure. *(Lesson 6-5)*

43.

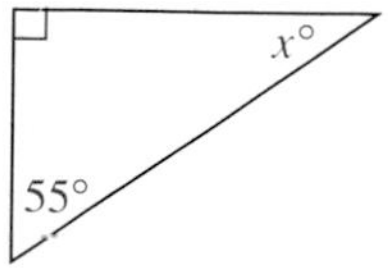

44.

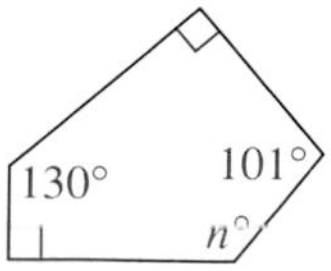

45.

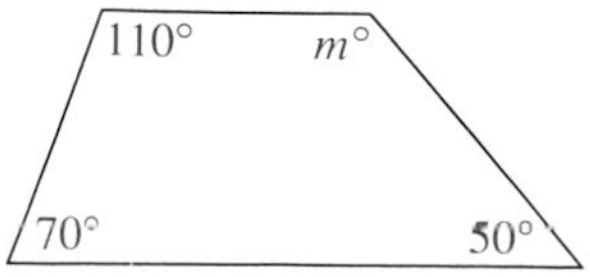

Tell whether line segments of the given lengths *can* or *cannot* be the sides of a triangle. If they can, tell whether the triangle would be *scalene*, *isosceles*, or *equilateral*. *(Lesson 6-6)*

46. 8 cm, 8 cm, 8 cm   **47.** 7 in., 6 in., 11 in.   **48.** 7 mm, 24 mm, 25 mm

The measures of two angles of a triangle are given. Tell whether the triangle is *acute*, *right*, or *obtuse*. *(Lesson 6-6)*

49. 15°, 52°   **50.** 45°, 45°   **51.** 38°, 67°   **52.** 75°, 29°

In the figure at the right, *ABCD* is a rhombus. Find each measure. *(Lesson 6-7)*

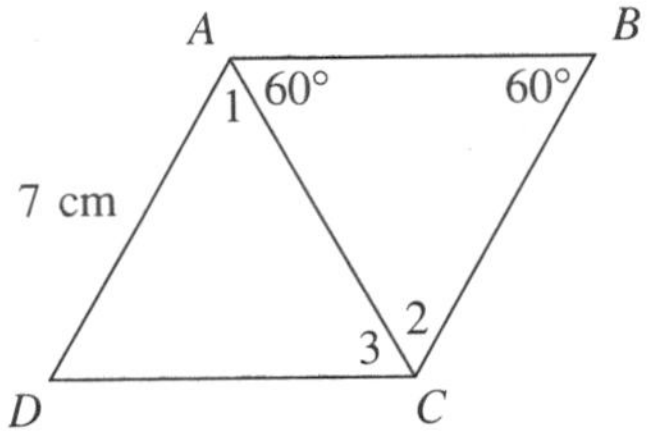

53. $m\angle 2$ **54.** $m\angle 3$

55. $m\angle 1$ **56.** the length of $\overline{BC}$

57. $m\angle D$ **58.** the length of $\overline{CD}$

Solve by identifying a pattern. *(Lesson 6-8)*

59. The figures below show the first four figures in a pattern of squares. How many small squares are in the ninth figure of this pattern?

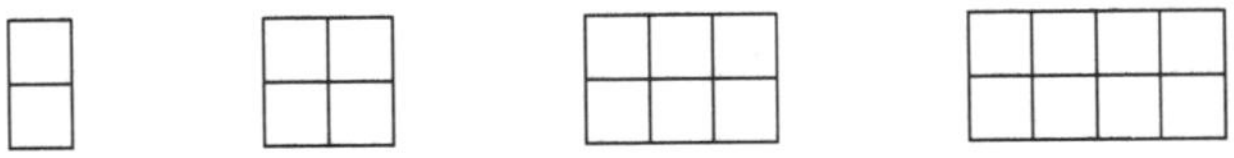

Is $\overleftrightarrow{XY}$ a line of symmetry? Write *Yes* or *No*. *(Lesson 6-9)*

60.   **61.** 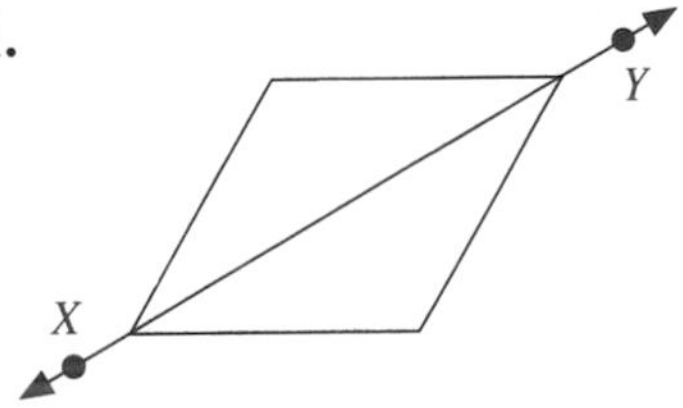   **62.**

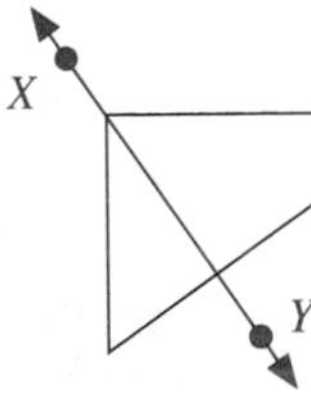

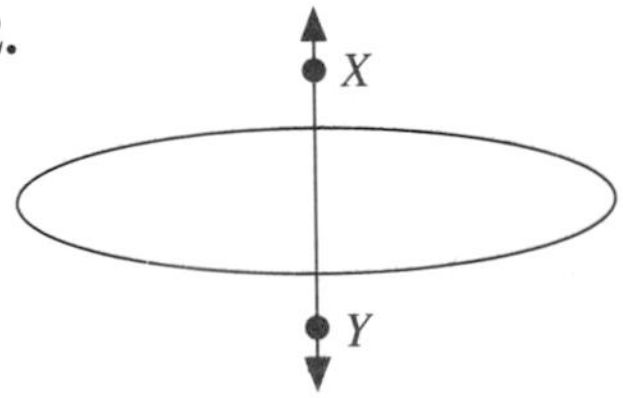

Trace each figure onto a piece of paper. Then draw all the lines of symmetry in each figure. If there are no lines of symmetry, write *none*. *(Lesson 6-9)*

63.   **64.**   **65.**

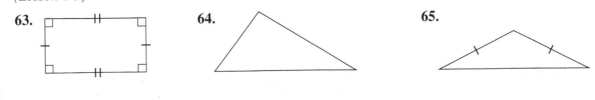

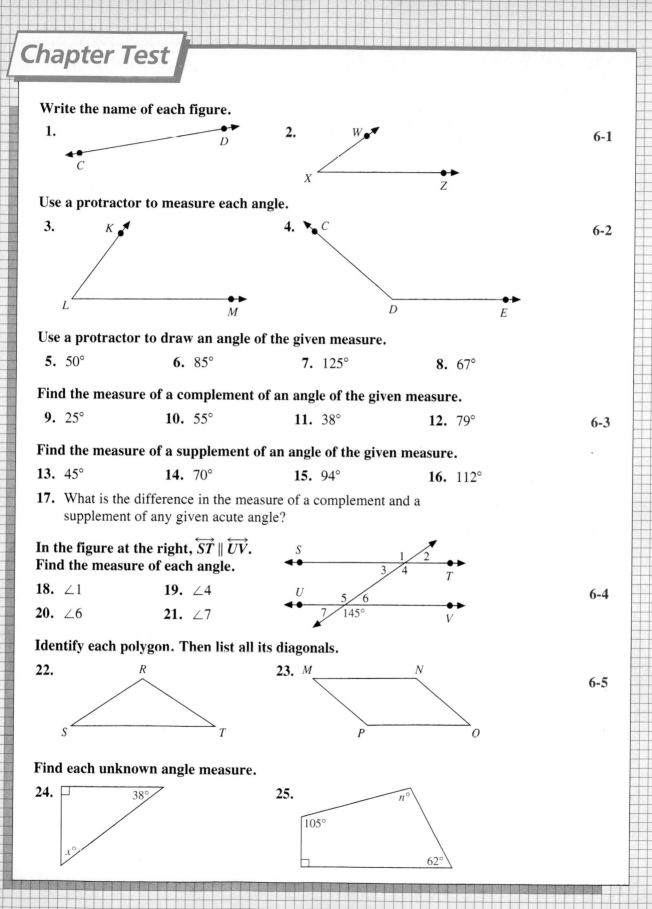

Chapter Test

Write the name of each figure.

1.

C *D*

2.

W *X* *Z*

6-1

Use a protractor to measure each angle.

3.

K *L* *M*

4.

C *D* *E*

6-2

Use a protractor to draw an angle of the given measure.

5. 50°   6. 85°   7. 125°   8. 67°

Find the measure of a complement of an angle of the given measure.

9. 25°   10. 55°   11. 38°   12. 79°

6-3

Find the measure of a supplement of an angle of the given measure.

13. 45°   14. 70°   15. 94°   16. 112°

17. What is the difference in the measure of a complement and a
supplement of any given acute angle?

**In the figure at the right, $\overleftrightarrow{ST} \parallel \overleftrightarrow{UV}$.
Find the measure of each angle.**

18. ∠1   19. ∠4

20. ∠6   21. ∠7

6-4

S 1 2 3 4 *T* *U* 5 6 7 145° *V*

Identify each polygon. Then list all its diagonals.

22.

R *S* *T*

23.

M *N* *P* *O*

6-5

Find each unknown angle measure.

24.

38° *x*°

25.

n° 105° 62°

Tell whether line segments of the given lengths *can* or *cannot* be the sides of a triangle. If they can, tell whether the triangle would be *scalene*, *isosceles*, or *equilateral*.

26. 3 ft, 5 ft, 9 ft   **27.** 6 m, 8 m, 10 m   **28.** 8 cm, 13 cm, 8 cm   **6-6**

The measures of two angles of a triangle are given. Tell whether the triangle is *acute*, *right*, or *obtuse*.

29. 26°, 75°   **30.** 31°, 44°   **31.** 62°, 28°

32. How many acute angles can there be in a triangle? how many obtuse angles? Explain.

In the figure at the right, *WXYZ* is a square. Find each measure.

33. $m\angle 1$ **34.** the length of $\overline{XY}$

35. $m\angle 2$ **36.** the length of $\overline{ZY}$

6-7

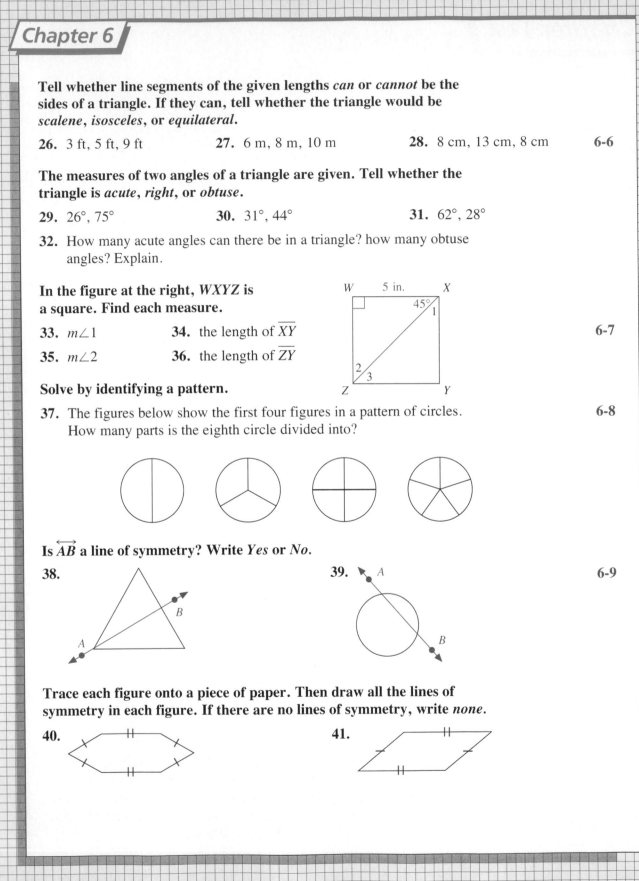

Solve by identifying a pattern.

37. The figures below show the first four figures in a pattern of circles. How many parts is the eighth circle divided into?   **6-8**

Is $\overleftrightarrow{AB}$ a line of symmetry? Write *Yes* or *No*.

38.   **39.**   **6-9**

Trace each figure onto a piece of paper. Then draw all the lines of symmetry in each figure. If there are no lines of symmetry, write *none*.

40.   **41.**

Tessellations

Objective: To recognize and name regular and semiregular tessellations. .

Terms to Know
• *tessellation*

A **tessellation** is a pattern in which identical copies of a figure cover a plane without gaps or overlaps. When the figure is a regular polygon, the pattern is a *regular* tessellation. A pattern formed by two or more types of regular polygons is a *semiregular* tessellation. For a pattern to be a regular or semiregular tessellation, the arrangement of regular polygons at every vertex must be identical.

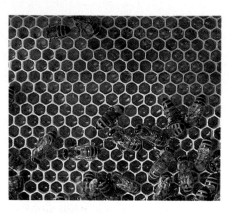

Example

Write the code for the tessellation in the honeycomb above.

Solution

The arrangement at every vertex is hexagon-hexagon-hexagon. Because a hexagon has 6 sides, the code is 6-6-6.

Exercises

Write the code for each tessellation.

1.

2.

Use the regular polygons at the right. Make a *template* for each figure by first tracing it onto heavy paper or cardboard, then cutting it out.

3. Draw a tessellation of equilateral triangles. Write the code for your tessellation.

4. Make a sketch that demonstrates why a tessellation of regular pentagons is impossible.

GROUP ACTIVITY

5. List the five combinations of regular polygons shown that will form a semiregular tessellation.

6. When you add a regular octagon to the set of regular polygons above, one other semiregular tessellation is possible. Create a template for a regular octagon. Then find the other tessellation.

Standardized Testing Practice

Choose the letter of the correct answer.

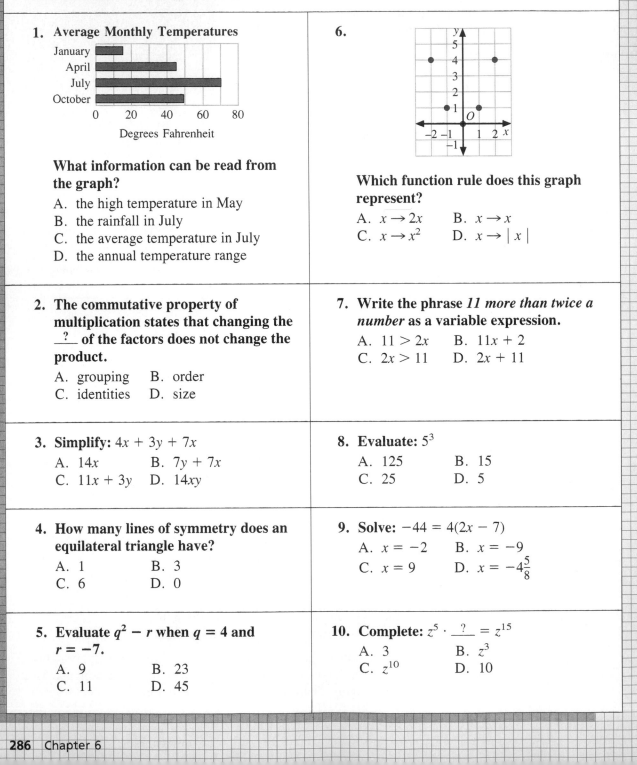

1. **Average Monthly Temperatures**

   What information can be read from the graph?
   A. the high temperature in May
   B. the rainfall in July
   C. the average temperature in July
   D. the annual temperature range

2. **The commutative property of multiplication states that changing the ___?___ of the factors does not change the product.**
   A. grouping   B. order
   C. identities   D. size

3. **Simplify:** $4x + 3y + 7x$
   A. $14x$ B. $7y + 7x$
   C. $11x + 3y$ D. $14xy$

4. **How many lines of symmetry does an equilateral triangle have?**
   A. 1   B. 3
   C. 6   D. 0

5. **Evaluate $q^2 - r$ when $q = 4$ and $r = -7$.**
   A. 9   B. 23
   C. 11   D. 45

6. **Which function rule does this graph represent?**
   A. $x \rightarrow 2x$ B. $x \rightarrow x$
   C. $x \rightarrow x^2$ D. $x \rightarrow |x|$

7. **Write the phrase *11 more than twice a number* as a variable expression.**
   A. $11 > 2x$ B. $11x + 2$
   C. $2x > 11$ D. $2x + 11$

8. **Evaluate:** 5^3
   A. 125   B. 15
   C. 25   D. 5

9. **Solve:** $-44 = 4(2x - 7)$
   A. $x = -2$ B. $x = -9$
   C. $x = 9$ D. $x = -4\frac{5}{8}$

10. **Complete:** $z^5 \cdot \underline{\quad?\quad} = z^{15}$
    A. 3   B. z^3
    C. z^{10} D. 10

11. What does $\overleftrightarrow{WX} \parallel \overleftrightarrow{YZ}$ mean?
  I. The lines are parallel.
  II. The lines are perpendicular.
  III. The lines do not intersect.
  A. I only   B. I and II
  C. I and III   D. II only

12. Evan started practicing piano at
  **3 P.M. and played for 2 h. How many
  songs did he practice?**
  A. 5   B. 6
  C. $1\frac{1}{2}$ D. not enough
                       information

13. At **6 P.M. the temperature is −11°F.
  By 8 P.M. it has fallen 7°F. What is the
  temperature at 8 P.M.?**
  A. −4°F   B. 4°F
  C. 18°F   D. −18°F

14. Choose the data most appropriate for
  a line graph.
  A. voters in areas of a city
  B. school enrollment over 10 years
  C. total sales of three magazines
  D. price per pound of apples in six
     grocery stores

15. The length of a book is about
  **230 mm. About how many centimeters
  long is the book?**
  A. 0.23   B. 2.3
  C. 2300   D. 23

16. Which figure has no lines of
  symmetry?
  A. regular octagon
  b. square
  C. scalene triangle
  D. rhombus

17.

What is a correct name for this figure?
  A. $\angle HZQ$ B. $\angle QHZ$
  C. $\angle 4$ D. $\angle ZQH$

18. The area of a parallelogram is 42 cm²
  and the height is 6 cm. Use the
  formula $A = bh$ to find the base.
  A. 7 cm   B. 9 cm
  C. 252 cm   D. 3 cm

19. A pictograph shows that 900 people
  bought tapes and 600 people bought
  CDs. If 6 symbols represent the people
  who bought tapes, how many people
  does one symbol represent?
  A. 250   B. 150
  C. 100   D. 300

20. Two parallel lines are intersected by a
  transversal. What is the relationship
  between the alternate interior angles?
  A. they are equal in measure
  B. they are complementary
  C. they are adjacent
  D. they are vertical

If you enter a number in the decimal (base 10) system into a computer, the computer will convert it to the binary (base 2) system. The computer will read the number as a combination of only two digits, 0 and 1. These *binary digits* are called *bits*. Computer programmers also use the hexadecimal (base 16) system to represent numbers. This system uses the digits 0, 1, 2, 3, 4, 5, 6, 7, 8, and 9 and the letters A, B, C, D, E, and F.

Representations of the Same Number	
decimal (base 10)	1219
binary (base 2)	10011000011
hexadecimal (base 16)	4C3

Computer Languages	
Name	Name Origin
Ada	Augusta **Ada** Lovelace (1815-1852)
APL	**A P**rogramming **L**anguage
BASIC	**B**eginner's **A**ll-purpose **S**ymbolic **I**nstruction **C**ode
COBOL	**Co**mmon **B**usiness-**O**riented **L**anguage
FORTRAN	**For**mula **Tran**slator
LISP	**Lis**t **P**rocessing
Pascal	Blaise **Pascal** (1623-1662)

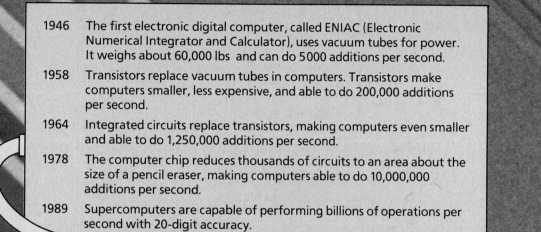

1946   The first electronic digital computer, called ENIAC (Electronic Numerical Integrator and Calculator), uses vacuum tubes for power. It weighs about 60,000 lbs and can do 5000 additions per second.

1958   Transistors replace vacuum tubes in computers. Transistors make computers smaller, less expensive, and able to do 200,000 additions per second.

1964   Integrated circuits replace transistors, making computers even smaller and able to do 1,250,000 additions per second.

1978   The computer chip reduces thousands of circuits to an area about the size of a pencil eraser, making computers able to do 10,000,000 additions per second.

1989   Supercomputers are capable of performing billions of operations per second with 20-digit accuracy.

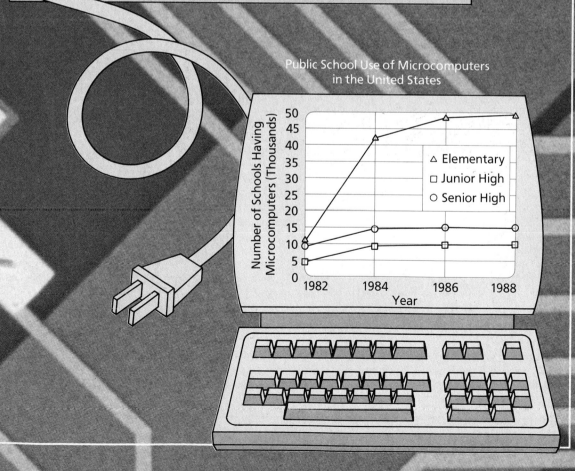

Public School Use of Microcomputers in the United States

Number of Schools Having Microcomputers (Thousands)

Year

△ Elementary
□ Junior High
○ Senior High

Factors and Prime Numbers

Objective: To find the prime factorization of a number.

Terms to Know
- *factor*
- *prime number*
- *composite number*
- *prime factorization*

1 At the right you see a 4 by 6 rectangular arrangement of 24 tiles. Describe all the other rectangular arrangements that can be made using exactly 24 tiles.

2 Describe all the rectangular arrangements that can be made if you have only 23 tiles.

A given number of tiles can be arranged in more than one way only when the given number has more than two *factors*. Recall that when one whole number is divisible by a second whole number, the second number is a **factor** of the first. A whole number greater than 1 with exactly two factors, 1 and the number itself, is called a **prime number**. A **composite number** has more than two factors.

Example 1

Tell whether each number is *prime* or *composite*.

  a. 11   **b.** 21   **c.** 31

Solution

  a. Prime. The only factors of 11 are 1 and 11.

  b. Composite. The factors of 21 are 1, 3, 7, and 21.

  c. Prime. The only factors of 31 are 1 and 31.

✔ Check Your Understanding

1. In Example 1, how many factors does each of the prime numbers have?

When you write a number as a product of prime numbers, you are writing the **prime factorization** of the number.

Example 2

Write the prime factorization of 140.

Solution

Use divisibility rules to help you make a *factor tree*.

$$140 = 2 \cdot 5 \cdot 2 \cdot 7$$
$$= 2 \cdot 2 \cdot 5 \cdot 7$$
$$= 2^2 \cdot 5 \cdot 7$$

Factor tree: 140 branches to 10 and 14; 10 branches to 2 and 5; 14 branches to 2 and 7.

✔ Check Your Understanding

2. In Example 2, how do you know that 140 is divisible by 10?

Guided Practice

Replace each __?__ with the correct phrase.

« **1.** A whole number whose only factors are 1 and itself is a(n) __?__.

« **2.** A number with more than two factors is a(n) __?__.

Rewrite each statement using exponents.

3. $450 = 2 \cdot 3 \cdot 3 \cdot 5 \cdot 5$ **4.** $2000 = 2 \cdot 2 \cdot 5 \cdot 2 \cdot 5 \cdot 2 \cdot 5$

Find all the factors of each number.

5. 48   **6.** 37   **7.** 64   **8.** 100

Tell whether each number is *prime* or *composite*.

9. 2   **10.** 10   **11.** 28   **12.** 17   **13.** 51   **14.** 63

Write the prime factorization of each number.

15. 49   **16.** 15   **17.** 18   **18.** 44   **19.** 90   **20.** 144

COMMUNICATION « *Discussion*

« **21.** List all the divisibility rules that you remember.

« **22.** How do the divisibility rules help you find the prime factorization of 5448?

Exercises

Tell whether each number is *prime* or *composite*.

1. 27   **2.** 32   **3.** 19   **4.** 41

5. 100   **6.** 61   **7.** 47   **8.** 96

Write the prime factorization of each number.

9. 27   **10.** 32   **11.** 22   **12.** 69

13. 20   **14.** 36   **15.** 28   **16.** 52

17. 108   **18.** 132   **19.** 500   **20.** 900

21. 1008   **22.** 624   **23.** 2625   **24.** 2808

25. List all the different ways to write 60 as a product of two whole-number factors. (*Hint:* There are six ways.)

26. What is the prime factorization of 1764?

Replace each __?__ with the expression that makes the statement true.

27. $y^5 = 1 \cdot$ __?__

28. $y^5 = y \cdot$ __?__

29. $y^5 = y^2 \cdot$ __?__

30. Use your answers to Exercises 27–29 to list all the factors of y^5.

31. Make a rule for finding all the factors of any given power a^n.

32. Use the rule you made in Exercise 31 to find all the factors of x^8.

GROUP ACTIVITY

33. Find all the prime numbers between 0 and 100. (*Hint:* There are 25 of these prime numbers.)

34. **a.** Find all the prime numbers between 100 and 200.
b. How many prime numbers are there in part (a)?
c. Are there *more than* or *fewer than* 25 prime numbers in part (a)?

35. **a.** How many prime numbers do you think there are between 200 and 300? Explain.
b. Find all the prime numbers between 200 and 300.
c. How many prime numbers are there in part (b)?
d. How does your answer to part (c) compare with your answer to part (a)?

36. **RESEARCH** The *sieve of Eratosthenes*, a method for finding prime numbers, has been used for more than 2000 years. Find how the sieve is used and report your findings to your group.

SPIRAL REVIEW

37. Find the mean, median, mode(s), and range: 84, 96, 72, 77, 91 (*Lesson 5-8*)

38. Find the sum: $\frac{9}{16} + \frac{3}{4}$ (*Toolbox Skill 17*)

39. Write the prime factorization of 96. (*Lesson 7-1*)

40. Draw a bar graph for the data below. (*Lesson 5-4*)

Cars Registered (Thousands)

State	Maine	Utah	Alaska	Idaho	Nevada
Number of Cars	698	760	225	625	624

41. Find the answer: $-12 - (-12)$ (*Lesson 3-4*)

42. Find the next three expressions in the pattern: $20n + 5$, $18n + 5$, $16n + 5$, __?__, __?__, __?__ (*Lesson 2-5*)

Greatest Common Factor

Objective: To find greatest common factors.

Terms to Know

- *common factor*
- *greatest common factor (GCF)*

John has 28 seedlings to plant and Henrietta has 40. They want to plant two rectangular gardens side by side using the same number of rows in each garden.

QUESTION What is the greatest number of rows they can have in their gardens?

This question can be answered using *common factors*. A number that is a factor of two numbers is called a **common factor** of those two numbers. The greatest number in a list of common factors is called the **greatest common factor (GCF).** The greatest number of rows that John and Henrietta can have in their gardens will equal the GCF of 28 and 40.

Example 1 Find the GCF of 28 and 40.

Solution 1 The factors of 28 are 1, 2, 4, 7, 14, and 28.
The factors of 40 are 1, 2, 4, 5, 8, 10, 20, and 40.
The common factors of 28 and 40 are 1, 2, and 4.
The GCF of 28 and 40 is 4.

Solution 2 Write the prime factorization of each number.

$$28 = 2^2 \cdot 7$$
$$40 = 2^3 \cdot 5$$
$$GCF = 2^2 = 4$$

The common prime factor is 2, ⟵ and the lesser power of 2 is 2^2.

Check Your Understanding

1. In Solution 1 of Example 1, why was 2 not chosen as the GCF?

2. In Solution 2 of Example 1, why were 5 and 7 not used as factors in the GCF?

ANSWER The greatest number of rows that John and Henrietta can have in their gardens is 4.

As shown in Example 1, you can use prime factorizations to find a GCF. To do this, find all the common prime factors. Then form a product using the least power that appears for each factor.

To find the GCF of variable expressions, include in it the least power that appears for each common variable factor.

Example 2	**Find the GCF.**
	a. $12a^4$ and $27a^6$
Solution	**a.** $12a^4 = 2^2 \cdot 3 \cdot a^4$

Find the GCF.

a. $12a^4$ and $27a^6$ **b.** $24xy$ and $84x$

a. $12a^4 = 2^2 \cdot 3 \cdot a^4$
   $27a^6 = 3^3 \cdot a^6$
   GCF $= 3 \cdot a^4 = 3a^4$

b. $24xy = 2^3 \cdot 3 \cdot x \cdot y$
   $84x = 2^2 \cdot 3 \cdot 7 \cdot x$
   GCF $= 2^2 \cdot 3 \cdot x = 12x$

✔ Check Your Understanding

3. In Example 2(a), why was the power a^4 used instead of a^6?
4. In Example 2(b), why was y not used as a factor in the GCF?

Guided Practice

«**1.** COMMUNICATION «*Reading* Replace each __?__ with the correct phrase.

A __?__ of two numbers is a number that is a factor of the two numbers. The __?__ is the greatest number in a list of common factors.

Identify the common prime factor(s).

2. $2^3 \cdot 5$	**3.** $2 \cdot 3^6$	**4.** $3^2 \cdot 5 \cdot 7$	**5.** $5 \cdot 7^2 \cdot 11$
$3 \cdot 5^4$	$3 \cdot 7^2$	$3 \cdot 7^3$	$5^3 \cdot 11^7 \cdot 13$

Identify the lesser power.

6. 2^3 and 2^8 **7.** 3^6 and 3^4 **8.** 11^7 and 11^3 **9.** 5 and 5^2

Find the GCF.

10. 24 and 56   **11.** 14 and 75   **12.** 45, 60, and 80

13. $36m^3$ and $45m^8$ **14.** $60r$ and 72   **15.** $20p^7$, $25p$, and $42p^4$

Exercises

Find the GCF.

1. 2 and 16   **2.** 7 and 14   **3.** 8 and 15

4. 54 and 99   **5.** 40 and 100   **6.** 50 and 81

7. 16, 24, and 72   **8.** 45, 72, and 108   **9.** $90x$ and $96x$

Find the GCF.

10. $48a$ and $51a$

11. $27bc$ and $49bd$

12. $75mn$ and $90m$

13. $12y$, 42, and $44y$

14. 15, $40k$, and 56

15. $32r^{12}$ and $36r^8$

16. $18y^5$ and $30y^2$

17. $16n^3$, $28n^2$, and $32n^5$

18. $20q^4$, $35q^5$, and $70q^{11}$

19. A 48-member band will be marching behind a 54-member band. They must both march in the same number of columns. What is the greatest number of columns in which they can march?

20. Mr. Liu's gym class has 24 students and Mr. Standish's gym class has 28 students. Each class is divided into teams. Teams from Mr. Liu's class play against teams from Mr. Standish's class. Each team must have the same number of students. What is the greatest number of students that each team can have?

THINKING SKILLS

Two whole numbers are said to be *relatively prime* if their only common factor is 1. Whole numbers do not have to be prime to be relatively prime.

Tell whether each pair of numbers is relatively prime. Write *Yes* or *No*.

21. 25 and 27

22. 12 and 55

23. 18 and 21

24. 40 and 99

25. Generate three pairs of composite numbers between 50 and 100 that are relatively prime.

26. Generate three pairs of composite numbers between 100 and 150 that are relatively prime.

LOGICAL REASONING

Classify each statement as *True* or *False*. If the statement is false, give a *counterexample* that demonstrates why it is false.

27. The GCF of an odd and an even number is always an odd number.

28. Two even numbers are never relatively prime.

29. The GCF of a prime number and an odd number is always odd.

30. The GCF of a prime number and an even number is always odd.

SPIRAL REVIEW

31. Solve: $5x - 13 = -3$ *(Lesson 4-5)*

32. An angle measures $37°$. Find the measure of an angle that is complementary to it.   *(Lesson 6-3)*

33. Estimate the sum: $2.8 + 3.4 + 2.9 + 3.2$ *(Toolbox Skill 4)*

34. Find the GCF: $4x^3$ and $22x$ *(Lesson 7-2)*

7-3 Least Common Multiple

Objective: To find least common multiples.

APPLICATION

Two nurses who work at the same hospital get together with their families every time they have the same weekend off. One of the nurses has every fourth weekend off. The other has every sixth weekend off.

QUESTION How often do the nurses and their families get together?

Terms to Know
• *multiple*
• *common multiple*
• *least common multiple (LCM)*

This question can be answered by finding the *least common multiple* of 4 and 6. When a number is multiplied by a nonzero whole number, the product is a **multiple** of the given number. A **common multiple** of two numbers is any number that is a multiple of both numbers. The **least common multiple (LCM)** of two numbers is the least number in the list of their common multiples.

Example 1 Find the LCM of 4 and 6.

Solution 1
the multiples of 4:   4, 8, 12, 16, 20, 24, 28, 32, 36, . . .
the multiples of 6:   6, 12, 18, 24, 30, 36, . . .
the common multiples of 4 and 6:   12, 24, 36, . . .
The LCM of 4 and 6 is 12.

Solution 2
Write the prime factorization of each number.
$4 = 2 \cdot 2 = 2^2$
$6 = 2 \cdot 3$
The prime factors are 2 and 3.

Form a product using the greatest power of each factor.
Greatest power of 2: 2^2
Greatest power of 3: 3
$LCM = 2^2 \cdot 3 = 12$

✔️ **Check Your Understanding**

1. In Solution 1 of Example 1, are all the multiples of 4 and 6 listed? Explain.

ANSWER The nurses and their families get together every 12 weeks.

As shown in Example 1, you can use prime factorizations to find an LCM. To do this, find all the prime factors that appear. Then multiply using the greatest power of each prime factor.

When you have to find the LCM of variable expressions, include in it the greatest power of each variable factor.

Example 2

Find the LCM.

a. $15a^2$ and $25a^4$

b. $4rs$ and $8s$

Solution

a. $15a^2 = 3 \cdot 5 \cdot a^2$
$25a^4 = 5^2 \cdot a^4$
$\text{LCM} = 3 \cdot 5^2 \cdot a^4 = 75a^4$

b. $4rs = 2^2 \cdot r \cdot s$
$8s = 2^3 \cdot s$
$\text{LCM} = 2^3 \cdot r \cdot s = 8rs$

✔️ **Check Your Understanding**

2. In Example 2(a), why was the power 5^2 used instead of 5?

3. In Example 2(b), why was r included as a factor in the LCM?

Guided Practice

COMMUNICATION « *Reading*

« **1.** Explain the meaning of the word *multiple* in the following sentence. *There were multiple-choice questions on the test.*

« **2.** Write a sentence using the word *multiple* in a mathematical context.

List four multiples of each number.

3. 7   **4.** 11   **5.** 8   **6.** 15

Identify all the prime factors that appear.

7. $3^2 \cdot 5$
$2 \cdot 5^7$

8. $5^2 \cdot 7$
$3 \cdot 7^2$

9. $2 \cdot 7^5 \cdot 11$
$7 \cdot 11^3 \cdot 13$

10. $2^2 \cdot 3^5 \cdot 11$
$5 \cdot 7 \cdot 13^2$

Identify the greater power.

11. 3^5; 3^2 **12.** 5^4; 5^3 **13.** 2^9; 2 **14.** 13^8; 13^{12}

Replace each __?__ with the correct number or expression.

15. $24 = 2^? \cdot \underline{\quad?\quad}$
$50 = \underline{\quad?\quad} \cdot 5^2$
$\text{LCM} = 2^? \cdot \underline{\quad?\quad} \cdot 5^2$
$= \underline{\quad?\quad}$

16. $9xy^3 = 3^? \cdot x \cdot \underline{\quad?\quad}$
$6x^2 = 2 \cdot \underline{\quad?\quad} \cdot x^2$
$\text{LCM} = 2 \cdot 3^? \cdot x^? \cdot \underline{\quad?\quad}$
$= \underline{\quad?\quad}$

Find the LCM.

17. 3 and 5

18. 4, 6, and 15

19. $6n$ and 9

20. $12x^9$ and $16x^7$

21. $6xy$ and $39yz$

22. $5y$, $12y^2$, and y^{10}

Find the LCM.

1. 2 and 5   **2.** 1 and 9   **3.** 9 and 27

4. 6 and 12   **5.** 8 and 10   **6.** 20 and 50

7. 4, 5, and 21   **8.** 9, 12, and 15   **9.** 2 and $3x$

10. $7r$ and s **11.** $6k$ and $4k$ **12.** $5x$ and $3x$

13. $6rs$ and $15st$ **14.** $8bz$ and $12bz$ **15.** $14a^9$ and $21a^5$

16. $16h^7$ and $24h^4$ **17.** $6n, 15n^4$, and 75   **18.** $16m^{10}, 18m$, and $30m^3$

19. Tracey Donovan can arrange her science class into lab groups of six or eight with no one left out. What is the least number of students Tracey Donovan can have in her science class?

20. Paper plates come in packages of 30, paper cups come in packages of 15, and paper napkins come in packages of 20. What is the least number of plates, cups, and napkins that Martha can buy to get an equal number of each?

COMPUTER APPLICATION

Sometimes you can use a BASIC computer program to find an answer quickly. Using a computer program can save you time when you are working with greater numbers. The program below computes the LCM of any two whole numbers.

```
10 PRINT "TO FIND LCM,"
20 PRINT "ENTER TWO NUMBERS."
30 INPUT A, B
40 FOR X = 1 TO B
50 LET A1 = A*X
60 LET Q = A1/B
70 IF Q = INT(Q) THEN 90
80 NEXT X
90 PRINT "LCM OF ";A;" AND ";B
100 PRINT "IS ";A1
110 END
```

Find the LCM. Use the BASIC program above if a computer is available.

21. 54 and 72   **22.** 56 and 70

23. 45 and 224   **24.** 98 and 180

25. 144 and 256   **26.** 168 and 196

27. Describe how you would explain to a friend the difference between a factor and a multiple.

28. For your notebook, outline the steps for finding a GCF and for finding an LCM. Explain any differences.

THINKING SKILLS

Exercises 29–34 require you to combine what you know about the GCF and the LCM.

Find both the GCF and the LCM.

29. 30 and 45

30. 63 and 84

31. 88 and 132

32. 12, 64, and 72

33. Find two numbers greater than 1 whose GCF is the lesser number and whose LCM is the greater number.

34. Is it possible for the GCF of two different numbers to equal the LCM of the numbers? Give a convincing argument to support your answer.

SPIRAL REVIEW

35. Find the product: $5\frac{3}{8} \times 2\frac{2}{3}$ *(Toolbox Skill 20)*

36. Find the LCM: $10mn$ and $12m$ *(Lesson 7-3)*

37. Write a variable expression that represents this phrase: seven more than a number y *(Lesson 4-6)*

38. Evaluate $|a - b|$ when $a = -2$ and $b = 5$. *(Lesson 3-6)*

Self-Test 1

Tell whether each number is *prime* or *composite*.

1. 23   **2.** 29   **3.** 33   **4.** 37   7-1

Write the prime factorization of each number.

5. 14   **6.** 12   **7.** 105   **8.** 400

Find the GCF.

9. 10 and 22   **10.** 18 and 45   7-2

11. $21bc$ and $66c$ **12.** $24x^5$ and $40x^2$

Find the LCM.

13. 5 and 15   **14.** 16 and 20   7-3

15. $6aw$ and $8ay$ **16.** $18n^7$ and $42n$

Modeling Fractions Using Fraction Bars

Objective: To use fraction bars to model fractions.

Materials

■ fraction bars: halves, thirds, fourths, sixths, twelfths

You can use fraction bars to help you visualize fractional parts of a whole. For example, let the entire fraction bar 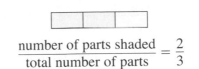 represent the whole. The diagram below then represents the fraction $\frac{2}{3}$.

$$\frac{\text{number of parts shaded}}{\text{total number of parts}} = \frac{2}{3}$$

Activity I   *Representing Fractions*

1 Write the fraction that is represented by each diagram.

a. **b.**

c. **d.**

2 Show how to use a fraction bar to represent each fraction.

a. $\frac{1}{4}$ **b.** $\frac{1}{3}$ **c.** $\frac{3}{4}$ **d.** $\frac{5}{6}$

Activity II   *Equivalent Fractions*

1 Describe the situation that you think is being represented by the fraction bars below.

2 Show how to use fraction bars to illustrate each statement.

a. $\frac{1}{2} = \frac{3}{6}$ **b.** $\frac{4}{12} = \frac{1}{3}$

c. $\frac{3}{4} = \frac{9}{12}$ **d.** $\frac{2}{12} = \frac{1}{6}$

3 Show two different fraction bars that represent the same amount as the given bar.

a. **b.** **c.**

4 Use your results from Step 3. Complete each statement.

a. $\frac{4}{6} = \underline{} = \underline{}$ **b.** $\frac{1}{3} = \underline{} = \underline{}$ **c.** $\frac{6}{12} = \underline{} = \underline{}$

Activity III *Comparing Fractions*

1 Which statement do you think the diagram at the right represents?

a. $\frac{3}{3} < \frac{4}{4}$ **b.** $\frac{3}{4} > \frac{4}{6}$

c. $\frac{3}{4} < \frac{4}{6}$ **d.** $\frac{3}{4} = \frac{4}{6}$

2 Show how to use fraction bars to compare each pair of fractions. Should the __?__ be replaced by $>$, $<$, or $=$?

a. $\frac{5}{6}$ __?__ $\frac{7}{12}$ **b.** $\frac{2}{3}$ __?__ $\frac{3}{4}$

c. $\frac{1}{4}$ __?__ $\frac{1}{3}$ **d.** $\frac{3}{4}$ __?__ $\frac{5}{12}$

3 Each diagram at the right represents a *unit fraction*.

  a. Describe what you think is meant by the term *unit fraction*.

  b. Write the fractions that the diagrams represent in order from least to greatest. What pattern do you see?

  c. Do you think $\frac{1}{50}$ is *greater than* or *less than* $\frac{1}{60}$? Explain.

Activity IV *Finding a Common Denominator*

1 Compare diagrams *A* and *B*, shown below.

  a. How are the diagrams alike?

  b. How are the diagrams different?

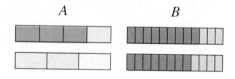

2 For each diagram, show an equivalent pair of fraction bars. The bars in the new pair should each have the same number of parts.

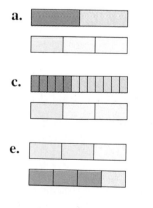

7-4 Equivalent Fractions

Objective: To find equivalent fractions and to write fractions in lowest terms.

Terms to Know
- *equivalent fractions*
- *lowest terms*

APPLICATION

Ali has a ruler that shows inches divided into fourths. While working on her art project, she draws a line that is three fourths of an inch long. The next day, she borrows Jared's ruler that shows inches divided into eighths. She finds that three fourths of an inch is the same as six eighths of an inch.

Fractions that represent the same amount are called **equivalent fractions.** So $\frac{3}{4}$ and $\frac{6}{8}$ are equivalent fractions.

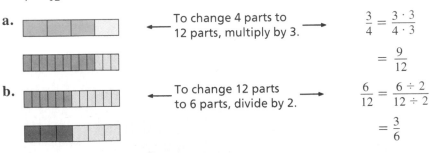

Example 1

Replace the $\underline{?}$ with the number that will make the fractions equivalent.

a. $\frac{3}{4} = \frac{?}{12}$ b. $\frac{6}{12} = \frac{?}{6}$

Solution

a.

To change 4 parts to 12 parts, multiply by 3.

$$\frac{3}{4} = \frac{3 \cdot 3}{4 \cdot 3}$$

$$= \frac{9}{12}$$

b.

To change 12 parts to 6 parts, divide by 2.

$$\frac{6}{12} = \frac{6 \div 2}{12 \div 2}$$

$$= \frac{3}{6}$$

✓ Check Your Understanding

1. How are the exercises in Example 1(a) and 1(b) alike? different?
2. In what way are the calculations in Example 1(b) similar to the calculations in Example 1(a)?

Generalization: *Finding Equivalent Fractions*

To find a fraction that is equivalent to a given fraction, multiply or divide both the numerator and the denominator of the given fraction by the same nonzero number.

$$\frac{a}{b} = \frac{a \cdot c}{b \cdot c} \qquad\qquad \frac{a}{b} = \frac{a \div c}{b \div c}, \quad b \neq 0, c \neq 0$$

A fraction is in **lowest terms** if the GCF of the numerator and the denominator is 1. You can write a fraction in lowest terms by using either the GCF or prime factorization.

Example 2	Write $\frac{12}{18}$ in lowest terms.
Solution 1	Divide both the numerator and the denominator by the GCF, which is 6.

$$\frac{12}{18} = \frac{12 \div 6}{18 \div 6} = \frac{2}{3}$$

$\longleftarrow$ You may find it easier to use this shortcut: $\dfrac{\overset{2}{\cancel{12}}}{\underset{3}{\cancel{18}}} = \dfrac{2}{3}$

Solution 2

$$\frac{12}{18} = \frac{\overset{1}{\cancel{2}} \cdot 2 \cdot \overset{1}{\cancel{3}}}{\underset{1}{\cancel{2}} \cdot \underset{1}{\cancel{3}} \cdot 3} = \frac{2}{3}$$

$\longleftarrow$ Write the prime factorization of the numerator and denominator and divide by all common factors.

Guided Practice

COMMUNICATION «*Reading*

«**1.** In a dictionary, find the word *equivalent*. Does it mean the same as *equal*?

«**2.** Explain the meaning of *equivalent* in the following sentence.
By working overtime, Mary earned the equivalent of two days' pay.

COMMUNICATION «*Writing*

Write the pair of equivalent fractions represented by each diagram.

«**3.** 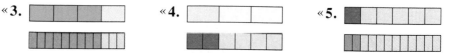 «**4.** «**5.**

Draw a diagram similar to those in Exercises 3–5 to represent each pair of equivalent fractions.

«**6.** $\frac{2}{4}, \frac{1}{2}$ «**7.** $\frac{2}{3}, \frac{4}{6}$ «**8.** $\frac{1}{4}, \frac{3}{12}$

Is each fraction written in lowest terms? Write *Yes* or *No*.

9. $\frac{10}{19}$ 10. $\frac{14}{25}$ 11. $\frac{9}{27}$ 12. $\frac{5}{45}$ 13. $\frac{28}{4}$ 14. $\frac{30}{17}$

Replace each __?__ with the number that will make the fractions equivalent.

15. $\frac{4}{5} = \frac{?}{25}$ 16. $\frac{1}{8} = \frac{?}{64}$ 17. $\frac{10}{?} = \frac{20}{18}$ 18. $\frac{5}{?} = \frac{35}{14}$

Write each fraction in lowest terms.

19. $\frac{8}{14}$ 20. $\frac{2}{12}$ 21. $\frac{25}{5}$ 22. $\frac{150}{60}$

Exercises

Replace each __?__ with the number that will make the fractions equivalent.

1. $\frac{1}{5} = \frac{?}{10}$ 2. $\frac{2}{7} = \frac{?}{21}$ 3. $\frac{22}{?} = \frac{44}{12}$ 4. $\frac{5}{?} = \frac{25}{75}$

5. $\frac{15}{4} = \frac{75}{?}$ 6. $\frac{60}{21} = \frac{20}{?}$ 7. $\frac{?}{6} = \frac{12}{36}$ 8. $\frac{?}{84} = \frac{17}{21}$

Write each fraction in lowest terms.

9. $\frac{3}{9}$ 10. $\frac{4}{8}$ 11. $\frac{6}{10}$ 12. $\frac{15}{24}$

13. $\frac{36}{27}$ 14. $\frac{35}{10}$ 15. $\frac{16}{18}$ 16. $\frac{36}{54}$

17. $\frac{72}{48}$ 18. $\frac{100}{86}$ 19. $\frac{284}{568}$ 20. $\frac{750}{850}$

21. The team won four tenths of its games. Write the fraction in lowest terms.

22. Find a fraction that has 16 as its numerator and is equivalent to four fifths.

THINKING SKILLS

Write each fraction in lowest terms.

23. $\frac{18}{3}$ 24. $\frac{36}{12}$ 25. $\frac{60}{15}$ 26. $\frac{45}{9}$

27. Compare your answers to Exercises 23–26. What do they have in common?

28. Make a list of ten different fractions that are equivalent to 4.

29. Determine whether this rule is *always*, *sometimes*, or *never* true: *If the numerator of a fraction is 0 and the denominator is not 0, then the fraction is equivalent to 0.*

30. Make up a rule similar to the one in Exercise 29 to help you determine if a fraction is equivalent to 1.

31. The equilateral triangle at the right is divided into equal parts by all its lines of symmetry. If one part is shaded, what fraction of the triangle is shaded?

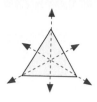

32. Suppose you divide a square into equal parts by drawing all its lines of symmetry. How many parts are there? If one part is shaded, what fraction of the square is shaded?

33. Using a regular polygon and all its lines of symmetry, model $\frac{1}{12}$.

The circles below are divided into equal parts by lines of symmetry. Name the fraction represented by each shaded part.

34.   35.

36.   37.

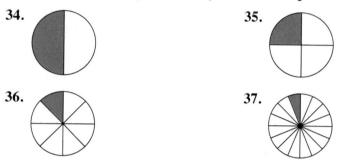

38. Show how you could use the circles in Exercises 34–37 to model the fact $\frac{3}{4} = \frac{12}{16}$.

SPIRAL REVIEW

39. Write $\frac{24}{150}$ in lowest terms.   *(Lesson 7-4)*

40. Estimate the quotient: $2502 \div 8$ *(Toolbox Skill 3)*

41. Name the figure: $\overset{\bullet\quad\bullet}{\underset{M\quad N}{\longrightarrow}}$ *(Lesson 6-1)*

42. Write an equation to represent this sentence: Eight less than a number x is 17.   *(Lesson 4-7)*

Challenge

1. Find a number for which all three statements below are true.
   a. When the number is divided by 3, the remainder is 2.
   b. When the number is divided by 4, the remainder is 3.
   c. When the number is divided by 5, the remainder is 4.

2. Find a number for which all three statements below are true.
   a. When the number is divided by 4, the remainder is 1.
   b. When the number is divided by 5, the remainder is 3.
   c. When the number is divided by 6, the remainder is 5.

Simplifying Algebraic Fractions

Objective: To simplify algebraic fractions.

CONNECTION

You have worked with fractions in arithmetic. You can apply what you know to work with fractions in algebra. A fraction that contains a variable is called an **algebraic fraction.**

Terms to Know

- *algebraic fraction*
- *simplify a fraction*
- *quotient of powers rule*

In Arithmetic		In Algebra
$\dfrac{3}{5}$ ←	← $\dfrac{\text{numerator}}{\text{denominator}}$ →	→ $\dfrac{a}{b}, b \neq 0$

Any fraction represents a division, so you know that the denominator of a fraction cannot be zero.

*Throughout this textbook, you may assume
that no denominator equals zero.*

You **simplify** a fraction or an algebraic fraction by writing it in lowest terms. To do this, write the prime factorization of both the numerator and the denominator, then divide by all the common factors.

Example 1

Simplify: **a.** $\dfrac{10xy}{8x}$ **b.** $\dfrac{n^5}{n^2}$

Solution

a. $\dfrac{10xy}{8x} = \dfrac{2 \cdot 5 \cdot \overset{1}{\cancel{x}} \cdot y}{2 \cdot 2 \cdot 2 \cdot \underset{1}{\cancel{x}}} = \dfrac{5y}{4}$

b. $\dfrac{n^5}{n^2} = \dfrac{\overset{1}{\cancel{n}} \cdot \overset{1}{\cancel{n}} \cdot n \cdot n \cdot n}{\underset{1}{\cancel{n}} \cdot \underset{1}{\cancel{n}}} = \dfrac{n^3}{1} = n^3$

In Example 1(b), the numerator and denominator of the fraction are powers of the same base. You may have noticed that the exponent in the simplified form is the same as the difference of the exponents in the given fraction.

Quotient of Powers Rule

To divide powers having the same base but different exponents, subtract the exponents.

$$\frac{a^m}{a^n} = a^{m-n}, a \neq 0$$

Example 2

Simplify: **a.** $\dfrac{5x^7}{x^3}$ **b.** $\dfrac{d^3}{4d}$

Solution

a. $\dfrac{5x^7}{x^3} = \dfrac{5x^{7-3}}{1} = 5x^4$

b. $\dfrac{d^3}{4d} = \dfrac{d^3}{4d^1} = \dfrac{d^{3\ 1}}{4} = \dfrac{d^2}{4}$

☑ **Check Your Understanding**

1. In Example 2(a), how would the solution be different if the denominator were x^4?

2. In Example 2(b), why is d rewritten as d^1?

Guided Practice

COMMUNICATION « *Reading*

Refer to the text on pages 306–307.

« **1.** In this textbook, what can you assume about the denominator of a fraction?

« **2.** State the quotient of powers rule in words.

Rewrite each fraction using the prime factorization of the numerator and the denominator.

3. $\dfrac{6a}{14}$ **4.** $\dfrac{10j}{15j}$ **5.** $\dfrac{12s}{18rs}$ **6.** $\dfrac{a^9}{a^5}$ **7.** $\dfrac{w^6}{7w}$

Simplify.

8. $\dfrac{4b}{16}$ **9.** $\dfrac{20x}{12x}$ **10.** $\dfrac{15m}{18mn}$ **11.** $\dfrac{n^{14}}{n^{11}}$ **12.** $\dfrac{a^4}{8a^3}$

Exercises

Simplify.

1. $\dfrac{21b}{28}$ **2.** $\dfrac{5}{10n}$ **3.** $\dfrac{36w}{45w}$ **4.** $\dfrac{18x}{4x}$ **5.** $\dfrac{6r}{3rs}$

6. $\dfrac{14vw}{14w}$ **7.** $\dfrac{c^{10}}{c^2}$ **8.** $\dfrac{r^7}{r}$ **9.** $\dfrac{6a^6}{a^2}$ **10.** $\dfrac{m^{12}}{13m^6}$

11. $\dfrac{48t}{4t}$ **12.** $\dfrac{6ab}{27b}$ **13.** $\dfrac{30}{6y}$ **14.** $\dfrac{n^9}{n^3}$ **15.** $\dfrac{8x^{24}}{x^8}$

16. Simplify the algebraic fraction whose numerator is $21n$ and whose denominator is $3n$.

17. Simplify the algebraic fraction whose numerator is b^6 and whose denominator is $6b$.

Simplify.

18. $\dfrac{9z^{15}}{3z^3}$　　**19.** $\dfrac{40n^{16}}{10n^{14}}$　　**20.** $\dfrac{10x^{21}}{15x^{13}}$　　**21.** $\dfrac{3v^{19}}{21v^{16}}$

22. $\dfrac{49a^3c^3}{14ac}$　　**23.** $\dfrac{16v^3w^3}{44v^2}$　　**24.** $\dfrac{12m^4n^2}{27m}$　　**25.** $\dfrac{35x^2y}{5xy}$

26. $\dfrac{18x^5y^5}{2y}$　　**27.** $\dfrac{20m^8}{24m^3n^2}$　　**28.** $\dfrac{3a^4b^3}{15a^2b}$　　**29.** $\dfrac{40s^6t^9}{16s^2t^7}$

LOGICAL REASONING

Exercises 30 and 31 refer to these statements.

I. $\dfrac{a^2}{a^3} = \dfrac{a \cdot a}{a \cdot a \cdot a} = \dfrac{1}{a}$　　　　**II.** $\dfrac{a^2}{a^3} = a^{2-3} = a^{-1}$

30. What logical conclusion can you make about the relationship between a^{-1} and $\frac{1}{a}$? Give a convincing argument to support your answer.

31. Use your results from Exercise 30. What do you think is another way to write a^{-5}?

SPIRAL REVIEW

32. Last baseball season, Ed had 38 hits, 27 runs batted in, and 4 home runs. How many singles did Ed hit?　*(Lesson 5-6)*

33. Simplify: $\dfrac{a^{12}}{a^5}$　*(Lesson 7-5)*

34. Complete: A ___?___ triangle has one 90° angle.　*(Lesson 6-6)*

35. Solve: $x - 17 = -9$　*(Lesson 4-3)*

36. Evaluate $a^2 + 2b - 15 \div c$ when $a = 4$, $b = 6$, and $c = 5$. *(Lesson 1-10)*

37. Find the sum: $-14 + 9 + (-3)$　*(Lesson 3-3)*

38. Write 12.5 cm in m.　*(Lesson 2-9)*

Historical Note

Recently, computers were used to find the three prime factors of a 155-digit number. Early computers could only do much simpler tasks. The first working computer, Mark I, went into operation at Harvard University in 1944. Rear Admiral Grace Hopper, U.S. Navy (Ret.), worked as a programmer on that computer. She had a long and distinguished career in the computer field. Among her many contributions was the development of computer languages.

Research

A computer language is named Ada in honor of Ada Lovelace. Find out who she was.

Comparing Fractions

Objective: To compare fractions.

Terms to Know

• *least common denominator (LCD)*

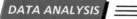 **DATA ANALYSIS**

The table below lists the lengths of different sizes of nails. For a project, a carpenter needs to use nails that are less than $1\frac{1}{3}$ in. long. The carpenter has some 3d nails.

QUESTION Are 3d nails short enough for the carpenter to use on the project?

To find the answer, first find the length of a 3d nail. The table shows that the length is $1\frac{1}{4}$ in. Then compare $1\frac{1}{4}$ to $1\frac{1}{3}$. Because both $1\frac{1}{4}$ and $1\frac{1}{3}$ have 1 as the whole-number part, you need to compare only the fractional parts of $1\frac{1}{4}$ and $1\frac{1}{3}$.

Penny Size (d)	2	3	4	5	6	7
Length (in.)	1	$1\frac{1}{4}$	$1\frac{1}{2}$	$1\frac{3}{4}$	2	$2\frac{1}{4}$

Example 1

Solution

Replace the __?__ with >, <, or =: $\frac{1}{4}$ __?__ $\frac{1}{3}$

To compare fractional parts of a whole, you can use shaded parts of bars to show the relationship.

◄── $\frac{1}{4}$ One fourth of a whole is less than

◄── $\frac{1}{3}$ one third of a whole, so $\frac{1}{4} < \frac{1}{3}$.

☑ **Check Your Understanding**

1. In Example 1, explain how you know that one fourth of a whole is less than one third of a whole.

ANSWER Because $\frac{1}{4} < \frac{1}{3}$, it follows that $1\frac{1}{4} < 1\frac{1}{3}$. So 3d nails are short enough to use.

You can also compare fractions whose denominators are different, but whose numerators are the same and greater than 1.

Example 2

Replace the __?__ with >, <, or =: $\dfrac{5}{6} \underline{} \dfrac{5}{12}$

Solution

$\longleftarrow \dfrac{5}{6}$

$\longleftarrow \dfrac{5}{12}$

One sixth of a whole is greater than one twelfth of a whole, so *five* sixths are greater than *five* twelfths.

$$\dfrac{5}{6} > \dfrac{5}{12}$$

There are times when the best way to compare fractions is to rewrite them as *equivalent fractions* with a common denominator. Then compare the numerators. The **least common denominator (LCD)** is the LCM of the denominators.

Example 3

Replace the __?__ with >, <, or =: $\dfrac{7}{12} \underline{} \dfrac{11}{18}$

Solution

The LCM of 12 and 18 is 36, so the LCD is 36.

$$\dfrac{7}{12} = \dfrac{7 \cdot 3}{12 \cdot 3} = \dfrac{21}{36} \qquad\qquad \dfrac{11}{18} = \dfrac{11 \cdot 2}{18 \cdot 2} = \dfrac{22}{36}$$

$$\dfrac{21}{36} < \dfrac{22}{36}, \text{ so } \dfrac{7}{12} < \dfrac{11}{18}.$$

☑ **Check Your Understanding**

2. In Example 3, how do you determine that the LCM is 36?

3. In Example 3, why is the number 3 used to multiply the numerator and denominator of $\dfrac{7}{12}$?

Guided Practice

COMMUNICATION « *Writing*

Write the statement represented by each diagram.

«**1.**

«**2.**

«**3.**

Draw a diagram similar to those in Exercises 1–3 to represent each statement.

«**4.** $\dfrac{1}{6} > \dfrac{1}{12}$ 　　 «**5.** $\dfrac{1}{4} < \dfrac{1}{3}$ 　　 «**6.** $\dfrac{3}{4} = \dfrac{9}{12}$ 　　 «**7.** $\dfrac{11}{12} > \dfrac{5}{6}$

Find the LCD of each pair of fractions.

8. $\dfrac{6}{7}, \dfrac{3}{14}$ 　　 **9.** $\dfrac{3}{25}, \dfrac{4}{5}$ 　　 **10.** $\dfrac{2}{3}, \dfrac{3}{5}$ 　　 **11.** $\dfrac{1}{6}, \dfrac{10}{21}$

Replace each __?__ with >, <, or =.

12. $\frac{1}{5}$ __?__ $\frac{1}{15}$ **13.** $\frac{7}{9}$ __?__ $\frac{7}{8}$ **14.** $\frac{4}{15}$ __?__ $\frac{3}{10}$ **15.** $\frac{4}{5}$ __?__ $\frac{3}{4}$

Exercises

Replace each __?__ with >, <, or =.

1. $\frac{1}{6}$ __?__ $\frac{1}{7}$ **2.** $\frac{1}{11}$ __?__ $\frac{1}{8}$ **3.** $\frac{15}{22}$ __?__ $\frac{15}{21}$ **4.** $\frac{11}{30}$ __?__ $\frac{11}{32}$

5. $\frac{5}{6}$ __?__ $\frac{11}{12}$ **6.** $\frac{4}{5}$ __?__ $\frac{7}{10}$ **7.** $\frac{2}{3}$ __?__ $\frac{5}{9}$ **8.** $\frac{1}{3}$ __?__ $\frac{2}{5}$

9. $\frac{16}{20}$ __?__ $\frac{4}{5}$ **10.** $\frac{21}{30}$ __?__ $\frac{7}{10}$ **11.** $\frac{5}{12}$ __?__ $\frac{3}{8}$ **12.** $\frac{3}{4}$ __?__ $\frac{7}{9}$

13. Marcia's history book is $8\frac{1}{4}$ in. wide. Her science book is $8\frac{1}{2}$ in. wide. Which book is wider?

14. Jon's graduation photograph is $3\frac{3}{8}$ in. high and $3\frac{3}{16}$ in. wide. Is the photograph wider than it is high? Explain.

Write in order from least to greatest. Use two inequality symbols.

15. $\frac{1}{5}, \frac{1}{7}, \frac{1}{4}$ **16.** $\frac{7}{8}, \frac{5}{8}, \frac{2}{3}$ **17.** $\frac{3}{8}, \frac{3}{40}, \frac{7}{16}$ **18.** $2\frac{8}{9}, 2\frac{2}{3}, 2\frac{4}{5}$

DATA ANALYSIS

Use the table at the right.

19. Is the winning height for 1948 *greater than* or *less than* for 1932?

20. Is the winning height for 1956 *greater than* or *less than* for 1952?

21. Would you display these data in a *bar graph* or a *line graph*? Explain.

22. In general, did the winning distances *increase* or *decrease* from 1932 to 1956?

23. Do you think the winning height for 1960 was *greater than*, *less than*, or *equal to* the winning height for 1956? Explain.

24. **RESEARCH** Find the winning height for 1960. Compare it to your answer to Exercise 23.

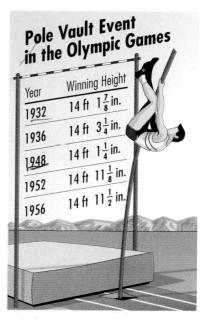

Pole Vault Event in the Olympic Games

Year	Winning Height
1932	14 ft $1\frac{7}{8}$ in.
1936	14 ft $3\frac{1}{4}$ in.
1948	14 ft $1\frac{1}{4}$ in.
1952	14 ft $11\frac{1}{8}$ in.
1956	14 ft $11\frac{1}{2}$ in.

A *professional photographer* takes pictures of many different subjects. To control the amount of time that film is exposed to light, the photographer changes the shutter speed. The speeds are marked on a camera with some numbers from 1 to 1000. These numbers represent denominators in fractions, as shown in the table at the right.

A photographer uses a faster shutter speed to photograph something moving quickly and a slower shutter speed for something standing still.

Use the table at the right.

25. List the shutter speeds in order from slowest to fastest: $\frac{1}{500}$, $\frac{1}{15}$, $\frac{1}{60}$

26. A photographer wants to take a "photo finish" of the cars in a race. Is a shutter speed of $\frac{1}{1000}$ or $\frac{1}{125}$ more appropriate?

27. A photographer has to take pictures of merchandise items for a catalogue. Is a shutter speed of $\frac{1}{250}$ or $\frac{1}{30}$ more appropriate?

28. On cloudy days, photographers reduce shutter speed one level to admit more light. A photographer is taking pictures on a sunny day, using a shutter speed of $\frac{1}{125}$. The day suddenly becomes cloudy. What speed should the photographer now use?

SPIRAL REVIEW

29. Draw the next two figures in the pattern below. Then write the first five numbers in the pattern.   *(Lesson 6-8)*

30. Replace the $\underline{\quad?\quad}$ with >, <, or =: $\frac{9}{11}\ \underline{\ ?\ }\ \frac{4}{5}$ *(Lesson 7-6)*

31. Find the GCF: 54 and 102   *(Lesson 7-2)*

32. Find the difference: $7\frac{4}{9} - 4\frac{5}{6}$ *(Toolbox Skill 19)*

33. Solve: $-24 = \frac{y}{-8}$ *(Lesson 4-4)*

34. Simplify: $7n^3 \cdot 9n^5$ *(Lesson 2-1)*

7-7

Fractions and Decimals

Objective: To write fractions as decimals and decimals as fractions.

EXPLORATION

1 Use a calculator to complete each statement.

a.  7. ÷ 20. = ?

b. 11. ÷ 15. = ?

2 Compare your answers to parts (a) and (b) of Step 1. How are they alike? How are they different?

3 Recall that a division bar acts like a division sign. Write each division in Step 1 using a fraction bar.

You can write any fraction as a decimal by dividing the numerator by the denominator. When the division results in a remainder of zero, the decimal is called a **terminating decimal.** When the remainder is not zero and a block of digits in the decimal repeats, the decimal is called a **repeating decimal.** You indicate that a block of digits repeats by putting a bar over those digits.

Example 1

Write each fraction or mixed number as a decimal.

a. $\dfrac{7}{11}$ b. $1\dfrac{3}{8}$

Solution

a.
$$
\begin{array}{r}
0.6363\ldots \\
11\overline{)7.0000} \\
\underline{6\ 6}\ \\
40 \\
\underline{33} \\
70 \\
\underline{66} \\
40 \\
\underline{33} \\
7
\end{array}
$$

$$\frac{7}{11} = 0.\overline{63}$$

b. First write $\dfrac{3}{8}$ as a decimal.
$$
\begin{array}{r}
0.375 \\
8\overline{)3.000} \\
\underline{2\ 4}\ \\
60 \\
\underline{56} \\
40 \\
\underline{40} \\
0
\end{array}
$$

$$\frac{3}{8} = 0.375$$

$$1\frac{3}{8} = 1.375$$

Check Your Understanding

1. In Example 1(a), why is the bar over both the 6 and the 3 instead of just over the 3?

2. If you used a calculator in Example 1(a), what number would be shown on an 8-digit display?

3. How could you use Example 1(b) to help you write $6\frac{3}{8}$ as a decimal?

In order to write a terminating decimal as a fraction in lowest terms, you first write the terminating decimal as a fraction whose denominator is a power of 10. Then you simplify the fraction.

Example 2	**Write each decimal as a fraction or mixed number in lowest terms.**
	a. 0.555 **b.** 4.24
Solution	**a.** $0.555 = \frac{555}{1000} = \frac{111}{200}$
	b. $4.24 = 4 + 0.24 = 4 + \frac{24}{100} = 4\frac{6}{25}$

☑ **Check Your Understanding**

4. In Example 2(a), how do you know that $\frac{111}{200}$ is written in lowest terms?

The chart below shows some of the most commonly used sets of equivalent decimals and fractions. You will find it helpful to become familiar with these.

Equivalent Decimals and Fractions

$0.2 = \frac{1}{5}$	$0.25 = \frac{1}{4}$	$0.125 = \frac{1}{8}$	$0.1\overline{6} = \frac{1}{6}$
$0.4 = \frac{2}{5}$	$0.5 = \frac{1}{2}$	$0.375 = \frac{3}{8}$	$0.\overline{3} = \frac{1}{3}$
$0.6 = \frac{3}{5}$	$0.75 = \frac{3}{4}$	$0.625 = \frac{5}{8}$	$0.\overline{6} = \frac{2}{3}$
$0.8 = \frac{4}{5}$		$0.875 = \frac{7}{8}$	$0.8\overline{3} = \frac{5}{6}$

Guided Practice

«**1.** COMMUNICATION «*Reading*   Replace each __?__ with the correct word or phrase.

When a fraction is written as a __?__, one of two types of decimals results. The two types of decimals are the __?__ and the repeating decimal. In a __?__ decimal, a pattern of digits repeats over and over.

Rewrite each repeating decimal with a bar over the repeating digits.

2. 0.41666 . . .   **3.** 1.825825 . . .   **4.** 0.32121 . . .

Write each decimal as a fraction whose denominator is a power of ten.

5. 0.18   **6.** 0.056   **7.** 0.475   **8.** 9.44   **9.** 4.6

Write each fraction or mixed number as a decimal.

10. $\frac{9}{20}$ **11.** $\frac{5}{8}$ **12.** $\frac{5}{11}$ **13.** $3\frac{8}{9}$ **14.** $7\frac{1}{3}$

Write each decimal as a fraction or mixed number in lowest terms.

15. 0.205 **16.** 0.81 **17.** $5.1\overline{6}$ **18.** 3.62 **19.** 7.75

Exercises

Write each fraction or mixed number as a decimal.

1. $\frac{7}{10}$ **2.** $\frac{3}{5}$ **3.** $\frac{9}{11}$ **4.** $\frac{1}{33}$

5. $6\frac{11}{20}$ **6.** $9\frac{13}{15}$ **7.** $10\frac{2}{3}$ **8.** $4\frac{4}{25}$

Write each decimal as a fraction or mixed number in lowest terms.

9. 0.432 **10.** 0.525 **11.** 0.19 **12.** 0.77

13. $0.1\overline{6}$ **14.** $0.\overline{3}$ **15.** 3.32 **16.** 5.95

17. 9.51 **18.** $7.8\overline{3}$ **19.** $2.\overline{6}$ **20.** 8.13

21. Jordan answered $\frac{7}{8}$ of the test questions correctly. Write this fraction as a decimal.

22. A soccer team won $\frac{6}{11}$ of its games. Write this fraction as a decimal.

ESTIMATION

Estimate each decimal as a fraction or mixed number.

23. 0.49 **24.** 0.42 **25.** 0.33 **26.** 0.748

27. 9.58 **28.** 4.19 **29.** 6.124 **30.** 5.168

PATTERNS

Write each fraction as a decimal.

31. $\frac{1}{9}$ **32.** $\frac{2}{9}$ **33.** $\frac{3}{9}$ **34.** $\frac{4}{9}$

35. Use your answers to Exercises 31–34. What pattern do you notice? Use this pattern to make a table of equivalent fractions and decimals for fractions between 0 and 1 that have 9 as a denominator.

Write each fraction as a decimal.

36. $\frac{1}{11}$ **37.** $\frac{2}{11}$ **38.** $\frac{3}{11}$ **39.** $\frac{4}{11}$

40. Use your answers to Exercises 36–39. What pattern do you notice? Use this pattern to make a table of equivalent fractions and decimals for fractions between 0 and 1 that have 11 as a denominator.

PROBLEM SOLVING/APPLICATION

At delicatessens, customers use fractions when ordering food. However, food items are usually weighed on scales that show weights as decimals.

41. A scale shows 0.52 lb. About what fraction of a pound is this?

42. Pete asks for $\frac{1}{4}$ lb of cheese. The deli clerk slices 0.26 lb. Is this *more than* or *less than* $\frac{1}{4}$ lb?

43. Martha asks for $\frac{3}{4}$ lb of roast beef. The deli clerk slices 0.73 lb. Is this *more than* or *less than* $\frac{3}{4}$ lb?

44. Vanessa asks for $1\frac{1}{2}$ lb of turkey that costs $4.99 per pound. The deli clerk slices 1.49 lb. What is the total cost of the turkey?

45. Lee asks for 1 lb of potato salad. The deli clerk first sets the scale to −0.03 lb. Next the clerk puts a container on the scale. Then the clerk puts potato salad into the container. The scale shows 1.01 lb.
    a. Why did the clerk set the scale to −0.03 lb before weighing?
    b. How much potato salad did the customer buy?

CALCULATOR

46. Using a calculator, find the decimal equivalents for these fractions.

$$\frac{1}{5}, \frac{1}{10}, \frac{1}{15}, \frac{1}{20}, \frac{1}{25}, \frac{1}{30}, \frac{1}{35}, \frac{1}{40}$$

Tell which decimals are repeating and which are terminating.

47. Use your answers to Exercise 46 to predict if $\frac{1}{45}$ and $\frac{1}{50}$ are repeating or terminating decimals. Use a calculator to check the predictions.

48. Using your answers to Exercises 46 and 47, make a rule for determining whether the decimal equivalent of a given fraction will terminate. (*Hint:* Consider the prime factors of the denominators.)

SPIRAL REVIEW

49. Measure the angle shown at the right. *(Lesson 6-2)*

50. Write $8\frac{5}{6}$ as a decimal. *(Lesson 7-7)*

51. Estimate the difference: $5827 - 1184$ *(Toolbox Skill 1)*

52. Solve: $3(m + 5) = -18$ *(Lesson 4-9)*

Strategy:
Drawing a Diagram

Objective: To solve problems by drawing a diagram.

Some problems may be easier to solve if you draw a diagram picturing the facts. Drawing a diagram while you read a problem can help you to understand the facts of the problem and determine what to find for the answer. For example, drawing grids or number lines might help you to solve problems involving distance. Drawing blocks might help solve problems involving lengths.

Problem

During a sightseeing tour, the tour bus travels 2 blocks due north, 3 blocks due east, 6 blocks due south, 11 blocks due west, and 4 blocks due north. At this point, where is the tour bus in relation to the starting point?

Solution

UNDERSTAND The problem is about a tour bus route.
Facts: travels 2 blocks due north,
     3 blocks due east,
     6 blocks due south,
     11 blocks due west,
     4 blocks due north
Find: the location of the bus relative to the starting point

PLAN You can use a piece of graph paper to draw a diagram. Let each square on the graph paper represent one block. Start near the middle of the graph paper and trace the route of the bus. Draw one arrow for each distance traveled. The first arrow will represent 2 blocks due north, the second arrow will represent 3 blocks due east, and so on. Be sure to identify the directions north, south, east, and west.

WORK

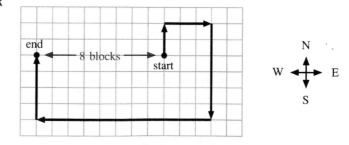

ANSWER Count the blocks between the starting point and the ending point. The tour bus is 8 blocks due west of its starting point.

Look Back What if the tour bus had traveled 11 blocks due east rather than 11 blocks due west? Where would the tour bus have been in relati... starting point?

Guided Practice

COMMUNICATION «Reading

Use this paragraph for Exercises 1–5.

A building has 30 floors, including two floors below ground level. An elevator in the building starts at ground level. It rises 20 floors, descends 11 floors, rises 5 floors, and descends 16 floors.

«**1.** How many floors does the building have in all?

«**2.** How many floors are above ground?

«**3.** Where does the elevator start?

«**4.** In what direction is the elevator traveling when it moves 16 floors?

«**5.** Draw a diagram to represent the situation.

6. What is the value of d in the diagram at the right?

6 m	◄─── d ───►
10 m	7 m

7. Solve by drawing a diagram: A softball field is located 12 mi due east of Jen's house. A bakery is located halfway between Jen's house and the softball field. The high school is located one third of the way from the bakery to the softball field. Where is the high school in relation to Jen's house?

Problem Solving Situations

Solve by drawing a diagram.

1. Bernie lives 8 blocks due east of Harry. Harry lives 3 blocks due west of Ian. Where does Ian live in relation to Bernie?

2. The capacities of 3 buckets are 3 gal, 5 gal, and 8 gal. How could you use these containers to measure exactly 4 gal of water?

Solve by drawing a diagram.

3. How many diagonals can be drawn in a hexagon?

4. The lengths of three steel rods are 8 cm, 14 cm, and 16 cm. How could you use these rods to mark off a length of 10 cm?

5. An elevator started at ground level. It rose 15 floors, descended 3 floors, rose 8 floors, descended 12 floors, and descended 2 floors. At this point, where was the elevator relative to ground level?

6. The exit for Greenville is 18 mi due west of the exit for Clairemont. The exit for Springfield is two thirds of the way from Clairemont to Greenville. The exit for Ashton is halfway between the exits for Springfield and Greenville. Where is the exit for Ashton in relation to the exit for Clairemont?

Solve using any problem solving strategy.

7. Seaside Gifts has postcards for $.25, $.35, and $.50. In how many different ways could a customer spend $1.35 on post-cards?

8. Sonia Garcia bought cans of soup for $1.19 each and cups of yogurt for $.89 each. She spent a total of $13.97. How many of each did Sonia buy?

> **PROBLEM SOLVING**
> ## CHECKLIST
> **Keep these in mind:**
> Using a Four-Step Plan
> Too Much or Not Enough
>   Information
> **Consider these strategies:**
> Choosing the Correct Operation
> Making a Table
> Guess and Check
> Using Equations
> Identifying a Pattern
> Drawing a Diagram

9. Sixteen players participated in a single-elimination tennis tournament. In such a tournament, each player is out after one loss. How many games did the winner play?

10. Mr. Cole's shoe size is one size less than four times his son's shoe size. Mr. Cole wears a size 11 shoe. What size shoe does his son wear?

11. David delivers meals. After he has picked up the meals, he drives 4 blocks due south to make his first delivery. To make his second delivery, he turns left and drives 6 blocks. From there he drives 4 blocks due north to make his third delivery. At this point, how far is David from where he picks up the meals?

12. What is the least number of magnets that you need to display twelve 4-inch by 6-inch postcards so that they can all be seen? Assume that each corner must have a magnet holding it up.

13. Rita bought two boxes of computer disks for $10.99 each and paid tax of $1.10. She gave the cashier two $20 bills. How much change did Rita receive?

Write a word problem that you could solve by using the given diagram. Then solve the problem.

14.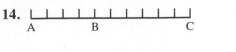
A   B   C

15.

SPIRAL REVIEW

16. Find the quotient: $2\frac{2}{9} \div 4\frac{1}{3}$ *(Toolbox Skill 21)*

17. **DATA,** *pages 92–93* Estimate the record high January temperature in Houston.   *(Lesson 5-2)*

18. Complete: A __?__ is a quadrilateral with exactly one pair of parallel sides.   *(Lesson 6-7)*

19. Karen lives 9 blocks due west of Julie and 6 blocks due west of Alice. Where does Alice live in relation to Julie?   *(Lesson 7-8)*

Self-Test 2

Write each fraction in lowest terms.

1. $\frac{9}{12}$ 2. $\frac{14}{16}$ 3. $\frac{20}{25}$ 4. $\frac{18}{48}$ 7-4

Simplify.

5. $\frac{8c}{24c}$ 6. $\frac{10x}{15xy}$ 7. $\frac{a^8}{a^2}$ 8. $\frac{8x^{14}}{x^5}$ 7-5

Replace each __?__ with >, <, or =.

9. $\frac{1}{8}$ _?_ $\frac{1}{7}$ 10. $\frac{7}{12}$ _?_ $\frac{7}{10}$ 11. $\frac{13}{16}$ _?_ $\frac{17}{24}$ 7-6

Write each fraction or mixed number as a decimal.

12. $\frac{11}{20}$ 13. $\frac{8}{9}$ 14. $2\frac{17}{25}$ 15. $4\frac{5}{6}$ 7-7

Write each decimal as a fraction or mixed number in lowest terms.

16. 0.78   17. 0.444   18. 3.56   19. $8.\overline{6}$

20. Solve by drawing a diagram: The lengths of three steel rods are 12 m, 26 m, and 19 m. How could you use these rods to mark off a length of 5 m?   7-8

Rational Numbers

Objective: To recognize rational numbers.

You have worked with whole numbers, integers, and fractions. All these numbers can be written in fractional form. For example, $5 = \frac{5}{1}$ and $-3 = \frac{-6}{2}$. Any number that can be written as a quotient of two integers $\frac{a}{b}$, where b does not equal zero, is called a **rational number.** All whole numbers, integers, and arithmetic fractions as well as many decimals are rational numbers.

Example 1

Express each rational number as a quotient of two integers.

a. -16 **b.** $2\frac{4}{7}$ **c.** $-1.\overline{3}$

Solution

a. $-16 = -\frac{16}{1}$

$\quad\quad = \frac{-16}{1}$

b. $2\frac{4}{7} = \frac{18}{7}$

c. $-1.\overline{3} = -1\frac{1}{3}$

$\quad\quad\quad = -\frac{4}{3}$

$\quad\quad\quad = \frac{-4}{3}$

Check Your Understanding

1. In Example 1(a), would the quotient $\frac{16}{-1}$ also represent -16? Explain.

2. In Example 1(a), would the quotient $\frac{-16}{-1}$ also represent -16? Explain.

Terms to Know

- *rational number*
- *irrational number*
- *real number*

Numbers that cannot be written as the quotient of two integers are called **irrational numbers.** These numbers are nonrepeating, nonterminating decimals. Any number that is either rational or irrational is called a **real number.** The chart below shows the parts of the *real number system*.

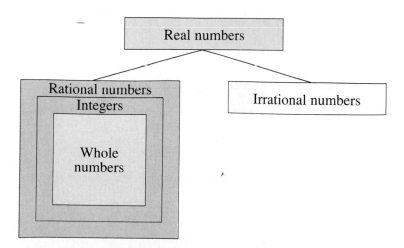

Every real number can be represented by a point on a number line.

Example 2

Use the number line at the right. Name the point that represents each number.

$$\overset{\longleftrightarrow}{\underset{-2\ \ AB\ -1\quad\ 0\quad\ \ 1\quad\ \ 2\ C\quad 3}{\mid\mid\mid\mid\quad\mid\quad\mid\quad\mid\quad\mid\mid\quad\mid}}$$

a. -1.5 b. -1.35 c. $2\frac{1}{3}$

Solution

a. -1.5 is halfway between -2 and -1. Point A represents -1.5.

b. -1.35 is between -2 and -1 and is greater than -1.5. Point B represents -1.35.

c. $2\frac{1}{3}$ is between 2 and 3. Point C represents $2\frac{1}{3}$.

Guided Practice

COMMUNICATION « Reading

Refer to the diagram on page 321. To which group(s) of numbers does each number belong?

«**1.** 3   «**2.** -10 «**3.** 6.23   **A.** whole numbers

«**4.** $\frac{7}{8}$ «**5.** $-\frac{14}{9}$ «**6.** $8.\overline{5}$ **B.** integers
     C. rational numbers

«**7.** $-0.\overline{27}$ «**8.** $5\frac{1}{3}$ «**9.** -18.3 **D.** irrational numbers
     E. real numbers

Express each rational number as a quotient of two integers.

10. 0.125   **11.** 0   **12.** -3 **13.** $-2\frac{1}{5}$ **14.** $0.1\overline{6}$

Use the number line at the right. Name the point that represents each number.

$$\overset{\longleftrightarrow}{\underset{-3\,D\quad\ -2\quad\ -1\ E\ F0\quad\ 1\quad G\ \ 2H\quad 3}{\mid\mid\quad\mid\quad\mid\mid\mid\quad\mid\quad\mid\mid\quad\mid}}$$

15. $1.\overline{5}$ **16.** -2.8 **17.** $2\frac{1}{4}$ **18.** $-\frac{2}{9}$ **19.** -0.654

Exercises

Express each rational number as a quotient of two integers.

1. $12\frac{3}{5}$ **2.** -50 **3.** $-4\frac{1}{4}$ **4.** $10\frac{5}{12}$

5. 20.2   **6.** -0.1 **7.** $-2.\overline{3}$ **8.** $8.\overline{6}$

Use the number line at the right. Name the point that represents each number.

$$\overset{\longleftrightarrow}{\underset{-3\,I\quad\ -2\,J\quad -1\quad\ K0\quad\ 1\quad LM\,2\ N\quad 3}{\mid\mid\quad\mid\mid\quad\mid\quad\mid\mid\quad\mid\quad\mid\mid\mid\quad\mid}}$$

9. $2.\overline{4}$ **10.** $-1.8\overline{3}$ **11.** 1.657   **12.** $-\frac{3}{13}$ **13.** $-\frac{14}{5}$ **14.** $1\frac{7}{8}$

15. On a horizontal number line, is the point that represents -1.5 to the *left* or *right* of the point that represents -1?

16. List five different ways to express 0 as a quotient of two integers.

Replace each __?__ with >, <, or =.

17. $-\frac{1}{2}$ __?__ $-\frac{1}{10}$ **18.** $-\frac{4}{9}$ __?__ -1 **19.** $-12.\overline{4}$ __?__ -12.4

20. 5.5 __?__ $\frac{16}{3}$ **21.** -4.315 __?__ -4.35 **22.** $3\frac{1}{7}$ __?__ $3.1\overline{6}$

NUMBER SENSE

To graph numbers on a number line, it is helpful to decide whether the number is close to a value whose graph is easily identifiable.

Tell whether each number is closest to -1, $-\frac{1}{2}$, 0, $\frac{1}{2}$, or 1.

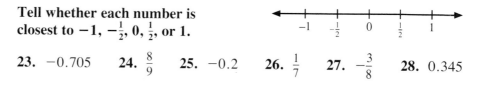

23. -0.705 **24.** $\frac{8}{9}$ **25.** -0.2 **26.** $\frac{1}{7}$ **27.** $-\frac{3}{8}$ **28.** 0.345

LOGICAL REASONING

Replace each __?__ with the word *All*, *Some*, or *No* to make the statement true.

29. __?__ whole numbers are integers.

30. __?__ integers are whole numbers.

31. __?__ rational numbers are integers.

32. __?__ real numbers are rational numbers.

SPIRAL REVIEW

33. Identify the polygon at the right. *(Lesson 6-7)*

34. Express 0.16 as a quotient of two integers. *(Lesson 7-9)*

35. Find the sum: $7\frac{3}{8} + 9\frac{11}{16}$ *(Toolbox Skill 18)*

36. Find the LCM: 24 and 36 *(Lesson 7-3)*

Graphing Open Sentences

Objective: To graph solutions of equations and inequalities on a number line.

CONNECTION

In both the English language and mathematics, there are many sentences that can be classified as either true or false. There are other sentences whose truth you cannot determine.

Terms to Know

- *open sentence*
- *solution of an open sentence*
- *graph of an open sentence*
- *inequality*

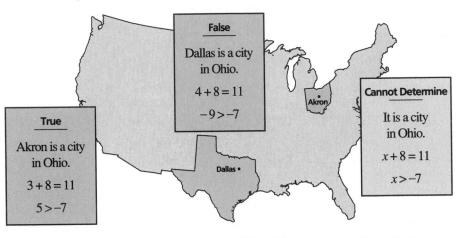

False
Dallas is a city in Ohio.
$4 + 8 = 11$
$-9 > -7$

Cannot Determine
It is a city in Ohio.
$x + 8 = 11$
$x > -7$

True
Akron is a city in Ohio.
$3 + 8 = 11$
$5 > -7$

You cannot determine the truth of the English sentence *It is a city in Ohio* because you do not know what city the word *it* represents. You cannot determine the truth of mathematical sentences like $x + 8 = 11$ and $x > -7$ because you do not know what number the variable x represents. For this reason, a mathematical sentence that contains a variable is called an **open sentence.**

An open sentence is by itself neither true nor false. When you substitute a real number for the variable, however, you can determine whether the result is true or false. Any value of the variable that results in a true sentence is called a **solution of the open sentence.** Because a solution is a real number, you can show the **graph of an open sentence** in one variable by graphing all the solutions on a number line.

Example 1

Solution

Graph the equation $8 = 12 + x$.

First solve the equation.
$$8 = 12 + x$$
$$8 - 12 = 12 + x - 12$$
$$-4 = x$$
The solution is -4.

Then graph the solution as a heavy dot on a number line.

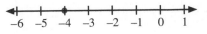

✓ Check Your Understanding

1. Would the graph in Example 1 be different if the equation were $x + 12 = 8$?

A mathematical sentence that has an inequality symbol between two numbers or quantities is an **inequality.** When an inequality is an open sentence, like $x > -4.2$, there are infinitely many real-number solutions. To graph an inequality like this on a number line, you use an open dot and an arrow.

Example 2

Solution

Graph the inequality $x > -4.2$.

Because -4.2 is not a solution of the inequality, you place an open dot at -4.2 on a number line. Then shade in a heavy arrow to the right to graph all numbers greater than -4.2.

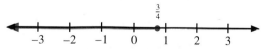

✎ **Check Your Understanding**

2. Why is -4.2 *not* a solution of $x > -4.2$?
3. Describe how the graph in Example 2 would be different if the inequality were $x < -4.2$.

Two other inequality symbols that are commonly used in mathematics are $\leq$ and $\geq$.

In Words	In Symbols
a is less than or equal to b.	$a \leq b$
a is greater than or equal to b.	$a \geq b$

On a number line, you graph an open sentence that contains one of these inequality symbols by using a closed dot and an arrow.

Example 3

Solution

Graph the inequality $y \leq \frac{3}{4}$.

Because $\frac{3}{4}$ is a solution of the inequality, you place a closed dot at $\frac{3}{4}$ on a number line. Then shade in a heavy arrow to the left to graph all numbers less than $\frac{3}{4}$.

✎ **Check Your Understanding**

4. Why is $\frac{3}{4}$ a solution of $y \leq \frac{3}{4}$?
5. Describe how the graph in Example 2 would be different if the inequality were $y \geq \frac{3}{4}$.

Guided Practice

For each open sentence, choose all the given numbers that are solutions.

1. $z + 8 = -1$ **a.** 7   **b.** -7 **c.** -9 **d.** 9

2. $n \geq -\frac{5}{8}$ **a.** $\frac{5}{8}$ **b.** -1.3 **c.** $-\frac{1}{2}$ **d.** $-\frac{5}{8}$

COMMUNICATION « *Writing*

Write each sentence in symbols.

«**3.** A number n is greater than or equal to -4.

«**4.** Fifteen is greater than a number x.

«**5.** A number q is less than -2.25.

«**6.** A number t is less than or equal to $-\frac{3}{5}$.

Write each inequality in words.

«**7.** $a < -14$ «**8.** $d \geq 3\frac{2}{9}$ «**9.** $k > -\frac{6}{7}$ «**10.** $m \leq -1.7$

Graph each open sentence.

11. $2r - 5 = 3$ **12.** $b < \frac{5}{6}$ **13.** $p > -7\frac{1}{3}$ **14.** $h \geq -3.5$

Exercises

Graph each open sentence.

1. $n + 4 = -5$ **2.** $c - 34 = -28$ **3.** $-11 = a - 8$

4. $69 = 62 + q$ **5.** $17 = 3m + 2$ **6.** $4z - 17 = -25$

7. $g < 2$ **8.** $y > -1$ **9.** $z > -3\frac{1}{3}$ **10.** $s < 5\frac{2}{5}$

11. $b \geq -6\frac{1}{7}$ **12.** $m \leq -\frac{2}{3}$ **13.** $a \leq 7.5$ **14.** $v \geq -2.25$

Match each statement with the correct graph.

15. $n \geq 25$

16. $m = 25$

17. $z < 25$

18. $p > 25$

19. $x \leq 25$

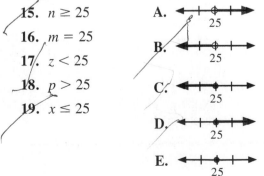

A.

B.

C.

D.

E.

The figure at the right shows the graph of $-3 < x < 1$. For Exercises 20–28, combine what you learned about this type of inequality in Chapter 1 with what you learned about graphing in this lesson.

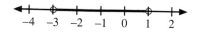

Tell how the graph of each inequality compares with the graph above.

20. $-3 \leq x \leq 1$ **21.** $1 > x > -3$ **22.** $-3 \leq x < 1$

Graph each inequality.

23. $-4 < t < 2$ **24.** $-1 > n > -5$ **25.** $0 \geq a \geq -3$

26. $1.5 \leq r \leq 6.5$ **27.** $-4 < x \leq 1$ **28.** $\frac{2}{3} \geq y \geq -\frac{2}{3}$

FUNCTIONS

There is a function of real numbers that is called the *greatest integer function*. This function pairs a real number with the greatest integer less than or equal to that number. The function rule is $x \rightarrow [x]$. For instance, when $x = 2\frac{1}{2}$, $[x] = 2$.

29. Complete the function table at the right.

30. List three numbers for which the greatest integer is 5.

31. Graph on a number line all the numbers for which the greatest integer is 0.

32. Use two inequality symbols to write a statement that represents all the numbers for which the greatest integer is 8. Use n as the variable.

x	$[x]$
$9\frac{1}{3}$	9
$7.\overline{6}$	?
4	?
$3.\overline{3}$	?
$\frac{7}{8}$	?

SPIRAL REVIEW

33. Estimate the sum: $157 + 389 + 438$ *(Toolbox Skill 2)*

34. Graph the inequality $x < -2.5$.   *(Lesson 7-10)*

35. DATA, *pages 288–289*   Write in scientific notation: the number of additions a computer could do in 1964   *(Lesson 2-10)*

36. *True* or *False?*: Two squares are always congruent.   *(Lesson 6-9)*

37. Use the formula $D = rt$. Let $r = 48$ mi/h and $t = 2.75$ h. Find D.   *(Lesson 4-10)*

38. Simplify: $\frac{30x}{42xy}$ *(Lesson 7-5)*

39. Evaluate $5n^4$ when $n = 3$.   *(Lesson 1-9)*

Negative and Zero Exponents

Objective: To simplify expressions involving negative and zero exponents, and to use negative exponents to write numbers in scientific notation.

EXPLORATION

1 Replace each ___?___ with the number that makes the statement true.

a. $\dfrac{a^2}{a^2} = \dfrac{\overset{1}{\cancel{a}} \cdot \overset{1}{\cancel{a}}}{\underset{1}{\cancel{a}} \cdot \underset{1}{\cancel{a}}} = \underline{\quad ? \quad}$

b. $\dfrac{a^2}{a^2} = a^{2-2} = a^?$

2 Use your results from Step 1. Complete: $1 = a^?$

3 Replace each ___?___ with the number that makes the statement true.

a. $\dfrac{a^3}{a^5} = \dfrac{\overset{1}{\cancel{a}} \cdot \overset{1}{\cancel{a}} \cdot \overset{1}{\cancel{a}}}{\underset{1}{\cancel{a}} \cdot \underset{1}{\cancel{a}} \cdot \underset{1}{\cancel{a}} \cdot a \cdot a} = \dfrac{1}{a \cdot a} = \dfrac{1}{a^?}$

b. $\dfrac{a^3}{a^5} = a^{3-5} = a^?$

4 Use your results from Step 3. Complete: $\dfrac{1}{a^2} = a^?$

If an expression has a negative exponent, you can write the expression as a fraction with a positive exponent.

If a is any nonzero number, then

$$a^{-n} = \frac{1}{a^n} \quad \text{and} \quad a^0 = 1.$$

Example 1

Simplify.

a. x^{-7} b. 3^{-3} c. $(-2)^{-2}$ d. $(5.67)^0$

Solution

a. $x^{-7} = \dfrac{1}{x^7}$ b. $3^{-3} = \dfrac{1}{3^3} = \dfrac{1}{27}$

c. $(-2)^{-2} = \dfrac{1}{(-2)^2} = \dfrac{1}{4}$ d. $(5.67)^0 = 1$

In Lesson 2-10 you learned to use positive powers of ten to express very large measures in *scientific notation*.

$$3.2 \times 10^{12} = 3.2 \times 1{,}000{,}000{,}000{,}000 = 3{,}200{,}000{,}000{,}000$$

Now that you know about negative exponents, you also can use scientific notation to express very small measures.

$$3.2 \times 10^{-4} = 3.2 \times \frac{1}{10^4} = 3.2 \times \frac{1}{10{,}000} = 3.2 \times 0.0001 = 0.00032$$

Example 2	**a.** Write 0.065 in scientific notation.
	b. Write 4.7×10^{-4} in decimal notation.
Solution	**a.** Move the decimal point to get a number that is at least 1, but less than 10.

$$0.065 = 6.5 \times 10^{-2}$$

2 places

b. Move the decimal point to the left.

$$4.7 \times 10^{-4} = 0.00047$$

4 places

✓ **Check Your Understanding**

1. How would Example 2(a) be different if the number were 0.65?
2. How would Example 2(b) be different if the number were 4.7×10^4?

Remember that you can use the [EE] or [EXP] key on your calculator to enter numbers in scientific notation. For instance, to enter 6.3×10^{-7} you use this key sequence.

| 6.3 | EE | | 7. | +/- |

Guided Practice

COMMUNICATION «*Reading*

Refer to the text on pages 328–329.

«**1.** What is the main idea of the lesson?

«**2.** List three major points that support the main idea.

Simplify.

3. 4^4 **4.** 6^3 **5.** $(-5)^2$ **6.** $(-3)^5$ **7.** $(-2)^7$

Write each number as a power of ten.

8. 0.01   **9.** 0.0001   **10.** 0.00001   **11.** $\dfrac{1}{1,000,000}$

Simplify.

12. x^{-4} **13.** 8^{-2} **14.** $(-3)^{-3}$ **15.** $(2,004,627)^0$

Write each number in scientific notation.

16. 0.0704   **17.** 0.00006   **18.** 0.00000057

Write each number in decimal notation.

19. 3.295×10^{-2} **20.** 1.7×10^{-5} **21.** 4.44×10^{-7}

Simplify.

1. y^{-12} 2. 7^{-2} 3. q^0 4. 324^0

5. 5^{-3} 6. w^{-200} 7. $(-4)^{-2}$ 8. $(-1)^{-13}$

Write each number in scientific notation.

9. 0.0072   10. 0.875   11. 0.0006012

12. 0.01234   13. 0.00000234   14. 0.0000005

Write each number in decimal notation.

15. 1.16×10^{-3} 16. 5.027×10^{-5} 17. 1×10^{-8}

18. 3.209×10^{-4} 19. 6.1×10^{-9} 20. 4×10^{-11}

21. When 0.00562 is written in scientific notation, what is the exponent in the power of 10?

22. Write the phrase *three to the negative fourth power* as a number without exponents.

CALCULATOR

Choose the key sequence you would use to enter each number on a calculator.

23. 2.18×10^{-6}

24. 2.18×10^6

A. [2.18] (+/_) (EE) [6]

B. [2.18] (EE) [6]

C. [2.18] (+/_) (EE) [6] (+/_)

D. [2.18] (EE) [6] (+/_)

CONNECTING MATHEMATICS AND SCIENCE

In biology you use a microscope to study organisms. Any organism viewed under a microscope is called a *specimen*. A specimen magnified at 100X appears 100 times as large as its actual size. Some specimens are too small to see with the human eye. The measurements of these specimens are usually written in scientific notation.

25. What is the meaning of 1500X?

26. The diameter of the smallest organism is 1.8×10^{-9} mm. Write this in decimal notation.

27. The width of a human hair is 0.02 cm. Write this in scientific notation.

28. A bacterial cell appears to be 0.24 cm wide when viewed under a microscope at 500X. What is the actual width of the cell? Write this width in scientific notation.

29. A virus cell with width 6.5×10^{-4} mm is viewed under a microscope at 20,000X. How wide does the cell appear to be?

30. RESEARCH Two basic types of microscopes are *optical microscopes* and *electron microscopes*. What method does each microscope use to magnify specimens? How powerful is each with regard to magnifying specimens?

SPIRAL REVIEW

31. Complete: Two lines in the same plane that do not intersect are __?__. *(Lesson 6-4)*

32. Write 0.0000074 in scientific notation. *(Lesson 7-11)*

33. Tell whether 111 is *prime* or *composite*. *(Lesson 7-1)*

34. DATA, pages 140–141 Draw a graph to represent the data in *Average Rates of Trains*. *(Lesson 5-5)*

Self-Test 3

Express each rational number as the quotient of two integers.

1. 3.4   **2.** $-4\frac{5}{8}$ **3.** -9 **4.** $1.\overline{2}$ **7-9**

Graph each open sentence.

5. $x + 5 = -2$ **6.** $3y - 4 = 14$ **7.** $22 = -5n + 7$ **7-10**

8. $p \le 4$ **9.** $x < -2.5$ **10.** $z \ge -\frac{1}{3}$

Simplify.

11. $(-8)^{-1}$ **12.** $(-3)^0$ **13.** w^{-3} **14.** 4^{-4} **7-11**

15. Write 0.00003 in scientific notation.

16. Write 2.43×10^{-7} in decimal notation.

Terms to Know

factor (p. 290)
prime number (p. 290)
composite number (p. 290)
prime factorization (p. 290)
common factor (p. 293)
greatest common factor (GCF)
  (p. 293)
multiple (p. 296)
common multiple (p. 296)
least common multiple (LCM)
  (p. 296)
equivalent fractions (p. 302)
lowest terms (p. 303)
algebraic fraction (p. 306)

simplify a fraction (p. 306)
quotient of powers rule (p. 306)
least common denominator (LCD)
  (p. 310)
terminating decimal (p. 313)
repeating decimal (p. 313)
rational number (p. 321)
irrational number (p. 321)
real number (p. 321)
open sentence (p. 324)
solution of an open sentence
  (p. 324)
graph of an open sentence (p. 324)
inequality (p. 325)

Choose the correct term from the list above to complete each sentence.

1. Fractions that represent the same amount are called __?__.

2. Any number that can be written as a quotient of two integers, where the divisor does not equal zero, is called a(n) __?__.

3. A(n) __?__ is a mathematical sentence that contains a variable.

4. A decimal in which a block of digits repeats is a(n) __?__ decimal.

5. A number with more than two factors is a(n) __?__ number.

6. The greatest number in a list of common factors is the __?__.

Tell whether each number is *prime* or *composite*. *(Lesson 7-1)*

7. 13   **8.** 81   **9.** 77   **10.** 43

Write the prime factorization of each number. *(Lesson 7-1)*

11. 75   **12.** 216   **13.** 350   **14.** 1200

Find the GCF. *(Lesson 7-2)*

15. 48 and 64   **16.** $66a$ and $84a$ **17.** $15xy$ and $24x$ **18.** $36z^2$ and $54z^5$

Find the LCM. *(Lesson 7-3)*

19. 6 and 15   **20.** 8 and 12   **21.** $4y$ and $5x$ **22.** $2c$ and $7c^2$

Write each fraction in lowest terms. *(Lesson 7-4)*

23. $\frac{6}{12}$ **24.** $\frac{6}{9}$ **25.** $\frac{8}{10}$ **26.** $\frac{3}{21}$ **27.** $\frac{48}{10}$ **28.** $\frac{27}{6}$

Chapter 7

Simplify. *(Lesson 7-5)*

29. $\dfrac{10ac}{5c}$ **30.** $\dfrac{12a}{15a}$ **31.** $\dfrac{v^{12}}{v^5}$ **32.** $\dfrac{x^{18}}{x^6}$ **33.** $\dfrac{4z^5}{z^2}$ **34.** $\dfrac{x^8}{2x}$

Replace each __?__ with >, <, or =. *(Lesson 7-6)*

35. $\dfrac{1}{5}$ _?_ $\dfrac{1}{6}$ **36.** $\dfrac{7}{9}$ _?_ $\dfrac{17}{18}$ **37.** $\dfrac{8}{11}$ _?_ $\dfrac{8}{13}$ **38.** $\dfrac{2}{3}$ _?_ $\dfrac{7}{12}$

Write each fraction or mixed number as a decimal. *(Lesson 7-7)*

39. $\dfrac{1}{5}$ **40.** $\dfrac{3}{10}$ **41.** $\dfrac{4}{9}$ **42.** $\dfrac{5}{11}$ **43.** $3\dfrac{7}{20}$ **44.** $5\dfrac{11}{50}$

Write each decimal as a fraction or mixed number in lowest terms. *(Lesson 7-7)*

45. 0.37   **46.** 0.255   **47.** 6.875   **48.** $4.\overline{6}$

Solve by drawing a diagram. *(Lesson 7-8)*

49. The capacities of three buckets are 10 gal, 4 gal, and 3 gal. How could you use each bucket once to measure exactly 9 gal of water?

50. How many diagonals can be drawn in a pentagon?

Express each rational number as the quotient of two integers. *(Lesson 7-9)*

51. 3.2   **52.** $6\dfrac{1}{5}$ **53.** -7 **54.** 0   **55.** 0.37   **56.** $-9.\overline{6}$

Use the number line below. Name the point that represents each number. *(Lesson 7-9)*

57. 2.25   **58.** $\dfrac{9}{2}$ **59.** -0.55 **60.** $-2\dfrac{3}{5}$ **61.** $-\dfrac{1}{3}$ **62.** 4.34

Graph each open sentence. *(Lesson 7-10)*

63. $x + 5 = 1$ **64.** $-4n + 5 = -3$ **65.** $z < -1.7$ **66.** $y \geq \dfrac{2}{3}$

Simplify. *(Lesson 7-11)*

67. m^{-5} **68.** 3^{-4} **69.** 12^0 **70.** $(-3)^{-2}$

Write each number in scientific notation. *(Lesson 7-11)*

71. 0.000074   **72.** 0.000821   **73.** 0.00000069   **74.** 0.005325

Write each number in decimal notation. *(Lesson 7-11)*

75. 2.7×10^{-3} **76.** 5.5×10^{-6} **77.** 9×10^{-9} **78.** 3.24×10^{-7}

Chapter Test

Write the prime factorization of each number.

1. 84
2. 150
3. 625
4. 1450

7-1

Find the GCF.

5. 36 and 64
6. $27a$ and $45a$
7. $48z^2$ and $112z^6$

7-2

Find the LCM.

8. 18 and 30
9. $3cd$ and $8d$
10. $5a^2$ and $2a^5$

7-3

Write each fraction in lowest terms.

11. $\frac{8}{12}$
12. $\frac{9}{15}$
13. $\frac{18}{8}$
14. $\frac{21}{6}$

7-4

Simplify.

15. $\frac{9x}{15}$
16. $\frac{24b}{18bc}$
17. $\frac{x^{15}}{x^5}$
18. $\frac{9n^8}{n^2}$

7-5

Replace each __?__ with >, <, or =.

19. $\frac{1}{4}$ __?__ $\frac{1}{6}$
20. $\frac{6}{13}$ __?__ $\frac{6}{7}$
21. $\frac{3}{8}$ __?__ $\frac{7}{16}$

7-6

Write each fraction or mixed number as a decimal.

22. $\frac{4}{5}$
23. $\frac{4}{15}$
24. $4\frac{1}{6}$
25. $7\frac{13}{20}$

7-7

Write each decimal as a fraction or mixed number in lowest terms.

26. 0.672
27. 0.27
28. $7.\overline{3}$
29. 2.88

30. Solve by using a diagram: Eight players participated in a single-elimination tennis tournament. In such a tournament, each player is out after one loss. How many games did the champion play?

7-8

Express each rational number as the quotient of two integers.

31. 7.4
32. -4.25
33. -8
34. $3.\overline{3}$

7-9

35. Write the number -1 as a quotient of two integers in three different ways.

36. Name three real numbers that are between $\frac{1}{3}$ and $\frac{1}{2}$.

Graph each open sentence.

37. $6 = x + 9$
38. $3w - 8 = 1$
39. $y \geq 5.5$
40. $m < \frac{7}{8}$

7-10

Write each number in scientific notation.

41. 0.00076
42. 0.00000335

Write each number in decimal notation.

43. 3×10^{-5}
44. 4.7×10^{-8}

7-11

Repeating Decimals as Fractions

Objective: To write repeating decimals as fractions.

You may have memorized fractional equivalents for some common repeating decimals, such as $0.\overline{3} = \frac{1}{3}$. Using algebra, however, it is possible to find the fractional equivalent of any repeating decimal.

Example

Write $0.\overline{27}$ as a fraction.

Solution

Let $n = 0.\overline{27}$.
Then $100n = 27.\overline{27}$. ◄── Multiply n by 10^2 because there are 2 digits in the repeating block.

Subtract n from $100n$.

$$
\begin{array}{r}
100n = 27.272727\ldots \\
- \quad n = -0.272727\ldots \\
\hline
99n = 27
\end{array}
$$

Solve for n.

$$\frac{99n}{99} = \frac{27}{99}$$

$$n = \frac{27}{99} = \frac{3}{11}$$

Exercises

Write each repeating decimal as a fraction.

1. $0.\overline{5}$ 2. $0.\overline{8}$ 3. $0.\overline{35}$ 4. $0.\overline{52}$

5. $2.\overline{3}$ 6. $6.\overline{7}$ 7. $3.\overline{21}$ 8. $9.\overline{45}$

9. $1.\overline{123}$ 10. $4.\overline{534}$ 11. $5.7\overline{2}$ 12. $2.3\overline{21}$

CALCULATOR

13. To multiply $\frac{1}{3} \times 3$ using his four-function calculator, Gustavo entered this key sequence.

The result was $\boxed{0.9999999}$. Francine entered the same key sequence on her scientific calculator and got the result $\boxed{1.}$. What conclusion can you make about $0.\overline{9}$ and 1?

14. Refer to Exercise 13. What do you think the results would be if Gustavo and Francine entered this key sequence on their calculators?

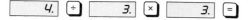

Standardized Testing Practice

Choose the letter of the correct answer.

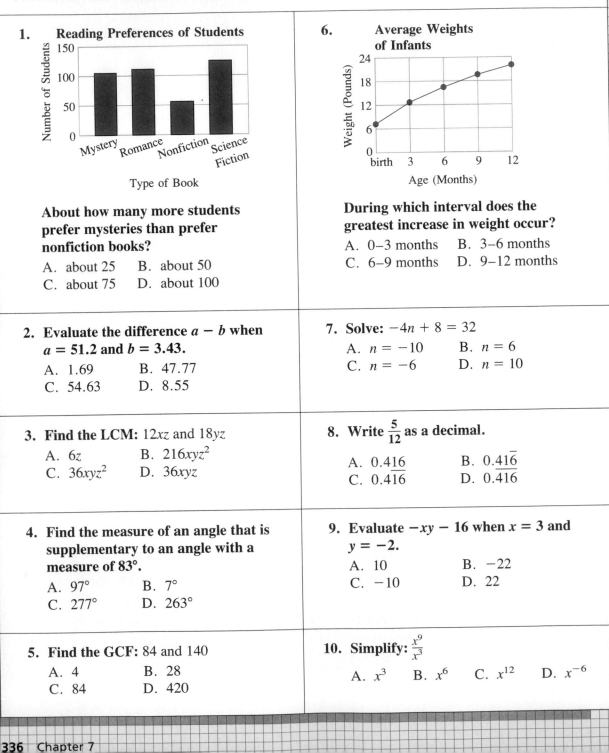

1. Reading Preferences of Students

About how many more students prefer mysteries than prefer nonfiction books?

A. about 25   B. about 50
C. about 75   D. about 100

2. Evaluate the difference $a - b$ when $a = 51.2$ and $b = 3.43$.

A. 1.69   B. 47.77
C. 54.63   D. 8.55

3. Find the LCM: $12xz$ and $18yz$

A. $6z$ B. $216xyz^2$
C. $36xyz^2$ D. $36xyz$

4. Find the measure of an angle that is supplementary to an angle with a measure of $83°$.

A. $97°$ B. $7°$
C. $277°$ D. $263°$

5. Find the GCF: 84 and 140

A. 4   B. 28
C. 84   D. 420

6. Average Weights of Infants

During which interval does the greatest increase in weight occur?

A. 0–3 months   B. 3–6 months
C. 6–9 months   D. 9–12 months

7. Solve: $-4n + 8 = 32$

A. $n = -10$ B. $n = 6$
C. $n = -6$ D. $n = 10$

8. Write $\frac{5}{12}$ as a decimal.

A. $0.41\overline{6}$ B. $0.41\overline{6}$
C. $0.41\overline{6}$ D. $0.\overline{416}$

9. Evaluate $-xy - 16$ when $x = 3$ and $y = -2$.

A. 10   B. -22
C. -10 D. 22

10. Simplify: $\dfrac{x^9}{x^3}$

A. x^3 B. x^6 C. x^{12} D. x^{-6}

11. **Choose the true statement about the data:** 42, 38, 51, 42, 72
    - A. The range is 30.
    - B. The mean is less than the mode.
    - C. The mode and the median are equal.
    - D. The mean and the median are equal.

12. **Write an equation for the situation.** Ella is 165 cm tall. She is 3 cm shorter than Deion. How tall is Deion?
    - A. $3d = 165$
    - B. $d + 165 = 3$
    - C. $d + 3 = 165$
    - D. $d - 3 = 165$

13. **At noon, the temperature was −9°C. During the next 5 h, it fell 4°C. What was the temperature at 5:00 P.M.?**
    - A. 5°C
    - B. 13°C
    - C. −5°C
    - D. −13°C

14. **Find the answer mentally:** 6(198)
    - A. 1212
    - B. 1248
    - C. 1188
    - D. 1152

15. **To which of the sets does the number 333 belong?**
    - A. whole numbers
    - B. integers
    - C. rational numbers
    - D. all of the above

16. **Eva did 10, 15, and 20 sit-ups each day during the first, second, and third weeks of a fitness program. If she continued the pattern, how many sit-ups did she do each day during the fifth week?**
    - A. 30
    - B. 40
    - C. 25
    - D. 35

17. **Choose the fraction that is *not* equivalent to $\frac{3}{4}$.**
    - A. $\frac{39}{52}$
    - B. $\frac{21}{28}$
    - C. $\frac{75}{100}$
    - D. $\frac{69}{96}$

18. **Find the sum of the measures of the angles in a 20-sided polygon.**
    - A. 3240°
    - B. 3600°
    - C. 360°
    - D. 3598°

19. **Simplify:** $8a^6 \cdot 5a^2$
    - A. $3a^4$
    - B. $40a^{12}$
    - C. $13a^8$
    - D. $40a^8$

20. **Describe the triangle.**
    - A. right, scalene
    - B. acute, scalene
    - C. equilateral
    - D. right, isosceles

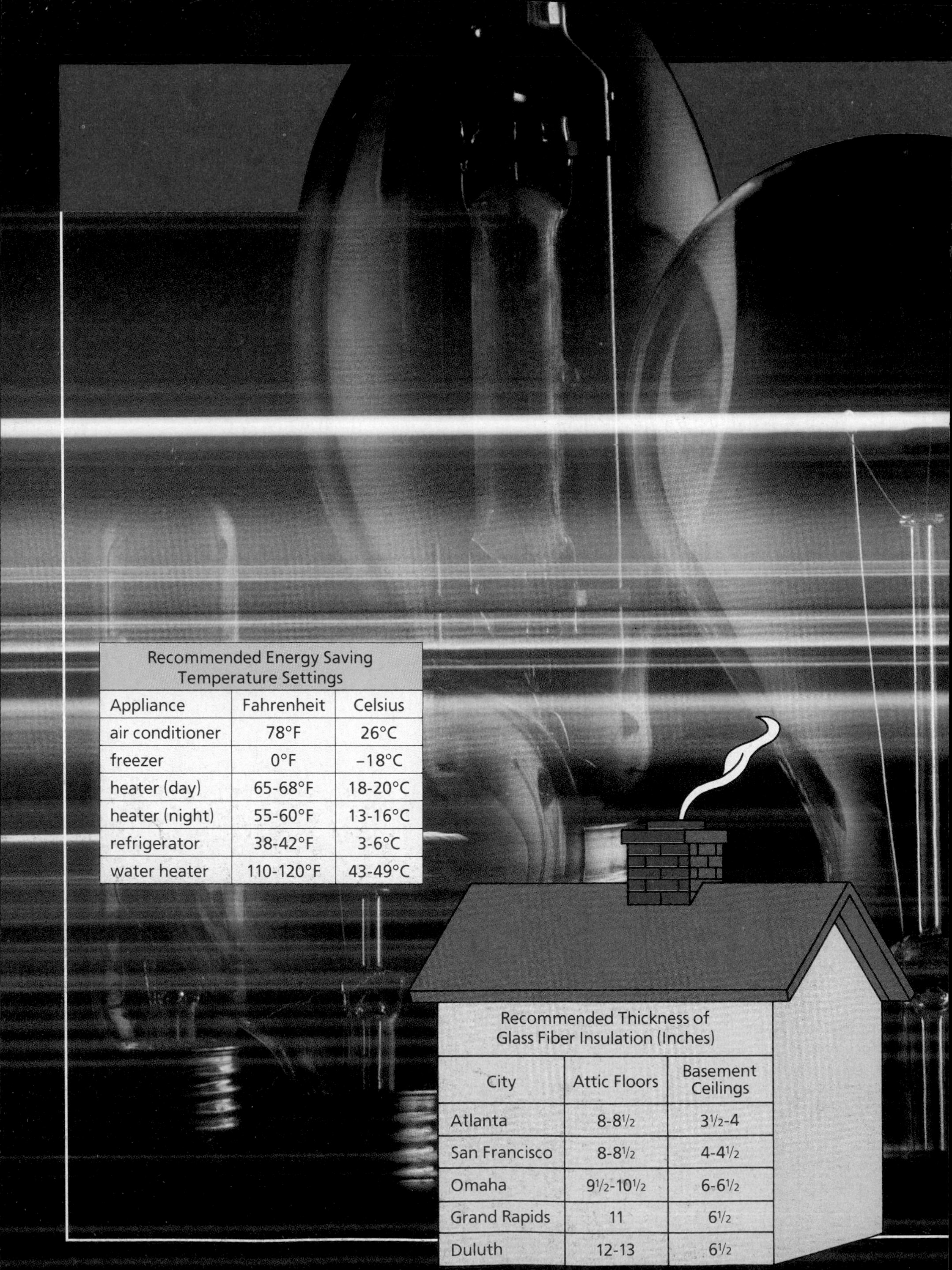

Recommended Energy Saving Temperature Settings		
Appliance	Fahrenheit	Celsius
air conditioner	78°F	26°C
freezer	0°F	−18°C
heater (day)	65-68°F	18-20°C
heater (night)	55-60°F	13-16°C
refrigerator	38-42°F	3-6°C
water heater	110-120°F	43-49°C

Recommended Thickness of Glass Fiber Insulation (Inches)		
City	Attic Floors	Basement Ceilings
Atlanta	8-8½	3½-4
San Francisco	8-8½	4-4½
Omaha	9½-10½	6-6½
Grand Rapids	11	6½
Duluth	12-13	6½

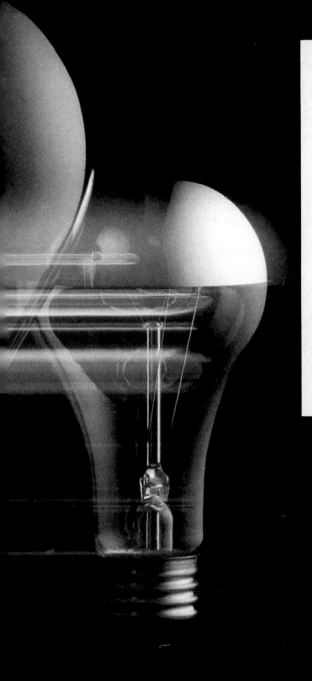

Water Heater
Electric
First hour rating: 58

ENERGY GUIDE

Estimates on the scale are based on a national average electric rate of 8.04¢ per kilowatt-hour.

Only models with first hour ratings of 56-64 gallons are used on this scale.

Model with lowest energy cost
$423
▼

$467

THIS ▼ MODEL

Model with highest energy cost
$547
▼

Estimated yearly energy cost

Your cost will vary depending on your local energy rate and how you use the product. This energy rate is based on U.S. Government standard tests.

Cost per kilowatt-hour		YEARLY COST
	4¢	$232
	6¢	$348
	8¢	$465
	10¢	$581
	12¢	$697
	14¢	$813

Ask your salesperson or local utility for the energy rate in your area.

The brightness of a light bulb is indicated by lumens, not wattage. Wattage indicates how much energy a bulb uses. It is most energy-efficient to use bulbs with high lumens and low wattage. Compared to an incandescent light bulb, a fluorescent bulb is 3 to 5 times more efficient. For example, a 40-watt fluorescent bulb gives off 3200 lumens, and a 60-watt incandescent bulb gives off only 882 lumens.

Multiplying Rational Numbers

Objective: To multiply rational numbers and algebraic fractions.

EXPLORATION

1 What part of the whole is shaded blue?

2 What part of the whole is shaded pink?

3 What part of the whole is shaded purple?

4 Use your results from Step 3. Complete: $\frac{1}{2} \cdot \frac{1}{3} = \frac{?}{?}$

5 Draw a model to represent each product.

    a. $\frac{1}{3} \cdot \frac{1}{4}$ **b.** $\frac{2}{3} \cdot \frac{1}{4}$ **c.** $\frac{3}{3} \cdot \frac{1}{4}$

To multiply with fractions, mixed numbers, and whole numbers, write all numbers as fractions and then multiply.

Generalization: *Multiplying Rational Numbers*

To multiply rational numbers, multiply the numerators and then multiply the denominators.

In Arithmetic

$$\frac{1}{3} \cdot \frac{4}{5} = \frac{4}{15}$$

In Algebra

$$\frac{a}{b} \cdot \frac{c}{d} = \frac{ac}{bd}, \ b \neq 0, \ d \neq 0$$

Two numbers, like $\frac{5}{9}$ and $\frac{9}{5}$, whose product is 1 are called **reciprocals.** The reciprocal of $-\frac{5}{9}$ is $-\frac{9}{5}$.

Multiplication Property of Reciprocals

The product of a number and its reciprocal is 1.

In Arithmetic

$$\frac{5}{9} \cdot \frac{9}{5} = 1$$

In Algebra

$$\frac{a}{b} \cdot \frac{b}{a} = 1, \ a \neq 0, \ b \neq 0$$

When two numbers are reciprocals of each other, they are also called **multiplicative inverses** of each other.

Example 1

Find each product. Simplify if possible.

a. $\frac{2}{3} \cdot \frac{7}{8}$

b. $-2\left(\frac{3}{4}\right)$

Solution

a. $\frac{2}{3} \cdot \frac{7}{8} = \frac{\overset{1}{2}}{3} \cdot \frac{7}{\underset{4}{8}} = \frac{1 \cdot 7}{3 \cdot 4} = \frac{7}{12}$

b. $-2\left(\frac{3}{4}\right) = \frac{-2}{1} \cdot \frac{3}{4}$ ⟵ Write each factor as a fraction.

$= \frac{\overset{-1}{\cancel{-2}}}{1} \cdot \frac{3}{\underset{2}{4}}$

$= \frac{-1 \cdot 3}{1 \cdot 2} = \frac{-3}{2} = -1\frac{1}{2}$

✓ Check Your Understanding

1. In Example 1(b), why is -2 rewritten as a fraction, $\frac{-2}{1}$?
2. Why is the answer to Example 1(b) negative?
3. How would the answer to Example 1(b) be different if $\frac{3}{4}$ were $-\frac{3}{4}$?

You can use the fraction key, $\boxed{a^{b/c}}$, on a calculator to perform operations with fractions and mixed numbers. To find the product $\left(3\frac{1}{8}\right)\left(\frac{1}{3}\right)$, you use the key sequence below. The result is $1\frac{1}{24}$.

$\boxed{1\_1\lrcorner\ 24.}$

$\boxed{3.}$ $\boxed{a^{b/c}}$ $\boxed{1.}$ $\boxed{a^{b/c}}$ $\boxed{8.}$ $\boxed{\times}$ $\boxed{1.}$ $\boxed{a^{b/c}}$ $\boxed{3.}$ $\boxed{=}$

Sometimes you may only need an estimate of a product, or you may want an estimate to check your work. To estimate, first round each factor and then multiply. If one factor is less than 1, you may want to use compatible numbers and then multiply.

Example 2

Estimate each product.

a. $\left(2\frac{5}{8}\right)\left(-3\frac{1}{3}\right)$

b. $\left(-\frac{5}{11}\right)\left(-5\frac{1}{12}\right)$

Solution

a. Round.

$\left(2\frac{5}{8}\right)\left(-3\frac{1}{3}\right) \rightarrow \underbrace{(3)(-3)}_{\text{about } -9}$

b. Use compatible numbers.

$\left(-\frac{5}{11}\right)\left(-5\frac{1}{12}\right) \rightarrow \underbrace{\left(-\frac{1}{2}\right)(-6)}_{\text{about } 3}$

✓ Check Your Understanding

4. In Example 2(a), $2\frac{5}{8}$ is rounded to 3 because $\frac{5}{8}$ is greater than or equal to $\frac{1}{2}$. Explain why $-3\frac{1}{3}$ is rounded to -3.
5. In Example 2(b), why is $-5\frac{1}{12}$ rewritten as -6?

Example 3

Simplify $\frac{5x}{3} \cdot \frac{7a}{10x}$.

Solution

$$\frac{5x}{3} \cdot \frac{7a}{10x} = \frac{\overset{1}{\cancel{5x}}}{3} \cdot \frac{7a}{\underset{2}{\cancel{10x}}}$$

$$= \frac{1 \cdot 7a}{3 \cdot 2} \quad \longleftarrow \quad \begin{array}{l}\text{Multiply the numerators,} \\ \text{and then the denominators.}\end{array}$$

$$= \frac{7a}{6}$$

When multiplying positive or negative rational numbers that are written as decimals, apply the rules you learned for multiplication of integers.

Guided Practice

« **1.** COMMUNICATION « *Reading* Replace each ___?___ with the correct word or number.

When you multiply a number and its reciprocal, the result is ___?___.
Another name for reciprocals is ___?___.
   To estimate the product of two numbers, one of which is less than 1, you can use ___?___ numbers and then multiply.

« **2.** COMMUNICATION « *Discussion* You know that $-\frac{7}{10} \cdot \frac{9}{10} = -\frac{63}{100}$.
Compare $-\frac{7}{10} \cdot \frac{9}{10}$ to $(-0.7)(0.9)$. Is the product $(-0.7)(0.9)$ *positive* or *negative*? Explain.

Find each product. Simplify if possible.

3. $\frac{5}{12} \cdot \frac{1}{5}$ **4.** $-\frac{3}{4} \cdot 3\frac{1}{3}$ **5.** $(-0.3)(0.8)$ **6.** $(-2.5)(-0.3)$

7. $\frac{4m}{7} \cdot \frac{n}{8m}$ **8.** $\frac{9a}{10} \cdot \frac{2}{3}$ **9.** $\frac{25}{c} \cdot \frac{6c}{5}$ **10.** $\frac{7w}{8} \cdot \frac{11}{14}$

Estimate each product.

11. $\left(5\frac{1}{8}\right)\left(-5\frac{7}{8}\right)$ **12.** $\left(-6\frac{7}{10}\right)\left(-3\frac{2}{5}\right)$ **13.** $\left(\frac{7}{20}\right)\left(8\frac{11}{12}\right)$ **14.** $\left(-\frac{6}{11}\right)\left(-4\frac{2}{11}\right)$

Exercises

Find each product. Simplify if possible.

1. $\frac{8}{9} \cdot \frac{4}{5}$ **2.** $\frac{2}{3} \cdot \frac{3}{2}$ **3.** $\frac{5}{6}\left(-2\frac{1}{4}\right)$ **4.** $\left(-4\frac{1}{3}\right)\left(-\frac{9}{10}\right)$

5. $(-56)\left(-\frac{1}{56}\right)$ **6.** $-\frac{3}{20} \cdot 16$ **7.** $5 \cdot 6\frac{7}{10}$ **8.** $3\frac{3}{4} \cdot 12$

9. $\left(-10\frac{1}{2}\right)\left(-5\frac{1}{3}\right)$ **10.** $\left(7\frac{1}{3}\right)\left(-1\frac{1}{5}\right)$ **11.** $(-0.7)(3.5)$ **12.** $\left(\frac{3n}{8}\right)\left(\frac{2}{n}\right)$

13. $\left(\frac{z}{10}\right)\left(\frac{5}{3z}\right)$ **14.** $(-2.8)(-0.9)$ **15.** $\left(\frac{15a}{4}\right)\left(\frac{b}{3}\right)$ **16.** $\left(\frac{b}{7}\right)\left(\frac{14a}{b}\right)$

Estimate each product.

17. $\left(-4\frac{1}{2}\right)\left(-1\frac{2}{3}\right)$ **18.** $\left(-17\frac{9}{10}\right)\left(\frac{2}{3}\right)$ **19.** $\left(-\frac{3}{11}\right)\left(7\frac{2}{3}\right)$ **20.** $\left(2\frac{5}{8}\right)\left(3\frac{3}{7}\right)$

21. Leigh made nine new window curtains. How many yards of material did Leigh use if each curtain required $2\frac{2}{3}$ yd?

22. The average weight of eight chickens was $7\frac{3}{4}$ lb. About how much was the total weight of the chickens?

Find each product. Simplify if possible.

23. $\frac{1}{2} \cdot 0.75$ **24.** $\frac{3}{8}(-0.6)$ **25.** $(-0.25)\left(-4\frac{1}{5}\right)$

26. $\left(3\frac{1}{7}\right)(10)\left(11\frac{1}{5}\right)$ **27.** $\left(-1\frac{1}{7}\right)\left(1\frac{3}{4}\right)(12)$ **28.** $\left(-5\frac{5}{8}\right)\left(\frac{3}{5}\right)(4)$

MENTAL MATH

You can use the distributive property to multiply a whole number by a mixed number mentally. For example, $6 \cdot 5\frac{2}{3} = 6(5) + 6\left(\frac{2}{3}\right) = 30 + 4 = 34$.

Multiply mentally.

29. $3 \cdot 9\frac{1}{4}$ **30.** $10 \cdot 5\frac{3}{5}$ **31.** $8 \cdot 4\frac{1}{4}$ **32.** $4 \cdot 5\frac{3}{8}$

⊞ CALCULATOR

Write the calculator key sequence you would use to find each product. Assume that the calculator has a fraction key.

33. $4\frac{2}{7} \cdot \frac{2}{7}$ **34.** $8\frac{11}{12} \cdot \frac{19}{20}$ **35.** $\frac{4}{5}\left(17\frac{3}{8}\right)$ **36.** $\left(1\frac{1}{2}\right)\left(111\frac{1}{2}\right)$

SPIRAL REVIEW

37. Estimate the product: 207×92 *(Toolbox Skill 1)*

38. Solve by using the guess-and-check strategy: Tim bought some shirts costing $19 each and some ties costing $14 each. He spent $127. How many of each did he buy? *(Lesson 4-2)*

39. Evaluate $7a - 3b$ when $a = -2$ and $b = -8$. *(Lesson 3-6)*

40. Simplify: $\left(\frac{5c}{9}\right)\left(\frac{d}{10c}\right)$ *(Lesson 8-1)*

41. Estimate the sum: $17 + 24 + 22 + 19$ *(Toolbox Skill 4)*

42. The weekly salaries at a store are $315, $295, $325, and $680. Does the mean describe these data well? Explain. *(Lesson 5-9)*

Dividing Rational Numbers

Objective: To find the quotients of rational numbers and of algebraic fractions.

1 How many halves are there in 3?

$$3 \div \frac{1}{2} = \underline{\quad?\quad}$$

A related multiplication is $3 \times \underline{\quad?\quad} = 6$.

2 How many two thirds are in 4?

$$4 \div \frac{2}{3} = \underline{\quad?\quad}$$

A related multiplication is
$4 \times \underline{\quad?\quad} = 6$.

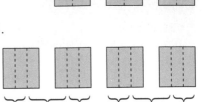

3 When you rewrite a division as a related multiplication, you multiply by the $\underline{\quad?\quad}$ of the divisor.

Generalization: *Dividing Rational Numbers*

To divide rational numbers, multiply by the reciprocal of the divisor.

In Arithmetic

$$\frac{1}{5} \div \frac{2}{3} = \frac{1}{5} \cdot \frac{3}{2} = \frac{3}{10}$$

In Algebra

$$\frac{a}{b} \div \frac{c}{d} = \frac{a}{b} \cdot \frac{d}{c} = \frac{ad}{bc}$$

$$b \neq 0, c \neq 0, d \neq 0$$

Example 1

Find each quotient. Simplify if possible.

a. $5 \div \frac{1}{2}$

b. $-\frac{5}{6} \div \left(-\frac{3}{4}\right)$

Solution

a. $5 \div \frac{1}{2} = \frac{5}{1} \cdot \frac{2}{1}$

$= 10$

b. $-\frac{5}{6} \div \left(-\frac{3}{4}\right) = \frac{-5}{6} \cdot \frac{-4}{3}$

$$= \frac{-5}{\overset{}{\underset{3}{6}}} \cdot \frac{\overset{-2}{\cancel{-4}}}{3}$$

$$= \frac{10}{9} = 1\frac{1}{9}$$

✓ Check Your Understanding

Would the answer to Example 1(a) be different if the quotient were $\frac{1}{2} \div 5$? Explain.

Some calculators have a reciprocal key, $\boxed{1/\text{x}}$. You can use this key to find a reciprocal, which the calculator displays as a decimal. To find the reciprocal of -8, you can use this key sequence.

$$\boxed{\quad 8.\quad}\ \boxed{+/_-}\ \boxed{1/\text{x}}$$

To find the quotient of variable expressions, write each expression as a fraction. Then multiply by the reciprocal of the divisor.

Example 2   **Simplify:**   **a.** $\dfrac{a}{2} \div \dfrac{a}{14}$ **b.** $\dfrac{3n}{2x} \div 6n$

Solution

a. $\dfrac{a}{2} \div \dfrac{a}{14} = \dfrac{a}{2} \cdot \dfrac{14}{a}$

$\quad = \dfrac{\overset{1}{\cancel{a}}}{\underset{1}{\cancel{2}}} \cdot \dfrac{\overset{7}{\cancel{14}}}{\underset{1}{\cancel{a}}}$

$\quad = 7$

b. $\dfrac{3n}{2x} \div 6n = \dfrac{3n}{2x} \div \dfrac{6n}{1}$

$\quad = \dfrac{3n}{2x} \cdot \dfrac{1}{6n}$

$\quad = \dfrac{\overset{1}{\cancel{3n}}}{2x} \cdot \dfrac{1}{\underset{2}{\cancel{6n}}}$

$\quad = \dfrac{1}{4x}$

When dividing positive or negative rational numbers that are written as decimals, apply the rules you learned for division of integers.

Guided Practice

«**1.** COMMUNICATION «*Discussion*   You know that $-\dfrac{1}{4} \div \dfrac{3}{4} = -\dfrac{1}{3}$. Compare $-\dfrac{1}{4} \div \dfrac{3}{4}$ to $(-0.25) \div (0.75)$. Is the quotient $(-0.25) \div (0.75)$ *positive* or *negative*? Explain.

Replace each __?__ with the number or variable expression that makes the statement true.

2. $\dfrac{4}{5} \div \left(-\dfrac{1}{8}\right) = \dfrac{4}{5} \cdot \underline{\ ?\ }$ **3.** $-\dfrac{6}{7} \div 7 = -\dfrac{6}{7} \cdot \underline{\ ?\ }$

4. $\dfrac{4z}{9} \div 18z = \dfrac{4z}{9} \cdot \underline{\ ?\ }$ **5.** $\dfrac{10}{a} \div \dfrac{5a}{6} = \dfrac{10}{a} \cdot \underline{\ ?\ }$

Find each quotient. Simplify if possible.

6. $-\dfrac{2}{3} \div \dfrac{1}{4}$ **7.** $8 \div \dfrac{2}{3}$ **8.** $-0.1 \div 10$ **9.** $1.5 \div (-0.75)$

10. $\dfrac{c}{5} \div \dfrac{c}{10}$ **11.** $\dfrac{4n}{5} \div 8n$ **12.** $\dfrac{m}{7} \div 7m$ **13.** $\dfrac{z}{8} \div \dfrac{z}{12}$

Find each quotient. Simplify if possible.

1. $-\dfrac{5}{8} \div \left(-\dfrac{5}{16}\right)$ 2. $\dfrac{3}{4} \div \dfrac{1}{5}$ 3. $18 \div \left(-\dfrac{1}{3}\right)$ 4. $-24 \div \dfrac{3}{4}$

5. $\dfrac{w}{5} \div \dfrac{w}{15}$ 6. $\dfrac{5n}{6} \div \dfrac{d}{5}$ 7. $\dfrac{2x}{3y} \div 3x$ 8. $\dfrac{m}{4} \div 8m$

9. $3\dfrac{1}{5} \div (-5)$ 10. $\dfrac{13}{6} \div \dfrac{6}{a}$ 11. $-1\dfrac{2}{3} \div \left(-\dfrac{1}{5}\right)$ 12. $-3\dfrac{1}{8} \div \left(-4\dfrac{1}{8}\right)$

13. $\dfrac{11}{5} \div 11y$ 14. $-\dfrac{14}{15} \div 7$ 15. $3.4 \div (-1.7)$ 16. $-0.56 \div (-0.7)$

17. To fasten tents to the ground, the campers cut pieces $5\frac{3}{4}$ ft long from a rope that was 46 ft long. How many pieces could be cut?

18. Al is working in a print shop on a page that is $8\frac{1}{2}$ in. wide. The page must be divided into four columns of equal width. What should be the width of each column?

CALCULATOR

Use a calculator reciprocal key $\boxed{1/x}$ to find the reciprocal of each number.

19. -2 20. -0.50 21. $\dfrac{4}{3}$ 22. 1.6

PROBLEM SOLVING/APPLICATION

APPLE MUFFINS

$1\frac{1}{2}$ c milk	2 tbsp baking soda
2 eggs	2 tsp cinnamon
$\frac{1}{2}$ c shortening	$\frac{2}{3}$ c raisins
$4\frac{1}{2}$ c flour	2 c chopped apple

Maria can make two dozen apple muffins using her recipe at the left. Sometimes she changes the amounts of the ingredients because she wants to make more than or less than the two dozen muffins.

23. Maria has $\frac{3}{4}$ c of raisins left in a box. Is this *more than* or *less than* the amount in the recipe?

24. To make 72 muffins, Maria triples the recipe. How much milk does she need?

25. Maria decides to make only one dozen muffins, so she divides each amount in the recipe by 2. Rewrite the recipe listing the new amounts.

26. **RESEARCH** Find a recipe in a cookbook or a magazine. How many items does the recipe make or how many people does it serve? Double the recipe and list the new amounts of the ingredients.

An *optician* makes and sells lenses and eyeglasses. A lens bends the rays of light entering the eye. The ability to bend light rays is measured in *diopters* (D) and depends on the *focal length* (f) of the lens.

To find the diopter measure, use the formula $D = \frac{1}{f}$, where f is in meters. If the focal length of a lens is 25 cm, first write 25 cm as 0.25 m. Then substitute for f.

$$D = \frac{1}{0.25} = 4$$ ⟵ Use a reciprocal key on a calculator to compute.

Use the formula $D = \frac{1}{f}$ to find each diopter measure.

27. $f = 400$ cm   **28.** $f = 15$ cm   **29.** $f = 0.02$ cm   **30.** $f = 800$ cm

SPIRAL REVIEW

31. Estimate the sum: $27 + 34 + 54 + 18$ *(Toolbox Skill 2)*

32. Find the quotient: $\frac{z}{9} \div \frac{z}{3}$ *(Lesson 8-2)*

33. Write an equation that represents the sentence: A number z divided by nine is fifteen.   *(Lesson 4-7)*

Self-Test 1

Find each product. Simplify if possible.

1. $(-4)\left(\frac{11}{12}\right)$ **2.** $(-0.6)(-1.8)$ **3.** $\left(\frac{6a}{7}\right)\left(\frac{c}{9a}\right)$ **4.** $\left(\frac{4n}{5}\right)\left(\frac{m}{2n}\right)$ **8-1**

Estimate each product.

5. $\left(6\frac{1}{5}\right)\left(-7\frac{7}{8}\right)$ **6.** $\left(2\frac{1}{2}\right)\left(10\frac{3}{4}\right)$

7. $\left(-\frac{6}{11}\right)\left(-8\frac{3}{8}\right)$ **8.** $\left(20\frac{13}{15}\right)\left(-\frac{6}{17}\right)$

Find each quotient. Simplify if possible.

9. $-\frac{1}{5} \div \frac{1}{8}$ **10.** $-6 \div -\frac{1}{3}$ **11.** $3.6 \div (-0.8)$ **8-2**

12. $\frac{5}{12} \div 2\frac{1}{12}$ **13.** $\frac{3n}{8} \div \frac{n}{16}$ **14.** $\frac{2c}{5} \div 3c$ **15.** $7 \div \frac{70}{z}$

8-3

Determining the Correct Form of an Answer

Objective: To determine the correct form of an answer when solving a problem.

When you solve a problem, you may have to decide which form of the answer is needed to make the answer appropriate for a given situation.

- A *whole number* is appropriate when the situation involves items that occur only as whole items, such as sweaters or posters.

- A *fraction* is appropriate when the situation involves items that can occur as fractions, such as yards of material.

- A *decimal* is appropriate when the situation involves items that can occur as decimals, such as amounts of money.

Example

Solve. Decide whether a *whole number*, a *fraction*, or a *decimal* is an appropriate answer.

a. Tami worked $22\frac{1}{2}$ h in three days. She worked the same number of hours each day. How many hours did she work each day?

b. Asparagus costs $2 per pound. How much does $\frac{2}{3}$ lb cost?

c. There will be 75 people at a luncheon. How many tables of 10 seats each are needed to seat everyone?

Solution

a. $22\frac{1}{2} \div 3 = 7\frac{1}{2}$

The answer is a number of hours, so a fraction is appropriate. Tami worked $7\frac{1}{2}$ h each day.

b. $\frac{2}{3} \cdot 2 = 1\frac{1}{3} = 1.\overline{3}$

The answer is an amount of money, so a decimal is appropriate. The cost of $\frac{2}{3}$ lb is $1.33.

c. $75 \div 10 = 7.5$

The answer is a number of tables, so a whole number is appropriate. Eight tables are needed.

✓ Check Your Understanding

1. In part (b), why is $1.\overline{3}$ written as $1.33?

2. In part (c), why is the answer *eight tables,* instead of *seven tables?*

Guided Practice

COMMUNICATION «*Discussion*

Decide whether a *whole number*, a *fraction*, or a *decimal* is the appropriate form for each quantity.

«**1.** a number of shirts

«**2.** a sale price

«**3.** a number of hours

«**4.** the number of people in class

Solve. Decide whether a *whole number,* **a** *fraction,* **or a** *decimal* **is an appropriate answer.**

5. A teacher wants to divide a class into groups of four. The class has 29 students. How many groups of four can there be?

6. Each week Raphael earns $26 and saves one fourth of his earnings. How much money does he save each week?

Exercises

Solve. Decide whether a *whole number,* **a** *fraction,* **or a** *decimal* **is an appropriate answer.**

1. The junior class is making 25 bows for prom decorations. Each bow uses $\frac{3}{4}$ yd of ribbon. How many yards of ribbon are needed?

2. A watercolor brush costs $3.75. How much do two brushes cost?

3. A tube of caulking can repair $2\frac{2}{3}$ classroom windows. How many tubes of caulking should be bought to repair 50 windows?

4. Jocelyn divides 5 lb of granola evenly between two bags. How many pounds of granola are in each bag?

5. Mark practices playing his guitar $\frac{3}{4}$ h each day. How many hours does he practice in seven days?

6. Ann has $7\frac{1}{4}$ cups of pecans. The recipe for one pecan pie calls for $1\frac{1}{4}$ cups of pecans. How many pecan pies can she bake?

7. Salmon costs $6 per pound. How much does $2\frac{1}{4}$ lb of salmon cost?

💻 COMPUTER APPLICATION

Many businesses use computers to keep a record of inventory and costs. A computer always shows rational numbers in decimal form. Decide whether a *whole number,* a *fraction,* or a *decimal* is the appropriate form of each quantity. Then write the correct form of the quantity.

8. The average number of cars sold in a week is 18.125.

9. The base price of a Model X1 car is 10995.54.

10. The number of hours a salesperson worked in a week is 43.75.

11. The average number of customers in a month is 159.41666667.

SPIRAL REVIEW

12. Make a sketch of $\overline{WY}$.   *(Lesson 6-1)*

13. A school bus seats 44 people. How many buses are needed to transport 204 people to the class outing?   *(Lesson 8-3)*

Modeling Fractions Using Fraction Tiles

Objective: To use fraction tiles to model addition and subtraction of fractions.

Materials

■ fraction tiles:
$1, \frac{1}{2}, \frac{1}{3}, \frac{1}{4}, \frac{1}{6}, \frac{1}{8}, \frac{1}{12}, \frac{1}{16}$

Fraction tiles are designed so that a square represents one whole unit. Fractions are represented by rectangular tiles that can be combined to form a whole square. You can use fraction tiles to add and subtract fractions. For example, you can use fraction tiles to show that the sum of $\frac{1}{3}$ and $\frac{1}{4}$ is $\frac{7}{12}$.

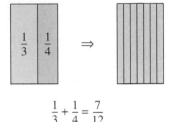

← Seven $\frac{1}{12}$ tiles completely cover both the $\frac{1}{3}$ and $\frac{1}{4}$ tiles.

$$\frac{1}{3} + \frac{1}{4} = \frac{7}{12}$$

Activity I *Finding Equivalent Fractions*

Suppose you want to cover a fraction tile representing $\frac{1}{2}$ and a fraction tile representing $\frac{1}{3}$. To do this, you want to use fraction tiles that are all the same size.

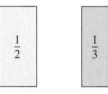

1 What is the largest size fraction tile you could use? How many of these fraction tiles cover the $\frac{1}{2}$ tile? How many cover the $\frac{1}{3}$ tile?

2 What is the smallest size fraction tile you could use? How many of these fraction tiles cover the $\frac{1}{2}$ tile? How many cover the $\frac{1}{3}$ tile?

3 Use your answers to Steps 1 and 2 to write two fractions equivalent to $\frac{1}{2}$ and two fractions equivalent to $\frac{1}{3}$.

Activity II *More Equivalent Fractions*

Arrange fraction tiles representing $\frac{1}{2}$ and $\frac{1}{4}$ as shown. Cover the tiles using fraction tiles that are all the same size.

1 What is the largest size fraction tile you can use?

2 What is the smallest size fraction tile you can use?

3 Use your answers to Steps 1 and 2 to complete the statement.

$$\frac{1}{2} + \frac{1}{4} = \underline{\quad?\quad} = \underline{\quad?\quad}$$

Activity III *Modeling Addition*

1 To complete one whole unit, fill the empty space using fraction tiles that are all the same size. Then replace each __?__ with a fraction in lowest terms that makes the statement true.

a.

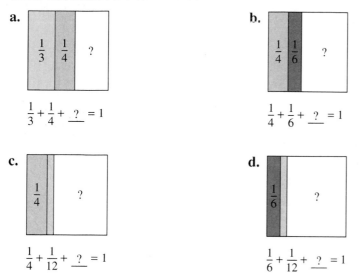

$$\frac{1}{3} + \frac{1}{4} + \underline{\ ?\ } = 1$$

b.

$$\frac{1}{4} + \frac{1}{6} + \underline{\ ?\ } = 1$$

c.

$$\frac{1}{4} + \frac{1}{12} + \underline{\ ?\ } = 1$$

d.

$$\frac{1}{6} + \frac{1}{12} + \underline{\ ?\ } = 1$$

2 In parts (c) and (d) of Step 2, is there more than one way to fill the empty space using fraction tiles that are all the same size? Explain.

Activity IV *Modeling Subtraction*

1 Show how to fill the empty space using fraction tiles that are all the same size. Express each answer as a fraction.

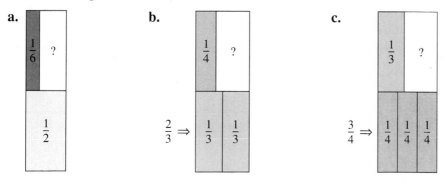

2 Use your answers to Step 1 to replace each __?__ with the correct fraction in lowest terms.

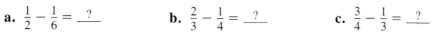

a. $\dfrac{1}{2} - \dfrac{1}{6} = \underline{\ ?\ }$ **b.** $\dfrac{2}{3} - \dfrac{1}{4} = \underline{\ ?\ }$ **c.** $\dfrac{3}{4} - \dfrac{1}{3} = \underline{\ ?\ }$

3 In Step 1, is there more than one way to fill each empty space using fraction tiles that are all the same size? Explain.

Adding and Subtracting Rational Numbers with Like Denominators

Objective: To add and subtract rational numbers and algebraic fractions with like denominators.

DATA ANALYSIS

Terry wants to make a bow tie and a sash. She refers to the back of the pattern to determine how many yards of material to buy. She plans to use fabric that is 60 in. wide. She will need $\frac{7}{8}$ yd for the tie and $\frac{5}{8}$ yd for the sash. Terry buys $\frac{12}{8}$ yd, or $1\frac{1}{2}$ yd, of fabric.

Tie	$\frac{7}{8}$ yd
Bow Tie	$\frac{7}{8}$ yd
Scarf	$1\frac{5}{8}$ yd
Sash:	
44/45 in.	$1\frac{1}{4}$ yd
58/60 in.	$\frac{5}{8}$ yd

Generalization: *Adding Rational Numbers with Like Denominators*

To add rational numbers with like denominators, add the numerators and write the sum over the denominator.

In Arithmetic

$$\frac{7}{8} + \frac{5}{8} = \frac{7 + 5}{8}$$

In Algebra

$$\frac{a}{c} + \frac{b}{c} = \frac{a + b}{c}, \ c \neq 0$$

Example 1

Find each sum. Simplify if possible.

a. $\frac{11}{12} + \frac{5}{12}$
b. $-1\frac{1}{5} + \left(-\frac{3}{5}\right)$

Solution

a. $\frac{11}{12} + \frac{5}{12} = \frac{11 + 5}{12}$

$= \frac{16}{12}$

$= \frac{4}{3} = 1\frac{1}{3}$

b. $-1\frac{1}{5} + \left(-\frac{3}{5}\right) = \frac{-6}{5} + \left(\frac{-3}{5}\right)$

$= \frac{-6 + (-3)}{5}$

$= \frac{-9}{5} = -1\frac{4}{5}$

✓ Check Your Understanding

1. Explain the first step in Example 1(b).

You can also subtract rational numbers with like denominators.

Example 2

Find each difference. Simplify if possible.

a. $2\frac{3}{10} - \frac{7}{10}$

b. $-3\frac{1}{4} - 6\frac{3}{4}$

Solution

a. $2\frac{3}{10} - \frac{7}{10} = \frac{23}{10} - \frac{7}{10}$

$$= \frac{23-7}{10}$$

$$= \frac{16}{10}$$

$$= \frac{8}{5} = 1\frac{3}{5}$$

b. $-3\frac{1}{4} - 6\frac{3}{4} = \frac{-13}{4} - \frac{27}{4}$

$$= \frac{-13-27}{4}$$

$$= \frac{-40}{4}$$

$$= -10$$

✅ **Check Your Understanding**

2. In Example 2(b), why does the numerator become -40?

3. What would the numerator in Example 2(b) have become if the problem had been $3\frac{1}{4} - 6\frac{3}{4}$ instead of $-3\frac{1}{4} - 6\frac{3}{4}$?

In Lesson 2-3 you learned to combine like terms. You can use those skills along with what you have just learned about adding and subtracting rational numbers to add and subtract algebraic fractions.

Example 3

Simplify.

a. $\frac{6}{x} + \frac{5}{x}$

b. $\frac{10n}{3} - \frac{4n}{3}$

Solution

a. $\frac{6}{x} + \frac{5}{x} = \frac{6+5}{x}$

$$= \frac{11}{x}$$

b. $\frac{10n}{3} - \frac{4n}{3} = \frac{10n-4n}{3}$

$$= \frac{6n}{3}$$

$$= \frac{2n}{1} = 2n$$

✅ **Check Your Understanding**

4. In Example 3(b), explain how $\frac{6n}{3}$ was simplified.

Guided Practice

COMMUNICATION «*Reading*

Refer to the text on pages 352–353.

«**1.** What is the main idea of the lesson?

«**2.** List two major points that support the main idea.

Express each rational number as a fraction.

3. $-2\frac{1}{4}$ **4.** $10\frac{5}{8}$ **5.** $7\frac{2}{3}$ **6.** $-1\frac{5}{6}$

Find each answer. Simplify if possible.

7. $1\frac{5}{9} + 3\frac{1}{9}$ **8.** $-\frac{4}{7} - \left(-5\frac{5}{7}\right)$ **9.** $-\frac{9}{10} + \left(-\frac{7}{10}\right) + \frac{1}{10}$

10. $2\frac{5}{8} + \left(-1\frac{1}{8}\right) + \frac{7}{8}$ **11.** $\frac{20}{15a} - \frac{8}{15a}$ **12.** $\frac{8b}{9} + \frac{2b}{9} + \frac{5b}{9}$

Exercises

Find each answer. Simplify if possible.

1. $\frac{14}{15} - \frac{6}{15}$ **2.** $\frac{1}{6} + \frac{5}{6}$ **3.** $-\frac{3}{5} - \left(-\frac{3}{5}\right)$

4. $\frac{3}{8} + \left(-\frac{5}{8}\right)$ **5.** $8\frac{1}{4} - 5\frac{3}{4}$ **6.** $\frac{3}{11} - 2\frac{1}{11}$

7. $-\frac{3}{10} - \left(-1\frac{7}{10}\right)$ **8.** $-8\frac{1}{10} - 2\frac{3}{10}$ **9.** $-1\frac{11}{14} + \left(-2\frac{3}{14}\right)$

10. $2\frac{3}{4} + \left(-3\frac{1}{4}\right)$ **11.** $-4\frac{5}{12} + \frac{11}{12} + 2\frac{7}{12}$ **12.** $-3\frac{2}{5} + \frac{3}{5} + \left(-2\frac{4}{5}\right)$

13. $\frac{8}{r} + \frac{4}{r}$ **14.** $\frac{12}{s} - \frac{5}{s}$ **15.** $\frac{11}{3m} - \frac{5}{3m}$

16. $\frac{3}{2a} + \frac{5}{2a}$ **17.** $\frac{7}{6x} + \frac{5}{6x}$ **18.** $\frac{13}{4r} + \frac{9}{4r}$

19. $\frac{9b}{2} + \frac{7b}{2}$ **20.** $\frac{18h}{5} - \frac{6h}{5}$ **21.** $\frac{18x}{25} - \frac{8x}{25}$

22. $\frac{16b}{5} - \frac{b}{5}$ **23.** $\frac{3c}{7} + \frac{5c}{7} + \frac{6c}{7}$ **24.** $\frac{8}{9n} + \frac{2}{9n} + \frac{11}{9n}$

25. Ken finished his mathematics assignment in $\frac{3}{4}$ h. Bert finished the same assignment in $1\frac{1}{4}$ h. How much longer did it take Bert to do the assignment than Ken?

26. Hazel spent $1\frac{3}{8}$ h mowing the lawn, $\frac{7}{8}$ h trimming the hedge, and $1\frac{5}{8}$ h weeding the garden. How many hours in all did Hazel spend doing this work?

Tell if each sum is *greater than* or *less than* 1. Do not find the sum.

27. $\frac{3}{5} + \frac{4}{5}$ **28.** $\frac{3}{7} + \frac{6}{7}$ **29.** $\frac{1}{8} + \frac{3}{8}$ **30.** $\frac{5}{6} + \frac{5}{6}$

CONNECTING MATHEMATICS AND INDUSTRIAL ARTS

Many measurements used in industrial arts involve fractions.

31. From a $33\frac{7}{8}$ in. long board, Jim cut a piece of wood that was $10\frac{5}{8}$ in. long. The saw blade shaved $\frac{1}{8}$ in. off the board. How long was the remaining piece?

32. A drawer front is larger than the drawer opening by $\frac{5}{8}$ in. on each edge. The front of the drawer is $26\frac{3}{8}$ in. long and $8\frac{3}{8}$ in. wide. What are the dimensions of the drawer opening?

33. Kate is designing a bookshelf. The top of the bookshelf and each of its 4 shelves will be $\frac{3}{4}$ in. thick. The bottom shelf will rest on the floor. The shelves will be $11\frac{1}{4}$ in. apart. Make a sketch of the bookshelf. What will be the overall height of the bookshelf?

CONNECTING ARITHMETIC AND PROBABILITY

The *probability* of an event is a number from 0 to 1 indicating the likelihood that the event will happen. The notation $P(E)$ indicates the probability of an event. The probability that an event will not happen is written $P(\text{not } E) = 1 - P(E)$.

Find the probability that each event will *not* happen.

34. $P(A) = \frac{1}{6}$ **35.** $P(B) = \frac{4}{9}$ **36.** $P(D) = \frac{3}{5}$ **37.** $P(C) = \frac{7}{16}$

38. Eight students run for class president. The probability of being elected is $\frac{1}{8}$. What is the probability of not being elected?

39. The probability of choosing a key that will open a door is $\frac{2}{11}$. What is the probability of choosing a key that will not open the door?

SPIRAL REVIEW

40. Write 0.0000047 in scientific notation. *(Lesson 7-11)*

41. Simplify: $\frac{14m}{5} - \frac{9m}{5}$ *(Lesson 8-4)*

8-5 Adding and Subtracting Rational Numbers with Unlike Denominators

Objective: To add and subtract rational numbers and algebraic fractions with unlike denominators.

EXPLORATION

1 Replace each ___?___ with a fraction that makes the statement true.

a. $\frac{4}{5} \cdot \underline{\ ?\ } = \frac{8}{10}$ b. $\frac{-3}{4} \cdot \underline{\ ?\ } = \frac{-15}{20}$ c. $\frac{7}{2} \cdot \underline{\ ?\ } = \frac{21}{6}$

2 What do all the answers in Step 1 have in common?

3 In Step 1, what is the relationship between $\frac{4}{5}$ and $\frac{8}{10}$, $\frac{-3}{4}$ and $\frac{-15}{20}$, and $\frac{7}{2}$ and $\frac{21}{6}$?

You may recall that the product of any number and 1 is the original number. When you multiply a fraction by a fractional form of 1, you obtain an equivalent fraction.

In order to add or subtract rational numbers with unlike denominators, you first write each rational number as a fraction. Then you write equivalent fractions having the LCD. You do this by multiplying each fraction by the form of 1 that will produce the LCD.

Example 1

Find each answer. Simplify if possible.

a. $\frac{1}{6} + \frac{3}{10}$ b. $-1\frac{1}{8} - 2\frac{3}{4}$

Solution

a. The LCD is 30.

$$\frac{1}{6} + \frac{3}{10} = \frac{1}{6} \cdot \frac{5}{5} + \frac{3}{10} \cdot \frac{3}{3}$$

$$= \frac{5}{30} + \frac{9}{30}$$

$$= \frac{5 + 9}{30}$$

$$= \frac{14}{30}$$

$$= \frac{7}{15}$$

b. The LCD is 8.

$$-1\frac{1}{8} - 2\frac{3}{4} = \frac{-9}{8} - \frac{11}{4}$$

$$= \frac{-9}{8} - \frac{11}{4} \cdot \frac{2}{2}$$

$$= \frac{-9}{8} - \frac{22}{8}$$

$$= \frac{-9 - 22}{8}$$

$$= \frac{-31}{8}$$

$$= -3\frac{7}{8}$$

Check Your Understanding

1. How would Example 1(b) be different if you used 16 as the common denominator?

The process demonstrated in Example 1 can also be used to add and subtract algebraic fractions.

Example 2   Simplify $\frac{5x}{12} + \frac{3x}{8}$.

Solution

$$\frac{5x}{12} + \frac{3x}{8} = \frac{5x}{12} \cdot \frac{2}{2} + \frac{3x}{8} \cdot \frac{3}{3} \qquad \longleftarrow \text{The LCD is 24.}$$

$$= \frac{10x}{24} + \frac{9x}{24}$$

$$= \frac{10x + 9x}{24} = \frac{19x}{24}$$

To find the sum or difference of rational numbers that are written as decimals, apply the rules that you learned for adding and subtracting integers. Remember to align decimal points.

Example 3   **Find each answer:**   **a.** $-4.25 + (-3.1)$ **b.** $-1.48 - (-3.8)$

Solution

a. $-4.25 + (-3.1) = -7.35$

b. $-1.48 - (-3.8) = -1.48 + 3.8 \qquad \longleftarrow$ Write the difference

$\qquad\qquad\qquad\qquad\ = 2.32$ as a related addition.

☑ **Check Your Understanding**

2. In Example 3(a), why is the answer negative?

3. In Example 3(b), why is the answer positive?

Guided Practice

«**1.** COMMUNICATION «*Reading*   Refer to the text on pages 356–357. Create an outline for your notebook describing how to add and subtract rational numbers with unlike denominators.

Find the LCD.

2. $\frac{2}{3}$ and $\frac{3}{5}$ **3.** $-\frac{n}{4}$ and $\frac{n}{6}$ **4.** $-1\frac{1}{8}$ and $-\frac{3}{16}$ **5.** 3 and $\frac{1}{2}$

Rewrite each sum or difference with equivalent fractions having the LCD.

6. $\frac{3}{4} + \frac{5}{8}$ **7.** $1\frac{11}{12} + 5\frac{3}{8}$ **8.** $-1\frac{5}{9} - 4$ **9.** $\frac{10m}{5} - \frac{4m}{11}$

Find each answer. Simplify if possible.

10. $\frac{5}{6} + \frac{3}{4}$ **11.** $1\frac{1}{4} - 5$ **12.** $-3\frac{1}{2} + \left(-4\frac{1}{6}\right) + \frac{2}{3}$

13. $\frac{4a}{5} - \frac{a}{4}$ **14.** $3x - \frac{x}{3}$ **15.** $\frac{4r}{5} + \frac{8r}{15} + \frac{r}{3}$

16. $-10 + 2.66$ **17.** $-9.7 - 2.8$ **18.** $5.4 + 0.19 + (-3.5)$

Exercises

Find each answer. Simplify if possible.

1. $\frac{2}{3} + \frac{5}{6}$

2. $\frac{7}{12} - \frac{5}{18}$

3. $6 - 2\frac{3}{5}$

4. $5\frac{1}{2} + \frac{1}{6}$

5. $\frac{4}{5} - \left(-\frac{3}{4}\right)$

6. $-3\frac{2}{3} + \frac{3}{4}$

7. $-3 + \left(-2\frac{1}{6}\right)$

8. $-4\frac{1}{2} - 1\frac{11}{18}$

9. $-1\frac{9}{10} - \left(-2\frac{1}{15}\right)$

10. $1\frac{1}{4} - 3\frac{1}{5}$

11. $\frac{3}{4} + \frac{5}{18} + \left(-\frac{4}{9}\right)$

12. $-3\frac{4}{7} + \left(-5\frac{1}{2}\right) + 4$

13. $\frac{5m}{3} - \frac{2m}{7}$

14. $\frac{8x}{5} + \frac{11x}{15}$

15. $3b + \frac{5b}{4}$

16. $7n - \frac{n}{2}$

17. $\frac{4h}{5} + \frac{3h}{2} + \frac{h}{4}$

18. $\frac{7a}{8} + \frac{5a}{6} + \frac{11a}{12}$

19. $-5.3 + 1.7$

20. $8.23 - 10.15$

21. $4.72 - (-11.8)$

22. $-15.4 - (-20)$

23. $-6 + 9.03 + (-7.4)$

24. $0.15 + (-4.8) + (-6.35)$

25. Darla needs $2\frac{3}{4}$ yd of material to make a dress, and $1\frac{2}{3}$ yd to make a matching jacket. How much material does she need in all?

26. Dana could jog $1\frac{1}{4}$ mi without stopping when she joined the track team. After two weeks she could jog $4\frac{1}{3}$ mi without stopping. How much farther could Dana jog after two weeks than when she began?

Find each answer. Simplify if possible.

27. $-10.5 - 9\frac{3}{4}$

28. $6.35 + 4\frac{3}{5}$

29. $7\frac{1}{9} - 4.\overline{3}$

30. $-3\frac{5}{6} + 1.\overline{6}$

31. $-\frac{1}{8} + 1.25 + \left(-3\frac{1}{2}\right)$

32. $0.6 + 5\frac{3}{8} + (-2.5)$

Evaluate each expression when $a = -2\frac{5}{6}$, $b = 3\frac{1}{2}$, $c = -4.8$, and $d = 0.6$.

33. $-ab$

34. $a \div b$

35. $2(a + b)$

36. $|a - b|$

37. $c \div d$

38. $5cd$

39. $d^2 - c$

40. $-c + d + 7$

DATA, *pages 338–339*

41. In Grand Rapids how much greater is the recommended thickness of insulation for attic floors than for basement ceilings?

42. How much greater is the recommended thickness of insulation for attic floors in Duluth than in Atlanta? (Use the figures for the upper end of each range.)

FUNCTIONS

Complete each function table. **Find each function rule.**

43.

x	$4x$
$2\frac{1}{6}$	$8\frac{2}{3}$
0.4	1.6
$-\frac{3}{8}$	?
$-1\frac{1}{4}$	?
-2.5	?

44.

x	$6x + \frac{1}{2}$
$-\frac{1}{2}$	$-2\frac{1}{2}$
$\frac{1}{2}$	$3\frac{1}{2}$
$\frac{1}{3}$	?
$\frac{1}{4}$	?
$\frac{1}{5}$	?

45.

x	?
-2	$-1\frac{2}{3}$
$-1\frac{2}{3}$	$-1\frac{1}{3}$
$-\frac{1}{3}$	0
$1\frac{1}{3}$	$1\frac{2}{3}$
$3\frac{2}{3}$	4

46.

x	?
-4	-1
-2	$-\frac{1}{2}$
$\frac{1}{2}$	$\frac{1}{8}$
1	$\frac{1}{4}$
5	$1\frac{1}{4}$

PATTERNS

The expression shown at the right is called a *series*. There is always a pattern to the numbers in a series.

$$\frac{1}{2} + \frac{1}{4} + \frac{1}{8} + \frac{1}{16} + \frac{1}{32} \cdots$$

47. What is the pattern of the fractions in the series?

48. Find the next two fractions in the series.

49. Find the sum of the fractions indicated.
  a. the first two fractions
  b. the first three fractions
  c. the first four fractions
  d. the first five fractions

50. Describe the relationship between the numerator and the denominator of each sum in Exercise 49.

51. Without adding, predict what will be the sum of the first six fractions in the series.

SPIRAL REVIEW

52. Write $\frac{12}{56}$ in lowest terms.   *(Lesson 7-4)*

53. Draw a pictograph to display the data.   *(Lesson 5-1)*

Populations of Large Cities (Millions)

City	New York	London	Tokyo	Mexico City	Beijing
Population	7.3	6.8	8.4	12.9	9.3

54. DATA, *pages 140–141*   What is the maximum speed of an elephant? *(Lesson 5-2)*

55. Simplify: $\frac{7a}{15} + \frac{4a}{5}$ *(Lesson 8-5)*

MIXED REVIEW

Operations with Rational Numbers

Jazz music is played all over the world today, but it has its origins in the experience of African-Americans. Music that would develop into jazz was first played in the early 1900s by African-Americans at memorial parades. Bands of musicians used brass instruments to create a sound that came to be called Dixieland jazz. Musicians have since developed many new forms of jazz, but Dixieland music is still popular today.

To find out what American city is often associated with Dixieland music and early jazz, complete the code boxes. Begin by solving Exercise 1. The answer to Exercise 1 is $\frac{35}{72}$, and the letter associated with Exercise 1 is W. Write W in the box above $\frac{35}{72}$ as shown below. Continue in this way with Exercises 2–22.

1. $\frac{7}{12} \cdot \frac{5}{6}$ (W)

2. $\frac{5}{8} \div \frac{3}{4}$ (Z)

3. $13\frac{4}{5} + 9\frac{4}{5}$ (N)

4. $4\frac{7}{10} - 2\frac{4}{5}$ (R)

5. $-6\left(\frac{7}{8}\right)$ (O)

6. $\left(-4\frac{2}{3}\right)\left(-2\frac{6}{7}\right)$ (L)

7. $\frac{5}{9} \div \left(-\frac{3}{4}\right)$ (E)

8. $-6\frac{3}{7} \div 9$ (S)

9. $\frac{3}{5} + \left(-3\frac{2}{5}\right)$ (R)

10. $-\frac{7}{9} + \left(-\frac{5}{12}\right)$ (O)

11. $-4\frac{1}{2} - \frac{5}{7}$ (N)

12. $9\frac{7}{8} - \left(-4\frac{3}{8}\right)$ (A)

13. $4\frac{4}{9} + \frac{3}{9} + \left(-2\frac{5}{9}\right)$ (C)

14. $-6 + \left(-2\frac{5}{6}\right) + \frac{11}{12}$ (D)

15. $(-0.4)(-0.9)$ (E)

16. $(5.6)(-0.5)$ (L)

17. $-15.19 \div 3.1$ (A)

18. $-36 \div (-1.8)$ (Z)

19. $-2.3 + (-3.8)$ (J)

20. $19 + (-14.6)$ (E)

21. $8.95 - 35.7$ (A)

22. $5.32 - (-0.74)$ (F)

?	?	W	?	?	?	?	?	?	?
$-5\frac{3}{14}$	0.36	$\frac{35}{72}$	$-1\frac{7}{36}$	$-2\frac{4}{5}$	-2.8	$-\frac{20}{27}$	-4.9	$23\frac{3}{5}$	$-\frac{5}{7}$

?	?	?	?	?	?	?	?
$2\frac{2}{9}$	$1\frac{9}{10}$	$14\frac{1}{4}$	$-7\frac{11}{12}$	$13\frac{1}{3}$	4.4	$-5\frac{1}{4}$	6.06

?	?	?	?
-6.1	-26.75	$\frac{5}{6}$	20

American music combines the musical traditions of many different peoples and places. To learn more about these musical traditions, solve each exercise. The word printed in blue following the correct answer matches the description printed in blue.

23. $\frac{2}{3} \cdot \frac{3}{2}$

Instrument created in America by African-Americans, based on African instruments.

a. 1   banjo   **b.** $2\frac{1}{6}$ harmonica

24. $\frac{1}{3}\left(-3\frac{1}{5}\right)$

City where country music is played at the *Grand Ole Opry*.

a. $-1\frac{1}{15}$ Nashville   **b.** $-1\frac{1}{5}$ Atlanta

25. $-\frac{11}{12} \div \frac{3}{4}$

Hawaiian instrument whose name means *leaping flea*.

a. $-1\frac{2}{9}$ ukulele   **b.** $-\frac{11}{16}$ holoku

26. $-3\frac{3}{5} \div 2\frac{6}{7}$

Kind of Mexican music usually played by a band of violins, guitars, and trumpets.

a. $-10\frac{2}{7}$ flamenco   **b.** $-1\frac{13}{50}$ mariachi

27. $\frac{15}{16} + \frac{9}{16} + \frac{13}{16}$

Instrument created by settlers in the Appalachians, based on similar European ones.

a. $2\frac{5}{16}$ dulcimer   **b.** $\frac{37}{48}$ mirliton

28. $-4\frac{2}{5} + \left(-\frac{3}{5}\right)$

Kind of jazz music that combines aspects of jazz with aspects of rock music.

a. $-3\frac{4}{5}$ synthesizer   **b.** -5 fusion

29. $\frac{1}{25} - \frac{1}{5}$

Kind of blues music for piano that was popular in the 1930s.

a. $\frac{1}{20}$ bebop   **b.** $-\frac{4}{25}$ boogie-woogie

30. $-\frac{1}{4} + \frac{2}{3} - \left(-1\frac{5}{6}\right)$

Decade sometimes called the Golden Age of Jazz.

a. $2\frac{1}{4}$ 1920s   **b.** $-1\frac{5}{12}$ 1940s

8-6 Estimating with Rational Numbers

Objective: To estimate the sum, difference, or quotient of rational numbers.

To replace some screens, the Okawas have a piece of screening that is $12\frac{1}{3}$ ft long. They cut off a piece that is $3\frac{3}{4}$ ft long and do not know if they have enough screening left for doors that require a total of 7 ft of screening. Since the Okawas do not need to know the exact footage, they estimate.

QUESTION About how many feet of screening are left?

Example 1

Estimate $12\frac{1}{3} - 3\frac{3}{4}$ by rounding.

Solution

$12\frac{1}{3} - 3\frac{3}{4}$ ◄—— Because $\frac{1}{3}$ is less than one half, round down.

Because $\frac{3}{4}$ is greater than one half, round up.

$\underbrace{12 - 4}$

about 8

✓ Check Your Understanding

How can you tell if a fraction is less than one half or greater than one half?

ANSWER About 8 ft of screening are left.

When estimating sums of rational numbers with the same sign, front-end estimation may give you a closer estimate than rounding. Add the integers and then adjust the answer.

Example 2

Estimate $\frac{5}{9} + 2\frac{4}{5} + 3\frac{1}{16}$ by using front-end-and-adjust estimation.

Solution

First, add the integers. $2 + 3 = 5$

Then adjust. $\frac{5}{9}$ is about $\frac{1}{2}$, $\frac{4}{5}$ is closer to 1, $\frac{1}{16}$ is closer to 0

$5 + \frac{1}{2} + 1 + 0 \rightarrow$ about $6\frac{1}{2}$

You can use compatible numbers to estimate the quotient of mixed numbers. You round the divisor to the nearest integer and then use a compatible dividend. To estimate the quotient when the divisor is a fraction, first write a simpler fraction. Then round the dividend and divide.

Example 3

Solution

Estimate: a. $13\frac{1}{3} \div 5\frac{9}{10}$ b. $-19\frac{7}{8} \div \frac{7}{15}$

a. $13\frac{1}{3} \div 5\frac{9}{10}$

$\downarrow \qquad \downarrow$

$\underbrace{12 \div 6}_{\text{about } 2}$

b. $-19\frac{7}{8} \div \frac{7}{15}$

$\downarrow \quad \downarrow$

$-20 \div \frac{1}{2} = \underbrace{(-20)(2)}_{\text{about } -40}$

Guided Practice

COMMUNICATION «*Reading*

« **1.** Using $8 - 5$ to estimate the sum $7\frac{2}{3} - 5\frac{1}{4}$ is an example of estimating by ___?___.

« **2.** Estimating by writing $9\frac{1}{16} \div 4\frac{1}{3}$ as $8 \div 4$ is an example of estimating by ___?___.

« **3.** Using $5 + 1 + 0$ to estimate the sum $3\frac{7}{8} + 2\frac{1}{6}$ is an example of estimating by ___?___.

Round to the nearest integer.

4. $\frac{3}{16}$ **5.** $2\frac{7}{12}$ **6.** $-3\frac{1}{4}$ **7.** $-15\frac{7}{8}$ **8.** $34\frac{5}{9}$

Estimate.

9. $4\frac{3}{4} - 2\frac{1}{2}$ **10.** $-9\frac{3}{11} + \left(-5\frac{1}{6}\right)$ **11.** $10\frac{7}{8} \div \left(-\frac{5}{14}\right)$ **12.** $18\frac{4}{5} \div 8\frac{3}{4}$

Exercises

Estimate.

1. $2\frac{8}{11} - 8\frac{1}{3}$ **2.** $7\frac{2}{3} - 2\frac{1}{2}$ **3.** $6\frac{5}{12} + 5\frac{1}{5}$

4. $-14\frac{1}{8} + \left(-10\frac{3}{5}\right)$ **5.** $-6\frac{3}{4} + 2\frac{2}{9}$ **6.** $-2\frac{3}{5} - 7\frac{3}{8}$

7. $\frac{9}{10} + 7\frac{3}{16} + 2\frac{4}{7}$ **8.** $3\frac{5}{6} + 9\frac{2}{5} + \frac{7}{8}$ **9.** $41\frac{1}{4} \div 11\frac{1}{8}$

10. $19\frac{7}{8} \div \left(-\frac{10}{29}\right)$ **11.** $-11\frac{3}{5} \div \frac{25}{101}$ **12.** $-28\frac{1}{4} \div \left(-5\frac{3}{4}\right)$

13. Gina makes quilts using stitches that are $\frac{9}{32}$ in. long. About how many stitches will there be in a row of stitches that is $8\frac{1}{2}$ in. long?

14. Melissa tutored for $2\frac{3}{4}$ h last week. This week she tutored for $7\frac{1}{3}$ h. About how many more hours did she tutor this week than last week?

15. Pedro cycled $9\frac{1}{4}$ mi one weekend, $10\frac{1}{4}$ mi the next weekend, and $13\frac{7}{8}$ mi the following weekend. About how many miles did he cycle altogether on the three weekends?

16. It takes Sam about $1\frac{2}{3}$ h to install a stereo system in an automobile at the plant where he is employed. About how many stereo systems can he plan to install in $8\frac{1}{4}$ h?

 CALCULATOR

Estimate to tell whether each calculator answer is *reasonable* or *unreasonable*. If an answer is unreasonable, find the correct answer.

17. $6\frac{1}{3} - 11\frac{12}{13}$ — $-18\ \_\ 10\ \lrcorner\ 39.$

18. $\frac{9}{10} + 4\frac{1}{8} + 3\frac{1}{6}$ — $8\ \_\ 23\ \lrcorner\ 120.$

19. $\left(-\frac{24}{49}\right)\left(8\frac{19}{20}\right)$ — $-4\ \_\ 94\ \lrcorner\ 245.$

20. $35\frac{1}{5} \div \frac{4}{23}$ — $6\ \_\ 14\ \lrcorner\ 115.$

SPIRAL REVIEW

21. Express 4.7 as the quotient of two integers.   *(Lesson 7-9)*

22. Find the sum: $-37 + (-19)$ *(Lesson 3-2)*

23. Graph $A(-3, -2)$, $B(4, -1)$, and $C(-2, -3)$ on a coordinate plane. *(Lesson 3-8)*

24. Estimate: $22\frac{1}{6} \div \frac{25}{49}$ *(Lesson 8-6)*

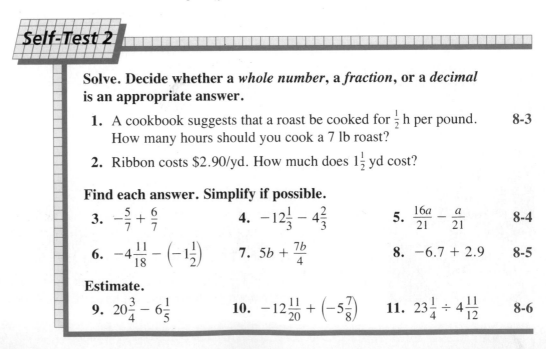

Self-Test 2

Solve. Decide whether a *whole number*, a *fraction*, or a *decimal* is an appropriate answer.

1. A cookbook suggests that a roast be cooked for $\frac{1}{2}$ h per pound. How many hours should you cook a 7 lb roast?   **8-3**

2. Ribbon costs \$2.90/yd. How much does $1\frac{1}{2}$ yd cost?

Find each answer. Simplify if possible.

3. $-\frac{5}{7} + \frac{6}{7}$ **4.** $-12\frac{1}{3} - 4\frac{2}{3}$ **5.** $\frac{16a}{21} - \frac{a}{21}$ **8-4**

6. $-4\frac{11}{18} - \left(-1\frac{1}{2}\right)$ **7.** $5b + \frac{7b}{4}$ **8.** $-6.7 + 2.9$ **8-5**

Estimate.

9. $20\frac{3}{4} - 6\frac{1}{5}$ **10.** $-12\frac{11}{20} + \left(-5\frac{7}{8}\right)$ **11.** $23\frac{1}{4} \div 4\frac{11}{12}$ **8-6**

8-7

Supplying Missing Facts

Objective: To solve problems by supplying missing facts.

Some problems involve quantities with different units of measure. To solve this type of problem, you may need to supply a fact that is not given in the problem. This fact might be the relationship between the two units of measure. Suppose a problem asks you to compare the number of ounces in two weights that are given in pounds. Before you can solve the problem, you will need to supply the fact that 1 lb = 16 oz.

Problem

Randy decides to spend an equal amount of time on each of his five homework subjects. If he plans to do homework for two hours altogether, how many minutes should he spend on each subject?

Solution

UNDERSTAND The problem is about the amount of time spent on different homework subjects.
Facts: equal time for each
subject
2 h for all 5 subjects
Find: number of minutes on
each subject

PLAN Divide the 2 h into 5 parts to find the time spent on each subject. This time will be in hours, but the question asks for minutes. The missing fact is 1 h = 60 min. Multiply the number of hours spent on each subject by 60 to find the number of minutes.

WORK

Divide: $2 \div 5 = \frac{2}{5}$ ⟵ $\frac{2}{5}$ h for each subject

Multiply: $\frac{2}{5} \cdot 60 = 24$ ⟵ 24 min for each subject

ANSWER Randy should spend 24 min on each subject.

Look Back An alternative method for solving this problem is to first find the total number of minutes in two hours, and then divide by five. Solve the problem again using this alternative method.

Guided Practice

Identify the missing fact. Do not solve the problem.

« **1.** The Evans family uses 21 quarts of milk per week. How many gallons do they use in 4 weeks?

« **2.** Ed is $67\frac{1}{2}$ in. tall and still growing. How many more inches must he grow before he is 6 ft tall?

Replace each __?__ with the correct number or word.

3. 1 day = __?__ hours

4. 1 __?__ = 128 fluid ounces

5. __?__ pounds = 1 ton

6. 1000 __?__ = 1 kilometer

Write a variable expression for each problem.

7. Amy went to camp for *c* weeks. How many days was she there?

8. There were *d* dimes in a parking meter. How many dollars is this?

9. Paul types *w* words in 1 h. How many words can he type in 1 min?

10. A classroom is *m* meters wide. How many centimeters is this?

Solve by supplying a missing fact.

11. Georgia bought 19 yards of rope and Mark bought 11 yards of rope. How many more feet of rope did Georgia buy?

Problem Solving Situations

Solve by supplying missing facts.

1. Norman drove due north for 85 min and then due east for 55 min. How many hours did he spend driving?

2. The band director needs 5 ft of fabric for each band costume. If there are nine members in the band, how many yards of the fabric should the director buy?

3. The town of Middleton celebrated its centennial 40 years ago. How many years old is the town now?

4. Walt divides 6 lb of birdseed among eight bird feeders. How many ounces of birdseed does he put in each bird feeder?

Solve using any problem solving strategy.

5. Manuel bought some used paperback books for $1.25 each and some used children's books for $.90 each. He paid a total of $7. How many of each type of book did he buy?

Solve using any problem solving strategy.

6. Jacob Kosofsky is six years older than five times his son's age. If Jacob Kosofsky is 31, how old is his son?

7. A baker plans to make 8 cheesecakes, each containing 10 oz of cream cheese. How many pounds of cream cheese are needed?

PROBLEM SOLVING
CHECKLIST
Keep these in mind:
  Using a Four-Step Plan
  Too Much or Not Enough
   Information
  Supplying Missing Facts
Consider these strategies:
  Choosing the Correct Operation
  Making a Table
  Guess and Check
  Using Equations
  Identifying a Pattern
  Drawing a Diagram

8. In the 1988 Olympics, the women's high jump event was won by Louise Ritter of the United States who jumped 6 ft 8 in. Guennadi Avdeenko of the USSR won the men's high jump event with a jump of 7 ft $9\frac{1}{2}$ in. How many inches higher did Avdeenko jump than Ritter?

9. Angela Bruegge works 40 h per week for 18 weeks. What is her total salary?

10. A fruit stand sells grapefruits for $.85 each, apples for $.25 each, and lemons for $.15 each. In how many different ways can a customer spend $3.30 on these fruits?

WRITING WORD PROBLEMS

Write a word problem that you would solve by supplying the missing fact. Then solve the problem.

11. 1 mile = 5280 feet

12. 10 years = 1 decade

SPIRAL REVIEW

13. Replace the __?__ with >, <, or =: $\frac{8}{9}$ __?__ $\frac{22}{27}$ *(Lesson 7-6)*

14. DATA, *pages 288–289* About how many more senior high schools than junior high schools had microcomputers in 1984? *(Lesson 5-3)*

15. Solve: $5x + 17 = 2$ *(Lesson 4-5)*

16. Sandy earns $2750 per month. How much does Sandy earn in two years? *(Lesson 8-7)*

Challenge

Find three different whole numbers a, b, and c so that $\frac{1}{a} + \frac{1}{b} + \frac{1}{c}$ is also a whole number.

8-8 Equations Involving Rational Numbers

Objective: To solve equations involving rational numbers.

CONNECTION

In Chapter 4 you solved equations involving integers. You can use inverse operations to solve questions involving fractions and decimals as well.

Example 1

Solve. Check each solution.

a. $7 = -3x + 5$ **b.** $2m + 5.1 = -1.3$

Solution

a.
$$7 = -3x + 5$$
$$7 - 5 = -3x + 5 - 5$$
$$2 = -3x$$
$$\frac{2}{-3} = \frac{-3x}{-3}$$
$$-\frac{2}{3} = x$$

The solution is $-\frac{2}{3}$.

✔️ **Check**
$$7 = -3x + 5$$
$$7 \stackrel{?}{=} (-3)\left(-\frac{2}{3}\right) + 5$$
$$7 \stackrel{?}{=} 2 + 5$$
$$7 = 7$$

b.
$$2m + 5.1 = -1.3$$
$$2m + 5.1 - 5.1 = -1.3 - 5.1$$
$$2m = -6.4$$
$$\frac{2m}{2} = \frac{-6.4}{2}$$
$$m = -3.2$$

The solution is -3.2.

✔️ **Check**
$$2m + 5.1 = -1.3$$
$$2(-3.2) + 5.1 \stackrel{?}{=} -1.3$$
$$-6.4 + 5.1 \stackrel{?}{=} -1.3$$
$$-1.3 = -1.3$$

Example 2

Solve $z - \frac{1}{2} = \frac{3}{4}$. Check the solution.

Solution

$$z - \frac{1}{2} = \frac{3}{4}$$
$$z - \frac{1}{2} + \frac{1}{2} = \frac{3}{4} + \frac{1}{2}$$
$$z = \frac{3}{4} + \frac{2}{4}$$
$$z = \frac{5}{4} = 1\frac{1}{4}$$

✔️ **Check** $z - \frac{1}{2} = \frac{3}{4}$
$$\frac{5}{4} - \frac{1}{2} \stackrel{?}{=} \frac{3}{4}$$
$$\frac{5}{4} - \frac{2}{4} \stackrel{?}{=} \frac{3}{4}$$
$$\frac{3}{4} = \frac{3}{4}$$

The solution is $1\frac{1}{4}$.

✔️ **Check Your Understanding**

1. In the Check for Example 1(a), why does $(-3)\left(-\frac{2}{3}\right)$ equal 2?
2. In Example 2, what equivalent fraction replaces $\frac{1}{2}$?
3. In Example 2, explain why the solution, $1\frac{1}{4}$, is substituted in the Check as a fraction, not as a mixed number.

Guided Practice

COMMUNICATION «*Reading*

Use this paragraph for Exercises 1–3.

Keith bought some fruit at \$.39 per pound. After using $2\frac{1}{2}$ lb of the fruit to make a salad, he had $1\frac{3}{4}$ lb left. How many pounds of fruit did Keith buy?

« **1.** Which equation could you use to solve the problem?

    a. $n + 2\frac{1}{2} = 1\frac{3}{4}$ **b.** $n - 2\frac{1}{2} = 1\frac{3}{4}$ **c.** $n + 1\frac{3}{4} = 2\frac{1}{2}$

« **2.** Identify any facts that are not needed to solve the problem.

« **3.** Suppose the question in the problem were "How much did it cost Keith for the fruit he used in his salad?" Identify any facts that are not needed to solve this problem.

Tell what operation(s) you would use to solve each equation. Do not solve.

    4. $4k + 18 = 11$ **5.** $\frac{6}{7} = \frac{3}{5} + a$ **6.** $-1.3 = 7b - 2.7$

Solve. Check each solution.

    7. $8 + 5j = 12$ **8.** $4.6 = 7n - 4.5$

    9. $x + \frac{3}{8} = -\frac{1}{2}$ **10.** $-1\frac{5}{6} + y = \frac{2}{3}$

Exercises

Solve. Check each solution.

    1. $6n + 3 = -1$ **2.** $9 + 4g = 12$ **3.** $x + \frac{2}{9} = \frac{1}{3}$

    4. $s - \frac{1}{8} = -\frac{3}{4}$ **5.** $\frac{1}{3} + n = -\frac{1}{2}$ **6.** $k + 2\frac{3}{4} = -\frac{5}{8}$

    7. $19.8 = 7.17 + 3x$ **8.** $m - 8 = \frac{3}{5}$ **9.** $-5c - 2.7 = 1.35$

    10. $3\frac{5}{8} + r = \frac{7}{8}$ **11.** $2a - 1.9 = -4.1$ **12.** $9.5 + 6j = 5.3$

Solve using an equation.

13. Joey trimmed the length of his new poster by $1\frac{1}{2}$ in. to fit the poster into a frame with length $28\frac{3}{4}$ in. What was the length of the poster before Joey trimmed it?

14. Isaac paid for two tickets to a foreign film with a \$20 bill. He received \$8.30 in change. What was the price of each ticket?

15. Laura bought three identical kites at Flights, Unlimited. She paid a total of $47.23, including $4.48 tax. What was the price of each kite?

16. Kate jogged $2\frac{1}{2}$ mi farther today than she had jogged yesterday. She jogged $7\frac{2}{3}$ mi today. How far did she jog yesterday?

CALCULATOR

Match each equation with the correct solution as it appears on a calculator with a fraction key.

17. $x + 2\frac{2}{3} = 4\frac{1}{6}$

18. $6\frac{3}{8} = 2q - 1\frac{7}{8}$

19. $-\frac{2}{3} = \frac{5}{6} + g$

20. $10\frac{1}{12} + y = 1\frac{5}{6}$

A. $-1 \lrcorner 1 \lrcorner 2.$

B. $-8 \lrcorner 1 \lrcorner 4.$

C. $4 \lrcorner 1 \lrcorner 8.$

D. $1 \lrcorner 1 \lrcorner 2.$

COMMUNICATION «*Writing*

Write an equation that represents each sentence.

«**21.** Three times a number q decreased by 11.3 is -8.9.

«**22.** The sum of a number n and two thirds is negative five ninths.

«**23.** Sixteen is a number t increased by three fifths.

«**24.** The difference of a number y and 4.9 is 12.7.

DATA ANALYSIS

Find the mean, median, mode(s), and range of each set of data.

25. 2.3, 0.8, -0.9, -1.1, 0.8, 1.4   **26.** $-\frac{3}{4}, 2\frac{7}{8}, 1\frac{5}{8}$

27. What equation would you use to find the greatest number in a group of numbers whose least number is $-\frac{3}{5}$ and whose range is $4\frac{2}{3}$? Find the missing number.

28. The mean of two numbers is -7.2. The greater number is 2.15. Use an equation to find the lesser number.

SPIRAL REVIEW

29. Complete: The measure of an acute angle is between __?__ and __?__. (*Lesson 6-3*)

30. Use the formula $D = rt$ to find r when $D = 147$ mi and $t = 3.5$ h. (*Lesson 4-10*)

31. Solve: $x - \frac{4}{5} = \frac{3}{10}$ (*Lesson 8-8*)

32. Find the product: $(-4)\left(\frac{5}{8}\right)$ (*Lesson 8-1*)

Using Reciprocals to Solve Equations

Objective: To use reciprocals to solve equations.

To solve equations in which the variable has been multiplied by a fraction, you will need to multiply by the reciprocal of the fraction. Remember that the product of a number and its reciprocal is 1.

Example 1

Solve $\frac{2}{3}x = -5$. Check the solution.

Solution

$$\frac{2}{3}x = -5$$

$$\frac{3}{2}\left(\frac{2}{3}x\right) = \frac{3}{2}(-5) \quad \longleftarrow \text{Multiply both sides by the reciprocal of } \frac{2}{3}.$$

$$1 \cdot x = \frac{3}{2} \cdot \frac{-5}{1}$$

$$x = \frac{-15}{2} = -7\frac{1}{2}$$

The solution is $-7\frac{1}{2}$.

✓ **Check**

$$\frac{2}{3}x = -5$$

$$\frac{2}{3}\left(\frac{-15}{2}\right) \overset{?}{=} -5$$

$$-5 = -5$$

Before you multiply by a reciprocal, you may need to use inverse operations to get the variable term alone on one side of the equals sign.

Example 2

Solve $4 = -\frac{5}{6}b + 1$. Check the solution.

Solution

$$4 = -\frac{5}{6}b + 1$$

$$4 - 1 = -\frac{5}{6}b + 1 - 1$$

$$3 = -\frac{5}{6}b$$

$$\left(-\frac{6}{5}\right)\left(\frac{3}{1}\right) = \left(-\frac{6}{5}\right)\left(-\frac{5}{6}b\right) \quad \longleftarrow \text{Multiply both sides by the reciprocal of } -\frac{5}{6}.$$

$$\frac{-18}{5} = 1 \cdot b$$

$$-3\frac{3}{5} = b$$

The solution is $-3\frac{3}{5}$.

✓ **Check**

$$4 = -\frac{5}{6}b + 1$$

$$4 \overset{?}{=} \frac{-5}{6}\left(\frac{-18}{5}\right) + 1$$

$$4 \overset{?}{=} 3 + 1$$

$$4 = 4$$

✓ **Check Your Understanding**

1. In Example 2, explain why $-\frac{6}{5}$ is used to multiply both sides of the equation.

2. In Example 2, explain why the number 3 is rewritten as $\frac{3}{1}$.

«**1. COMMUNICATION** «*Reading* Replace each __?__ with the correct word.

To solve $-\frac{4}{5}t = 12$, you __?__ each side of the equation by the __?__ of $-\frac{4}{5}$.

2. Replace each __?__ with the number that makes the statement true.

$$\frac{2}{5}w - 3 = -7$$

$$\frac{2}{5}w - 3 + \underline{\ ?\ } = -7 + \underline{\ ?\ }$$

$$\frac{2}{5}w = \underline{\ ?\ }$$

$$\frac{?}{?}\left(\frac{2}{5}w\right) = \frac{?}{?}(-4)$$

$$1 \cdot w = \frac{5}{2} \cdot \frac{-4}{?}$$

$$w = \underline{\ ?\ }$$

Solve. Check each solution.

3. $-15 = \frac{2}{3}m$ **4.** $\frac{9}{10}j + 11 = 5$ **5.** $-\frac{8}{9}q - 5 = -13$

Solve. Check each solution.

1. $\frac{4}{11}j = 16$ **2.** $\frac{3}{4}x = -8$ **3.** $-\frac{5}{6}k = 9$

4. $-\frac{8}{9}c = 12$ **5.** $\frac{1}{2}v - 3 = 8$ **6.** $\frac{8}{15}y + 9 = 15$

7. $12 + \frac{2}{5}f = 5$ **8.** $-1 = \frac{7}{12}z + 6$ **9.** $25 = \frac{14}{25}r + 4$

10. $-\frac{2}{9}b + 4 = 1$ **11.** $\frac{6}{7}t - 2 = -10$ **12.** $-18 = -\frac{5}{11}x - 8$

13. Jennifer carried home a number of one-half pound packages of sliced deli meats and a 10 lb bag of oranges. Her total load was $13\frac{1}{2}$ lb. How many packages of deli meats did she carry?

14. After Marty had eaten some apples from a basket, he still had 14 apples left. This is seven eighths of the number of apples Marty picked. How many apples did Marty pick?

You can use reciprocals to solve equations involving fractions when the variable has been multiplied by an integer other than 1. For example, to solve $2x = \frac{4}{5}$, multiply both sides by $\frac{1}{2}$.

Solve. Check each solution.

15. $3y = \frac{3}{8}$

16. $-5z = \frac{15}{16}$

17. $\frac{3}{5} = -2m + \frac{1}{5}$

18. $3a - \frac{2}{5} = 2$

19. $14b + \frac{5}{8} = -\frac{1}{4}$

20. $-9d + 1\frac{1}{2} = 4\frac{1}{5}$

WRITING ABOUT MATHEMATICS

21. Write a paragraph describing how you solve an equation involving rational numbers written as fractions.

22. Write a paragraph comparing the methods for solving equations that involve integers, equations that involve decimals, and equations that involve fractions.

MENTAL MATH

Use mental math to solve each equation.

23. $\frac{1}{2}y = 2$

24. $-\frac{7}{8}b = \frac{7}{8}$

25. $\frac{3}{4}x = 1$

26. $m + \frac{6}{11} = 1\frac{6}{11}$

27. $r - \frac{1}{2} = \frac{1}{2}$

28. $4 = \frac{5}{6}g + 4$

CALCULATOR

Solve. Check the solution. Use a calculator with a fraction key if you have one available to you.

29. $\frac{1}{3}g = \frac{4}{5}$

30. $\frac{2}{3}v = -\frac{5}{8}$

31. $-\frac{1}{4}x + \frac{2}{3} = -\frac{3}{5}$

32. $\frac{4}{9} + \frac{11}{6}n = \frac{5}{6}$

33. $\frac{1}{2} = -\frac{7}{12}p - \frac{3}{4}$

34. $-\frac{3}{5} = \frac{3}{4} + \frac{15}{2}w$

COMMUNICATION « *Writing*

Write an equation for each sentence.

« **35.** Two thirds of a number t is negative fourteen.

« **36.** The sum of 11 and three fifths of a number z is ten.

« **37.** The difference of one half of a number g and three is five sixths.

« **38.** One fourth of a number n is nine and two thirds.

Write each equation in words.

« **39.** $\frac{1}{3}g = 14$

« **40.** $\frac{4}{5}c = 16$

« **41.** $-2 = \frac{1}{3}x + 9$

« **42.** $\frac{3}{4}m - 4 = 11$

« **43.** $10 = \frac{3}{8}s + \frac{1}{2}$

« **44.** $5d = -\frac{7}{12}$

THINKING SKILLS

The example at the right shows an *alternative method* for solving an equation involving rational numbers.

For Exercises 45–48, analyze the example at the right.

45. How does the number 20 relate to the denominators 4, 2, and 5?

46. What happened to the denominators when each side was multiplied by 20?

47. Which property allows you to simplify the left side of the equation from $20\left(\frac{3}{4}n - \frac{1}{2}\right)$ to $15n - 10$?

48. Create a plan for solving an equation using the method illustrated above. Use your plan to solve each equation.

   a. $4 = -\frac{2}{3}y + \frac{1}{2}$ **b.** $\frac{1}{2}m - \frac{3}{5} = \frac{2}{3}$ **c.** $\frac{1}{2} + \frac{3}{4}x = -\frac{1}{8}$

$$\frac{3}{4}n - \frac{1}{2} = \frac{2}{5}$$

$$20\left(\frac{3}{4}n - \frac{1}{2}\right) = 20\left(\frac{2}{5}\right)$$

$$20\left(\frac{3}{4}n\right) - 20\left(\frac{1}{2}\right) = 20\left(\frac{2}{5}\right)$$

$$15n - 10 = 8$$

$$15n = 18$$

$$n = \frac{18}{15} = 1\frac{1}{5}$$

SPIRAL REVIEW

49. Solve by drawing a diagram: A telephone installer left the telephone company office and drove 8 blocks due west, 10 blocks due north, 7 blocks due east, and 10 blocks due south. At this point, how far was the installer from the telephone company office? *(Lesson 7-8)*

50. Graph $x \geq 2.5$ on a number line. *(Lesson 7-10)*

51. The measure of an angle is 42°. Find its supplement. *(Lesson 6-3)*

52. Solve: $\frac{3}{4}z = 15$ *(Lesson 8-9)*

53. Find the quotient: $\frac{8}{15} \div \left(-\frac{2}{3}\right)$ *(Lesson 8-2)*

F annie Merritt Farmer (1857–1915) introduced the idea of exact measurement of ingredients in her cookbook first published in 1896. She advocated the use of measuring spoons and cups and directed that dry ingredients be leveled off with a knife blade. Earlier cookbooks used only vague terms such as "a pinch of salt." She championed the study of nutrition and gave courses to nurses on preparing diets for sick people. She also lectured on dietetics at Harvard Medical School.

Research

Find out how recipe ingredients are measured in other countries.

Transforming Formulas

Objective: To evaluate and transform formulas containing rational numbers.

Alex Gibbons is an exchange student living in France for the school year. He wants to borrow his host-family's kitchen and cook his favorite recipe from home. The recipe suggests a baking temperature of 325°F, but the ovens in France use the Celsius temperature scale. In order to use this recipe, Alex needs to know that 325°F is equivalent to 163°C.

A formula that relates Fahrenheit and Celsius temperatures is $F = \frac{9}{5}C + 32$, where F represents a Fahrenheit temperature and C represents the equivalent Celsius temperature. To find a value for C, you may want to *solve* the formula for C. To solve for a particular variable, you **transform** the formula by using inverse operations to get that variable alone on one side of the equals sign.

> **Terms to Know**
> • *transform a formula*

Example 1

Solve $F = \frac{9}{5}C + 32$ for C.

Solution

$$F = \frac{9}{5}C + 32$$

$$F - 32 = \frac{9}{5}C + 32 - 32 \qquad \longleftarrow \text{Subtract 32 from both sides.}$$

$$F - 32 = \frac{9}{5}C$$

$$\frac{5}{9}(F - 32) = \frac{5}{9}\left(\frac{9}{5}C\right) \qquad \longleftarrow \begin{array}{l}\text{Multiply both sides by}\\ \text{the reciprocal of } \frac{9}{5}.\end{array}$$

$$\frac{5}{9}(F - 32) = C$$

$$C = \frac{5}{9}(F - 32)$$

Example 2

Find the equivalent Celsius temperature for 325°F.

Solution

Use the formula $C = \frac{5}{9}(F - 32)$.

$$C = \frac{5}{9}(325 - 32) \qquad \longleftarrow \text{Substitute 325 for } F.$$

$$C = \frac{5}{9}(293) = 162.\overline{7} \approx 163$$

325°F is approximately 163°C.

✔️ **Check Your Understanding**

1. In Example 1, why is the *difference* $F - 32$ multiplied by $\frac{5}{9}$?
2. In Example 2, why is 325 substituted for F and not for C?

Guided Practice

« **1.** Explain the meaning of the word *transformed* in the following sentence.

    A caterpillar is transformed into a butterfly.

« **2.** The prefix *trans-*, as in transform, may mean ''across,'' ''beyond,'' ''from one place to another,'' or ''change.'' List three other words that begin with this prefix. Give a definition of each word.

Choose the letter of a correct transformation of each formula.

3. $P = C - r$ **a.** $C = Pr$ **b.** $C = P + r$ **c.** $C = \dfrac{P}{r}$

4. $D = \dfrac{1}{f}$ **a.** $f = D$ **b.** $f = 1 - D$ **c.** $f = \dfrac{1}{D}$

Solve for the variable shown in color.

5. $C = \dfrac{22}{7}d$ **6.** $V = \dfrac{1}{3}Bh$ **7.** $A = \dfrac{1}{2}bh$ **8.** $D = rt$

9. Use the formula $d = \dfrac{7}{22}C$ to find d (in inches) when $C = 11$ in.

10. Use the formula $h = \dfrac{3V}{B}$ to find h (in feet) when $V = 12$ ft^3 and $B = 4$ ft^2.

Exercises

Solve for the variable shown in color.

1. $A = bh$ **2.** $P = 4s$ **3.** $P = a + b + c$

4. $m = \dfrac{a + b}{2}$ **5.** $F = \dfrac{W}{d}$ **6.** $6.28r = C$

7. $I = Prt$ **8.** $c = np$ **9.** $W = IE$

10. $l = \dfrac{A}{w}$ **11.** $P = \dfrac{a}{2} + 110$ **12.** $P = 2l + 2w$

13. Use the formula $F = \dfrac{W}{d}$ to find F (in pounds) when $W = 40$ ft-lb and $d = 8$ lb.

14. Use the formula $A = \dfrac{1}{2}bh$ to find A (in square feet) when $b = 12$ ft and $h = 2\dfrac{1}{2}$ ft.

15. Use the formula $C = \dfrac{5}{9}(F - 32)$ to find the equivalent Celsius temperature for each Fahrenheit temperature.
     a. 0°F   **b.** 32°F   **c.** 212°F

16. Use the formula $F = \dfrac{9}{5}C + 32$ to find the equivalent Fahrenheit temperature for 37°C.

Ohm's Law is used in science. It states that in an electric circuit $E = IR$, where E represents the force (in volts), I represents the current (in amperes), and R represents the resistance (in ohms).

17. a. Solve Ohm's Law for I.

   b. What current does a force of 6 volts produce in a circuit with 0.15 ohms of resistance?

18. a. Solve Ohm's Law for R.

   b. Calculate the resistance of an electric steam iron that takes 8 amperes of current from a source supplying 110 volts.

GROUP ACTIVITY

A plumber's fee is a minimum of $30 plus an hourly rate of $22.

19. Write a formula that relates the plumber's fee (f) and the number of hours (h) worked. Put f alone on one side of the equals sign.

20. Make a table showing both the fees and the number of hours worked for a plumber who works 1, 2, 3, 4, and 5 h.

21. Solve the formula you wrote in Exercise 19 for h.

22. A plumber's fee is $184. How many hours did the plumber work?

23. RESEARCH   Find three formulas that are used in everyday life. (*Hint*: taxicab rates, telephone rates, electric or water rates)

SPIRAL REVIEW

24. Complete: A rhombus has four equal __?__.   (*Lesson 6-7*)

25. Solve $A = \frac{1}{2}ps$ for s.   (*Lesson 8-10*)

26. Find the sum: $-\frac{3}{4} + \frac{5}{16}$ (*Lesson 8-5*)

Self-Test 3

1. Heidi swam 20 ft farther than Kimberly. Kimberly swam 25 yd. How many yards did Heidi swim?   **8-7**

Solve. Check each solution.

2. $13 = 9 + 20z$ **3.** $4.8n - 7 = 16.4$ **4.** $v - \frac{1}{2} = 1\frac{6}{7}$ **8-8**

5. $-\frac{2}{9}x = 24$ **6.** $14 = \frac{4}{7}g - 6$ **7.** $-1 = 5 + \frac{3}{8}a$ **8-9**

8. Solve $V = lwh$ for h.   **8-10**

9. Use the formula you found in Exercise 8 to find h (in meters) when $V = 30$ m^3, $l = 5$ m, and $w = 3$ m.

FOCUS ON APPLICATIONS

Using a Stock Price Table

Objective: To read and interpret a stock price table.

A shareholder is a person who owns one or more shares of stock. Each share represents part ownership in a corporation.

Stock price tables like the one shown below can be found in the business section of a newspaper. They summarize one day of activity on a given stock exchange. Stock prices in these tables are usually given in eighths, fourths, and halves of a dollar. A share of stock listed at $27\frac{3}{4}$ is worth \$27.75.

Stock	Sym	Div	%	PE	Sales 100s	Hi	Lo	Close	Chg
BayOil	BOIL	.50	3.7	10	278	14	$13\frac{5}{8}$	$13\frac{5}{8}$	$-\frac{3}{8}$
BookClb	BKCL	1.20	4.1	8	1161	$29\frac{5}{8}$	$27\frac{3}{4}$	$29\frac{3}{8}$	$+1\frac{5}{8}$
BusiServ	BUSV	.72	2.4	16	71	$30\frac{1}{8}$	$29\frac{7}{8}$	30	...
ByteSys	BTSY	2.80	5.3	13	335	$53\frac{3}{8}$	$52\frac{7}{8}$	53	$-1\frac{1}{4}$

The last four columns in a stock price table display information about the various prices paid for one share of stock on a given day. The columns marked *Hi* and *Lo* display the highest and lowest prices paid per share that day. The column marked *Close* displays the price paid per share in the last transaction of the day for that stock. The column marked *Chg* displays the change in the closing price from the previous day.

Example

How much did the value of 100 shares of BayOil stock change from the close of the previous day to the close of the day shown?

Solution

The *Chg* column for BayOil shows $-\frac{3}{8}$.

This means that the closing price for one share of stock was $\$\frac{3}{8}$ lower than the closing price on the previous day.

Multiply to find the change in the value of 100 shares of BayOil stock.

$$-\frac{3}{8} \cdot 100 = \frac{-75}{2} = -37\frac{1}{2}$$

Write $-37\frac{1}{2}$ as a decimal: -37.5

The value of 100 shares of BayOil stock decreased by \$37.50 from the close of the previous day to the close of the day shown.

Stocks are sometimes *split*. When a stock splits "*n* for 1," the number of shares is multiplied by *n*, and the price is divided by *n*, so that the investment keeps its value. For example, suppose a shareholder owns 10 shares of stock worth $80 per share. The total value is $800. If the stock splits 2 for 1, the shareholder then owns 10 · 2, or 20, shares of stock at $80 ÷ 2, or $40, per share. The total value is still $800.

Exercises

Use the stock price table on page 378. For Exercises 1–6, name the stock or the category corresponding to the given information.

1. Hi: 29⅝
2. Lo: 13⅝
3. Close: 13⅝
4. BusiServ: 29⅞
5. ByteSys: 53⅜
6. BookClb: 29⅜

7. Write the high price for BusiServ stock as a decimal.

8. Did the price of BookClb stock *increase* or *decrease* from the previous day's price?

9. Marion bought 100 shares of BusiServ stock at its lowest price during the day. How much did she pay in all for the stock?

10. Find the previous day's closing price for BayOil stock.

11. Find the previous day's closing price for BookClb stock.

12. How much did the value of 400 shares of BookClb stock change from the close of the previous day to the close of the day shown?

13. Isabel bought 50 shares of BayOil stock at its highest price. How much would she have saved if she had bought it at its lowest price?

14. Jen owned 60 shares of BusiServ stock at the close of the day. Then the stock split 3 for 1. How many shares did she own after the split? How much was each share worth?

15. Dave owned 200 shares of ByteSys stock at the close of the day. Then the stock split 4 for 1. How many shares did he own after the split? How much was each share worth?

16. **RESEARCH** Find out what the columns marked *Div*, *%*, *PE*, and *Sales 100s* represent.

Terms to Know

reciprocal (p. 340)
multiplicative inverse (p. 340)

multiplication property of reciprocals
(p. 340)
transform a formula (p. 375)

Choose the correct term from the list above to complete each sentence.

1. Another name for the reciprocal of a number is the ___?___.

2. The ___?___ states that the product of a number and its reciprocal is one.

Find each answer. Simplify if possible. *(Lessons 8-1, 8-2, 8-4, 8-5)*

3. $\frac{5}{8} \cdot \frac{3}{4}$

4. $\frac{5}{6} \div \frac{2}{3}$

5. $11\frac{5}{6} + 2\frac{5}{6}$

6. $2\frac{1}{6} - \frac{1}{3}$

7. $-6\left(\frac{3}{4}\right)$

8. $\left(-5\frac{1}{4}\right)\left(-2\frac{2}{3}\right)$

9. $\frac{2}{9} \div \left(-\frac{3}{4}\right)$

10. $-4\frac{4}{9} \div 10$

11. $\frac{4}{5} + \left(-4\frac{2}{5}\right)$

12. $-\frac{8}{9} + \left(-\frac{11}{12}\right)$

13. $\left(-4\frac{3}{8}\right) \div \left(-2\frac{3}{16}\right)$

14. $1\frac{3}{10} \div \left(-3\frac{3}{5}\right)$

15. $-3\frac{1}{2} - \frac{2}{3}$

16. $-3\frac{5}{6} - \left(-7\frac{1}{6}\right)$

17. $3\frac{2}{7} + \frac{1}{7} + \left(-1\frac{3}{7}\right)$

18. $-3 + \left(-1\frac{3}{5}\right) + \frac{9}{10}$

19. $(-0.6)(-0.9)$

20. $(4.8)(-0.5)$

21. $-6.09 \div 2.9$

22. $-48 \div (-1.2)$

23. $-1.6 + (-4.7)$

24. $12 + (-8.1)$

25. $3.75 - 62.5$

26. $2.16 - (-0.54)$

27. $\left(\frac{2a}{5}\right)\left(\frac{3}{8a}\right)$

28. $\left(\frac{x}{4}\right)\left(\frac{8z}{x}\right)$

29. $\frac{3c}{2} \div \frac{7c}{3}$

30. $\frac{6a}{11} \div 4a$

31. $\frac{7}{x} + \frac{4}{x}$

32. $\frac{a}{6} + \frac{7a}{12}$

33. $\frac{3r}{10} - \frac{2r}{15}$

34. $\frac{19m}{12} - \frac{m}{12}$

35. $8n - \frac{7n}{9}$

36. $4c + \frac{c}{7}$

37. $\frac{3}{10n} + \frac{9}{10n} + \frac{13}{10n}$

38. $\frac{3m}{8} + \frac{m}{4} + \frac{4m}{3}$

Solve. Decide whether a *whole number*, a *fraction*, or a *decimal* is an appropriate answer. *(Lesson 8-3)*

39. Margaret spends a total of $2\frac{1}{3}$ h commuting to work every day. How many hours does she spend commuting to work in 5 days?

40. A music practice room costs $15 per hour to rent. What is the cost to rent the room for $3\frac{1}{2}$ h?

41. John's mother sent him to the grocery store to buy yogurt. She gave him $5.53. Each container of yogurt cost $.79. How many containers of yogurt was John able to buy?

42. The total weight of 12 identical packages is 15 lb. What is the weight of each package?

Chapter 8

Estimate. *(Lessons 8-1, 8-6)*

43. $\left(3\frac{7}{8}\right)\left(2\frac{1}{9}\right)$ **44.** $\left(-6\frac{1}{3}\right)\left(-4\frac{3}{4}\right)$ **45.** $\left(\frac{6}{13}\right)\left(-9\frac{1}{8}\right)$ **46.** $\left(\frac{8}{27}\right)\left(8\frac{2}{5}\right)$

47. $13\frac{6}{7} - 8\frac{1}{12}$ **48.** $4\frac{5}{8} + 3\frac{1}{9} + 2\frac{9}{10}$ **49.** $-12\frac{5}{6} \div \left(-\frac{7}{16}\right)$ **50.** $-23\frac{9}{11} \div \frac{4}{13}$

51. $-5\frac{5}{6} + \left(-10\frac{2}{9}\right)$ **52.** $-9\frac{3}{10} - \left(-\frac{4}{5}\right)$ **53.** $18\frac{1}{2} \div \left(-3\frac{2}{3}\right)$ **54.** $-16\frac{1}{7} \div \left(-5\frac{1}{4}\right)$

Solve by supplying missing facts. *(Lesson 8-7)*

55. When Steve reached the age of 30, he was $6\frac{1}{2}$ ft tall. At birth he was 21 in. long. How many inches did Steve grow in 30 years?

56. Kiko runs $4\frac{1}{4}$ mi each day. How far does she run in a week?

57. Lisa bought $3\frac{1}{2}$ dozen bran muffins for a party. How many muffins did she buy?

58. Marty swam for $\frac{1}{2}$ h on Monday and $\frac{3}{5}$ h on Tuesday. For how many more minutes did Marty swim on Tuesday than on Monday?

Solve. Check each solution. *(Lessons 8-8, 8-9)*

59. $5x + 9 = 13$ **60.** $3a - 8 = -9$ **61.** $-4c - 8.1 = 3.9$

62. $10p - 0.5 = 2.35$ **63.** $4 + j = -\frac{2}{5}$ **64.** $\frac{1}{8} + h = 4$

65. $17 + 10x = 5$ **66.** $-8b + 13 = 7$ **67.** $-7y + 4.8 = 9$

68. $3n - 3.7 = -4.3$ **69.** $m - \frac{5}{6} = \frac{2}{3}$ **70.** $\frac{7}{8} + c = \frac{1}{4}$

71. $-1.2 = 5b - 4.7$ **72.** $-5.9 = -6n + 2.5$ **73.** $3\frac{1}{2} + a = 2\frac{4}{5}$

74. $p + \frac{3}{4} = 2\frac{1}{3}$ **75.** $\frac{5}{8}x = 30$ **76.** $-\frac{2}{5}n = 5$

77. $-\frac{5}{6}a + 9 = 7$ **78.** $\frac{2}{3}y - 3 = -13$ **79.** $\frac{3}{4}c = -8$

80. $-\frac{1}{3}p = -7$ **81.** $19 = -\frac{7}{8}k + 5$ **82.** $11 + \frac{4}{5}m = 5$

Solve for the variable shown in color. *(Lesson 8-10)*

83. $d = 2r$ **84.** $F = ma$ **85.** $A = \frac{1}{2}ps$ **86.** $a = \frac{M - m}{2}$

87. Use the formula $C = \frac{5}{9}(F - 32)$ to find the equivalent Celsius temperature for 86°F.

88. Use the formula $P = 2l + 2w$ to find P (in feet) when $l = 5\frac{1}{4}$ ft and $w = 3\frac{1}{2}$ ft.

Chapter Test

Find each answer. Simplify if possible.

1. $\left(-\dfrac{5}{9}\right)\left(-\dfrac{3}{10}\right)$ 2. $\left(-2\dfrac{1}{2}\right)\left(4\dfrac{2}{3}\right)$ 3. $(2.6)(-3.4)$ 4. $\left(\dfrac{3a}{5b}\right)\left(\dfrac{10}{9a}\right)$ 8-1

5. $\dfrac{15}{16} \div (-3)$ 6. $-0.846 \div (-1.8)$ 7. $\dfrac{2n}{9} \div \dfrac{4n}{5}$ 8. $\dfrac{12x}{13} \div 3x$ 8-2

Solve. Decide whether a *whole number*, a *fraction*, or a *decimal* is an appropriate answer.

9. Lucia has 24 yd of rope to cut into pieces measuring $2\dfrac{1}{4}$ yd each. How many pieces of this size can Lucia cut from the rope?   8-3

10. Mr. Richards charges $20 per hour for tutoring. How much will he earn for $5\dfrac{3}{4}$ h of tutoring?

Find each answer. Simplify if possible.

11. $\dfrac{7}{12} + \dfrac{11}{12}$ 12. $-2\dfrac{5}{6} - \dfrac{5}{6}$ 13. $\dfrac{9}{a} + \dfrac{7}{a}$ 14. $\dfrac{14x}{9} - \dfrac{8x}{9}$ 8-4

15. $1\dfrac{5}{6} - \dfrac{7}{9}$ 16. $-\dfrac{5}{8} - \left(-\dfrac{2}{3}\right)$ 17. $-5\dfrac{1}{4} + 2\dfrac{11}{12}$ 18. $-5 + 2\dfrac{1}{2} + \left(-\dfrac{3}{4}\right)$ 8-5

19. $\dfrac{5a}{9} - \dfrac{5a}{18}$ 20. $7c + \dfrac{8c}{11}$ 21. $8.6 + (-9.9)$ 22. $-42.6 - 0.58$

Estimate.

23. $17\dfrac{1}{10} - 5\dfrac{9}{11}$ 24. $8\dfrac{11}{13} + 4\dfrac{1}{7} + 5\dfrac{7}{12}$ 25. $\left(-9\dfrac{7}{8}\right)\left(\dfrac{7}{16}\right)$ 8-6

Solve by supplying missing facts.

26. Leroy earns $350 per week. How much does Leroy earn in a year?   8-7

27. Identify any missing facts and state where you might find the facts needed to solve this problem: The Amazon river is 3912 mi long. The Nile river is the longest river in the world. How much longer is the Nile river than the Amazon river?

Solve. Check each solution.

28. List the steps you would use to solve the equation $-2x - 7.5 = -6.2$.   8-8

29. $3z - 8 = -10$ 30. $-2.6 = 4a + 6.6$ 31. $4\dfrac{3}{5} + m = \dfrac{7}{10}$

32. $-\dfrac{3}{5}x = 4$ 33. $\dfrac{5}{8}c - 17 = 3$ 34. $11 = \dfrac{6}{7}b + 2$ 8-9

Solve for the variable shown in color.

35. $D = \dfrac{m}{v}$ 36. $A = \dfrac{1}{2}ap$ 37. $a + b + c = 180$ 8-10

38. Use the formula $V = \dfrac{1}{3}Bh$ to find V (in cubic inches) when $B = 4$ in.^{2} and $h = 6$ in.

Adding and Subtracting More Algebraic Fractions

Objective: To add and subtract algebraic fractions with unlike denominators that contain variables.

Sometimes the unlike denominators in algebraic fractions contain variables. In order to add and subtract such fractions, you have to find the LCD using the skills you learned in Lesson 7-3 for finding an LCM.

Example

Simplify $\dfrac{2x}{3y^2} + \dfrac{5x}{6y^2}$.

Solution

$\dfrac{2x}{3y^2} + \dfrac{5x}{6y^2} = \dfrac{2x}{3y^2} \cdot \dfrac{2}{2} + \dfrac{5x}{6y^2}$ ⟵ The LCM of $3y^2$ and $6y^2$ is $6y^2$.

$\qquad\qquad = \dfrac{4x}{6y^2} + \dfrac{5x}{6y^2}$

$\qquad\qquad = \dfrac{4x + 5x}{6y^2}$

$\qquad\qquad = \dfrac{9x}{6y^2} = \dfrac{3x}{2y^2}$

Exercises

Simplify.

1. $\dfrac{2}{x} + \dfrac{1}{5x}$

2. $\dfrac{5}{mn} - \dfrac{5}{6mn}$

3. $\dfrac{9}{2a^2} - \dfrac{3}{a^2}$

4. $\dfrac{3b}{c} + \dfrac{8b}{7c}$

5. $\dfrac{10x}{3yz} + \dfrac{4x}{yz}$

6. $\dfrac{8m}{7n^3} + \dfrac{2m}{n^3}$

7. $\dfrac{5}{6xy} - \dfrac{7}{12xy}$

8. $\dfrac{9}{5a} - \dfrac{7}{15a}$

9. $\dfrac{3}{10n^3} + \dfrac{6}{5n^3}$

10. $\dfrac{2b}{9ac} + \dfrac{5b}{3ac}$

11. $\dfrac{5r}{4s} - \dfrac{7r}{8s}$

12. $\dfrac{7u}{4v^2} - \dfrac{u}{2v^2}$

13. $\dfrac{7}{3m} - \dfrac{3}{4m}$

14. $\dfrac{7}{2m^2} + \dfrac{4}{3m^2}$

15. $\dfrac{11}{4rs} + \dfrac{1}{5rs}$

16. $\dfrac{5m}{2n} - \dfrac{2m}{5n}$

17. $\dfrac{4b}{5c^3} + \dfrac{8b}{3c^3}$

18. $\dfrac{3m}{2np} - \dfrac{m}{7np}$

19. $\dfrac{7}{10r} - \dfrac{1}{6r}$

20. $\dfrac{9}{8uv} + \dfrac{3}{10uv}$

21. $\dfrac{5r}{4s^3} + \dfrac{17r}{6s^3}$

22. $\dfrac{11x}{12y} + \dfrac{5x}{9y}$

23. $\dfrac{13u}{6vw} - \dfrac{7u}{8vw}$

24. $\dfrac{4x}{9y^2} - \dfrac{2x}{15y^2}$

25. $\dfrac{9}{mn} + \dfrac{3}{2mn} + \dfrac{7}{6mn}$

26. $\dfrac{5u}{2v} + \dfrac{9u}{10v} + \dfrac{3u}{5v}$

27. $\dfrac{7}{12x^2} + \dfrac{1}{6x^2} + \dfrac{9}{8x^2}$

28. $\dfrac{b}{2c^2} + \dfrac{2b}{5c^2} + \dfrac{2b}{3c^2}$

29. **THINKING SKILLS** A student simplified $\dfrac{5x + 12y}{8y^2}$ as $\dfrac{5x + 3}{2y}$. Compare the two expressions. Did the student simplify correctly? Explain.

Cumulative Review

Standardized Testing Practice

Choose the letter of the correct answer.

1. **Identify a pair of vertical angles.**

   A. $\angle 2$ and $\angle 7$ B. $\angle 4$ and $\angle 5$
   C. $\angle 3$ and $\angle 5$ D. $\angle 6$ and $\angle 1$

2. **What is the prime factorization of 80?**
   A. $8 \cdot 10$ B. $2 \cdot 5 \cdot 8$
   C. $2^4 \cdot 5$ D. $5 \cdot 16$

3. **Decide which statistical measure best describes the data.**
   4, 3.9, 12.8, 3.7, 3.6, 4.1
   A. mean   B. median
   C. mode   D. range

4. **Decide which is the appropriate form of the answer.**
   Vans hold 12 students each. If 54 students plan to travel in vans, how many vans will be needed?
   A. decimal   B. fraction
   C. dollars   D. whole number

5. **Which equation could you use to solve the problem?**
   Gia's home run total is six more than three times Kelly's home run total. If Gia's total is 15, what is Kelly's total?
   A. $6x + 3 = 15$ B. $6x - 3 = 15$
   C. $3x + 15 = 6$ D. $3x + 6 = 15$

6. **Identify the figure.**

   A. rhombus   B. rectangle
   C. trapezoid   D. hexagon

7. **Find the quotient:** $\frac{t}{12} \div \frac{2t}{3}$
   A. $\frac{1}{2}$ B. 8

   C. $\frac{1}{8}$ D. $\frac{t^2}{18}$

8. **Simplify:** s^{-6}
   A. $-6s$ B. $\frac{s}{6}$
   C. $\frac{1}{s^6}$ D. $s - 6$

9. **A marching band can march in rows of either 8 or 14 with no one left out. What is the least number of members the band could have?**
   A. 22   B. 112
   C. 56   D. 2

10. **Choose the correct relationship:**
    $m = |x - y|, n = |y - x|$
    A. $m = n$
    B. $m > n$
    C. $m < n$
    D. cannot determine

11. **Solve:** $9x + 2 + 4x = 41$

    A. $x = 3$

    B. $x = 3\frac{4}{13}$

    C. $x = 2$

    D. $x = 507$

12. **Phillip bought $4\frac{1}{2}$ yd of cotton and $1\frac{3}{4}$ yd of wool. How many yards of fabric did he buy altogether?**

    A. $6\frac{1}{4}$ yd   B. $5\frac{4}{6}$ yd

    C. $5\frac{2}{3}$ yd   D. $2\frac{3}{4}$ yd

13. **An elevator goes up 6 floors, goes down 8 floors, and then goes down 3 more floors. How much lower is the elevator than it was at its starting point?**

    A. 11 floors   B. 2 floors

    C. 5 floors   D. 17 floors

14. **Write 0.0000498 in scientific notation.**

    A. 4.98×10^{-4}

    B. 4.98×10^{4}

    C. 4.98×10^{5}

    D. 4.98×10^{-5}

15. **On a trip, Carolina plans to spend 10 days camping, and then 11 days at a resort. How many weeks long is her trip?**

    A. 21   B. 2

    C. 7   D. 3

16. **Complete:** A triangle with sides that measure 4 cm, 4 cm, and 6 cm is a(n) _?_ triangle.

    A. scalene   B. isosceles

    C. equilateral   D. rhomboid

17. **Complete:** An angle that measures 79° is a(n) _?_ angle.

    A. obtuse   B. acute

    C. right   D. adjacent

18.
    Which of the following inequalities does this graph represent?

    I. $x \geq -1$ II. $x \leq -1$ III. $x > -1$

    A. I only   B. II only

    C. III only   D. I and II

19. **Which of the following can make a bar graph visually misleading?**

    A. drawing it horizontally

    B. not shading the bars

    C. putting a gap in the scale

    D. using a different color

20. **Solve:** $14 + \frac{15}{16}x = 11$

    A. $x = -3\frac{1}{5}$ B. $x = 26\frac{2}{3}$

    C. $x = -3$ D. $x = -2\frac{13}{16}$

Construction of a Spiral Based
on the Golden Rectangle

Golden Rectangle

$$\frac{x+y}{x} = \frac{x}{y} = 1.618$$

Ratio, Proportion, and Percent 9

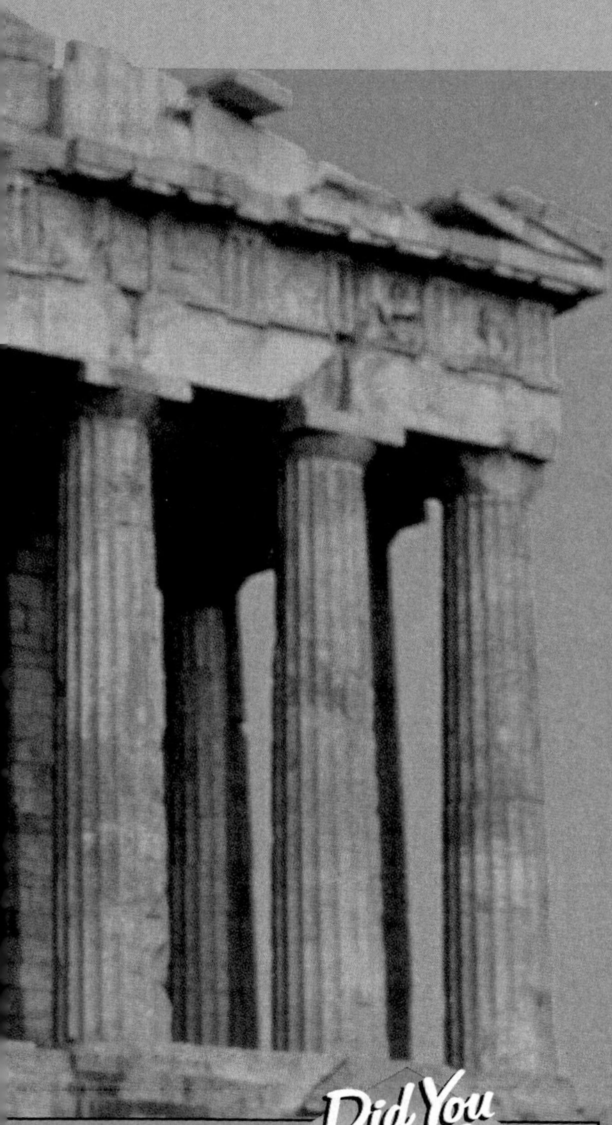

Preference for Rectangle Proportions			
rectangle shape	ratio of length to width	people rating most pleasing	people rating least pleasing
	1.00	3.0%	27.8%
	1.20	0.2%	19.7%
	1.25	2.0%	9.4%
	1.33	2.5%	2.5%
	1.44	7.7%	1.2%
	1.50	20.6%	0.4%
	1.62	35.0%	0.0%
	1.75	20.0%	0.8%
	2.00	7.5%	2.5%
	2.50	1.5%	35.7%

Did You Know?

Ancient philosophers believed, and modern psychologists have demonstrated, that certain shapes are more pleasing to the eye than others. One such shape is the **golden rectangle**, whose ratio of length to width is about 1.618. For thousands of years, the golden rectangle has been used in art and architecture. The front of the Parthenon, a Greek temple built between 447 and 432 B.C., was designed to fit inside a golden rectangle.

9-1

Ratio and Rate

Objective: To write ratios as fractions in lowest terms and to write unit rates.

Terms to Know
- *ratio*
- *rate*
- *unit rate*

DATA ANALYSIS

The table at the right lists the win-loss records of four teams in a football league. You can use *ratios* to describe information in the chart.

A **ratio** is a comparison of two numbers by division. The ratio of two numbers a and b ($b \neq 0$) can be written in three ways.

a to b $a:b$ $\dfrac{a}{b}$

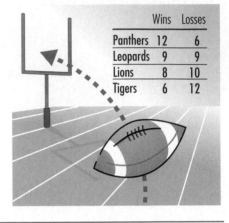

	Wins	Losses
Panthers	12	6
Leopards	9	9
Lions	8	10
Tigers	6	12

Example 1

Use the table above. Write each ratio in lowest terms.

a. Tiger wins to Tiger losses

b. Lion losses : Lion wins

c. $\dfrac{\text{Panther wins}}{\text{Panther losses}}$

Solution

To write a ratio in lowest terms, write it as a fraction in lowest terms.

a. $\dfrac{6}{12} = \dfrac{1}{2}$ or $1:2$ or   1 to 2

b. $\dfrac{10}{8} = \dfrac{5}{4}$ ◄—— Do not write as a mixed number.

c. $\dfrac{12}{6} = \dfrac{2}{1}$ ◄—— Keep the 1 in the denominator.

✓ Check Your Understanding

1. In Example 1, is the ratio of Panther wins to losses the same as the ratio of Panther losses to wins? Explain.

When the quantities being compared are in different units, first write the quantities in the same unit and then write the ratio.

Example 2
Solution

Write the ratio as a fraction in lowest terms: 21 days to 6 weeks

Change 21 days to weeks: 7 days = 1 week, so 21 days = 3 weeks

$\dfrac{3 \text{ weeks}}{6 \text{ weeks}} \rightarrow \dfrac{3}{6} = \dfrac{1}{2}$ The ratio of 21 days to 6 weeks is $\dfrac{1}{2}$.

✓ Check Your Understanding

2. In Example 2, is the answer the same when you write the number of weeks as a number of days? Explain.

A ratio that compares two unlike quantities is called a **rate**. A **unit rate** is a rate for *one* unit of a given quantity. An example of a unit rate is *miles per hour*, which indicates the number of miles for one hour.

Example 3

Write the unit rate.

a. 150 mi in 3 h   **b.** $42 for 4 shirts

Solution

a. $\dfrac{\text{miles}}{\text{hours}} \rightarrow \dfrac{150}{3} = \dfrac{50}{1}$ The rate is 50 mi in 1 h, or 50 mi/h.

b. $\dfrac{\text{dollars}}{\text{shirts}} \rightarrow \dfrac{42}{4} = \dfrac{10.5}{1}$ The rate is $10.50 for 1 shirt, or $10.50/shirt.

Guided Practice

COMMUNICATION « *Writing*

«**1.** Write the ratio 5 *to* 6 in two other ways.

«**2.** Write the unit rate in words: 22 mi/gal

«**3.** Write the unit rate in symbols: 53 words per minute

Replace each __?__ with the correct number.

4. 2 ft = __?__ in.   **5.** 3 h = __?__ min

6. __?__ years = 730 days   **7.** __?__ yd = 21 ft

Write each ratio as a fraction in lowest terms.

8. $\dfrac{14}{7}$ **9.** $\dfrac{36}{15}$ **10.** 2 h:1 day

11. 5 in. to 2 ft   **12.** $2 to $.30   **13.** 4 yd:4 ft

Write the unit rate.

14. $45 for 3 lb   **15.** 150 words in 3 min   **16.** 160 mi in 4 h

Exercises

Write each ratio as a fraction in lowest terms.

1. 18 to 24   **2.** 2:32   **3.** 35:21   **4.** 49 to 14

5. $\dfrac{36}{18}$ **6.** $\dfrac{14}{14}$ **7.** 40 to 25   **8.** 100:75

9. 28 days to 2 weeks   **10.** 2 m to 40 cm   **11.** 9 in.:2 ft

12. $.50:$3   **13.** 12 oz to 1 lb   **14.** 5 yd:10 ft

Write the unit rate.

15. 300 mi in 6 h   **16.** 205 words in 5 min   **17.** 250 mi on 10 gal

18. $42 for 5 h   **19.** 33 m in 15 s   **20.** $6 for 12 L

21. Sean's chemistry book describes salt water that is made by combining 1 c salt with 9 c water. Write the ratio of water to salt as a fraction in lowest terms.

22. DATA, *pages 2–3*   Write the ratio of Houston Symphony Pops concerts to Mozart concerts as a fraction in lowest terms.

Write each ratio as a fraction in lowest terms.

23. $10x$ to $20y$ **24.** $15y$ to $5w$ **25.** $12n : 4n$ **26.** $\dfrac{8t}{6t}$

DATA ANALYSIS

Use the pictograph at the right.

27. How many cars were sold in January?

28. How many more full size cars were sold than luxury cars?

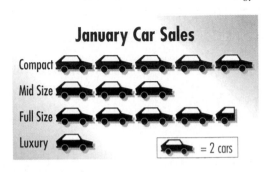

January Car Sales

29. Write the ratio of *compact cars sold* to *mid size cars sold* as a fraction in lowest terms.

30. Write the ratio of *full size cars sold* to *luxury cars sold* as a fraction in lowest terms.

PROBLEM SOLVING/APPLICATION

When shopping, you may find the same items packaged in various sizes or quantities. The item with the lesser unit price is usually the better buy.

31. Find the better buy: 8 pens for $3.60 or 12 pens for $4.80.

32. Find the better buy: 3 shirts for $25 or 4 shirts for $32.50.

33. Describe a situation in which it may *not* be preferable to buy an item with the lesser unit price.

34. Go to a grocery store and find the prices for different quantities of the same item. Determine the better buy.

SPIRAL REVIEW

35. Solve: $\frac{2}{3}x = 16$ *(Lesson 8-9)*

36. Write the unit rate: $62.50 for 5 h   *(Lesson 9-1)*

37. Write the prime factorization of 240.   *(Lesson 7-1)*

38. Solve: $2x + 7 = -17$ *(Lesson 4-5)*

9-2 Proportions

Objective: To solve proportions.

Terms to Know
- *proportion*
- *terms*
- *cross products*
- *solve a proportion*

EXPLORATION

1 Is the statement $\frac{3}{4} = \frac{9}{12}$ *true* or *false*?

2 Complete: $3(12) =$ __?__ ; $4(9) =$ __?__ $\frac{3}{4} \times \frac{9}{12}$

3 Is the statement $\frac{3}{12} = \frac{4}{9}$ *true* or *false*?

4 Complete: $3(9) =$ __?__ ; $12(4) =$ __?__ $\frac{3}{12} \times \frac{4}{9}$

5 Compare your answers in Steps 1 and 2 with your answers in Steps 3 and 4. Complete.

Two ratios are equal if the product of the first numerator and the second denominator equals the product of the first __?__ and the second __?__.

A **proportion** is a statement that two ratios are equal.

You write: $\frac{3}{4} = \frac{9}{12}$ or $3:4 = 9:12$

You read: *3 is to 4 as 9 is to 12*.

The numbers 3, 4, 9, and 12 are the **terms** of the proportion. If a statement is a true proportion, the **cross products** of the terms are equal.

In Arithmetic	In Algebra
$\frac{3}{4} = \frac{9}{12}$	$\frac{a}{b} = \frac{c}{d}$
$3(12) = 4(9)$	$ad = bc, b \neq 0, d \neq 0$
$36 = 36$	

Example 1

Tell whether each proportion is *True* or *False*.

a. $\frac{6}{9} = \frac{8}{12}$

b. $\frac{4}{6} = \frac{0.9}{1.3}$

Solution

a. $\frac{6}{9} \times \frac{8}{12}$

$6(12) \stackrel{?}{=} 9(8)$

$72 \stackrel{?}{=} 72$

True

b. $\frac{4}{6} \times \frac{0.9}{1.3}$

$4(1.3) \stackrel{?}{=} 6(0.9)$

$5.2 \stackrel{?}{=} 5.4$

False

☑ **Check Your Understanding**

1. Would the proportion in Example 1(b) be true if 1.3 were 1.35? Explain.

Sometimes one of the terms of a proportion is a variable. To **solve a proportion,** you find the value of the variable that makes the proportion true. You can use cross products when you solve a proportion.

Example 2

Solve each proportion.

a. $\dfrac{n}{6} = \dfrac{3}{2}$

b. $\dfrac{6}{4} = \dfrac{b}{10}$

Solution

a.  $\dfrac{n}{6} \bowtie \dfrac{3}{2}$

$2n = 6(3)$

$2n = 18$

$\dfrac{2n}{2} = \dfrac{18}{2}$

$n = 9$

b. $\dfrac{6}{4} \bowtie \dfrac{b}{10}$

$6(10) = 4b$

$60 = 4b$

$\dfrac{60}{4} = \dfrac{4b}{4}$

$15 = b$

✔️ **Check Your Understanding**

2. In Example 2(b), would the solution be different if the proportion were $\dfrac{4}{6} = \dfrac{10}{b}$? Explain.

A calculator may be helpful in solving proportions. For instance, to solve the proportion $\dfrac{35}{10} = \dfrac{n}{7}$, you can use this key sequence.

$$\boxed{35.} \; \boxed{\times} \; \boxed{7.} \; \boxed{\div} \; \boxed{10.} \; \boxed{=}$$

This sequence solves the equation $10n = 35(7)$, which results from the cross products of the proportion. First multiply 35 by 7. Then divide by 10 to find n.

Guided Practice

COMMUNICATION «*Reading*

«**1.** What is the main idea of the lesson?

«**2.** List two major points that support the main idea of the lesson.

COMMUNICATION «*Writing*

«**3.** Write *five is to six as ten is to twelve* in symbols.

«**4.** Write *three is to one as fifteen is to five* in symbols.

«**5.** Write the proportion $\dfrac{6}{9} = \dfrac{2}{3}$ in words.

«**6.** Write the proportion $\dfrac{c}{4} = \dfrac{10}{40}$ in words.

Write the cross products of the terms.

7. $\dfrac{32}{8} = \dfrac{20}{5}$

8. $\dfrac{24}{18} = \dfrac{40}{30}$

9. $\dfrac{25}{40} = \dfrac{n}{8}$

10. $\dfrac{c}{2} = \dfrac{168}{12}$

Tell whether each proportion is *True* or *False*.

11. $\dfrac{5}{8} = \dfrac{15}{24}$ **12.** $\dfrac{12}{4} = \dfrac{4}{12}$ **13.** $\dfrac{18}{9} = \dfrac{6}{1}$ **14.** $\dfrac{10}{15} = \dfrac{2.2}{3.3}$

Solve each proportion.

15. $\dfrac{3}{8} = \dfrac{b}{24}$ **16.** $\dfrac{16}{8} = \dfrac{6}{n}$ **17.** $\dfrac{3}{a} = \dfrac{5}{10}$ **18.** $\dfrac{y}{4} = \dfrac{2.1}{1.4}$

Exercises

Tell whether each proportion is *True* or *False*.

1. $\dfrac{21}{24} = \dfrac{7}{8}$ **2.** $\dfrac{10}{12} = \dfrac{3}{4}$ **3.** $\dfrac{8}{6} = \dfrac{6}{8}$ **4.** $\dfrac{8}{10} = \dfrac{12}{15}$

5. $\dfrac{12}{6} = \dfrac{5}{3}$ **6.** $\dfrac{25}{6} = \dfrac{8}{2}$ **7.** $\dfrac{1.5}{3} = \dfrac{10}{20}$ **8.** $\dfrac{50}{100} = \dfrac{3.2}{6.4}$

Solve each proportion.

9. $\dfrac{3}{2} = \dfrac{9}{n}$ **10.** $\dfrac{12}{b} = \dfrac{4}{6}$ **11.** $\dfrac{22}{10} = \dfrac{m}{5}$ **12.** $\dfrac{a}{5} = \dfrac{9}{3}$

13. $\dfrac{1.2}{1.5} = \dfrac{y}{10}$ **14.** $\dfrac{2.5}{5} = \dfrac{c}{8}$ **15.** $\dfrac{6}{z} = \dfrac{4}{6}$ **16.** $\dfrac{7}{p} = \dfrac{1}{7}$

17. Is $\dfrac{4}{9} = \dfrac{36}{81}$ a true proportion? Explain.

18. A number n is to 200 as 5 is to 4. Write a proportion and solve for n.

Solve each proportion.

19. $\dfrac{14}{2a} = \dfrac{6}{9}$ **20.** $\dfrac{3w}{5} = \dfrac{24}{10}$ **21.** $\dfrac{n}{25} = \dfrac{4}{n}$ **22.** $\dfrac{d}{9} = \dfrac{4}{d}$

23. Solve $\dfrac{a}{b} = \dfrac{c}{d}$ for c when $a = 8$, $b = 12$, and $d = 18$.

24. Solve $\dfrac{w}{x} = \dfrac{y}{z}$ for z when $w = 20$, $x = 8$, and $y = 15$.

Solve each proportion. (Use the distributive property.)

25. $\dfrac{3}{2} = \dfrac{9}{n + 2}$ **26.** $\dfrac{z - 5}{8} = \dfrac{3}{4}$ **27.** $\dfrac{6}{y - 4} = \dfrac{3}{5}$ **28.** $\dfrac{2}{3} = \dfrac{6 + n}{12}$

CALCULATOR

Choose the key sequence you would use to solve each proportion.

29. $\dfrac{30}{19} = \dfrac{m}{57}$

30. $\dfrac{30}{57} = \dfrac{m}{19}$

31. $\dfrac{30}{57} = \dfrac{19}{m}$

A. $\boxed{30.}\ \boxed{\times}\ \boxed{19.}\ \boxed{\div}\ \boxed{57.}$

B. $\boxed{30.}\ \boxed{\times}\ \boxed{19.}\ \boxed{\times}\ \boxed{57.}$

C. $\boxed{30.}\ \boxed{\times}\ \boxed{57.}\ \boxed{\div}\ \boxed{19.}$

D. $\boxed{57.}\ \boxed{\times}\ \boxed{19.}\ \boxed{\div}\ \boxed{30.}$

CONNECTING ARITHMETIC AND ALGEBRA

A *direct variation* is a relationship between two variables in which one variable changes at the same rate as the other. The ratio of the two variables is always the same. If the variables are x and y, you can write this relationship as the proportion

$$\frac{y_1}{x_1} = \frac{y_2}{x_2},$$

where y_1 and y_2 are different values of y, and x_1 and x_2 are different values of x. You say that y is *directly proportional* to x.

Use the proportion above.

32. Find y_2 when $x_1 = 10$, $x_2 = 25$, and $y_1 = 16$.

33. Find y_1 when $x_1 = 2$, $x_2 = 3$, and $y_2 = 27$.

34. Find x_2 when $x_1 = 14$, $y_1 = 24$, and $y_2 = 60$.

35. Find x_1 when $x_2 = 8$, $y_1 = 3.9$, and $y_2 = 6$.

36. An employee's wages are directly proportional to the number of hours worked. One employee earns $120 for working 5 h. How much will that employee earn for working 8 h?

37. The length of the shadow of a tree is directly proportional to the height of the tree. A bonsai tree that is 5 in. tall casts a shadow 3.5 in. long. How long is the shadow of a tree that is 20 ft tall?

SPIRAL REVIEW

38. Find the complement of an angle with measure 62°.   *(Lesson 6-3)*

39. Simplify: $\dfrac{8x^2y^3}{2xyz}$ *(Lesson 7-5)*

40. Solve the proportion: $\dfrac{9}{16} = \dfrac{a}{32}$ *(Lesson 9-2)*

41. Find the product: $-\dfrac{8}{27} \cdot \dfrac{3}{4}$ *(Lesson 8-1)*

42. Estimate the quotient: $3404 \div 43$ *(Toolbox Skill 3)*

Bonsai is the art of growing small, ornamentally shaped trees in shallow containers.

Scale Drawings

Objective: To interpret and use scale drawings.

APPLICATION

Scale drawings are drawings that represent real objects. The **scale** is the ratio of the size of the drawing to the actual size of the object. You can use proportions to find the actual measurements of an object when you have a scale drawing.

Below is the scale drawing of a school cafeteria. The scale is 1 in. : 16 ft.

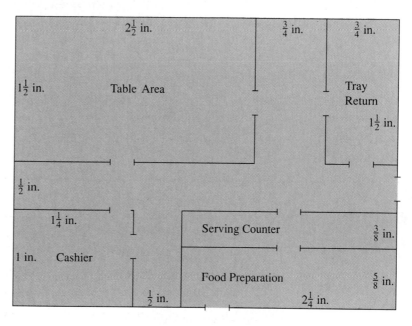

Example 1

Solution

Find the actual length of the tray return area in the scale drawing above.

Let a represent the actual length of the tray return area in feet. Use the scale to write a proportion.

$$\frac{\text{scale drawing length (in.)}}{\text{actual length (ft)}} \; \begin{array}{l} \rightarrow \\ \rightarrow \end{array} \; \frac{1}{16} = \frac{1\frac{1}{2}}{a}$$

$$1a = 16\left(1\frac{1}{2}\right)$$

$$a = 16\left(\frac{3}{2}\right) = 24$$

The actual length of the tray return area is 24 ft.

✓ Check Your Understanding

1. In Example 1, why is the numerator $1\frac{1}{2}$, not $\frac{3}{4}$?

Manufacturers often build models of their products before making the actual product. A **scale model** has the same shape but is usually smaller than the actual object it represents. You can use proportions to find the dimensions of scale models.

Example 2

A car model is being built with a scale of 2 in.:5 ft. The actual length of the car is 12 ft. What is the length of the model?

Solution

Let m represent the length of the model. Use the scale to write a proportion.

$$\frac{\text{model length (in.)}}{\text{actual length (ft)}} \quad \begin{array}{c} \rightarrow \\ \rightarrow \end{array} \quad \frac{2}{5} = \frac{m}{12}$$

$$2(12) = 5m$$

$$24 = 5m$$

$$\frac{24}{5} = \frac{5m}{5}$$

$$4\frac{4}{5} = m$$

The length of the model is $4\frac{4}{5}$ in.

✏️ **Check Your Understanding**

2. In Example 2, could the proportion have been written $\frac{5}{2} = \frac{12}{m}$? Explain.

Guided Practice

COMMUNICATION «*Reading*

«**1.** Explain what it means when a scale drawing shows a scale of 1 in.:10 ft.

«**2.** The following shows the scale on a drawing: What scale does this represent?

$$\vdash\!\!-\!\!-\!\!-\!\!-\!\!-\!\!\dashv$$
1 ft

COMMUNICATION «*Writing*

«**3.** For a scale drawing of a house, write a scale that shows ''every 2 in. represents 15 ft.''

«**4.** For a scale model of a statue, write a scale that shows ''every centimeter represents 12 m.''

Use the scale drawing of the school cafeteria on page 395. Write a proportion you could use to find each actual dimension. Do not solve.

5. the length of the cashier's room

6. the width of the tray return area

7. the width of the serving counter

8. the length of the table area

Use the scale drawing of the school cafeteria on page 395. Find each actual dimension.

9. the width of the corridor outside the serving counter

10. the length of the cashier's room

11. the length of the serving counter

The scale of a car model is 3 in. : 10 ft. Find each dimension of the model.

12. The actual width of the seat is 2 ft.

13. The actual height of the tire is 3 ft.

14. The actual height of the window is $1\frac{1}{4}$ ft.

Exercises

Use the scale drawing below of a basketball court. The scale is 1 in. : 25 ft. Find each actual dimension.

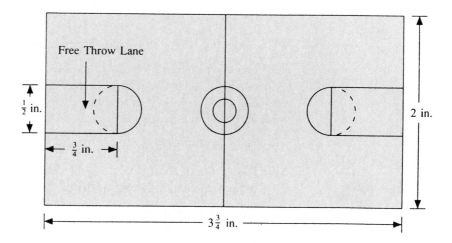

1. the width of the court
2. the width of the free throw lane
3. the length of the free throw lane
4. the length of the court

5. A kitchen is 13 ft wide and 14 ft long. The scale of a floor plan of the house containing this kitchen is 1 in. : 8 ft. Find the width and the length of the kitchen on the floor plan.

6. A scale model of a building is constructed using the scale 1 in. : 20 ft. The actual height of the building is 240 ft. Find the height of the scale model.

7. A model train is built using the scale 1 : 60. The actual length of the dining car is 2100 cm. Find the length of the dining car of the model train.

8. The model of a jet plane is $2\frac{1}{2}$ ft long. The scale of the model is 1:40. What is the actual length of the plane?

9. The length of the kitchen in a floor plan is 6 in. The actual length of the kitchen is 15 ft. What is the scale of the floor plan?

10. The wheelbase of a new car is 7 ft. The wheelbase on a model of the car is 2 in. What is the scale of the model?

GROUP ACTIVITY

11. **a.** Measure a room and record the results.
    b. Choose a convenient scale and make a scale drawing of the room. Draw all the furniture in the room as it would be seen from above.
    c. Sketch a new table that will fit in an empty part of the room. If you were to shop for the table, what size would you look for?

12. What scale might you choose to make a scale model of a ship? Explain your choice.

13. A statue of a person is 21 ft high. Decide what the scale might be. Explain your choice.

RESEARCH

14. Find what HO and O scales are in railroad modeling.

15. The northeastern face of Mount Rushmore in South Dakota has a large carving of four Presidents' faces. What scale did the sculptor use?

SPIRAL REVIEW

16. Find the sum: $\frac{5}{9} + \frac{7}{12}$ *(Lesson 8-5)*

17. The model of a car is $6\frac{1}{2}$ in. long. The scale of the model is 1:20. What is the actual length of the car?   *(Lesson 9-3)*

18. Solve by drawing a diagram: On a nature walk, Ed hiked 3 km north, 7 km east, 12 km south, 7 km west, 2 km north. Where was Ed in relation to his original position?   *(Lesson 7-8)*

9-4

Strategy:
Using Proportions

Objective: To use proportions to solve problems.

You can solve many problems that involve equal ratios or equal rates by using proportions.

Problem

The Daily Gazette charges $7.20 for 3 weeks of home newspaper delivery. At this rate, what is the cost of 8 weeks of home delivery?

Solution

UNDERSTAND The problem is about home newspaper delivery.
Facts: $7.20 for 3 weeks of home delivery
Find: cost of 8 weeks of home delivery

PLAN Let c = the cost of 8 weeks of home delivery. Write a proportion using the ratio of the number of weeks to the cost of home delivery. Then solve the proportion using cross products.

WORK
$$\frac{3}{7.2} = \frac{8}{c}$$ ← number of weeks
← cost of home delivery

$$3c = (7.2)(8)$$
$$3c = 57.6$$
$$\frac{3c}{3} = \frac{57.6}{3}$$
$$c = 19.2$$

ANSWER The cost of 8 weeks of home newspaper delivery is $19.20.

Look Back An alternative method is to solve this problem without using a proportion. Describe how you could do this.

Guided Practice

COMMUNICATION « *Writing*

Write the proportion you would use to solve each problem. Use the variable *n*. Do not solve.

« **1.** The ratio of students to parents at a play was 5 to 6. There were 150 parents at the play. How many students were at the play?

« **2.** Last week Mary drove her car 440 mi and used 20 gal of gasoline. At the same rate, how many miles could she drive her car using 35 gal of gasoline?

« **3.** The cost of 3 oranges is $1.68. What is the cost of 7 oranges?

Solve using a proportion.

4. Josefina sells helium balloons. She charges $9 for 12 balloons. At this rate, what will Josefina charge for 50 balloons?

5. A photocopy machine copied 50 pages in 1.5 min. At this rate, how long will the machine take to copy 90 pages?

«6. **COMMUNICATION** «*Discussion* Could you solve the Problem on page 399 by using the proportion $\frac{7.2}{3} = \frac{c}{8}$? Explain.

Problem Solving Situations

Solve using a proportion.

1. In Plainsville, 5 out of every 7 town meeting members voted to approve the school budget. A total of 190 members voted to approve the school budget. How many town meeting members are there in all?

2. Maria spent 3 h addressing 50 wedding invitations. At this rate, how long will it take Maria to address 125 wedding invitations?

3. Travel Rent-a-Car charges customers $135 to rent a compact car for 3 days with unlimited mileage. At this rate, what will it cost to rent a compact car for 5 days?

4. Shirley paid $12.72 for 8 cans of frozen orange juice to make a punch for a party. Elliot bought 10 cans of the same brand of orange juice to make punch for another party. How much did Elliot pay for the juice?

Solve using any problem solving strategy.

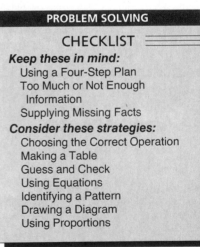

> **PROBLEM SOLVING**
> ## CHECKLIST
> **Keep these in mind:**
> Using a Four-Step Plan
> Too Much or Not Enough Information
> Supplying Missing Facts
> **Consider these strategies:**
> Choosing the Correct Operation
> Making a Table
> Guess and Check
> Using Equations
> Identifying a Pattern
> Drawing a Diagram
> Using Proportions

5. Centerville's movie theater is 11 blocks due south of the grocery store. The bank is 5 blocks due north of the movie theater. Where is the bank in relation to the grocery store?

6. Shelly distributed 6 lb of nuts equally among 32 gift baskets. How many ounces of nuts did Shelly put in each basket?

7. There are twice as many geese as ducks in a lake. In all, there are 51 geese and ducks. How many geese are in the lake?

8. There are 6 counselors for every 50 campers at a summer camp. The camp has 21 counselors. Find the number of campers.

9. The figures below show the first four figures in a pattern of squares. How many squares are in the ninth figure in this pattern?

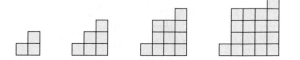

WRITING WORD PROBLEMS

Write a word problem that you could solve using each proportion. Then solve the problem.

10. $\dfrac{3}{35} = \dfrac{x}{140}$

11. $\dfrac{4}{n} = \dfrac{6}{\$1.44}$

SPIRAL REVIEW

12. Solve using a proportion: In the Springfield Chorus, 2 out of every 7 singers are altos. There are 112 singers in the chorus. How many are altos? *(Lesson 9-4)*

13. Graph the inequality $x < 2.5$. *(Lesson 7-10)*

14. Find the mean, median, mode(s), and range: 11, 2, 17, 13, 13, 17, 13 *(Lesson 5-8)*

Self-Test 1

Write each ratio as a fraction in lowest terms.

1. $\dfrac{18}{30}$ **2.** 30 min : 2 h   **3.** 1 lb to 8 oz   **4.** 21 : 14   **9-1**

Write the unit rate.

5. 110 mi in 2 h   **6.** $16 for 4 lb   **7.** $1.95 for 3 combs

Tell whether each proportion is *True* or *False*.

8. $\dfrac{5}{10} = \dfrac{20}{40}$ **9.** $\dfrac{6.8}{1.7} = \dfrac{8}{2}$ **10.** $\dfrac{3}{4} = \dfrac{20}{15}$ **11.** $\dfrac{9}{10} = \dfrac{10}{11}$ **9-2**

Solve each proportion.

12. $\dfrac{2}{8} = \dfrac{b}{40}$ **13.** $\dfrac{17}{a} = \dfrac{17}{4}$ **14.** $\dfrac{10}{3} = \dfrac{d}{10}$ **15.** $\dfrac{9.2}{n} = \dfrac{4}{5}$

16. The model of a truck is $4\frac{3}{4}$ in. wide. The scale of the model is 3 in. : 5 ft. What is the actual width of the truck?   **9-3**

17. Kathy pays $41.25 for 5 dancing lessons. At this rate, how much does she pay for 12 dancing lessons?   **9-4**

Using a Ten-by-Ten Grid

Objective: To use ten-by-ten grids to model percents.

Materials

■ graph paper

A percent is a ratio that compares a number to 100. For example, 1 percent, written 1%, is the same as the ratio 1 : 100.

Each square in the ten-by-ten grid at the right represents 1%. You can use such a grid to model ratios and percents. For example, the ratio 48 : 100, or 48%, is modeled at the right.

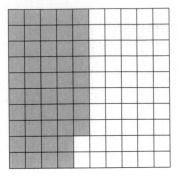

Activity I *Interpreting Models of Percents*

1 What percent of the ten-by-ten grid at the left below is red?

2 What percent of the ten-by-ten grid at the left below is not red?

3 What percent of the ten-by-ten grid at the right below is blue?

4 What percent of the ten-by-ten grid at the right below is not blue?

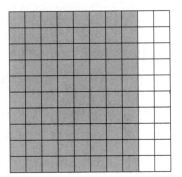
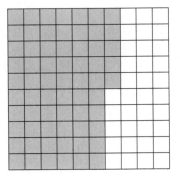

Activity II *Drawing Models of Percents*

Draw a ten-by-ten grid on a piece of graph paper.

1 Draw $'s in squares to represent 12%.

2 Draw •'s in empty squares to represent 35%.

3 Draw #'s in empty squares to represent 10%.

4 What percent of the grid does not contain any $'s?

5 What percent of the grid does not contain any •'s?

6 What percent of the grid does not contain any #'s?

7 What percent of the grid contains a symbol?

8 What percent of the grid does not contain a symbol?

Activity III   *Percents Less than 1 and Greater than 100*

Use the grids below for Steps 1 and 3.

i.
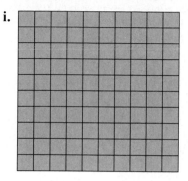

ii.
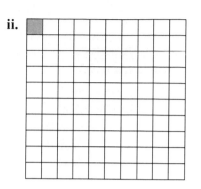

iii.

iv.

1 Choose the number of the grid that represents each percent.

   a. $\frac{1}{2}\%$ **b.** 1%   **c.** 50%   **d.** 100%

2 Tell how much of a ten-by-ten grid you would have to shade to represent each percent.

   a. $\frac{1}{5}\%$ **b.** 20%   **c.** $\frac{2}{3}\%$ **d.** $66\frac{2}{3}\%$

3 What percent would you obtain if you added the percents represented by grids (i) and (ii)? grids (i) and (iii)? grids (i) and (iv)?

Activity IV   *Converting Fractions and Decimals to Percents*

1 Use the grid at the right. Write each ratio as a fraction in simplest form, as a decimal, and as a percent.

   a. red squares : all squares
   b. blue squares : all squares
   c. blank squares : all squares

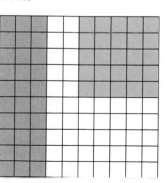

2 Model each number by shading squares on four separate ten-by-ten grids.

   a. $\frac{3}{4}$ **b.** $\frac{2}{5}$ **c.** 0.15   **d.** 0.08

3 Use your models to write each number in Step 2 as a percent.

Percent

Objective: To write fractions and decimals as percents and to write percents as fractions and decimals.

Terms to Know
• *percent*

Jenine attempted twenty shots during the last basketball game. The school paper reported that she made 85% of her shots.

The symbol % is read "percent." A **percent** is a ratio that compares a number to 100. Percent means "per hundred," "hundredths," or "out of every hundred."

$$1\% = \frac{1}{100} = 0.01$$

Jenine's Basketball Record

Shots made	17
Shots attempted	20

$$\frac{\text{Shots made}}{\text{Shots attempted}} \quad \frac{17}{20}$$

Example 1

Solution

Write each fraction as a percent: **a.** $\frac{17}{20}$ **b.** $\frac{3}{8}$

a. Write $\frac{17}{20}$ as a fraction whose denominator is 100.

$$\frac{17}{20} = \frac{17 \cdot 5}{20 \cdot 5} = \frac{85}{100} = 85\%$$

b. The denominator of $\frac{3}{8}$ is not a factor of 100, so use division.

$$\frac{3}{8} = 0.37\tfrac{1}{2} \quad \longleftarrow \text{Divide } 8\overline{)3.00} \text{ to the } \textit{hundredths} \text{ place.}$$

$$= 37\tfrac{1}{2}\% \quad \longleftarrow 0.37\tfrac{1}{2} \text{ means } 37\tfrac{1}{2} \textit{ hundredths.}$$

✓ **Check Your Understanding**

1. When changing $\frac{17}{20}$ to a percent in Example 1(a), why do you write the ratio as a fraction with a denominator of 100?
2. If you used a calculator in Example 1(b), what number would be displayed for $0.37\tfrac{1}{2}$?

Notice that to write a decimal as a percent, as in Example 1(b), you move the decimal point two places to the right and insert the symbol %.

$$0.37\tfrac{1}{2} = 3\,7\underbrace{.}\tfrac{1}{2}\%$$

Example 2

Solution

Write each decimal as a percent: **a.** 0.43   **b.** 0.09   **c.** 2.4

a. $0.43 = 0.\underbrace{43}\%$ **b.** $0.09 = 0.\underbrace{09}\%$ **c.** $2.4 = 2.\underbrace{40}\%$

       $= 43\%$ $= 9\%$ $= 240\%$

You can also write percents as decimals and as fractions. To write a percent as a decimal, you move the decimal point two places to the left and drop the % symbol.

Notice in Examples 3 and 4 that a percent may be less than 1 or greater than 100.

Example 3

Write each percent as a decimal.

a. 74%   **b.** 0.25%   **c.** 110%   **d.** $8\frac{1}{2}\%$

Solution

a. $74\% = 74\% = 0.74$

b. $0.25\% = 0\ 0.25\% = 0.0025$

c. $110\% = 1\ 1\ 0\ \% = 1.1$

d. $8\frac{1}{2}\% = 0\ 8.5\% = 0.085$

Example 4

Write each percent as a fraction or a mixed number in lowest terms.

a. 125%   **b.** 3.6%   **c.** $\frac{2}{5}\%$

Solution

a. $125\% = \frac{125}{100} = \frac{5}{4} = 1\frac{1}{4}$

b. $3.6\% = \frac{3.6}{100} = \frac{(3.6)(10)}{(100)(10)} = \frac{36}{1000} = \frac{9}{250}$

c. $\frac{2}{5}\% = \frac{\frac{2}{5}}{100} = \frac{2}{5} \div 100 = \frac{2}{5} \cdot \frac{1}{100} = \frac{1}{250}$

☑ Check Your Understanding

3. Describe how the position of the decimal point is changed in writing a percent as a decimal.

4. In Example 3(d), why is $8\frac{1}{2}$ rewritten as 8.5?

5. Why are the numerator and denominator in Example 4(b) multiplied by 10?

Guided Practice

COMMUNICATION « *Reading*

« **1.** What is a percent?

« **2.** Describe two methods for changing a fraction to a percent.

Write each ratio as a percent.

3. 18 out of 100   **4.** $2:100$

5. 56 to 100   **6.** $\frac{100}{100}$

7. To write $1\frac{5}{8}$ as a percent, replace each __?__ with the number that makes the statement true.

$$1 = \frac{?}{100} = \underline{\ ?\ }\%$$

$$\frac{5}{8} = 0.625 = \underline{\ ?\ }\%$$

$$1\frac{5}{8} = 1 + \frac{5}{8} = \underline{\ ?\ }\% + \underline{\ ?\ }\% = \underline{\ ?\ }\%$$

Write each fraction or decimal as a percent.

8. $\frac{11}{50}$ **9.** $\frac{3}{5}$ **10.** $\frac{7}{8}$ **11.** $\frac{1}{200}$

12. 0.45   **13.** 0.002   **14.** 1   **15.** 2.78

Write each percent as a decimal.

16. 65%   **17.** 4%   **18.** 9.2%   **19.** $7\frac{1}{4}\%$

Write each percent as a fraction or a mixed number in lowest terms.

20. 36%   **21.** 0.5%   **22.** $\frac{1}{4}\%$ **23.** 375%

Exercises

Write each fraction, mixed number, or decimal as a percent.

1. $\frac{2}{25}$ **2.** $\frac{21}{50}$ **3.** $\frac{7}{10}$ **4.** $\frac{1}{5}$ **5.** $\frac{5}{8}$

6. $\frac{9}{16}$ **7.** 0.21   **8.** 0.08   **9.** 0.099   **10.** 1.7

11. $0.15\frac{1}{5}$ **12.** $\frac{7}{12}$ **13.** $\frac{15}{32}$ **14.** $1\frac{4}{15}$ **15.** $2\frac{2}{7}$

Write each percent as a decimal.

16. 6%   **17.** 49%   **18.** 0.3%   **19.** 587%   **20.** $24\frac{1}{3}\%$

Write each percent as a fraction or mixed number in lowest terms.

21. 50%   **22.** $\frac{3}{4}\%$ **23.** 2.3%   **24.** 475%   **25.** 0.15%

26. Christina made 22 out of 25 shots in last week's basketball game. What percent of the shots Christina attempted did she make?

27. The Gray Company receives about 40 letters each day. Sixteen of these letters have postage stamps. The remainder of the letters have been passed through a postage meter. What percent of the letters received each day have postage stamps on them?

28. Steve got 15 out of 20 problems correct on a math test. What percent of the problems did he get wrong?

29. The Robins won 30 games and lost 20. What percent of the games they played did they win?

PATTERNS

30. Write each percent as a decimal.
    a. 200%   **b.** 20%   **c.** 2%

31. Use the results of Exercise 30 to find the first six decimals in the pattern.

200%	20%	2%	0.2%	0.02%	0.002%
_?_	_?_	_?_	_?_	_?_	_?_

32. Write each decimal as a percent.
    a. 9   **b.** 0.9   **c.** 0.09

33. Use the results of Exercise 32 to find the first six percents in the pattern.

9	0.9	0.09	0.009	0.0009	0.00009
_?_ %	_?_ %	_?_ %	_?_ %	_?_ %	_?_ %

THINKING SKILLS

Complete to find what percent the first number is of the second number.

34. 97; 100 $\longrightarrow \dfrac{?}{100} = \underline{\;?\;}\%$ **35.** 19; 19 $\longrightarrow \dfrac{19}{?} = \underline{\;?\;}\%$

36. 100; 200 $\longrightarrow \dfrac{?}{200} = \underline{\;?\;}\%$ **37.** 200; 100 $\longrightarrow \dfrac{200}{?} = \underline{\;?\;}\%$

38. Analyze your answers to Exercises 34–37. Create a plan for finding what percent one number is of another number.

SPIRAL REVIEW

39. Write 7.5% as a fraction and as a decimal.   *(Lesson 9-5)*

40. Find the GCF: 108 and 60   *(Lesson 7-2)*

41. Find the difference: $\dfrac{29}{37} - \dfrac{18}{37}$ *(Lesson 8-4)*

42. Use a protractor to draw an angle with measure 123°.   *(Lesson 6-2)*

Challenge

A photograph is reduced to four fifths of its original size. What percent of the reduced size is the original size?

Finding a Percent of a Number

Objective: To find a percent of a number and to estimate percents of numbers.

APPLICATION

Luis Mendez wants to buy a compact disc player. The regular price is $280 but the disc player is on sale today at a 30% discount.

QUESTION How much will Luis save if he buys the compact disc player today?

Example 1

Solution

Find 30% of 280.

To find a percent of a number, you multiply.

$$30\% \text{ of } 280$$
$$\downarrow \quad \downarrow \quad \downarrow$$
$$0.3 \quad \cdot \quad 280 = 84$$

✓ Check Your Understanding

1. In Example 1, 30% was written as a decimal. What is an alternate way to write 30% in order to multiply?

ANSWER Luis will save $84 if he buys the disc player today.

You can use equations to find a percent of a number.

Example 2

Solution

What number is 64% of 75?

Write an equation. Let n be the missing number.

Using a fraction:

What number is 64% of 75?
$$\downarrow \quad \downarrow \quad \downarrow \quad \downarrow$$
$$n \quad = \quad \frac{64}{100} \quad \cdot \quad 75$$
$$n \quad = \quad \frac{16}{25} \quad \cdot \quad 75$$
$$n \quad = \quad 48$$

Using a decimal:

What number is 64% of 75?
$$\downarrow \quad \downarrow \quad \downarrow \quad \downarrow$$
$$n \quad = 0.64 \cdot 75$$
$$n \quad = 48$$

So 48 is 64% of 75.

✓ Check Your Understanding

2. Explain why 64% can be written as either $\frac{64}{100}$ or 0.64.

3. What is $\frac{64}{100}$ as a fraction in lowest terms?

Most calculators have a percent key, $\boxed{\%}$. The way a calculator uses the percent key depends on the type of calculator. To find 64% of 75, you may be able to use one of these two key sequences.

$\boxed{64.}$ $\boxed{\%}$ $\boxed{\times}$ $\boxed{75.}$ $\boxed{=}$ *or*

$\boxed{75.}$ $\boxed{\times}$ $\boxed{64.}$ $\boxed{\%}$

Consult the user's manual to determine how your calculator handles percents.

Even if you have a calculator, you may find it helpful to memorize the equivalent percents, decimals, and fractions in the chart below.

Equivalent Percents, Decimals, and Fractions

$20\% = 0.2 = \frac{1}{5}$	$25\% = 0.25 = \frac{1}{4}$	$12\frac{1}{2}\% = 0.125 = \frac{1}{8}$	$16\frac{2}{3}\% = 0.1\overline{6} = \frac{1}{6}$
$40\% = 0.4 = \frac{2}{5}$	$50\% = 0.5 = \frac{1}{2}$	$37\frac{1}{2}\% = 0.375 = \frac{3}{8}$	$33\frac{1}{3}\% = 0.\overline{3} = \frac{1}{3}$
$60\% = 0.6 = \frac{3}{5}$	$75\% = 0.75 = \frac{3}{4}$	$62\frac{1}{2}\% = 0.625 = \frac{5}{8}$	$66\frac{2}{3}\% = 0.\overline{6} = \frac{2}{3}$
$80\% = 0.8 = \frac{4}{5}$		$87\frac{1}{2}\% = 0.875 = \frac{7}{8}$	$83\frac{1}{3}\% = 0.8\overline{3} = \frac{5}{6}$
$100\% = 1$			

You can use equivalent values from the chart above to estimate percents of numbers.

Example 3

Estimate.

a. 19% of 126   **b.** $16\frac{2}{3}\%$ of $358.75   **c.** 65% of 299

Solution

a. 19% of 126   ⟵ 19% is about 20%.
     ↓   ↓   ↓   You can use the equivalent
     $\frac{1}{5}$ · 125   fraction for 20%.

     about 25

b. $16\frac{2}{3}\%$ of $358.75   **c.** 65% of 299
     ↓   ↓   ↓   ↓   ↓   ↓
     $\frac{1}{6}$ · $360 $\frac{2}{3}$ · 300

     about $60   about 200

☑ **Check Your Understanding**

4. In Example 3(c), explain why 65% is about $\frac{2}{3}$.

Guided Practice

« **1.** **COMMUNICATION** « *Reading*   Replace each __?__ with the correct word or number.
To find what number is 48% of 64, you __?__ 64 by __?__.

Choose the letter of the equation you would use to find each answer.

2. 6% of 180
   a. $n = 6 \cdot 180$
   b. $n = 0.6 \cdot 180$
   c. $n = 0.06 \cdot 180$

3. 25% of 488
   a. $n = 25 \cdot 488$
   b. $n = \frac{1}{4} \cdot 488$
   c. $n = 2.5 \cdot 488$

Find each answer.

4. 5% of 320 is what number?

5. 50% of 98 is what number?

6. $12\frac{1}{2}\%$ of 84 is what number?

7. What number is 36% of 70?

Estimate.

8. 89% of 51

9. 17% of $6\frac{1}{4}$

10. $33\frac{1}{3}\%$ of $59.99

Exercises

Find each answer.

1. What number is 40% of $236?

2. What number is $37\frac{1}{2}\%$ of 120?

3. What number is 2.5% of 9600?

4. 50% of 528 is what number?

5. 100% of 456 is what number?

6. 8% of $24 is what number?

7. Mei bought an electric guitar priced at $350. The sales tax rate is 4.5%. How much sales tax did she pay?

8. Kwasi scored 85% on a 60-item test. How many items did he get correct?

Estimate.

9. 88% of 41

10. 60% of 62

11. 33% of 147

12. $83\frac{1}{3}\%$ of 123

13. 25% of $23.95

14. 47% of $505

CALCULATOR

Write two different calculator key sequences you could use to find each answer.

15. 5% of 312

16. 2.65% of 180

17. 16% of 975

18. 300% of 159

19. 0.7% of 34.9

20. 26.42% of 100

MENTAL MATH

Finding 10% of a number is the same as multiplying the number by 0.1. You mentally move the decimal point one place to the left. Knowing 10% of a number can help you find other percents of the same number. For example, 5% of a number is one half of 10% of the same number.

Find each answer mentally.

21. a. 10% of $24   **b.** 5% of $24   **c.** 40% of $24

22. a. 10% of 320   **b.** 20% of 320   **c.** 30% of 320

23. a. 10% of 4.6   **b.** 5% of 4.6   **c.** 20% of 4.6

24. Sam's dinner at a restaurant cost $18. Use the fact that 15% = 10% + 5% to find the 15% tip Sam paid on his dinner.

NUMBER SENSE

Replace each __?__ with >, <, or =.

25. 40% of 48 __?__ 48

26. 120% of 52 __?__ 52

27. n __?__ 200% of n, $n > 0$

28. x __?__ 90% of x, $x > 0$

29. 50 __?__ 100% of 50

30. 100% of y __?__ y, $y > 0$

31. 175% __?__ $1\frac{3}{4}$

32. $\frac{1}{2}$% __?__ $\frac{1}{2}$

 **COMPUTER APPLICATION**

A department store staff uses a spreadsheet to give final prices of items when they are marked up by a given percent. The percents of markup are in column A, the original prices of the items are in row 1, and the final prices appear below the original prices.

	A	B	C	D	E	F
1		$10.00	$20.00	$30.00	$40.00	$50.00
2	40%	$14.00	$28.00	$42.00	$56.00	$70.00
3	50%	$15.00				
4	60%	$16.00				
5	70%	$17.00				

Find the correct value for each cell in the spreadsheet above.

33. C3   **34.** D3   **35.** E3   **36.** F3   **37.** C4   **38.** D4

39. E4   **40.** F4   **41.** C5   **42.** D5   **43.** E5   **44.** F5

SPIRAL REVIEW

45. Laura pays $75 for four guitar lessons. At this rate, how much would she pay for ten lessons? *(Lesson 9-4)*

46. Solve $P = 2l + 2w$ for w. *(Lesson 8-10)*

47. What number is 15% of 24? *(Lesson 9-6)*

Finding the Percent One Number Is of Another

Objective: To find the percent one number is of another.

You can use an equation to find out what percent one number is of another number.

Example

a. What percent of 75 is 21?

b. What percent of 16 is 22?

c. 12.5 is what percent of 500?

Solution

Let n = the percent.

a. What percent of 75 is 21?

$$n \cdot 75 = 21$$
$$75n = 21$$
$$\frac{75n}{75} = \frac{21}{75}$$
$$n = \frac{21}{75} = \frac{7}{25} = \frac{28}{100} = 28\%$$

So 28% of 75 is 21.

b. What percent of 16 is 22?

$$n \cdot 16 = 22$$
$$16n = 22$$
$$\frac{16n}{16} = \frac{22}{16}$$
$$n = \frac{22}{16} = \frac{11}{8} = 1.37\frac{1}{2} = 137\frac{1}{2}\%$$

So $137\frac{1}{2}\%$ of 16 is 22.

c. 12.5 is what percent of 500?

$$12.5 = n \cdot 500$$
$$12.5 = 500n$$
$$\frac{12.5}{500} = \frac{500n}{500}$$
$$n = \frac{12.5}{500} = 0.025 = 2.5\%$$

So 12.5 is 2.5% of 500.

Check Your Understanding

1. In Example 1(a), why was $\frac{7}{25}$ written with a denominator of 100?

2. Explain how you know that $\frac{12.5}{500} = 0.025$ in Example 1(c).

Guided Practice

Write an equation that represents each question. Let *n* represent the unknown percent.

«**1.** What percent of 100 is 37?

«**2.** What percent of 25 is 19.5?

«**3.** 28 is what percent of 5.6?

«**4.** 6 is what percent of 40?

Write a question involving percents that could be represented by each equation.

«**5.** $34n = 17$

«**6.** $\frac{1}{2}n = \frac{3}{4}$

«**7.** $23.7 = 60n$

«**8.** $78 = 12.5n$

Write each fraction as a percent.

9. $\frac{16.2}{60}$

10. $\frac{72}{64.8}$

11. $\frac{9.3}{12.4}$

Find each answer.

12. What percent of 50 is 37?

13. 120 is what percent of 150?

14. What percent of 75 is 100?

15. 2.4 is what percent of 120?

Exercises

Find each answer.

1. What percent of 20 is 16?

2. What percent of 200 is 68?

3. What percent of 40 is 28?

4. What percent of 32 is 8?

5. 24 is what percent of 25?

6. 3.5 is what percent of 50?

7. 18 is what percent of 18?

8. What percent of 66 is 44?

9. 12.2 is what percent of 36.6?

10. 12 is what percent of 2400?

11. 80 is what percent of 3200?

12. What percent of 48 is 2.4?

13. Mike got 35 items correct on a 40-item test. What percent of the items did he get correct?

14. The sale price of the luggage Caroline bought was $160. The sales tax she paid was $8.80. What was the sales tax rate?

15. The goal of a local charity is to raise $400,000. At the present time, it has raised $250,000. What percent of the goal has been reached?

16. A local charity wishes to raise $500,000. At the end of the fund drive, $600,000 was raised. What percent of its goal did the charity reach?

17. Last November, it snowed on 6 out of 30 days. What percent of the days did it not snow last November?

18. Jenna's March telephone bill showed $21.50 for local charges and $8.50 for long distance charges. What percent of her total bill was for long distance charges?

19. DATA, *pages 338–339* In Grand Rapids, the recommended basement ceiling insulation is what percent of the attic floor insulation?

20. DATA, *pages 140–141* The maximum speed of a grizzly bear is what percent of the maximum speed of a chicken?

Interest is money paid for the use of money. Money deposited in a bank, on which interest is paid, is called the *principal*. The simple interest formula is $I = Prt$, where I is the interest earned, P is the principal, r is the interest rate per year, and t is the time in years.

Use the formula $I = Prt$.

21. Find I when $P = \$1000$, $r = 5\%$ per year, and $t = 1$ year.

22. Find I when $P = \$500$, $r = 7.2\%$ per year, and $t = 2$ years.

23. Find I when $P = \$2500$, $r = 6\frac{1}{2}\%$ per year, and $t = 3$ years.

24. Find I when $P = \$2000$, $r = 5\%$ per year, and $t = 6$ months.

25. Find r when $I = \$57.50$, $P = \$1250$, and $t = 1$ year.

26. Find t when $I = \$1147.50$, $P = \$6000$, and $r = 8\frac{1}{2}\%$ per year.

SPIRAL REVIEW

27. Replace the __?__ with $>$, $<$, or $=$: $\frac{7}{9}$ _?_ $\frac{7}{10}$ *(Lesson 7-6)*

28. What percent of 40 is 48? *(Lesson 9-7)*

29. Find the LCM: 15 and 40 *(Lesson 7-3)*

30. Evaluate $-12 + 5n^2$ when $n = -4$. *(Lesson 3-6)*

Finding a Number When a Percent of It Is Known

Objective: To find a number when a percent of it is known.

APPLICATION

In an election for senior class president, 126 students voted for Takashi Noma. Takashi received 60% of the total vote.

QUESTION How many students voted in the election?

You can use an equation to find a number when a percent of it is known. In the situation above, 60% of the total vote is 126 students.

Example 1
Solution

60% of what number is 126?

Let n = the missing number.
60% of what number is 126?

$$0.6 \cdot n = 126 \qquad \longleftarrow \text{Write 60\% as a decimal.}$$
$$0.6n = 126$$
$$\frac{0.6n}{0.6} = \frac{126}{0.6}$$
$$n = 210 \qquad \text{So 60\% of 210 is 126.}$$

✓ **Check Your Understanding**

1. If 60% of a number is 126, is the missing number *more than* or *less than* 126? Explain.

ANSWER 210 students voted in the election.

Example 2
Solution

120 is $66\frac{2}{3}\%$ of what number?

Let n = the missing number.

120 is $66\frac{2}{3}\%$ of what number?

$$120 = \frac{2}{3} \cdot n \qquad \longleftarrow \text{Write } 66\frac{2}{3}\% \text{ as a fraction.}$$

$$120 = \frac{2}{3}n$$

$$\left(\frac{3}{2}\right)\left(\frac{120}{1}\right) = \left(\frac{3}{2}\right)\left(\frac{2}{3}n\right)$$

$$180 = n \qquad \text{So 120 is } 66\frac{2}{3}\% \text{ of 180.}$$

✓ **Check Your Understanding**

2. In Example 2, why is $66\frac{2}{3}\%$ rewritten as the fraction $\frac{2}{3}$?

3. In Example 2, why is $\frac{3}{2}$ used to multiply both sides of the equation?

1. Which equation would you use to solve the following problem?
   A shoe salesperson earns a commission of 12% of all sales. How much must the salesperson sell to earn $288?
   a. $0.12 \cdot n = 288$ **b.** $0.12 \cdot 288 = n$ **c.** $0.12 = 288 \cdot n$

COMMUNICATION « *Writing*

Write an equation to represent each question. Use n for the missing number. Do not solve.

« **2.** 24 is 50% of what number?   « **3.** 45 is 25% of what number?

« **4.** $16\frac{2}{3}\%$ of what number is 88?   « **5.** $18\frac{1}{2}\%$ of what number is 160?

Solve each equation.

6. $0.08n = 60$ 7. $45 = \frac{3}{5}n$ 8. $3.4n = 17$

Find each answer.

9. 75% of what number is 24?   **10.** 25% of what number is 36?

11. 2000 is $83\frac{1}{3}\%$ of what number?   **12.** 48 is 120% of what number?

Find each answer.

1. 10% of what number is 42?   **2.** 30 is 60% of what number?

3. 17 is 50% of what number?   **4.** 40% of what number is 114?

5. 130 is 6.5% of what number?   **6.** 3% of what number is 54?

7. 75% of what number is 60?   **8.** 216 is 90% of what number?

9. 15 is $33\frac{1}{3}\%$ of what number?   **10.** 250% of what number is 50?

11. 180 is $\frac{1}{2}\%$ of what number?   **12.** $37\frac{1}{2}\%$ of what number is 72?

13. Rozlyn got a score of 80% on her last French test. She got 40 items correct. How many items were on the test?

14. There are 4230 season ticket holders for the Cougars football games. This represents $56\frac{2}{3}\%$ of the seating capacity at the stadium. What is the seating capacity at the stadium?

15. Kent earns an 8% commission on all furniture sales. How much must he sell to earn a commission of $336?

16. In last year's Freshman Class election, one candidate received 36 votes. This result represents 45% of the total class vote. How many students voted in the Freshman Class election last year?

The table at the right shows the percent of the Recommended Daily Allowance (RDA) of nutrients contained in a 1 oz serving of a certain cereal.

Use the table at the right.

17. How many calories are in one serving of cereal with whole milk?

18. How many more milligrams of potassium are in one serving of cereal with skim milk than with whole milk?

19. What percent of the RDA for Vitamin C does one serving of plain cereal give?

20. How much less is the percent of the RDA for Vitamin D for plain cereal than for cereal with skim milk?

NUTRITION INFORMATION
SERVING SIZE: 1 OZ (284 g, ABOUT 1 CUP)
ALONE OR WITH 1/2 CUP SKIM MILK OR WHOLE MILK
SERVINGS PER PACKAGE: 12

	CEREAL	WITH SKIM MILK	WITH WHOLE MILK
CALORIES	110	150	180
PROTEIN	2 g	6 g	6 g
CARBOHYDRATE	24 g	30 g	30 g
FAT	0 g	0 g	4 g
CHOLESTEROL	0 mg	0 mg	15 mg
SODIUM	290 mg	350 mg	350 mg
POTASSIUM	35 mg	240 mg	220 mg

PERCENTAGE OF U.S. RECOMMENDED DAILY ALLOWANCES (U.S. RDA)

	CEREAL	WITH SKIM MILK	WITH WHOLE MILK
PROTEIN	4	15	15
VITAMIN A	100	100	100
VITAMIN C	100	100	100
THIAMIN	100	100	100
RIBOFLAVIN	100	110	110
NIACIN	100	100	100
CALCIUM	*	15	15
IRON	100	100	100
VITAMIN D	50	60	60
VITAMIN E	100	100	100
VITAMIN B_6	100	100	100
FOLIC ACID	100	100	100
VITAMIN B_{12}	100	110	110
PHOSPHORUS	4	15	15
MAGNESIUM	2	6	6
ZINC	100	100	100
COPPER	2	4	10

*CONTAINS LESS THAN 2% OF THE U.S. RDA OF THIS NUTRIENT

21. a. How many grams of protein are in one serving of cereal with skim milk?
b. What percent of the RDA for protein does one serving of cereal with skim milk give?
c. Use your answers to parts (a) and (b) to find the RDA for protein.

22. The RDA for calcium for teenagers is 1200 mg. How many milligrams of calcium are in one serving of cereal with skim milk?

23. The table shows that one serving of cereal with whole milk provides 110% of the RDA for Vitamin B_{12}. Explain what this percent means.

24. RESEARCH Compare the nutrient percents in the table above with those given on another box of cereal. Do the two cereals provide the same percents? If not, which brand seems more nutritious to you?

CAREER/APPLICATION

A *retail salesperson* sells goods or services directly to consumers. Retail salespeople work in many businesses, such as department stores, car dealerships, insurance agencies, and supermarkets. The work of retail salespeople often involves percents.

25. A salesperson at a clothing store marks an item priced at $32.50 for a 40% discount. What is the final selling price of the item?

26. A supermarket cashier charges a 5% sales tax on taxable items. A customer buys $14.69 worth of nontaxable items and $8.40 worth of taxable items. What was the total cost of the purchases?

27. A salesperson at Central Autos earned a $765 commission on the sale of a $9000 car. What is the percent of commission?

28. A real estate agent determined that the value of a house had increased by 25%. The owners sold the house for $22,000 more than the price they had originally paid. What did the owners pay for the house originally?

SPIRAL REVIEW

29. Write $\frac{40}{64}$ in lowest terms.   *(Lesson 7-4)*

30. 36% of what number is 18?   *(Lesson 9-8)*

31. Find the quotient: $\frac{14}{15} \div \left(-\frac{7}{10}\right)$ *(Lesson 8-2)*

32. Solve the proportion: $\frac{8}{3} = \frac{n}{9}$ *(Lesson 9-2)*

Self-Test 2

1. Write $\frac{13}{20}$ as a percent.
2. Write 0.45 as a percent.   9-5
3. Write 130% as a decimal.
4. Write 28% as a fraction.

Find each answer.

5. 65% of 29 is what number?
6. What number is $4\frac{1}{2}$% of 150?   9-6
7. What percent of 99 is 16.5?
8. 34 is what percent of 68?   9-7
9. 57 is 15% of what number?
10. 2.5% of what number is 4?   9-8

Percent of Increase or Decrease

Objective: To find a percent of increase or decrease.

DATA ANALYSIS

After two years on the soccer team, Matt and Terry compared their records.

QUESTION Which player improved more?

To find the answer, find the *percent of increase* in scoring for each player. The **percent of increase** tells what percent the amount of increase is of the original amount.

$$\text{percent of increase} = \frac{\text{amount of increase}}{\text{original amount}}$$

Soccer Goals Scored

	Matt	Terry
Last Year	10	6
This Year	14	9

Example 1

Find the percent of increase.

a. original amount: 10
new amount: 14

b. original amount: 6
new amount: 9

Solution

a. amount of increase: $14 - 10 = 4$

$$\text{percent of increase} = \frac{\text{amount of increase}}{\text{original amount}}$$

$$= \frac{4}{10} = \frac{40}{100} = 40\%$$

b. amount of increase: $9 - 6 = 3$

$$\text{percent of increase} = \frac{\text{amount of increase}}{\text{original amount}}$$

$$= \frac{3}{6} = \frac{1}{2} = 50\%$$

✓ Check Your Understanding

1. In part (a) of Example 1, why is the fraction $\frac{4}{10}$ used, not $\frac{4}{14}$?

ANSWER Matt's percent of increase was 40%, and Terry's percent of increase was 50%. So Terry improved more.

You can use a similar method to find a *percent of decrease*. The **percent of decrease** tells what percent the amount of decrease is of the original amount. When you refer to a sale price of an item, you often call the percent of decrease the *rate of discount*.

Example 2   The original price of a stereo is $150, and the new price is $100. Find the percent of decrease.

Solution   amount of decrease: $150 − $100 = $50

$$\text{percent of decrease} = \frac{\text{amount of decrease}}{\text{original amount}}$$

$$= \frac{50}{150} = \frac{1}{3} = 33\frac{1}{3}\%$$

Guided Practice

« **1.** COMMUNICATION « *Reading* Replace each ___?___ with the correct word.

When you find the percent a price has been lowered, you find the rate of ___?___.
When you find the percent a price has been raised, you find the percent of ___?___.

Write the fraction you would use to find the percent of increase or decrease. Do not find the percent of increase or decrease.

original amount	new amount
2. $40	$30
3. 125 cm	100 cm
4. 7.5 ft	12.5 ft
5. 250 calculators	200 calculators

Find the percent of increase.

6. original cost: $200; new cost: $230

7. original population: 1000; new population: 1750

Find the percent of decrease.

8. original price: $20; new price: $19

9. original distance: 84 ft; new distance: 56 ft

Exercises

Find the percent of increase.

1. original score: 100; new score: 118

2. original record: 25 s; new record: 35 s

3. original weight: 150 lb; new weight: 180 lb

4. original price: $15; new price: $25

5. original length: 23 m; new length: 46 m

Find the percent of decrease.

6. original number of boxes: 40; new number of boxes: 14

7. original cost: $100; new cost: $79

8. original population: 500; new population: 460

9. original number of students: 144; new number of students: 20

10. original fare: $250; new fare: $240

11. Last week Suzanne earned $15 doing errands. This week she earned $40. Find the percent of increase in Suzanne's pay.

12. Kyle bought a painting for $600 and later sold it for $800. Find the percent of increase in the value of the painting.

13. The original price of a jacket is $144. The new price is $54. Find the rate of discount.

14. A runner finished a race in 10.5 s. In the next race the runner finished in 9.8 s. Find the percent of decrease in the running time.

Tell whether there is an *increase* or a *decrease*. Then find the percent of increase or decrease.

15. original number of acres: 40; new number of acres: 90

16. original weight: 16 kg; new weight: 21 kg

17. original population: 3.6 million; new population: 3 million

18. original length: 2.5 cm; new length: 1.25 cm

19. **WRITING ABOUT MATHEMATICS** Write a paragraph to a friend describing how to find the percent of decrease from $15 to $13.80.

THINKING SKILLS

Decide whether each statement is *True* or *False*. Give a convincing argument to support your answer.

20. Twice an amount is the same as an increase of 200%.

21. Half an amount is the same as a decrease of 50%.

SPIRAL REVIEW

22. Solve: $1.4x + 0.4 = 20$ *(Lesson 8-8)*

23. Write the ratio 3 min : 10 s as a fraction in lowest terms.   *(Lesson 9-1)*

24. Write $\frac{17}{20}$ as a decimal.   *(Lesson 7-7)*

25. The original price of a bracelet is $3. The new price is $2.40. Find the percent of decrease.   *(Lesson 9-9)*

MIXED REVIEW
Working with Percent

As early as 1709, American colonists were using the word *barbecue* to refer to meat roasted over an open flame. The word *barbecue* came from a Spanish word *barbacoa*. Spanish explorers had learned about barbecuing meat from a group of people native to the West Indies. The Spaniards then borrowed the word *barbacoa*, which meant a stick framework used to roast or smoke meat.

To discover the name of the people from whom the Spanish explorers learned the word *barbacoa*, complete the code boxes below. Begin by solving Exercise 1. The answer to Exercise 1 is 20, and T is associated with Exercise 1. Write T in the box above 20 as shown below. Continue in this way with Exercises 2–21.

1. What number is 25% of 80? (T)

2. What number is 74% of 50? (A)

3. What percent of 160 is 56? (F)

4. What percent of 325 is 13? (O)

5. $66\frac{2}{3}\%$ of what number is 80? (H)

6. 70% of what number is 294? (L)

7. 60% of 125 is what number? (T)

8. 15% of 320 is what number? (O)

9. 42 is what percent of 150? (I)

10. 32 is what percent of 256? (H)

11. $37\frac{1}{2}\%$ of what number is 24? (O)

12. 3% of what number is 6? (P)

13. What number is 65% of 270? (I)

14. What number is 82% of 200? (A)

15. What percent of 75 is 66? (E)

16. What percent of 228 is 76? (E)

17. 50% of what number is 457? (T)

18. 396 is 75% of what number? (N)

19. 18 is $\frac{1}{2}\%$ of what number? (I)

20. $33\frac{1}{3}\%$ of 87 is what number? (E)

21. What percent of 420 is 378? (P)

?	?	?	T	?	?	?	?
914	120	88	20	164	28	528	4

?	?	?	?	?	?	?	?	?	?	?	?	?
90	29	48	200	420	$33\frac{1}{3}$	64	35	12.5	37	175.5	75	3600

Many words in the English language have their origins in other languages. By solving Exercises 22–29, you can learn about the origins of some common English words. Solve each exercise. The word printed in blue following the correct answer matches the description printed in blue.

22. 65 is 13% of what number?

Word that comes from an Arabic word meaning *calendar*.

a. 845   schedule   **b.** 500   almanac

23. What percent of 225 is 75?

Word that comes from a Hindi word for a tie-dyeing process.

a. $33\frac{1}{3}$ bandanna   **b.** 168   batik

24. 80% of what number is 900?

Word that comes from a French word referring to a type of *cloak*.

a. 720   poncho   **b.** 1125   limousine

25. 40 is 5% of what number?

Word that comes from two Chinese words referring to a fish sauce.

a. 2   tartar   **b.** 800   ketchup

26. What number is $12\frac{1}{2}$% of 88?

Word that comes from an Ojibwa word meaning *head first*.

a. 11   chipmunk   **b.** 1100   woodchuck

27. $37\frac{1}{2}$% of what number is 3000?

Word that comes from an Aztec word meaning *bitter water*.

a. 8000   chocolate   **b.** 1125   coffee

28. What percent of 120 is 90?

Word that comes from a Spanish word meaning *lizard*.

a. 108   crocodile   **b.** 75   alligator

29. 16% of what number is 400?

Word that comes from a Senegalese word meaning *to eat*.

a. 2500   yam   **b.** 64   okra

9-10 Percents and Proportions

Objective: To use proportions to solve percent problems.

CONNECTION

You have learned to solve proportions and you know that a percent is a ratio. You can use proportions to solve percent problems. To do this, you write the problem as a proportion.

$$\frac{\text{part}}{\text{whole}} = \frac{\text{part}}{\text{whole}}$$

One of the ratios in the proportion will represent the percent. The *whole* part of this ratio will always be 100.

Example 1

Solution

What number is 25% of 56?

The ratio represented by 25% is $\frac{25}{100}$. The part is 25. The whole is 100. You are looking for a percent of 56, so n is the part and 56 is the whole.

Write the proportion and solve.

$$\underset{\text{whole}}{\underset{\longrightarrow}{\textit{unknown} \text{ part}}} \quad \frac{n}{56} = \frac{25}{100} \quad \overset{\text{part}}{\underset{\text{whole}}{\longleftarrow}}$$

$$100n = 56(25)$$

$$\frac{100n}{100} = \frac{1400}{100}$$

$$n = 14$$

So 14 is 25% of 56.

Example 2

Solution

What percent of 25 is 20?

You are looking for the percent. So in the ratio representing the percent, n is the part and 100 is the whole. In the other ratio, 20 is the part and 25 is the whole.

$$\underset{\text{whole}}{\underset{\longrightarrow}{\text{part}}} \quad \frac{20}{25} = \frac{n}{100} \quad \overset{\textit{unknown} \text{ part}}{\underset{\text{whole}}{\longleftarrow}}$$

$$20(100) = 25n$$

$$\frac{2000}{25} = \frac{25n}{25}$$

$$80 = n$$

So 80% of 25 is 20.

✓ Check Your Understanding

1. In Example 1, will the result change if you write $\frac{25}{100}$ in simplest form before you find the cross products? Explain.

2. In Example 2, why is the answer 80% rather than 80?

Example 3

Solution

48.6 is 60% of what number?

The ratio represented by 60% is $\frac{60}{100}$. The part is 60. The whole is 100. The number 48.6 is a percent of another number. So in the other ratio, 48.6 is the part and n is the whole.

$$\underset{unknown \text{ whole}}{\frac{part}{}} \quad \longrightarrow \quad \frac{48.6}{n} = \frac{60}{100} \quad \longleftarrow \quad \frac{part}{whole}$$

$$(48.6)(100) = 60n$$

$$\frac{4860}{60} = \frac{60n}{60}$$

$$81 = n$$

So 48.6 is 60% of 81.

Guided Practice

COMMUNICATION « *Reading*

Tell whether the number shown in color is a *part* or a *whole*.

«**1.** What number is 70% of 220?   «**2.** 25% of what number is 45?

«**3.** 30% of 12.4 is what number?   «**4.** 24 is what percent of 36?

«**5.** What percent of 51 is 34?   «**6.** 0.28 is 80% of what number?

COMMUNICATION « *Writing*

«**7.** In symbols, write the proportion described by the statement: 10 is to 25 as an unknown number is to 100.

«**8.** In symbols, write a proportion that represents the situation: 50% of 236 is what number?

«**9.** In words, write the question that is represented by the proportion: $\frac{9.5}{38} = \frac{n}{100}$

«**10.** In words, write the proportion that represents the situation: 36 is 25% of what number?

Find each answer using a proportion.

11. What number is 45% of 6?   **12.** 8 is what percent of 32?

13. 140% of what number is 21?   **14.** 90% of 450 is what number?

15. What percent of 24 is 9?   **16.** 2.8 is 5% of what number?

17. COMMUNICATION «*Discussion* In Example 2 on p. 424, how can you tell that 25 is the whole?

Find each answer using a proportion.

1. What number is 4% of 236?
2. 18 is 40% of what number?
3. 6% of what number is 210?
4. 18% of 350 is what number?
5. 18 is what percent of 120?
6. What percent of 40 is 25?
7. 10% of what number is 24.5?
8. 7% of what number is 84?
9. What number is 48% of 2200?
10. 11% of what number is 1.21?
11. What percent of 60 is 4.2?
12. 95 is what percent of 95?

13. The Weston Library has 8000 books. Fiction books make up 35% of the total. How many fiction books does the library have?

14. Martin sells stereo equipment and earns a commission on all sales. Last month he earned $840 in commissions. His total monthly sales were $14,000. What percent of Martin's sales was his commission?

15. Last month, General Cable Company made service calls to 135 customers. This represents 3% of all the customers subscribing to General Cable Company. How many customers does General Cable Company have?

PROBLEM SOLVING/APPLICATION

One way that state and local governments raise money is through sales taxes. Tables like the one shown at the right are used to find the sales tax for various purchase prices.

Use the tax table at the right for Exercises 16–20.

16. Find the sales tax for an item priced at $8.73.

17. What interval has a sales tax of $.07.

18. What is the total amount, including tax, that you would pay for an item priced at $5.54?

19. Including tax, how much more would you pay for an item priced at $6.00 than for one priced at $5.00?

5.5% Sales Tax Table			
Amount of Sale	Tax	Amount of Sale	Tax
0.00–0.09	0.00	4.96–5.13	0.28
0.10–0.27	0.01	5.14–5.31	0.29
0.28–0.45	0.02	5.32–5.49	0.30
0.46–0.63	0.03	5.50–5.67	0.31
0.64–0.81	0.04	5.68–5.85	0.32
0.82–0.99	0.05	5.86–6.03	0.33
1.00–1.17	0.06	6.04–6.21	0.34
1.18–1.35	0.07	6.22–6.39	0.35
1.36–1.53	0.08	6.40–6.57	0.36
1.54–1.71	0.09	6.58–6.75	0.37
1.72–1.89	0.10	6.76–6.93	0.38
1.90–2.07	0.11	6.94–7.11	0.39
2.08–2.25	0.12	7.12–7.29	0.40
2.26–2.43	0.13	7.30–7.47	0.41
2.44–2.61	0.14	7.48–7.65	0.42
2.62–2.79	0.15	7.66–7.83	0.43
2.80–2.97	0.16	7.84–8.01	0.44
2.98–3.15	0.17	8.02–8.19	0.45
3.16–3.33	0.18	8.20–8.37	0.46
3.34–3.51	0.19	8.38–8.55	0.47
3.52–3.69	0.20	8.56–8.73	0.48
3.70–3.87	0.21	8.74–8.91	0.49
3.88–4.05	0.22	8.92–9.09	0.50
4.06–4.23	0.23	9.10–9.27	0.51
4.24–4.41	0.24	9.28–9.45	0.52
4.42–4.59	0.25	9.46–9.63	0.53
4.60–4.77	0.26	9.64–9.81	0.54
4.78–4.95	0.27	9.82–9.99	0.55

20. Jen has a coupon for 10% off the cost of two mugs. The price of one mug is $4.95. Sales tax is taken on the discounted price. Including tax, what will Jen pay for two mugs?

21. The price of a compact disc is $13.87. The cost, including tax, is $14.77. What is the sales tax rate to the nearest tenth of a percent?

22. RESEARCH Find the general sales tax rate for each of the 50 states.

WRITING ABOUT MATHEMATICS

23. Write an outline for your notebook entitled *How to Solve a Percent Problem Using a Proportion*.

24. a. Use the method you learned in Lesson 9-8 to solve this problem.
7.5% of what number is 112.5?
Then solve the problem again using a proportion.

    b. Write a paragraph describing how these two methods are alike, and how they are different.

SPIRAL REVIEW

25. Write 0.943 as a percent. *(Lesson 9-5)*

26. Solve by supplying missing facts: Will spent 37 min doing mathematics homework and 48 min doing science homework. How many hours did Will spend doing homework? *(Lesson 8-7)*

27. Write 0.000058 in scientific notation. *(Lesson 7-11)*

28. Solve: $3(x + 5) = 9$ *(Lesson 4-9)*

29. Find the answer using a proportion: 7 is what percent of 8?
(Lesson 9-10)

Self-Test 3

Find the percent of increase or decrease.

1. original score: 14; new score: 49   9-9

2. original price: $25; new price: $22.50

3. original length: 75 km; new length: 60 km

4. original population: 3000; new population: 4000

Find the answer using a proportion.

5. What number is 20% of 16?   6. 9% of what number is 108?   9-10

7. 2.4 is 6% of what number?   8. What percent of 45 is 18?

Chapter Review

ratio (p. 388)
rate (p. 389)
unit rate (p. 389)
proportion (p. 391)
terms (p. 391)
cross products (p. 391)
solve a proportion (p. 392)

scale drawings (p. 395)
scale (p. 395)
scale model (p. 396)
percent (p. 404)
percent of increase (p. 419)
percent of decrease (p. 419)

Choose the correct term from the list above to complete each sentence.

1. The __?__ is the ratio of the size of a drawing to the actual size of the object it represents.

2. A __?__ is a rate for one unit of a given quantity.

3. A comparison of two numbers by division is a __?__.

4. A __?__ is a statement that two ratios are equal.

5. A __?__ represents a ratio that compares a number to 100.

6. In the proportion $\frac{6}{9} = \frac{10}{15}$, the __?__ of the proportion are 6, 9, 10, and 15.

Write each ratio as a fraction in lowest terms. *(Lesson 9-1)*

7. $\frac{2}{10}$

8. 60 cm to 3 m

9. 8:44

10. $\frac{48}{32}$

Write the unit rate. *(Lesson 9-1)*

11. 21 h in 3 days

12. 108 words in 4 min

13. $21.25 for 5 h

Tell whether each proportion is *True* or *False*. *(Lesson 9-2)*

14. $\frac{12}{12} = \frac{6}{5}$

15. $\frac{12}{9} = \frac{4.8}{3.6}$

16. $\frac{6}{32} = \frac{3}{48}$

17. $\frac{5}{7} = \frac{6}{8}$

Solve each proportion. *(Lesson 9-2)*

18. $\frac{n}{5} = \frac{2}{10}$

19. $\frac{14}{7} = \frac{8}{y}$

20. $\frac{w}{1.6} = \frac{5}{8}$

21. $\frac{2}{9.2} = \frac{5}{z}$

Solve. *(Lesson 9-3)*

22. The scale on the floor plan of a museum is 1 in. : 16 ft. On the floor plan, the Far Eastern Art room is $1\frac{1}{4}$ in. wide. Find the actual width of the room.

23. The scale of a model stadium is 3 in. : 8 ft. The actual height of the stadium is 40 ft. Find the height of the scale model.

Chapter 9

Solve using a proportion. *(Lesson 9-4)*

24. Tom can read 34 pages of a novel in 50 min. At this rate, how many minutes will it take Tom to read a 272-page novel?

25. A department store charges $6.50 for 4 blank cassette tapes. What will the department store charge for 10 blank cassette tapes?

Write each fraction or mixed number as a percent. *(Lesson 9-5)*

26. $\frac{11}{20}$ **27.** $\frac{7}{16}$ **28.** $2\frac{3}{4}$ **29.** $\frac{1}{3}$ **30.** $\frac{5}{8}$

Write each decimal as a percent. *(Lesson 9-5)*

31. 0.65   **32.** 0.03   **33.** 1.2   **34.** 0.428   **35.** 0.0095

Write each percent as a fraction or mixed number in lowest terms and as a decimal. *(Lesson 9-5)*

36. 75%   **37.** 250%   **38.** $5\frac{1}{4}$%   **39.** 19.5%   **40.** 0.05%

Estimate. *(Lesson 9-6)*

41. $33\frac{1}{3}$% of 272   **42.** 51% of 79   **43.** 13% of 64   **44.** 24% of 258

Find each answer. *(Lessons 9-6, 9-7, 9-8)*

45. What number is 20% of 95?

46. 4% of 350 is what number?

47. What percent of 18 is 9?

48. What percent of 225 is 180?

49. 22 is what percent of 550?

50. 0.17 is what percent of 6.8?

51. 15% of what number is 90?

52. 28 is $66\frac{2}{3}$% of what number?

53. 16.8 is 24% of what number?

54. 75% of what number is 273?

Find the percent of increase or percent of decrease. *(Lesson 9-9)*

55. original distance: 50 km; new distance: 200 km

56. original weight: 75 lb; new weight: 80 lb

57. original time: 3.2 s; new time: 1.6 s

58. original cost: $50; new cost: $40

Find each answer by using a proportion. *(Lesson 9-10)*

59. What number is 48% of 150?

60. 60% of what number is 2.7?

61. What percent of 104 is 39?

62. 12% of 24 is what number?

Write each ratio as a fraction in lowest terms.

1. $\frac{4}{16}$ **2.** 50:75   **3.** $1 to $.40   **4.** 20 in. to 1 yd   9-1

Write the unit rate.

5. 80 mi on 2 gal   **6.** $51 for 6 h   **7.** 135 mi in 3 h

Solve each proportion.

8. $\frac{20}{a} = \frac{5}{6}$ **9.** $\frac{7}{8} = \frac{y}{16}$ **10.** $\frac{2.5}{c} = \frac{10}{7}$ **11.** $\frac{1.8}{6} = \frac{m}{4}$ 9-2

12. Explain the difference between a ratio and a proportion.

13. The scale of a model house is 1 in. : 6 ft. The actual house is 48 ft   9-3
wide. Find the width of the model house.

Solve using a proportion.

14. Isaac Holmes can grade 20 exams in 3 h. At this rate, how long will it   9-4
take Isaac Holmes to grade 72 exams?

Write each fraction or decimal as a percent.

15. 0.829   **16.** $\frac{2}{5}$ **17.** $\frac{7}{25}$ **18.** 0.06   **19.** 1.04   9-5

Write each percent as a fraction in lowest terms and as a decimal.

20. 64%   **21.** 90%   **22.** 7.2%   **23.** 350%   **24.** $6\frac{1}{2}$%

Find each answer.

25. What number is 18% of 70?   **26.** 80% of 32.5 is what number?   9-6

27. What percent of 128 is 38.4?   **28.** 104 is what percent of 78?   9-7

29. 35 is $87\frac{1}{2}$% of what number?   **30.** 60% of what number is 390?   9-8

Find the percent of increase or percent of decrease.

31. original number of passengers: 40; new number of passengers: 10   9-9

32. original length: 12.5 m; new length: 17.5 m

33. Give two situations that reflect a 100% increase.

Find each answer using a proportion.

34. 700 is 35% of what number?   **35.** What percent of 75 is 50?   9-10

The Compound Interest Formula

Objective: To calculate compound interest using the compound interest formula.

Banks pay interest for the use of principal that you deposit in your account. If you leave the interest in your account, you will earn **compound interest**, which is interest on the principal *and* any interest previously earned.

Banks pay interest to accounts at specific **interest periods**. Interest that is paid to an account once a year is *compounded annually*.

The total amount (A) in an account at the end of a period of time (t) in years is given by the compound interest formula: $A = P\left(1 + \frac{r}{n}\right)^{nt}$, where P is the principal, r is the annual interest rate, and n is the number of interest periods per year.

Example

Let $P = \$6000$, $r = 8\%$, $n = 4$, and $t = 3$. Find A.

Solution

Use the compound interest formula. $\qquad A = P\left(1 + \frac{r}{n}\right)^{nt}$

$$A = 6000\left(1 + \frac{0.08}{4}\right)^{4(3)} = 6000(1 + 0.02)^{12}$$
$$= 6000(1.02)^{12}$$
$$\approx 6000(1.2682418) \approx 7609.4508 \qquad \longleftarrow \text{Round.}$$

The value of A is $\$7609.45$. (This represents the amount at the end of 3 years when an account earns 8% interest compounded 4 times a year.)

You can use this key sequence on a scientific calculator to find the answer to the Example.

Exercises

Find each answer using the compound interest formula. Assume that no deposits or withdrawals are made after the first deposit.

1. Let $P = \$2400$, $r = 12\%$, $n = 2$, and $t = 4$. Find A.

2. Let $P = \$9000$, $r = 7.2\%$, $n = 4$, and $t = 5$. Find A.

3. Let $P = \$4500$, $r = 9\%$, $n = 12$, and $t = 0.5$. Find A.

4. Cari deposited $2000 in a savings account. The account earned 8% annual interest compounded semiannually (twice a year). How much money was in the account at the end of 3 years?

5. Thrift Savings Bank pays 10% annual interest compounded quarterly (four times a year) on its two year certificates. How much would a $5000 certificate be worth at the end of two years?

Cumulative Review

Standardized Testing Practice

Choose the letter of the correct answer.

1.

Which three points lie on the same line?

A. P, Q, and R B. P, Q, and T

C. T, Q, and S D. S, R, and P

2. A health club charges an initial fee of $90, plus $30 per month. Doug paid $360 to the health club. For how many months did he sign up to be a member?

A. 12   B. 9   C. 30   D. 15

3. Which is *not* a rational number?

A. 1.7   B. $\frac{12}{0}$ C. $\frac{2}{7}$ D. $0.\overline{3}$

4. Find the sum of the measures of the angles of a pentagon.

A. 900   B. 1260

C. 540   D. 360

5. Five rulers cost $1.95. What is the cost of 12 rulers?

A. $4.68   B. $23.64

C. $9.75   D. $3.90

6.

Hours Spent on Homework	Tally	Frequency
0	ＨＨＴ	5
1	ＩＩＩＩ	4
2	ＨＨＴ ＩＩ	7
3	ＩＩＩＩ	4
	Total	20

What is the mean of these data?

A. 1 h   B. 3 h

C. 2 h   D. 1.5 h

7. Corrine exercises for about $\frac{3}{4}$ h every day. About how long does she exercise in a week?

A. 3 h   B. 18 h

C. $\frac{3}{4}$ h   D. 6 h

8. What is 45% of 120?

A. 45   B. 54

C. $2.\overline{6}$ D. 0.45

9. Which fraction is equivalent to $\frac{2}{5}$?

A. $\frac{12}{15}$ B. $\frac{5}{2}$ C. $\frac{24}{60}$ D. $2\frac{1}{5}$

10. Solve: $z + \frac{1}{3} = -\frac{2}{5}$

A. $z = -\frac{11}{15}$ B. $z = -\frac{1}{15}$

C. $z = -\frac{2}{15}$ D. $z = -1\frac{1}{2}$

11. Which property does this statement illustrate?

$\frac{1}{4}(284) = \frac{1}{4} \cdot 200 + \frac{1}{4} \cdot 84$

A. identity property of multiplication
B. additive property of addition
C. associative property of addition
D. distributive property

12. Find the sum. Simplify if possible.

$-2\frac{7}{12} + 3\frac{11}{12}$

A. $-6\frac{1}{2}$ B. $1\frac{1}{3}$

C. $5\frac{1}{2}$ D. $6\frac{1}{2}$

13. A furniture store manager buys lamps for $60 and sells them for $80. What is the percent of increase?

A. 20%   B. $33\frac{1}{3}\%$

C. 25%   D. 140%

14. Which of the following has the greatest value?

$1, |-2|, -14, |0|, |3|-|8|$

A. $|-2|$ B. -14
C. 1 D. $|3| - |8|$

15. Which number is most likely to be estimated?

A. the hourly wage of a cashier
B. the postage for a package
C. the number of frames on a roll of film
D. the number of people who visit an airport in one year

16. A mail carrier walks 18 blocks due east, 26 blocks due west, and 5 blocks due east. Where is the mail carrier in relation to the starting point?

A. 3 blocks due west
B. 3 blocks due east
C. 5 blocks due west
D. 5 blocks due east

17. Solve: $\frac{10}{15} = \frac{x}{36}$

A. $x = \frac{2}{3}$ B. $x = 12$

C. $x = 15$ D. $x = 24$

18. Find the product. Simplify if possible.

$\frac{8c}{15} \cdot \frac{10}{c}$

A. $\frac{18c}{15c}$ B. $5\frac{1}{3}$

C. $\frac{4c^2}{75}$ D. $\frac{18c}{15c}$

19. Write the ratio in lowest terms.
16 in. to 4 ft

A. $\frac{4}{1}$ B. $\frac{1}{3}$

C. $\frac{16}{4}$ D. $\frac{1}{4}$

20. What should be true of a double line graph?

A. the lines may not cross
B. one line should show an increase and one should show a decrease
C. the intervals on the horizontal axis should be equal
D. the bottom line should be red

Archery Target

48 in.

9.6 in.

48 in.

Playing Field Dimensions

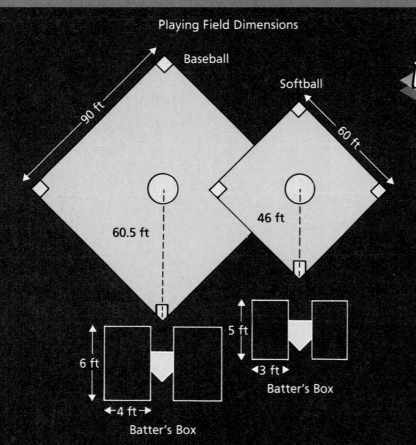

Baseball

Softball

90 ft

60 ft

60.5 ft

46 ft

5 ft

◄3 ft►

Batter's Box

6 ft

◄4 ft→

Batter's Box

Did You Know?

Compared to ice skating, roller skating is a very young sport. Ice skates made from bones were used as early as 50 B.C. In contrast, roller skates were not invented until the 18th century. The first roller skates were merely a refinement of ice skates, having wooden rollers attached to iron blades. It wasn't until 1863 that a four-wheeled skate was introduced.

Standard Dimensions of Playing Fields

Game	Length	Width
American football	120 yd	53 yd 1 ft
basketball	94 ft	50 ft
Canadian football	160 yd	65 yd
field hockey	100 yd	60 yd
ice hockey	200 ft	85 ft
roller hockey	120-140 ft	60-70 ft
men's lacrosse	110 yd	$53\frac{1}{3}$-60 yd
women's lacrosse	120 yd	70 yd
polo	300 yd	200 yd
soccer	110 yd	80 yd

Perimeter

Objective: To find the perimeter of a polygon.

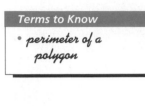

Terms to Know

- perimeter of a polygon

APPLICATION

The athletic field at the new Jay City High School needs to be enclosed by fencing. The figure below is a sketch of the field.

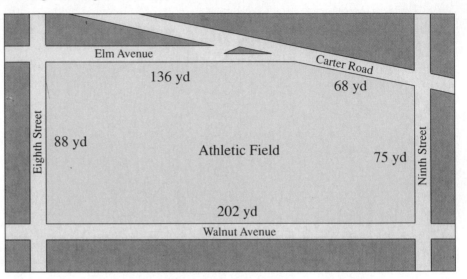

QUESTION What amount of fencing is needed to enclose the field?

The shape of the athletic field is a pentagon. To find the amount of fencing, you need to find the *perimeter* of this pentagon. The **perimeter** of a polygon is the sum of the lengths of all its sides.

Example 1

Find the perimeter of the pentagon in the sketch above.

Solution

Let the variables a, b, c, d, and e represent the lengths of the sides. To find the perimeter, substitute the lengths of the sides into the formula $P = a + b + c + d + e$.

$$P = a + b + c + d + e$$
$$= 88 + 136 + 68 + 75 + 202$$
$$= 569$$

The perimeter of the pentagon is 569 yd.

✓ Check Your Understanding

1. How would the formula in Example 1 be different if the field were shaped like a hexagon?

ANSWER The amount of fencing needed is 569 yd.

To find the perimeter of some figures, you need to apply the properties of polygons that you studied in Chapter 6.

Example 2

Solution

The length of one side of a regular pentagon is 4.9 cm. Find the perimeter.

First make a sketch of the figure. Because it is a regular pentagon, all sides can be marked with the same measure.

Let n represent the length of one side. To find the perimeter, substitute the length of one side into the formula $P = 5n$.

$$P = 5n = 5(4.9) = 24.5$$

The perimeter of the pentagon is 24.5 cm.

4.9 cm   4.9 cm

4.9 cm   4.9 cm

4.9 cm

Example 3

Solution

The length of a rectangle is $5\frac{1}{2}$ ft and the width is $4\frac{3}{4}$ ft. Find the perimeter.

First make a sketch of the figure. Because it is a rectangle, opposite sides can be marked with the same measure.

Let l and w represent the length and width, respectively. To find the perimeter, substitute the length and width into the formula $P = 2l + 2w$.

$$P = 2l + 2w = 2(5\tfrac{1}{2}) + 2(4\tfrac{3}{4}) = 11 + 9\tfrac{1}{2} = 20\tfrac{1}{2}$$

The perimeter is $20\frac{1}{2}$ ft.

$4\frac{3}{4}$ ft

$5\frac{1}{2}$ ft $5\frac{1}{2}$ ft

$4\frac{3}{4}$ ft

☑ **Check Your Understanding**

2. How are Examples 1 and 2 alike? How are they different?

3. How would the solution of Example 2 be different if the figure were a regular octagon?

Guided Practice

COMMUNICATION « *Writing*

Write a formula that can be used to find the perimeter of each figure.

« **1.**

x

y y

x

« **2.**

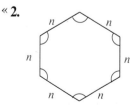

« **3.** a rhombus with one side that measures s

« **4.** a triangle with sides that measure a, b, and c

Find the perimeter of each figure.

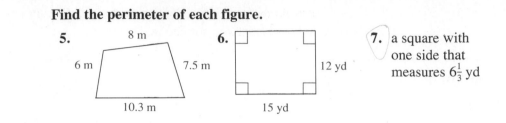

5. 8 m, 6 m, 7.5 m, 10.3 m

6. 12 yd, 15 yd

7. a square with one side that measures $6\frac{1}{3}$ yd

«**8.** COMMUNICATION «*Discussion* The sides of a triangle measure 1 yd, 60 in., and 4 ft. Describe three ways to find the perimeter.

Exercises

Find the perimeter of each figure.

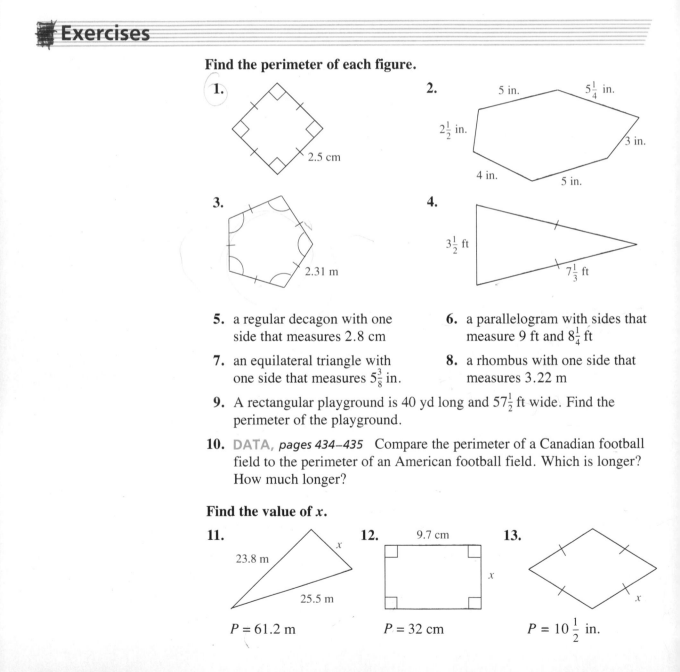

1. 2.5 cm

2. 5 in., $5\frac{1}{4}$ in., $2\frac{1}{2}$ in., 3 in., 4 in., 5 in.

3. 2.31 m

4. $3\frac{1}{2}$ ft, $7\frac{1}{3}$ ft

5. a regular decagon with one side that measures 2.8 cm

6. a parallelogram with sides that measure 9 ft and $8\frac{1}{4}$ ft

7. an equilateral triangle with one side that measures $5\frac{3}{8}$ in.

8. a rhombus with one side that measures 3.22 m

9. A rectangular playground is 40 yd long and $57\frac{1}{2}$ ft wide. Find the perimeter of the playground.

10. DATA, *pages 434–435* Compare the perimeter of a Canadian football field to the perimeter of an American football field. Which is longer? How much longer?

Find the value of x.

11. 23.8 m, x, 25.5 m

$P = 61.2$ m

12. 9.7 cm, x

$P = 32$ cm

13. x, x

$P = 10\frac{1}{2}$ in.

Solve, if possible. If there is not enough information, tell what additional facts are needed.

14. The perimeter of a regular octagon is 19.6 m. What is the length of one side?

15. The perimeter of a rectangle is 25 cm. What is its width?

16. What is the length of one side of a square with perimeter 14 ft?

17. Four sides of a pentagon each measure $10\frac{1}{2}$ ft. What is the length of the fifth side?

18. One side of a trapezoid measures 32 in., two other sides each measure 14 in., and the perimeter is 108 in. What is the length of the fourth side?

PATTERNS

In the following patterns, the length of one side of a small square (☐) is 1 unit. Find the perimeter of the ninth figure in each pattern.

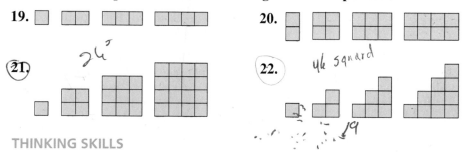

19. ☐ ☐☐ ☐☐☐ ☐☐☐☐

20.

21. 24°

22. 46 squared

19

THINKING SKILLS

23. In triangle ABC, the length of $\overline{BC}$ is 15 mm. The length of $\overline{AC}$ is 4 mm less than the length of $\overline{BC}$, and the length of $\overline{AB}$ is half the length of $\overline{BC}$. What is the perimeter of the triangle?

24. The length of a rectangle is 5 in. less than twice the width. The perimeter is 62 in. Find the width and the length.

SPIRAL REVIEW

25. Find the mean, median, mode(s), and range of the data: 18, 25, 19, 22, 18, 20, 22, 24, 18   (Lesson 5-8)

26. Simplify: $\frac{4x}{5} + \frac{11x}{5}$ (Lesson 8-4)

27. Find the perimeter of a regular hexagon with one side that measures $9\frac{1}{2}$ in.   (Lesson 10-1)

28. Write 0.00000084 in scientific notation.   (Lesson 7-11)

29. Find the measure of an angle supplementary to a 30° angle. (Lesson 6-3)

30. Solve the proportion: $\frac{15}{45} = \frac{a}{6}$ (Lesson 9-2)

444

Circles and Circumference

Objective: To identify and find the radius, diameter, and circumference of a circle.

Terms to Know

- circle
- center
- radius (radii)
- chord
- diameter
- circumference

The set of all points in a plane that are a given distance from a given point in the plane is called a **circle.** The given point is the **center** of the circle. For instance the rim of the Ferris wheel in the picture at the right is a circle. A circle is named by its center.

A line segment whose two endpoints are the center of a circle and a point on the circle is called a **radius.** A circle has an infinite number of *radii*, all with the same length.

A line segment whose endpoints are both on a circle is called a **chord** of the circle. Any chord that passes through the center of a circle is a **diameter.** The diameter, d, of any circle is twice its radius, r.

$$d = 2r$$

Example 1

Use the circle at the right.
a. Name the circle.
b. Identify as many radii, chords, and diameters as are shown.
c. Find the radius.
d. Find the diameter.

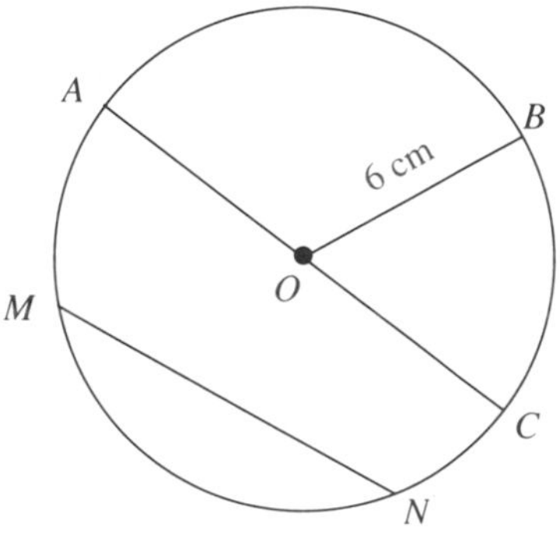

Solution

a. circle O
b. radii: $\overline{OA}$, $\overline{OB}$, and $\overline{OC}$
   chords: $\overline{MN}$ and $\overline{AC}$
   diameter: $\overline{AC}$
c. The length of radius $\overline{OB}$ is labeled as 6 cm. The radius is 6 cm.
d. $r = 6$, so $d = 2r = 2(6) = 12$.
   The diameter is 12 cm.

✏️ **Check Your Understanding**

1. In Example 1, why is $\overline{AC}$ identified as both a chord and a diameter of circle O?

The distance around a circle is called its **circumference,** C. In all circles, the ratio of the circumference to the diameter is equal to the same number, represented by the Greek letter π (*pi*).

$$\frac{C}{d} = \pi$$

Formulas: *Circumference of a Circle*

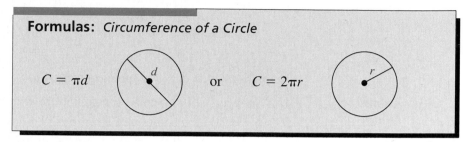

$C = \pi d$ \qquad or \qquad $C = 2\pi r$

Because the number represented by π is irrational, there is no common fraction or decimal that you can use to name it. However, π is so important in mathematics that the following two *approximations* are often used.

$$\pi \approx 3.14 \qquad \text{and} \qquad \pi \approx \frac{22}{7}$$

In general, use 3.14 as an approximation for π. You can use $\frac{22}{7}$ when multiples of 7 make the calculations easier. When you use 3.14 for π, you generally round your answer to the tenths' place.

Example 2

The diameter of a circle is 21 cm. Find the circumference.

Solution

Use the formula $C = \pi d$.

$$C \approx \frac{22}{7} \cdot 21 \qquad \longleftarrow \quad \text{Substitute } \frac{22}{7} \text{ for } \pi \text{ and 21 for } d.$$

$$C \approx 66$$

The circumference of the circle is approximately 66 cm.

Check Your Understanding

2. Why was $\frac{22}{7}$ and not 3.14 substituted for π in Example 2?
3. How would the solution of Example 2 be different if the radius of the circle were 21 cm?

Many calculators have a ⬚π key. When you press this key, the calculator recalls from its memory a decimal approximation of π that has several decimal places. For instance, this key sequence might be used to find the circumference in Example 2.

To the nearest tenth, the answer in the display rounds to 66.0. Notice that this is the same as the result when you substitute $\frac{22}{7}$ for π.

Example 3	The circumference of a circle is 25.12 m. Find the radius.
Solution	Use the formula $C = 2\pi r$.

$$25.12 \approx 2(3.14)r \quad \longleftarrow \quad \text{Substitute 25.12 for } C$$
$$25.12 \approx 6.28r \qquad\qquad \text{and 3.14 for } \pi.$$

$$\frac{25.12}{6.28} \approx \frac{6.28r}{6.28}$$

$$4 \approx r$$

The radius of the circle is approximately 4 m.

 **Check Your Understanding**

4. Why was 3.14 and not $\frac{22}{7}$ substituted for π in Example 3?
5. How would the solution of Example 3 be different if you were asked to find the diameter of the circle instead of the radius?

Guided Practice

COMMUNICATION « *Reading*

Refer to the text on pages 440–442.

«**1.** Explain how *a radius* is different from *a diameter*.

«**2.** What characteristic of a circle is represented by the number π?

«**3.** Name two commonly used approximations for π.

«**4.** Identify two frequently used formulas for circumference.

Exercises 5 and 6 show a series of steps that apply one of the circumference formulas to the given figure. In each step, replace the __?__ with the number that makes the statement true.

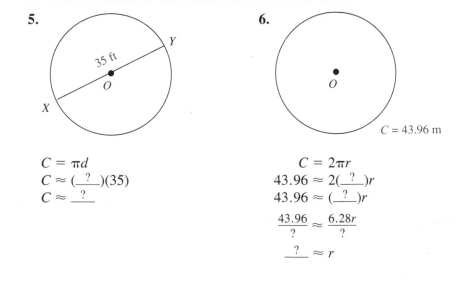

5.

$$C = \pi d$$
$$C \approx (\underline{\ ?\ })(35)$$
$$C \approx \underline{\ ?\ }$$

6.

$C = 43.96 \text{ m}$

$$C = 2\pi r$$
$$43.96 \approx 2(\underline{\ ?\ })r$$
$$43.96 \approx (\underline{\ ?\ })r$$
$$\frac{43.96}{?} \approx \frac{6.28r}{?}$$
$$\underline{\ ?\ } \approx r$$

Use the figure at the right.

7. Name the circle.

8. Identify as many radii, chords, and diameters as are shown.

9. Find the diameter.

10. Find the radius.

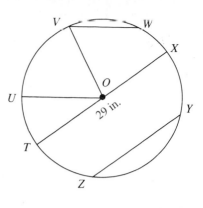

Find the circumference of a circle with the given measure.

11. $d = 14$ yd 12. $r = 8$ m

Find the radius and diameter of a circle with the given circumference.

13. 5.2 m 14. 220 ft

Exercises

Use the figure at the right.

1. Name the circle.

2. Identify as many radii, chords, and diameters as are shown.

3. Find the radius.

4. Find the diameter.

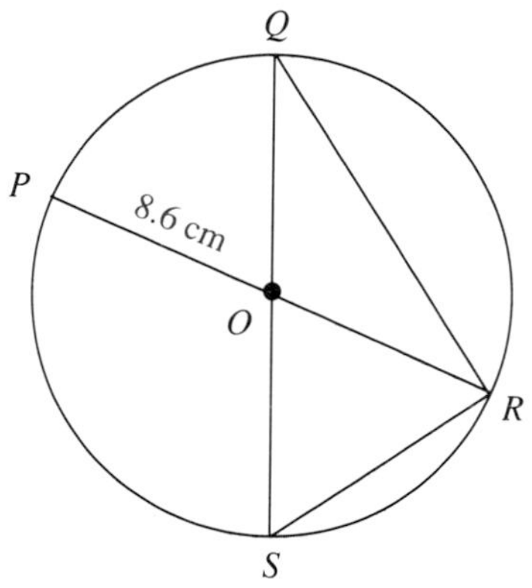

Find the circumference of a circle with the given measure.

5. $d = 2$ m 6. $d = 10.5$ cm

7. $r = 2\frac{1}{3}$ yd 8. $r = 2.5$ cm

9. $r = 7$ mm 10. $d = 3.5$ m

11. $d = 5.2$ m 12. $r = 14$ in.

Find the radius and diameter of a circle with the given circumference.

13. 56.52 m 14. 33 ft 15. 44 yd 16. 34.54 cm

17. In a public park, the length of the fence that encloses a circular garden is 157 m. What is the approximate diameter of the garden?

18. The radius of the front wheel of a tricycle is 14 in. The radius of each back wheel is 6 in. How much larger is the circumference of the front wheel than the circumference of a back wheel?

Replace each __?__ with *exactly* or *approximately* to make a true statement.

19. The product π(3) is __?__ equal to 9.42.

20. The product 2(π)(6) is __?__ equal to the product π(12).

21. The circumference of a circle with diameter 7 in. is __?__ 22 in.

22. The circumference of a circle with radius 5 cm is __?__ 10(π) cm.

WRITING ABOUT MATHEMATICS

Write an essay entitled *A World without Circles*. Here are two questions that you should address in this essay.

23. How would nature be different if there were no circular shapes?

24. How would your house and school be different if no circles were used in manufacturing the items that you use?

PROBLEM SOLVING/APPLICATION

Exercises 25–28 refer to a monocycle wheel whose radius is 35 in.

25. What is the circumference of the wheel?

26. Approximately how far does the wheel travel in five complete turns?

27. How many complete turns does the wheel make in traveling 1 mi?

28. If the wheel has turned one million times, how many miles has it traveled?

SPIRAL REVIEW

29. Michiko jogs 4.5 mi/day. How many miles does she jog per week? *(Lesson 8-7)*

30. Write the prime factorization of 150. *(Lesson 7-1)*

31. Find the radius of a circle with circumference 785 cm. *(Lesson 10-2)*

32. What percent of 60 is 15? *(Lesson 9-7)*

33. The measures of two angles of a triangle are 30° and 60°. Tell whether the triangle is *acute*, *right*, or *obtuse*. *(Lesson 6-6)*

34. Evaluate $-3a^2$ when $a = 5$. *(Lesson 3-6)*

10-3

Area of Polygons

Objective: To find the area of rectangles, parallelograms, triangles, and trapezoids.

Terms to Know

• *area*
• *base*
• *height*

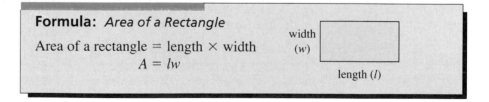

EXPLORATION

The length of one side of a square on a sheet of grid paper can be thought of as one *unit*. A square is then one *square unit*. The number of square units enclosed by a figure on grid paper is the *area* of the figure.

1 Draw and cut out a rectangle like the one shown below.
  a. Find the area of the rectangle by counting the square units.
  b. Can you describe a simpler way to find the area?

2 Draw and cut out a square that is three units on each side.
  a. Find the area of the square by counting the square units.
  b. Can you describe a simpler way to find the area?

The region enclosed by a plane figure is called the **area** (*A*) of the figure. When possible, it is far more efficient to find an area using a formula than counting units within a figure.

One area formula that may be familiar to you is the formula for the area of a rectangle.

Formula: *Area of a Rectangle*

Area of a rectangle = length × width

$$A = lw$$

When the length and width of a rectangle are equal, the figure is a square. To find the area of a square, you need to know only one measure, the length of a side.

Formula: *Area of a Square*

Area of a square = (length of side)²

$$A = s^2$$

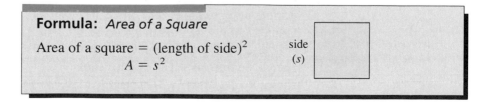

Example 1

Solution

Find the area of a rectangle with width $3\frac{1}{2}$ ft and length $4\frac{1}{2}$ ft.

Use the formula $A = lw$.

$$A = \left(4\frac{1}{2}\right)\left(3\frac{1}{2}\right)$$ ← Substitute $4\frac{1}{2}$ for l and $3\frac{1}{2}$ for w.

$$A = \left(\frac{9}{2}\right)\left(\frac{7}{2}\right) = \frac{63}{4} = 15\frac{3}{4}$$ The area is $15\frac{3}{4}$ ft².

✔ **Check Your Understanding**

1. How would Example 1 be different if the figure were a square with the length of one side equal to $3\frac{1}{2}$ ft?

Any parallelogram can be "rearranged" to form a rectangle. For this reason, the area formula for a parallelogram is closely related to the rectangle formula.

To use the parallelogram formula, either pair of parallel sides can be the **bases.** The **height** is the perpendicular distance between the bases.

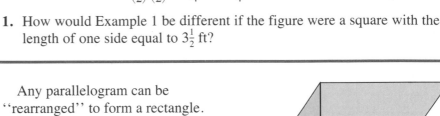

Formula: *Area of a Parallelogram*

Area of a parallelogram = base × height
$$A = bh$$

height (h)
base (b)

Example 2

Solution

Find the height of a parallelogram with base 6 yd and area 46 yd².

Use the formula $A = bh$.
$$46 = 6h$$ ← Substitute 46 for A and 6 for b.

$$\frac{46}{6} = \frac{6h}{6}$$

$$h = \frac{46}{6} = 7\frac{4}{6} = 7\frac{2}{3}$$ The height is $7\frac{2}{3}$ yd.

A diagonal of a parallelogram separates it into identical triangles. You can find the area of a triangle by thinking of it as *one half* the area of a parallelogram with the same base and height.

To use the triangle formula, let any side of the triangle be the **base.** The **height** is the perpendicular distance between the base and the opposite vertex.

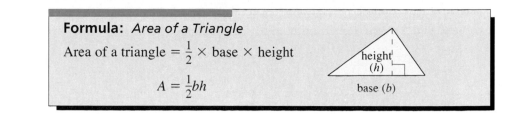

Formula: *Area of a Triangle*

Area of a triangle $= \frac{1}{2} \times$ base $\times$ height

$$A = \frac{1}{2}bh$$

height (h)

base (b)

Example 3

Solution

Find the area of a triangle with base 11 cm and height 50 mm.

All measures that are substituted into a formula must be expressed in the same unit, so change 50 mm to 5 cm.

Use the formula $A = \frac{1}{2}bh$.

$$A = \frac{1}{2}(11)(5) \qquad \longleftarrow \text{Substitute 11 for } b \text{ and 5 for } h.$$

$$A = \frac{1}{2}(55) = 27\frac{1}{2}$$

The area is 27.5 cm².

☑ **Check Your Understanding**

2. How would the solution to Example 3 be different if 11 cm had been changed to millimeters? Which answer is correct?

3. Would the answer to Example 3 be different if the base were 50 mm and the height were 11 cm?

Guided Practice

COMMUNICATION « *Writing*

Make a sketch of each figure.

« **1.** a rectangle with length 5 cm and width 2.2 cm

« **2.** a triangle with base 15 in. and height 1 ft

« **3.** a parallelogram with base $8\frac{1}{2}$ in. and height 6 in.

« **4.** a square with a side that measures *t* inches

Find the area of each figure.

5.

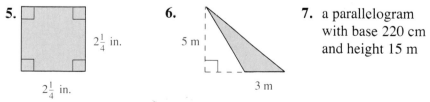

$2\frac{1}{4}$ in.

$2\frac{1}{4}$ in.

6.

5 m

3 m

7. a parallelogram with base 220 cm and height 15 m

8. Find the height of a triangle with area 36 ft² and base 8 ft.

« **9.** COMMUNICATION « *Discussion* How many square inches are in a square foot? How many square centimeters are in a square meter?

Find the area of each figure.

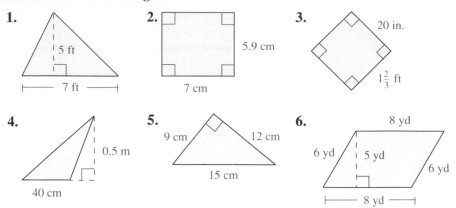

1. 5 ft, 7 ft

2. 5.9 cm, 7 cm

3. 20 in., $1\frac{2}{3}$ ft

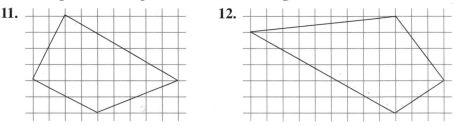

4. 0.5 m, 40 cm

5. 9 cm, 12 cm, 15 cm

6. 8 yd, 6 yd, 5 yd, 6 yd, 8 yd

7. a square with length of one side equal to $2\frac{1}{2}$ ft

8. a parallelogram with height $1\frac{1}{3}$ yd and base $2\frac{1}{2}$ ft

9. John Lano's garden is 7 yd wide and 27 ft long. Find the area.

10. DATA, *pages 434–435* Which of the playing fields listed has the greatest area?

SPATIAL SENSE

Find the area in square units of each quadrilateral. (*Hint:* Use the grid lines to separate each quadrilateral into triangles.)

11.

12.

THINKING SKILLS

Use the figure at the right. Replace each ___?___ with the variable or expression that makes the statement true.

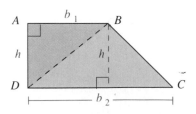

13. The area of triangle *ABD* is $\frac{1}{2}h$ ___?___ .

14. The area of triangle *BCD* is $\frac{1}{2}$ ___?___ b_2.

15. Use the results of Exercises 13 and 14. The area of trapezoid *ABCD* is ___?___ + ___?___ .

16. By the distributive property, $\frac{1}{2}hb_1 + \frac{1}{2}hb_2 = \frac{1}{2}h(\underline{\;?\;} + \underline{\;?\;})$.

Apply the formula $A = \frac{1}{2}h(b_1 + b_2)$. Find the area of each trapezoid.

17.

10 m

5 m

12 m

18.

11 ft

$3\frac{1}{4}$ ft

5 ft

CAREER/APPLICATION

An *interior designer* plans a room to make it attractive and functional. A designer often calculates and compares the costs of materials.

Family Room

windows

12 ft

windows

door

door

15 ft

Use the floor plan above.

19. A hardwood floor costs $9.50 per square foot to install, including materials and labor. Find the cost of installing a floor in this room.

20. Wall-to-wall carpeting costs $28.50 per square yard, including installation. How much does it cost to carpet this room?

21. Square ceramic tiles that measure 8 in. on each side cost $.89 each. The installation charge is $250. What is the total cost of installing the tiles in this room?

22. GROUP ACTIVITY   Using a newspaper, find the costs of several types of flooring. Which type is most expensive? least expensive? Calculate the costs of installing three types of flooring in your classroom. What factors should you consider when choosing a flooring?

SPIRAL REVIEW

23. Find the product: $\frac{2x}{3} \cdot \frac{9}{16}$ *(Lesson 8-1)*

24. Find the area of a triangle with base $1\frac{3}{4}$ yd and height 48 ft. *(Lesson 10-3)*

Exploring Area Using Tangrams

Objective: To explore the
concept of area using
tangrams.

Materials

▪ tangrams

The **tangram** is an ancient Chinese
puzzle. It consists of the seven puzzle
pieces shown in the figure at the right.
 To help you with the activities in this
lesson, each tangram piece in the figure
has been identified by one of the letters
from *A* through *G*. If you have a manufactured set of tangram pieces, you will
not see these letters on the pieces.

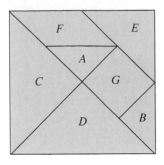

Activity I *Identifying the Tangram Pieces*

1 List the letters of the tangram pieces that are triangles. Next to each
letter, classify the triangle by its sides and by its angles.

2 List the letters of the tangram pieces that are quadrilaterals. Next to each
letter, give the most appropriate name for the quadrilateral.

Activity II *Calculating Areas*

Suppose that the area of triangle
A is one square unit. Because
triangle *B* is identical to *A*, its
area also would be one square
unit. The area of triangles *A* and
B together must be the same as
the area of square *G*, so the area
of *G* would be two square units.

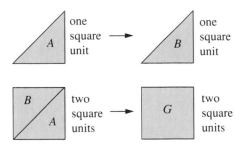

1 Use the reasoning described above. Copy and complete this table.

Tangram piece	A	B	C	D	E	F	G
Area in square units	1	1	?	?	?	?	2

2 Now suppose that the area of square *G* is one square unit. Find the areas
of the other pieces. Copy and complete this table.

Tangram piece	A	B	C	D	E	F	G
Area in square units	?	?	?	?	?	?	1

3 Next, suppose that the area of triangle *C* is one square unit. Find the
areas of the other pieces. Copy and complete this table.

Tangram piece	A	B	C	D	E	F	G
Area in square units	?	?	1	?	?	?	?

Activity III *Taking a Different Look at Area*

This time, suppose that the area of the entire
tangram puzzle is one square unit. It follows
that the area of each tangram piece has to be
some fractional part of one square unit. For
instance, the figure at the right shows that
the area of triangle C is one fourth the area
of the entire tangram puzzle. Therefore, its
area is $\frac{1}{4}$ square unit.

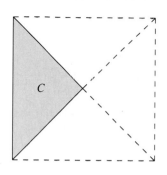

1 Use the reasoning described above to find
the areas of the other six tangram pieces.
Copy and complete this table.

Tangram piece	A	B	C	D	E	F	G
Area in square units	?	?	$\frac{1}{4}$	?	?	?	?

2 Show how to use four of the tangram pieces to form a square with area
equal to $\frac{1}{2}$ square unit.

3 Show how to use triangles A, B, C, and E to form each of the following
quadrilaterals. What is the area of each quadrilateral?

a. a square

b. a rectangle that is not square

c. a parallelogram that is not a rectangle

d. a trapezoid

Activity IV *Creating Tangram Designs*

For many centuries, people have enjoyed the
challenge of arranging the tangram pieces to form
familiar shapes. The Chinese produced entire books
illustrating these designs.

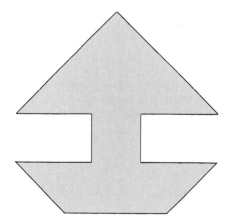

1 Show how the seven tangram pieces can be arranged to form the "sailboat" design shown in the
figure at the right.

2 Arrange the seven tangram pieces to form your
own design. Trace the outline of your design on a
sheet of paper and trade outlines with a partner.
Determine how the tangram pieces were arranged
to form your partner's design. How does the area
of your partner's design compare to the area of
your design?

10-4 Area of Circles

Objective: To find the area of circles.

No matter what the shape of a figure is, its area is measured in square units. Although at first it may seem impossible to count the number of square units in a circular region, the following method gives you a fairly good *estimate* of the area of a circle.

1 Estimate the area of a circle with radius 4 units.

    a. Count the square units that contain any part of the circular region. Call this number a.

    b. Count the square units that lie entirely within the circle. Call this number b.

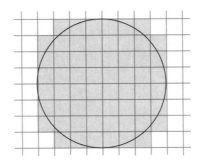
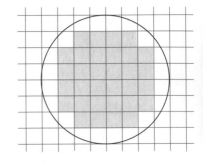

    c. Find the average of a and b: $\dfrac{a+b}{2} = \underline{\ ?\ }$.

    d. Complete: The area of the circle is about $\underline{\ ?\ }$ square units.

2 Use the method from Step 1 to estimate the area of these circles.

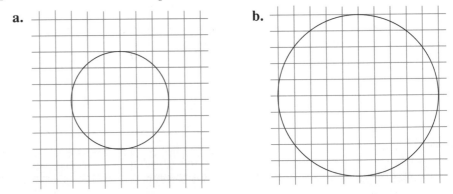

    a.   **b.**

In Lesson 10-3, you found the areas of polygons by using their bases and heights. It may seem that finding the area of a circle is totally unrelated, because you describe the size of a circle by its radius or diameter rather than its base and height. However, it is possible to use what you know about polygons to develop an area formula for circles.

Suppose that you could separate a circle into equal parts, then rearrange the parts as shown below. The new figure looks very much like a parallelogram.

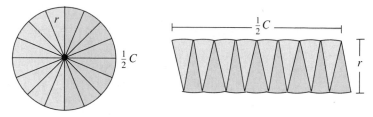

Notice that the base of this "parallelogram" is equal to one half the circumference of the circle, or $\frac{1}{2}C$. The height of the "parallelogram" is just about equal to the radius of the circle, which is r. Now you can use the area formula for a parallelogram to find the area of this figure.

$$A = b \cdot h$$
$$A = \frac{1}{2}C \cdot r$$

You know that $C = 2\pi r$, so replace the C in the formula with $2\pi r$.

$$A = \frac{1}{2}(2\pi r) \cdot r$$
$$A = \pi r^2$$

Formula: *Area of a Circle*

$$A = \pi r^2$$

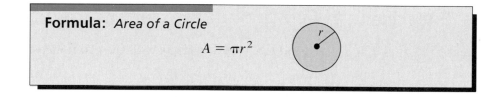

Example

The radius of a circle is 4.4 m. Find the area.

Solution

Use the formula $A = \pi r^2$.   ⟵ Substitute 3.14 for π
$$A \approx 3.14(4.4)^2$$ and 4.4 for r.
$$A \approx 3.14(19.36) \approx 60.7904$$
To the nearest tenth, the area of the circle is 60.8 m^2.

☑ **Check Your Understanding**

1. Why was 3.14 and not $\frac{22}{7}$ substituted for π in the Example?
2. How would the Example be different if the *diameter* were 4.4 m?

60.821234

%on off

If you have a calculator with the 🔲 and 🔲 keys, you can use this key sequence to find the area of the circle in the Example.

[π] [×] [4.4] [x²] [=]

To the nearest tenth, the answer in the display rounds to 60.8.

Guided Practice

COMMUNICATION «*Reading*

Refer to the text on pages 440–442 and on pages 452–453.

«**1.** Explain how the circumference of a circle is different from its area.

«**2.** Compare the circumference formulas to the area formula for a circle. How are they alike? How are they different?

Find each answer.

3. $(3.14)(8)(8)$

4. $\frac{22}{7}(14)(14)$

5. $\frac{22}{7}\left(\frac{7}{10}\right)^2$

6. $(3.14)(6.5)^2$

Find the area of a circle with the given measure.

7. $d = 2$ mm

8. $r = 2.5$ cm

9. $r = 21$ ft

10. $d = 3\frac{1}{2}$ in.

«**11.** **COMMUNICATION** «*Discussion*   Calculate both the circumference and the area of a circle with radius 2 cm. Are the answers the same or different? Explain.

Exercises

Find the area of a circle with the given measure.

1. $r = 10$ in.

2. $r = 7$ mm

3. $d = 6$ yd

4. $r = 1\frac{1}{2}$ ft

5. $d = 4\frac{2}{3}$ mi

6. $d = 3.4$ km

7.

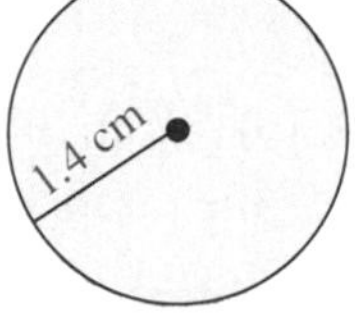

8.
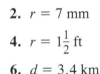

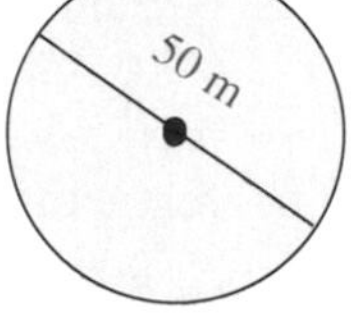

9. Sometimes the beam from a lighthouse can be seen for 30 mi in all directions. Over how many square miles of area can the beam be seen?

10. The diameter of a circular field is 600 ft. One fourth of the field is to be planted with corn. How many square yards are to be planted with corn?

Match each problem with the calculator key sequence that can be used to solve it. (Not every key sequence will be used.)

11. Find the area of a circle with radius 6 cm.

12. Find the circumference of a circle with diameter 6 cm.

13. Find the circumference of a circle with radius 6 cm.

14. Find the area of a circle with diameter 6 cm.

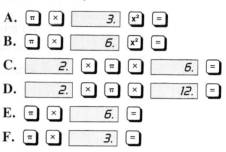

A. π × [$3.$] x^2 =

B. π × [$6.$] x^2 =

C. [$2.$] × π × [$6.$] =

D. [$2.$] × π × [$12.$] =

E. π × [$6.$] =

F. π × [$3.$] =

PROBLEM SOLVING/APPLICATION

A **composite** figure is made up, or *composed*, of two or more familiar geometric figures. For instance, the figure at the right is composed of a rectangle and a half-circle. The half-circle is called a **semicircle**.

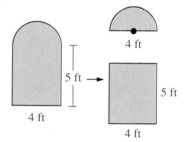

15. How do you think you can find the area of a semicircle?

16. What is the area of the semicircle in the figure above?

17. What is the area of the rectangle in the figure above?

18. What is the total area of the figure above?

Find the area of each composite figure.

19. 20. 21.

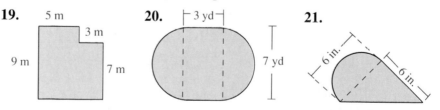

Finding the area of some composite figures involves subtracting areas rather than adding. In each figure, find the area of the shaded region.

22. 23. 24.

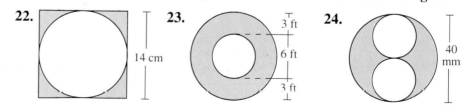

FUNCTIONS

Because the area and circumference of a circle are quantities that depend on the radius of the circle, each quantity is a *function* of the radius.

Copy and complete each function table.

25.

Radius (r)	Circumference (2πr)
1	?
2	?
3	6π
4	?
5	10π

26.

Radius (r)	Area (πr²)
1	?
2	4π
3	?
4	?
5	25π

27. Refer to the tables that you completed for Exercises 25 and 26. Describe the pattern in each table in words.

28. Refer to the tables that you completed for Exercises 25 and 26. Using r as the variable, write a function rule for each table.

THINKING SKILLS

29. Determine what happens to the circumference of a circle when its radius is doubled.

30. Determine what happens to the area of a circle when its radius is doubled.

SPIRAL REVIEW

31. Write 26.7% as a decimal.   *(Lesson 9-5)*

32. Find the area of a circle with diameter 10.6 m.   *(Lesson 10-4)*

33. Estimate: $1\frac{7}{8} + 7\frac{1}{6}$ *(Lesson 8-6)*

34. Solve: $2x + 9 = 17$ *(Lesson 4-5)*

Historical Note

Claudius Ptolemy, a Greek-Egyptian who lived in the second century A.D., wrote a book on astronomy that was the standard in its field for centuries. Much of Ptolemy's work was based on dividing the circumference of the circle into 360 parts. He developed this idea from a Babylonian practice originated nearly 2000 years earlier.

Research

What number was the base of the Babylonian number system? Why is it believed that this number was used?

Choosing Perimeter, Circumference, or Area

Objective: To decide whether perimeter, circumference, or area must be calculated to solve a given problem.

In everyday life, problems that you encounter seldom instruct you to "find the perimeter" or "find the circumference" or "find the area." Usually you must make a decision about which measure is needed.

- Calculate *perimeter* or *circumference* when you need to find the distance around a figure— perimeter for polygons, circumference for circles.

- Calculate *area* when you need to find the measure of the region enclosed by a figure.

Example

Tell whether you would need to calculate *perimeter*, *circumference*, or *area* to find each measure.

a. the amount of artificial turf needed to cover a football field

b. the length of the chalk line around a football field

Solution

a. The artificial turf covers a region enclosed by a rectangle. You need to calculate the *area* of the rectangle.

b. The chalk line outlines the sides of a rectangle. You need to calculate the distance around the rectangle, or its *perimeter*.

✓ **Check Your Understanding**

In what type of situation do you need to find the *circumference* of a figure?

Guided Practice

COMMUNICATION «*Reading*

Refer to the text above.

«**1.** How do you know when you need to calculate either a perimeter or a circumference? What is the example given?

«**2.** How do you know when you need to calculate area? What is the example given?

Tell whether you would need to calculate *perimeter*, *circumference*, or *area* to find each measure.

3. the amount of grazing space in a square pasture

4. the amount of fencing needed for a circular garden

«5. COMMUNICATION «*Discussion* Describe an everyday situation in which you needed to choose perimeter, circumference, or area to solve a problem.

Exercises

Tell whether you would need to calculate *perimeter*, *circumference*, or *area* to find each measure.

1. the amount of wood needed to frame a circular mirror

2. the amount of wallpaper needed for a rectangular wall

3. the amount of decorative molding needed around a square ceiling

4. the amount of floor space covered by a circular rug

Use the figure at the right. Tell whether you need to calculate *perimeter*, *circumference*, or *area* to solve each problem. Then solve.

5. How much space is there for a swimmer to float in the pool?

6. What amount of fencing is needed to enclose the concrete deck?

7. How much fencing is needed to enclose the entire recreation area?

8. What amount of concrete is needed to cover the play area?

9. How much space does the concrete deck provide for chairs?

10. One package of lawn seed covers 100 ft². How many packages are needed to seed the lawn?

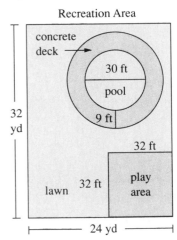

Recreation Area

concrete deck

30 ft

pool

9 ft

32 yd

32 ft

32 ft

lawn   play area

24 yd

SPIRAL REVIEW

11. Solve: $4x + 17 + x = -8$ (*Lesson 4-9*)

12. Find the LCM of 154 and 75.   (*Lesson 7-3*)

13. Tell whether you need to calculate *perimeter*, *circumference*, or *area* to find the distance around a circular track.   (*Lesson 10-5*)

14. Solve $A = \frac{1}{2}bh$ for h.   (*Lesson 8-10*)

10-6

Problems with No Solution

Objective: To recognize when a problem has no solution.

In solving a problem, you must evaluate the answer to your calculations to be sure that it makes sense in a real-world context. You will find that some problems have no solution, even though you have arrived at a correct answer to your calculations.

Problem

The perimeter of a triangular field is 64 ft. The measures of two sides are 16 ft and 14 ft. What is the length of the third side?

Solution

UNDERSTAND The problem is about the perimeter of a field.
Facts: triangular shape
        perimeter 64 ft
        two sides measure 16 ft and 14 ft
Find: length of the third side

PLAN Use the perimeter formula $P = a + b + c$. Substitute 64 for P, 16 for a, and 14 for b. Solve the resulting equation for c.

WORK
$$P = a + b + c$$
$$64 = 16 + 14 + c$$
$$64 = 30 + c$$
$$64 - 30 = 30 + c - 30$$
$$34 = c$$

ANSWER The solution of the equation is 34. However, the sum of the lengths of any two sides of a triangle must be greater than the length of the third side, and $16 + 14 < 34$. The problem has no solution.

Look Back What if the given information were that the lengths of the two sides are 18 ft and 14 ft? Does the problem have a solution?

Guided Practice

COMMUNICATION « *Reading*

Explain why each answer has no meaning in the given context.

«**1.** *Context:* How many sweaters were bought? *Answer:* $1\frac{3}{4}$

«**2.** *Context:* What are the measures of the three angles of a triangle? *Answer:* 35°, 65°, and 90°

«**3.** *Context:* What is the perimeter of a pentagon? *Answer:* −60

«**4.** *Context:* How many people attended? *Answer:* 156.3

Explain why each problem has no solution.

5. A collection of seven coins consists of dimes and quarters and has a total value of $1.20. How many of each type of coin are there?

6. The perimeter of a rectangular auditorium is 306 ft. If the length is 68 yd, what is the width?

«7. COMMUNICATION «Discussion   Describe how the *problems with no solution* in this lesson are different from the *problems with not enough information* that you studied in Lesson 5-6.

Problem Solving Situations

Explain why each problem has no solution.

1. Books at the sale cost $.75 each, including tax. Janeen said she spent $8 at the sale. How many books did she buy?

2. The measure of each of two sides of a triangle is 16 in. The perimeter is measured to be 30 in. What is the measure of the third side?

3. Jake says he has $.60 in dimes and nickels. He has twelve coins altogether. How many of each type of coin does he have?

4. Admission to the school play was $3 for adults and $2 for students. The receipts were recorded as $1400. If 237 adults attended the play, how many students attended?

Solve using any problem solving strategy.

5. The cost of carpeting a rectangular room with width 12 ft and length 15 ft is $640. What is the cost of the carpeting per square yard?

6. The perimeter of a square is 32 cm. What is its area?

7. Roast beef costs $4.98 per pound. How many pounds can you buy for $12.45?

8. When you double a number and add five, the result is three less than twice the number. What is the number?

9. The perimeter of an isosceles triangle is 60 m. The measure of one side is 15 m. What are the measures of the other two sides?

10. The perimeter of a rectangle is 40 cm. What is the greatest possible area that this rectangle could have?

> **PROBLEM SOLVING**
> ## CHECKLIST
> **Keep these in mind:**
> Using a Four-Step Plan
> Too Much or Not Enough
>   Information
> Supplying Missing Facts
> Problems with No Solution
>
> **Consider these strategies:**
> Choosing the Correct Operation
> Making a Table
> Guess and Check
> Using Equations
> Identifying a Pattern
> Drawing a Diagram
> Using Proportions

For each exercise, use the collection of coins that is pictured at the right.

11. Write a word problem about the coins that has exactly one solution.

12. Write a word problem about the coins that has no solution.

SPIRAL REVIEW

13. Write a variable expression that represents the phrase *the sum of a number x and nine*.   *(Lesson 4-6)*

14. Explain why this problem has no solution: Four coins have a total value of $.48. What are the coins?   *(Lesson 10-6)*

15. Solve: $\frac{5}{8}a = 24$ *(Lesson 8-9)*

16. 18% of 70 is what number?   *(Lesson 9-6)*

Self-Test 1

Find the perimeter of each figure.

1. a rhombus with one side that measures 11.2 cm   **10-1**

2. a rectangle with sides that measure $6\frac{1}{2}$ in. and $9\frac{1}{4}$ in.

3. A circle has a radius of 7 m.   **10-2**
   a. Find the diameter.   **b.** Find the circumference.

Find the area of each figure.

4. a square with one side that measures $5\frac{1}{2}$ in.   **10-3**

5. a triangle with base 12 cm and height 6 cm

6. a circle with radius 8.4 cm   **10-4**

7. Tell whether you would need to calculate *perimeter*, *circumference*, or *area* to find the amount of fencing needed to enclose a circular circus tent.   **10-5**

8. Explain why this problem has no solution: A triangle has a perimeter of 27 in. Two of the sides measure 18 in. and 9 in. Find the length of the third side.   **10-6**

10-7

Congruent Polygons

Objective: To identify corresponding parts of congruent figures.

Terms to Know
- *congruent*
- *corresponding angles*
- *corresponding sides*

CONNECTION

In arithmetic, you learned that fractions and decimals that represent the same number are called *equivalent*. For instance, you learned that there are infinitely many ways to represent the number $\frac{1}{2}$.

$$\frac{1}{2} = \frac{2}{4} = \frac{3}{6} = \frac{4}{8} = \dots \quad \text{and} \quad \frac{1}{2} = 0.5 = 0.50 = 0.500 = \dots \ .$$

Similarly, in geometry there are infinitely many ways to picture a given size and shape. For example, although the four triangles below have different names and are positioned differently, they are identical in size and shape.

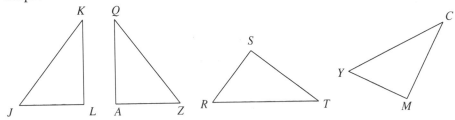

Geometric figures that have the same size and shape are called **congruent** figures. The symbol for congruent is $\cong$.

Line segments are congruent when they have the same length. In the figure at the right, for example, line segments *AB* and *CD* each measure $1\frac{1}{4}$ in., so they are congruent.

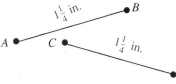

In Words	In Symbols
Line segment AB is congruent to line segment CD.	$\overline{AB} \cong \overline{CD}$

Angles are congruent when they have the same degree measure. In the figure at the right, angles *P* and *Q* are congruent because they each measure 127°.

In Words	In Symbols
Angle P is congruent to angle Q.	$\angle P \cong \angle Q$

Polygons are congruent when there is a way to match up their vertices so that all pairs of **corresponding angles** and all pairs of **corresponding sides** are congruent. In the triangles at the right, the red markings indicate that these corresponding parts are congruent.

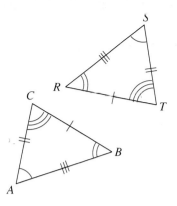

$$\angle A \cong \angle S \qquad \overline{AB} \cong \overline{SR}$$
$$\angle B \cong \angle R \qquad \overline{BC} \cong \overline{RT}$$
$$\angle C \cong \angle T \qquad \overline{AC} \cong \overline{ST}$$

To state that the triangles are congruent, you list the vertices of each triangle in the same order as the corresponding angles.

In Words	In Symbols
Triangle ABC is congruent to triangle SRT.	$\triangle ABC \cong \triangle SRT$

Example

Use the figures at the right to complete each statement.

a. $\overline{SP} \cong \underline{\ ?\ }$

b. $\angle B \cong \underline{\ ?\ }$

c. quadrilateral $PQRS \cong$ quadrilateral $\underline{\ ?\ }$

d. If the length of $\overline{DB}$ is 15 cm, then the length of $\underline{\ ?\ }$ is 15 cm.

e. If $m\angle R = 98°$, then $m\ \underline{\ ?\ } = 98°$.

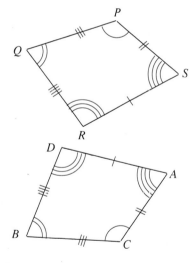

Solution

Use the red markings to match up the vertices as follows.

P and C	Q and B
R and D	S and A

a. $\overline{SP} \cong \overline{AC}$ **b.** $\angle B \cong \angle Q$

c. quadrilateral $PQRS \cong$ quadrilateral $CBDA$

d. $\overline{DB} \cong \overline{RQ}$, so the length of $\overline{RQ}$ is 15 cm.

e. $\angle R \cong \angle D$, so $m\angle D = 98°$.

✔️ Check Your Understanding

1. List all the pairs of corresponding angles and corresponding sides in the figure for the Example.

2. In part (c) of the Example, why is the answer not given as "quadrilateral $PQRS \cong$ quadrilateral $ABCD$"?

Guided Practice

COMMUNICATION « *Writing*

«**1.** Write the following statement in symbols: Angle *PQR* is congruent to angle *XYZ*.

«**2.** Write the following statement in words: $\triangle PQR \cong \triangle XYZ$

Use the figure at the right. Suppose you are given the information that $\triangle DEF \cong \triangle ABC$.

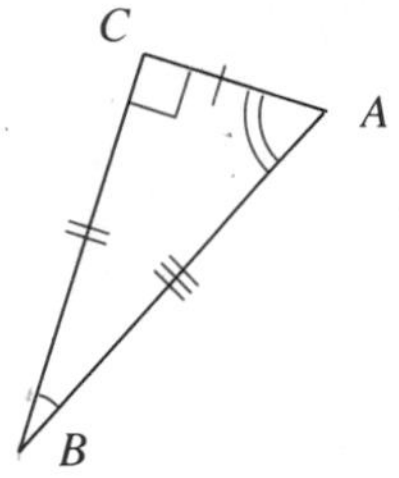

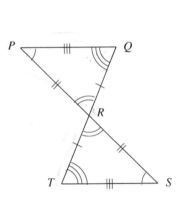

«**3.** Which side of $\triangle DEF$ would have three marks?

«**4.** Which angle of $\triangle DEF$ would have two marks?

«**5.** Which angle of $\triangle DEF$ would be a right angle?

«**6.** Make a sketch of $\triangle DEF$.

Use the figure at the right to complete each statement.

7. $\angle Q \cong$ __?__ **8.** $\overline{TR} \cong$ __?__

9. $\triangle PQR \cong$ __?__

10. If the length of $\overline{RQ}$ is 22 in., then the length of __?__ is 22 in.

11. If $m\angle S = 45°$, then m __?__ $= 45°$.

12. If $m\angle PRQ = 68°$, then m __?__ $= 68°$.

Exercises

Use the figure at the right to complete each statement.

1. $\angle F \cong$ __?__ **2.** $\angle CBE \cong$ __?__

3. $\overline{BC} \cong$ __?__ **4.** $\overline{FE} \cong$ __?__

5. quadrilateral *ABEF* $\cong$ quadrilateral __?__

6. If the length of $\overline{CD}$ is 4.6 m, then the length of __?__ is 4.6 m.

7. If $m\angle FEB = 70°$, then m __?__ $= 70°$.

8. Which line segment forms a side of both quadrilaterals?

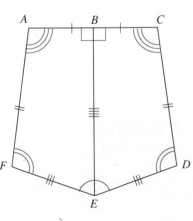

9. If △PQR ≅ △MNO, which side of △MNO is congruent to $\overline{RP}$?

10. Rectangle JKLM is congruent to rectangle WXYZ. List all the segments that are congruent to $\overline{MJ}$.

LOGICAL REASONING

Tell whether each statement is *true* or *false*. If the statement is false, give a counterexample to show why it is false.

11. Two rectangles are congruent if they have the same perimeter.

12. Two squares are congruent if they have the same perimeter.

13. If two circles have the same radius, then they are congruent.

14. If two circles have the same area, then they are congruent.

15. If two rectangles have the same area, then they are congruent.

16. If two rectangles are congruent, then they have the same area.

CONNECTING GEOMETRY AND ALGEBRA

17. Give a set of coordinates for point S so that quadrilateral ABCD ≅ quadrilateral PQRS.

18. Give two different sets of coordinates for point Z so that △MLN ≅ △XYZ.

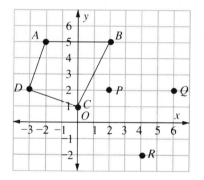

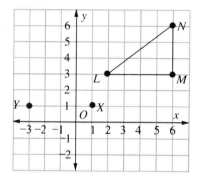

SPIRAL REVIEW

19. Simplify: $\frac{2a}{5} + \frac{3a}{4}$ *(Lesson 8-5)*

20. Solve: $x - 18 = -4$ *(Lesson 4-3)*

21. Solve: $\frac{x}{6} = \frac{20}{3}$ *(Lesson 9-2)*

22. Use the figures below to complete this statement: If the length of $\overline{WY}$ is 8.3 cm, then the length of __?__ is 8.3 cm.   *(Lesson 10-7)*

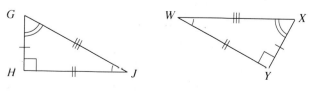

Similar Polygons

Objective: To recognize similar polygons and use corresponding sides to find unknown lengths.

Terms to Know
• *similar*

When you see a new type of car in an automobile showroom, do you realize that the plans for the car were actually begun several years ago? In creating these plans, engineers make a scale model of the car that has the same shape as the actual car but is much smaller in size. Scale models like these are examples of *similar* geometric figures.

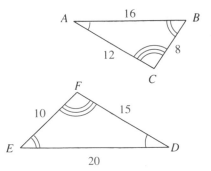

Geometric figures that have the same shape but not necessarily the same size are called **similar** figures. For instance, the circles shown at the right are similar. The symbol for similar is ~.

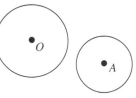

In Words	In Symbols
Circle O is similar to circle A.	$\odot O \sim \odot A$

Polygons are similar when there is a way to match up their vertices so that all pairs of corresponding angles are congruent and all pairs of corresponding sides are *in proportion*. In the triangles at the right, for example, the following pairs of angles are congruent.

$\angle A \cong \angle D \qquad \angle B \cong \angle E \qquad \angle C \cong \angle F$

Using the corresponding angles as a guide, you can identify the corresponding sides.

$$\overline{AB} \text{ and } \overline{DE} \qquad \overline{BC} \text{ and } \overline{EF} \qquad \overline{AC} \text{ and } \overline{DF}$$

Now you can check that each ratio between the lengths of corresponding sides is equal to the same number, $\frac{4}{5}$.

$$\frac{\overline{AB}}{\overline{DE}} \to \frac{16}{20} = \frac{4}{5} \qquad \frac{\overline{BC}}{\overline{EF}} \to \frac{8}{10} = \frac{4}{5} \qquad \frac{\overline{AC}}{\overline{DF}} \to \frac{12}{15} = \frac{4}{5}$$

The corresponding sides are in proportion, so the triangles are similar. You write $\triangle ABC \sim \triangle DEF$, being sure to list the vertices in the same order as the corresponding angles.

When you know that polygons are similar, you can use the fact that corresponding sides are in proportion to find unknown measures.

Example

In the figures at the right, quadrilateral $WXYZ \sim$ quadrilateral $QPRS$. Find the length of $\overline{PR}$.

Solution

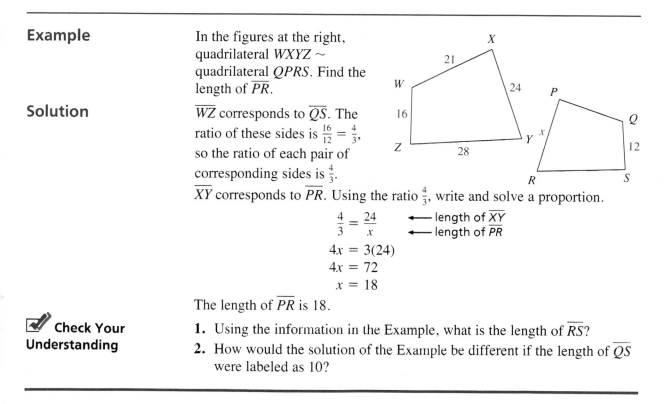

$\overline{WZ}$ corresponds to $\overline{QS}$. The ratio of these sides is $\frac{16}{12} = \frac{4}{3}$, so the ratio of each pair of corresponding sides is $\frac{4}{3}$.

$\overline{XY}$ corresponds to $\overline{PR}$. Using the ratio $\frac{4}{3}$, write and solve a proportion.

$$\frac{4}{3} = \frac{24}{x} \qquad \begin{array}{l} \longleftarrow \text{length of } \overline{XY} \\ \longleftarrow \text{length of } \overline{PR} \end{array}$$
$$4x = 3(24)$$
$$4x = 72$$
$$x = 18$$

The length of $\overline{PR}$ is 18.

Check Your Understanding

1. Using the information in the Example, what is the length of $\overline{RS}$?
2. How would the solution of the Example be different if the length of $\overline{QS}$ were labeled as 10?

Guided Practice

COMMUNICATION « *Reading*

Refer to the text on pages 462–463 and on pages 466–467. Copy and complete each outline in your notebook.

« **1.** *Congruent Geometric Figures*
   ___?___ size
   ___?___ shape

Congruent Polygons
corresponding angles are ___?___
corresponding sides are ___?___

Refer to the text on pages 462–463 and on pages 466–467. Copy and complete each outline in your notebook.

« **2.** *Similar Geometric Figures*
  __?__ size
  __?__ shape

Similar Polygons
corresponding angles are __?__
corresponding sides are __?__

Use the figures at the right to write each ratio.

3. $\dfrac{\text{length of } \overline{LM}}{\text{length of } \overline{PQ}}$

4. $\dfrac{\text{length of } \overline{MN}}{\text{length of } \overline{QR}}$

5. $\dfrac{\text{length of } \overline{NK}}{\text{length of } \overline{RS}}$

6. $\dfrac{\text{length of } \overline{KL}}{\text{length of } \overline{SP}}$

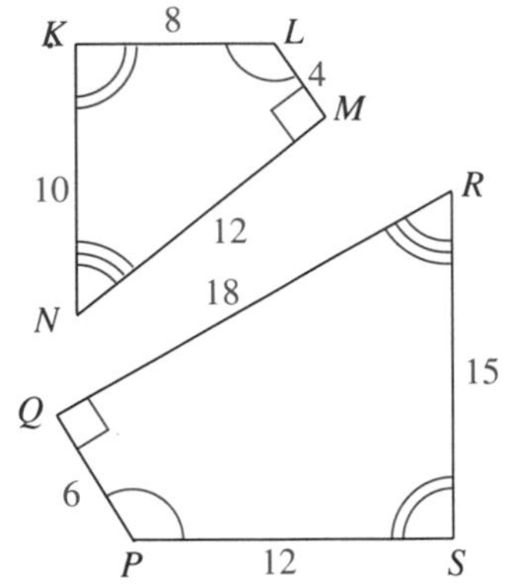

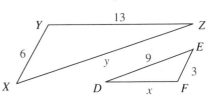

Use the figures at the right to complete.

7. Written in lowest terms, the ratio of each pair of corresponding sides is __?__.

8. quadrilateral *KLMN* ~ quadrilateral __?__

9. Explain why △*RST* is *not* similar to △*ABC*.

10. Given that △*XYZ* ~ △*EFD*, find the values of *x* and *y*.

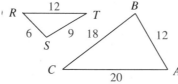

« **11.** COMMUNICATION « *Discussion* If two polygons are congruent, are they also similar? Explain.

Exercises

Find the values of *x* and *y*.

1. △*QRS* ~ △*PNM*

2. △*XYZ* ~ △*CBA*

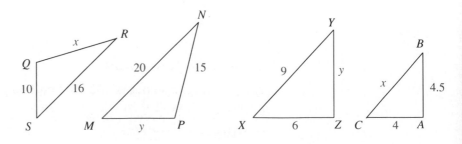

3. quad. *WXYZ* ~ quad. *JKGH*   **4.** quad. *DCBA* ~ quad. *PQRS*

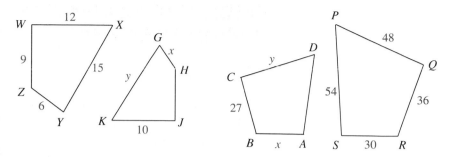

5. The measures of three sides of a triangle are 16 cm, 10 cm, and 12 cm. The shortest side of a similar triangle measures 15 cm. What is the perimeter of the similar triangle?

6. The length of a rectangle is 12 ft and its width is 8 ft. The width of a similar rectangle is 18 ft. What is the area of the similar rectangle?

7. Suppose that $\triangle PQR \cong \triangle MNO$, and the length of $\overline{PQ}$ is 10 cm. Can you determine the length of any side of $\triangle MNO$? Explain.

8. Suppose that $\triangle PQR \sim \triangle MNO$, and the length of $\overline{PQ}$ is 10 cm. Can you determine the length of any side of $\triangle MNO$? Explain.

Exercises 9–12 refer to the figures at the right. In the figures, $\triangle RST \sim \triangle LMN$.

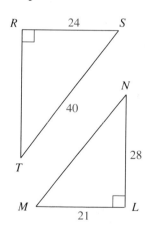

9. Find the length of $\overline{RT}$.

10. Find the length of $\overline{NM}$.

11. Write the ratio $\dfrac{\text{perimeter of } \triangle RST}{\text{perimeter of } \triangle LMN}$ in lowest terms. How is this ratio related to the ratio of the corresponding sides?

12. Suppose that $\triangle ABC \sim \triangle LMN$. You know that the length of $\overline{AB}$ is 4. What is the perimeter of $\triangle ABC$?

LOGICAL REASONING

Tell whether two figures of the given type are *always*, *sometimes*, or *never* similar.

13. two squares

14. two rectangles

15. two isosceles triangles

16. two equilateral triangles

17. two regular hexagons

18. two congruent polygons

19. a right triangle and an acute triangle

20. an obtuse triangle and an equilateral triangle

COMPUTER APPLICATION

The **midpoint** of a line segment is the point that separates it into two congruent segments. In the figure at the right, point D is the midpoint of $\overline{AB}$ and point E is the midpoint of $\overline{BC}$.

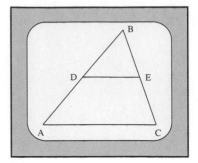

21. Draw the figure at the right, making $\triangle ABC$ any size that you choose. Use geometric drawing software if it is available.

22. Copy and complete the chart.

$\triangle ABC$	$\triangle DBE$
length of $\overline{AB} = $?	length of $\overline{DB} = $?
length of $\overline{BC} = $?	length of $\overline{BE} = $?
length of $\overline{AC} = $?	length of $\overline{DE} = $?
$m\angle BAC = $?	$m\angle BDE = $?
$m\angle B = $?	$m\angle B = $?
$m\angle BCA = $?	$m\angle BED = $?
perimeter $= $?	perimeter $= $?

23. Refer to the chart that you completed for Exercise 22. What is the relationship between $\triangle ABC$ and $\triangle DBE$?

24. **THINKING SKILLS** Repeat Exercises 21–23 for three different triangles. Compare the results. Make a generalization by completing this statement: When a line segment is drawn between the midpoints of two sides of a triangle, _____?_____.

SPIRAL REVIEW

25. Solve: $2.4x - 10 = 6.2$ *(Lesson 8-8)*

26. In the figures at the right, $\triangle YJT \sim \triangle ACB$. Find the values of x and y. *(Lesson 10-8)*

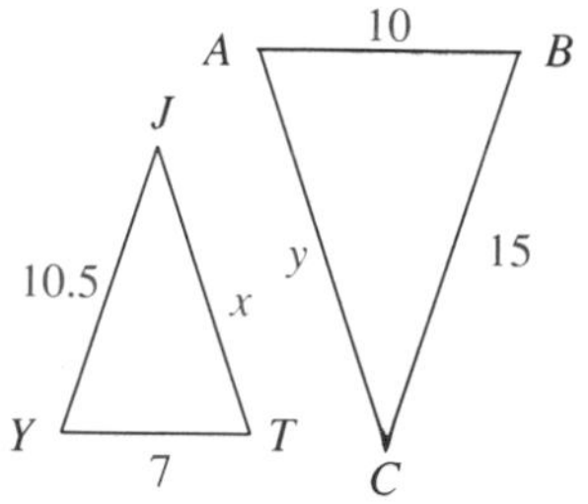

27. Find the area of a circle with radius 14 cm. *(Lesson 10-4)*

28. The scale of the floor plan of a theater is 1 in. : 12 ft. On the floor plan, the stage is $2\frac{3}{4}$ in. wide. Find the actual width of the stage. *(Lesson 9-3)*

29. Write 6.89×10^{-4} in decimal notation. *(Lesson 7-11)*

30. Tell whether line segments of 5 m, 5 m, and 8 m *can* or *cannot* be the sides of a triangle. If they can, tell whether the triangle would be *scalene*, *isosceles*, or *equilateral*. *(Lesson 6-6)*

Indirect Measurement

Objective: To use similar triangles to find an unknown measurement that cannot be measured directly.

Jed Carter was standing near a tree on a sunny day. Jed cast a shadow of 12 ft at the same time that the tree cast a shadow of 18 ft. Jed is 6 ft tall.

QUESTION How tall is the tree?

Clearly it would be difficult to find the height of the tree by making a *direct measurement* with a ruler or tape measure. In cases like this, it often is possible to use similar triangles to make an **indirect measurement.**

Terms to Know
- *indirect measurement*

Example 1

In the figure below, the triangles are similar. Find the unknown height *h*.

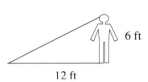
6 ft
12 ft

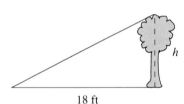
h
18 ft

Solution

Write a proportion involving the corresponding sides of the triangles.

$\text{length of Jed's shadow} \longrightarrow$ $\quad \dfrac{12}{18} = \dfrac{6}{h} \quad$ $\longleftarrow \text{Jed's height}$
$\text{length of tree's shadow} \longrightarrow$ $\qquad\qquad\qquad \longleftarrow \text{tree's height}$

Write $\dfrac{12}{18}$ as $\dfrac{2}{3}$. $\qquad \dfrac{2}{3} = \dfrac{6}{h}$

Solve the proportion $\qquad 2h = 3(6)$
using cross products. $\qquad 2h = 18$
$\qquad\qquad\qquad\qquad h = 9$

The unknown height is 9 ft.

✓ Check Your Understanding

1. Using the solution of Example 1, write the ratio of Jed's height to the tree's height in lowest terms. How does it compare to the ratio of the length of Jed's shadow to the length of the tree's shadow?

2. In Example 1, if the length of Jed's shadow were 18 ft, what would be the length of the shadow cast by the tree at the same time?

ANSWER The tree is 9 ft tall.

Example 2

In the figure below, $\triangle ABE \sim \triangle ACD$. Find the unknown height h.

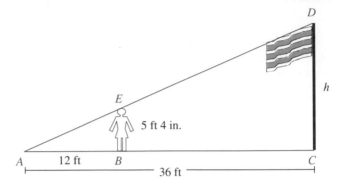

Solution

Write a proportion involving the corresponding sides of the triangles.

$\dfrac{\text{length of person's shadow} \longrightarrow}{\text{length of flagpole's shadow} \longrightarrow} \quad \dfrac{12}{36} = \dfrac{5\ \text{ft}\ 4\ \text{in.}}{h} \quad \begin{array}{l} \longleftarrow \text{person's height} \\ \longleftarrow \text{flagpole's height} \end{array}$

Write 5 ft 4 in. as $5\frac{1}{3}$ ft.

$\dfrac{12}{36} = \dfrac{5\frac{1}{3}}{h}$

Write $\frac{12}{36}$ as $\frac{1}{3}$.

$\dfrac{1}{3} = \dfrac{5\frac{1}{3}}{h}$

Solve the proportion using cross products.

$h = 3\left(5\frac{1}{3}\right)$

$h = \left(\dfrac{3}{1}\right)\left(\dfrac{16}{3}\right) = 16$

The unknown height is 16 ft.

Check Your Understanding

3. How would the solution of Example 2 be different if you wrote 5 ft 4 in. as 64 in. and solved the proportion using this number?

4. How would the solution of Example 2 be different if the person's height were 5 ft 6 in.?

Guided Practice

COMMUNICATION « *Reading*

Refer to the text on pages 471–472.

« **1.** What is meant by *direct measurement*?

« **2.** Describe the process of indirect measurement that is presented in this lesson.

« **3.** How is the figure for Example 2 different from the figure for Example 1?

« **4.** How are the measurements involved in Example 2 different from the measurements involved in Example 1?

COMMUNICATION « *Writing*

«**5.** Draw a diagram to represent this situation: A fence post that is 2 m tall casts a shadow that is 6 m long. At the same time, a nearby tree that is 6 m tall casts a shadow 18 m long.

«**6.** Write a description of the situation that is pictured in the figures at the right.

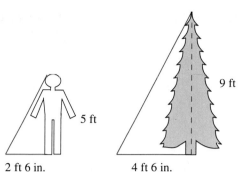

5 ft
9 ft
2 ft 6 in.
4 ft 6 in.

Exercises 7–10 refer to the figure at the right. In this figure, the triangles are similar.

7. Write the ratio of the length of the person's shadow to the length of the building's shadow in lowest terms.

8. Write 5 ft 3 in. as a number of feet.

9. Write a proportion involving the corresponding sides of the triangles.

10. Find the unknown height h.

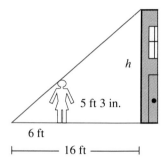

h
5 ft 3 in.
6 ft
16 ft

Exercises

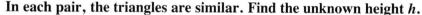

In each pair, the triangles are similar. Find the unknown height h.

1.

h
5 m
12 m
4 m

2.

5 ft 10 in.
h
5 ft
12 ft

3.

STOP
h
6 ft 3 in.
10 ft
16 ft

4.

h
1.8 m
3 m
10 m

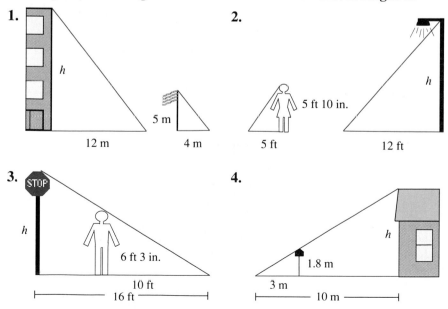

5. A student who is 5 ft 6 in. tall casts a shadow that is 15 ft long. At the same time a nearby tree casts a shadow that is 45 ft long. What is the height of the tree?

6. A man who is 1.6 m tall casts a shadow that is 50 cm long. At the same time a nearby television tower casts a shadow that is 8 m long. How tall is the television tower?

PROBLEM SOLVING/APPLICATION

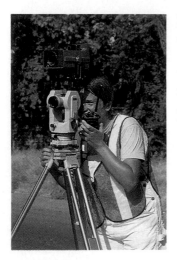

The figure at the right illustrates a method of indirect measurement that is used by surveyors. In this case, the unknown distance is the distance across Chosen Lake, represented by the length of line segment AB. The surveyors create similar triangles by marking off $\overline{DE}$ parallel to $\overline{AB}$.

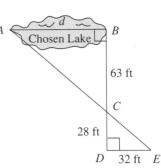

7. Why is it true that $m\angle ACB = m\angle ECD$?

8. Which two angles are equal in measure as a result of the fact that $\overline{DE}$ is parallel to $\overline{AB}$? Explain.

9. Complete this statement: $\triangle ABC \sim \triangle \underline{\ ?\ }$

10. Write a proportion involving corresponding sides of the triangles.

11. What is the distance across Chosen Lake?

12. The figure below shows how surveyors laid out similar triangles along the banks of the Sage River. Use these triangles to calculate the distance across the river.

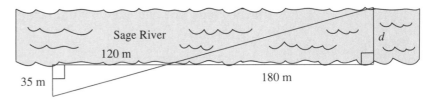

SPIRAL REVIEW

13. Find the perimeter of a rectangle with length 16 cm and width 7 cm. *(Lesson 10-1)*

14. In the figure at the right, the triangles are similar. Find h. *(Lesson 10-9)*

15. The sum of three times a number and four is 52. Find the number. *(Lesson 4-8)*

16. Solve by drawing a diagram: Sara hiked 4 mi south, 2 mi east, 8 mi north, and 2 mi west. Where was she then in relation to her starting point? *(Lesson 7-8)*

Square Roots

Objective: To find the square root of a number using a table or a calculator.

EXPLORATION

Terms to Know

• *square root*
• *perfect square*

1 Evaluate x^2 when x has the given value.
   a. 10   **b.** -10 **c.** 15   **d.** -15

2 If $x^2 = 144$ and x is a positive number, what is the value of x?

3 If $x^2 = 81$ and x is a negative number, what is the value of x?

4 If $x^2 = 0$, what is the value of x?

5 Do you think there is any real number x for which $x^2 = -100$? Give a convincing argument to support your answer.

If $a^2 = b$, the number a is called a **square root** of b. For instance, 7 is a square root of 49 because $7^2 = 7 \cdot 7 = 49$. Notice that -7 is also a square root of 49 because $(-7)^2 = (-7)(-7) = 49$. The symbol $\sqrt{}$ is used to indicate the positive square root.

In Words	**In Symbols**
The positive square root of 49 is 7.	$\sqrt{49} = 7$
The negative square root of 49 is -7.	$-\sqrt{49} = -7$

In the real number system, the square root of a negative number does not exist. This is true because there is no real number a for which a^2 is a negative number. Therefore, an expression like $\sqrt{-49}$ has no meaning in the real number system.

Example 1

Find each square root.
a. $-\sqrt{16}$ **b.** $\sqrt{2500}$ **c.** $\sqrt{0}$ **d.** $\sqrt{\dfrac{16}{49}}$ **e.** $-\sqrt{0.16}$

Solution

a. $4^2 = (4)(4) = 16$, so $-\sqrt{16} = -4$.

b. $50^2 = 50 \cdot 50 = 2500$, so $\sqrt{2500} = 50$.

c. $0^2 = 0$, so $\sqrt{0} = 0$.

d. $\left(\dfrac{4}{7}\right)^2 = \dfrac{4}{7} \cdot \dfrac{4}{7} = \dfrac{16}{49}$, so $\sqrt{\dfrac{16}{49}} = \dfrac{4}{7}$.

e. $(0.4)^2 = (0.4)(0.4) = 0.16$, so $-\sqrt{0.16} = -0.4$.

✔️ **Check Your Understanding**

1. How would the answer to part (e) of Example 1 be different if you were asked to find $\sqrt{-0.16}$?

When $\sqrt{n}$ is an integer, the number n is called a **perfect square.** You could never list all the perfect squares, of course, but you can indicate the set of perfect squares by showing this pattern.

$$\{0^2, 1^2, 2^2, 3^2, 4^2, 5^2, \ldots\} = \{0, 1, 4, 9, 16, 25, \ldots\}$$

If an integer is not a perfect square, then its square root is an irrational number. Therefore, numbers like $\sqrt{2}$, $\sqrt{3}$, $\sqrt{5}$, and $\sqrt{6}$ are irrational. Because an irrational number cannot be expressed as a quotient of two integers, there is no common fraction or decimal that indicates the exact value of these square roots. However, you can *approximate* these numbers either by using a table of squares and square roots like the one on page 480 or by using a calculator.

Example 2	Approximate $\sqrt{46}$ to the nearest thousandth.
Solution 1	Use the table on page 480. Find 46 in the *Number* column. Read the number across from 46 in the *Square Root* column. $\sqrt{46} \approx 6.782$
Solution 2	Use a calculator. Enter $\boxed{46.}$ $\boxed{\sqrt{x}}$. The calculator may show $\boxed{6.7823299}$ or $\boxed{6.782329983}$, depending on the number of digits in the display. To the nearest thousandth, either number rounds to 6.782, and $\sqrt{46} \approx 6.782$.

Check Your Understanding

2. In Example 2, why is it incorrect to write $\sqrt{46} = 6.782$?
3. Use a calculator to find $(6.782)^2$. Explain the result.

Notice that, in the order of operations, the square root symbol is a grouping symbol.

$$\sqrt{9 + 16} = \sqrt{25} = 5$$

Guided Practice

COMMUNICATION « *Writing*

Write each phrase or sentence in symbols.

« **1.** the positive square root of one and sixty-nine hundredths

« **2.** the square of negative nine

« **3.** The square of sixty is three thousand, six hundred.

« **4.** The negative square root of four ninths is negative two thirds.

Write each expression or statement in words.

« **5.** $-\sqrt{196}$ « **6.** $\left(\frac{1}{16}\right)^2$ « **7.** $(0.2)^2 = 0.04$ « **8.** $-\sqrt{1} = -1$

Tell whether each expression represents a *rational number*, represents an *irrational number*, or is *not a real number*.

9. $\sqrt{81}$ **10.** $\sqrt{75}$ **11.** $\sqrt{\dfrac{9}{49}}$ **12.** $-\sqrt{0.04}$

13. $\sqrt{-36}$ **14.** $-\sqrt{36}$ **15.** $\sqrt{1000}$ **16.** $\sqrt{10,000}$

Find each square root.

17. $-\sqrt{64}$ **18.** $\sqrt{121}$ **19.** $\sqrt{\dfrac{4}{25}}$

20. $\sqrt{0.09}$ **21.** $-\sqrt{1.44}$ **22.** $-\sqrt{\dfrac{1}{36}}$

Use the table on page 480 or use a calculator. Approximate each square root to the nearest thousandth.

23. $\sqrt{116}$ **24.** $-\sqrt{8}$ **25.** $\sqrt{40}$

26. $-\sqrt{45}$ **27.** $\sqrt{132}$ **28.** $-\sqrt{72}$

«**29.** COMMUNICATION «*Discussion* Explain the difference between the expressions $\sqrt{25} + \sqrt{144}$ and $\sqrt{25 + 144}$.

Exercises

Find each square root.

1. $\sqrt{25}$ **2.** $-\sqrt{4}$ **3.** $\sqrt{1}$ **4.** $-\sqrt{81}$ **5.** $-\sqrt{900}$

6. $\sqrt{12,100}$ **7.** $\sqrt{\dfrac{9}{64}}$ **8.** $-\sqrt{\dfrac{4}{81}}$ **9.** $-\sqrt{0.49}$ **10.** $\sqrt{6.25}$

Use the table on page 480 or use a calculator. Approximate each square root to the nearest thousandth.

11. $-\sqrt{89}$ **12.** $\sqrt{50}$ **13.** $\sqrt{140}$ **14.** $-\sqrt{131}$

Find the exact square root if possible. Otherwise, approximate the square root to the nearest thousandth.

15. $\sqrt{400}$ **16.** $\sqrt{40}$ **17.** $\sqrt{0.04}$ **18.** $\sqrt{\dfrac{1}{25}}$

19. $-\sqrt{33}$ **20.** $\sqrt{225}$ **21.** $\sqrt{1.44}$ **22.** $-\sqrt{84}$

Find each answer. If necessary, round to the nearest thousandth.

23. $\sqrt{36} + \sqrt{64}$ **24.** $\sqrt{36 + 64}$ **25.** $\sqrt{24 + 25}$

26. $\sqrt{24} + \sqrt{25}$ **27.** $\sqrt{9} \cdot \sqrt{16}$ **28.** $\sqrt{9 \cdot 16}$

29. Name three perfect squares between 200 and 300.

30. Using the table of squares and square roots on page 480, find a number between 5000 and 5100 that is a perfect square.

CONNECTING ALGEBRA AND GEOMETRY

The symbol $\sqrt{}$ is read *the square root of* because evaluating $\sqrt{x}$ gives you the length of the side of a square with area x.

31. Suppose that the area of a square is 196 m². What is its perimeter?

32. The area of a square is 45 ft². Approximate its perimeter to the nearest tenth of a foot.

Area = 25 square units
Side = $\sqrt{25}$ units
  = 5 units

MENTAL MATH

Most people agree that it is a good idea to memorize these simple square roots in order to do calculations more quickly and accurately.

$\sqrt{1} = 1$	$\sqrt{16} = 4$	$\sqrt{49} = 7$	$\sqrt{100} = 10$
$\sqrt{4} = 2$	$\sqrt{25} = 5$	$\sqrt{64} = 8$	$\sqrt{121} = 11$
$\sqrt{9} = 3$	$\sqrt{36} = 6$	$\sqrt{81} = 9$	$\sqrt{144} = 12$

Find each answer mentally.

33. $5 \cdot \sqrt{121}$

34. $29 - \sqrt{49}$

35. $(0.1)(\sqrt{36})$

36. $\sqrt{100} + 87$

37. $\frac{1}{2}(\sqrt{25})$

38. $\frac{\sqrt{16}}{8}$

39. $(\sqrt{81} + 1)(\sqrt{144} - \sqrt{4})$

40. $\sqrt{1} + \sqrt{64} - \sqrt{9}$

ESTIMATION

When a square root is irrational, you have learned to approximate its value using a table or calculator. However, to check your answers for reasonableness, you also should be able to *estimate* the value by locating it between consecutive integers. Here is an example.

> 76 is between 64 and 81, two perfect squares.
>
> $\sqrt{76}$ is between $\sqrt{64}$ and $\sqrt{81}$.
>
> $\sqrt{76}$ is between 8 and 9.

Between which two consecutive integers does each square root lie?

41. $\sqrt{38}$

42. $\sqrt{130}$

43. $-\sqrt{19}$

44. $-\sqrt{105}$

Choose the best estimates *without using a table or a calculator*.

45. $\sqrt{18}$ **a.** about 3.8   **b.** about 4.2   **c.** about 4.9

46. $\sqrt{56}$ **a.** about 7.5   **b.** about 7.9   **c.** about 8.3

47. $\sqrt{8.1}$ **a.** about 0.9   **b.** about 2.8   **c.** about 9.0

48. $\sqrt{0.26}$ **a.** about 0.51   **b.** about 1.3   **c.** about 5.1

Some artists and architects have created designs based on the *golden rectangle*. In a **golden rectangle,** the ratio $\frac{\text{length}}{\text{width}}$ equals $\frac{1 + \sqrt{5}}{2}$. People seem to find the rectangle's proportions pleasing to the eye. The Triumphal Arch of Constantine in Rome approximates two golden rectangles.

49. Approximate $\frac{1 + \sqrt{5}}{2}$ to the nearest thousandth.

50. Use your answer to Exercise 49. If the width of a rectangle is 1 m, about how long must it be in order to be a golden rectangle?

51. Which of these rectangles is most nearly a golden rectangle?

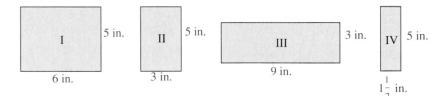

52. DATA ANALYSIS Make models of the rectangles in Exercise 51. Then ask each of 30 people to choose the rectangle with the shape that seems most pleasant to look at. Display the results in a frequency table. Did most people choose the golden rectangle?

SPIRAL REVIEW

53. Find the GCF of 96 and 16.   *(Lesson 7-2)*

54. Find $\sqrt{0.0081}$.   *(Lesson 10-10)*

> **Challenge**
>
> The perimeter of a rectangle is 18 cm. What is its greatest possible area?
> The area of a rectangle is 36 cm². What is its least possible perimeter?

Table of Squares and Square Roots

NO.	SQUARE	SQUARE ROOT	NO.	SQUARE	SQUARE ROOT	NO.	SQUARE	SQUARE ROOT
1	1	1.000	51	2,601	7.141	101	10,201	10.050
2	4	1.414	52	2,704	7.211	102	10,404	10.100
3	9	1.732	53	2,809	7.280	103	10,609	10.149
4	16	2.000	54	2,916	7.348	104	10,816	10.198
5	25	2.236	55	3,025	7.416	105	11,025	10.247
6	36	2.449	56	3,136	7.483	106	11,236	10.296
7	49	2.646	57	3,249	7.550	107	11,449	10.344
8	64	2.828	58	3,364	7.616	108	11,664	10.392
9	81	3.000	59	3,481	7.681	109	11,881	10.440
10	100	3.162	60	3,600	7.746	110	12,100	10.488
11	121	3.317	61	3,721	7.810	111	12,321	10.536
12	144	3.464	62	3,844	7.874	112	12,544	10.583
13	169	3.606	63	3,969	7.937	113	12,769	10.630
14	196	3.742	64	4,096	8.000	114	12,996	10.677
15	225	3.873	65	4,225	8.062	115	13,225	10.724
16	256	4.000	66	4,356	8.124	116	13,456	10.770
17	289	4.123	67	4,489	8.185	117	13,689	10.817
18	324	4.243	68	4,624	8.246	118	13,924	10.863
19	361	4.359	69	4,761	8.307	119	14,161	10.909
20	400	4.472	70	4,900	8.367	120	14,400	10.954
21	441	4.583	71	5,041	8.426	121	14,641	11.000
22	484	4.690	72	5,184	8.485	122	14,884	11.045
23	529	4.796	73	5,329	8.544	123	15,129	11.091
24	576	4.899	74	5,476	8.602	124	15,376	11.136
25	625	5.000	75	5,625	8.660	125	15,625	11.180
26	676	5.099	76	5,776	8.718	126	15,876	11.225
27	729	5.196	77	5,929	8.775	127	16,129	11.269
28	784	5.292	78	6,084	8.832	128	16,384	11.314
29	841	5.385	79	6,241	8.888	129	16,641	11.358
30	900	5.477	80	6,400	8.944	130	16,900	11.402
31	961	5.568	81	6,561	9.000	131	17,161	11.446
32	1,024	5.657	82	6,724	9.055	132	17,424	11.489
33	1,089	5.745	83	6,889	9.110	133	17,689	11.533
34	1,156	5.831	84	7,056	9.165	134	17,956	11.576
35	1,225	5.916	85	7,225	9.220	135	18,225	11.619
36	1,296	6.000	86	7,396	9.274	136	18,496	11.662
37	1,369	6.083	87	7,569	9.327	137	18,769	11.705
38	1,444	6.164	88	7,744	9.381	138	19,044	11.747
39	1,521	6.245	89	7,921	9.434	139	19,321	11.790
40	1,600	6.325	90	8,100	9.487	140	19,600	11.832
41	1,681	6.403	91	8,281	9.539	141	19,881	11.874
42	1,764	6.481	92	8,464	9.592	142	20,164	11.916
43	1,849	6.557	93	8,649	9.644	143	20,449	11.958
44	1,936	6.633	94	8,836	9.695	144	20,736	12.000
45	2,025	6.708	95	9,025	9.747	145	21,025	12.042
46	2,116	6.782	96	9,216	9.798	146	21,316	12.083
47	2,209	6.856	97	9,409	9.849	147	21,609	12.124
48	2,304	6.928	98	9,604	9.899	148	21,904	12.166
49	2,401	7.000	99	9,801	9.950	149	22,201	12.207
50	2,500	7.071	100	10,000	10.000	150	22,500	12.247

The Pythagorean Theorem

Objective: To use the Pythagorean Theorem to find unknown lengths.

Terms to Know
- *hypotenuse*
- *legs*
- *Pythagorean Theorem*

 **DATA ANALYSIS**

In a right triangle, the side opposite the right angle is always the longest side. It is called the **hypotenuse.** The two shorter sides are called the **legs.** In ancient times, people gathered data about the hypotenuse and legs of many right triangles and began to see a pattern forming. The table below lists these data for several right triangles. Can you see the pattern?

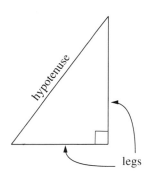

Some Common Measures of Right Triangles

Lengths of Legs	3	5	6	7	8	9
	4	12	8	24	15	12
Length of Hypotenuse	5	13	10	25	17	15

A good way to find the pattern is to work with the simplest of the triangles listed, the "3-4-5" right triangle, and observe what happens when you build a square on each of the legs and on the hypotenuse.

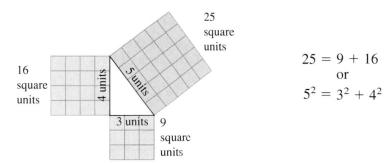

$$25 = 9 + 16$$
or
$$5^2 = 3^2 + 4^2$$

As you can see, the area of the square on the hypotenuse is equal to the sum of the areas of the squares on the two legs. More than 2500 years ago, the Greek mathematician Pythagoras proved that this relationship applies to all right triangles, not just the 3-4-5 right triangle. In his honor, this property is called the **Pythagorean Theorem.**

> ### The Pythagorean Theorem
> If the length of the hypotenuse of a right triangle is c and the lengths of the legs are a and b, then the following relationship holds true.
> $$c^2 = a^2 + b^2$$

If you know the lengths of two sides of a right triangle, you can use the Pythagorean theorem to find the unknown length of the third side.

Example 1

Find the unknown length. If necessary, round to the nearest tenth.

a.

5 cm

c

8 cm

b.

28 ft

a

35 ft

Solution

a. $c^2 = a^2 + b^2$
$c^2 = 5^2 + 8^2$
$c^2 = 25 + 64$
$c^2 = 89$
$c = \sqrt{89}$
$c \approx 9.434$

The length of the hypotenuse is about 9.4 cm.

b. $c^2 = a^2 + b^2$
$35^2 = a^2 + 28^2$
$1225 = a^2 + 784$
$441 = a^2$
$\sqrt{441} = a$
$21 = a$

The length of the leg is 21 ft.

✔️ **Check Your Understanding**

1. Why were the squares of the lengths added in part (a) of Example 1, but subtracted in part (b)?

It also has been proved that the *converse* of the Pythagorean theorem is a true statement. You obtain the converse of a statement by interchanging the "if" and "then" parts of the statement.

Converse of the Pythagorean Theorem

If the sides of a triangle have lengths a, b, and c such that $c^2 = a^2 + b^2$, then the triangle is a right triangle.

Example 2

Is a triangle with sides of the given lengths a right triangle?
a. 12 ft, 16 ft, 20 ft
b. 9 m, 15 m, 13 m

Solution

a. $c^2 = a^2 + b^2$
$20^2 \overset{?}{=} 12^2 + 16^2$
$400 \overset{?}{=} 144 + 256$
$400 = 400$

Yes, it is a right triangle.

b. $c^2 = a^2 + b^2$
$15^2 \overset{?}{=} 9^2 + 13^2$
$225 \overset{?}{=} 81 + 169$
$225 \neq 250$

No, it is not a right triangle.

✔️ **Check Your Understanding**

2. In Example 2, how do you know which length to substitute for c in the equation $c^2 = a^2 + b^2$?

Guided Practice

COMMUNICATION « *Reading*

« **1.** What is the main idea of the lesson?

« **2.** Explain the difference between the Pythagorean theorem and the *converse* of the Pythagorean theorem.

Replace each __?__ with = or ≠ to make a true statement.

3. $7^2 + 3^2$ __?__ 10^2

4. $5^2 + 12^2$ __?__ 13^2

5. $6^2 + 8^2$ __?__ 10^2

6. $4^2 + 8^2$ __?__ 9^2

Find the unknown length. If necessary, round to the nearest tenth. Use the table on page 480 or a calculator as needed.

7. c, 36 in., 15 in.

8. b, 6 m, 12 m

Is a triangle with sides of the given lengths a right triangle? Write *Yes* or *No*.

9. 5 in., 8 in., 11 in.

10. 6 cm, 10 cm, 8 cm

Exercises

Find the unknown length. If necessary, round to the nearest tenth. Use the table on page 480 or a calculator as needed.

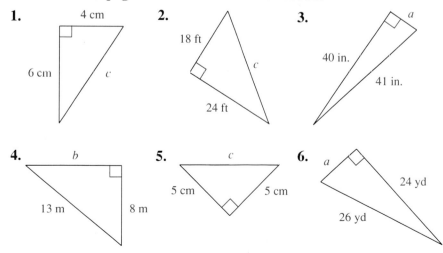

1. 4 cm, 6 cm, c

2. 18 ft, 24 ft, c

3. a, 40 in., 41 in.

4. b, 13 m, 8 m

5. c, 5 cm, 5 cm

6. a, 24 yd, 26 yd

Is a triangle with sides of the given lengths a right triangle? Write *Yes* or *No*.

7. 3 yd, 4 yd, 5 yd

8. 8 mm, 10 mm, 12 mm

9. 7 m, 24 m, 25 m

10. 9 ft, 12 ft, 14 ft

11. 13 cm, 5 cm, 12 cm

12. 15 in., 25 in., 20 in.

PROBLEM SOLVING/APPLICATION

Often the Pythagorean Theorem is used as a method of indirect measurement. In the following figures, use the Pythagorean Theorem to find the unknown distance. If necessary, round to the nearest tenth.

13.

support wire — l, 5 yd, 9 yd

14.

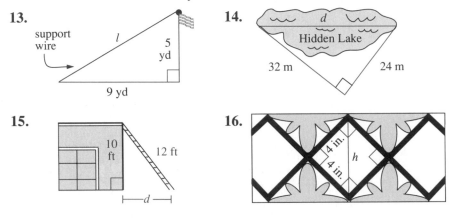

d

Hidden Lake

32 m   24 m

15.

10 ft   12 ft d

16.

4 in.   4 in. h

Solve. If necessary, round the answer to the nearest tenth.

17. Martine Lancois left her house and walked 2 km due east and then 6 km due north. At this point, how far was Martine from her house?

18. A ladder that is 6 m long leans against a wall. The bottom of the ladder rests 2 m from the base of the wall. How far up the wall does the ladder reach?

19. A public garden is to be shaped like a rectangle with length 80 yd and width 60 yd. What is the length of a diagonal walkway across this garden?

20. A square gate is reinforced by a wooden brace across its diagonal. The perimeter of the gate is 12 ft. What is the length of the brace?

Any three positive integers a, b, and c such that $c^2 = a^2 + b^2$ are said to form a **Pythagorean triple.** For instance, all the integers in the chart on page 481 form Pythagorean triples. To *generate* triples like these, you substitute positive integers x and y, $x > y$, into these formulas.

$$a = x^2 - y^2 \qquad b = 2xy \qquad c = x^2 + y^2$$

21. Find the triple that is generated by the values $x = 2$, $y = 1$.

22. Find the values of x and y that generate the triple 8–15–17.

23. Make an organized list of all the Pythagorean triples generated by values of x from 2 through 6.

24. COMPUTER APPLICATION Write a BASIC program that generates a Pythagorean triple when values are input for x and y.

SPIRAL REVIEW

25. Solve: $\frac{1}{4} + n = -\frac{5}{8}$ *(Lesson 8-8)*

26. Find the unknown length in the figure at the right.   *(Lesson 10-11)*

27. Graph the inequality $x \le 3.5$ *(Lesson 7-10)*

28. Find the area of a square with side 7 in. *(Lesson 10-3)*

(figure: right triangle with legs 8 cm and 14 cm, hypotenuse b)

Self-Test 2

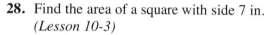

Use the congruent quadrilaterals in the Example on page 463 to complete each statement.

1. $\overline{QR} \cong$ ___?___   **2.** $\angle A \cong$ ___?___   10-7

3. Use the similar quadrilaterals in the Example on page 467 to find the length of $\overline{PQ}$.   10-8

4. A person 6 ft tall casts a shadow 15 ft long. At the same time, a building casts a shadow 50 ft long. How tall is the building?   10-9

Find each square root. Use the table on page 480 if necessary.

5. $\sqrt{8100}$ **6.** $\sqrt{\dfrac{25}{64}}$ **7.** $\sqrt{125}$ 10-10

8. Is a triangle with sides of 9 m, 12 m, and 13 m a right triangle? Write *Yes* or *No*.   10-11

9. Find the unknown length in the triangle at the right. If necessary, round to the nearest tenth.

(figure: right triangle ABC with right angle at B, leg $AB = 4$, leg $BC = 7$)

Chapter Review

perimeter (p. 436)
circle (p. 440)
center (p. 440)
radius (p. 440)
chord (p. 440)
diameter (p. 440)
circumference (p. 441)
area (p. 445)
base (p. 446)
height (p. 446)

congruent (p. 462)
corresponding angles (p. 463)
corresponding sides (p. 463)
similar (p. 466)
indirect measurement (p. 471)
square root (p. 475)
perfect square (p. 476)
hypotenuse (p. 481)
legs (p. 481)
Pythagorean Theorem (p. 481)

Choose the correct term from the list above to complete each sentence.

1. The distance around a circle is called its __?__.

2. The __?__ is the longest side of a right triangle.

3. Two figures that have the same shape and size are called __?__.

4. A __?__ of a circle is a line segment whose endpoints are the center and a point on the circle.

5. A __?__ has a square root that is an integer.

6. The region enclosed by a plane figure is called its __?__.

Find the perimeter of each figure.   *(Lesson 10-1)*

7. a regular hexagon with one side that measures 6.2 cm

8. a parallelogram with sides that measure 7 ft and $10\frac{1}{4}$ ft

Use the circle at the right.   *(Lesson 10-2)*

9. Name the circle.

10. Identify as many radii, chords, and diameters as shown.

11. Find the radius.

12. Find the diameter.

13. Find the circumference.

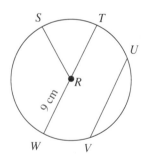

Find the area of each figure.   *(Lessons 10-3 and 10-4)*

14. a square with length of one side equal to $5\frac{1}{2}$ ft

15. a rectangle with length 4.8 m and width 2.6 m

16. a triangle with base 12 in. and height 9 in.

17. a circle with radius 16 cm

Tell whether you would need to calculate *perimeter*, *circumference*, **or** *area* **to find each measure.** *(Lesson 10-5)*

18. the amount of asphalt needed to tar a square playground

19. the fence around the playground

Explain why each problem has no solution. *(Lesson 10-6)*

20. A collection of five coins has a value of $1 and consists of dimes and quarters. How many of each type of coin are there?

21. The perimeter of a rectangle is 24 cm. The length is 12 cm. Find the width.

Use the figures at the right to complete each statement. *(Lesson 10-7)*

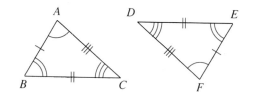

22. $\overline{AB} \cong$ __?__ **23.** $\angle C \cong$ __?__

24. If the length of $\overline{BC}$ is 8 m, then the length of __?__ is 8 m.

25. If $m\angle B = 65°$, then m __?__ $= 65°$.

In the figures at the right
ABCDE ~ UVRST.
(Lesson 10-8)

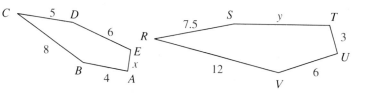

26. Find x.   **27.** Find y.

Solve. *(Lesson 10-9)*

28. In the figures at the right, the triangles are similar. Find the unknown height h.

Find the exact square root if possible. Otherwise, approximate each square root to the nearest thousandth. *(Lesson 10-10)*

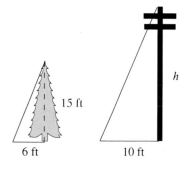

15 ft

6 ft   10 ft

29. $\sqrt{900}$ **30.** $\sqrt{\dfrac{100}{121}}$

31. $\sqrt{76}$ **32.** $\sqrt{142}$

Use the Pythagorean theorem. *(Lesson 10-11)*

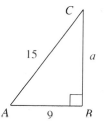

33. Is a triangle with sides of 10 ft, 24 ft, and 26 ft a right triangle? Write *Yes* or *No*.

34. Find the length of the third side of the triangle at the right.

1. Find the perimeter of a square with one side that measures $7\frac{1}{4}$ in.   **10-1**

2. Find the circumference of a circle with radius 21 in.   **10-2**

3. Find the area of a triangle with base 1 yd and height 4 ft.   **10-3**

4. What is the greatest possible area for a rectangle with a perimeter of 16 ft?

5. Find the area of a circle with radius 2.3 m.   **10-4**

6. When the radius of a circle is tripled, what happens to the circumference and area of the circle?

7. Tell whether you would need to calculate *perimeter*, *circumference*, or *area* to find the amount of sod needed to cover a soccer field.   **10-5**

8. Explain why this problem has no solution: La Toya has $1.10 in nickels and dimes. She has ten coins altogether. How many of each type of coin does she have?   **10-6**

Use the figures at the right to complete each statement.

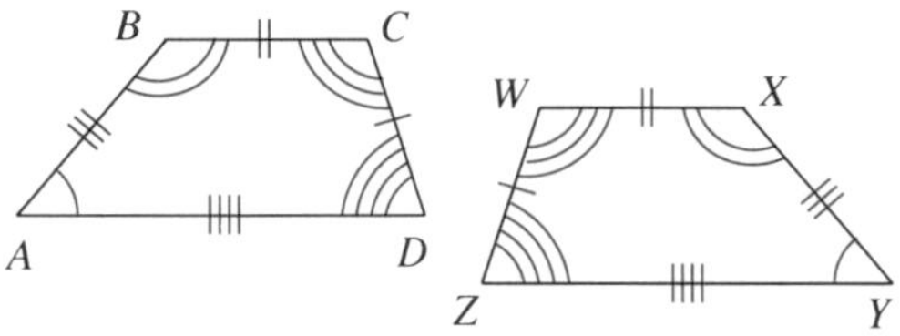

9. $\overline{AB} \cong$ ___?___   **10-7**

10. $\angle C \cong$ ___?___

11. In the figures at the right $\triangle ABC \sim \triangle ZYX$. Find the length of $\overline{YZ}$.   **10-8**

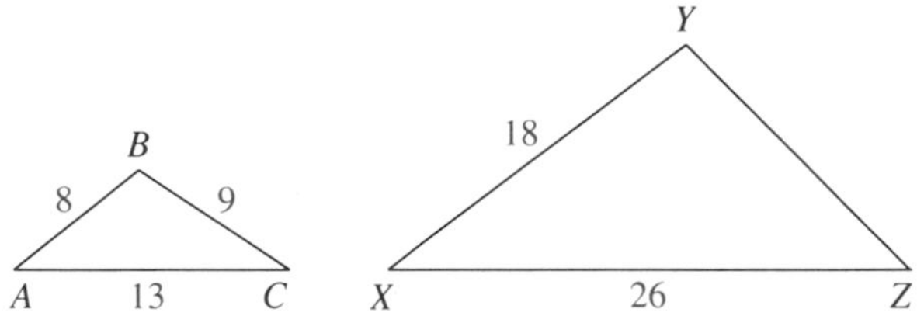

12. In the figure at the right, the triangles are similar. Find the unknown height h.   **10-9**

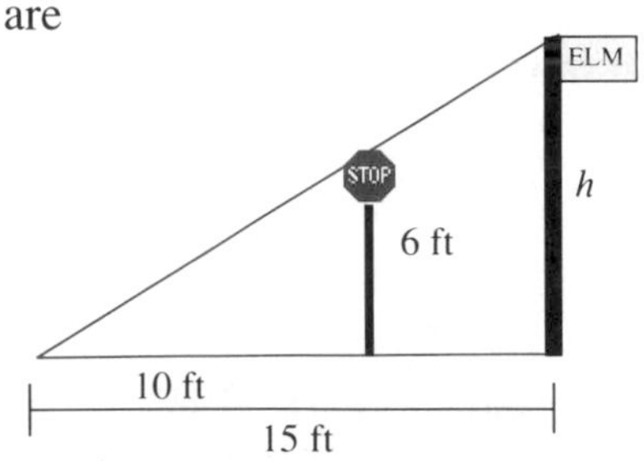

Find each square root if possible. Otherwise, approximate the square root to the nearest thousandth.

13. $\sqrt{3600}$ 14. $\sqrt{\dfrac{9}{49}}$ 15. $\sqrt{91}$ 16. $\sqrt{147}$ **10-10**

17. Is a triangle with sides of 16 ft, 30 ft, and 34 ft a right triangle? Write *Yes* or *No*.   **10-11**

Simplifying Square Roots

Objective: To simplify square roots using the product property of square roots.

You probably know that $\sqrt{144} = 12$ because you remember that $12 \cdot 12 = 144$. However, you could also find $\sqrt{144}$ as follows.

$$\sqrt{144} = \sqrt{4 \cdot 36} = \sqrt{4} \cdot \sqrt{36} = 2 \cdot 6 = 12$$

This method illustrates the following property of square roots.

The Product Property of Square Roots

The square root of the product of two nonnegative real numbers is equal to the product of their square roots.

$$\sqrt{ab} = \sqrt{a} \cdot \sqrt{b} \qquad a \geq 0, b \geq 0$$

You can use the product property to simplify square roots.

Example

Solution

Simplify $\sqrt{80}$.

First write 80 as the product of two factors, one of which is the greatest perfect square possible: $80 = 16 \cdot 5$

Then use the product property of square roots.

$$\sqrt{80} = \sqrt{16 \cdot 5} = \sqrt{16} \cdot \sqrt{5} = 4\sqrt{5}$$

Exercises

Simplify each square root.

1. $\sqrt{12}$ 2. $\sqrt{45}$ 3. $\sqrt{50}$ 4. $\sqrt{24}$

5. $\sqrt{8}$ 6. $\sqrt{27}$ 7. $\sqrt{40}$ 8. $\sqrt{63}$

9. $\sqrt{28}$ 10. $\sqrt{44}$ 11. $\sqrt{32}$ 12. $\sqrt{48}$

13. $\sqrt{124}$ 14. $\sqrt{125}$ 15. $\sqrt{60}$ 16. $\sqrt{200}$

17. The Example above showed that $\sqrt{80} = 4\sqrt{5}$.
    a. Approximate $\sqrt{80}$ to the nearest thousandth.
    b. Approximate $\sqrt{5}$ to the nearest thousandth, then multiply by 4.
    c. Compare your answers to (a) and (b). What do you conclude?

18. **THINKING SKILLS** Do you think there is a *Quotient Property of Square Roots*? Formulate a possible statement of this property, similar to the statement of the *Product Property* given above. Then create an example to show how this property might be used.

Standardized Testing Practice

Choose the letter of the correct answer.

1.

Find the coordinates of point Q.

A. $(-2, -1)$ B. $(-2, 1)$
C. $(1, -2)$ D. $(-1, 2)$

2. Complete: 35% of __?__ is 63.

A. 180   B. 55.6
C. 63   D. 1.8

3. Write the phrase *the sum of three times a number t and seven* **as a variable expression.**

A. $3t - 7$ B. $3t + 7$
C. $t^3 + 7$ D. $3(t + 7)$

4. The lengths of the legs of a right triangle are 9 in. and 12 in. Find the hypotenuse.

A. 54 in.   B. 15 in.
C. 21 in.   D. 7.9 in.

5. Find the next expression in the pattern: $x + 2, 2x + 4, 3x + 6,$ __?__

A. $4x + 7$ B. $3x + 8$
C. $4x + 8$ D. $4x + 12$

6.

Identify a diameter of this circle.

A. $\overline{TR}$ B. $\overline{PQ}$
C. $\overline{SQ}$ D. $\overline{TQ}$

7. Add: $-36 + (-17)$

A. -53 B. 53
C. 19   D. -19

8. Write 475% as a fraction or mixed number in lowest terms.

A. $\frac{475}{100}$ B. $47\frac{1}{2}$
C. $\frac{19}{40}$ D. $4\frac{3}{4}$

9. Evaluate $|t + 3| - s$ **when** $t = -8$ **and** $s = 4$.

A. 15   B. -9
C. 7   D. 1

10. Simplify: $\sqrt{4900}$

A. 60   B. 2450
C. 70   D. 700

11.

6 cm

10 cm

Find the area of this triangle.

A. 9.5 cm^2 B. 30 cm^2
C. 60 cm^2 D. 19 cm^2

12. Simplify: $(7x^2)(2x)$

A. $14x^2$ B. $14x^3$
C. 7   D. 1

13. Which type of quadrilateral has exactly two lines of symmetry?

A. rhombus
B. square
C. trapezoid
D. none of the above

14. The price of a book is marked down from $18.95 to $15.16. Find the percent of decrease.

A. 25%   B. 20%
C. 37.9%   D. 80%

15. Use the formula $m = \dfrac{a+b}{2}$ to find m when $a = 17$ and $b = 27$.

A. 5   B. 22
C. 27   D. 44

16.

Complete: $\angle JKG \cong \underline{\ ?\ }$

A. $\angle LKG$ B. $\angle LGK$
C. $\angle KGM$ D. $\angle JKL$

17. Solve: $15 + 2q = 9$

A. $q = \dfrac{9}{17}$ B. $q = -3$

C. $q = 12$ D. $q = -12$

18. Which of the following is equivalent to 4%?

I. $\dfrac{4}{100}$ II. 0.04   III. $\dfrac{1}{25}$

A. I only
B. I and II only
C. I, II, and III
D. II and III only

19. A strip of molding is $\frac{4}{5}$ the length of a room. The room is $12\frac{1}{2}$ ft long. Find the length of the molding.

A. $13\dfrac{3}{10}$ ft   B. 10 ft

C. $11\dfrac{7}{10}$ ft   D. $15\dfrac{5}{8}$ ft

20. Which fraction is greater than $-\frac{3}{4}$?

A. $-\dfrac{15}{20}$ B. $-\dfrac{7}{8}$

C. $-\dfrac{13}{14}$ D. $-\dfrac{1}{2}$

Statistics and Circle Graphs 11

Kinds of Wood

Type of Wood	Color of Wood	Typical Use
white ash	light brown	baseball bats
lodgepole pine	light yellow	poles
incense cedar	reddish brown	pencils
white oak	grayish brown	furniture
silver maple	reddish brown	construction
paper birch	light brown	toys
white basswood	yellowish brown	wooden utensils

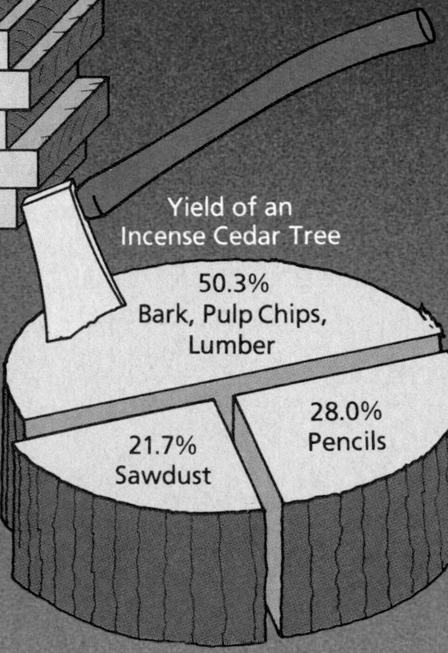

Did You Know?

Most people think of the sassafras tree in connection with the tea made from its bark. But its wood is also extremely useful. Early Native Americans prized the wood for its resistance to shock and decay and used it for making dugout canoes. Today, sassafras wood is still used for making small boats.

Yield of an Incense Cedar Tree

50.3%
Bark, Pulp Chips, Lumber

21.7%
Sawdust

28.0%
Pencils

493

Frequency Tables with Intervals

Objective: To organize data in a frequency table with intervals.

CONNECTION

In Chapter 5 you learned that statistics is the branch of mathematics that deals with collecting, organizing, and analyzing data. You also learned that data can be organized in a frequency table. When data are scattered, a frequency table that has the data grouped in equal intervals may be easier to interpret.

Bowling Scores

156	131	124	115	143
158	158	131	121	131
143	152	130	137	143
124	137	131	121	130
152	158	131	124	143

Example

Solution

Make a frequency table for the data above. Use intervals such as 111–120.

- Make a table with three columns.
- List the intervals in the first column.
- Make a tally mark in the second column next to each interval for every score that falls within that interval.
- Record the total number of tally marks for each interval in the third column.

Bowling Scores

Score	Tally	Frequency
111–120	I	1
121–130	⊬⊬ II	7
131–140	⊬⊬ II	7
141–150	IIII	4
151–160	⊬⊬ I	6

✓ Check Your Understanding

What would be the frequency for the last interval in the Example if that interval were 155–160?

Guided Practice

«**1. COMMUNICATION** «*Reading*   Replace each __?__ with the correct word or phrase.

Data can be organized in a __?__ table. A large set of data can be condensed by grouping the data into equal __?__. The number of times that an item occurs is called the __?__ of that item.

2. The first interval in a frequency table is 50–54. What will be the next four intervals?

3. Make a frequency table for the data. Use intervals such as 61–70.

Scores on a Mathematics Test

65	70	70	85	69	85	100	81	70	83	81	90	82	87
96	68	94	100	95	81		83	100	66	81	83	96	84	85

« **4.** COMMUNICATION «*Discussion* Can you tell what the individual data items are in a frequency table with intervals? without intervals, as you saw in lesson 5-10? Explain.

Exercises

Make a frequency table for the data.

1. **Compact Disc Prices (Dollars)**

7.99	7.99	10.99	12.99	10.99	12.99	14.99	15.49	14.99	15.49
14.99	14.49	12.49	12.99	12.99	11.99	11.99	13.99	13.99	13.49

(Use intervals such as 7.00–8.99.)

2. **Ages of Hospital Nurses**

23	25	23	29	23	34	31	47	55	57	56	57	45
37	24	41	59	60	61	45	45	52	39	50	39	

(Use intervals such as 20–29.)

3. **Annual Tuition at Private Colleges (Dollars)**

7800	12000	9550	14000	13750	7500	6400
10900	11200	9600	13400	13200	7000	6900
12300	11200	8500	7400	11500	12200	7600

(Use intervals such as 6000–7999.)

4. **Total Points Scored in Basketball Games**

79	89	75	71	80	83	85	88
93	90	86	74	96	98	105	91
107	82	110	91	88	95	78	86

(Use intervals such as 70–79.)

5. Make a frequency table for the data. Choose reasonable intervals.

Cost of a One Week Vacation (Dollars)

890	945	1200	1050	1450	1300
1200	1100	950	1425	1680	1650
1450	1500	1050	1750	1695	1200
1250	1370	980	1575	1625	1300

6. DATA, *pages 186–187* Using intervals such as 1–5, make a frequency table with intervals for the data in the frequency table.

7. THINKING SKILLS
Examine the frequency table at the right. Do you notice anything wrong with this table? Determine what is wrong and explain.

Scores on a Biology Test

Scores	Tally	Frequency
60–70	IIII	4
70–80	HHT II	7
80–90	HHT	5
90–100	HHT HHT	10

SPIRAL REVIEW

8. The original price is $40 and the new price is $45. Find the percent of increase. *(Lesson 9-9)*

9. Write a variable expression that represents this phrase: three less than twice a number x *(Lesson 4-6)*

10. Find $\sqrt{324}$. *(Lesson 10-10)*

11. Make a frequency table for the data. Use intervals such as 20–24. *(Lesson 11-1)*

Number of Hours Worked per Week

20	26	20	35	29	26	40	20	40
29	40	37	24	20	29	25	24	35
20	20	29	37	37	40	37	37	40
37	24	35	24	26	37	20	37	29

Challenge

The exact *percent* frequency for each data item is shown in the table below. What is the least number of total data items that there could be?

Data Item	0	1	2	3	4
Frequency	6.25%	12.5%	12.5%	43.75%	25%

11-2

Stem-and-Leaf Plots

Objective: To make and interpret stem-and-leaf plots.

A frequency table with intervals will not show you the value of each piece of data. An alternative display in which no information is lost is the **stem-and-leaf plot.** In this type of display, each number is represented by a *stem* and a *leaf.* The **leaf** is the last digit on the right of the number. The **stem** is the digit or digits of the number remaining when the leaf is dropped.

Terms to Know

- *stem-and-leaf plot*
- *stem*
- *leaf*

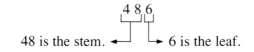

48 is the stem. ← ← → 6 is the leaf.

Example 1

Make a stem-and-leaf plot for the data.

Scores on a Science Test

78	74	80	84	99	65	81	80
62	95	100	91	75	82	76	92
80	75	100	82	77	70	79	85
90	95	86	82	75	69	83	97

Solution

Write the stems in order from least to greatest to the left of a vertical line. Write each leaf to the right of its stem.

6	5 2 9	← This row contains 65, 62, and 69.
7	8 4 5 6 5 7 0 9 5	
8	0 4 1 0 2 0 2 5 6 2 3	
9	9 5 1 2 0 5 7	
10	0 0	

↑
Stems   Leaves

Rearrange the leaves for each stem in order from least to greatest. Title the stem-and-leaf plot.

Scores on a Science Test

6	2 5 9
7	0 4 5 5 5 6 7 8 9
8	0 0 0 1 2 2 2 3 4 5 6
9	0 1 2 5 5 7 9
10	0 0

✓ Check Your Understanding

1. How would Example 1 be different if a score of 96 were included?

Example 2

Use the stem-and-leaf plot at the right.

a. What is the best free throw average?

b. How many teams have a free throw average of 0.735?

c. Find the median of the free throw averages.

Team Free Throw Averages

0.69	2　4　4
0.70	0　1
0.71	
0.72	5　5
0.73	1　5　5　8　9
0.74	
0.75	2　4　5

Solution

a. The greatest stem is 0.75. The greatest leaf for 0.75 is 5. The best free throw average is 0.755.

b. The stem is 0.73. The number 5 appears twice beside this stem. Two teams have a free throw average of 0.735.

c. The order of the 15 leaves reflects the order of the averages. The middle leaf is 1. The stem corresponding to this leaf is 0.73. The median of the free throw averages is 0.731.

✓ **Check Your Understanding**

2. Describe how Example 2(b) would be different if you were asked to find the number of teams with a free throw average greater than 0.735.

3. Describe how Example 2(c) would be different if you were asked to find the range of the free throw averages.

Guided Practice

COMMUNICATION «*Reading*

List the data that are represented by each stem-and-leaf plot.

«**1.**
3.4	0　5　5　8
3.5	6　8
3.6	0　0　1　5

«**2.**
23	0　8
24	6
25	5　5　8　9

Identify the stem and the leaf of each number.

3. 19　　　　　　**4.** 2.78　　　　　　**5.** 20.57

6. 1228　　　　　**7.** 0.253　　　　　**8.** 400

«**9.** COMMUNICATION «*Discussion* Refer to the stem-and-leaf plot in Example 2. Does the stem-and-leaf plot include a free throw average of 0.710? of 0.740? Explain.

10. Make a stem-and-leaf plot for the data.

Prices of Compact Disc Players (Dollars)

190	206	210	228	230	194
209	215	229	236	195	215
238	195	210			

Use the stem-and-leaf plot at the right.

11. Find the least capacity.

12. How many theaters have a capacity of 220?

13. How many theaters have a capacity less than 200?

14. Find the mean, median, mode(s), and range of the data.

Capacities of Movie Theaters

16	5 9
17	5 8
18	0 0
19	0 8
20	0 0 5
21	5 8 9
22	
23	4 6 8 8

Exercises

Make a stem-and-leaf plot for the data.

1. **Earned Run Averages of Leading Pitchers**

2.56	2.80	2.78
2.92	2.62	2.20
2.67	2.90	2.90
2.32	2.25	2.82
2.77	2.84	2.82
2.62		

2. **Ages of Company Presidents**

45	58	60	62
48	50	42	60
56	58	55	48
38	55	47	
39	50	65	
39	35	44	

Use the stem-and-leaf plot at the right.

3. What is the least cost?

4. What is the greatest cost?

5. How many customers pay $41 per month?

Monthly Cost of Cable (Dollars)

2	5 5 6 8 9 9
3	0 0 2 5 5 6 7 8
4	0 1 1 1 2 5 5
5	0 2 6

6. How many customers pay $27 per month?

7. How many customers pay more than $36 per month?

8. How many customers pay less than $44 per month?

9. How many customers pay from $35 through $50 per month?

10. Find the mean, median, mode(s), and range of the data.

CONNECTING STATISTICS AND SOCIAL STUDIES

11. Make a stem-and-leaf plot for the data below.

Age at First Inauguration of United States Presidents

57	61	57	57	58	57	61	54	68	51	49	64
50	48	65	52	56	46	54	49	50	47	55	54
42	51	56	55	51	54	51	60	62	43	55	56
61	52	69	64								

Use the stem-and-leaf plot that you made for Exercise 11 for Exercises 12–15.

12. How many presidents were 56 years old at their inauguration?

13. How many presidents were less than 55 years old at their inauguration?

14. How many presidents were 55–64 years old at their inauguration?

15. How many presidents were 40–49 years old at their inauguration?

SPIRAL REVIEW

16. Simplify: $\frac{7x}{4} + \frac{9x}{4}$ *(Lesson 8-4)*

17. PATTERNS Find the next three expressions in the pattern:
$2n + 4, 3n + 4, 4n + 4,$ __?__, __?__, __?__ *(Lesson 2-5)*

18. Solve: $4(n + 5) = -24$ *(Lesson 4-9)*

19. Use the stem-and-leaf plot on page 499 entitled *Capacities of Movie Theaters*. How many theaters have a capacity greater than 200? *(Lesson 11-2)*

20. Solve using an equation: On Sunday, the number of visitors at a crafts fair was 43 less than twice the number of visitors on Saturday. There were 279 visitors on Sunday. How many visitors were there on Saturday? *(Lesson 4-8)*

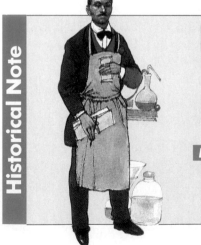

George Washington Carver (1864–1943), an African-American agricultural chemist, developed ways to diversify crops to prevent soil depletion. Through his research at Tuskegee Institute, he also synthesized hundreds of new products based on sweet potatoes, peanuts, pecans, soybeans, wood shavings, and cotton stalks. Statistics played a role in his research.

Research

Find out what some of the products were that George Washington Carver synthesized.

Historical Note

Histograms and Frequency Polygons

Objective: To interpret histograms and frequency polygons.

Terms to Know
- *histogram*
- *frequency polygon*

CONNECTION

Data displayed in a frequency table can also be displayed in a *histogram*. A **histogram** is a bar graph like the one shown at the right that is used to show frequencies. In this special type of bar graph, there are no spaces between consecutive bars.

Example 1

Use the histogram above.

a. How many employees work 36–40 h per week?

b. Between which two consecutive intervals does the greatest increase in frequency occur? What is the increase?

c. How many employees work fewer than 31 h per week?

Solution

a. There are 20 employees who work 36–40 h per week.

b. The greatest increase occurs between the intervals of 31–35 and 36–40. The frequencies for these intervals are 12 and 20. The increase is 20 − 12 = 8.

c. There are 12 employees who work 26–30 h, 8 employees who work 21–25 h, and 4 employees who work 16–20 h. Since 12 + 8 + 4 = 24, 24 people work fewer than 31 h.

✓ Check Your Understanding

How would Example 1(c) have been different if you were asked to find how many employees work 31 or more hours per week?

Frequencies can also be displayed in a line graph called a **frequency polygon.** One way to construct a frequency polygon is to connect the midpoints of the tops of the bars of a histogram, as shown at the right. You also connect the first point to the origin and the last point to the other end of the horizontal axis.

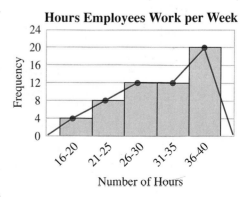

Example 2

Use the frequency polygon at the right.

a. How many students are 14 years old?

b. Exactly 3 students are a certain age. What is this age?

c. How many students are there in all?

Ages of Students

Solution

a. There are 5 students who are 14 years old.

b. The age for which there are exactly 3 students is 15.

c. Find the sum of all the frequencies: $2 + 5 + 3 + 4 + 6 = 20$
There are 20 students in all.

Guided Practice

COMMUNICATION «*Reading*

Refer to the text on pages 501–502.

« **1.** How are an ordinary bar graph and a histogram different?

« **2.** How are an ordinary line graph and a frequency polygon different?

« **3.** COMMUNICATION «*Writing* On one set of axes, draw a histogram and a frequency polygon for the data in the frequency table below.

Exam Scores

Score	Tally	Frequency
60–69	I	1
70–79	IIII	4
80–89	HHT II	7
90–99	HHT IIII	9

Use the histogram at the top of page 501.

4. How many employees work more than 20 h per week?

5. Which interval includes the number of hours worked per week by exactly 8 employees?

Use the frequency polygon at the top of this page.

6. How many students are more than 15 years old?

7. How many students are 13–16 years old?

Exercises

Use the histogram at the right.

1. How many students ran 9–16 laps?

2. How many students ran 9 laps or more?

3. How many students ran 16 laps or less?

4. How many students were there in all?

5. Did the frequency increase or decrease between the intervals 9–16 and 17–24?

6. What was the greatest decrease between consecutive intervals?

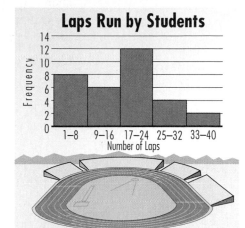

Laps Run by Students

Use the frequency polygon at the right.

7. How many band members are 64 in. tall?

8. What is the height of only 4 band members?

9. How many band members are more than 66 in. tall?

10. How many band members are less than 66 in. tall?

11. How many band members are 65–68 in. tall?

12. How many band members are there in all?

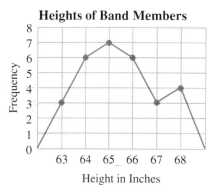

Heights of Band Members

PROBLEM SOLVING/APPLICATION

There are specific ranges of test scores associated with each letter grade. For example, a grade of B is sometimes given for any score in the interval 80–89. For this reason, histograms are often used to display test scores.

13. Make a histogram for the data below using intervals such as 60–69.

Scores on a Mathematics Test

87	66	74	74	88	97	89	78	68	84	75	85	77
69	93	65	73	76	83	85	79	80	96	84	88	

Use the histogram that you made for Exercise 13 for Exercises 14–17.

14. What letter grade corresponds to each interval in the histogram?

15. How many students received a grade of C on the test?

16. What percent of the students earned a grade of B or better on the test?

17. What was the mode of the letter grades on the test?

GROUP ACTIVITY

Use the data below for Exercises 18 and 19.

Hourly Wages of Part-time Baby Sitters (Dollars)

1.00	1.50	2.25	0.75	0.75	1.50	2.00	2.50	1.75	1.25	0.75
0.75	3.00	2.50	1.75	1.50	1.00	2.50	2.75	2.00	0.75	1.50

18. Make a frequency polygon without intervals for the data.

19. Make a frequency polygon for the data using intervals such as 0.75–1.24.

20. Which of the graphs that you drew for Exercises 18 and 19 displays the individual data items? Explain.

21. Which of the graphs that you drew for Exercises 18 and 19 gives a better picture of the overall trend in the hourly wages of part-time baby sitters? Explain.

SPIRAL REVIEW

22. Use the frequency polygon on page 503. How many band members are 63 in. tall? *(Lesson 11-3)*

23. Write $\frac{13}{20}$ as a decimal. *(Lesson 7-7)*

24. Solve: $\frac{3}{4}x = 18$ *(Lesson 8-9)*

25. What number is 4.5% of 2000? *(Lesson 9-6)*

26. Find the mean, median, mode(s), and range:
7.9, 7.3, 6.5, 8.4, 7.3, 6.4, 5.7, 6.9 *(Lesson 5-8)*

Box-and-Whisker Plots

Objective: To make and interpret box-and-whisker plots.

EXPLORATION

1 What is the median of the data below?

3   4   10   11   11   14   20   23   23   23   25   27

2 What is the median of the items in the above list that are less than the median in Step 1? greater than the median in Step 1?

3 Into how many sections do the medians you found in Steps 1 and 2 divide the data?

The median divides a set of data into two parts. The **first quartile** is the median of the lower part. The **third quartile** is the median of the upper part. You can display these measures of a set of data along with the least and greatest values of the data in a **box-and-whisker plot.**

Example 1

Make a box-and-whisker plot for the data.

Gasoline Mileage (Miles per Gallon)

24   20   18   25   22   31   30   20   28   29   35   24   38

Solution

Arrange the data in order from least to greatest. Find the median, first quartile, third quartile, least value, and greatest value.

18   20   20   22   24   24   25   28   29   30   31   35   38

| ↑ | | ↑ | | | | ↑ | | | ↑ | | | ↑ |

least value   first quartile   median   third quartile   greatest value

$$\frac{20 + 22}{2} = 21 \qquad\qquad \frac{30 + 31}{2} = 30.5$$

Display the five numbers as points below a number line.

```
15        20        25        30        35        40
├┼┼┼┼┼┼┼┼┼┼┼┼┼┼┼┼┼┼┼┼┼┼┼┼┼┤
     ●      ●       ●        ●              ●
```

Draw a box with ends at the quartiles. Draw a vertical line through the box at the median. Draw *whiskers* from the ends of the box to the points representing the least and greatest values. Title the graph.

Gasoline Mileage (Miles per Gallon)

```
15        20        25        30        35        40
├┼┼┼┼┼┼┼┼┼┼┼┼┼┼┼┼┼┼┼┼┼┼┼┼┼┤
     ●──┌────┬───────┐──────────●
        │    │       │
     ●──└────┴───────┘
```

Box-and-whisker plots are often used to compare sets of data.

Example 2

Use the box-and-whisker plot below.

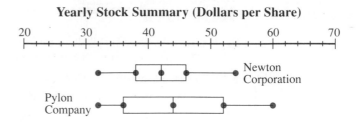

Yearly Stock Summary (Dollars per Share)

a. Find the median, the first quartile, the third quartile, and the lowest and highest price per share for Newton stock.

b. Which stock had the greater range in price during the year?

Solution

a. Locate the box-and-whisker for Newton Corporation.
   The vertical line through the box is at the median, $42.
   The left end of the box is at the first quartile, $38.
   The right end of the box is at the third quartile, $46.
   The lowest price per share is $32.
   The highest price per share is $54.

b. The overall length of the box-and-whisker plot is greater for the Pylon Company than for the Newton Corporation. Pylon Company stock had the greater range in price during the year.

Check Your Understanding

Describe a different way to find the answer to Example 2(b).

Guided Practice

« **1.** COMMUNICATION « *Reading* In statistics, the suffix *-ile* means "a division of a specified size."
   a. What is the meaning of the word *quartile*?
   b. What is the meaning of the word *percentile*?

« **2.** COMMUNICATION « *Discussion* Is the median counted when you find the first and third quartiles of a set of data? Explain.

Identify the median and the first and third quartiles for each set of data.

3. 12.4, 12.6, 12.8, 13.0, 13.4, 13.7, 14.0, 14.6, 14.8

4. 127, 130, 136, 138, 142, 150, 155, 159, 166, 168

5. Draw a box-and-whisker plot using the data in Exercise 3.

Use the box-and-whisker plot on page 506.

6. Which stock had the lesser median price per share for the year?

7. Which stock had the greatest price per share for the year?

Exercises

Make a box-and-whisker plot for each set of data.

1. **Box Seat Prices at Baseball Games (Dollars)**

   12.50   12.00   13.25   11.00   11.50   11.25   14.00   9.00
   11.75   13.50   14.50

2. **Prices for 13-inch Color Televisions (Dollars)**

   229   299   349   215   250   375   399   415   400   240
   269   329   345   379   411

Use the box-and-whisker plot below for Exercises 3–8.

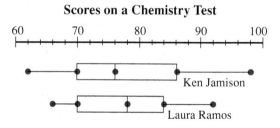

Scores on a Chemistry Test

3. Find the median, the first quartile, the third quartile, and the lowest and highest scores for Ken Jamison's class.

4. Find the median, the first quartile, the third quartile, and the lowest and highest scores for Laura Ramos's class.

5. Which teacher's class had the student with the greatest score?

6. Which teacher's class had the student with the least score?

7. Which teacher's class had the greater range of scores on the test?

8. Which teacher's class had the greater median score on the test?

LOGICAL REASONING

Tell whether each statement is *always*, *sometimes*, or *never* true.

9. A box-and-whisker plot displays at least two numbers from a set of data.

10. When the number of items in a set of data is odd, the median is one of the data items.

11. Exactly half of the items in a set of data are less than the median.

12. WRITING ABOUT MATHEMATICS The box-and-whisker plot below shows data about theater attendance. Use the box-and-whisker plot to write a paragraph describing the relationship between the attendance at plays and musicals.

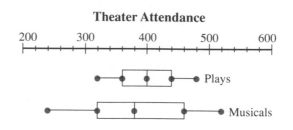

Theater Attendance

SPIRAL REVIEW

13. Graph the inequality $x \geq -3$. *(Lesson 7-10)*

14. Find the sum of the measures of the angles of a pentagon. *(Lesson 6-5)*

15. Make a box-and-whisker plot to display the following data. *(Lesson 11-4)*

Number of Field Goals Scored

12	20	19	15	24	26
27	23	21	18	14	29

16. Find the quotient: $3\frac{1}{3} \div \left(-\frac{2}{5}\right)$ *(Lesson 8-2)*

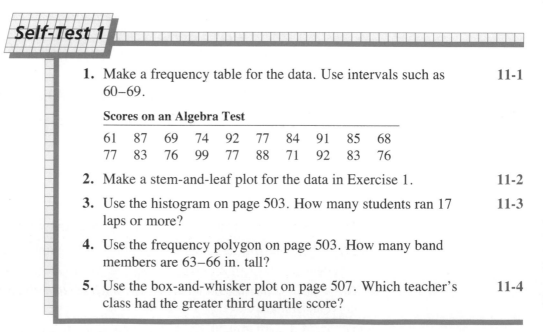

Self-Test 1

1. Make a frequency table for the data. Use intervals such as 60–69.   **11-1**

Scores on an Algebra Test

61	87	69	74	92	77	84	91	85	68
77	83	76	99	77	88	71	92	83	76

2. Make a stem-and-leaf plot for the data in Exercise 1.   **11-2**

3. Use the histogram on page 503. How many students ran 17 laps or more?   **11-3**

4. Use the frequency polygon on page 503. How many band members are 63–66 in. tall?

5. Use the box-and-whisker plot on page 507. Which teacher's class had the greater third quartile score?   **11-4**

11-5 Scattergrams

Objective: To read and interpret scattergrams.

DATA ANALYSIS

People who analyze data sometimes use a **scattergram** such as the one at the right to display the relationship between two sets of data. In a scattergram, the data are represented by points, but the points are not connected like those on a line graph. Furthermore, a scattergram may have more than one point for a given number on either scale.

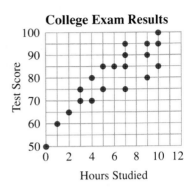

College Exam Results

Example 1

Use the scattergram above.

a. What was the test score of the student who studied for 6 h?

b. How many students studied for 7 h?

c. Find the mode(s) of the test scores.

Solution

a. Locate 6 on the horizontal axis. Move up to the red point, then left to the vertical axis. The student's score was 85.

b. Locate 7 on the horizontal axis. Move up and count the points. A total of 4 students studied for 7 h.

c. Locate the test score with the greatest number of points beside it. The mode of the test scores is 85.

✓ Check Your Understanding

1. In Example 1(b), where on the horizontal axis does the vertical line corresponding to *7 h studied* begin?

2. How would Example 1(c) have been different if you had been asked to find the range of the test scores?

The points on a scattergram usually do not lie on a line. However, on some scattergrams the points lie *near* a line. A line that can be drawn near the points on a scattergram is called the **trend line.** It can be used to make predictions about the data.

If the trend line slopes upward to the right, there is a **positive correlation** between the sets of data. If the trend line slopes downward to the right, there is a **negative correlation.**

Example 2

Use the scattergram at the right.

a. Is there a *positive* or *negative* correlation between the hours studied and the test scores?

b. Predict the test score for a student who studies for 8 h.

College Exam Results

Test Score vs *Hours Studied*

Solution

a. The trend line slopes upward to the right. As the number of hours studied increases, the test scores tend to increase. There is a positive correlation between the hours studied and the test scores.

b. Locate 8 on the horizontal axis. Move up to the trend line, then left to the vertical axis. The predicted score is about 85.

Guided Practice

COMMUNICATION « *Reading*

Refer to the text on pages 509–510.

« **1.** What is the main idea of the lesson?

« **2.** List two major points that support the main idea.

Identify each scattergram as having a *positive correlation*, a *negative correlation*, or *no correlation*.

3. 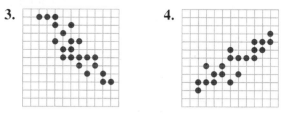 **4.** **5.**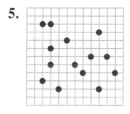

Use the scattergram in Example 2 above.

6. What was the test score of the student who studied for 2 h?

7. How many students had a score of 75?

8. Find the range of the number of hours studied.

9. Predict the test score for a student who studied for 12 h.

10. What was the test score of the student who studied for 1 h?

Exercises

Use the scattergram at the right.

1. How much does the woman who is 71 in. tall weigh?

2. How tall is the woman who weighs 140 lb?

3. How many women are 64 in. tall?

4. How many women weigh 155 lb?

5. Find the mode(s) of the heights.

6. Find the range of the weights.

7. Two women are 67 in. tall. How much does each woman weigh?

8. As the heights increase, do the weights tend to *increase* or *decrease*?

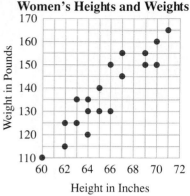

Women's Heights and Weights

Use the scattergram at the right.

9. Is there a *positive* or a *negative* correlation between the number of movies attended and a person's age?

10. Who will probably attend more movies in a year, a 15-year-old or a 25-year-old?

11. Predict the number of movies a 23-year-old will attend in a year.

12. Predict the age of a person who attends 18 movies per year.

13. Draw a scattergram to display the data.

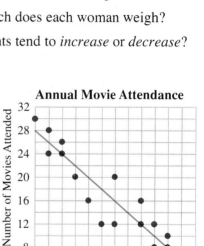

Annual Movie Attendance

Jump Shots Made in 25 Attempts from Various Distances

Distance from Basket (ft)	25	15	10	20	15	5	10	5	25
Jump Shots Made	6	10	15	7	9	18	12	20	5

14. Use the scattergram you made in Exercise 13. Tell whether there is a *positive correlation*, a *negative correlation*, or *no correlation* between the two sets of data.

DATA, *pages 492–493*

15. How many trees have branches that spread 32 ft?

16. What is the spread of the branches of the tree that is 60 ft tall?

Car Registrations and Passenger Train Cars over Ten Years

Cars (Millions)	117	119	122	123	124	126	128	132	135	137
Train Cars	4493	4241	4347	3945	3736	2610	2580	2502	2307	2350

17. Using statistical graphing software or graph paper, draw a scattergram displaying the data in the table above.

18. Draw the trend line for the scattergram you drew in Exercise 17.

19. Use the trend line that you drew in Exercise 18. Is there a *positive* or a *negative* correlation between the number of passenger car registrations and the number of passenger train cars?

20. Use the trend line that you drew in Exercise 18. If the number of passenger cars registered continues to increase over time, what will probably happen to the number of passenger train cars?

RESEARCH

21. Use an almanac to find the area and the average depth of the world's ten largest bodies of salt water besides the four oceans. Draw a scattergram showing the relationship between these two sets of data.

22. Use the scattergram that you drew in Exercise 21. Tell whether there is a *positive correlation*, a *negative correlation*, or *no correlation* between the two sets of data.

SPIRAL REVIEW

23. Find the mean, median, mode(s), and range of the data: 12, 14, 16, 18, 23, 26, 19, 10, 12, 28, 15, 12   *(Lesson 5-8)*

24. Make a stem-and-leaf plot for the data.   *(Lesson 11-2)*

Ages of Chorus Members

26	27	28	34	32	36	41	45	48	49	25	29
19	17	16	19	20							

25. Solve: $2x - 2.4 = 3.8$ *(Lesson 8-8)*

26. Use the scattergram on page 511 entitled *Annual Movie Attendance.* How many movies did the 19-year-old attend?   *(Lesson 11-5)*

27. What number is 5% of 720?   *(Lesson 9-6)*

28. Solve by making a table: Michael asks the attendant at Village Laundry to give him change for a quarter. In how many different ways could the attendant give Michael change using only dimes and nickels? *(Lesson 3-7)*

Interpreting Circle Graphs

Objective: To read and interpret circle graphs.

 **APPLICATION**

The athletic director at State University analyzed how the annual football budget was spent. The results were presented to the budget director in a *circle graph* like the one at the right.

A **circle graph** is used to represent data expressed as parts of a whole. The entire circle represents the whole, or 100%, of the data. Each wedge or **sector** represents a part of the data.

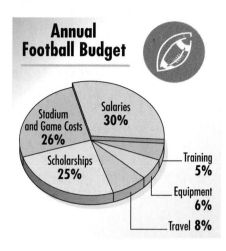

Annual Football Budget

Salaries 30%
Stadium and Game Costs 26%
Scholarships 25%
Training 5%
Equipment 6%
Travel 8%

Example 1

Use the circle graph above.

a. If the total budget was $3,500,000, how much was spent on scholarships?

b. If $1,500,000 was spent on salaries, how much was the total budget?

Solution

a. 25% of the budget was spent on scholarships.

Let n = the amount spent on scholarships.

n is 25% of $3,500,000. $\Rightarrow$ $n = 0.25(3,500,000)$
$= 875,000$

The amount spent on scholarships was $875,000.

b. 30% of the budget was spent on salaries.

Let n = the total budget.

30% of n is $1,500,000. $\Rightarrow$ $0.3n = 1,500,000$
$n = 5,000,000$

The total budget was $5,000,000.

✓ **Check Your Understanding**

1. In Example 1(b), what had to be done in order for the equation $0.3n = 1,500,000$ to be rewritten as $n = 5,000,000$?

The sectors in a circle graph are sometimes labeled with the data items themselves, rather than with percents.

Example 2

Use the circle graph at the right.

a. What was the total budget?

b. What percent of the budget was spent on utilities?

c. What percent of the budget was not spent on salaries?

Library Budget
(Thousands of Dollars)

Salaries **325**

New Purchases **90**

Computer Costs **25**

Utilities **37.5**

Maintenance **12.5**

Supplies **10**

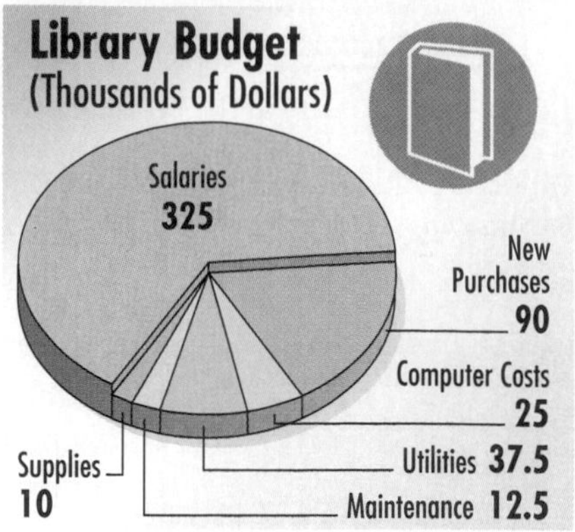

Solution

a. Add the amounts budgeted for all the expenses.
$325 + 10 + 12.5 + 37.5 + 25 + 90 = 500$
The total budget was $500,000.

b. The amount spent on utilities was $37,500. The total budget was $500,000. Let n = the percent of the budget spent on utilities.

n is $37,500 out of $500,000. $\Rightarrow$ $n = \dfrac{37,500}{500,000}$
$= 0.075$

7.5% of the budget was spent on utilities.

c. Subtract the amount spent on salaries from the total budget.
$500,000 - 325,000 = 175,000$ ◄── amount not spent on salaries
Let n = the percent of the budget not spent on salaries.

n is $175,000 out of $500,000. $\Rightarrow$ $n = \dfrac{175,000}{500,000}$
$= 0.35$

35% of the total budget was not spent on salaries.

✓ **Check Your Understanding**

2. In Example 2(a), why is the total budget $500,000 instead of $500?

Guided Practice

COMMUNICATION « *Reading*

Refer to the text on pages 513–514.

« **1.** What percent is represented by the entire circle in a circle graph?

« **2.** What are two ways that you can label the sectors in a circle graph?

Find each answer.

3. What number is 24% of 4500?   **4.** 30% of what number is 18?

Use the circle graph on page 513.

5. If the total budget was $4,000,000, how much was spent on travel?

6. If $228,000 was spent on equipment, how much was spent on stadium and game costs?

Use the circle graph on page 514.

7. What percent of the budget was spent on new purchases?

8. What percent of the budget was spent on maintenance?

Exercises

Use the circle graph at the right.

1. What percent of the events were not concerts?

2. If there was a total of 300 events at the stadium, how many were basketball events?

3. If there was a total of 240 events at the stadium, how many were baseball events?

4. If there were 36 concerts at the stadium, how many events were there in all?

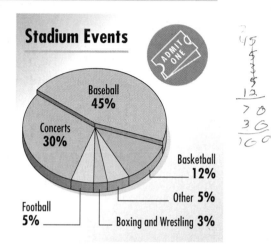

5. If there were 12 boxing and wrestling events at the stadium, how many events were there in all?

6. If there were 11 football events at the stadium, how many baseball events were there?

Use the circle graph at the right.

7. What is the total budget for a month?

8. What percent of the budget is spent on transportation?

9. What percent of the budget is saved?

10. What percent of the budget is not spent on housing?

11. What percent of the budget is not spent on clothing?

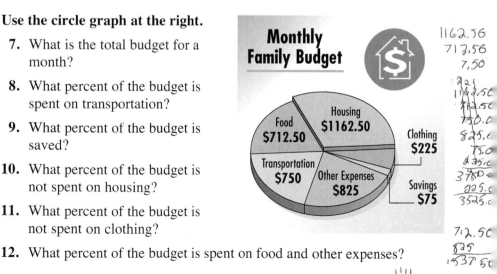

12. What percent of the budget is spent on food and other expenses?

ESTIMATION

13. Estimate the fraction of the monthly budget spent on housing.

14. Estimate the fraction of the monthly budget that is used for savings and other expenses.

15. What fraction of music buyers are 25–29 years old?

16. What fraction of music buyers are less than 35 years old?

17. What fraction of music buyers are 20–29 years old?

18. Out of 3000 customers, how many customers would probably be 15–24 years old?

MENTAL MATH

Use the circle graph at the right. Find each answer mentally.

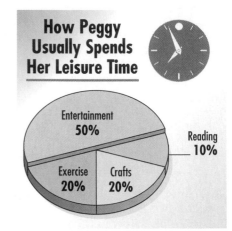

How Peggy Usually Spends Her Leisure Time

19. If Peggy had 90 h of leisure time last month, how many hours did she probably spend reading?

20. If Peggy used 10.5 h of her leisure time last month for entertainment, about how many hours of leisure time did she probably have?

21. About how many hours of leisure time did Peggy probably have last week if she spent 6 h exercising?

22. How many hours will Peggy probably spend on crafts if she has 15 h of leisure time?

SPIRAL REVIEW

23. Make a box-and-whisker plot for the data. *(Lesson 11-4)*

    Test Scores

65	68	72	75
78	80	82	84
86	91	98	

24. Refer to the circle graph on page 513. If the total budget was $3,250,000, how much was spent on travel? *(Lesson 11-6)*

25. Find the area of a triangle with base 14 in. and height 9 in. *(Lesson 10-3)*

26. What percent of 30 is 25? *(Lesson 9-7)*

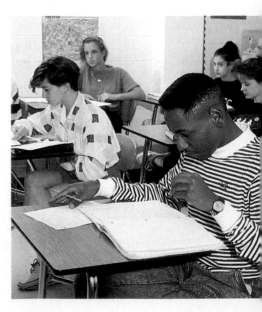

11-7

Constructing Circle Graphs

Objective: To construct circle graphs to display data.

When you construct a circle graph, you divide the circle into sectors. Each sector represents a percent of the data.

The sum of the angles formed by the sectors of a circle is 360°. These angles are called **central angles** because they each have a vertex at the center of the circle. Angle *AOB* in the diagram at the right is a central angle.

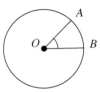

Example

Draw a circle graph to display the data.

Sales of Compact Discs

Type	rock	country	jazz	classical
Number Sold	240	180	85	35

Solution

• Find the percent of the total number sold made up by each type of music. Write each percent as a decimal. Then multiply each decimal by 360° to find the number of degrees for each sector.

Find the total number sold: $240 + 180 + 85 + 35 = 540$

rock: $\frac{240}{540} \approx 44.4\%$ $\quad 0.444(360°) \approx 160°$

country: $\frac{180}{540} \approx 33.3\%$ $\quad 0.333(360°) \approx 120°$

jazz: $\frac{85}{540} \approx 15.7\%$ $\quad 0.157(360°) \approx 57°$

classical: $\frac{35}{540} \approx 6.5\%$ $\quad 0.065(360°) \approx 23°$

• Use a template or a compass to draw a circle. Draw a radius. Then use a protractor to draw the central angle for each sector.

• Label each sector with the corresponding type of music and the number of compact discs sold.

• Include a title on the circle graph.

Compact Disc Sales

Rock 240

Classical 35

Jazz 85

Country 180

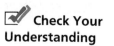
Check Your Understanding

In the Example, why must you find the total number sold?

You can use a calculator to find the number of degrees in each sector when you are making a circle graph. For example, this key sequence would be used to find the number of degrees for the rock sector in the Example on page 517.

$$\boxed{240.} \quad \boxed{\div} \quad \boxed{540.} \quad \boxed{\times} \quad \boxed{360.} \quad \boxed{=}$$

Guided Practice

COMMUNICATION « Reading

« **1.** Explain the meaning of the word *sectors* in the following sentence. *At the end of World War II, Berlin was divided into four sectors.*

« **2.** Write a sentence using the word *sector* in a mathematical context.

Write each fraction as a percent.

3. $\dfrac{450}{2000}$ **4.** $\dfrac{75}{375}$ **5.** $\dfrac{300}{900}$ **6.** $\dfrac{30}{513}$

Write each percent as a decimal.

7. 18%   **8.** 79%   **9.** 8.5%   **10.** 39.2%

11. Copy and complete the chart below as if you were going to draw a circle graph to display the data. Do not draw the graph.

The Movies People Rent

	Type of Movie	Percent	Measure of Central Angle for the Sector
a.	comedy	20%	?
b.	mystery	40%	?
c.	musical	12%	?
d.	action/adventure	25%	?
e.	other	?	?

Draw a circle graph to display each set of data.

12. Protein Source Preferences

Source	pork	chicken	fish	beef	other
Percent	10%	40%	15%	24%	11%

13. Election Results

Candidate	Tyrall	Little	Clemmons	Evans
Number of Votes	1,050,000	1,260,000	510,000	180,000

Draw a circle graph to display each set of data.

1. **Library Inventory**

Type of Book	adult fiction	children's fiction	nonfiction	reference	other
Percent	35%	20%	25%	15%	5%

2. **Annual Budget for a Four-Year Public University**

Budget Item	room and board	tuition and fees	books	travel	other
Percent	41%	24%	11%	7%	17%

3. **Cars Rented from Yoko's Rent-A-Car**

Type	compact	mid size	full size	mini-vans
Number Rented	90	110	150	50

4. **Survey of Favorite Chicken Restaurants**

Restaurant	Chicken Hut	Millie's Chicken	Chicken-to-Go	Country Chicken
Number of Votes	50	48	42	60

 CALCULATOR

For Exercises 5–8, choose the letter of the calculator key sequence you would use to find the number of degrees for each sector described.

A. [700.] [÷] [1600.] [×] [360.] [=]

B. [550.] [÷] [2000.] [×] [360.] [=]

C. [700.] [÷] [2000.] [×] [360.] [=]

D. [2000.] [÷] [700.] [×] [360.] [=]

E. [550.] [÷] [1600.] [×] [360.] [=]

F. [360.] [÷] [550.] [×] [1600.] [=]

5. Out of a monthly budget of $2000, Kaya spends $700 on rent and household expenses.

6. Out of a monthly budget of $1600, Faraj spends $550 on rent and household expenses.

7. Out of a monthly budget of $2000, Anthony spends $550 on car expenses.

8. Out of a monthly budget of $1600, Mary Kate spends $700 on car expenses.

A *Hotel-Restaurant-Travel Administrator* manages hospitality services, such as hotels and motels, amusement parks, campgrounds, and food services. One aspect of an administrator's job is monitoring receipts. The table below shows the annual receipts of a mid-sized motel.

9. What percent of the total receipts came from room rentals?

10. Which category contributed the lowest percent of the receipts?

Annual Receipts	
rooms:	$490,560
food:	$236,785
beverages:	$93,584
telephone:	$24,500
other:	$9,499
total:	$854,928

11. Estimate the fraction of the total income that came from food and beverage sales.

12. Draw a circle graph to display the data.

CONNECTING CIRCLE GRAPHS AND GEOMETRY

Finding the area of a sector is like finding the number of degrees in a central angle of a circle graph. You find the fraction of 360° made up by the central angle, and multiply this fraction by the total area. So the area of a sector with a central angle of $n°$ and radius r is $\frac{n}{360}\pi r^2$.

Find the area of each sector.

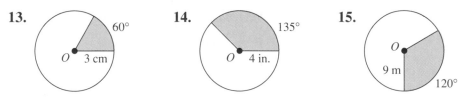

13. 60° O 3 cm

14. 135° O 4 in.

15. O 9 m 120°

SPIRAL REVIEW

16. Write $\frac{5}{8}$ as a percent.   *(Lesson 9-5)*

17. Solve: $2(n + 6) = -14$ *(Lesson 4-9)*

18. Draw a circle graph to display the data.   *(Lesson 11-7)*

Commuting Methods of Acme Employees

Method	car	car pool	walk	train	bus
Percent	40%	7%	8%	25%	20%

Using Statistical Graphing Software

Objective: To use statistical graphing software to display and analyze data.

Most large companies have a department whose purpose is to analyze data about the company's performance. People working in such departments have to present summaries to managers in a form that can be quickly and easily understood. Graphs are often included in the summaries, and **statistical graphing software** is used to produce them.

Statistical graphing software can be very flexible. For example, the data below could be presented in either a bar or a circle graph. Graphs developed using statistical graphing software look very professional and can have a strong visual impact.

Personal Computer Sales

Region	Sales (Dollars)
Northeast	45,225,000
North-Central	33,675,000
Northwest	19,950,000
Southeast	31,475,000
South-Central	24,500,000
Southwest	25,175,000
Total	180,000,000

Exercises

Use statistical graphing software to create a circle graph displaying the data in the table above.

1. How many sectors does the circle graph have?

2. How are the sectors labeled?

3. If each region had the same amount of sales, what fraction of the sales would this be?

4. How many regions, if any, had less than 17% of the sales?

5. How many regions, if any, had at least 25% of the sales?

6. What percent of the sales did the southern regions have?

7. What percent of the sales did the eastern regions have?

8. Suppose that the regions are reorganized so that $10,225,000 in sales from the northeast region are counted as sales for the southeast region. How will this affect the sectors for these two regions? How will this affect the sectors for the other regions?

9. What can you visualize in the circle graph of this set of data that you could not visualize in a bar graph of the data?

Choosing an Appropriate Data Display

Objective: To decide which method for displaying data is appropriate.

When you have data to display, you must decide which type of display is appropriate.

*   A *histogram* or *frequency polygon* may be most appropriate when you want to visually compare the frequencies at which specific data items or groups of data items occur.

*   A *circle graph* may be most appropriate when you want to visually compare parts of a set of data to the whole.

*   A *scattergram* may be most appropriate when you want to display the correlation between two sets of data.

*   A *stem-and-leaf plot* may be most appropriate when you want to display the individual data items in an ordered and concise manner.

*   A *box-and-whisker plot* may be most appropriate when you want to display the median, first and third quartiles, and least and greatest values of a set of data, or compare these aspects of two sets of data.

Example

Decide whether it would be appropriate to draw a *scattergram* or a *histogram* to display the data. Then make the display.

Class Grade Point Averages

3.4	2.9	2.9	2.3	3.2	2.5	3.1	3.2	2.9	2.8
2.2	3.8	3.3	2.8	2.7	3.0	2.4	2.6	3.4	2.5
2.9	3.1	4.0	3.7	2.6					

Solution

Histogram. A scattergram is used to compare two sets of data. Only one set of data is given. For this reason, a scattergram is not appropriate.

A histogram is appropriate. It will show how the grade point averages for the class are clustered.

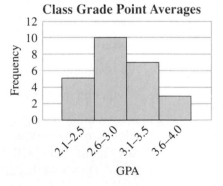

Check Your Understanding

In the Example, would it be appropriate to draw a scattergram if you were also given the grade that each student received in English? Explain.

Guided Practice

«1. **COMMUNICATION** «*Discussion* If you are given a set of data in the form of a frequency table with intervals, which displays will automatically be inappropriate?

2. Decide whether it would be appropriate to draw a *circle graph* or a *stem-and-leaf plot* to display the data. Then make the display.

Students in Activities after School

Activity	sports	job	tutoring	other
Students	50	75	34	16

Exercises

Tell which of the two types of displays would be appropriate for each set of data. Then make the display.

1. *box-and-whisker plot* or *circle graph*

   Scores on an English Test

100	78	76	78	68	80	89	85	97	75	70	100
98	78	95	84	83	58	93	84	80	90	100	78

2. *scattergram* or *stem-and-leaf plot*

   Weekly Grocery Budgets (Dollars)

125	86	75	52	88	97	78	85	85	116
97	104	88	93	84	71	107	76	96	74

3. *histogram* or *circle graph*

   Survey of the Number of Women in Certain Occupations

Occupation	professional	technical	service	other
Women	125	225	90	60

4. *scattergram* or *box-and-whisker plot*

   Managers' Salaries at Thorn's Department Store

Experience (Years)	1	11	5	6	7	12	6	10
Salary (Thousands)	18	32	24	24	30	35	26	32

SPIRAL REVIEW

5. 18 is 20% of what number? *(Lesson 9-8)*

6. Decide whether it would be appropriate to draw a *circle graph* or a *stem-and-leaf plot* to display the data. Then make the display. *(Lesson 11-8)*

   Price per Gallon of Gasoline (Dollars)

1.29	1.26	1.34	1.38	1.32	1.38	1.35
1.27	1.31	1.40	1.27	1.43	1.41	

11-9

Strategy:
Using Logical Reasoning

Objective: To solve problems using logical reasoning.

Some problems require you to make conclusions based on the relationships in the facts of the problem. Using a table to organize the facts can help you to reason logically and find the answer to the problem.

Problem

Bob, Frank, and Tran play baseball, football, and tennis. No one plays a sport that begins with the same letter as his name. Tran and the baseball player are cousins. Who plays each sport?

Solution

UNDERSTAND The problem is about three people who play sports.
Facts: first letter of name is not first letter of sport
         the baseball player's cousin is Tran
Find: who plays each sport

PLAN Make a table to show all the possibilities. As you read each fact of the problem, put an × in a space to eliminate a possibility. Write "yes" to indicate a match.

WORK The first letter of the name does not match the first letter of the sport. So eliminate *baseball* from Bob's row, *football* from Frank's row, and *tennis* from Tran's row.

Tran and the baseball player are cousins, so Tran does not play baseball. Eliminate *baseball* from Tran's row. Tran must play football.

Because neither Bob nor Tran plays baseball, Frank must play baseball.

	Baseball	Football	Tennis
Bob	×		
Frank	yes	×	
Tran	×	yes	×

Frank plays baseball, so eliminate *tennis* from Frank's row. Tran plays football, so eliminate *football* from Bob's row. Bob must play tennis.

	Baseball	Football	Tennis
Bob	×	×	yes
Frank	yes	×	×
Tran	×	yes	×

ANSWER Bob plays tennis, Frank plays baseball, and Tran plays football.

Look Back What if the baseball player were Frank's cousin, instead of Tran's cousin? What sport would each person play?

Guided Practice

COMMUNICATION « *Reading*

Use the problem below for Exercises 1–6.

Art Klein, Jim Pierce, and Sy Fischer teach art, gym, and science. No teacher teaches a subject that sounds like his first name. Art is not interested in athletics. Which teacher teaches each subject?

« **1.** Can Sy Fischer be the science teacher? Explain.

« **2.** Can Art Klein be the art teacher? Explain.

« **3.** What does the second sentence tell you about a subject Jim Pierce does not teach?

« **4.** What does the third sentence tell you about a subject Art Klein does not teach?

5. Use your answers to Exercises 1–4 to complete a table of the problem.

6. Use your table from Exercise 5 to write an answer to the problem.

Solve using logical reasoning.

7. Jason, Sharon, and Alicia study painting, piano, and ballet. The painter painted a portrait of Jason. Sharon and the dancer are sisters. Who studies each art form?

8. Adebayo, Botan, Carl, and Dennis were the first four people to finish a marathon. Adebayo did not finish first or second. Botan did not finish second or third. Carl did not finish first or fourth. Dennis did not finish second. Adebayo finished before Dennis. In what order did the runners finish?

Problem Solving Situations

Solve using logical reasoning.

1. Ann Fernandez, Ellen Taylor, and Sonia Morris teach French, typing, and mathematics. No one teaches a subject that begins with the same letter as her last name. Ellen Taylor speaks only English. Who teaches each subject?

2. Abdul, Ben, and Carmen placed first, second, and third in a student council election. Abdul was not first, Carmen was not third, and Ben was not second. Carmen finished ahead of Ben. In what order did the students finish?

3. Marv, Shirley, and Stan are the officers in a bowling league. The president and Stan are in-laws. Shirley and the treasurer are married. Marv and the secretary are brothers. Who serves in each position?

Solve using logical reasoning.

4. In a civil case, Jeffries, Daniels, Powers, and Watkins are the judge, defendant, plaintiff, and witness. No one has a role that begins with the first letter of his or her last name. Only Powers went to law school. Jeffries, Watkins, and the plaintiff live in the same town. Who has each role?

Solve using any problem solving strategy.

5. Maura has four $5 bills, four $10 bills, and four $20 bills in a box. She takes two bills from the box at random and finds the total value of the bills. How many different totals are possible?

6. A typist can type 100 words in 2 min. How long would it take the typist to type 475 words?

7. Aaron, Bianca, and Chris live in Atlanta, Boston, and Chicago. No one lives in a city that begins with the first letter of his or her name. Bianca has never been to the South. Who lives in Boston?

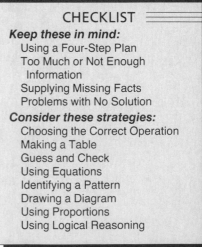

PROBLEM SOLVING

CHECKLIST

Keep these in mind:
   Using a Four-Step Plan
   Too Much or Not Enough
    Information
   Supplying Missing Facts
   Problems with No Solution

Consider these strategies:
   Choosing the Correct Operation
   Making a Table
   Guess and Check
   Using Equations
   Identifying a Pattern
   Drawing a Diagram
   Using Proportions
   Using Logical Reasoning

8. Marisa is $62\frac{1}{2}$ in. tall. Marty is $6\frac{1}{2}$ ft tall. Which person is taller? By how many inches?

9. Roger was charged $20.75 for some pumpkins. The price of a pumpkin was $4.95. How many pumpkins did Roger buy?

Write a word problem that you could solve using logical reasoning. Then solve the problem.

10.

	Cat	Dog	Gerbil
Cindy	×		
Doug	×	×	
Grace			×

11.

	Blue	Brown	Green
Mary			×
Fran		×	
Jim	×		×

SPIRAL REVIEW

12. Solve: $4 = -\frac{3}{5}n + 16$ *(Lesson 8-9)*

13. Torres, Wong, and Lipshaw are a doctor, a dentist, and a journalist. The dentist went to college with Torres. Wong and the journalist are neighbors. Lipshaw and Wong are both patients of the doctor. What is each person's profession?   *(Lesson 11-9)*

14. The lengths of the legs of a right triangle are 7 cm and 24 cm. Find the length of the hypotenuse.   *(Lesson 10-11)*

Self-Test 2

Use the scattergram at the top of page 511.

1. How many women are 70 in. tall?   **11-5**

2. Is there a *positive* or a *negative* correlation between a woman's weight in pounds and height in inches?

3. Use the circle graph on page 513. If the total budget was $4,200,000, how much was spent on training?   **11-6**

4. Draw a circle graph to display the data.   **11-7**

City Budget (Millions of Dollars)

Budget Item	Schools	Police	Fire	Other
Cost	22	11	7	10

5. Decide whether it would be appropriate to draw a *circle graph* or a *scattergram* to display the data. Then make the display.   **11-8**

Allowance Record (Dollars)

Savings	10	5	2	8	5	3	6
Entertainment	2	7	11	2	5	7	4

6. Harry, Matt, and Phyllis are majoring in history, mathematics, and physics. No one majors in a subject that begins with the same letter as his or her first name. Phyllis and the history major take English together. Who majors in each subject?   **11-9**

Analyzing Census Data

Objective: To analyze census data.

Every ten years, the Census Bureau conducts a population census to determine the number of individuals living in the United States. The 1990 census marked the 200th anniversary of the first census.

The census is important because the distribution of billions of tax dollars and the redistricting of local and state governments depend on the results.

The table below shows some preliminary figures for the 1990 census. Cities are ranked according to population.

1990 Rank	1990 Population	Change from 1980	1980 Rank
1. New York	7,033,179	−0.5%	1
2. Los Angeles	3,420,000	15.2%	3
3. Chicago	2,725,979	−9.3%	2
4. Houston	1,609,723	0.9%	5
5. Philadelphia	1,543,313	−8.6%	4
6. San Diego	1,094,524	25%	8
7. Dallas	990,957	9.5%	7
8. Phoenix	971,565	23%	9
9. Detroit	970,156	−19.3%	6
10. San Antonio	926,558	17.9%	11

Example

a. What was the population of Phoenix in 1980?

b. Suppose by the year 2000 the population of San Diego increases from the 1990 level by another 25%. Predict what will be the population of San Diego by the year 2000.

Solution

a. Let n = the population of Phoenix in 1980.
Then $0.23n$ = the increase in the population from 1980 to 1990, and $n + 0.23n$ = the total population of Phoenix in 1990.

$$n + 0.23n = 971{,}565$$
$$1.23n = 971{,}565$$
$$\frac{1.23n}{1.23} = \frac{971{,}565}{1.23}$$
$$n \approx 789{,}890$$

The population of Phoenix in 1980 was about 789,890.

b. Find the amount of increase: $0.25(1{,}094{,}524) = 273{,}631$
Add the amount of increase to the original amount.

$$1{,}094{,}524 + 273{,}631 = 1{,}368{,}155$$

The population of San Diego will be about 1,368,155 by the year 2000.

Use the table on page 528 for Exercises 1–10.

1. Which city had the greatest decrease in population from 1980 to 1990?

2. Which city had the least increase in population from 1980 to 1990?

3. In which cities did the population increase from 1980 to 1990? In what part of the country are these cities located?

4. What was the mean of the populations of the top 10 cities in 1990? How many cities had a population greater than the mean?

5. What was the median of the populations of the top 10 cities in 1990?

6. What was the range of the populations of the top 10 cities in 1990?

7. What was the population of Los Angeles in 1980?

8. What was the population of Detroit in 1980?

9. Suppose by the year 2000 the population of the city of New York decreases from the 1990 level by another 0.5%. Predict what will be the population of the city of New York by the year 2000.

10. Suppose by the year 2000 the population of Dallas increases from the 1990 level by another 9.5%. Predict what will be the population of Dallas by the year 2000.

RESEARCH

11. Find the population of the United States according to every census taken since the census began in 1790. Draw a line graph of the data. Describe the trends you see in the data.

12. Find the 1980 census and 1990 census populations for your state. Did the population increase or decrease from 1980 to 1990? Predict what will be your state's population by the year 2000.

Chapter Review

stem-and-leaf plot (p. 497)
stem (p. 497)
leaf (p. 497)
histogram (p. 501)
frequency polygon (p. 501)
first quartile (p. 505)
third quartile (p. 505)
box-and-whisker plot (p. 505)

scattergram (p. 509)
trend line (p. 509)
positive correlation (p. 509)
negative correlation (p. 509)
circle graph (p. 513)
sector (p. 513)
central angle (p. 517)

Choose the correct term from the list above to complete each sentence.

1. A(n) __?__ is a bar graph used to show frequencies.

2. A(n) __?__ is used to represent data expressed as parts of a whole.

3. If the trend line slopes upward to the right there is a(n) __?__ between the sets of data.

4. The __?__ is the median of the upper part of the data.

5. The __?__ is the last digit on the right of a number.

6. To construct a(n) __?__, you can connect the midpoints of the tops of the bars of a histogram.

Use the data at the right for Exercises 7 and 8. *(Lessons 11-1, 11-2)*

7. Make a frequency table for the data. Use intervals such as 16–20.

8. Make a stem-and-leaf plot for the data.

Points Scored per Game

17	24	18	31	22	16	22
18	23	24	30	18	25	33
27	23	18	20	22	24	31

Use the stem-and-leaf plot at the right. *(Lesson 11-2)*

9. What is the greatest test score?

10. How many test scores of 84 are there?

11. How many test scores are less than 90?

12. Find the mean, median, mode(s), and range of the test scores.

Learner's Permit Test Scores

6	0 2 4 8
7	2 6 6 6 8 8
8	4 4 6 8 8 8 8
9	0 0 2 4 4 6 6 6 8 8
10	0 0

Make a box-and-whisker plot for the data. *(Lesson 11-4)*

13. **Ages of Workers in an Office (Years)**

31	29	34	19	24	42	33
31	27	28	30	24	30	

Use the histogram below. *(Lesson 11-3)*

14. How many students are 72–75 in. tall?

15. How many students are less than 68 in. tall?

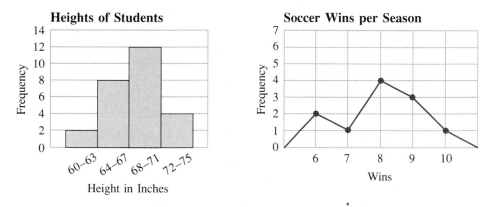

Use the frequency polygon above. *(Lesson 11-3)*

16. How many times did the soccer team win 9 games per season?

17. How many seasons has the soccer team played?

Use the box-and-whisker plot below for Exercises 18–20.
(Lesson 11-4)

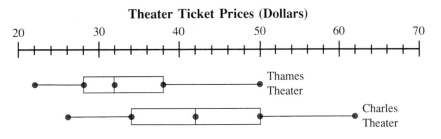

18. Find the median, the first quartile, the third quartile, and the highest and lowest ticket prices for the Thames Theater.

19. Which theater has the lesser range in ticket prices?

20. Which theater has the greater median ticket price?

Draw a circle graph to display the data. *(Lesson 11-7)*

21. Colors of Cars Sold

Color	Blue	White	Red	Gray	Black
Number Sold	60	80	40	20	40

Use the scattergram at the right. *(Lesson 11-5)*

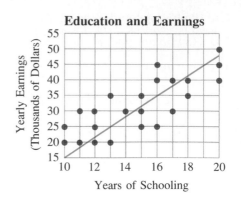

Education and Earnings

22. How much money does the person who went to school for 14 years earn?

23. How many people earn $30,000 per year?

24. Predict the yearly earnings for someone who has had 19 years of schooling.

25. Is there a *positive* or a *negative* correlation between the number of years of schooling and earnings per year?

Use the circle graph at the left below. *(Lesson 11-6)*

26. If the Jones family's total annual budget is $45,000, how much do they spend annually on housing?

27. If $3500 is spent on entertainment, how much is the total budget?

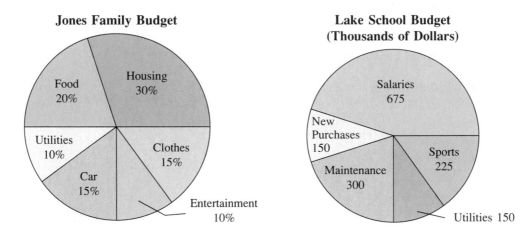

Jones Family Budget

Food 20%
Housing 30%
Utilities 10%
Clothes 15%
Car 15%
Entertainment 10%

Lake School Budget
(Thousands of Dollars)

Salaries 675
New Purchases 150
Maintenance 300
Sports 225
Utilities 150

Use the circle graph at the right above. *(Lesson 11-6)*

28. What was the total budget?

29. What percent of the budget was spent on sports?

30. Decide whether it would be appropriate to draw a *scattergram* or a *stem-and-leaf plot* to display the data at the right. Then make the display. *(Lesson 11-8)*

Leading Batting Averages

0.322	0.308	0.366	0.304	0.301
0.312	0.311	0.356	0.306	0.305
0.303	0.307	0.312	0.304	0.306

31. Ed, Alice, and Will are an editor, an artist, and a writer. No one has a name that starts with the same letter as his or her job. Alice and the editor had lunch together. Who has each job? *(Lesson 11-9)*

Chapter Test

Use the data below for Exercises 1, 3, and 4.

Hours Worked per Week

32	28	40	38	26	42	35	40	38
28	40	41	35	36	30	38	29	44
68	40	37	38	35	32	26	30	40

1. Make a frequency table for the data. Use intervals such as 21–30.   **11-1**

2. What are appropriate intervals for a frequency table that displays these prices: $1.49, $1.99, $2.99, $3.50, $3.99, $4.99?

3. Make a stem-and-leaf plot for the data, *Hours Worked per Week*.   **11-2**

4. Use the stem-and-leaf plot you made for Exercise 3 to find the median number of hours worked per week.

Use the histogram at the right for Exercises 5 and 6.

5. How many employees are 31–40 years old?   **11-3**

6. How many employees are more than 40 years old?

7. Draw a box-and-whisker plot for the data below.   **11-4**

Prices of Stereo Systems (Dollars)

800	400	550	700	1200
750	700	600	600	850
750	650	600	700	550

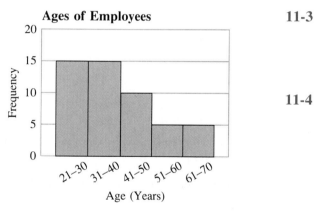

Use the box-and-whisker plot below for Exercises 8–10.

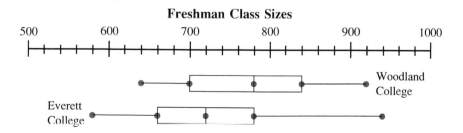

8. Which college had the largest freshman class?

9. Which college had the greater median freshman class size?

10. Find the median, the first quartile, the third quartile, and the highest and lowest freshman class size for Woodland College.

Use the scattergram below.

11. How many people missed 12 days of work due to illness?

11-5

12. Is there a *positive* or a *negative* correlation between the two sets of data?

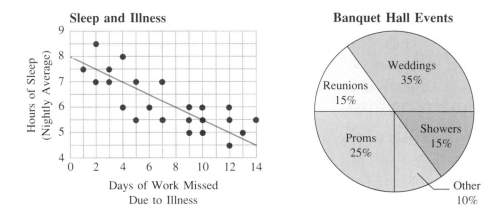

13. What is the main purpose for using a scattergram to display data?

Use the circle graph above.

14. If there was a total of 200 events at the banquet hall, how many events were weddings?

11-6

15. If there were 36 reunions at the banquet hall, how many events were there in all?

16. Draw a circle graph to display the data.

11-7

Animals in a Zoo

Class	Mammals	Birds	Reptiles	Amphibians
Animals	210	60	30	20

17. Decide whether it would be appropriate to draw a *circle graph* or a *box-and-whisker plot* for the data below. Then make the display.

11-8

Monthly Commuting Expenses

96   36   40   89   36   64   96   25   10   25   50   36   40   89

18. Hilda, Wendy, and Cara are a hostess, a waitress, and a chef. No one works at a job that begins with the first letter of her name. Hilda and the chef are sisters. Who has each job?

11-9

Percentiles

Objective: To find a given percentile.

When you take a standardized test such as the SAT or the ACT, your score is often reported in terms of percentiles. A *percentile* shows a person's score in relation to other scores in a particular group. For example, if your score on a test is at the 65th percentile, this means approximately 65% of the people who took the test had a lesser score than you while approximately 35% of the people had a greater score.

Example

The following are the test scores of 28 students arranged in order from lowest position (1) to highest position (28).

a. Find the 65th percentile.   **b.** Find the 75th percentile.

Position	1	2	3	4	5	6	7	8	9	10	11	12	13	14
Score	62	65	67	69	71	71	72	74	74	75	77	77	78	79

Position	15	16	17	18	19	20	21	22	23	24	25	26	27	28
Score	81	82	84	84	85	88	88	89	90	93	95	96	98	98

Solution

a. Find 65% of 28: $0.65(28) = 18.2$

The answer is not a whole number. When this is the case, round up to the next whole number: $18.2 \rightarrow 19$

Read the score in the 19th position: 85

The 65th percentile is a score of 85.

b. Find 75% of 28: $0.75(28) = 21$

The answer is a whole number. When this is the case, use that number and the next number: 21 and 22

Find the average of the scores in those positions.

$$\frac{88 + 89}{2} = 88.5 \longleftarrow \text{The scores in positions 21 and 22 are 88 and 89, respectively.}$$

The 75th percentile is a score of 88.5.

Exercises

Use the data consisting of the 28 test scores above. Find each percentile.

1. 90th   **2.** 77th   **3.** 40th   **4.** 95th

5. 62nd   **6.** 25th   **7.** 82nd   **8.** 80th

9. 60th   **10.** 55th   **11.** 50th   **12.** 85th

Standardized Testing Practice

Choose the letter of the correct answer.

1. A swimmer, a volleyball player, and a skier are named Arthur, Bernardo, and Carol. Carol's sport only happens outdoors. What can you conclude?
   A. Carol is the skier.
   B. Carol is not the swimmer.
   C. Arthur is not the skier.
   D. all of the above

2. Simplify: $(36x^2)^0$
   A. 1   B. 0
   C. $36x^2$ D. $36x^{20}$

3. Helen has two thirds as many pens as Dave has. Helen has 6 pens. How many pens does Dave have?
   A. $5\frac{1}{3}$ B. $6\frac{2}{3}$
   C. 4   D. 9

4. The radius of a circle is 10 in. What is the area of the circle?
   A. 31.4 in.^{2} B. 62.8 in.^{2}
   C. 100 in.^{2} D. 314 in.^{2}

5. A frequency table shows that 8 out of a total of 20 students are 12–15 years old. How many students are 16–19 years old?
   A. 8   B. 12
   C. 20   D. not enough information

6. The lengths of the legs of a right triangle are 0.3 cm and 0.4 cm. What is the length of the hypotenuse?
   A. 0.84 cm   B. 0.25 cm
   C. 0.5 cm   D. 0.7 cm

7. Solve: $\frac{x}{7} = -21$
   A. $x = 3$ B. $x = -147$
   C. $x = -3$ D. $x = 147$

8. There is exactly one pair of parallel sides in a quadrilateral. What special type of quadrilateral is it?
   A. rhombus   B. trapezoid
   C. rectangle   D. parallelogram

9. Find the range of the data.
   $3, 7, -2, 0, 7, -1, 8, -5, 2$
   A. 2   B. 8
   C. 13   D. 7

10. A sector of a circle graph shows that 35% of consumers prefer wheat bread. Find the number of degrees for that sector.
    A. 35°   B. 126°
    C. 63°   D. 145°

11.

Ages of Car Buyers

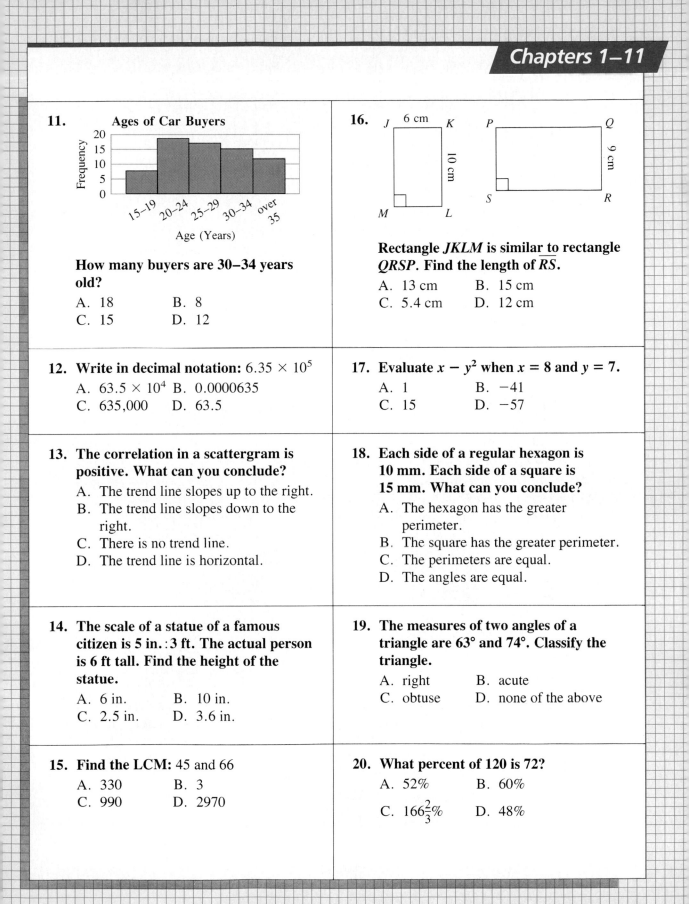

How many buyers are 30–34 years old?

A. 18 B. 8

C. 15 D. 12

16.

Rectangle *JKLM* is similar to rectangle *QRSP*. Find the length of $\overline{RS}$.

A. 13 cm B. 15 cm

C. 5.4 cm D. 12 cm

12. Write in decimal notation: 6.35×10^5

A. 63.5×10^4 B. 0.0000635

C. 635,000 D. 63.5

17. Evaluate $x - y^2$ **when** $x = 8$ **and** $y = 7$.

A. 1 B. -41

C. 15 D. -57

13. The correlation in a scattergram is positive. What can you conclude?

A. The trend line slopes up to the right.

B. The trend line slopes down to the right.

C. There is no trend line.

D. The trend line is horizontal.

18. Each side of a regular hexagon is 10 mm. Each side of a square is 15 mm. What can you conclude?

A. The hexagon has the greater perimeter.

B. The square has the greater perimeter.

C. The perimeters are equal.

D. The angles are equal.

14. The scale of a statue of a famous citizen is 5 in. : 3 ft. The actual person is 6 ft tall. Find the height of the statue.

A. 6 in. B. 10 in.

C. 2.5 in. D. 3.6 in.

19. The measures of two angles of a triangle are 63° and 74°. Classify the triangle.

A. right B. acute

C. obtuse D. none of the above

15. Find the LCM: 45 and 66

A. 330 B. 3

C. 990 D. 2970

20. What percent of 120 is 72?

A. 52% B. 60%

C. $166\frac{2}{3}\%$ D. 48%

Winter Bald Eagle Count In Southern New Jersey

Delaware Bay
Atlantic Coast

Number of Eagles

Year

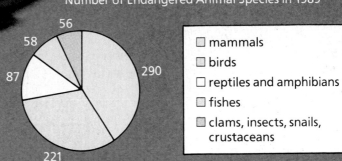

Number of Endangered Animal Species in 1989

56
58
290
87
221

- ☐ mammals
- ☐ birds
- ☐ reptiles and amphibians
- ☐ fishes
- ☐ clams, insects, snails, crustaceans

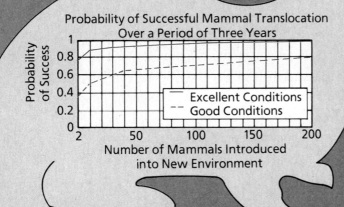

Probability of Successful Mammal Translocation Over a Period of Three Years

Probability of Success

1
0.8
0.6
0.4
0.2
0

— Excellent Conditions
-- Good Conditions

2   50   100   150   200

Number of Mammals Introduced into New Environment

Did You Know?

One method scientists use to preserve endangered species from extinction is **translocation**. Moving, or translocating, a group of animals to a new environment sometimes gives animals a better chance of survival. Translocation is most likely to be successful if the animals are moved to an environment in which they have thrived in the past. The probability that the animals will survive increases with the number of animals translocated.

Rolling a Number Cube

Objective: To explore the ratios that occur when number cubes are rolled.

Materials

■ number cubes

A *number cube* has six sides numbered from 1 through 6. When you roll a number cube, any one of the six sides is equally likely to land face up.

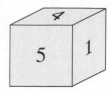

Activity I

1 Roll a number cube 30 times. Copy and complete the table below to record the number of 1's, 2's, 3's, 4's, 5's, and 6's that land face up.

1's	2's	3's	4's	5's	6's	Total number of rolls
?	?	?	?	?	?	30

2 For each of the six numbers, write a ratio comparing the number of times the number landed face up to the total number of rolls, 30. For example, if twelve 1's are rolled, the ratio is $\frac{12}{30}$. (Do not simplify the fractions.)

3 Find the sum of the six ratios.

Activity II

1 Use your results from Activity I to predict about how many 2's would land face up if a number cube were rolled 60 times.

2 Now roll a number cube 60 times and count the number of 2's that land face up.

3 How close was your prediction to the actual number of 2's that landed face up?

Activity III

1 Suppose you rolled two number cubes and recorded the sum of the numbers that landed face up. For example, if 3 lands face up on one cube and 5 lands face up on the second cube, you would record the number 8. Make a table similar to the one above, listing all the possible sums in order from least to greatest.

2 Roll a pair of number cubes 30 times. Complete the table that you made in Step 1.

3 For each sum, write a ratio comparing the number of times the sum occurred to the total number of rolls, 30.

4 Find the sum of all the ratios.

Probability and Odds

Objective: To find the probability of an event and to determine the odds in favor of an event.

Terms to Know

- *outcome*
- *equally likely*
- *event*
- *probability*
- *odds in favor*

When the spinner at the right is spun, there are six possible **outcomes.** The spinner may stop on any one of the six numbered sectors of the circle. (Assume that the spinner will not stop on the line between two sectors.) Each outcome is **equally likely** to happen.

An **event** is any group of outcomes. When all possible outcomes are equally likely, the **probability** of an event E, written $P(E)$, is given by the formula below.

$$P(E) = \frac{\text{number of favorable outcomes}}{\text{number of possible outcomes}}$$

A probability is a number from 0 through 1 and is often written as a fraction in lowest terms.

Example 1

Use the spinner shown above. Find each probability.
a. $P(5)$ **b.** $P(\text{red})$ **c.** $P(7)$
d. $P(\text{not white})$ **e.** $P(\text{red, blue, or white})$

Solution

There are 6 possible outcomes when the spinner is spun. Each outcome is equally likely to occur. Use the formula above.

a. $P(5) = \frac{1}{6}$ ⟵ The one favorable outcome is landing on 5.

b. $P(\text{red}) = \frac{3}{6} = \frac{1}{2}$ ⟵ Three sectors are red, so there are 3 favorable outcomes.

c. $P(7) = \frac{0}{6} = 0$ ⟵ The probability of an *impossible* event is 0.

d. $P(\text{not white}) = \frac{5}{6}$ ⟵ Five sectors are not white, so there are 5 favorable outcomes.

e. $P(\text{red, blue, or white}) = \frac{6}{6} = 1$ ⟵ The probability of a *certain* event is 1.

✓ Check Your Understanding

1. Why is it impossible for the event in Example 1(c) to occur?
2. In Example 1(d), describe the five favorable outcomes.
3. Why is the event in Example 1(e) certain to occur?

An event may have favorable outcomes and unfavorable outcomes. The **odds in favor** of an event E are given by the formula below.

$$\text{odds in favor of event } E = \frac{\text{number of favorable outcomes}}{\text{number of unfavorable outcomes}}$$

Example 2

Use the spinner at the right. Find the odds in favor of each event.
  a. white   **b.** number > 2

Solution

a. odds $= \dfrac{1}{5}$ ⟵ 1 favorable outcome
                   ⟵ 5 unfavorable outcomes

The odds in favor of spinning white are 1 to 5.

b. odds $= \dfrac{4}{2} = \dfrac{2}{1}$

The odds in favor of spinning a number greater than 2 are 2 to 1.

✔️ **Check Your Understanding**

4. Describe the favorable and unfavorable outcomes for the event in Example 2(a).

5. Find the odds in favor of the event *not red*.

Guided Practice

COMMUNICATION « *Writing*

Write each expression in symbols.

« **1.** The probability that a spinner lands on a number less than 5 is $\frac{2}{3}$.

« **2.** The probability that a spinner lands on a color that is not red is $\frac{1}{2}$.

Write each expression in words. Assume that a spinner has been spun.

« **3.** $P(\text{white or red})$ « **4.** $P(\text{number} > 2)$ « **5.** $P(\text{not } 4)$

6. The spinner shown is spun. List all the possible outcomes.

7. Describe the favorable outcomes for the event the spinner shown lands on red. Describe the unfavorable outcomes.

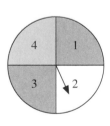

Find each probability for the spinner shown.

  8. $P(2)$ **9.** $P(\text{red})$ **10.** $P(5)$

11. $P(\text{white or blue})$ **12.** $P(\text{even number})$ **13.** $P(\text{number} > 0)$

Find the odds in favor of each event for the spinner shown.

14. red   **15.** 1 or 2   **16.** not blue

Exercises

Use the spinner at the right. Find each probability.

1. P(red)

2. P(2 or 3)

3. P(odd number)

4. P(blue)

5. P(not red)

6. P(number divisible by 3)

7. P(not white)

8. P(5)

9. P(9)

10. P(number > 0)

Use the spinner at the right. Find the odds in favor of each event.

11. W

12. red

13. W, X, or Y

14. blue

15. not white

16. not V

17. red or U

18. white or red

19. X or Y

20. white or not red

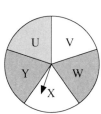

21. Assume that there is an equal chance that a coin will show heads or tails facing up when it has been tossed.
  a. Find P(heads).
  b. Find the odds in favor of heads.

A number cube that has six sides numbered 1 through 6 is rolled. Find each probability.

22. P(1)

23. P(6)

24. P(odd number)

25. P(prime number)

26. P(3, 4, 5, or 6)

27. P(number < 6)

28. P(number > 8)

29. P(number < 9)

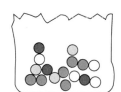

The bag shown at the right contains 3 red, 6 blue, 4 white, and 3 green marbles. One marble is selected *at random*, that is, purely by chance. Find each probability.

30. P(blue)

31. P(not blue)

32. P(red or green)

33. P(red, green, or white)

Use the bag of marbles shown above. One marble is selected at random. Find the odds in favor of each event.

34. red

35. blue or white

36. white

37. not red

38. not blue

39. white or green

You can look at a sample of items chosen at random from a large group to get an idea of what the large group is like. The probability that an event occurs in a sample can be used as an estimate of the probability that the event occurs in the entire group.

40. A sample of 18 film packages is chosen at random from a bin containing 100-speed and 200-speed film packages. Five of the film packages chosen are 100-speed film. Estimate the probability that any package chosen from the entire carton is 100-speed film.

41. A sample of 15 pairs of baby socks is chosen at random from a table of blue pairs of socks and pink pairs of socks at a department store. Six of the pairs of socks in the sample are blue. Estimate the probability that any pair of socks chosen from the entire table is blue.

THINKING SKILLS

42. A number cube is rolled. The probability that a 4 lands face up is $\frac{1}{6}$. What is the probability that a 4 does not land face up?

43. A bag contains 9 marbles. Two marbles are red, 4 marbles are blue, and 3 marbles are green. What is the probability that a marble drawn from the bag is blue? What is the probability that a marble drawn from the bag is not blue?

44. **a.** In Exercise 42, what is $P(4) + P(\text{not } 4)$?
    b. In Exercise 43, what is $P(\text{blue}) + P(\text{not blue})$?

45. Use your answers to Exercise 44.
    a. Complete the statements to formulate generalizations about the probability that an event E occurs or does not occur.
    For any event E, $P(E) + P(\text{not } E) = \underline{\ ?\ }$.
    For any event E, $P(\text{not } E) = 1 - \underline{\ ?\ }$.
    b. A spinner is spun and the probability that the arrow lands on red is $\frac{1}{3}$. Use the generalizations in part (a) to find the probability that the spinner does not land on red.

SPIRAL REVIEW

46. A number cube that has sides numbered 1 through 6 is rolled. Find $P(0)$.   *(Lesson 12-1)*

47. Find the circumference of a circle with radius 9 in.   *(Lesson 10-2)*

48. Solve: $12x - 15x = -15$ *(Lesson 4-9)*

Challenge

A box contains 40 pens. Some are red and some are blue. If the odds of selecting a red pen are 3 to 5, how many red pens are in the box?

Sample Spaces and Tree Diagrams

Objective: To use a tree diagram to find sample spaces and probabilities.

A list of all possible outcomes is called the **sample space**. A **tree diagram** is one way of showing a sample space and of finding the number of all possible outcomes.

In the previous lesson, you found probabilities when you rolled one number cube or spun one spinner. Sometimes you perform more than one such activity at the same time or consecutively.

Example

Three coins are tossed. Make a tree diagram to show the sample space. Then find each probability.

a. *P*(exactly 2 heads)
b. *P*(at least one tail)

Solution

When each coin is tossed, there are two possibilities, either heads (H) or tails (T).

Terms to Know
- *sample space*
- *tree diagram*

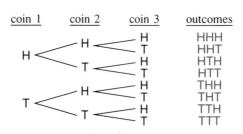

a. There are 3 outcomes that give exactly 2 heads: HHT, HTH, and THH. There are 8 possible outcomes.

$$P(\text{exactly 2 heads}) = \frac{3}{8}$$

b. *At least one tail* means one tail, two tails, or three tails. There are 7 favorable outcomes.

HHT, HTH, THH ⟵ 1 tail
HTT, THT, TTH ⟵ 2 tails
TTT ⟵ 3 tails

$$P(\text{at least one tail}) = \frac{7}{8}$$

 **Check Your Understanding**

1. How would the sample space in the Example be different if only two coins were tossed?

2. If part (b) were changed to *P*(at most one tail), list the favorable outcomes and then find the probability.

Guided Practice

« **1.** Replace each __?__ with the correct word.

The list of all possible outcomes of an experiment is called the __?__. A __?__ diagram can be used to show such a list.

« **2.** Choose the phrase that means the same as *at least one*.
  a. more than one   **b.** one or more   **c.** not more than one

3. The spinner at the right is spun and a coin is tossed. Make a tree diagram to show the sample space.

Use the tree diagram from Exercise 3 to find each probability.

4. P(red, heads)

5. P(not blue, tails)

6. P(green, heads)

7. P(red or white, heads)

8. P(red or blue, tails)

9. P(white, heads or tails)

Exercises

1. Four coins are tossed. Make a tree diagram to show the sample space.

Use the tree diagram from Exercise 1 to find each probability.

2. Find P(exactly 3 tails).

3. Find P(at least 2 heads).

4. Find P(5 heads).

5. Find P(no tails).

6. The number cube below numbered 1 through 6 is rolled and the spinner is spun. Make a tree diagram to show the sample space.

Use the tree diagram from Exercise 6 to find each probability.

7. P(2, blue)

8. P(odd number, red)

9. P(1 or 3, green)

10. P(7, purple)

11. P(number < 6, blue)

12. P(not 5, red)

13. P(even number, red or blue)

14. P(3, not green)

15. P(8, green)

16. P(not 3, not red)

Make a tree diagram and solve.

Dee's Diner offers customers the lunch choices shown at the right. Customers may order a serving of meat, one vegetable, and one soup for one low price. If a customer chooses a meal at random, find each probability.

Meat	chicken, fish, beef
Vegetable	peas, corn
Soup	tomato, onion

17. *P*(chicken, peas, onion soup)

18. *P*(beef, peas or corn, tomato soup)

19. *P*(fish, corn, not tomato soup)

20. *P*(not beef, not corn, tomato soup or onion soup)

Whitney can choose an outfit from the following clothes: two pairs of slacks (navy or gray), four blouses (red, blue, striped, or paisley), and three pairs of shoes (white, blue, or black). If Whitney chooses an outfit at random, find each probability.

21. *P*(navy slacks, paisley blouse, white shoes)

22. *P*(gray slacks, not a red blouse, not white shoes)

SPIRAL REVIEW

23. Solve the proportion: $\frac{3}{x} = \frac{8}{16}$ *(Lesson 9-2)*

24. The radius of a circle is 14 cm. Find the area. *(Lesson 10-4)*

25. Write 68.4% as a decimal. *(Lesson 9-5)*

26. The spinner in Exercise 6 on page 546 is spun and a coin is tossed. Make a tree diagram to show the sample space. Find *P*(blue, tail). *(Lesson 12-2)*

12-3

The Counting Principle

Objective: To use the counting principle to find the number of possible outcomes and to find a probability.

Terms to Know

• *counting principle*

Susan plans to buy a music system. The table below shows the brands of stereo components that she can afford to buy.

Receiver	Speakers	Compact Disc (CD) player
Supra	Ultra	Star
Majestic	Excel	Carlyle
Heritage	Lobo	
Dynamic		

QUESTION How many different systems can Susan choose?

To answer the question, you can make a tree diagram and count the number of favorable outcomes. You may not find this method convenient when the number of final branches in the tree diagram is too great. Example 1 shows another method you can use to answer the question.

Example 1

Find the number of different music systems that could be chosen from the components in the table above.

Solution

There are three stages in choosing a music system from the table above: choosing a receiver, choosing speakers, and choosing a CD player.

Receiver	*Speakers*	*CD player*
4 choices	3 choices	2 choices

You find the number of possible music systems by multiplying the number of choices for each component.

$$4 \cdot 3 \cdot 2 = 24$$

✓ Check Your Understanding

1. Suppose you made a tree diagram to represent the sample space of possible music systems. How many final branches would there be?
2. Suppose there were a fifth brand of receiver listed in the table above. How many different music systems could be chosen?

ANSWER Susan can choose 24 different stereo systems.

Although a tree diagram lists each outcome in a sample space, it is often easier to use the **counting principle** to find the total number of possible outcomes.

The Counting Principle

The number of outcomes for an event is found by multiplying the number of choices for each stage of the event.

Example 2

A whole number from 1 through 20 is chosen at random, then a letter from A through W is chosen at random. Use the counting principle to find P(odd number, C or D).

Solution

First use the counting principle to find the number of favorable outcomes.

number of favorable   number of favorable
choices for an odd number   choices for a C or D
10   ·   2 $= 20$

There are 20 favorable outcomes.

Then find the number of possible choices.

number of choices   number of choices
for a number   for a letter
20   ·   23 $= 460$

There are 460 possible outcomes.

$$P(\text{odd number, C or D}) = \frac{20}{460} = \frac{1}{23}$$

✓ **Check Your Understanding**

3. What would P(odd number, C or D) have been in Example 2 if the letter had been chosen at random from A through T?

Guided Practice

COMMUNICATION «*Reading*

Refer to pages 548–549 of the text.

«**1.** Describe the counting principle in words.

«**2.** Complete: According to the counting principle, if there are m ways to do one activity, n ways to do another activity, and p ways to do a third activity, then there are _?_ ways to do the three activities.

Use the counting principle to find the number of possible outcomes.

3. rolling a number cube twice   **4.** tossing a coin five times

Use the counting principle to find each probability.

5. A coin is tossed five times. Find P(all tails).

6. A number cube is rolled twice. Find $P(4, 4)$.

Exercises

Use the counting principle to find the number of possible outcomes.

1. selecting one hour of the day and one minute of the hour

2. choosing a letter from A through X and a digit from 0 through 9

3. rolling a number cube 3 times

4. choosing an outfit from 3 sport coats, 4 shirts, 5 ties, and 3 pairs of pants

5. choosing a dinner from 6 main courses, 4 vegetables, 2 salads, and 3 beverages

Use the counting principle to find each probability.

6. A coin is tossed four times. Find P(all heads).

7. A number cube is rolled twice. Find P(odd number, 4).

8. A coin is tossed six times. Find P(all tails).

9. One day of the week and one month of the year are selected at random. Find P(Monday or Tuesday, month ending in ''r'').

Exercises 10–17 refer to the number cube, spinners, and cards at the right. Determine the number of possible outcomes.

10. spin spinner A and spin spinner B

11. draw a card and roll the number cube

12. spin spinner A, spin spinner B, and select a card

Find each probability.

13. You spin spinner A twice. Find P(red, red).

14. You draw a card and spin spinner B. Find P(V, not blue).

15. You spin spinner A and spin spinner B.
    a. Find P(yellow or white, blue).
    b. Find P(not red, not red).

16. You roll the number cube, spin spinner A, and select a card.
    a. Find P(2, red, X).
    b. Find P(even number; not white; X, Y, or Z).

17. You roll the number cube, spin spinner A, spin spinner B, and select a card. Find P(number < 5, red, red, not W).

A

B

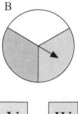

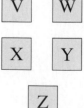

Use this paragraph for Exercises 18–23.

License plates for motor vehicles usually contain six symbols. The first three symbols are often letters from A through Z and the last three symbols are often digits from 0 through 9.

18. How many letters are there from A through Z?

19. How many digits are there from 0 through 9?

In Exercises 20 and 21, assume that letters and digits can repeat.

20. Use the counting principle to find the number of different license plates possible.

21. Find the probability that the first digit on a license plate is a 5.

In Exercises 22 and 23, assume that no letters and no digits repeat.

22. Use the counting principle to find the number of different license plates possible.

23. Find the probability that the third letter on a license plate is a T.

24. A number cube is rolled and a letter from A through M is chosen at random. Find P(odd number, A or M).   *(Lesson 12-3)*

25. Solve the proportion: $\frac{5}{15} = \frac{x}{12}$ *(Lesson 9-2)*

26. Tell whether 15 is *prime* or *composite*.   *(Lesson 7-1)*

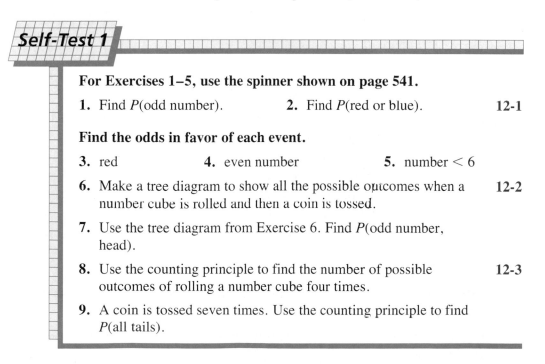

Self-Test 1

For Exercises 1–5, use the spinner shown on page 541.

1. Find P(odd number).   **2.** Find P(red or blue).   **12-1**

Find the odds in favor of each event.

3. red   **4.** even number   **5.** number < 6

6. Make a tree diagram to show all the possible outcomes when a   **12-2**
number cube is rolled and then a coin is tossed.

7. Use the tree diagram from Exercise 6. Find P(odd number, head).

8. Use the counting principle to find the number of possible   **12-3**
outcomes of rolling a number cube four times.

9. A coin is tossed seven times. Use the counting principle to find P(all tails).

Independent and Dependent Events

Objective: To find the probability of two independent or two dependent events.

EXPLORATION

Write the letter *A* on each of 3 index cards, the letter *B* on each of 2 index cards, and the letter *C* on 1 index card. Place all 6 index cards in a large envelope or a box.

1 Draw a card at random. What is the probability of drawing an *A*?

2 Replace the card you selected and draw again. What is the probability of drawing an *A* again?

3 Remove an *A*. Do not replace the card, then draw again. What is the probability of drawing an *A*?

4 Was the probability of drawing an *A* different when you replaced the first card than when you did not? Explain.

Two events may occur either at the same time or one after the other. The two events are **independent** if the occurrence of the first event does not affect that of the second.

Probability of Independent Events

If events *A* and *B* are independent, the probability of both events occurring is found by multiplying the probabilities of the events.

$$P(A \text{ and } B) = P(A) \cdot P(B)$$

Example 1

A bag contains 3 red, 5 white, and 4 blue marbles. Two marbles are drawn at random with replacement. Find *P*(white, then red).

Solution

With replacement means that the first marble is replaced before the second is drawn. The two events are independent.

First find the probability of each event.

$$P(\text{white}) = \frac{5}{12} \qquad P(\text{red}) = \frac{3}{12} = \frac{1}{4}$$

Then multiply. $P(\text{white, then red}) = P(\text{white}) \cdot P(\text{red})$

$$= \frac{5}{12} \cdot \frac{1}{4} = \frac{5}{48}$$

✓ **Check Your Understanding**

1. Explain why the events in Example 1 are independent.
2. In Example 1, why is $P(\text{white}) = \frac{5}{12}$?

Two events are **dependent** if the occurrence of the first event affects that of the second.

Example 2

Use the bag of marbles described in Example 1. Two marbles are drawn at random without replacement. Find $P(\text{white, then red})$.

Solution

Without replacement means that the first marble is not replaced before the second one is drawn. The events are dependent.
$P(\text{white, then red}) = P(\text{white}) \cdot P(\text{red after removing a white})$
First find the probability of each event.

$P(\text{white}) = \dfrac{5}{12}$ ⟵ There are 5 white marbles.
  ⟵ There are 12 marbles altogether.

$P(\text{red after white}) = \dfrac{3}{11}$ ⟵ There are still 3 red marbles.
  ⟵ There are 11 marbles left.

Then multiply. $P(\text{white, then red}) = P(\text{white}) \cdot P(\text{red after white})$

$$= \frac{5}{12} \cdot \frac{3}{11}$$

$$= \frac{15}{132} = \frac{5}{44}$$

☑ Check Your Understanding

3. Explain why the events described in Example 2 are dependent.
4. Consider Examples 1 and 2. Explain why $P(\text{white, then red})$ is different when the marbles are drawn without replacement than when the marbles are drawn with replacement.

Guided Practice

COMMUNICATION «*Discussion*

«**1.** Explain in your own words the difference between independent events and dependent events.

«**2.** Give an example of two events that are independent.

«**3.** Give an example of two events that are dependent.

«**4.** COMMUNICATION «*Reading* Replace the __?__ with the correct expression.
If A and B are independent events, then $P(A \text{ and } B) = $ __?__ .

5. A number from 1 through 10 is chosen. Find $P(7)$.

6. A wallet contains three $5 bills, two $10 bills, and one $20 bill. Two bills are selected at random.
a. Find $P(\$10, \text{ then } \$5)$ if the bills are selected with replacement.
b. Find $P(\$10, \text{ then } \$5)$ if the bills are selected without replacement.

Exercises

A drawer contains 10 blue socks, 15 white socks, and 5 black socks. Two socks are selected at random with replacement. Find each probability.

1. *P*(blue, then white) **2.** *P*(white, then white) **3.** *P*(black, then white)

4. *P*(both blue) **5.** *P*(both black) **6.** *P*(not blue, white)

Exercises 7–16 refer to the cards shown below.

Two cards are drawn at random with replacement. Find each probability.

7. *P*(M, then A) **8.** *P*(not M, then A)

9. *P*(I, then T) **10.** *P*(vowel, then vowel)

Two cards are drawn at random without replacement. Find each probability.

11. *P*(M, then A) **12.** *P*(A, then A)

13. *P*(M, then M) **14.** *P*(E, then T)

15. *P*(vowel, then vowel) **16.** *P*(vowel, then M)

A number cube is rolled three times in a row. Find each probability.

17. *P*(all 4's)

18. *P*(odd number, then odd number, then even number)

A drawer contains 10 white socks and 6 blue socks. Khaled reaches in the drawer without looking and selects 2 socks.

19. What is the probability that he selects two blue socks?

20. What is the probability that he selects first one blue sock and then one white sock?

21. What is the probability that he selects no blue socks?

A bucket contains 10 yellow, 8 white, and 2 orange tennis balls. Without looking, Trina selects 3 tennis balls.

22. What is the probability that she selects three yellow tennis balls?

23. What is the probability that she selects no white tennis balls?

24. Trina selects a ball and replaces it. She then selects two more balls without looking. Find *P*(yellow, then orange, then yellow).

Six boys and four girls are finalists in a contest. All of their names are placed in a hat and three names are drawn at random. The prizes are awarded in the order the names are drawn. Find each probability.

25. P(3 girls)

26. P(3 boys)

27. P(boy, then girl, then boy)

28. P(girl, then boy, then girl)

The combination for a school locker consists of three one-digit numbers. For example, (0, 5, 9) and (3, 4, 1) are possible combinations. In Exercises 29–32, digits may repeat. If combinations are randomly chosen, find each probability.

29. P(2, 2, 2)

30. P(4, 3, 9)

31. P(odd, odd, even)

32. P(all digits are the same)

33. P(all digits are different but none of the digits is 9)

SPIRAL REVIEW

34. In the figures at the right, $\triangle ABC \sim \triangle XYZ$. Find the values of m and n.
(Lesson 10-8)

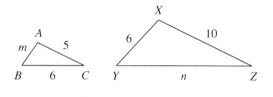

35. A number cube is rolled. Find P(6).
(Lesson 12-1)

36. The Tigers won 40 home games last year and 50 home games this year. Find the percent of increase. *(Lesson 9-9)*

37. A bag contains 4 red marbles and 3 blue marbles. A marble is drawn at random with replacement. Find P(red, then blue). *(Lesson 12-4)*

38. MENTAL MATH   Use mental math to find the solution:
$13 = 2a - 7$ *(Lesson 4-1)*

Hilda Geiringer (1893–1973) received her doctorate in mathematics from the University of Vienna in 1917. She did research and taught at universities in Berlin, Brussels, and Istanbul. She came to the United States in 1938 and taught at Bryn Mawr College and Wheaton College. She made important contributions to probability theory based on the biological studies of Gregor Johann Mendel. She was an avid mountain climber and had a wide knowledge of literature, poetry, and classical music.

Research

Find out why Gregor Johann Mendel is famous.

12-5

Strategy:
Using a Venn Diagram

Objective: To use Venn diagrams to solve problems.

The English mathematician John Venn used diagrams to show the relationships between collections of objects. These diagrams are known today as *Venn diagrams*. You can use Venn diagrams as a strategy for solving problems.

Problem

Of 160 students in the senior class, 55 take French, 90 take History, and 25 take both subjects.

a. How many take only French? How many take only History?

b. How many take either French or History?

c. How many seniors take neither of those subjects?

Solution

UNDERSTAND The problem is about the number of students who take different subjects.

Facts: 160 in class; 55 in French, 90 in History, 25 in both French and History

Find: number who take only one, either one, or neither

PLAN A Venn diagram illustrates the situation and can be used to answer the questions. To draw a Venn diagram, first draw a rectangle that represents the 160 students. Then, inside the rectangle, draw 2 overlapping circles. Label one circle *F* for French and the other *H* for History. The area where the circles overlap represents the 25 students who take both classes.

WORK

a. number taking only French
$55 - 25 = 30$
number taking only History
$90 - 25 = 65$

b. number taking either subject
$30 + 65 + 25 = 120$

c. number taking neither subject
$160 - 120 = 40$

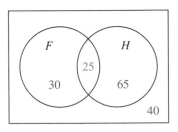

ANSWER **a.** 30 students take only French and 65 take only History.

b. 120 students take either French or History.

c. 40 students take neither subject.

Look Back What if the total number of students in the class were 175 instead of 160. Which answer(s) would change? Which answer(s) would remain the same?

Guided Practice

COMMUNICATION « *Reading*

Use this paragraph for Exercises 1–4.

At Marlin High School, 80 juniors have a part-time job, 110 juniors drive to school, and 35 juniors both drive and have part-time jobs. There are 280 juniors at the school.

« **1.** How many juniors are there at the school?

« **2.** How many juniors drive to school?

« **3.** Draw a Venn diagram to represent the situation.

« **4.** How many juniors neither drive nor have part-time jobs?

Solve using a Venn diagram.

5. Of 50 people at a concert, 32 like rock music, 14 like country music, and 8 like both types of music. How many people like neither rock nor country music?

Problem Solving Situations

Solve using a Venn diagram.

Of 250 Wayland residents surveyed, 140 read the *Daily Times*, 80 read the *Evening Tribune*, and 45 read both newspapers.

1. How many residents read only the *Daily Times*?

2. How many residents read at least one of the newspapers?

3. How many residents read neither newspaper?

In a class of 42 people, 23 wear a watch, 12 wear tennis shoes, and 8 wear both tennis shoes and a watch.

4. How many wear tennis shoes but not a watch?

5. How many do not wear tennis shoes and do not wear a watch?

6. If one person in the class were chosen at random, what is the probability that the person is not wearing a watch?

Of the 154 student musicians at Sweetwater High School, 72 are in the marching band, 35 are in the jazz band, and 20 are in both.

7. How many are in neither band?

8. How many play in the jazz band, but not the marching band?

9. How many play only in the marching band?

Solve using any problem solving strategy.

10. A dentist schedules 15 min for each patient. How many patients can the dentist see in one day?

11. At one point in a race, Julie was $\frac{1}{2}$ mi behind Marvin, Marvin was $\frac{3}{4}$ mi ahead of Lance, and Julie was $\frac{1}{4}$ mi behind Kim. List the order of the runners at that point.

12. How many diagonals can be drawn in an octagon?

13. A spinner has a red, a blue, and a yellow sector, each numbered 1, 2, or 3. The number of the blue sector is an odd number. The red sector is not numbered 1. The number of the yellow sector is one more than the number of the red sector. What number is the red sector?

14. Of 24 students in a gym class, 12 liked volleyball, 8 liked softball, and 6 liked both. How many students liked neither?

WRITING WORD PROBLEMS

Write a word problem that you could solve using each Venn diagram. Then solve the problem.

15.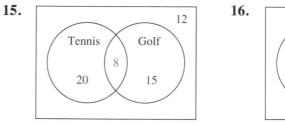

16.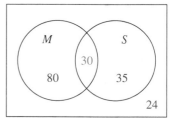

SPIRAL REVIEW

17. Solve the proportion: $\frac{6}{9} = \frac{10}{n}$ *(Lesson 9-2)*

18. Kinta's math scores were 78, 84, 96, 82, and 80. Find the mean, the median, the mode(s), and the range. *(Lesson 5-8)*

19. Find the next three numbers in the pattern: 8, 9, 11, 14, __?__, __?__, __?__ *(Lesson 2-5)*

20. There are 29 students in a history class. Of these, 16 students take geometry, 18 take Spanish, and 11 take both geometry and Spanish. How many students take neither geometry nor Spanish? *(Lesson 12-5)*

Permutations and Combinations

Objective: To find the number of permutations and the number of combinations of a group of items.

Terms to Know
- *permutation*
- *factorial*
- *combination*

The personnel director of a computer software company plans to interview 3 people for a job opening. The director can choose any one of the 3 people for the first interview, either of the other 2 for the second, and the remaining person for the last. By the counting principle, the director can arrange the interviews in $3 \cdot 2 \cdot 1 = 6$ different orders.

An arrangement of a group of items in a particular order is called a **permutation**. The number of permutations of 3 items is $3 \cdot 2 \cdot 1 = 6$. The expression $3 \cdot 2 \cdot 1$ can be written 3!, which is read "3 **factorial.**"

Example 1

Find the number of permutations of the letters in the word MATH.

Solution

Use the counting principle. As a letter is chosen for each position in the "word," there is one less letter available for the next position.

position 1		position 2		position 3		position 4	
4	$\cdot$	3	$\cdot$	2	$\cdot$	1	= 24

There are 24 permutations of the letters in the word MATH.

✓ **Check Your Understanding**

1. In Example 1, list the 6 permutations that start with the letter M.
2. Explain how you could use a factorial to solve Example 1.

You can also make an arrangement using only part of a group of items.

Example 2

In how many different ways can first and second prizes be awarded in a contest among 5 people?

Solution

Find the number of permutations of *5 people taken 2 at a time*.

number of choices for first place		number of choices for second place	
5	$\cdot$	4	= 20

There are 20 ways to award first and second prizes.

Permutations can be described by a brief notation. The number of permutations of 5 items can be written as $_5P_5$. Similarly, the number of permutations of 5 items taken 2 at a time can be written as $_5P_2$.

$$_5P_5 = 5! = 5 \cdot 4 \cdot 3 \cdot 2 \cdot 1 = 120 \qquad _5P_2 = 5 \cdot 4 = 20$$

A **combination** is a group of items chosen from within a group of items. The difference between one combination and another is the items in the combination, not the order of the items.

To understand the difference between a permutation and a combination, consider this example. Suppose there are three candidates, Ling, Jackson, and Garcia, running for two positions on a committee. There are six possible orders in which the candidates can finish in the voting, but only three different possible results of the election.

Possible Orders	Possible Election Results
Ling, Jackson, Garcia	
Jackson, Ling, Garcia	Ling and Jackson are elected.
Ling, Garcia, Jackson	
Garcia, Ling, Jackson	Ling and Garcia are elected.
Jackson, Garcia, Ling	
Garcia, Jackson, Ling	Jackson and Garcia are elected.

The list on the left contains all the permutations for this situation. The list on the right contains all the combinations for this situation.

Notice that there are 6 permutations of 3 people taken 2 at a time.

$$_3P_2 = 3 \cdot 2 = 6$$

To find the number of combinations of items taken 2 at a time, divide by $_2P_2$ to eliminate answers that are the same except for order.

$$_3P_2 \div {_2P_2} = 6 \div 2 = 3$$

There are 3 combinations of 3 people taken 2 at a time.

Example 3

A committee of 3 is to be chosen from a group of 7 people. Find the number of combinations.

Solution

Find the number of combinations of 7 people taken 3 at a time. First find $_7P_3$. Then divide by $_3P_3$.

$$\frac{_7P_3}{_3P_3} = \frac{7 \cdot 6 \cdot 5}{3 \cdot 2 \cdot 1} = \frac{210}{6} = 35$$

There are 35 ways that a committee of 3 can be chosen.

Combinations can also be described by a brief notation. The number of combinations of 7 items taken 3 at a time is written as $_7C_3$.

$$_7C_3 = \frac{_7P_3}{_3P_3} = 35$$

You can use a calculator to find a number of permutations. If your calculator has a factorial key, you can use this calculator key sequence to find $_4P_4$.

If your calculator does not have a factorial key, you can use this calculator key sequence to find $_4P_4$.

Guided Practice

Tell whether each phrase refers to a *permutation* or a *combination*.

1. the number of ways to arrange 5 books on a shelf

2. the number of ways to select 3 books from a choice of 12 books

COMMUNICATION « *Writing*

Write each permutation or combination in words.

«**3.** $_2P_2$ «**4.** $_5P_1$ «**5.** $_9C_3$ «**6.** $_7C_4$

Use notation as in Exercises 3–6 to express each permutation or combination.

«**7.** the number of permutations of 8 out of 15 items

«**8.** the number of combinations of 7 items taken 4 at a time

Find the value of each expression.

9. 6!   **10.** 7!   **11.** $_5P_4$ **12.** $_4C_3$

Find the number of permutations.

13. the letters in the word QUIET   **14.** 3 out of 6 students

Find the number of combinations.

15. 5 scarves taken 2 at a time   **16.** 4 out of 6 committee members

Exercises

Find the number of permutations.

1. 2 posters on 2 walls

2. the digits 1, 2, 3, 4, and 5

3. the letters in the word MARBLE

4. the letters in the word COMPUTE

5. a president and a secretary from 12 students

6. first, second, and third prizes among 9 contestants

Find the number of combinations.

7. 6 cassette tapes taken 3 at a time

8. a committee of 3 selected from a group of 8

9. 3 colors selected from 10 choices

10. 2 videotapes selected from a group of 6

Find the value of each expression.

11. $_4P_4$ 12. $_4C_4$ 13. $_6C_2$ 14. $_8P_2$

15. In how many different ways can 5 people sit in a row of 5 seats?

16. In how many different ways can 5 books be chosen from a collection of 5 books?

17. In how many different ways can 4 paintings be selected from a group of 10 paintings?

18. There are 25 members in the journalism club. In how many different ways can a chairperson and a treasurer be chosen?

19. **WRITING ABOUT MATHEMATICS** Write a paragraph describing the difference between a permutation and a combination.

CALCULATOR

Use a calculator to find the number of permutations.

20. 7 shirts hanging in a closet

21. the letters in the word PHONE

22. 10 charms on a bracelet

23. 9 hats on 9 heads

24. A calculator displays greater numbers in scientific notation. What is the least whole number with a factorial great enough that it appears in scientific notation on your calculator display?

CONNECTING MATHEMATICS AND PHYSICAL EDUCATION

25. A school system belongs to a league of 8 cross-country teams. A school plays each of the other schools one time during the season. How many cross-country meets does the league have in a season?

26. How many different basketball squads with 5 players can a physical education teacher choose from a class of 12 students?

27. In how many different ways can a coach arrange the batting order of the 9 starting players on a baseball team?

THINKING SKILLS

28. Find $_7C_2$.   29. Find $_7C_5$.

30. Compare the results of Exercises 28 and 29. What do you notice?

31. Write a formula comparing $_nC_r$ and $_nC_{n-r}$.

CONNECTING ALGEBRA AND GEOMETRY

Through any 2 points there is one and only one line. That is, 2 points determine a line. The number of lines determined by a group of points, no three of which lie on the same line, is the number of combinations of the group taken 2 at a time.

Find the number of lines determined by the given number of points. Assume that no three points lie on the same line.

32. 3 **33.** 4 **34.** 5 **35.** 6

36. PATTERNS Use your answers to Exercises 32–35.
   a. What is the pattern that describes the increase in the number of possible lines?
   b. Use the pattern to predict the number of lines determined by 7 points, by 8 points, and by 9 points.

SPIRAL REVIEW

37. 25% of what number is 20? *(Lesson 9-8)*

38. The length of a rectangle is 12 cm. The width of the rectangle is 8.5 cm. Find the perimeter. *(Lesson 10-1)*

39. Find the number of permutations of the letters in the word BRIGHT. *(Lesson 12-6)*

40. A coin is tossed 4 times. Find P(all heads). *(Lesson 12-4)*

41. DATA, *pages 538–539* In 1989, what percent of endangered species were birds? *(Lesson 11-6)*

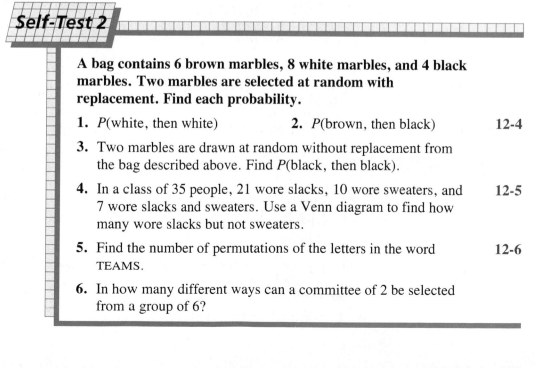

Self-Test 2

A bag contains 6 brown marbles, 8 white marbles, and 4 black marbles. Two marbles are selected at random with replacement. Find each probability.

1. P(white, then white) **2.** P(brown, then black) 12-4

3. Two marbles are drawn at random without replacement from the bag described above. Find P(black, then black).

4. In a class of 35 people, 21 wore slacks, 10 wore sweaters, and 7 wore slacks and sweaters. Use a Venn diagram to find how many wore slacks but not sweaters. 12-5

5. Find the number of permutations of the letters in the word TEAMS. 12-6

6. In how many different ways can a committee of 2 be selected from a group of 6?

Experimental Probability

Objective: To find the experimental probability of an event.

Terms to Know
- *experimental probability*

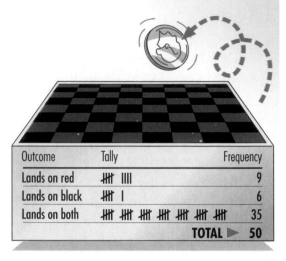

A nickel is tossed onto a checkerboard 50 times. If the nickel does not land on the checkerboard, the toss is not counted. The results are shown in the frequency table at the right.

In many cases, the possible outcomes are not equally likely to happen. In the situation described above, the nickel is not equally likely to land on red, on black, or on both. So you cannot use the probability formula given on page 541 to determine the probability of the given event. Instead, you find the **experimental probability** based on the collection of actual data.

Outcome	Tally	Frequency
Lands on red	ⅢⅢ ⅢⅠ	9
Lands on black	ⅢⅢ Ⅰ	6
Lands on both	ⅢⅢ ⅢⅢ ⅢⅢ ⅢⅢ ⅢⅢ ⅢⅢ ⅢⅢ	35
	TOTAL ▶	**50**

Example

Use the results in the table above to find the experimental probability that the nickel lands on both colors.

Solution

$P(\text{both red and black}) = \dfrac{35}{50} = \dfrac{7}{10}$

✔️ Check Your Understanding

1. What does 35 represent in the fraction in the Example?
2. If a dime were used instead of a nickel, do you think the results would have been the same? Explain.

Guided Practice

COMMUNICATION « *Reading*

«**1.** What is the objective of this lesson? How is it different from the objective of Lesson 12-1?

«**2.** When should experimental probability be used rather than the probability formula?

Refer to the frequency table on page 564. Find each probability.

3. *P*(only black)   **4.** *P*(only red)

5. *P*(orange)   **6.** *P*(only red or only black)

7. *P*(only red, only black, or both red and black)

Exercises

Matt, Flavio, and Juana each tossed a paper cup 40 times. They recorded whether the cup landed up, down, or on its side. Use their results in the table below to find each experimental probability.

	Up	Down	Side
Matt	8	12	20
Flavio	6	10	24
Juana	7	11	22

1. *P*(up) for Matt   **2.** *P*(up) for Flavio   **3.** *P*(up) for Juana

4. *P*(down) for Matt   **5.** *P*(side) for Flavio   **6.** *P*(side) for Juana

7. *P*(down) for Flavio   **8.** *P*(up) for the whole group

9. *P*(down) for the whole group   **10.** *P*(side) for the whole group

The table below shows Stan's results when he tossed a bottle cap 80 times. Use the results to find each experimental probability.

Outcome	Frequency	Tally
up	||||| ||||| ||||| ||||| ||||| ||||| ||||| ||||| |||||	45
down	||||| ||||| ||||| ||||| ||||| ||||| |||||	35

11. *P*(up)   **12.** *P*(down)   **13.** *P*(up or down)

14. Penny also tossed a bottle cap 80 times. It landed down 56 times. Find the probability that the bottle cap does not land down.

15. Killian tossed a paper cup. It landed up 20 times, down 26 times, and on its side 34 times. Find *P*(up), *P*(down), and *P*(side).

16. A baseball player got a hit 36 times in the last 150 times at bat. What is the probability that the player will get a hit the next time at bat? Write the probability as a decimal to the nearest thousandth.

17. DATA, *pages 538–539* The chances of successful mammal translocation are determined by experimental probability. Find the probability that the translocation of 50 mammals into a new environment under excellent conditions will be successful over a period of three years.

Randomly select a page in this textbook. Count the number of words that contain 1, 2, 3, 4, 5, 6, or more than 6 letters. (*Note:* Numbers should be counted as words. For example, the number 387 should be counted as a 3-letter word. Do not count operation symbols.)

18. Copy and complete the chart to record your data.

Outcome	Tally	Frequency
1 letter		
2 letters		
3 letters		
4 letters		
5 letters		
6 letters		
more than 6 letters		
		Total:

Use the results recorded in your chart in Exercise 18 to find each experimental probability.

19. $P(4)$ 　　　　**20.** $P(3)$ 　　　　**21.** $P(2)$

22. $P(\text{more than 6})$ 　　**23.** $P(\text{fewer than 5})$ 　　**24.** $P(\text{more than 4})$

25. Which word length appeared most often on the page you chose for your experiment? Which word length appeared least often?

26. Would you expect the same results if you chose another page?

27. Solve using a proportion: *The Weekly News* charges $14 for delivery for 8 weeks. At this rate, what is the charge for 26 weeks? (*Lesson 9-4*)

28. Graph the inequality $x < -1.5$. (*Lesson 7-10*)

29. Alexis tossed a paper cup. It landed up 20 times, down 30 times, and on its side 40 times. Find $P(\text{side})$. (*Lesson 12-7*)

30. Find the measure of an angle complementary to 80°. (*Lesson 6-3*)

31. Draw a circle graph to display the data. (*Lesson 11-7*)

Tickets Sold by Grade Level

Grade	Freshman	Sophomore	Junior	Senior
Number Sold	60	30	90	120

32. Solve by supplying the missing fact: On three separate days, Angela spent 45 min, 55 min, and 50 min studying biology. How many hours in all did she spend studying biology? (*Lesson 8-7*)

Using BASIC to Simulate an Experiment

Objective: To estimate the waiting time for an event to happen.

Car license plates in many states have three numbers and three letters, for example 169 SPZ or 343 HAT. A typical *waiting time* problem about license plates is "About how many license plates would you have to inspect to find one whose letter section has the form 'consonant,' 'vowel,' 'consonant'?" A *waiting time* problem asks for the number of experiments you have to perform until you get a particular event.

The computer program below randomly selects three letters and tells you if each letter is a vowel or a consonant. It does this by picking a decimal at random between 0 and 1. If the decimal is less than $\frac{5}{26}$ (the probability of getting a vowel), the word VOWEL is printed. If the random decimal is greater than or equal to $\frac{5}{26}$, the word CONSONANT is printed. The program performs twenty experiments.

```
10   FOR E = 1 TO 20
20   FOR N = 1 TO 3
30   LET D = RND (1)
40   IF D < 5/26 THEN PRINT "VOWEL ";
50   IF D >= 5/26 THEN PRINT "CONSONANT ";
60   NEXT N
70   PRINT
80   NEXT E
90   END
```

Exercises

Use the computer program above.

1. Run the program. Count the number of experiments needed until you first see a CONSONANT VOWEL CONSONANT outcome.

2. Run the program 10 times. Find the average waiting time until you see a VOWEL CONSONANT VOWEL outcome.

3. Find the average waiting time until you see at least one VOWEL.

Change lines 40 and 50 in the program above. Use the new program for Exercises 4 and 5.

```
40   IF D < 1/2 THEN PRINT "ODD ";
50   IF D >= 1/2 THEN PRINT "EVEN ";
```

4. Run the program 10 times. Find the average waiting time until you see the numbers in the order ODD EVEN ODD.

5. Find the average waiting time until you see at least one ODD.

Using Samples to Make Predictions

Objective: To make predictions using sampling and probability concepts.

Terms to Know
• *random sample*

The Fairtown Tribune conducted a poll before a primary election. The poll takers chose a group of 400 people at random from among the 220,000 registered voters. Each person in this **random sample,** or sample chosen at random from a larger group, was asked, "If the election were held today, which candidate would receive your vote?" The results are shown in the table below.

Candidate	Number of Supporters
Perez	115
Say	90
Wong	109
Wiley	86

Example

How many votes might Say expect to receive in the election?

Solution

Use the sample results to find P(Say). There were 400 people in the random sample. Ninety people in the sample would vote for Say.

$$P(\text{Say}) = \frac{90}{400} = 0.225$$

The number of votes expected is the product of this probability and the total number of registered voters.

P(Say) × number of registered voters = expected number of votes

0.225 × 220,000 = 49,500

Say might expect to receive about 49,500 votes.

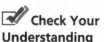 **Check Your Understanding**

1. How does a poll taker determine the number of votes a candidate can expect?
2. About how many voters are not expected to vote for Say?

Guided Practice

« **1.** COMMUNICATION «*Reading*   If the sample described above included 400 of Say's relatives, friends, and business associates, would that be a random sample? Explain.

« **2.** COMMUNICATION «*Discussion*   Explain how you would select a random sample to predict the outcome of a class election.

A television ratings service asked a random sample of 300 of the 44,000 residents of Millsburg to name their favorite program. Of those polled, 60 chose *The Hendersons*, 44 chose *The Lawyers*, 74 chose *Win That Vacation*, 80 chose *Comedy Hour*, and 42 chose *The Detectives*. Use these sample results for Exercises 3–6.

3. Find P(The Hendersons).

4. Find P(The Detectives).

If the whole town were questioned, about how many residents would be expected to select each program as their favorite?

5. The Hendersons

6. The Detectives

Exercises

Refer to the text on page 568. About how many votes might each candidate expect to receive?

1. Perez

2. Wiley

3. Wong

Solve. In Exercises 4–7, assume that each sample is random.

4. When 50 seniors were questioned, 40 said they plan to attend college. About how many of the 480 seniors are expected to attend college?

5. When 75 residents of a senior citizen apartment were questioned, 35 said they favor construction of a new shopping center across the street. About how many of the 450 residents are expected to favor construction of the new shopping center?

6. A random sample of 150 owners of a new Brandon automobile were questioned to determine their satisfaction. Of those polled, 120 stated they were very satisfied, 25 stated they were satisfied, and 5 stated they were dissatisfied. About how many of the 12,000 owners of new Brandon automobiles are expected to be dissatisfied?

7. A random survey, or poll, of 400 people showed that 120 listen to radio station KMAT in the mornings and 80 people listen to station KEFA in the evenings. The survey was conducted in an area with a population of about 18,000 people. About how many people in the area are expected to listen to KMAT in the mornings? About how many are expected to listen to KEFA in the evenings?

DATA ANALYSIS

A cable television company polled a random sample of 200 of their 42,000 customers. The customers were asked to rate their service as excellent, good, fair, or poor. The results are shown in the table at the right.

Quality of Service	Number Who Voted
Excellent	104
Good	58
Fair	34
Poor	4

Find the probability that a person polled feels the quality of service is at each level.

8. excellent
9. good
10. fair
11. poor

Use the results of the cable company poll to predict how many of the 42,000 subscribers feel their cable service is at each level.

12. excellent
13. good
14. fair
15. poor

CAREER/APPLICATION

A *quality control inspector* checks random samples of certain manufactured goods to find the number of defective goods in the sample. The inspector then predicts the number of defective goods in a larger quantity of the item. Depending on the seriousness of the defect or on the number of defective goods predicted, the manufacturer may decide not to sell certain *production lots* of the item.

16. A quality control inspector selected 40 out of a lot of 6000 light bulbs to inspect and found 3 to be defective. Predict the number of defective light bulbs that might be found in the entire lot.

17. Ten out of a sample of 100 goods inspected were defective. It was later found that 45 goods in the entire production lot were defective. Approximate the number of goods in this production lot.

18. A manufacturer of computer diskettes will not sell a production lot that contains more than 5% defective diskettes. A quality control inspector checks a sample of 50 diskettes in a certain lot and finds that 2 are defective. Would the manufacturer sell this production lot?

19. Suppose 30 out of the 2400 cartons of milk in a production lot have leaks. What percent of the cartons in the lot have leaks?

Tell whether it is appropriate to use a poll to predict each of the following. If it is not appropriate, describe a more appropriate method.

20. the winner of a spelling contest

21. the winner of a golf tournament

22. the favorite television program among students in a school

23. the favorite book read by students in a school during the summer vacation

SPIRAL REVIEW

24. Find the area of a triangle with base 14 in. and height 10 in. *(Lesson 10-3)*

25. When 40 tenth graders, selected randomly, were questioned, 12 said they walk to school. About how many of the 520 tenth graders in the town walk to school? *(Lesson 12-8)*

26. Simplify: x^{-3} *(Lesson 7-11)*

27. Estimate: $-15\frac{1}{3} \div \frac{5}{14}$ *(Lesson 8-6)*

28. Tell whether you need to calculate *perimeter, circumference,* or *area* to find the amount of grazing land in a rectangular field. *(Lesson 10-5)*

Self-Test 3

Charlita tossed a paper cup 150 times. It landed up 84 times, down 27 times, and on its side 39 times. Find each experimental probability.

1. *P*(up) **2.** *P*(down) **3.** *P*(side) 12-7

4. The *Daily Newsreporter* polled a random sample of 500 registered voters before the local school board election. The results are shown below. 12-8

Candidate	Number of Supporters
Bernedez	150
Caldwell	180
Stewart	70
Aluga	100

There are 180,000 registered voters. From about how many voters does Aluga expect support in the election?

Using a Table of Random Numbers

Objective: To use a table of random numbers to select a random sample.

Marie DeForge is a marketing executive for a publishing company. She plans to survey 50 students from a university with 20,000 students. She will use a table of random numbers to select the 50 students.

A table of random numbers is a list of numbers in which the digits 0, 1, 2, 3, 4, 5, 6, 7, 8, and 9 are arranged randomly.

Example

a. How can Marie use a table of random numbers to select 50 students at random?

b. What are the first 5 numbers selected by this method?

Solution

a. Marie assigns each student a five-digit number from 00001 to 20000. To select a random sample of 50 students, she randomly selects a starting point and a direction from the table on page 573. She will discard any number that is greater than 20,000. Suppose she starts on column 4, line 1, and reads down the column. When she completes column 4, she will read down column 5, and so on.

b. The first number in column 4, line 1, is 02011. The first student selected has number 2011. The next number is 85393. That number is discarded. The next four numbers Marie selects are 16656, 07972, 10281, and 03427.

Exercises

Use the random number table on page 573 to answer each question. Start with the indicated column and line and read down the column.

1. The mayor's office intends to interview a random sample of 15 of the town's 15,000 households. Each household is assigned a number from 00001 to 15000. (column 4, line 1)

   a. What are the first five numbers that will be discarded?

   b. Why will the numbers in part (a) be discarded?

   c. What are the 15 numbers selected by this method?

2. A magazine plans to celebrate its twentieth anniversary by randomly selecting 20 of its 97,000 subscribers and awarding them a free lifetime subscription. (column 2, line 6)

   a. To use the table of random numbers, what group of numbers can be assigned to the subscribers?

   b. What are the first and last numbers chosen in the random sample?

   c. How many numbers will be discarded?

3. The customer service department of a bank plans to interview a random sample of 25 of the bank's 85,056 bank card holders to get their reactions to the benefits of the card. Describe how to use a table of random numbers to select 25 card holders at random.

Table of Random Numbers

COL. LINE	1	2	3	4	5	6	7	8	9	10
1	10480	15011	01536	02011	81647	91646	69179	14194	62590	36207
2	22368	46573	25595	85393	30995	89198	27982	53402	93965	34095
3	24130	48360	22527	97265	76393	64809	15179	24830	49340	32081
4	42167	93093	06243	61680	07856	16376	39440	53537	71341	57004
5	37570	39975	81837	16656	06121	91782	60468	81305	49684	60672
6	77921	06907	11008	42751	27756	53498	18602	70659	90655	15053
7	99562	72905	56420	69994	98872	31016	71194	18738	44013	48840
8	96301	91977	05463	07972	18876	20922	94595	56869	69014	60045
9	89579	14342	63661	10281	17453	18103	57740	84378	25331	12566
10	85475	36857	53342	53988	53060	59533	38867	62300	08158	17983
11	28918	69578	88231	33276	70997	79936	56865	05859	90106	31595
12	63553	40961	48235	03427	49626	69445	18663	72695	52180	20847
13	09429	93969	52636	92737	88974	33488	36320	17617	30015	08272
14	10365	61129	87529	85689	48237	52267	67689	93394	01511	26358
15	07119	97336	71048	08178	77233	13916	47564	81056	97735	85977
16	51085	12765	51821	51259	77452	16308	60756	92144	49442	53900
17	02368	21382	52404	60268	89368	19885	55322	44819	01188	65255
18	01011	54092	33362	94904	31273	04146	18594	29852	71585	85030
19	52162	53916	46369	58586	23216	14513	83149	98736	23495	64350
20	07056	97628	33787	09998	42698	06691	76988	13602	51851	46104
21	48663	91245	85828	14346	09172	30168	90229	04734	59193	22178
22	54164	58492	22421	74103	47070	25306	76468	26384	58151	06646
23	32639	32363	05597	24200	13363	38005	94342	28728	35806	06912
24	29334	27001	87637	87308	58731	00256	45834	15398	46557	41135
25	02488	33062	28834	07351	19731	92420	60952	61280	50001	67658
26	81525	72295	04839	96423	24878	82651	66566	14778	76797	14780
27	29676	20591	68086	26432	46901	20849	89768	81536	86645	12659
28	00742	57392	39064	66432	84673	40027	32832	61362	98947	96067
29	05366	04213	25669	26422	44407	44048	37937	63904	45766	66134
30	91921	26418	64117	94305	26766	25940	39972	22209	71500	64568
31	00582	04711	87917	77341	42206	35126	74087	99547	81817	42607
32	00725	69884	62797	56170	86324	88072	76222	36086	84637	93161
33	69011	65795	95876	55293	18988	27354	26575	08625	40801	59920
34	25976	57948	29888	88604	67917	48708	18912	82271	65424	69774
35	09763	83473	73577	12908	30883	18317	28290	35797	05998	41688
36	91567	42595	27958	30134	04024	86385	29880	99730	55536	84855
37	17955	56349	90999	49127	20044	59931	06115	20542	18059	02008
38	46503	18584	18845	49618	02304	51038	20655	58727	28168	15475
39	92157	89634	94824	78171	84610	82834	09922	25417	44137	48413
40	14577	62765	35605	81263	39667	47358	56873	56307	61607	49518
41	98427	07523	33362	64270	01638	92477	66969	98420	04880	45585
42	34914	63976	88720	82765	34476	17032	87589	40836	32427	70002
43	70060	28277	39475	46473	23219	53416	94970	25832	69975	94884
44	53976	54914	06990	67245	68350	82948	11398	42878	80287	88267
45	76072	29515	40980	07391	58745	25774	22987	80059	39911	96189
46	90725	52210	83974	29992	65831	38857	50490	83765	55657	14361
47	64364	67412	33339	31926	14883	24413	59744	92351	97473	89286
48	08962	00358	31662	25388	61642	34072	81249	35648	56891	69352
49	95012	68379	93526	70765	10592	04542	76463	54328	02349	17247
50	15664	10493	20492	38391	91132	21999	59516	81652	27195	48223

Chapter Review

Terms to Know

outcome (p. 541)
equally likely (p. 541)
event (p. 541)
probability (p. 541)
odds in favor (p. 542)
sample space (p. 545)
tree diagram (p. 545)
counting principle (p. 549)

independent events (p. 552)
dependent events (p. 553)
permutation (p. 559)
factorial (p. 559)
combination (p. 560)
experimental probability (p. 564)
random sample (p. 568)

Choose the correct term from the list above to complete each sentence.

1. Two events are __?__ if the occurrence of the first affects the occurrence of the second.

2. A(n) __?__ is any group of outcomes.

3. A(n) __?__ is a list of all possible outcomes.

4. An arrangement of a group of items in a particular order is a(n) __?__.

5. The __?__ of an event is based on the collection of actual data.

6. The expression 5! is an example of a(n) __?__.

Use the spinner at the right. Find each probability. *(Lesson 12-1)*

7. $P(4)$ 8. $P(8)$ 9. P(blue)

Use the spinner at the right. Find the odds in favor of each event. *(Lesson 12-1)*

10. odd number   11. red   12. not blue

13. The spinner at the right is spun and a coin is tossed. Make a tree diagram to show the sample space. *(Lesson 12-2)*

Use the tree diagram from Exercise 13 to find each probability. *(Lesson 12-2)*

14. $P(4, \text{heads})$ 15. $P(2 \text{ or } 5, \text{tails})$

16. $P(\text{not } 1, \text{not heads})$ 17. $P(6, \text{tails})$

Use the counting principle. *(Lesson 12-3)*

18. Teruko is buying a couch. She has a choice of 3 styles, 5 colors, and 4 different fabrics. Find the number of possible choices.

19. A whole number from 1 through 15 is chosen at random and a coin is tossed. Find $P(\text{even}, \text{heads})$.

Exercises 20–23 refer to a box that contains 7 red plastic chips, 5 blue plastic chips, and 3 white plastic chips.

Two chips are drawn at random with replacement. Find each probability. *(Lesson 12-4)*

20. P(blue, then red)

21. P(red, then white)

Two chips are drawn at random without replacement. Find each probability. *(Lesson 12-4)*

22. P(red, then red)

23. P(blue, then white)

24. In a mathematics class of 28 students, 14 are also taking physics, 12 are taking biology, and 5 are taking both physics and biology. Use a Venn diagram to find how many students in the mathematics class are not taking either physics or biology. *(Lesson 12-5)*

Find the number of permutations. *(Lesson 12-6)*

25. the letters in the word NUMBER

26. the letters in the word TRUCK taken three at a time

Find the number of combinations. *(Lesson 12-6)*

27. 4 books selected from a group of 9

28. 7 students selected 2 at a time

29. A baseball cap is tossed onto a floor made of brown and white tiles. The results are shown in the table at the right. Find P(brown). *(Lesson 12-7)*

Outcome	Frequency
Lands on white	12
Lands on brown	10
Lands on both	28

In a poll taken before a recent election, 250 voters were asked their reasons for favoring candidates. *(Lesson 12-8)*

Do you favor a candidate because of the candidate's	Number polled responding "Yes"
television appearance?	65
decisions on key issues?	92
past political voting record?	56
political party affiliation?	37

30. Find the probability that a person polled favors a candidate because of the candidate's television appearance.

31. The district where the poll was taken includes 400,000 registered voters. About how many of these voters are expected to favor a candidate because of the candidate's television appearance?

Chapter Test

Use the spinner at the right. Find each probability.

1. $P(4)$ **2.** $P(0)$ **3.** $P(\text{white})$ **12-1**

Use the spinner at the right. Find the odds in favor of each event.

4. even number   **5.** blue   **6.** not white

7. Explain why it is impossible for the probability of an event to be greater than 1.

8. A coin is tossed and the spinner above is spun. Make a tree diagram to show the sample space.   **12-2**

Use the tree diagram from Exercise 8 to find each probability.

9. $P(\text{heads, 5})$ **10.** $P(\text{tails, odd number})$

11. Ed is buying a suit. He has a choice of 6 colors, 4 styles, and 2 fabrics. Use the counting principle to find the number of possible outcomes.   **12-3**

12. A whole number from 1 through 10 is chosen at random and a letter from M through Z is chosen at random. Use the counting principle to find $P(\text{prime number, X or Y})$.

A bag contains 10 red marbles, 5 blue marbles, and 3 white marbles.

13. Two marbles are chosen with replacement. Find $P(\text{blue, then white})$.   **12-4**

14. Two marbles are chosen without replacement. Find $P(\text{red, then red})$.

15. Write a word problem that involves dependent events. Then solve the problem.

16. Of 100 people, 55 like basketball, 45 like hockey, and 20 like both. Use a Venn diagram to find how many people do not like either hockey or basketball.   **12-5**

17. Find the number of permutations of the letters in the word HISTORY.   **12-6**

18. Find the number of permutations of the letters in the word LAWYER taken four at a time.

19. A committee of 5 people is to be chosen from a group of 8 people. Find the number of combinations.

20. A paper cup is tossed 60 times. It lands on its side 38 times, on its top 8 times, and on its bottom 14 times. Find $P(\text{bottom})$.   **12-7**

21. A random survey of a class of 50 students showed that 12 of the students have dogs as pets. The survey was conducted in a school that has 750 students. About how many students in the school are expected to have dogs as pets?   **12-8**

Expected Value

Objective: To find the expected value of an event.

Many decisions are based on a knowledge of **expected value.** The expected value of an experiment is the sum of the products of each outcome and its probability. Expected value tells you the result that is obtained *on average* when an experiment is performed repeatedly.

Example

An outdoor theater loses $4000 on concert days when it rains and earns $12,000 on concert days when it does not rain. The probability that it will rain on a concert day is 0.2. Find the amount that the outdoor theater can expect to earn on average for a concert.

Solution

There are two possible outcomes: it will rain and the theater will lose $4000; it will not rain and the theater will earn $12,000.

$P(\text{rain}) = 0.2; P(\text{no rain}) = 1 - P(\text{rain}) = 0.8$

$$\begin{aligned}
\text{expected value} &= -4000 \cdot P(\text{rain}) + 12{,}000 \cdot P(\text{no rain}) \\
&= -4000(0.2) + 12{,}000(0.8) \\
&= -800 + 9600 = 8800
\end{aligned}$$

The outdoor theater can expect to earn $8800 on average for a concert.

Exercises

1. An appliance store offers each customer an envelope that contains a discount. The discounts and their probabilities are shown in the table. Find the discount that a customer can expect on average.

Discount	Probability
$100	0.02
$50	0.06
$20	0.32
$10	0.60

2. A building contractor bids on a job to build a new school. On such a project, the contractor has a 75% chance of making a profit of $48,000 and a 25% chance of losing $10,000. What can the contractor expect to earn on average on this job?

3. A game is played by tossing a coin. If you toss heads, you win 10 points. If you toss tails, you lose 5 points. Find the number of points that you can expect to win on average for a given toss.

4. The players who finish first, second, third, and fourth in a tennis tournament earn $120,000, $80,000, $50,000, and $20,000, respectively. How much can one of the four finalists expect to win on average if all four finalists have an equal chance of winning?

Standardized Testing Practice

Choose the letter of the correct answer.

1. There are 5 math teachers, 4 science teachers, 3 administrators, and 7 school board members. A committee must be formed that consists of one member of each group. How many different committees are possible?
   A. 19   B. 120
   C. 96   D. 420

2. Find the quotient: $\frac{k}{3} \div k$

   A. $\frac{k^2}{3}$ B. 3

   C. $\frac{1}{3}$ D. $\frac{2k}{3}$

3. The price of ABC Company's stock is $6.00 more than twice the price of XYZ Incorporated's stock. ABC Company's stock price is $15.00. What is XYZ Incorporated's stock price?
   A. $13.50   B. $18.00
   C. $42.00   D. $4.50

4. A number cube is rolled. What is the probability of rolling an even number?

   A. 0   B. $\frac{1}{3}$

   C. $\frac{1}{2}$ D. 1

5. Write $\frac{13}{65}$ as a percent.
   A. 5%   B. 500%
   C. 2%   D. 20%

6. Prices of Audio Tapes (Dollars)

   What is the median audio tape price?
   A. $6.25   B. $7.00
   C. $4.75   D. $8.50

7. Simplify: $9q + 3 + 4q$
   A. $12q + 4$ B. $9 + 7q$
   C. $16q$ D. $13q + 3$

8. The number x is 15% of 60. The number y is 75% of 12.

   Which of the following is true?
   A. $x > y$ B. $x < y$
   C. $x = y$ D. cannot be determined

9. A coin is flipped 3 times. What is P(heads, then heads, then tails)?

   A. $\frac{1}{2}$ B. $\frac{1}{3}$

   C. $\frac{1}{8}$ D. $\frac{1}{16}$

10. The number 14.38 is to be displayed in a stem-and-leaf plot. What is its stem?
    A. 1   B. 14
    C. 14.3   D. 14.38

11.

How many students study both mathematics and music?

A. 48   B. 12
C. 83   D. 95

12.

What is the correct name for this figure?

A. $\overrightarrow{PQ}$ B. $\overrightarrow{QP}$
C. $\overleftrightarrow{PQ}$ D. $\overline{PQ}$

13. The base of a parallelogram is 14 cm and the height is 6 cm. Find the area.

A. 40 cm^2 B. 20 cm^2
C. 84 cm^2 D. cannot be
                         determined

14. To compare the incomes from five different sales regions to the total income, which data display is most appropriate?

A. scattergram
B. box-and-whisker plot
C. stem-and-leaf plot
D. circle graph

15. Find the GCF: 210 and 315

A. 3   B. 630
C. 105   D. 7

16. Specialty Books Monthly Sales

About what fraction of the sales are from fiction books?

A. $\frac{1}{4}$ B. $\frac{1}{3}$ C. $\frac{2}{3}$ D. $\frac{1}{6}$

17. Brad has 5 dimes, 1 quarter, and 6 nickels. In how many different ways can he make $.50?

A. 7   B. 3
C. 5   D. 4

18. James plans to put a wallpaper border around his rectangular bedroom. Which measurement of the room should he find?

A. area   B. circumference
C. diameter   D. perimeter

19. In a random sample of 300 students in a school, 80 rated science as their favorite subject. Predict how many students in the school's total population of 1650 would rate science as their favorite subject.

A. about 80   B. about 440
C. about 1430 D. about 220

20. Simplify: $12 + 4(3 - 5)^2$

A. 64   B. 28
C. -1 D. -64

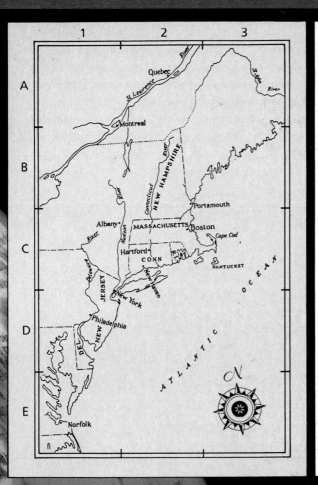

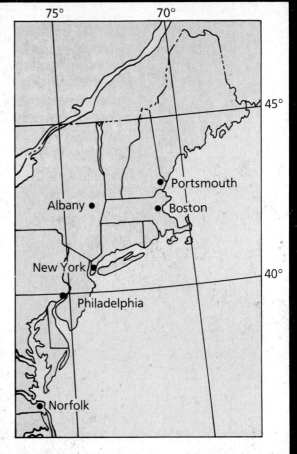

The latitude and longitude lines on a map or globe represent a coordinate system. Latitude is given in degrees north or south of the equator, and longitude is given in degrees east or west of Greenwich, England. By knowing the latitude and longitude of a place, you can locate it on a map or globe. For example, Philadelphia, Pennsylvania, is located at approximately 40° north 75° west.

Equations with Two Variables

Objective: To find solutions of equations with two variables.

Terms to Know

* *solution of an equation with two variables*

CONNECTION

In previous chapters you worked with many different formulas. For example, you learned that the formula for the circumference of a circle is $C = 2\pi r$. This formula is an example of an equation with two variables.

A **solution of an equation with two variables** is an ordered pair of numbers that make the equation true. For example, two solutions of the equation $y = \frac{1}{2}x$ are (12, 6) and (20, 10). An equation with two variables may have *infinitely many* solutions. Using a table can help you find some of these solutions.

Example 1

Find solutions of the equation $y = 5x + 4$. Use $-2, -1, 0, 1,$ and 2 as values for x.

Solution

Make a table.

$y = 5x + 4$		
x	y	(x, y)
-2	$5(-2) + 4 = -6$	$(-2, -6)$
-1	$5(-1) + 4 = -1$	$(-1, -1)$
0	$5(0) + 4 = 4$	$(0, 4)$
1	$5(1) + 4 = 9$	$(1, 9)$
2	$5(2) + 4 = 14$	$(2, 14)$

☑ **Check Your Understanding**

1. In Example 1, what is the value of y when $x = 1$?
2. In Example 1, what is the value of x when $y = -6$?

Sometimes it is easier to find solutions of an equation with two variables if you first get one variable alone on one side of the equation.

Example 2

Find solutions of the equation $y - 4x = 7$. Use $-2, -1, 0, 1,$ and 2 as values for x.

Solution

Transform the equation so that y is alone on one side.

$$y - 4x = 7$$
$$y - 4x + 4x = 7 + 4x$$
$$y = 7 + 4x$$

Then make a table.

$y = 7 + 4x$		
x	y	(x, y)
-2	$7 + 4(-2) = -1$	$(-2, -1)$
-1	$7 + 4(-1) = 3$	$(-1, 3)$
0	$7 + 4(0) = 7$	$(0, 7)$
1	$7 + 4(1) = 11$	$(1, 11)$
2	$7 + 4(2) = 15$	$(2, 15)$

Guided Practice

Refer to the text on page 582.

« **1.** What is a solution of an equation with two variables?

« **2.** How should you write solutions of an equation with two variables x and y?

Choose the letter of the correct value of y for the given value of x.

3. $y = 3x + 9; x = -2$ **a.** 3   **b.** 9   **c.** 15

4. $y = -2x + 4; x = -1$ **a.** -2 **b.** 2   **c.** 6

5. $y = -x + 5; x = 5$ **a.** 0   **b.** -5 **c.** 10

6. $y = -\frac{1}{3}x + 8; x = -9$ **a.** 5   **b.** 11   **c.** 35

Transform each equation to get y alone on one side.

7. $x + y = 8$ **8.** $y - x = -2$

9. $4x + y = 4$ **10.** $y + 5 = x$

11. Find solutions of the equation $y = \frac{3}{4}x + 1$. Use $-8, -1, 0, 2$, and 4 as values for x.

« **12. COMMUNICATION** *« Discussion*   In Exercise 11, for which values of x was the corresponding value of y an integer? In general, for which values of x will the corresponding value of y in the equation $y = \frac{3}{4}x + 1$ be an integer?

Find solutions of each equation. Use $-2, -1, 0, 1$, and 2 as values for x.

13. $y = x - 1$ **14.** $y = -2x + 3$ **15.** $y = \frac{1}{2}x - 2$

16. $x + y = 4$ **17.** $2x + y = 5$ **18.** $y - 3x = 0$

Exercises

Find solutions of each equation. Use $-6, -2, 0, 2$, and 6 as values for x.

1. $y = 4x + 5$ **2.** $y = -x - 1$ **3.** $y = -\frac{1}{3}x + 5$

4. $y = -6x$ **5.** $y = 5x - 9$ **6.** $y = \frac{5}{6}x - 3$

7. $x + y = -4$ **8.** $x + y = 8$ **9.** $2x + y = -1$

10. $y - 4x = 3$ **11.** $7x + y = 0$ **12.** $y - 2 = 3x$

13. An electrician charges a $20 travel charge and $36 per hour.
    a. Write an equation in two variables that relates what the electrician charges (C) to the number of hours worked (h).
    b. What is the charge for a job that takes 3 h?
    c. An electrician charges $182. How many hours did the electrician work?

14. A telephone call from Darby to Weston costs 22¢ for the first minute and 14¢ for each additional minute. The equation $C = 14(m - 1) + 22$ relates the total cost of a call in cents (C) to the length of the call in minutes (m).
    a. Find the cost, in dollars, of a call that lasts 12 min.
    b. Find the length of a call that costs 78¢.

ESTIMATION

For each equation, estimate a value of y for the given value of x.

15. $y = 1.3x + 6; x = 2.8$ 16. $y = 4.6x - 9; x = 6.2$

17. $y = 0.49x + 6.8; x = 8$ 18. $y = -0.34x - 1.3; x = 6$

DATA, *pages 580–581*

19. Which city has map coordinates A2?

20. Name two cities with map coordinates C2.

21. What are the map coordinates of Norfolk?

22. Which city is located at approximately 42.5° north, 74° west?

23. Approximate the latitude and longitude of Portsmouth.

24. RESEARCH   Find the latitude and longitude of your town or city.

SPIRAL REVIEW

25. Find the number of permutations of the letters in the word RADIO. *(Lesson 12-6)*

26. Graph the points $A(3, 1)$, $B(2, -3)$, $C(0, 5)$, and $D(4, 0)$ on a coordinate plane.   *(Lesson 3-8)*

27. Find $\sqrt{3600}$.   *(Lesson 10-10)*

28. Find solutions of the equation $2x + y = -5$. Use $-2, -1, 0, 1$, and 2 as values for x.   *(Lesson 13-1)*

13-2

Graphing Equations with Two Variables

Objective: To graph equations with two variables.

Terms to Know

• *graph of an equation with two variables*

1 Tell whether the coordinates of points A and B are solutions of the equation $y = -2x + 4$.

2 Find two points on the line other than points A and B whose coordinates are solutions of the equation $y = -2x + 4$.

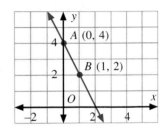

3 Predict whether it is possible to find a point on the line whose coordinates are not a solution of the equation $y = -2x + 4$.

4 How is the line in the figure above related to the equation $y = -2x + 4$?

The **graph of an equation with two variables** is all the points whose coordinates are solutions of the equation. The graph of an equation such as $y = -2x + 4$ is a line on a coordinate plane.

Example 1

Solution

Graph the equation $y = \frac{2}{3}x - 2$.

• Make a table.

• Find at least three solutions of the equation. Choose reasonable values for x. Use -3, 0, and 3 as values for x.

	$y = \frac{2}{3}x - 2$	
x	y	(x, y)
-3	$\frac{2}{3}(-3) - 2 = -4$	$(-3, -4)$
0	$\frac{2}{3}(0) - 2 = -2$	$(0, -2)$
3	$\frac{2}{3}(3) - 2 = 0$	$(3, 0)$

• Graph each solution as a point on a coordinate plane. Label each point with its coordinates.

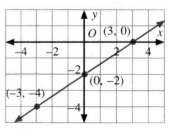

• Conncct the points with a straight line.

✓ Check Your Understanding

1. In Example 1, why are -3 and 3 good choices as values for x?

2. In Example 1, if you had used other values for x besides -3, 0, and 3, would the graph of the equation have been the same line?

Example 2

Graph the equation $x + y = 3$.

Solution

Transform the equation. Then find at least three solutions. Use -1, 0, and 1 as values for x.

$$x + y = 3$$
$$x + y - x = 3 - x$$
$$y = 3 - x$$

$y = 3 - x$		
x	y	(x, y)
-1	4	$(-1, 4)$
0	3	$(0, 3)$
1	2	$(1, 2)$

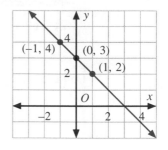

You can use a graphing calculator to graph equations with two variables once you have gotten y alone on one side. Key sequences vary depending on the type of graphing calculator. On some graphing calculators you would use the key sequence below to graph $y = 3x - 2$.

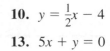

Guided Practice

« **1. COMMUNICATION** « *Reading* Replace each _?_ with the correct word.

The graph of each solution of an equation with two variables is a(n) _?_ on a coordinate plane. The graph of all the solutions of an equation such as $3x - y = 6$ is a(n) _?_ on a coordinate plane.

Tell whether the graph of each ordered pair is on the line shown in the graph at the right. Write *Yes* or *No*.

2. $(0, -3)$ **3.** $(-3, 0)$ **4.** $(1, 2)$

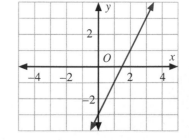

Use the line in the graph at the right to complete each ordered pair.

5. $(2, \underline{\ ?\ })$ **6.** $(3, \underline{\ ?\ })$ **7.** $(\underline{\ ?\ }, -1)$

Graph each equation.

8. $y = x - 3$

9. $y = -3x + 2$

10. $y = \frac{1}{2}x - 4$

11. $x + y = 2$

12. $4x + y = 8$

13. $5x + y = 0$

Exercises

Graph each equation.

1. $y = 4x + 2$ **2.** $y = 3x - 1$ **3.** $y = -4x - 2$

4. $y = -x + 4$ **5.** $y = \frac{2}{3}x$ **6.** $y = -\frac{1}{3}x + 2$

7. $x + y = 1$ **8.** $x + y = -5$ **9.** $2x + y = 6$

10. $3x + y = -1$ **11.** $y - 3 = x$ **12.** $y + 2 = \frac{3}{4}x$

13. Graph the line that passes through the points $(-4, 0)$ and $(0, 1)$.

14. Graph the line that passes through the points $(2, 0)$ and $(0, 3)$.

 CALCULATOR

Choose the letter of the graphing calculator key sequence you would use to graph each equation.

15. $y = 6x - 5$

16. $y = 5x - 6$

A. GRAPH Y= [ ] 6 X|T [-] [ ] 5

B. Y= [ ] 5 X|T [-] [ ] 6 GRAPH

C. Y= X|T [ ] 5 [-] [ ] 6 GRAPH

D. Y= [ ] 6 X|T [-] [ ] 5 GRAPH

THINKING SKILLS

17. Think of the equation $y = 2$ as meaning $y = 0x + 2$. Complete each ordered pair.

  a. $(-1, \underline{\ ?\ })$ **b.** $(0, \underline{\ ?\ })$ **c.** $(3, \underline{\ ?\ })$ **d.** $(5, \underline{\ ?\ })$

18. Use the ordered pairs from Exercise 17 to graph the equation $y = 2$ on a coordinate plane. Describe the graph in words.

19. Think of the equation $x = -3$ as meaning $0y - 3 = x$. Complete each ordered pair.

  a. $(\underline{\ ?\ }, -4)$ **b.** $(\underline{\ ?\ }, -2)$ **c.** $(\underline{\ ?\ }, 0)$ **d.** $(\underline{\ ?\ }, 1)$

20. Use the ordered pairs from Exercise 19 to graph the equation $x = -3$ on a coordinate plane. Describe the graph in words.

SPIRAL REVIEW

21. Solve $mx + y = b$ for y. *(Lesson 8-10)*

22. Evaluate $-3a + b$ when $a = -5$ and $b = 7$. *(Lesson 3-6)*

23. Find the area of a circle with radius 4 m. *(Lesson 10-4)*

24. Graph the equation $y = 2x - 5$. *(Lesson 13-2)*

Slope and Intercepts

Objective: To find the slope and intercepts of a line and to write the equation of a line using the slope and *y*-intercept.

Terms to Know

• *slope*
• *y-intercept*
• *x-intercept*
• *slope-intercept form of an equation*

As you move from one point to another on a line, the vertical movement is called the *rise*, and the horizontal movement is called the *run*. The **slope** of a line is the ratio of the *rise* to the *run*. The slope of a line describes the line's steepness and direction. The line at the right has slope $\frac{3}{2}$.

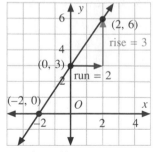

$$\text{slope} = \frac{\text{rise}}{\text{run}} = \frac{3}{2}$$

The slope is the same between any two points on a given line.

The *y*-coordinate of the point where a graph crosses the *y*-axis is called the **y-intercept.** Notice that the value of *x* at this point is 0. The *y*-intercept of the line above is 3.

The *x*-coordinate of the point where a graph crosses the *x*-axis is called the **x-intercept.** Notice that the value of *y* at this point is 0. The *x*-intercept of the line above is -2.

The **slope-intercept form of an equation** is $y = mx + b$. In this equation, *m* represents the slope of the line, and *b* represents the *y*-intercept. The slope-intercept form of the equation for the line above is $y = \frac{3}{2}x + 3$.

Example 1

Find the slope and the *y*-intercept of the line at the right. Then use them to write an equation for the line.

Solution

- Use two points on the line to find the slope. The rise is -4 and the run is 2.

  $$\text{slope} = \frac{\text{rise}}{\text{run}} = \frac{-4}{2} = -2$$

  The slope is -2, or $m = -2$.

- The *y*-coordinate of the point where the line crosses the *y*-axis is 5. The *y*-intercept is 5, or $b = 5$.

- Use the equation $y = mx + b$.
  An equation for the line is $y = -2x + 5$.

✓ Check Your Understanding

1. In Example 1, why is the rise negative?

2. Which two points on the graph in Example 1 were used to find the slope of the line? Would the slope have been the same if two other points on the line had been used?

When an equation is not written in slope-intercept form, you can transform the equation to find the slope and y-intercept.

Example 2
Solution

Find the slope, the y-intercept, and the x-intercept of the line $y + 6 = -3x$.

• Write the equation in slope-intercept form.

$$y + 6 = -3x$$
$$y + 6 - 6 = -3x - 6$$
$$y = -3x - 6$$
$$y = -3x + (-6)$$

Compare $y = -3x + (-6)$ to $y = mx + b$. $m = -3$ and $b = -6$.

• To find the x-intercept, let $y = 0$. Solve for x.

$$y + 6 = -3x$$
$$0 + 6 = -3x$$
$$6 = -3x$$
$$-2 = x \qquad \longleftarrow \text{Both sides were divided by } -3.$$

The slope is -3, the y-intercept is -6, and the x-intercept is -2.

☑️ **Check Your Understanding**

3. In Example 2, why was the equation $y = -3x - 6$ written as $y = -3x + (-6)$?

4. Where will a graph of the equation in Example 2 cross the y-axis?

By looking at the graph of a line, you can tell easily if its slope is *positive* or *negative*. A line with *positive slope* rises as the value of x increases. A line with *negative slope* falls as the value of x increases.

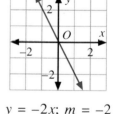

$y = 2x; \; m = 2 \qquad y = -2x; \; m = -2$

Guided Practice

COMMUNICATION «*Reading*

«**1.** Explain the meaning of the word *intercepted* in the sentence below.
The quarterback's pass was intercepted.

«**2.** Explain the meaning of the word *slope* in the sentence below.
The ski instructor brought the class to the beginner slope.

«**3.** Write a sentence using each word in a mathematical context.
a. intercept   **b.** slope

Write an equation for the line with the given slope and y-intercept.

4. slope $= 2$
y-intercept $= 0$

5. slope $= -1$
y-intercept $= \frac{1}{2}$

6. slope $= \frac{3}{4}$
y-intercept $= -1$

Use the graph at the right.

7. What are the coordinates of the point where the line crosses the x-axis?

8. What are the coordinates of the point where the line crosses the y-axis?

9. Find the slope and the y-intercept of the line. Then use them to write an equation for the line.

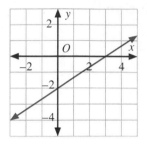

Find the slope, the y-intercept, and the x-intercept of each line.

10. $y = 3x - 6$

11. $y = -x - 5$

12. $y = \frac{4}{5}x - 8$

13. $y = \frac{2}{3}x$

14. $y - 3x = 6$

15. $4 + y = x$

Exercises

Find the slope and the y-intercept of each line. Then use them to write an equation for each line.

1. **2.** **3.**

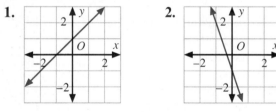
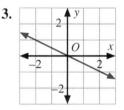

Find the slope, the y-intercept, and the x-intercept of each line.

4. $y = 2x - 10$

5. $y = -3x + 9$

6. $y = x + 5$

7. $y = -x - 1$

8. $y = 6x + 5$

9. $y = -2x + 1$

10. $y = -\frac{1}{3}x + 2$

11. $y = x - 4\frac{1}{2}$

12. $y = -\frac{2}{5}x$

13. $y = 3x$

14. $y - 8 = 2x$

15. $3 + y = -3x$

16. $y - x = 2$

17. $y - 4x = 6$

18. $y - 7 = -\frac{1}{4}x$

19. The equation of a line is $y = -7x + 21$. What are the coordinates of the point where the line crosses the x-axis? the y-axis?

20. The equation of a line is $y - 5x = 10$. What are the coordinates of the point where the line crosses the x-axis? the y-axis?

CALCULATOR

Choose the letter of the correct graphing calculator display for each equation.

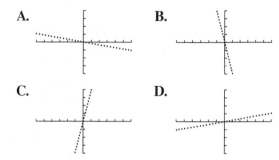

A.   B.

C.   D.

21. $y = 4x$

22. $y = -4x$

23. $y = \frac{1}{4}x$

24. $y = -\frac{1}{4}x$

CAREER/APPLICATION

Slopes are used in the planning and construction of the roof of a building. The slope of a roof is called the *pitch*. A carpenter must know the pitch of a roof to cut rafters of the correct length.

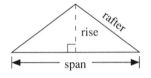

$$\text{Pitch} = \frac{\text{the rise of the roof}}{\text{half the span of the roof}}$$

25. The rise of a roof is 12 ft and the span is 24 ft. Find the pitch.

26. The pitch of the roof Tony is building is 5 to 12. Half of the span is 16 ft. What is the rise?

27. About how long should a rafter for the roof described in Exercise 25 be?

LOGICAL REASONING

Tell whether each statement is *always*, *sometimes*, or *never* true.

28. The slope of a line with positive x- and y-intercepts is positive.

29. The slope of a line with negative x- and y-intercepts is negative.

SPIRAL REVIEW

30. Solve: $4x + 5.1 = -2.9$ *(Lesson 8-8)*

31. In a poll of 30 people, 20 people preferred So-Brite toothpaste. In a similar poll of 87 people, about how many people would be expected to prefer So-Brite toothpaste?   *(Lesson 12-8)*

32. Find the slope, the y-intercept, and the x-intercept of the line $y = 3x - 1$.   *(Lesson 13-3)*

Using Geoboards

Objective: To use a geoboard to explore slopes and intercepts.

Materials

- 5 × 5 peg geoboard
- rubber bands

In the previous lesson you learned that the slope of a line is represented by the vertical change divided by the horizontal change between any two points on the line. Geoboards can help you visualize the slope of a line.

The geoboard pictured at the right simulates a coordinate plane. The peg at the center represents the origin, and the two rubber bands represent the axes.

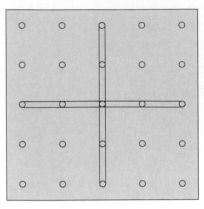

Activity I *Finding Slopes*

Construct a coordinate plane on a geoboard as shown in the diagram above.

1. Stretch a rubber band between the points (2, 2) and (−2, −2). Between these two points, what is the rise? the run? What is the slope of the line?

2. Stretch another rubber band between the points (2, 2) and (0, 0). Between these two points, what is the rise? the run? What is the slope of the line?

3. Compare the slopes in Steps 1 and 2. What can you conclude about the slope between two different pairs of points on the same line?

Activity II *Finding Intercepts*

1. Stretch a rubber band between each pair of points. Name the *x*-intercept and the *y*-intercept of each line.
   a. (−1, 2), (1, −2)   **b.** (2, 1), (−1, −2)
   c. (2, 2), (−2, −2)   **d.** (2, −1), (−1, 2)

2. You can stretch a rubber band between (0, −2) and two other points to produce a line with an *x*-intercept of −1. Give the coordinates of these two other points.

3. You can stretch a rubber band between (0, −2) and two other points to produce a line with an *x*-intercept of 1. Give the coordinates of these two other points.

4. You can stretch a rubber band between (−2, 0) and two other points to produce a line with a *y*-intercept of −1. Give the coordinates of these two other points.

Activity III *Parallel and Perpendicular Lines*

1 The diagram at the right shows one rubber band stretched between the points $(-2, 0)$ and $(0, 2)$, and another stretched between the points $(-1, -2)$ and $(2, 1)$. What is the relationship between the two lines?

2 Find the slope of each line in Step 1. Compare the slopes.

3 Stretch one rubber band between the points $(-2, 2)$ and $(0, -2)$, and another between the points $(0, 2)$ and $(2, -2)$. What is the relationship between the two lines?

4 Find the slope of each line in Step 3. Compare the slopes.

5 Use your answers to Steps 2 and 4 to predict the relationship between the slopes of any two parallel lines.

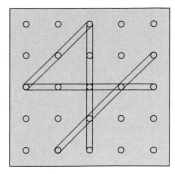

6 The diagram at the right shows one rubber band stretched between the points $(-1, -2)$ and $(2, 1)$, and another stretched between the points $(-2, 1)$ and $(1, -2)$. What is the relationship between the two lines?

7 Find the slope of each line in Step 6, and their product.

8 Stretch one rubber band between the points $(-1, 2)$ and $(1, -2)$, and another between the points $(-2, -1)$ and $(2, 1)$. What is the relationship between the two lines?

9 Find the slope of each line in Step 8, and their product.

10 Use your answers to Steps 7 and 9 to predict the relationship between the slopes of any two perpendicular lines.

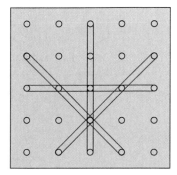

Activity IV *Vertical and Horizontal Lines*

1 The diagram at the right shows a red rubber band stretched between the points $(-2, 1)$ and $(2, 1)$. Between these two points, what is the rise? the run? the slope?

2 Stretch a rubber band between the points $(-2, -2)$ and $(2, -2)$. Between these two points, what is the rise? the run? the slope?

3 Use your answers to Steps 1 and 2 to predict what will be the slope of any horizontal line.

4 The diagram at the right shows a blue rubber band stretched between the points $(1, 2)$ and $(1, -2)$. Between these two points, what is the rise? the run? the slope?

5 Stretch a rubber band between the points $(-2, 2)$ and $(-2, -2)$. Between these two points, what is the rise? the run? the slope?

6 Use your answers to Steps 4 and 5 to predict what will be the slope of any vertical line.

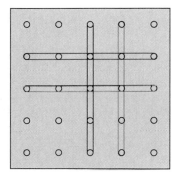

Systems of Equations

Objective: To solve systems of equations by graphing.

APPLICATION

Terms to Know
- *system of equations*
- *solution of a system*

Downtown Delivery charges $2 per pound to deliver a package, plus a service fee of $6. ABC Delivery charges $3 per pound, but only a $4 service fee. To find out how much to charge, the companies use the equations $y = 2x + 6$ and $y = 3x + 4$, where y is what a company charges to deliver a package, and x is the weight of a package.

QUESTION For what weight package will the charges be the same?

To answer this question, you find a solution common to both equations.

Two equations with the same variables form a **system of equations.** An ordered pair that is a solution of *both* equations is called a **solution of the system.** You can solve a system of equations by graphing.

Example 1

Solve the system by graphing: $\quad y = 2x + 6$
$\qquad\qquad\qquad\qquad\qquad\qquad y = 3x + 4$

Solution

Make a table for each equation. Then graph both equations on one coordinate plane.

$y = 2x + 6$		
x	y	(x, y)
-1	4	$(-1, 4)$
0	6	$(0, 6)$
1	8	$(1, 8)$

$y = 3x + 4$		
x	y	(x, y)
-1	1	$(-1, 1)$
0	4	$(0, 4)$
1	7	$(1, 7)$

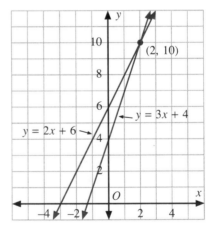

The solution is $(2, 10)$.

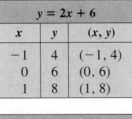 **Check**

$$y = 2x + 6 \qquad\qquad y = 3x + 4$$
$$10 \overset{?}{=} 2(2) + 6 \qquad 10 \overset{?}{=} 3(2) + 4$$
$$10 \overset{?}{=} 4 + 6 \qquad\quad 10 \overset{?}{=} 6 + 4$$
$$10 = 10 \qquad\qquad 10 = 10$$

ANSWER Each company will deliver a 2-lb package for $10.

Example 2	Solve the system by graphing:	$y = -\frac{1}{2}x + 2$
		$y = -\frac{1}{2}x - 1$

Solution

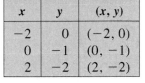

$y = -\frac{1}{2}x + 2$		
x	y	(x, y)
-2	3	$(-2, 3)$
0	2	$(0, 2)$
2	1	$(2, 1)$

$y = -\frac{1}{2}x - 1$		
x	y	(x, y)
-2	0	$(-2, 0)$
0	-1	$(0, -1)$
2	-2	$(2, -2)$

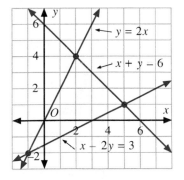

The lines are parallel. They do not intersect. The system has no solution.

 **Check Your Understanding**

If you were to use different values of x to graph the system in Example 2, would the lines still be parallel?

Guided Practice

COMMUNICATION «*Reading*

Refer to the text on pages 594–595.

«**1.** When does a system of equations have no solution?

«**2.** How can you check if an ordered pair is a solution of the system?

Use the graph to name the solution of each system.

3. $x + y = 6$
$\quad\ y = 2x$

4. $x +\ \ y = 6$
$\quad\ x - 2y = 3$

5. $x - 2y = 3$
$\quad\quad\ y = 2x$

«**6.** COMMUNICATION «*Discussion* When graphing the equations from a system of equations, do you need to use the same values of x in each table of values? Explain.

Solve each system by graphing.

7. $y = 3x - 1$
$y = -x - 5$

8. $y = -3x - 2$
$y = -3x + 1$

9. $y = \frac{1}{3}x$
$y = 2x - 5$

Exercises

Solve each system by graphing.

1. $y = 4x - 2$
$y = -3x + 5$

2. $y = -x + 5$
$y = x + 3$

3. $y = 3x + 6$
$y = 3x - 2$

4. $y = 4x - 6$
$y = -2x - 6$

5. $y = 3x + 3$
$y = -2x - 2$

6. $y = 6x - 4$
$y = 2x$

7. $y = -2x$
$y = -2x + 3$

8. $y = -x - 4$
$y = -x + 3$

9. $y = -x - 4$
$y = -2x - 7$

10. $y = -2x + 7$
$y = 2x - 1$

11. $y = x - 1$
$y = -\frac{1}{2}x + 2$

12. $y = \frac{1}{3}x - 2$
$y = \frac{1}{3}x + 3$

13. $y - 3x = -2$
$y - x = 0$

14. $y + 7 = 2x$
$y + 2 = -\frac{1}{2}x$

15. $y + 2 = -\frac{1}{3}x$
$y - 1 = \frac{2}{3}x$

CONNECTING ALGEBRA AND GEOMETRY

Use this system of equations for Exercises 16–20:
$y = -2x + 8$
$y = 3x + 3$

16. Graph the system.

17. What is the solution of the system?

18. Where does each line intersect the x-axis?

19. What figure is formed by the two lines and the x-axis?

20. Find the area of the figure in Exercise 19.

COMPUTER APPLICATION

Use function graphing software or graph paper.

21. Graph this system of three equations on one coordinate plane:
$y = 3x - 5, y = -x + 3, y = x - 1$

22. Describe the graph of the system in Exercise 21. Does the system of three equations have a solution? Explain.

23. Graph this system of three equations on one coordinate plane:
$y = \frac{1}{2}x + 4, y = \frac{3}{2}x - 4, y = \frac{1}{2}x - 4$

24. Describe the graph of the system in Exercise 23. Does the system of three equations have a solution? Explain.

When you graph two equations with two variables on a coordinate plane, the lines will intersect, coincide, or be parallel. Two lines *coincide* when they contain all the same points.

25. Tell whether two lines with the same slope and different *y*-intercepts will *intersect*, *coincide*, or be *parallel*.

26. Tell whether two lines with the same slope and the same *y*-intercept will *intersect*, *coincide*, or be *parallel*.

27. Tell whether two lines with different slopes will *intersect*, *coincide*, or be *parallel*.

Use your answers to Exercises 25–27. Without graphing, tell whether the two lines *intersect*, *coincide*, or are *parallel*.

28. $y = 5x + 2$, $y = 5x + 7$ **29.** $y = -2x$, $y = 4x + 9$

30. $y = 3x + 6$, $y - 3x = 6$ **31.** $y - x = 8$, $y + 11 = x$

32. The lengths of the legs of a right triangle are 10 cm and 24 cm. Find the length of the hypotenuse. *(Lesson 10-11)*

33. Simplify: $\dfrac{10y^3}{y}$ *(Lesson 7-5)*

34. Solve the system by graphing: $y = -2x + 3$, $y = -4x + 7$ *(Lesson 13-4)*

Self-Test 1

Find solutions of each equation. Use −2, −1, 0, 1, and 2 as values for *x*.

1. $y = 6x - 3$ **2.** $y - 2x = 5$ 13-1

Graph each equation.

3. $y = \frac{3}{4}x + 2$ **4.** $x + y = 4$ 13-2

Find the slope, the *y*-intercept, and the *x*-intercept of each line.

5. $y - 8 = 5x$ **6.** $y = 4x$ 13-3

Solve by graphing.

7. $y = x + 1$
$y = 3x - 5$ **8.** $y = \frac{2}{3}x + 6$ 13-4

$y = \frac{2}{3}x - 2$

Using Function Graphing Software

Objective: To use function graphing software to find solutions to systems of equations.

A computer can help you solve a system of equations, especially when the lines do not intersect at a point whose coordinates are integers.

Function graphing software enables you to graph a system of equations and then magnify or zoom in on the solution in order to get closer and closer estimates of the coordinates. The box in the diagram below indicates the section of a graph to be magnified.

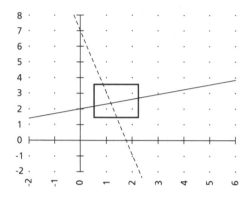

Exercises

The equations of the lines in the diagram above are $y = \frac{3}{10}x + 2$ and $y = -4x + 7$. Use function graphing software to graph the system of equations shown in the diagram.

1. Estimate each coordinate of the solution to the nearest integer.

2. Zoom in on the graph using a box whose center is at the solution of the system. Continue to zoom in until you can accurately estimate each coordinate of the solution to the nearest tenth. What are the coordinates of the solution to the nearest tenth?

3. Zoom in on the graph again until you can accurately estimate each coordinate of the solution to the nearest hundredth. What are the coordinates of the solution to the nearest hundredth?

Use function graphing software to graph the system of equations $y = -\frac{1}{2}x + 2$ and $y = 3x + 8$.

4. In what quadrant do the lines intersect?

5. Estimate each coordinate of the solution to the nearest integer.

6. Zoom in on the graph until you can accurately estimate each coordinate of the solution to the nearest hundredth. What are the coordinates of the solution to the nearest hundredth?

Strategy:
Solving a Simpler Problem

Objective: To solve problems by using simpler problems.

You can solve some complicated problems by solving a simpler problem or by using simpler numbers. Use strategies such as making a table and identifying a pattern to solve the simpler problem, and then solve the original.

Problem

The 10 houses on one side of a street are numbered with the even numbers from 2 to 20. What is the sum of these house numbers?

Solution

UNDERSTAND The problem is about the sum of house numbers.

Facts: houses are numbered with the even numbers from 2 to 20
Find: the sum of the house numbers

PLAN Make a table of partial sums and identify a pattern. Then use the pattern to solve the original problem.

WORK

Number of Houses	House Numbers	Sum	Pattern
1	2	2	$1^2 + 1 = 2$
2	2 + 4	6	$2^2 + 2 = 6$
3	2 + 4 + 6	12	$3^2 + 3 = 12$
4	2 + 4 + 6 + 8	20	$4^2 + 4 = 20$

The pattern shows that the sum of the house numbers is the sum of the number of houses and the square of the number of houses.
The number of houses is 10. So $10^2 + 10 = 110$.

ANSWER The sum of the house numbers is 110.

Look Back In the Problem, you made a pattern. You can *generalize* this pattern by using the expression $n^2 + n$. This expression will give you the sum of the first n even whole numbers. Use this expression to find the sum of the first 50 even whole numbers.

Guided Practice

Use the problem below for Exercises 1–5.

The rooms in a hotel are numbered consecutively from 1 to 185. How many plastic digits are needed to number all the rooms?

« **1.** List the 1-digit numbers that appear on the rooms. How many of them are there?

« **2.** What are the least and greatest 2-digit numbers that appear on the rooms? How many 2-digit numbers are there in all?

« **3.** What are the least and greatest 3-digit numbers that appear on the rooms? How many 3-digit numbers are there in all?

« **4.** Use your answers to Exercises 1–3. How many plastic digits are needed to form the 1-digit numbers? the 2-digit numbers? the 3-digit numbers?

5. Use your answers to Exercise 4 to write an answer to the problem.

Solve using a simpler problem.

6. A grocer stacks cans in a display using the pattern shown at the right. Each row has 1 fewer can than the row below it. How many cans are there in a stack with 20 rows? (*Hint*: Pair the rows beginning with the top and bottom rows.)

7. Find the sum of the first 50 odd whole numbers. (*Hint*: Find partial sums and identify a pattern.)

Problem Solving Situations

Solve using a simpler problem.

1. How many triangles are there in the figure at the right? (*Hint*: There are four different sized triangles in the figure.)

2. A marching band is marching in a triangular formation. There is 1 band member in the first row. Each of the other rows contains 2 more band members than the row in front of it. There are 11 rows in all. How many band members are there?

3. Ellen sold tickets to a play. On the first day she sold two tickets. On each of the next twelve days she sold two more tickets than she sold the day before. How many tickets did Ellen sell in all?

4. The lockers on one side of a hallway are numbered consecutively from 1 to 100. What is the sum of these locker numbers?

Solve using any problem solving strategy.

PROBLEM SOLVING CHECKLIST

Keep these in mind:
Using a Four-Step Plan
Too Much or Not Enough Information
Supplying Missing Facts
Problems with No Solution

Consider these strategies:
Choosing the Correct Operation
Making a Table
Guess and Check
Using Equations
Identifying a Pattern
Drawing a Diagram
Using Proportions
Using Logical Reasoning
Using a Venn Diagram
Solving a Simpler Problem

5. Sean has two more dimes than quarters. The coins are worth $5.10. How many of each does he have?

6. Admission to the science museum is $7 for adults and $3 for children. One day, the receipts were recorded as $724. If 80 adults went to the science museum that day, how many children went to the science museum that day?

7. There are 128 sophomores at Lincoln High School. A survey showed that 51 play an instrument, 68 play a sport, and 23 play an instrument and a sport. How many sophomores do neither?

8. The houses on River Road are numbered consecutively from 10 to 132. How many brass digits are needed to form all the house numbers?

WRITING WORD PROBLEMS

Using the information given, write a word problem that you could solve using a simpler problem. (You might need to add more information.) Then solve the problem.

9. The 84 houses on Oak Street are numbered consecutively.

10. Each row in a chorus line has 2 more dancers than the row in front of it.

SPIRAL REVIEW

11. Find the circumference of a circle with radius $5\frac{1}{2}$ in. *(Lesson 10-2)*

12. Find the sum of the first 200 even whole numbers. *(Lesson 13-5)*

13. Graph the inequality $x \geq 4\frac{3}{4}$. *(Lesson 7-10)*

Challenge

What type of figure do you get when you graph the equation $|x| + |y| = 4$ on the coordinate plane?

Inequalities with One Variable

Objective: To solve and graph inequalities with one variable.

1 You know that $12 > 8$. Replace each ___?___ with $>$ or $<$.

a. $12 + 2$ ___?___ $8 + 2$

b. $12 - 2$ ___?___ $8 - 2$

c. $2(12)$ ___?___ $2(8)$

d. $\dfrac{12}{2}$ ___?___ $\dfrac{8}{2}$

e. $-2(12)$ ___?___ $-2(8)$

f. $\dfrac{12}{-2}$ ___?___ $\dfrac{8}{-2}$

2 Use your answers to Step 1. What are the two situations where performing a given operation on both sides of an inequality changes the inequality symbol?

You solve an inequality involving addition or subtraction the same way you solve an equation involving addition or subtraction.

> **Generalization:** *Solving Inequalities Using Addition or Subtraction*
>
> If a number has been *added* to the variable, subtract that number from both sides of the inequality.
> If a number has been *subtracted* from the variable, add that number to both sides of the inequality.

Example 1

Solve and graph each inequality.

a. $x + 4 \le 6$

b. $a - 5 > -2$

Solution

a.
$$x + 4 \le 6$$
$$x + 4 - 4 \le 6 - 4$$
$$x \le 2$$

✓ **Check**

You cannot check every point on the graph, but you should check at least one. Choose $x = 0$.

$$x + 4 \le 6$$
$$0 + 4 \overset{?}{\le} 6$$
$$4 \le 6$$

b.
$$a - 5 > -2$$
$$a - 5 + 5 > -2 + 5$$
$$a > 3$$

✓ **Check**

Choose $a = 5$.

$$a - 5 > -2$$
$$5 - 5 \overset{?}{>} -2$$
$$0 > -2$$

Inequalities involving multiplication or division are solved in much the same way as equations involving multiplication or division. The only difference concerns multiplying or dividing by a negative number.

> **Generalization:** *Solving Inequalities Using Multiplication or Division*
>
> If the variable has been multiplied or divided by a *positive* number, solve as you would solve an equation.
> If the variable has been multiplied or divided by a *negative* number, solve as you would solve an equation, and *reverse the direction of the inequality symbol*.

Example 2

Solve and graph each inequality.

a. $5k < -10$

b. $\dfrac{n}{-3} \geq 1$

Solution

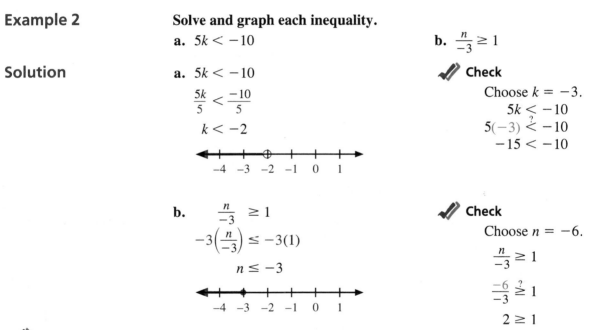

a. $5k < -10$

$\dfrac{5k}{5} < \dfrac{-10}{5}$

$k < -2$

✅ **Check**

Choose $k = -3$.

$5k < -10$

$5(-3) \overset{?}{<} -10$

$-15 < -10$

b. $\dfrac{n}{-3} \geq 1$

$-3\left(\dfrac{n}{-3}\right) \leq -3(1)$

$n \leq -3$

✅ **Check**

Choose $n = -6$.

$\dfrac{n}{-3} \geq 1$

$\dfrac{-6}{-3} \overset{?}{\geq} 1$

$2 \geq 1$

✅ **Check Your Understanding**

1. In Example 2(a), why do you not reverse the inequality symbol?
2. In Example 2(b), why do you reverse the inequality symbol?

Guided Practice

COMMUNICATION «*Reading*

Refer to the text on pages 602–603.

«**1.** What is the main idea of the lesson?

«**2.** List three major points that support the main idea.

Tell whether you would *add*, *subtract*, *multiply*, or *divide* to solve each inequality. Do not solve.

3. $a - 5 > 3$ **4.** $\frac{1}{2}k \le 5$ **5.** $8p \ge -8$ **6.** $4 + b < 9$

Tell whether you would reverse the inequality symbol when solving. Write *Yes* or *No*. Do not solve.

7. $-2x \le 16$ **8.** $5y < 30$ **9.** $\frac{m}{8} \ge 40$ **10.** $n - 9.4 > 3$

Choose the letter of the graph that matches each inequality.

11. $x > 0.6$

12. $x < 0.6$

13. $x \ge 0.6$

14. $x \le 0.6$

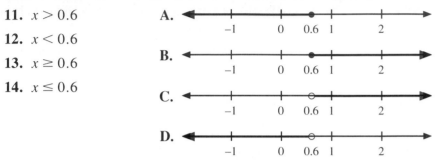

Solve and graph each inequality.

15. $y - 3 < 1$ **16.** $6v \ge 12$ **17.** $2x \le -1$

18. $r + 2.7 > 1.5$ **19.** $-\frac{2}{3}t \le -2$ **20.** $\frac{s}{-4} < 1$

Exercises

Solve and graph each inequality.

1. $n - 5 > 12$ **2.** $b + 9 < 11$ **3.** $3c < 18$

4. $\frac{a}{2} \le -4$ **5.** $\frac{b}{-3} < -7$ **6.** $r - 2\frac{2}{3} \ge 4\frac{2}{3}$

7. $p + 2.5 > -5$ **8.** $5x \le -15$ **9.** $-4g < 16$

10. $-\frac{2}{3}y \ge -6$ **11.** $-\frac{1}{6}y > 3$ **12.** $y + 7\frac{3}{4} \ge 3$

13. $h - 1.8 \le -2.6$ **14.** $-3x > 6$ **15.** $-5h > -30$

16. $\frac{d}{5} \ge -2$ **17.** $w + 1 < -1$ **18.** $m + 1\frac{2}{3} < -3\frac{1}{2}$

Write an inequality representing each statement. Then solve each inequality.

19. The sum of a number n and 2 is greater than or equal to 4.

20. The product of 3 and a number n is less than or equal to 15.

21. The quotient of a number n divided by -6 is less than 12.

ESTIMATION

To estimate the solution of an inequality, first decide which inverse operation to use. Then use an appropriate estimation method for that operation.

Choose the letter of the best estimate for the given inequality.

22. $x - 3.02 < 1.8$ **a.** $x < 5$ **b.** $x < -1$

23. $31t > 894$ **a.** $t > 3$ **b.** $t > 30$

24. $\frac{a}{6.2} \leq 0.47$ **a.** $a \leq 3$ **b.** $a \leq 12$

25. $m + 12.2 > 5.46$ **a.** $m > 17$ **b.** $m > -7$

DATA, *pages 140–141*

Write an inequality to represent each statement. Then use the appropriate data to solve each inequality for g.

26. The maximum speed, g, of a giraffe is greater than 3 times the maximum speed, k, of a chicken.

27. Twice the maximum speed, g, of a giraffe is less than the maximum speed, c, of a cheetah.

28. Use your answers to Exercises 26 and 27 to complete this statement with the correct numbers: $\underline{\ ?\ } < g < \underline{\ ?\ }$. Then write the statement in words.

SPIRAL REVIEW

29. Write 345% as a fraction in lowest terms. *(Lesson 9-5)*

30. Solve: $3x + 11 = 29$ *(Lesson 4-5)*

31. Solve: $\frac{3}{4}z - 8 = 16$ *(Lesson 8-9)*

32. Solve and graph $-9b \leq -81$. *(Lesson 13-6)*

33. Write the unit rate: 225 mi in 5 h *(Lesson 9-1)*

Two-Step Inequalities with One Variable

Objective: To solve and graph inequalities with one variable using two steps.

APPLICATION

Derek tunes his own car. The engine's danger zone in revolutions per minute (rpm) is 5000 to 6000. As Derek works on the car, he realizes that the idle speed in rpm can be multiplied by 6 and still be more than 200 rpm below the danger zone.

QUESTION What is the idle speed of Derek's car's engine?

To answer the question, write and solve an inequality. Let $r =$ the idle speed in rpm.

six times the rpm	plus	200	is less than	5000
$6r$	$+$	200	$<$	5000

Solving an inequality like this involves using two steps.

Example 1
Solution

Solve the inequality $6r + 200 < 5000$.

$$6r + 200 < 5000$$
$$6r + 200 - 200 < 5000 - 200 \quad \longleftarrow \text{Subtract 200 from both sides of the inequality.}$$
$$6r < 4800$$

$$\frac{6r}{6} < \frac{4800}{6} \quad \longleftarrow \text{Divide both sides of the inequality by 6.}$$

$$r < 800$$

☑ **Check Your Understanding**

1. In Example 1, which operation was undone first?
2. Was the inequality symbol in Example 1 reversed? Explain.

ANSWER The idle speed of Derek's car's engine is less than 800 rpm.

Generalization: *Solving Two-Step Inequalities*

First undo the addition or subtraction. Then undo the multiplication or division. Remember to reverse the direction of the inequality symbol when you multiply or divide by a negative number.

Example 2

Solution

Solve and graph the inequality $-16a - 12 \leq 36$.

$$-16a - 12 \leq 36$$
$$-16a - 12 + 12 \leq 36 + 12 \qquad \longleftarrow \text{Add 12 to both sides.}$$
$$-16a \leq 48$$

$$\frac{-16a}{-16} \geq \frac{48}{-16} \qquad \longleftarrow \begin{array}{l}\text{Divide both sides by } -16. \\ \text{Reverse the inequality symbol.}\end{array}$$
$$a \geq -3$$

✔ Check

Choose $a = 0$.
$$-16a - 12 \leq 36$$
$$-16(0) - 12 \overset{?}{\leq} 36$$
$$0 - 12 \overset{?}{\leq} 36$$
$$-12 \leq 36$$

✔ Check Your Understanding

3. In Example 2, what is happening at the step where the inequality symbol gets reversed?

4. Describe how the graph in Example 2 would be different if the inequality were $-16a - 12 < 36$.

Guided Practice

COMMUNICATION « *Writing*

Write an inequality that represents each sentence. Do not solve.

« **1.** The sum of twice a number and 5 is less than 17.

« **2.** The difference of 9 subtracted from three times a number is greater than or equal to 6.

« **3.** The sum of 3 and the quotient of a number divided by 2 is less than or equal to 1.

Write each inequality in words.

« **4.** $4n - 8 > 10$ « **5.** $5n + 2 \leq -4$ « **6.** $\frac{n}{-5} + 7 < 3$

Tell whether you would first *add* or *subtract* to solve each inequality. Then tell what number you would add or subtract.

7. $3n + 7 < 16$ **8.** $4k - 6 > 14$ **9.** $-5a + 1 \leq -24$

Solve and graph each inequality.

10. $9b - 6 > 21$ **11.** $7h + 5 \geq -9$ **12.** $-4n - 9 < -1$

13. $-\frac{2}{3}x - 2 > 1$ **14.** $4 + \frac{n}{-3} \leq 5$ **15.** $2a - 1.6 < 4.8$

Exercises

Solve and graph each inequality.

1. $5t - 8 > 22$
2. $7h - 4 \leq 24$
3. $11 + 2a \geq 5$
4. $5 + 6x \geq 7$
5. $\frac{n}{2} - 1 \leq 1$
6. $\frac{z}{3} - 9 < -7$
7. $2b - 3 < -5$
8. $5t + 4 \geq 4$
9. $2 + \frac{k}{5} > -1$
10. $5 + \frac{x}{-4} \leq 6$
11. $-8a + 6 \leq 30$
12. $-2q - 7 \leq -9$
13. $\frac{3}{4}y - 2 \geq -4$
14. $-\frac{4}{5}m + 3 < 7$
15. $-4m - 2.7 \geq 4.1$
16. $3b + 4.5 > 6$
17. $\frac{a}{-3} + 2 < 9$
18. $\frac{t}{-2} - 4 \leq -4$

19. The sum $-2v + 5$ is less than -23. What can you conclude about the value of v?

20. The difference $9j - 6$ is at least 48. What can you conclude about the value of j?

Solve each inequality.

21. $2(3a - 1) < 16$
22. $15 \leq 3(t - 4)$
23. $8 > 4(2x + 5)$

SPIRAL REVIEW

24. A number cube that has six sides numbered 1 through 6 is rolled. Find $P(\text{number} > 4)$. *(Lesson 12-1)*

25. Find solutions of the equation $y = 3x - 5$. Use $-2, -1, 0, 1,$ and 2 as values for x. *(Lesson 13-1)*

26. Quadrilateral $ABCD$ is congruent to quadrilateral $WXYZ$. Which side of $WXYZ$ is congruent to $\overline{AD}$? *(Lesson 10-7)*

27. Solve and graph the inequality $-2y + 8 < 10$. *(Lesson 13-7)*

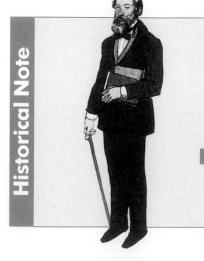

Historical Note

Jakob Steiner (1796–1863) has been called the greatest geometer since Apollonius of Perga (3rd century B.C.). Yet Steiner did not learn to read or write until he was 14 years old. Born into a poor family in Switzerland, Steiner left home at the age of 18 to study mathematics. He eventually became a professor at the University of Berlin, where he taught many other famous mathematicians including Leopold Kronecker and Bernhard Riemann.

Research

Both Apollonius of Perga and Jakob Steiner were interested in conic sections. Find out what conic sections are.

Inequalities with Two Variables

Objective: To determine solutions of inequalities with two variables and to graph inequalities with two variables.

The line shown in the figure at the right is the graph of $y = -2x + 1$. The line divides the coordinate plane into two parts. The red shaded part *above* the line shows the points for which $y > -2x + 1$. The blue shaded part *below* the line shows the points for which $y < -2x + 1$.

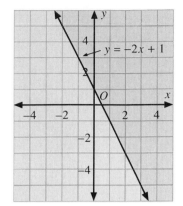

A **solution of an inequality with two variables** is an ordered pair of numbers that make the inequality true. Since the point $(-2, 1)$ is in the blue part, $(-2, 1)$ is a solution of the inequality $y < -2x + 1$.

Terms to Know

• *solution of an inequality with two variables*

Example 1

Tell whether $(-1, 6)$ is a solution of the inequality $y \geq 3x + 8$. Write *Yes* or *No*.

Solution

Substitute -1 for x and 6 for y.

$$y \geq 3x + 8$$
$$6 \overset{?}{\geq} 3(-1) + 8$$
$$6 \overset{?}{\geq} -3 + 8$$
$$6 \geq 5$$

Yes, $(-1, 6)$ is a solution of the inequality $y \geq 3x + 8$.

Every inequality with two variables has a related equation. For example, the inequality $y < x + 4$ has the related equation $y = x + 4$. The graph of the related equation forms the *boundary* of the graph of the inequality.

> **Generalization:** *Graphing an Inequality with Two Variables*
>
> If an inequality is of the form $y > mx + b$, make a dashed boundary line and shade above.
> If an inequality is of the form $y \geq mx + b$, make a solid boundary line and shade above.
> If an inequality is of the form $y < mx + b$, make a dashed boundary line and shade below.
> If an inequality is of the form $y \leq mx + b$, make a solid boundary line and shade below.

Example 2
Solution

Graph the inequality $y > 2x - 4$.

- Begin with the equation $y = 2x - 4$. Three points on this line are $(0, -4)$, $(1, -2)$, and $(2, 0)$. Draw a dashed line through these points to show that points on the line are *not* solutions of the inequality.

- To graph $y > 2x - 4$, shade the region *above* the line.

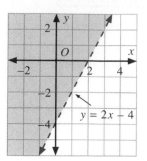

$y = 2x - 4$

✔ Check

Choose the point $(0, 0)$.

$$y > 2x - 4$$
$$0 \overset{?}{>} 2(0) - 4$$
$$0 \overset{?}{>} 0 - 4$$
$$0 > -4$$

✔ Check Your Understanding

1. Describe how the graph in Example 2 would be different if the inequality were $y < 2x - 4$.

2. If the inequality in Example 2 were $y \geq 2x - 4$, would the points on the boundary line be solutions of the inequality?

Guided Practice

COMMUNICATION « *Reading*

Refer to the text on pages 609–610.

« **1.** If an inequality contains the inequality symbol ''$\geq$,'' do you use a *solid* or a *dashed* boundary line on the graph?

« **2.** If an inequality is of the form $y < mx + b$, do you shade *above* or *below* the boundary line?

« **3.** When the graph of an inequality has a dashed boundary line, are the points on that line solutions of the inequality?

Use the graph at the right. Tell whether each ordered pair is a solution of the inequality $y \leq -\frac{1}{2}x - 2$. Write *Yes* or *No*.

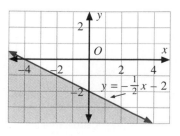

$y = -\frac{1}{2}x - 2$

4. $(0, -2)$ **5.** $(1, -1)$

6. $(2, 1)$ **7.** $(-5, 0)$

8. $(-3, -1)$ **9.** $(4, -4)$

Tell whether each ordered pair is a solution of each inequality. Write
Yes or *No*.

10. $(3, -2); y > -3x - 2$ **11.** $(-4, 2); y < \frac{1}{2}x + 4$

Graph each inequality.

12. $y < -x + 3$ **13.** $y \geq 2x$ **14.** $y > \frac{1}{3}x + 1$

Exercises

Tell whether each ordered pair is a solution of each inequality. Write
Yes or *No.*

1. $(2, 1); y > x - 3$ **2.** $(6, 5); y > x + 2$

3. $(-1, 2); y < -2x$ **4.** $(-2, 1); y \leq -x - 1$

5. $(-3, -2); y \geq 2x + 5$ **6.** $(3, -4); y \leq -3x + 7$

7. $(10, -5); y < \frac{3}{5}x - 9$ **8.** $(-8, 1); y > \frac{1}{4}x + 3$

Graph each inequality.

9. $y > x$ **10.** $y < -x$ **11.** $y \leq 3x - 2$

12. $y \geq -2x + 4$ **13.** $y > -x - 1$ **14.** $y \geq x - 2$

15. $y < 3x + 1$ **16.** $y > -2x$ **17.** $y \leq 4x - 3$

18. $y < \frac{1}{3}x + 2$ **19.** $y > -\frac{1}{4}x + 2$ **20.** $y \geq -\frac{2}{3}x$

Tell whether each ordered pair is a solution of the inequality
$y > -x + 3$. Write *Yes* or *No.*

21. $(0, 2)$ **22.** $(2, 1)$ **23.** $(4, -1)$ **24.** $(-3, 1)$

25. $(-2, 4.9)$ **26.** $(1, 2.1)$ **27.** $(0, 3\frac{1}{3})$ **28.** $(2, \frac{3}{4})$

GROUP ACTIVITY

29. Graph the inequalities $y > 2x + 3$ and $y \leq -3x - 2$ on the same
coordinate plane.
  a. Name three points that are solutions of both inequalities.
  b. Rewrite one of the two inequalities so that the points $(-1, -4)$,
     $(0, -6)$, and $(-3, -3)$ are solutions of both inequalities.

30. Graph the inequalities $y \leq x + 3$ and $y \geq x - 3$ on the same
coordinate plane.
  a. Describe the region common to both inequalities.
  b. Rewrite the two inequalities so that the graphs have no points in
     common and the boundary line of each region is dashed.

COMMUNICATION « *Writing*

Write an inequality with two variables to represent each statement.

« **31.** The value of a number y is less than or equal to the sum of five times the number x and three.

« **32.** The value of a number y is greater than or equal to the difference of seven subtracted from the number x.

« **33.** The amount of time that Jamie spends doing homework (j) is greater than twice the amount of time that her younger sister Anna spends doing homework (a).

« **34.** The cost of a house in the Midwest (m) is less than one-third the cost of a similar house in the Northeast (n).

Write each inequality in words.

« **35.** $y < 3x$ « **36.** $y > 6x + 8$ « **37.** $k \geq m + 1$ « **38.** $c \leq \frac{1}{2}t$

SPIRAL REVIEW

39. Complete the function table at the right.
(*Lesson 2-6*)

40. Graph the equation $y = 4x - 2$.
(*Lesson 13-2*)

41. Draw an angle that measures 48°.
(*Lesson 6-2*)

42. Tell whether $(-1, 4)$ is a solution of the inequality $y < -x + 5$.
(*Lesson 13-8*)

x	$2x + 4$
1	6
4	12
5	?
7	?
10	?

Self-Test 2

1. The 128 apartments in a building are numbered consecutively with metal numbers beginning with the number 1. How many metal digits are needed to number the apartments?   **13-5**

Solve and graph each inequality.

2. $x - 7 > -4$ **3.** $-2a \geq 8$ **13-6**

4. $\frac{c}{5} + 7 < 18$ **5.** $-3b - 4 \leq -5$ **13-7**

Graph each inequality.

6. $y \leq 2x - 3$ **7.** $y > 4x + 1$ **13-8**

13-9 **Function Rules**

Objective: To use a function rule to find values of a function.

 CONNECTION

Terms to Know

- *domain*
- *range*
- *values of a function*
- *function notation*

A function pairs a number in one set of numbers with exactly one number in another set of numbers. In previous lessons you learned to represent a function using function tables, function rules, ordered pairs, and graphs of ordered pairs. Function rules were stated using arrow notation.

Another way to state a function rule is by using an equation with two variables. For example, the function rule $x \rightarrow 4x + 2$ can also be written as $y = 4x + 2$. The set of all possible values of x is called the **domain** of the function, and the set of all possible values of y is called the **range** of the function. The values of y are also referred to as the **values of the function.**

Example 1

Solution

Find values of each function. Use −2, 1, and 5 as values for x.

a. $y = -6x - 5$ **b.** $y = x^2 + 1$

Make function tables.

a.

$y = -6x - 5$	
x	y
−2	$-6(-2) - 5 = 7$
1	$-6(1) - 5 = -11$
5	$-6(5) - 5 = -35$

When $x = -2$, $y = 7$.
When $x = 1$, $y = -11$.
When $x = 5$, $y = -35$.

b.

$y = x^2 + 1$	
x	y
−2	$(-2)^2 + 1 = 5$
1	$(1)^2 + 1 = 2$
5	$(5)^2 + 1 = 26$

When $x = -2$, $y = 5$.
When $x = 1$, $y = 2$.
When $x = 5$, $y = 26$.

 **Check Your Understanding**

1. In Example 1(a), do you first *multiply* or *subtract* when finding each value of the function?

2. In Example 1(b), do you first *find the power* or *add* when finding each value of the function?

A third type of notation used to state a function rule is **function notation**. In function notation, the function in Example 1(b) would be written $f(x) = x^2 + 1$, which is read "f of x equals $x^2 + 1$" or "the value of f at x is $x^2 + 1$."

Function notation is a useful shorthand for showing how values are paired. Instead of writing "when $x = 1$, $y = -11$" in Example 1(a), you could use function notation to write $f(1) = -11$.

Example 2

Solution

Find $f(-3)$, $f(0)$, and $f(5)$ for the function $f(x) = |x - 2|$.

Make a function table.

$f(x) = |x - 2|$	
x	$f(x)$
-3	$|-3 - 2| = 5$
0	$|0 - 2| = 2$
5	$|5 - 2| = 3$

$f(-3) = 5$, $f(0) = 2$, and $f(5) = 3$

☑ **Check Your Understanding**

3. In Example 2, why is the answer $f(-3) = 5$ instead of $f(-3) = -5$?

Guided Practice

COMMUNICATION « *Writing*

Write each sentence using function notation.

« **1.** The function f pairs x with $7x - 1$.

« **2.** The function f pairs x with $24x$.

« **3.** The value of the function f at $x = -1$ is 0.

« **4.** The value of the function f at $x = 3$ is 9.

Write each statement in words.

« **5.** $f(x) = 25x$ « **6.** $f(x) = \frac{x}{2}$ « **7.** $f(2) = 6$ « **8.** $f(1) = -1$

Find each answer.

9. $3(-4)^2$ **10.** $(-6 + 3)^2$ **11.** $|5 - 8|$ **12.** $|-7| + 9$

Find values of each function. Use -1, 3, and 6 as values for x.

13. $y = 3x - 7$ **14.** $y = |x - 5|$ **15.** $y = (x + 4)^2$

Find $f(-2)$, $f(0)$, and $f(4)$ for each function.

16. $f(x) = 4x + 1.5$ **17.** $f(x) = \frac{1}{2}x$ **18.** $f(x) = x^2 - 1$

Exercises

Find values of each function. Use -6, 2, and 3 as values for x.

1. $y = \frac{5}{6}x$ **2.** $y = -x + 10$ **3.** $y = 2x^2 + 1$

4. $y = (x - 7)^2$ **5.** $y = |x| + 2.5$ **6.** $y = |x + 6|$

Find $f(-4)$, $f(-2)$, and $f(5)$ for each function.

7. $f(x) = -2x$

8. $f(x) = \frac{1}{2}x - 2$

9. $f(x) = (x + 1)^2$

10. $f(x) = x^2 + 3$

11. $f(x) = 5x^2$

12. $f(x) = |2x - 1|$

CONNECTING MATHEMATICS AND SCIENCE

Functions are used in physics to describe relationships between physical quantities. The function $P(s) = \frac{3}{200}s^3$ shows how the power, in watts, generated by a windmill is related to wind speed (s).

13. Find the power generated by a windmill when the wind speed is 2 mi/h.

14. Find the power generated by a windmill when the wind speed is 4 mi/h.

15. The wind speed in Exercise 14 is twice the wind speed in Exercise 13. Is the power generated by the windmill in Exercise 14 likewise twice the power generated by the windmill in Exercise 13? Explain.

PROBLEM SOLVING/APPLICATION

Engineers use *pulse functions* to make machines perform tasks at specific time intervals. A pulse function can have a value of either 0 or 1. Depending on the machine, the task will be performed either when the value is 1 or when the value is 0.

Let $p(t) = 1$ if t is a multiple of 4, and let $p(t) = 0$ if t is any other number. Use this pulse function for Exercises 16–18.

16. Suppose a machine stamps a package when $p(t) = 1$. Will the machine stamp a package when $t = 48$?

17. Suppose a machine tightens a bolt when $p(t) = 1$. Will the machine tighten a bolt when $t = 38$?

18. Suppose that a machine punches a hole in a metal plate when $p(t) = 0$. How many holes will the machine punch during the interval $1 \leq t \leq 60$?

SPIRAL REVIEW

19. Graph the function $x \rightarrow 2x + 3$ when $x = -4, -3, -1, 0$, and 1. *(Lesson 3-9)*

20. Find the absolute value: $|-4|$ *(Lesson 3-1)*

21. Find the slope, the y-intercept, and the x-intercept of the line $y - 2 = -5x$.   *(Lesson 13-3)*

22. Find $f(-1)$, $f(0)$, and $f(2)$ for the function $f(x) = |x + 8|$. *(Lesson 13-9)*

Function Graphs

Objective: To graph functions on a coordinate plane.

 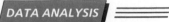

The graph at the right shows the function $d(t) = 85t$. This function represents the distance that a car moving at a steady rate of 85 km/h will travel in t hours.

You can graph a function by graphing the equation specified by the function rule. In doing so, you may want to express the rule in terms of x and y, rather than x and $f(x)$. For example, the rule for the graph at the right would be expressed as $y = 85x$.

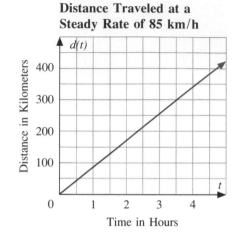

Distance Traveled at a Steady Rate of 85 km/h

Example

Solution

Graph each function: **a.** $y = x^2$ **b.** $y = |x|$

Make a function table for each function. Then graph the ordered pairs and connect the points.

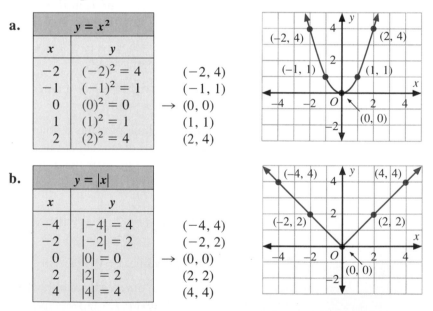

a.

$y = x^2$	
x	y
-2	$(-2)^2 = 4$
-1	$(-1)^2 = 1$
0	$(0)^2 = 0$
1	$(1)^2 = 1$
2	$(2)^2 = 4$

$(-2, 4)$
$(-1, 1)$
$\rightarrow (0, 0)$
$(1, 1)$
$(2, 4)$

b.

$y = |x|$	
x	y
-4	$|-4| = 4$
-2	$|-2| = 2$
0	$|0| = 0$
2	$|2| = 2$
4	$|4| = 4$

$(-4, 4)$
$(-2, 2)$
$\rightarrow (0, 0)$
$(2, 2)$
$(4, 4)$

✓ **Check Your Understanding**

How many values did you find in each function table in the Example before making each graph?

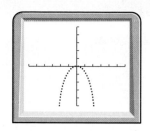

You can use a graphing calculator to graph a variety of functions. On some graphing calculators you would use the key sequence below to graph the function $y = -x^2$.

Guided Practice

COMMUNICATION « *Reading*

Refer to the text on pages 616–617.

«**1.** Describe in words the graph of the function $y = x^2$.

«**2.** Describe in words the graph of the function $y = |x|$.

«**3.** Describe in words the graph of the function $y = -x^2$.

«**4.** Is the graph of a function always a straight line?

Tell whether each ordered pair is on the graph of the function shown at the right. Write *Yes* or *No*.

5. $(1, 0)$ **6.** $(2, 2)$ **7.** $(-3, 2)$

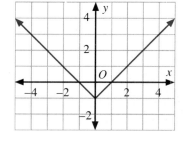

Use the graph at the right to complete each ordered pair.

8. $(3, \underline{\ ?\ })$ **9.** $(-2, \underline{\ ?\ })$

10. $(\underline{\ ?\ }, -1)$ **11.** $(2, \underline{\ ?\ })$

Graph each function.

12. $y = \frac{1}{2}x - 1$ **13.** $y = 3x$ **14.** $y = \frac{1}{3}x^2$

15. $y = (x + 2)^2$ **16.** $y = |x - 1|$ **17.** $y = -|x|$

Exercises

Graph each function.

1. $y = 2x + 1$ **2.** $y = -4x - 3$ **3.** $y = -3x^2$

4. $y = 3x^2$ **5.** $y = |x - 2|$ **6.** $y = |x| - 2$

7. $y = \frac{3}{5}x$ **8.** $y = -\frac{2}{3}x$ **9.** $y = -\frac{1}{3}x^2$

10. $y = \frac{1}{4}x^2$ **11.** $y = x^2 - 1$ **12.** $y = (x - 1)^2$

13. $y = x^2 + 3$ **14.** $y = |x| + 2$ **15.** $y = -|x| + 1$

 CALCULATOR

Choose the letter of the correct graphing calculator display for each function.

16. $y = 2x^2$

17. $y = -2x^2$

18. $y = \frac{1}{2}x^2$

19. $y = -\frac{1}{2}x^2$

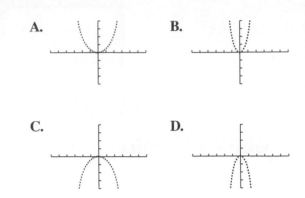

DATA ANALYSIS

The graph at the top of page 616 shows a relationship between time and distance traveled. Many functions represent a relationship between a physical quantity and time, and these relationships can be displayed in graphs.

Use the graphs below. Write the letter of the graph that most likely represents each situation relating temperature (T) to time (t).

20. Natu turned on the heat in his apartment.

21. Andrea turned on the air conditioner in her house.

22. Julieta heated and then sipped a cup of soup for lunch.

23. Jason put ice in a glass of fruit juice and then drank it while he began his homework.

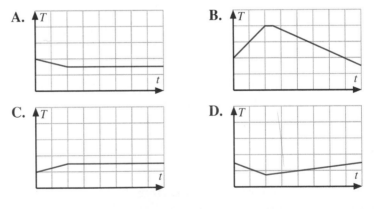

24. **WRITING ABOUT MATHEMATICS** Graph the functions $d(x) = 4x$ and $C(x) = 4x$, where $d(x)$ represents the distance traveled by a person jogging x mi at 4 mi/h, and $C(x)$ represents the cost of buying x yd of material at $4 per yard. Write a paragraph that discusses the similarities and differences in the graphs.

MENTAL MATH

A graph on a coordinate plane represents a function only if each value of x is paired with *exactly one* value of y. You can mentally check whether a graph represents a function. To do this, imagine drawing vertical lines through the graph at various values of x. If any of these lines contain two or more points on the graph, then the graph does not represent a function. This process is called a *vertical-line test*.

Use the vertical-line test to tell mentally whether each graph represents a function. Write *Yes* or *No*.

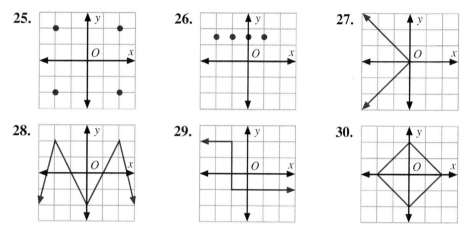

25.   26.   27.

28.   29.   30.

SPIRAL REVIEW

31. What number is 55% of 250?   *(Lesson 9-6)*

32. Find the sum of the measures of the angles of a nonagon. *(Lesson 6-5)*

33. Graph the function $y = -\frac{2}{3}x^2$.   *(Lesson 13-10)*

Self-Test 3

Find values of each function. Use -2, 0, and 2 as values for x.

  1. $y = x^2 - 4$ **2.** $y = |x| - 5$ 13-9

Find $f(-4)$, $f(0)$, and $f(8)$ for each function.

  3. $f(x) = 4x + 7$ **4.** $f(x) = \frac{1}{4}x^2$

Graph each function.

  5. $y = 2x - 1$ **6.** $y = x^2 - 2$ 13-10

Chapter Review

Terms to Know

solution of an equation with two
   variables (p. 582)
graph of an equation with two
   variables (p. 585)
slope (p. 588)
y-intercept (p. 588)
x-intercept (p. 588)
slope-intercept form of an equation
   (p. 588)

system of equations (p. 594)
solution of a system (p. 594)
solution of an inequality with two
   variables (p. 609)
domain of a function (p. 613)
range of a function (p. 613)
values of a function (p. 613)
function notation (p. 613)

**Choose the correct term or phrase from the list above to complete
each sentence.**

1. The __?__ is the set of all possible x-values of the function.

2. An ordered pair that is a solution of both equations in a system of
   equations is a(n) __?__.

3. All points whose coordinates are solutions of an equation with
   two variables represent the __?__.

4. The __?__ of a line is the ratio of the rise to the run.

5. $y = mx + b$ is called the __?__.

6. The y-coordinate of the point where a graph crosses the y-axis is
   the __?__.

**Find solutions of each equation. Use $-2, -1, 0, 1,$ and 2 as values
for x.** *(Lesson 13-1)*

7. $y = 7x + 4$ 8. $y = \frac{1}{2}x + 3$ 9. $y + 5x = -4$

Graph each equation. *(Lesson 13-2)*

10. $y = 2x$ 11. $y = \frac{1}{3}x - 3$ 12. $y - 2x = 4$

**Find the slope and the y-intercept of each line. Then use them to write
an equation for each line.** *(Lesson 13-3)*

13.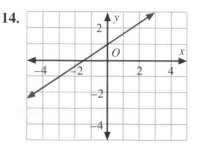

14.

Find the slope, the y-intercept, and the x-intercept of each line.
(Lesson 13-3)

15. $y - 7 = 8x$

16. $y - 6x = 18$

17. $y = -4x$

Solve each system by graphing. *(Lesson 13-4)*

18. $y = x + 5$
$y = -2x - 1$

19. $y = -3x$
$y = -x - 2$

20. $y = 2x - 2$
$y = x - 2$

21. $y = \frac{3}{4}x + 5$
$y = \frac{3}{4}x - 2$

Solve using a simpler problem. *(Lesson 13-5)*

22. How many squares are in the figure at the right?

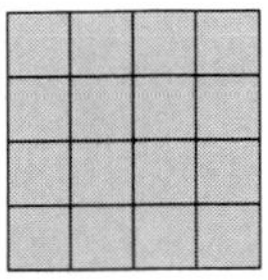

23. Ellen decided to save dimes. On the first day she saved one dime. Every day after that she saved two more dimes than she had saved the day before. How many dimes had Ellen saved after 30 days?

Solve and graph each inequality. *(Lessons 13-6, 13-7)*

24. $x + 1 \leq 7$

25. $c - 8 > -9$

26. $4z \geq 16$

27. $\frac{m}{-2} < 2$

28. $\frac{a}{2} - 1 > -2$

29. $\frac{2}{3}x - 4 \leq 4$

30. $3a + 7 < 4$

31. $-4m - 6 \geq 14$

32. $-5y + 4.8 < 1.3$

Tell whether each ordered pair is a solution of each inequality. Write *Yes* or *No*. *(Lesson 13-8)*

33. $(2, -4); y \geq 3x + 1$

34. $(5, 7); y < 2x - 3$

35. $(4, 2); y > -\frac{1}{2}x$

Graph each inequality. *(Lesson 13-8)*

36. $y > 4x + 3$

37. $y \leq -2x - 4$

38. $y \geq \frac{1}{3}x - 2$

Find values of each function. Use -3, 0, and 4 as values for x.
(Lesson 13-9)

39. $y = -7x + 11$

40. $y = |x| - 6$

41. $y = (x - 1)^2$

Find $f(-2)$, $f(0)$, $f(3)$ for each function. *(Lesson 13-9)*

42. $f(x) = 5x - 8$

43. $f(x) = x^2 + 5$

44. $f(x) = |x + 4|$

Graph each function. *(Lesson 13-10)*

45. $y = x + 2$

46. $y = x^2 - 3$

47. $y = |x| + 3$

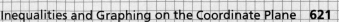

Chapter Test

Find solutions of each equation. Use −2, −1, 0, 1, and 2 as values for x.

1. $y = 3x - 2$
2. $y = \frac{1}{2}x + 1$
3. $y - 2x = 5$

13-1

Graph each equation.

4. $y = 2x + 3$
5. $y = \frac{3}{4}x$
6. $3x + y = 4$

13-2

Find the slope, the y-intercept, and the x-intercept of each line.

7. $y = -4x + 12$
8. $y - 5 = -\frac{1}{2}x$

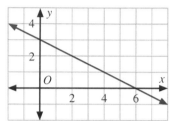

13-3

9. Find the slope and the y-intercept of the line at the right. Then use them to write an equation for the line.

10. **a.** Draw a line with positive slope.
    b. Draw a line with negative slope.

Solve each system by graphing.

11. $y = 2x + 4$
    $y = -x + 7$

12. $y = -5x + 2$
    $y = -5x - 2$

13-4

13. On a coordinate plane, what represents the solution of a system of two equations?

14. Solve using a simpler problem: Find the sum of the first 100 odd whole numbers.

13-5

Solve and graph each inequality.

15. $c - 3 \le 5$
16. $4m > -2$
17. $x + 2\frac{1}{2} < 3\frac{3}{4}$

13-6

18. $\frac{m}{-3} + 2 \ge 1$
19. $5a + 11 \ge 21$
20. $-3n - 7 < 2$

13-7

Graph each inequality.

21. $y > 3x + 1$
22. $y \le -2x - 3$
23. $y < -x + 2$

13-8

Find $f(5)$, $f(0)$, and $f(-3)$ for each function.

24. $f(x) = -3x + 9$
25. $f(x) = |x - 3|$

13-9

Graph each function.

26. $y = x^2 + 3$
27. $y = |x| - 1$

13-10

Graphing Inequalities with One Variable on the Coordinate Plane

Objective: To graph inequalities with one variable on the coordinate plane.

In Lesson 13-8, you graphed inequalities with two variables. To graph inequalities with one variable on the coordinate plane, you follow a similar procedure. You graph the boundary line and then shade the appropriate region.

A boundary line of the form $y = k$ is a horizontal line. A boundary line of the form $x = k$ is a vertical line.

Example

Graph $x > 1$ on a coordinate plane.

Solution

- Begin with the equation $x = 1$. Three points on this line are $(1, 0)$, $(1, -3)$, and $(1, 6)$. Draw a dashed boundary line through these points.

- To graph $x > 1$, shade the region to the *right* of the line, where every value of x is greater than 1.

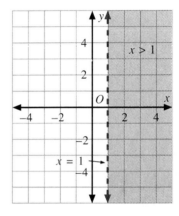

✔️ **Check**

Choose the point $(2, 0)$.
$$x > 1$$
$$2 \overset{?}{>} 1$$
$$2 > 1$$

Exercises

Tell whether you shade to the *right* or to the *left* of the boundary line.

1. $x > 4$ **2.** $x \geq -2$ **3.** $x \leq 3$

4. $x < -1$ **5.** $x \geq 5$ **6.** $x \leq -7$

Tell whether you shade *above* or *below* the boundary line.

7. $y > 0$ **8.** $y < 4$ **9.** $y \leq -6$

10. $y \geq -5$ **11.** $y < -2$ **12.** $y > 1$

Graph each inequality on a coordinate plane.

13. $x < -4$ **14.** $x > 5$ **15.** $y \geq 2$ **16.** $y \leq -2$

17. $y + 3 \leq 6$ **18.** $2x \geq 4$ **19.** $3x - 4 < -7$ **20.** $-2y - 5 > 3$

Cumulative Review

Standardized Testing Practice

Choose the letter of the correct answer.

1. Which ordered pair is a solution of $2x + y = 4$?
   I. (2, 2)   II. (7, −10)   III. (0, 4)
   A. I only   B. II only
   C. I and II   D. II and III

2. The central angle for one sector of a circle graph is 54°. What percent of the total amount is represented by this sector?
   A. 54%   B. 15%
   C. 36%   D. 30%

3. Simplify: $7x + 9z − 4z + 6x$
   A. $18xz$ B. $13x + 5z$
   C. $16x + 2z$ D. $x + 5z$

4. Two letters are chosen without replacement from the word MAGNET. What is $P(\text{N, then T})$?
   A. $\frac{1}{30}$ B. $\frac{1}{36}$

   C. $\frac{11}{30}$ D. $\frac{2}{11}$

5. Find the x-intercept: $y = \frac{2}{3}x − 4$

   A. $\frac{2}{3}$ B. −4

   C. 6   D. 4

6. Two cocaptains are chosen from a team of 6 swimmers. In how many different ways can this be done?
   A. 36   B. 12
   C. 15   D. 24

7. The scale of a model house is 2 in. : 15 ft. The actual width of the house is 30 ft. Find the width of the model house.
   A. 4 in.   B. 30 in.
   C. 1 in.   D. 2.25 in.

8. Solve: $5x + 3x + 9 = 33$
   A. $x = 8$ B. $x = 3$

   C. $x = 1\frac{16}{17}$ D. $x = 5\frac{1}{4}$

9.   4.3 | 0 1 1 3 8
        4.4 | 1 4 5 5 5 9
        4.5 | 0 2 3

   How many data items in this stem-and-leaf plot are less than 4.45?
   A. 7   B. 10
   C. 9   D. 11

10. Simplify: $\frac{15n^2}{n^6}$

    A. $\frac{15}{n^3}$ B. $15n^4$

    C. $15n^8$ D. $\frac{15}{n^4}$

11.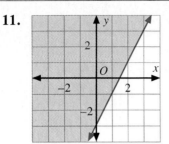

Which inequality does this graph represent?

A. $y > 2x - 3$
B. $y \leq 2x - 3$
C. $y < 2x - 3$
D. $y \geq 2x - 3$

16.

Outcome	Frequency
red	36
yellow	18
blue	26

A three-colored spinner was spun 80 times. The outcomes are recorded in the table above. Find the experimental probability of the spinner landing on red.

A. $\frac{1}{3}$
B. $\frac{9}{20}$
C. $\frac{11}{20}$
D. $\frac{2}{3}$

12. Solve: $17 + j = 6$

A. $j = 9$
B. $j = -9$
C. $j = 11$
D. $j = -11$

17. Simplify: $-5(-4) + 6(-2)$

A. 8
B. 20
C. -32
D. -52

13. A bag contains 5 red, 1 white, and 2 blue marbles. One marble is chosen at random from the bag. Find the odds in favor of selecting a blue marble.

A. $\frac{1}{4}$
B. $\frac{1}{8}$
C. $\frac{1}{3}$
D. $\frac{5}{8}$

18. The length of one side of a regular octagon is 15 cm. Find the perimeter.

A. 90 cm
B. 23 cm
C. 120 cm
D. cannot be determined

14. Write the phrase *the total of four times a number z and fifteen* as a variable expression.

A. $4(z + 15)$
B. $z + 4 \cdot 15$
C. $4z + 15$
D. $z^4 + 15$

19. Find the sum: $\frac{3x}{8} + \frac{5x}{12}$

A. $\frac{11x}{20}$
B. $\frac{x}{3}$
C. $\frac{19x}{24}$
D. $\frac{17x}{12}$

15. Solve: $-4x - 10 < 2$

A. $x < 2$
B. $x > {}^-3$
C. $x > -2$
D. $x < -3$

20. What percent of 120 is 78?

A. 78%
B. 65%
C. 93.6%
D. 154%

Dimensions of the Statue of Liberty	
pedestal height	154 ft
statue height	151 ft 1 in.
right arm length	45 ft
hand length	16 ft 5 in.
index finger length	8 ft
nail width	1 ft 6 in.
nose length	4 ft 6 in.
mouth width	3 ft
eye width	2 ft 6 in.

When the Statue of Liberty was dedicated in 1886, Grover Cleveland was the President of the United States. When the Statue was rededicated in 1986, Ronald Reagan was the President. The restoration project that preceded the rededication required 5000 yd^3 of concrete.

Surface Area and Volume 14

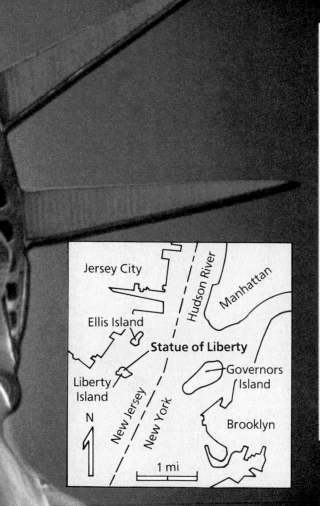

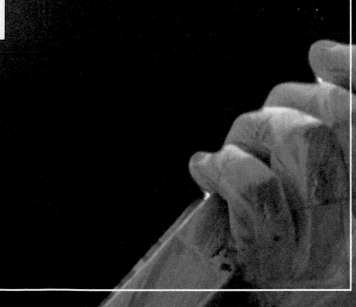

1834	Frédéric Auguste Bartholdi, designer of the Statue of Liberty, born in France
1870	Bartholdi makes first signed and dated model of the Statue of Liberty
1876	Statue's hand and torch exhibited in Philadelphia
1878	Statue's head and shoulders exhibited in Paris
1884	Statue completed in Paris
1885	Statue dismantled and shipped to United States
1886	Statue unveiled and dedicated on Bedloe's Island in Upper New York Bay
1924	Statue declared a national monument
1933	Statue placed under jurisdiction of the National Park Service
1936	Fiftieth anniversary restoration
1984	Restoration begins in preparation for Centennial celebration
1985	New torch and gilded flame are installed
1986	Statue's centennial rededication

Space Figures

Objective: To identify space figures.

CONNECTION

Terms to Know

- *space figure*
- *face*
- *edge*
- *vertex*
- *polyhedron*
- *base*
- *prism*
- *cube*
- *pyramid*
- *cylinder*
- *cone*
- *sphere*

In Chapters 6 and 10 you studied polygons. Polygons are sometimes referred to as *plane figures* because they lie in a plane. **Space figures** are three-dimensional figures that enclose part of space.

Some space figures have flat surfaces called **faces.** A line segment on a space figure where two faces intersect is called an **edge.** A point where edges intersect is called a **vertex.**

A **polyhedron** is a space figure whose faces are polygons. *Prisms* and *pyramids* are polyhedrons. They are identified by the number and shape of their **bases.**

A **prism** has *two* parallel congruent bases. The other faces of the prisms you will study in this book are rectangles. A **cube** is a rectangular prism whose faces are all squares.

A **pyramid** has one base. Its other faces are triangles.

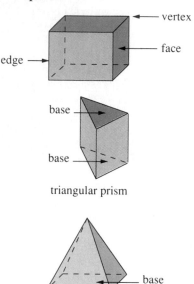

triangular prism

rectangular pyramid

Example 1

Identify each space figure.

a.

b.

c.

Solution

a. Triangular pyramid. The figure has one triangular base.

b. Hexagonal prism. The figure has two hexagonal bases that are congruent and parallel.

c. Rectangular prism. The figure has two rectangular bases that are congruent and parallel.

☑ **Check Your Understanding**

1. In Example 1(a), how can you tell that the figure is not a prism?

2. Give two reasons why the figure in Example 1(b) is not a pyramid.

Space figures that have curved surfaces are not polyhedrons. Three such figures are a *cylinder*, a *cone*, and a *sphere*.

A **cylinder** has two circular bases that are congruent and parallel.

A **cone** has one circular base and a vertex.

A **sphere** is the set of all points in space that are the same distance from a given point called the center.

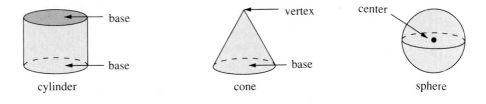

| cylinder | cone | sphere |

Example 2

Identify each space figure.

a.   b.   c.

Solution

a. Cone. The figure has one circular base.

b. Sphere.

c. Cylinder. The figure has two circular bases that are congruent and parallel.

Guided Practice

Replace each __?__ with the correct word.

« **1.** A polyhedron with two congruent parallel bases is a(n) __?__.

« **2.** A polyhedron with one base is a(n) __?__.

« **3.** A space figure with a curved surface and two circular bases that are congruent and parallel is a(n) __?__.

« **4.** A space figure with a curved surface, one circular base, and a vertex is a(n) __?__.

Name one property that is shared by the two types of figures.

« **5.** cylinders and prisms   « **6.** cones and pyramids

Identify each space figure.

7.

8.

9.

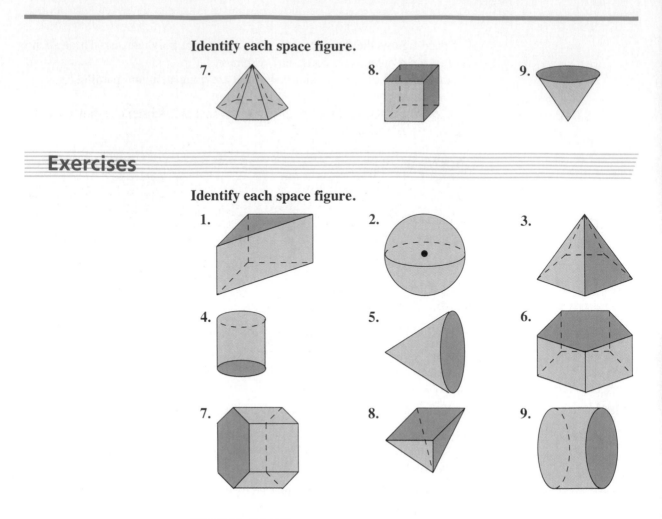

Exercises

Identify each space figure.

1.

2.

3.

4.

5.

6.

7.

8.

9.

SPATIAL SENSE

Learning to draw space figures can help you visualize them better. The diagrams below shows the steps involved in drawing a rectangular prism. First draw two congruent rectangles, one to the right of and above the other. Next draw lines connecting the corresponding vertices of the rectangles. Then make dashes for the edges of the prism that should not be visible.

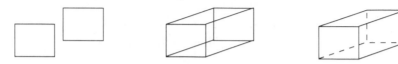

10. Draw a triangular prism.

11. Draw a cylinder. (*Hint:* Make the bases look like ovals.)

12. Draw a cone. (*Hint:* Draw an oval for the base and a point above the center of the base for the vertex.)

13. Draw a rectangular pyramid. (*Hint:* Draw a parallelogram for the base and a point above the base for the vertex.)

CAREER/APPLICATION

An *industrial designer* makes drawings of a product before it is manufactured. The drawings include front, top, and side views, as shown below.

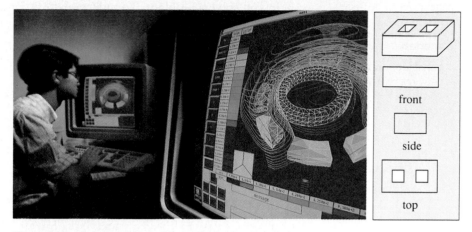

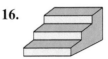

Sketch front, top, and side views of each object.

14.

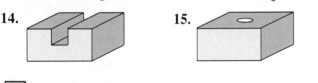

15.

16.

COMPUTER APPLICATION

Suppose you have drawn a cylinder resting on one of its circular bases. A side view of the cylinder is shown in the diagram at the right. You can use drawing software to rotate it around a horizontal axis or a vertical axis through its center.

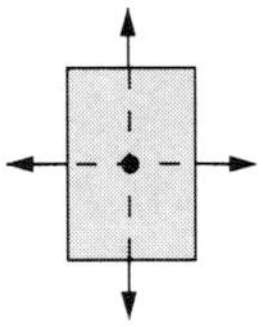

Use computer aided design (CAD) software or spatial sense.

17. Rotate the cylinder 90° around the vertical axis. Will the resulting figure look different from the original?

18. Rotate the cylinder 90° around the horizontal axis. Make a sketch of the resulting figure.

19. Different amounts of rotation around the horizontal axis will result in different views of the cylinder. Choose another rotation and sketch the resulting figure.

SPIRAL REVIEW

20. Identify the space figure at the right. *(Lesson 14-1)*

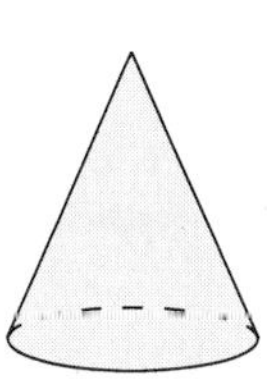

21. Solve and graph $x - 3 < 2$. *(Lesson 13-6)*

22. Find the area of a triangle with base 10 in. and height 7 in. *(Lesson 10-3)*

Paper Folding and Cutting

Materials

- construction paper
- centimeter ruler
- compass
- scissors

Later in this chapter you may want to make a model of a space figure to help you solve a problem. To make a pattern for a model, you must know the number of faces of the space figure and their shapes.

Activity I *Recognizing a Pattern*

1 A pattern for a polyhedron is shown below. How many rectangular faces does the polyhedron have? How many triangular faces? Do you think the polyhedron is a *prism* or a *pyramid*?

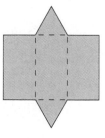

2 On a sheet of construction paper, construct a larger version of the pattern. Include the dashed lines.

3 Cut out the figure you have constructed. Fold along the dashed lines until the edges meet. Tape edges together where they meet.

4 You have built a model of a space figure. Identify it. Was it what you expected?

5 If the triangles in the pattern were squares, you would have a pattern for a common object. Name a common object that the figure would resemble.

Activity II *Making a Pattern*

1 Draw a square with sides 10 cm long.

2 On each side of the square, construct an equilateral triangle with sides 10 cm long.

3 Cut out the figure you have constructed. Fold along the edges of the square until the edges of the triangles meet. Tape the edges together.

4 You have built a model of a space figure. Identify the space figure.

5 Construct an equilateral triangle with sides 4 cm long. On each side of that triangle, construct an isosceles triangle whose two equal sides are 5 cm long. Cut out, fold, and tape the pattern. Identify the space figure.

Now that you have cut and folded paper patterns to make models of space figures, you may be able to look at a pattern and visualize the space figure that it would make. You may then want to copy the pattern and construct the model as a check.

Activity III *Visual Thinking*

1 Which pattern(s) could be folded to form a cube?

a.   **b.**   **c.**

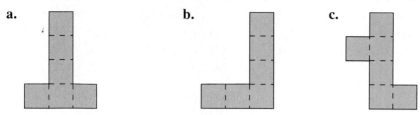

2 Which pattern(s) could be folded to make a cylinder?

a.   **b.**   **c.**

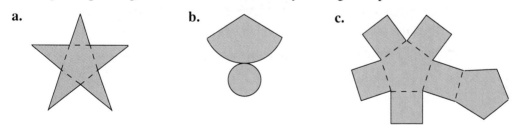

3 Identify the space figure that could be formed by folding each pattern.

a.   **b.**   **c.**

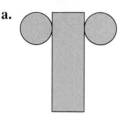 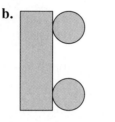 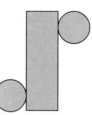

You can also develop your visual thinking by looking at a drawing of a space figure and figuring out a pattern that can be used to construct it.

Activity IV *More Visual Thinking*

Make a pattern for each space figure.

a.   **b.**   **c.**

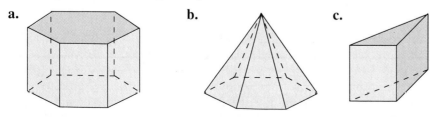

Surface Area of Prisms

Objective: To find the surface area of a prism.

Terms to Know
- *surface area*

Megan needs to wrap the box containing her brother John's present. She has a piece of wrapping paper measuring 2000 cm². The dimensions of the box are given in the diagram below.

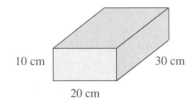

10 cm   30 cm

20 cm

QUESTION Does Megan have enough wrapping paper?

You need to find the *surface area* of the box, or prism.

The **surface area** (S.A.) of a prism is the sum of the areas of the bases and faces of the prism. Surface area is expressed in square units.

Example 1

Solution

Find the surface area of the rectangular prism shown above.

Make a sketch of the rectangular faces and label the dimensions.

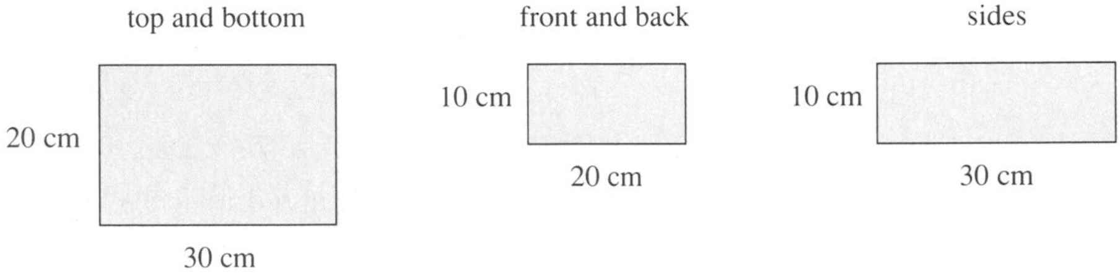

top and bottom   front and back   sides

20 cm

30 cm

10 cm

20 cm

10 cm

30 cm

S.A. = top and bottom + front and back + sides
S.A. = 2(30 · 20)   + 2(20 · 10)   + 2(30 · 10)
S.A. = 1200   + 400   + 600
S.A. = 2200

The surface area of the rectangular prism is 2200 cm².

ANSWER No, Megan does not have enough wrapping paper. She needs more than 2200 cm² of wrapping paper, and she has only 2000 cm².

Example 2	Find the surface area of the triangular prism at the right.

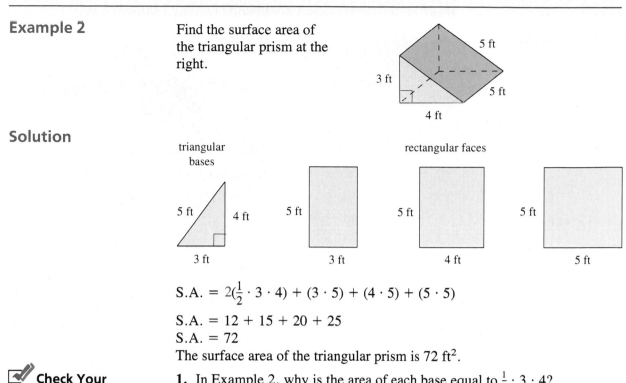

Solution

triangular bases   rectangular faces

S.A. = $2(\frac{1}{2} \cdot 3 \cdot 4) + (3 \cdot 5) + (4 \cdot 5) + (5 \cdot 5)$

S.A. = $12 + 15 + 20 + 25$
S.A. = 72

The surface area of the triangular prism is 72 ft^2.

✓ Check Your Understanding

1. In Example 2, why is the area of each base equal to $\frac{1}{2} \cdot 3 \cdot 4$?
2. In Example 2, why can you not triple the area of one rectangular face to find the area of the three rectangular faces?

Guided Practice

COMMUNICATION « *Reading*

«1. Explain the meaning of the word *surface* in the following sentence.
   Clues to the mystery began to surface after some investigation.

«2. The prefix *sur-*, as in *surface*, means "over." List two other words that begin with this prefix. Write the definition of each word.

Refer to the text on pages 634–635.

«3. State the main idea of the lesson.

«4. List two major points that support the main idea.

5. The cube at the right has edges 2 cm long.
   a. What is the area of each face of the cube?
   b. How many faces does the cube have?
   c. What is the surface area of the cube?

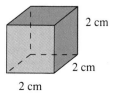

Draw all the faces of each prism and label their dimensions.

6. 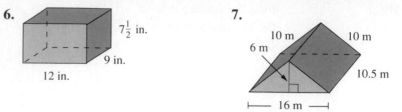 $7\frac{1}{2}$ in.

9 in.

12 in.

7. 10 m   10 m

6 m

10.5 m

├── 16 m ──┤

8. Find the surface area of the prism in Exercise 6 above.

9. Find the surface area of the prism in Exercise 7 above.

Exercises

Find the surface area of each prism.

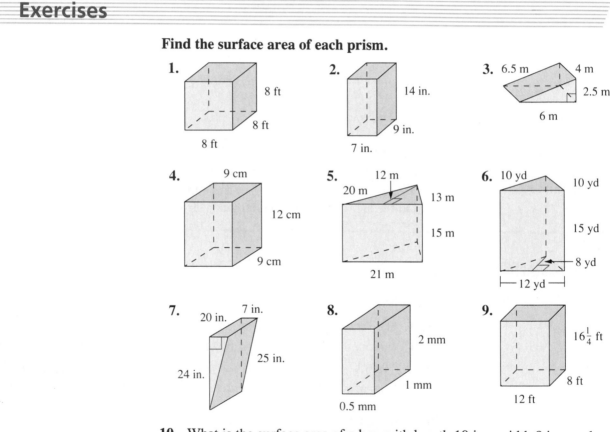

1. 8 ft

8 ft

8 ft

2. 14 in.

9 in.

7 in.

3. 6.5 m   4 m

2.5 m

6 m

4. 9 cm

12 cm

9 cm

5. 12 m

20 m   13 m

15 m

21 m

6. 10 yd   10 yd

15 yd

8 yd

├── 12 yd ──┤

7. 20 in.   7 in.

25 in.

24 in.

8. 2 mm

1 mm

0.5 mm

9. $16\frac{1}{4}$ ft

8 ft

12 ft

10. What is the surface area of a box with length 10 in., width 8 in., and height $4\frac{1}{2}$ in.?

11. What is the surface area of a compact disc case with length 14.3 cm, width 12.5 cm, and height 1 cm?

12. The length of a box of granola is 5 in. The width is 2 in. The height is $7\frac{3}{4}$ in. Find the surface area of the box of granola.

13. A perfume bottle is shaped like a rectangular prism with length 4 cm, width 2 cm, and height 7.5 cm. Find the surface area of the bottle of perfume.

14. What is the surface area of a cube with edges of length 1 cm?

15. What is the surface area of a cube with edges of length 2 ft?

16. One cube has edges of length 5 in. Another cube has edges of length 10 in. Find the ratio of their surface areas.

17. One cube has edges of length 10 m. Another cube has edges of length 30 m. Find the ratio of their surface areas.

CONNECTING GEOMETRY AND ALGEBRA

The height of each triangular face of the pyramid at the right is l. The base is a square with a side of length n. If you know the values of l and n, you can find the surface area of this pyramid.

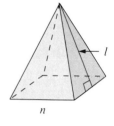

Use the figure above.

18. Choose the letter of the formula for the area of the base.

   a. nl **b.** $4n$ **c.** $\frac{1}{2}nl$ **d.** n^2

19. Choose the letter of the formula for the area of a triangular face.

   a. $\frac{1}{2}n^2$ **b.** $\frac{1}{2}l^2$ **c.** $\frac{1}{2}nl$ **d.** l^2

20. Use your answers to Exercises 18 and 19 to write a formula for the surface area of the pyramid.

21. Use the formula that you wrote in Exercise 20 to find the surface area of the pyramid when $n = 12$ ft and $l = 8$ ft.

SPIRAL REVIEW

22. Find the surface area of the prism at the right. *(Lesson 14-2)*

23. Solve: $4x - 7.9 = -1.1$ *(Lesson 8-8)*

24. Find $f(-2)$, $f(0)$, and $f(4)$ for the function $f(x) = \frac{1}{2}x + 3$. *(Lesson 13-9)*

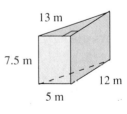

25. Draw an angle that measures $54°$. *(Lesson 6-2)*

26. DATA, *pages 626–627* In lowest terms, write the ratio of the length of the nose on the Statue of Liberty to the width of the mouth. *(Lesson 9-1)*

Challenge

Number the 8 vertices of a cube from 1 to 8 so that the sum of the 4 numbers at the vertices of each face is 18.

Surface Area of Cylinders

Objective: To find the surface area of a cylinder.

EXPLORATION

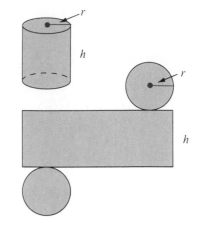

1 How many bases does the cylinder in the diagram at the right have?

2 What is the formula for the area of a base of the cylinder?

3 What is the formula for the circumference of a base of the cylinder?

4 The curved surface of a cylinder flattens out into a rectangle. How does the length of the rectangle compare with the circumference of a base of the cylinder?

The surface area of a cylinder consists of the areas of a rectangle and two congruent circles. The length of the rectangle is the circumference of a base of the cylinder, and the width is the height of the cylinder.

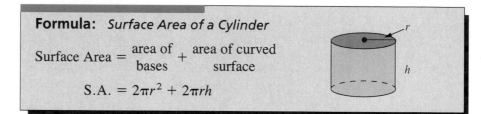

Formula: *Surface Area of a Cylinder*

Surface Area = $\dfrac{\text{area of bases}}{}$ + $\dfrac{\text{area of curved surface}}{}$

S.A. $= 2\pi r^2 + 2\pi rh$

Example

Find the surface area of the cylinder at the right.

Solution

S.A. $= 2\pi r^2 + 2\pi rh$

S.A. $\approx 2\left(\dfrac{22}{7}\right)(7)^2 + 2\left(\dfrac{22}{7}\right)(7)(30)$

S.A. $\approx 308 + 1320 \approx 1628$

The surface area of the cylinder is approximately 1628 cm².

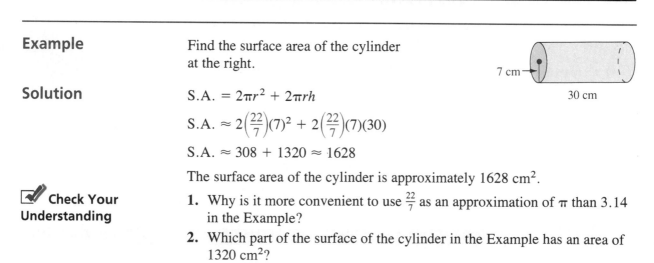

7 cm

30 cm

✓ **Check Your Understanding**

1. Why is it more convenient to use $\frac{22}{7}$ as an approximation of π than 3.14 in the Example?

2. Which part of the surface of the cylinder in the Example has an area of 1320 cm²?

You can use the π and x^2 keys on a calculator to find the surface area of a cylinder. You would use this key sequence to find the surface area of a cylinder with radius 4 and height 15.

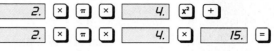

Guided Practice

COMMUNICATION « *Reading*

Refer to the text on page 638.

« **1.** Which two measurements must you know to compute the surface area of a cylinder?

« **2.** What is the shape of a pattern for the curved surface of a cylinder?

« **3.** State the formula for the surface area of a cylinder.

Use the cylinder at the right.

4. Find the area of a base.

5. Find the circumference of the base.

6. Find the area of the curved surface.

7. Use your answers to Exercises 4 and 6 to find the surface area.

Find the surface area of each cylinder.

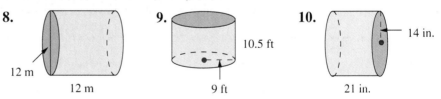

Exercises

Find the surface area of each cylinder.

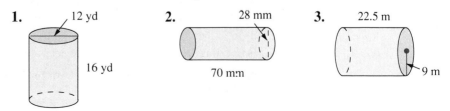

Find the surface area of each cylinder.

4. 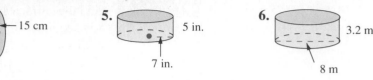 ←— 15 cm

20 cm

5. 5 in.

7 in.

6. 3.2 m

8 m

7. The diameter of a dime is about 18 mm. The height is about 1.3 mm. Find the approximate surface area of a stack of 15 dimes.

8. A drum is closed on the top and the bottom. The diameter of the drum is 20 in. The height is 27 in. Find the approximate surface area of the drum.

9. Find the approximate surface area of the paperweight below.

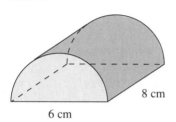

8 cm

6 cm

10. Compare the areas of the two labels. What do you notice?

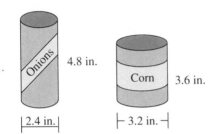

Onions 4.8 in.

Corn 3.6 in.

2.4 in. 3.2 in.

CALCULATOR

For Exercises 11 and 12, choose the letter of the key sequence you would use to find each value.

A. 2. × π × 3. × 10. =

B. 2. × π × 3. x^2 =

C. 2. × π × 10. x^2 =

11. the area of the two bases of a cylinder with radius 3 and height 10

12. the area of the curved surface of a cylinder with radius 3 and height 10

PROBLEM SOLVING/APPLICATION

The formula for the surface area of the curved surface of a cone, or *lateral area*, is L.A. $= \pi r l$, where r is the radius of the base of the cone, and l is the *slant height*.

l

r

13. For a play, a costume maker has to make a cone-shaped hat with radius 3 in. and slant height 12 in. What will be the lateral area of the hat to the nearest whole number?

14. Suppose the costume maker plans to decorate the hat described in Exercise 13 using 3 sequins per square inch. How many sequins will the costume maker use?

15. Suppose the costume maker plans to put ribbon around the bottom edge of the hat described in Exercise 13. Rounding up to the nearest inch, how much ribbon will the costume maker use?

THINKING SKILLS

16. Decide which of the following will change when the height of a cylinder is doubled.

    a. area of a base   **b.** area of curved surface   **c.** surface area

17. Decide which of the following will change when the radius of a can is doubled.

    a. area of a base   **b.** area of curved surface   **c.** surface area

SPIRAL REVIEW

18. Find the surface area of the cylinder at the right. *(Lesson 14-3)*

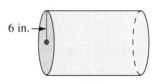

6 in.

18 in.

19. Find the area of a rectangle with length 9 cm and width 5 cm. *(Lesson 10-3)*

20. Graph the function $y = |x| - 3$. *(Lesson 13-10)*

21. A number cube that has sides numbered 1 through 6 is rolled. Find $P(\text{less than 3})$. *(Lesson 12-1)*

22. Estimate the quotient: $12\frac{1}{2} \div 3\frac{7}{8}$ *(Lesson 8-6)*

Self-Test 1

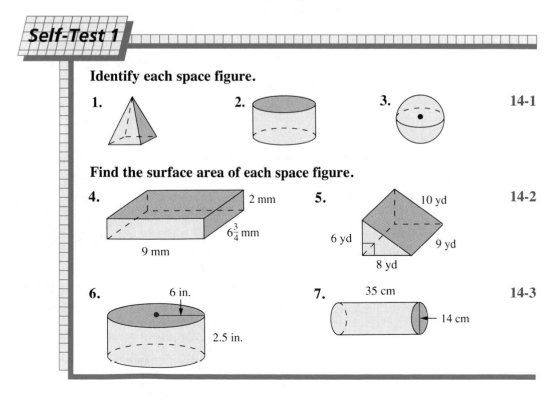

Identify each space figure.

1.   **2.**   **3.**   14-1

Find the surface area of each space figure.

4.   2 mm $6\frac{3}{4}$ mm   9 mm   **5.**   10 yd   6 yd   9 yd   8 yd   14-2

6.   6 in.   2.5 in.   **7.**   35 cm   14 cm   14-3

Volumes of Prisms and Pyramids

Objective: To find the volumes of prisms and pyramids.

When you pack a suitcase or pour a glass of water, the amount of clothing or the amount of liquid your container can hold depends on its *volume*. The **volume** V of a space figure is the amount of space it encloses. To measure volume you use cubic units: for instance, cubic centimeters (cm^3), cubic inches ($in.^3$), and cubic yards (yd^3).

You can find the volume of a rectangular prism by counting the number of unit cubes that can fit inside the prism.

Terms to Know
- *volume*

Example 1

Find the volume of the prism at the right.

Solution

To measure the volume, think of the prism as layers of unit cubes that measure 1 cm on each edge.

Number of cubes in 1 layer: $4 \cdot 2 = 8$
Number of cubes in 3 layers: $3 \cdot 8 = 24$

The volume is 24 cm^3.

☑ **Check Your Understanding**

1. How would the volume change if the height were 5 cm instead of 3 cm?
2. How would the volume change if the length of the base were 6 cm instead of 4 cm?

In Example 1, notice that you multiplied the area of a base of the prism by the height of the prism to find the volume. In fact, the volume of any prism is the product of the area of a base (B) and the height (h). The volume of any pyramid is one third the product of the area of the base and the height.

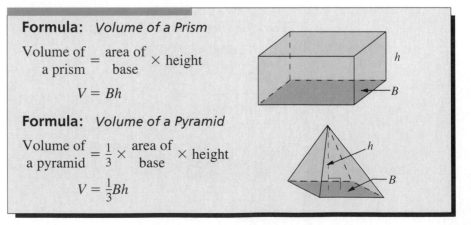

Formula: *Volume of a Prism*

$$\text{Volume of a prism} = \text{area of base} \times \text{height}$$

$$V = Bh$$

Formula: *Volume of a Pyramid*

$$\text{Volume of a pyramid} = \frac{1}{3} \times \text{area of base} \times \text{height}$$

$$V = \frac{1}{3}Bh$$

Example 2

Solution

Find the volume of the pyramid at the right.

The base is a rectangle.
$B = (5.5)(6) = 33$

$V = \frac{1}{3}Bh$

$V = \frac{1}{3}(33)(8) = 88$

The volume is 88 cm³.

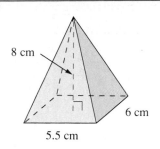

8 cm

6 cm

5.5 cm

Guided Practice

COMMUNICATION «*Reading*

Refer to the text on page 642.

«**1.** What does B represent in the formula $V = Bh$?

«**2.** The volume of a space figure is represented by the formula $V = \frac{1}{3}Bh$. What figure is it?

Write the area formula needed to find the base area of each space figure.

3. rectangular prism   **4.** triangular prism   **5.** square pyramid

6. The lengths of the sides of the base of a given pyramid and the height of the pyramid are measured in feet. What is the most convenient unit of measure for the volume?

The area of the base and the height of each space figure are given. Find the volume.

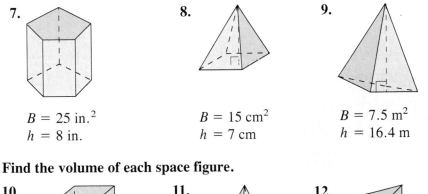

7.

$B = 25$ in.²
$h = 8$ in.

8.

$B = 15$ cm²
$h = 7$ cm

9.

$B = 7.5$ m²
$h = 16.4$ m

Find the volume of each space figure.

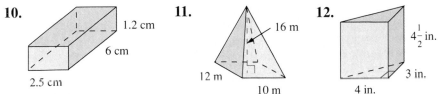

10.

1.2 cm

6 cm

2.5 cm

11.

16 m

12 m

10 m

12.

$4\frac{1}{2}$ in.

3 in.

4 in.

Exercises

Find the volume of each space figure with the given base and height.

a. a prism   **b. a pyramid**

1. $B = 22$ in.^{2} **2.** $B = 19$ cm^2 **3.** $B = 48$ ft^2 **4.** $B = 25$ m^2
$h = 6$ in. $h = 12$ cm $h = 7$ ft $h = 9$ m

Find the volume of each space figure.

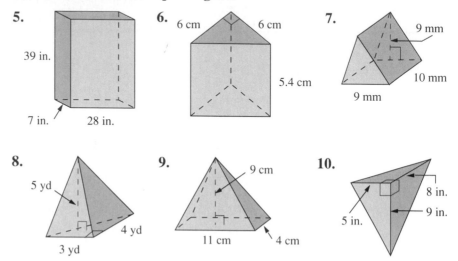

5. 39 in., 7 in., 28 in.

6. 6 cm, 6 cm, 5.4 cm

7. 9 mm, 10 mm, 9 mm

8. 5 yd, 4 yd, 3 yd

9. 9 cm, 11 cm, 4 cm

10. 8 in., 9 in., 5 in.

Find the volume of each space figure described in Exercises 11–14.

11. a cube with sides of length 8 m

12. a rectangular prism with base length $10\frac{1}{2}$ units, base width 8 units, and height 6 units

13. a square pyramid with base edges 15 ft and height 8 ft

14. a square pyramid with base edges 11 m and height 7.2 m

15. The length of a pancake mix box is 15 cm, the width is 5 cm, and the height is 21 cm. What is the volume of the box?

16. A tent is shaped like a square pyramid with height 2 m. If the bottom edges of the tent are 3 m long, find the volume of the tent.

Find the area of the base of each prism described.

17. The volume is 102 yd^3.   **18.** The volume is 216 cm^3.
The height is 12 yd.   The height is 15 cm.

Find the area of the base of each pyramid described.

19. The volume is 98 m^3.   **20.** The volume is 108 yd^3.
The height is 6 m.   The height is 9 yd.

DATA ANALYSIS

Many structures are in the shape of rectangular prisms or pyramids. The table below shows the base edge lengths and heights of some structures with square bases.

Structure	Type of figure	Length of each base edge (ft)	Height (ft)
World Trade Center, each tower, New York City	Prism	209	1350
Transamerica Building, San Francisco	Pyramid	145	853
Great Pyramid of Cheops, Giza, Egypt	Pyramid (original measurements)	756	480

Find the volume of each structure to the nearest cubic foot.

21. a tower at the World Trade Center

22. the Transamerica Building

23. ESTIMATION Estimate to decide which is larger, the volume of a World Trade Center tower or the volume of the Great Pyramid.

 CALCULATOR

For Exercises 24 and 25, use the fact that a cubic foot of water is approximately 7.481 gallons. Use a calculator to find the number of gallons of water needed to fill each swimming pool.

24.

5 ft
20 ft
50 ft

25.

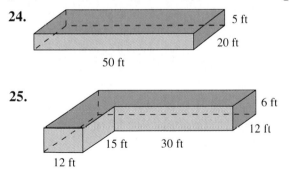

6 ft
12 ft
15 ft   30 ft
12 ft

SPIRAL REVIEW

26. Find the volume of the prism shown at the right.   *(Lesson 14-4)*

27. What number is 64% of 350?   *(Lesson 9-6)*

28. Solve and graph: $2x - 7 \geq 5$ *(Lesson 13-7)*

29. Find the area of a circle with radius 5.6 cm.   *(Lesson 10-4)*

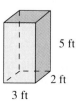

5 ft
2 ft
3 ft

Volumes of Cylinders and Cones

Objective: To find the volumes of cylinders and cones.

The formulas for the volumes of a cylinder and a cone are similar to those for a prism and a pyramid. The base of a cylinder or a cone is a circle, so use πr^2 for the area of the base, B, in the formulas.

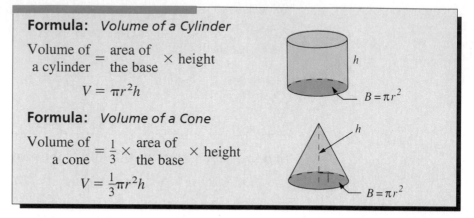

Formula: *Volume of a Cylinder*

$$\text{Volume of a cylinder} = \text{area of the base} \times \text{height}$$

$$V = \pi r^2 h$$

$$B = \pi r^2$$

Formula: *Volume of a Cone*

$$\text{Volume of a cone} = \frac{1}{3} \times \text{area of the base} \times \text{height}$$

$$V = \frac{1}{3}\pi r^2 h$$

$$B = \pi r^2$$

Example 1

The diameter of a cylinder is 30 m and the height is 11 m. Find the volume of the cylinder.

Solution

The radius, r, is $\frac{1}{2}(30) = 15$

$V = \pi r^2 h$
$V \approx 3.14(15^2)(11)$
$V \approx 7771.5$

The volume is approximately 7771.5 m³.

Example 2

Find the volume of a cone with radius 14 in. and height 12 in.

Solution

$V = \frac{1}{3}\pi r^2 h$

$V \approx \frac{1}{3}\left(\frac{22}{7}\right)(14^2)(12)$

$V \approx 2464$

The volume is approximately 2464 in.³.

✓ Check Your Understanding

1. How would Example 1 be different if the diameter were 30 in. and the height were 11 in.?
2. How would Example 1 be different if the object were a cone?
3. Why was $\frac{22}{7}$ used for π in Example 2?

Guided Practice

Replace each __?__ with the correct word, phrase, or expression.

« **1.** To find the volume of a cylinder or cone, you need to know the area of a __?__ and the __?__.

« **2.** The base area of a cone or cylinder with radius r is $B = $ __?__.

« **3.** The volume of a cone is __?__ the volume of a cylinder with the same radius and height.

Match each figure with the formula for its volume.

4. cone

5. prism

6. pyramid

7. cylinder

A. $V = Bh$

B. $V = \pi r^2 h$

C. $V = \frac{1}{3}Bh$

D. $V = \frac{1}{3}\pi r^2 h$

Replace each __?__ with the correct number.

8.

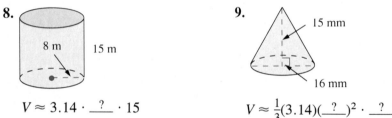

$V \approx 3.14 \cdot$ __?__ $\cdot 15$

9.

$V \approx \frac{1}{3}(3.14)(\underline{\ ?\ })^2 \cdot$ __?__

Find the volume of each space figure.

10.

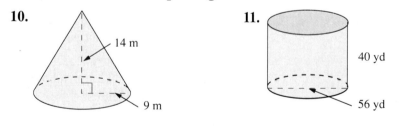

11.

« **12.** Suppose you know the volume, radius, and height of a cylinder.
 a. How would the volume change if you were to double the height?
 b. How would the volume change if you were to double the radius?

« **13.** Which has a greater effect on the volume of a cylinder, doubling the radius or doubling the height?

Exercises

Find the volume of each space figure.

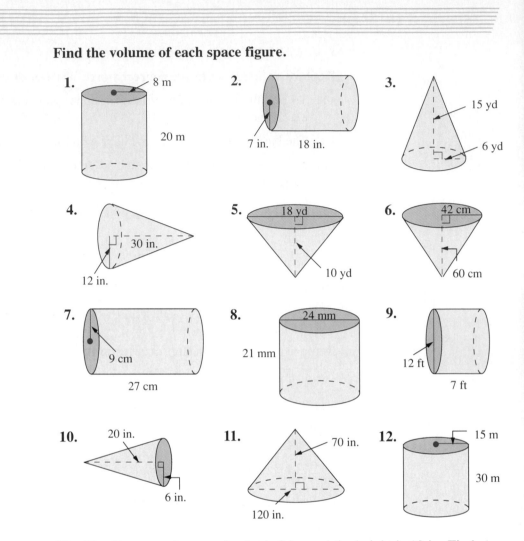

1. 8 m, 20 m

2. 7 in., 18 in.

3. 15 yd, 6 yd

4. 30 in., 12 in.

5. 18 yd, 10 yd

6. 42 cm, 60 cm

7. 9 cm, 27 cm

8. 24 mm, 21 mm

9. 12 ft, 7 ft

10. 20 in., 6 in.

11. 70 in., 120 in.

12. 15 m, 30 m

13. The diameter of a can of paint is 8 in. and the height is 10 in. Find the volume.

14. The height of a funnel is 12 cm and the radius of the base is 7 cm. Find the volume of the funnel.

15. The volume of a cylinder with radius 8 cm is approximately 1004.8 cm^3. Find the height of the cylinder. Use $\pi \approx 3.14$.

16. The volume of a cone with radius 7 yd is approximately 462 yd^3. Find the height of the cone. Use $\pi \approx \frac{22}{7}$.

17. WRITING ABOUT MATHEMATICS Write a paragraph describing how you would explain to a friend the difference between the surface area and the volume of a space figure.

CONNECTING GEOMETRY AND SCIENCE

The relationship between the surface area and the volume of an animal's body is an important factor in the animal's survival. An animal that has more skin area per unit of volume loses body heat faster, so an animal in a warmer climate would benefit from greater skin area.

18. For a cylinder with $r = 5$ cm and $h = 10$ cm, the ratio of surface area to volume is $\frac{3}{5}$. For a cylinder with $r = 15$ cm and $h = 30$ cm, this ratio is $\frac{1}{5}$. Do you think that a larger or a smaller animal has the greater ratio of surface area to volume?

19. The area of the curved surface of a cylinder is S.A. $= 2\pi rh$. Find this area for the following cylinders.
   a. $r = 1.5$ cm and $h = 15$ cm
   b. $r = 1.5$ cm and $h = 10$ cm
   c. Do you think it would be better for an arctic animal to have long or short legs?

20. **RESEARCH** Find out how the long ears and legs of a black-tailed jack rabbit help it survive.

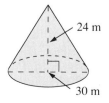

 **CALCULATOR**

Find the volume of the shaded space figure. Use a calculator with a ⬚π key if you have one available to you.

21.

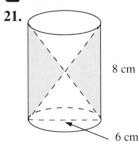

8 cm

6 cm

22.

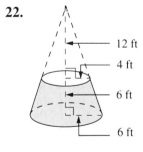

12 ft

4 ft

6 ft

6 ft

SPIRAL REVIEW

23. Find the volume of the cone shown at the right. *(Lesson 14-5)*

24 m

30 m

24. Find $\sqrt{8100}$. *(Lesson 10-10)*

25. Solve by graphing: $y = 5x - 3$
$y = x + 1$ *(Lesson 13-4)*

26. Solve: $\frac{x}{15} = \frac{9}{5}$ *(Lesson 9-2)*

14-6 Spheres

Objective: To find the surface area and volume of a sphere.

APPLICATION

Suppose you were to cut open a sphere and lay it flat. The area of the figure formed is the *surface area* of the sphere. This surface area would be four times the area of a circle with the same radius as the sphere.

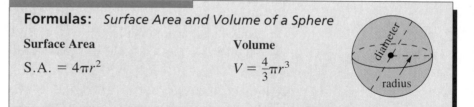

Use these formulas to find the surface area and the volume of a sphere.

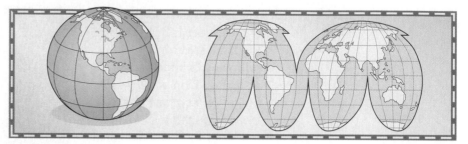

Formulas: *Surface Area and Volume of a Sphere*

Surface Area

$$S.A. = 4\pi r^2$$

Volume

$$V = \frac{4}{3}\pi r^3$$

Example

Find the surface area and volume of a sphere with diameter 12 m.

Solution

$r = \frac{1}{2}d = \frac{1}{2} \cdot 12 = 6$

Substitute 6 for r in each formula.

S.A. $= 4\pi r^2$ $V = \frac{4}{3}\pi r^3$

S.A. $\approx 4(3.14)(6^2)$

S.A. $\approx 4(3.14)(36)$ $V \approx \frac{4}{3}(3.14)(6^3)$

S.A. ≈ 452.16

$V \approx \frac{4}{3}(3.14)(216)$

$V \approx 904.32$

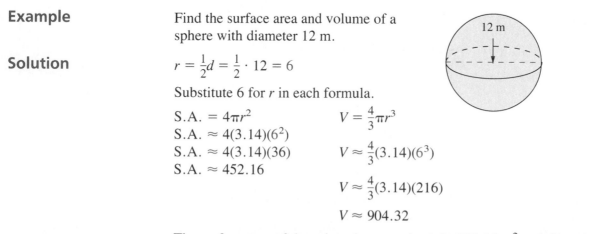

The surface area of the sphere is approximately 452.16 m^2 and the volume is approximately 904.32 m^3.

✓ Check Your Understanding

1. Why was it necessary to find one half the diameter in the Example?
2. What does 216 represent in the Example?

Guided Practice

Explain the meaning of the word *sphere* or *hemisphere* in each sentence.

« **1.** The courtroom is a judge's sphere.

« **2.** The right hemisphere of the brain largely controls a person's musical ability.

« **3.** Chile is in the southern hemisphere.

Replace each __?__ with the correct number.

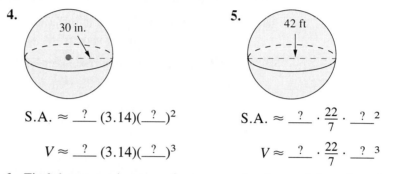

4.

30 in.

$$\text{S.A.} \approx \underline{\;?\;}\,(3.14)(\underline{\;?\;})^2$$

$$V \approx \underline{\;?\;}\,(3.14)(\underline{\;?\;})^3$$

5.

42 ft

$$\text{S.A.} \approx \underline{\;?\;} \cdot \frac{22}{7} \cdot \underline{\;?\;}^2$$

$$V \approx \underline{\;?\;} \cdot \frac{22}{7} \cdot \underline{\;?\;}^3$$

6. Find the approximate surface area and volume of the sphere in Exercise 4.

7. Find the approximate surface area and volume of the sphere in Exercise 5.

Exercises

Find the surface area of each sphere.

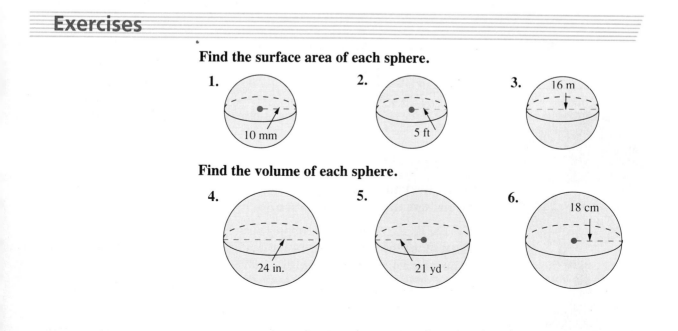

1.

10 mm

2.

5 ft

3.

16 m

Find the volume of each sphere.

4.

24 in.

5.

21 yd

6.

18 cm

The radius of a basketball is 12 cm.

7. Find the surface area. **8.** Find the volume.

The diameter of a soccer ball is 22 cm.

9. Find the surface area. **10.** Find the volume.

Find the surface area and the volume of the sphere described.

11. The radius is 27 m. **12.** The radius is 63 in.

13. The diameter is 84 cm. **14.** The diameter is 90 ft.

The ratio of the surface area to the volume of any sphere with radius r is $3 : r$. Use this information for Exercises 15–17.

15. The ratio of the surface area to the volume of a given sphere is $3 : 7$. What is the radius of the sphere?

16. What is the ratio of the surface area to the volume of a sphere with radius 39 in.?

17. The surface area of a sphere is 1017.38 cm^2. The volume is 3052.14 cm^3. Using the ratio above, write a proportion to find the radius r.

The hemisphere shown below is one half a sphere with the same radius. The volume of the hemisphere is one half the volume of the sphere.

Volume of a hemisphere $= \frac{1}{2} \cdot$ volume of the sphere

$$V = \frac{1}{2}\left(\frac{4}{3}\pi r^3\right)$$

$$V = \frac{2}{3}\pi r^3$$

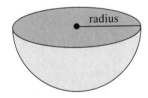

Find the volume of each hemisphere.

18.

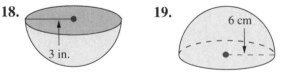

3 in.

19.

6 cm

20.

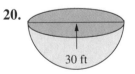

30 ft

🖩 CALCULATOR

Use a calculator with a ⌨ key if you have one available to you. Write the answers in scientific notation.

21. The diameter of Earth is approximately 7927 mi.
   a. Find the surface area of Earth.
   b. Find the volume of Earth.

22. The radius of Jupiter is approximately 44,431 mi.
   a. Find the surface area.
   b. Find the volume of Jupiter.

**A sphere with radius 3 cm fits inside a
cylinder with height 6 cm, as shown.**

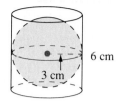

6 cm

3 cm

23. Show that the surface area of the sphere
   is equal to the area of the curved surface
   of the cylinder.

24. Show that the volume of the sphere is
   two thirds the volume of the cylinder.

25. Find the volume that is inside the cylinder, but outside the sphere.

26. A sphere with radius 3 cm fits inside a cube with edges 6 cm. Find the
   volume that is inside the cube, but outside the sphere.

27. A hemisphere with radius 3 in. fits inside a rectangular prism with
   length 6 in., width 6 in., and height 3 in. Find the volume that is
   inside the prism, but outside the hemisphere.

SPIRAL REVIEW

28. Find the volume of the sphere shown
   at the right.   *(Lesson 14-6)*

9 ft

29. Write 0.0000081 in scientific
   notation.   *(Lesson 7-11)*

30. The lengths of the legs of a right triangle are 7 cm and 24 cm. Find the
   hypotenuse.   *(Lesson 10-11)*

31. DATA, *pages 434–435*   The length of a field hockey field is how many
   feet greater than the width of the field?   *(Lesson 8-7)*

32. Find the slope and the y-intercept of the line $y - 5x = 3$.
   (Lesson 13-3)

33. Find the number of permutations of 7 objects taken 3 at a time.
   (Lesson 12-6)

Historical Note

Archimedes, a Greek mathematician who lived during the third
century B.C., is considered to be one of the greatest mathematicians of all time. He discovered many important relationships in geometry. Among them is the fact that the volume of a sphere that fits snugly
inside a cylinder is two thirds the volume of the cylinder. He also discovered that a body floating in liquid loses in weight an amount equal to
that of the liquid displaced.

Research

Archimedes was also an inventor. Find out about some of his
inventions.

14-7

Strategy:
Making a Model

Objective: To solve problems by making a model.

In 1990, scientists had difficulty opening a solar panel on the Hubble Space Telescope. To investigate, they built a model. They were able to locate and solve the problem and then fix the telescope.

Sometimes making a model of a problem situation can help you to solve the problem.

Problem

A children's museum has display stands like the one at the right. Each stand consists of 12 cubes. The top and all four sides of the stand are painted red. The bottom is unpainted. The cubes are unpainted on the faces that cannot be seen. How many of the cubes in each stand have exactly 2 red faces and how many have exactly 3 red faces?

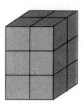

Solution

UNDERSTAND The problem is about the color of the faces of cubes in a display stand.
Facts: 5 sides of a stack of cubes are painted red
Find: the number of cubes with exactly 2 red faces
the number of cubes with exactly 3 red faces

PLAN Make a stack of 12 cubes like the one shown in the figure. Color the appropriate faces red. Take the stack apart and count the number of cubes with 2 red faces, and the number with 3 red faces.

WORK The stack has 3 layers of 4 cubes.

Top layer: Middle layer: Bottom layer:

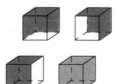   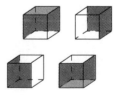

Each cube has 3 red faces.   Each cube has 2 red faces.   Each cube has 2 red faces.

ANSWER In each stand, 8 cubes have exactly 2 red faces and 4 cubes have exactly 3 red faces.

Look Back Suppose the bottom of each stand were also painted red. How would your answer be different?

Guided Practice

Use this problem for Exercises 1–7.

A fence is built using 24 sections, each 1 unit long. What is the greatest possible rectangular area that can be enclosed?

« **1.** What is the shape of the area enclosed by the fence?

« **2.** What is the perimeter of the area enclosed by the fence?

« **3.** What convenient object could you use to represent a section of the fence?

« **4.** If the length of the rectangle enclosed by the fence is 1 unit, what is the width? the area?

« **5.** How many different rectangles can be built using the 24 sections of the fence?

« **6.** Make a sketch of a model showing the arrangement of the sections that has the greatest possible area.

7. Use your model from Exercise 6 to write an answer to the problem.

8. Solve by making a model: A rectangular prism with volume 20 cubic units is formed using cubes with edges 1 unit long.
 a. How many such prisms are possible?
 b. What are the dimensions of the prism with the greatest surface area?

Problem Solving Situations

Solve by making a model.

1. All 6 faces of a wooden cube are painted blue. The cube is then cut into 27 smaller cubes. Tell how many of the smaller cubes have the number of painted faces indicated.

 a. exactly 3 blue faces   **b.** exactly 2 blue faces
 c. exactly 1 blue face   **d.** no blue faces

2. Suppose a cube is painted so that the top and bottom are painted blue and the other 4 faces are painted red. The cube is then cut into 27 smaller cubes. How many of the smaller cubes have at least 1 red face and 1 blue face?

3. A rectangular prism is to be formed using exactly 30 cubes. How many different prisms can be formed? Which prism has the least surface area that can be painted? Which prism has the greatest surface area?

Solve using any problem solving strategy.

4. Of 150 students, 60 take biology, 50 take Spanish, and 40 take both subjects. How many students take neither subject?

5. If 3 serving-size bottles of Tree Ripe apple juice cost $1.47, how much will 10 bottles cost?

6. The figures below show the number of sectors resulting when 1, 2, 3, and 4 diameters are drawn in a circle. How many sectors result when 10 diameters are drawn in a circle?

PROBLEM SOLVING

CHECKLIST

Keep these in mind:
   Using a Four-Step Plan
   Too Much or Not Enough
     Information
   Supplying Missing Facts
   Problems with No Solution

Consider these strategies:
   Choosing the Correct Operation
   Making a Table
   Guess and Check
   Using Equations
   Identifying a Pattern
   Drawing a Diagram
   Using Proportions
   Using Logical Reasoning
   Using a Venn Diagram
   Solving a Simpler Problem
   Making a Model

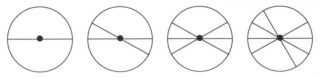

7. An isosceles triangle has perimeter 54 in. The length of one side is 27 in. Find the lengths of the other two sides.

8. A rectangular flower bed is marked off using 72 sections of border fencing. What is the largest possible area of the flower bed?

9. Ted reads 30 min each day after dinner. In a seven-day week, how many hours does he read after dinner?

10. Ming, Damian, and Cesar are the presidents of the Math, Drama, and Chess Clubs. No person's club begins with the same letter as his or her name. Damian's best friend is president of the Math Club. Which person is president of each club?

A drama club presents a play using signing for hearing-impaired audience.

WRITING WORD PROBLEMS

Using any or all of the figures below, write a word problem that you could solve by making a model. Then solve the problem.

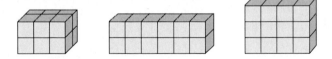

11. Write a problem involving volume.

12. Write a problem involving surface area.

SPIRAL REVIEW

13. A box is formed to hold 12 cubes with edges 1 unit long. Find the dimensions of the box with the least surface area. *(Lesson 14-7)*

14. Find the volume of a cube with sides of 1.5 m. *(Lesson 14-4)*

15. A number cube is tossed. Find $P(4)$. *(Lesson 12-1)*

16. Write 0.546 as a fraction in lowest terms. *(Lesson 7-7)*

Self-Test 2

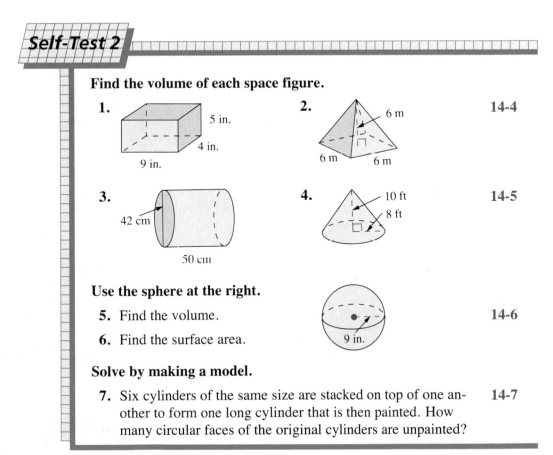

Find the volume of each space figure.

1. 5 in. / 4 in. / 9 in.   **2.** 6 m / 6 m / 6 m   14-4

3. 42 cm / 50 cm   **4.** 10 ft / 8 ft   14-5

Use the sphere at the right.

5. Find the volume.   14-6

6. Find the surface area.   9 in.

Solve by making a model.

7. Six cylinders of the same size are stacked on top of one another to form one long cylinder that is then painted. How many circular faces of the original cylinders are unpainted?   14-7

Packaging a Consumer Product

Objective: To explore how surface area and volume are used in making decisions about packaging consumer products.

Surface area and volume are important considerations in packaging a consumer product. Among the many factors affecting a manufacturer's decision in choosing packaging are cost, convenience, storage capacity, and visual appeal.

Example

Family Farm Foods is considering the three metal cans shown for a new Hearty Stew.

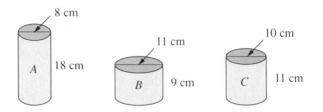

a. Which can holds the most stew?

b. Which can uses the least amount of metal?

Solution

a. The can that holds the most stew is the one with the greatest volume. Find the volume of each can. Use the formula for the volume of a cylinder, $V = \pi r^2 h$.

Can A: $V = \pi r^2 h \approx (3.14)(4)^2(18) \approx 904.32$

Can B: $V = \pi r^2 h \approx (3.14)(5.5)^2(9) \approx 854.865$

Can C: $V = \pi r^2 h \approx (3.14)(5)^2(11) \approx 863.5$

Can A holds the most stew.

b. The can that uses the least amount of metal is the can with the least surface area. Find the surface area of each can. Use the formula for the surface area of a cylinder, S.A. $= 2\pi r^2 + 2\pi rh$.

Can A: S.A. $= 2\pi r^2 + 2\pi rh$
$\approx 2(3.14)(4)^2 + 2(3.14)(4)(18)$
≈ 552.64

Can B: S.A. $= 2\pi r^2 + 2\pi rh$
$\approx 2(3.14)(5.5)^2 + 2(3.14)(5.5)(9)$
≈ 500.83

Can C: S.A. $= 2\pi r^2 + 2\pi rh$
$\approx 2(3.14)(5)^2 + 2(3.14)(5)(11)$
≈ 502.4

Can B uses the least amount of metal.

Exercises

Exercises 1–4 refer to the boxes shown below. Family Farm Foods is considering the three cardboard boxes below for a powdered product.

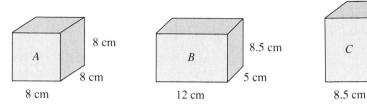

1. Which box uses the least amount of cardboard?

2. If the label covers the sides of the box, but not the top and bottom, which box label uses the least amount of paper?

3. Which box takes up the least area on a shelf?

4. Which box can hold the most?

Cans like the one at the right are to be shipped in boxes of 12.

5. Find the length, width, and height of the box needed to ship the 12 cans in one layer of three rows, four cans each.

6. Find the dimensions of the box needed to ship the 12 cans in two layers of two rows, three cans each.

7. Find the dimensions of the box needed to ship the 12 cans in one layer of two rows, six cans each.

8. Which box uses the least cardboard?

GROUP ACTIVITY

Suppose you had to design a cylindrical container for a new type of juice.

9. Make a sketch of your container. Label the dimensions.

10. What materials might you use to make the container? Why?

11. Find the volume of your container.

12. How many containers would fit on a shelf that is 2 ft wide, 3 ft deep, and 2 ft high?

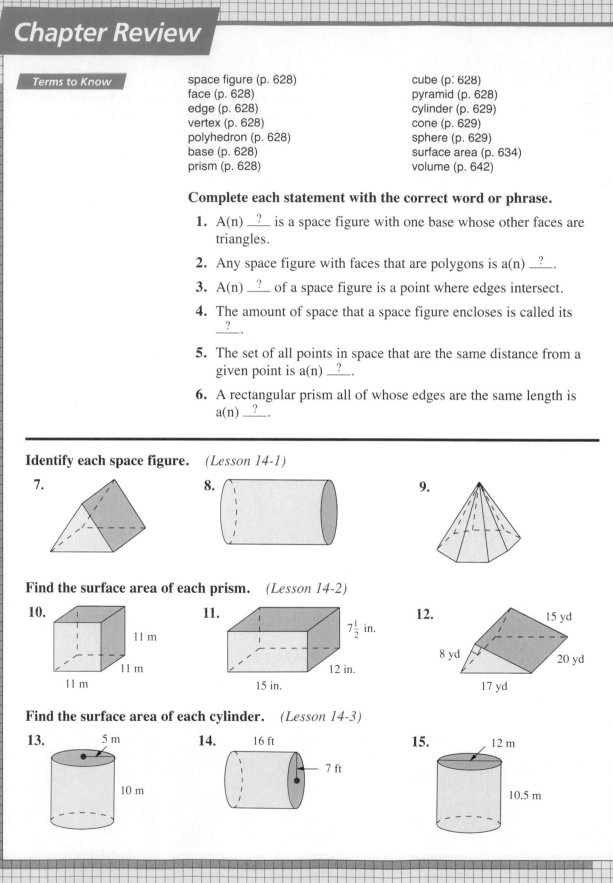

space figure (p. 628)
face (p. 628)
edge (p. 628)
vertex (p. 628)
polyhedron (p. 628)
base (p. 628)
prism (p. 628)

cube (p. 628)
pyramid (p. 628)
cylinder (p. 629)
cone (p. 629)
sphere (p. 629)
surface area (p. 634)
volume (p. 642)

Complete each statement with the correct word or phrase.

1. A(n) _?_ is a space figure with one base whose other faces are triangles.

2. Any space figure with faces that are polygons is a(n) _?_.

3. A(n) _?_ of a space figure is a point where edges intersect.

4. The amount of space that a space figure encloses is called its _?_.

5. The set of all points in space that are the same distance from a given point is a(n) _?_.

6. A rectangular prism all of whose edges are the same length is a(n) _?_.

Identify each space figure. *(Lesson 14-1)*

7.

8.

9.

Find the surface area of each prism. *(Lesson 14-2)*

10. 11 m, 11 m, 11 m

11. $7\frac{1}{2}$ in., 12 in., 15 in.

12. 15 yd, 8 yd, 20 yd, 17 yd

Find the surface area of each cylinder. *(Lesson 14-3)*

13. 5 m, 10 m

14. 16 ft, 7 ft

15. 12 m, 10.5 m

Chapter 14

Find the volume of each prism or pyramid. *(Lesson 14-4)*

16.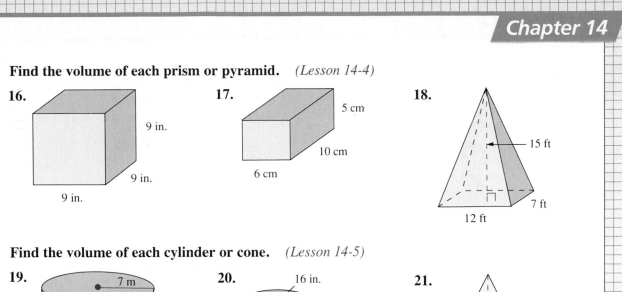

9 in.
9 in.
9 in.

17.

5 cm
10 cm
6 cm

18.

15 ft
7 ft
12 ft

Find the volume of each cylinder or cone. *(Lesson 14-5)*

19.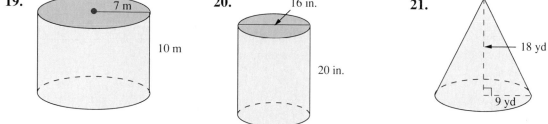

7 m
10 m

20.

16 in.
20 in.

21.

18 yd
9 yd

Find the surface area and volume of each sphere. *(Lesson 14-6)*

22.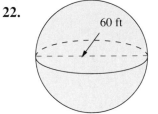

60 ft

23.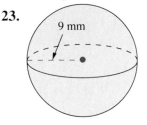

9 mm

24.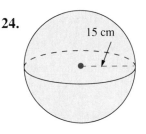

15 cm

Solve by making a model. *(Lesson 14-7)*

25. Λ rectangular prism like the one shown at the right is painted so that the top and bottom are red and the other 4 faces are blue. The prism is then cut into 18 cubes. How many of the cubes have exactly 1 red face and 1 blue face?

26. A rectangular prism with volume 8 cubic units is formed using cubes with edges 1 unit long. How many such prisms are possible? What are the dimensions of the prism with the greatest surface area?

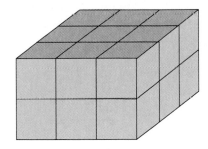

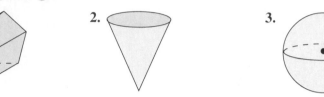

Chapter Test

Identify each space figure.

1.
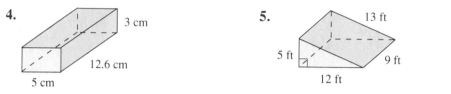

2.

3.

14-1

Find the surface area of each space figure.

4.
3 cm
12.6 cm
5 cm

5.
13 ft
5 ft
9 ft
12 ft

14-2

6. A cube has edges *n* cm long. A second cube has edges 2*n* cm long. What is the ratio of the surface area of the first cube to that of the second cube?

Find the surface area of each space figure.

7.
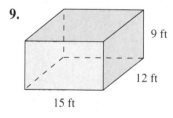
10 yd
15 yd

8.
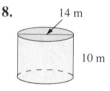
14 m
10 m

14-3

Find the volume of each space figure.

9.
9 ft
12 ft
15 ft

10.
12 yd
6 yd
8 yd

14-4

11.
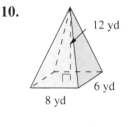
50 cm
50 cm

12.
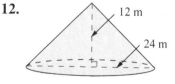
12 m
24 m

14-5

13. Find the surface area and volume of a sphere with radius 24 in.

14-6

14. When the radius of a sphere is doubled, what happens to the surface area and volume of the sphere?

15. A fence is built using 36 sections, each one unit long. What is the greatest possible rectangular area that can be enclosed?

14-7

Volume of Composite Space Figures

Objective: To determine volumes of composite space figures.

A **composite space figure** is a combination of two or more space figures. The thumbtack at the right is a composite space figure composed of two cylinders and a cone. The top of the thumbtack is a wide cylinder; the post is a narrow cylinder. The point is a cone.

(diagram: 5 mm, 1 mm, 6 mm, radius 0.5 mm, 3 mm)

Example

Find the volume of the thumbtack pictured above. Round your answer to the nearest unit.

Solution

The volume of the thumbtack is the sum of the volumes of the three parts.

Top: $V = \pi r^2 h \approx 3.14(5^2)(1) \approx 78.5$
Post: $V = \pi r^2 h \approx 3.14(0.5)^2(6) \approx 4.71$
Point: $V = \frac{1}{3}\pi r^2 h \approx \frac{1}{3}(3.14)(0.5^2)(3) \approx 0.785$

Total Volume $\approx 78.5 + 4.71 + 0.785$
≈ 83.995

The volume is about 84 mm^3.

Exercises

Find the volume of each composite space figure. Round your answer to the nearest unit.

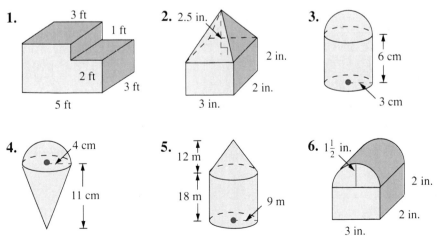

1. 3 ft, 1 ft, 2 ft, 3 ft, 5 ft

2. 2.5 in., 2 in., 2 in., 3 in.

3. 6 cm, 3 cm

4. 4 cm, 11 cm

5. 12 m, 18 m, 9 m

6. $1\frac{1}{2}$ in., 2 in., 2 in., 3 in.

Standardized Testing Practice

Choose the letter of the correct answer.

1.

What is the volume of this cone?

A. 22 m^3 B. 198 m^3

C. 44 m^3 D. 66 m^3

2. Simplify: $\sqrt{25 + 144}$

A. 17   B. 13

C. 84.5   D. 14

3. Write $4y - x = 12$ in slope-intercept form.

A. $4y = x + 12$ B. $x = 4y - 12$

C. $y = \frac{1}{3}x - 4$ D. $y = \frac{1}{4}x + 3$

4. Solve: $r - 13 = -2$

A. $r = 15$ B. $r = -15$

C. $r = 11$ D. $r = -11$

5. Write 0.35 as a fraction.

A. 35% B. $\frac{9}{20}$

C. $\frac{7}{20}$ D. $\frac{7}{10}$

6.

Which equation does this graph represent?

A. $y = -3x + 1$ B. $y = \frac{1}{3}x + 1$

C. $y = 3x + 1$ D. $y = -\frac{1}{3}x + 1$

7. Solve: $8.9 + 3a = 14.6$

A. $a = 7.8$ B. $a = 1.9$

C. $a = 17.1$ D. $a = 1.2$

8. Simplify: $12(3t - 5)$

A. $15t - 17$ B. $36t - 5$

C. $3t - 60$ D. $36t - 60$

9. The measure of an angle is 73°. What is the measure of its complement?

A. $17°$ B. $27°$

C. $107°$ D. $117°$

10. Solve: $-4w - 9 < -1$

A. $w < 2$ B. $w < -2$

C. $w > -2$ D. $w < -32$

11.

4 cm

10 cm

5 cm

What is the surface area of this prism?

A. 200 cm^2 B. 180 cm^2

C. 220 cm^2 D. 190 cm^2

12. Doris buys $13\frac{1}{2}$ ft of chain and uses $6\frac{7}{8}$ ft for a dog leash. Estimate the amount of chain she has left.

A. about 20 ft   B. about 2 ft

C. about 14 ft   D. about 7 ft

13. In a pictograph, one symbol represents 24 sheep. How many symbols are needed to represent 84 sheep?

A. $3\frac{1}{2}$ B. 60

C. 61   D. 7

14. The height of a cylinder is 8 in. and the radius is 6 in. Find the surface area of the cylinder.

A. 301.44 in.^{2} B. 527.52 in.^{2}

C. 414.48 in.^{2} D. 904.32 in.^{2}

15. The base of a pyramid is a square with sides of length 12 cm. The height of the pyramid is 8 cm. Find the volume of the pyramid.

A. 384 cm^3 B. 32 cm^3

C. 1152 cm^3 D. 128 cm^3

16.

-2   -1   0   1   2

Which inequality does this graph represent?

A. $m + 4 \geq 3$ B. $m - 1 > -2$

C. $m - 2 \geq -1$ D. $4m \leq -4$

17. The price of a carpet is discounted by 15%. The amount of the discount is $39.60. What is the original price of the carpet?

A. $264   B. $45.54

C. $5.94   D. $594

18. The radius of a sphere is 6 mm. What is the surface area of the sphere?

A. 904.32 mm^2 B. 452.16 mm^2

C. 75.36 mm^2 D. 113.04 mm^2

19. There are 3 teachers for every 50 students in a school. The school has 750 students. How many teachers are there at the school?

A. 125   B. 50

C. 703   D. 45

20. One letter from A through Z and one digit from 0 through 9 are selected. How many different outcomes are possible?

A. 234   B. 676

C. 260   D. 36

First Nine Rows of Pascal's Triangle

Polynomials 15

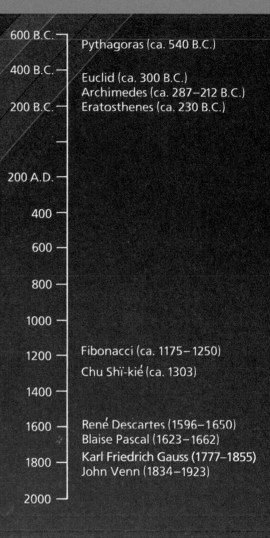

600 B.C. — Pythagoras (ca. 540 B.C.)

400 B.C. — Euclid (ca. 300 B.C.)
Archimedes (ca. 287–212 B.C.)
200 B.C. — Eratosthenes (ca. 230 B.C.)

200 A.D. —

400 —

600 —

800 —

1000 —

1200 — Fibonacci (ca. 1175–1250)
Chu Shï-kié (ca. 1303)

1400 —

1600 — René Descartes (1596–1650)
Blaise Pascal (1623–1662)

1800 — Karl Friedrich Gauss (1777–1855)
John Venn (1834–1923)

2000 —

Pascal's triangle is a triangular array of numbers. One of its applications is in the multiplication of binomials. The triangle is named for the French mathematician Blaise Pascal, who wrote about it in 1653. Earlier mathematicians were also familiar with the triangle. In 1303, the Chinese mathematician Chu Shï-kié included an illustration of it in a book.

Contributions to Mathematics

Archimedes	Archimedes' spiral
René Descartes	Cartesian coordinates
Eratosthenes	sieve of Eratosthenes
Euclid	Euclidean geometry
Leonardo Fibonacci	Fibonacci sequence
Karl Friedrich Gauss	Gaussian distribution
Blaise Pascal	Pascal's triangle
Pythagoras	Pythagorean theorem
John Venn	Venn diagram

Simplifying Polynomials

Objective: To simplify polynomials.

CONNECTION

Since Chapter 1, you have been writing variable expressions and simplifying some expressions. In this lesson you will simplify *polynomials*. Each expression below is a **polynomial,** a variable expression consisting of one or more *terms*.

$$3x^2 \qquad -4t \qquad 2a^2 - 3ab + 2b^2 \qquad x^3 - 1$$

Some polynomials have special names.

Terms to Know

- *polynomial*
- *monomial*
- *binomial*
- *trinomial*
- *standard form*

A **monomial** has one term. *Examples*: $3x^2$ and $-4t$
A **binomial** has two terms. *Example*: $x^3 - 1$
A **trinomial** has three terms. *Example*: $2a^2 - 3ab + 2b^2$

When you are working with a polynomial, it is often helpful to write the polynomial in **standard form.** To do this, write the terms in order from the highest to the lowest power of one of the variables.

Example 1

Write each polynomial in standard form.

  a. $4x^2 + x - 3x^3$ **b.** $8c^3 + 7 - 9c + 2c^4$

Solution

  a. $4x^2 + x - 3x^3$ **b.** $8c^3 + 7 - 9c + 2c^4$
       $= -3x^3 + 4x^2 + x$ $= 2c^4 + 8c^3 - 9c + 7$

✔️ **Check Your Understanding**

1. In Example 1(a) and in Example 1(b), list the powers in order.
2. In Example 1(b), why is 7 the last term?

Like terms have the same variables raised to the same powers. To *simplify* a polynomial, you combine like terms and write the resulting polynomial in standard form.

Example 2

Simplify $12c^3 - 4c^2 - 8c^3 - 5 + 7c^2 - 4c$.

Solution

$12c^3 - 4c^2 - 8c^3 - 5 + 7c^2 - 4c$
$= (12c^3 - 8c^3) + (-4c^2 + 7c^2) - 5 - 4c$ ⟵ Group like terms.
$= 4c^3 + 3c^2 - 5 - 4c$ ⟵ Combine like terms.
$= 4c^3 + 3c^2 - 4c - 5$ ⟵ Write in standard notation.

✔️ **Check Your Understanding**

3. In Example 2, why does $4c^3$ come before $3c^2$?
4. In Example 2, why are $4c^3$, $3c^2$, and $4c$ not combined?

Guided Practice

Replace each __?__ with the correct word or expression.

«**1.** A __?__ is a variable expression with one or more terms. A __?__ has one term, a __?__ has two terms, and a __?__ has three terms.

«**2.** A polynomial is in __?__ form when its terms are arranged in order from highest to lowest powers. To simplify a polynomial, you __?__ and write the resulting polynomial in standard form.

Is the polynomial a *monomial*, a *binomial*, or a *trinomial*?

3. $ab + 3$ **4.** $x + y - 2xy$ **5.** 5 **6.** $-t^6 + s^4$

Tell whether the terms are *like terms* or *unlike terms*.

7. $3m^3, 5m^3$ **8.** $7x^4, 4x^7$ **9.** xy^3, xy **10.** $3ab^2, 5ab^2$

Write each polynomial in standard form.

11. $3g^3 + 4g^4 - 3g + 8 - 7g^2$ **12.** $4k - 8k^4 + 7k^2 - 9k^3$

13. $5a + 8a^7 - 2a^3 + 9a^5 - 6$ **14.** $7x^3 + 2x - 5x^8 + 9 - x^5$

Simplify.

15. $5x^3 + 6x - 2x^3 + 8x - x^2 + 5$

16. $7c^2 + 8c + 2c^2 - 9c^3 - 5c - 7$

17. $4 - 5a^3 - 2a^2 + 5a - 4a^3 + 8a^2$

18. $3n - 7 + 8n^2 + 5n^3 - 3n^2 - 8n$

Exercises

Write each polynomial in standard form.

1. $x^2 - 2 + x$ **2.** $5 - 3a^2 + 6a$

3. $m - 3 + m^3 + 3m^2$ **4.** $-x^2 + 8x - 1 + x^3$

5. $4z^2 - 6z + 3z^4 - 7 + 8z^3$ **6.** $2a^4 - 3 + 2a^2 - 13a - 7a^3$

7. $c^3 - 5c - 2 + c^4$ **8.** $-z^7 + 5z^3 + 2z - z^5$

9. $k^2 - 8k^4 - 6 + 3k^3$ **10.** $n^3 - 1 + 2n^6 + 4n^4 + 3n^7$

Simplify.

11. $2x^2 + x + 2 + 3x - x^2 + 5$ **12.** $-9d^3 - 4d^2 - d^3 + 4 - 2d^2$

13. $4e - 3e^2 + 4e^2 + 7e - 20$ **14.** $r^3 - 7r^2 - 6r^3 - 5 + 7$

15. $x^2 + 6x + 7 - 5x^2 + 5 + 3x$ **16.** $2w^3 - 6w^2 + 7w^3 - 7$

Simplify.

17. $-3x^2 - 6x - 2 - 5x - 4x^2 + 3$

18. $2g^2 - 4 + 7g + 5g^2 + 1$

19. $8w^3 + 6w^2 - 12w - 2w^2 + 1$

20. $-x^3 + 9x - 3x^3 - 6x^2 + 2$

21. $x^2 - 2x + 1 + 3x^2 + 4x + x^5$

22. $-x^2 + 8 + 3x^4 - 9x + x^4$

23. $2a^4 - 6a^3 - 2a^2 - 4a^3 - 5a^2$

24. $x^3 - x^2 + 7 + 3x^3 - 11 + x^2$

25. $-x^3 + 2x - 3 - 4x^3 - 2x + 3$

26. $6 - 2a^4 + a^2 - 1 + 6a^4 - a^2$

Write each polynomial in standard form for the variable in color.

27. $a^2b + ab^3 + a^2b^2 + 4$; b

28. $m^2n^3 - mn + mn^4 + m^2n^2 - 1$; n

29. $x^3y^3 - 4 + x^2y - xy^2$; y

30. $c^4d - cd^4 + c^2d^2 + 6 - c^3d^3$; d

31. $a^3b^2 + ab^3 - a^2b - 4$; a

32. $xz^3 - x^3z + 9 - x^2z^2$; x

33. $a^4c^2 + a^3c^4 - a^2c + ac^3$; c

34. $mn^4 - m^2n^3 - m^3n^2 + m^4n$; m

CONNECTING MATHEMATICS AND LANGUAGE ARTS

Explain what the prefix of the word means. For example, the prefix *uni-* in *unicorn* means *one*. Then give an everyday word with the same prefix as the given word.

35. polynomial

36. monomial

37. binomial

38. trinomial

THINKING SKILLS

39. Write in standard form in three different ways: $x^3yz^2 + xy^3z^3 + x^2y^2z$

40. Create a polynomial that can be written in standard form in four different ways.

SPIRAL REVIEW

41. Find the volume of a sphere whose radius is 9 cm.   *(Lesson 14-6)*

42. Simplify: $c^3 - 3c^2 + 8c - 9c^2 + 3c^3 - 6$ *(Lesson 15-1)*

43. Solve and graph: $3x \le -18$ *(Lesson 13-6)*

44. Find $\sqrt{529}$.   *(Lesson 10-10)*

Modeling Polynomials Using Tiles

Objective: To use algebra tiles to model polynomials.

Materials
■ algebra tiles

You can use algebra tiles to model polynomials. For example, let the

tile 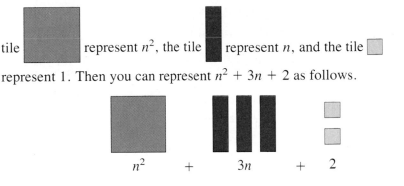 represent n^2, the tile represent n, and the tile represent 1. Then you can represent $n^2 + 3n + 2$ as follows.

$$n^2 \quad + \quad 3n \quad + \quad 2$$

Activity I *Representing Polynomials*

1 Write the polynomial represented by the diagram below.

2 Show how to use algebra tiles to represent $3n^2 + 4n + 5$.

Activity II *Adding Polynomials*

1 Name the polynomial represented by each set of tiles.

2 How many tiles representing n^2 are there in all?
3 How many tiles representing n are there in all?
4 How many tiles representing 1 are there in all?
5 Write the sum of the polynomials represented by the diagrams.
6 Show how to use algebra tiles to represent the sum.

Activity III *Subtracting Polynomials*

1 Show how to use algebra tiles to represent $2n^2 + 3n + 5$.
2 Remove or cross out one tile representing n^2, two tiles representing n, and three tiles representing 1.
3 Write the polynomial represented by the tiles that are left.
4 Complete: When you subtract the polynomial __?__ from the polynomial $2n^2 + 3n + 5$, the difference is __?__.

Adding Polynomials

Objective: To add polynomials.

CONNECTION

In previous chapters you performed the four basic operations with rational numbers. You can also add, subtract, multiply, and divide polynomials.

Example 1
Solution

Add: $(n^2 + 3n + 1) + (2n^2 + 2n + 4)$

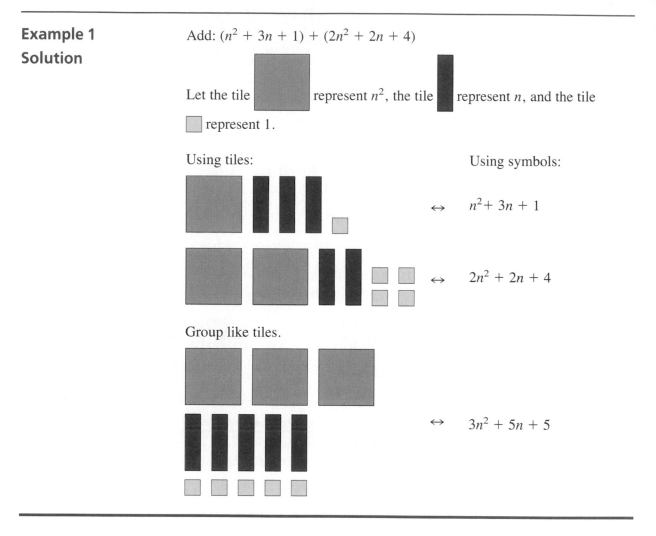

Let the tile ▢ represent n^2, the tile ▮ represent n, and the tile ▢ represent 1.

Using tiles: Using symbols:

$\leftrightarrow$ $n^2 + 3n + 1$

$\leftrightarrow$ $2n^2 + 2n + 4$

Group like tiles.

$\leftrightarrow$ $3n^2 + 5n + 5$

Using algebra tiles can help you see how to add polynomials.

Generalization: *Adding Polynomials*
To add polynomials, combine like terms.

When adding polynomials, insert a zero term for a missing power.

| Example 2 | Add: $(6x^4 - 2x^3 + 7x^2 + x - 6) + (-7x^4 + 2x^3 - 5x + 7)$ |

Solution 1

Line up like terms vertically.

$$\begin{array}{r} 6x^4 - 2x^3 + 7x^2 + x - 6 \\ -7x^4 + 2x^3 + 0x^2 - 5x + 7 \\ \hline -1x^4 + 0x^3 + 7x^2 - 4x + 1 \end{array}$$ ⟵ Insert $0x^2$ since there is no x^2 term.

The sum is written $-x^4 + 7x^2 - 4x + 1$.

Solution 2

$(6x^4 - 2x^3 + 7x^2 + x - 6) + (-7x^4 + 2x^3 - 5x + 7)$
$= (6x^4 - 7x^4) + (-2x^3 + 2x^3) + 7x^2 + (x - 5x) + (-6 + 7)$
$= -1x^4 + 0x^3 + 7x^2 + (-4x) + 1$
$= -x^4 + 7x^2 - 4x + 1$

✓ **Check Your Understanding**

1. In Solution 2, why does $x + (-5x) = -4x$?

2. In Solutions 1 and 2, why is there no x^3 term in the answer?

Guided Practice

COMMUNICATION « *Reading*

Refer to the text on pages 672–673.

«**1.** Describe two ways to add polynomials.

«**2.** Why would you insert a zero term for a missing power?

COMMUNICATION « *Writing*

«**3.** Write the polynomial that is represented by each group of tiles. Then write the sum of the polynomials. Use n as the variable.

«**4.** Draw a diagram similar to that of Exercise 3 to represent the addition $(3n^2 + 2n + 3) + (2n^2 + 4n + 5)$.

Replace each __?__ with the term that makes the statement true.

5. $(5x^2 + 3x + 2) + (2x^2 + 6x + 6)$
   $= 7x^2 + \underline{?} + 8$

6. $(3a^3 + 4a^2 - 5a + 1) + (5a^3 - a^2 - 2a + 3)$
   $= 8a^3 + \underline{?} - \underline{?} + 4$

7. $(-c^4 + 2c^3 + 7c) + (2c^4 + c^3 - 5c^2 + 2c)$
   $= \underline{?} + 3c^3 - \underline{?} + 9c$

8. $(3m^3 - m^2 + 5m + 2) + (m^3 + m^2 - 3m + 6)$
   $= 4m^3 + \underline{?} + 8$

Add.

9. $2x^2 + 3x + 4$
   $\underline{5x^2 - 4x - 3}$

10. $y^2 + 8y - 2$
    $\underline{7y^2 - 5y + 8}$

11. $-3c^3 - 5c^2 - 11c - 5$
    $\underline{2c^3 - 6c^2 + 9c - 7}$

12. $z^4 - z^3 - 2z^2 - 3z$
    $\underline{4z^4 + 5z^2 + 2z - 1}$

13. $(6c^3 + 8c^2 - 12c - 4) + (9c^3 + c^2 + 3c + 7)$

14. $(8w^3 - 9w^2 - 6w + 9) + (5w^3 - 3w^2 - 8w - 11)$

15. $(z^5 - 2z^3 - 9z + 3) + (-z^4 + 2z^3 - 10)$

16. $(-6r^4 + 4r^2 + 1) + (-5r^2 - 7r + 4)$

Exercises

Add.

1. $(2a^2 + 3a + 5) + (3a^2 + a + 5)$

2. $(4c^2 + 3c + 2) + (5c^2 + 8c + 4)$

3. $(3x^2 + 5x + 9) + (4x^2 + 6x + 2)$

4. $(5a^2 + 8a + 1) + (6a^2 + 2a + 5)$

5. $(7v^2 + 3v + 6) + (2v^2 + 3v + 1)$

6. $(9n^2 + 4n + 3) + (8n^2 + 3n + 8)$

7. $(8y^3 + 6y^2 + 3y + 2) + (6y^3 + 6y^2 + 6y + 6)$

8. $(8a^4 + 6a^3 + 5a^2 + 2a + 4) + (4a^4 + 2a^3 + 3a^2 + 6a + 2)$

9. $(2b^4 + 3b^3 + 4b^2 + 5b + 1) + (6b^4 + 2b^3 + 8b^2 + 2b + 6)$

10. $(2d^4 + 5d^3 + 4d^2 + d + 3) + (5d^4 + 2d^3 + 3d^2 + 3d + 2)$

11. $(7y^2 - 6y + 8) + (-6y^2 + 2y + 1)$

12. $(-7a^2 - 4a + 5) + (9a^2 + 2a - 7)$

13. $(4x^2 + 6x + 9) + (2x^2 - 3x - 4)$

14. $(5r^3 + 9r^2 - 3r - 6) + (3r^3 - 7r^2 - 6r - 5)$

15. $(3k^4 - k^2 + 4k - 6) + (-4k^4 - 7k^3 - 2k^2 - 3k + 4)$

16. $(8k^5 - 3k^3 + 2) + (2k^4 - 3k^2 + 5)$

17. $(4k^5 - 1) + (k^5 - k^4 + 1)$

18. $(k^2 - k - 1) + (k^2 - 1)$

19. Find the sum of $x^5 + 3x^4 - 5x^2 + 4$ and $4x^5 + 3x^3 + 6x^2 - 9$.

20. Find the sum of $n^6 - 5n^4 + 6n^2 - 4$ and $-5n^6 + n^5 - n^4 - n^2 + 2$.

Add. Write the answer in standard form.

21. $(a^4 + 6a^2 + 4a^3 + 6 + 7a) + (a^3 + 3a^2 + 5a^4 + 4a + 1)$

22. $(x^2 + 3x^3 + 7x + 8x^4 + 1) + (6x^4 + 3x^3 + 5x + 6 + 3x^2)$

23. $(3d^2 - 6d - 2d^3 + 1) + (-d^3 + 2d + 5d^2 - 5)$

24. $(b - 4b^2 - b^3 + 1) + (b^2 + 2b^3 - 3b - 5)$

25. $(b^4 - b + 4b^2 - 3) + (-2b^3 - b^2 + b^4 + 2)$

26. $(a^3 - 8a^2 - 2a + 1) + (a^3 + 5a^2 + 3a^4 - 7)$

27. WRITING ABOUT MATHEMATICS   Write a paragraph describing two methods of adding polynomials. Give advantages of each.

THINKING SKILLS

To write the *opposite* of a polynomial, you write the polynomial with all its signs changed to their opposites. For instance, the opposite of the polynomial $2x + 1$ is the polynomial $-2x - 1$.

28. Write the opposite of $2x^2 - 5x - 1$ in standard form.

29. Find the sum of $2x^2 - 5x - 1$ and its opposite.

30. Make a generalization about the sum of a polynomial and its opposite.

FUNCTIONS

31. Complete the function table below.

x	$x^2 - 4x$	$x^3 + 4x + 4$	$x^3 + x^2 + 4$
-2	?	?	?
-1	?	?	?
0	?	?	?
1	?	?	?
2	?	?	?

32. Use the table in Exercise 31 to complete the following statement: The sum of the values of two polynomial functions is equal to __?__.

SPIRAL REVIEW

33. Graph $y = 2x - 4$ on a coordinate plane.   *(Lesson 13-2)*

34. Find the surface area of a cube whose edge is 8 in.   *(Lesson 14-2)*

35. Add: $(7a^4 + 3a^3 + 2a^2 + 5) + (-6a^3 + 4a^2 - 8)$ *(Lesson 15-2)*

36. Make a stem-and-leaf plot for the given data.   *(Lesson 11-2)*

Scores on a Mathematics Test

77   68   84   82   79   91   73   88   94   79

15-3 Subtracting Polynomials

Objective: To subtract polynomials.

EXPLORATION

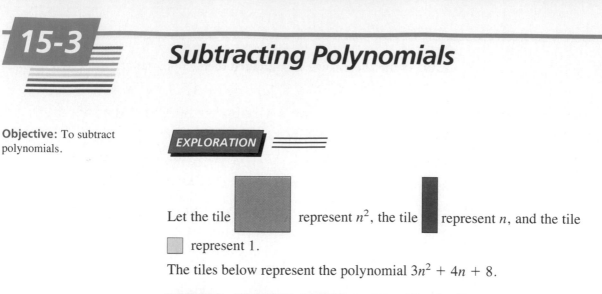

Let the tile [ ] represent n^2, the tile [ ] represent n, and the tile [ ] represent 1.

The tiles below represent the polynomial $3n^2 + 4n + 8$.

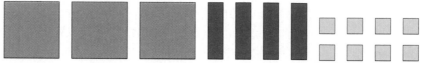

■ Copy the diagram or use algebra tiles. Remove 2 tiles representing n^2, one tile representing n, and five tiles representing 1.

■ What polynomial is represented by the tiles that you removed?

■ What polynomial is represented by the tiles that are left?

■ Complete: $(3n^2 + 4n + 8) - (2n^2 + n + 5) = \underline{\ ?\ }$

In the Exploration you removed tiles to represent the polynomial that was being subtracted. Recall that subtraction may also be thought of as addition of the opposite. To subtract a rational number, you add the opposite of that number. To subtract a polynomial, you use a similar procedure.

> **Generalization:** *Subtracting Polynomials*
>
> To subtract a polynomial, add the opposite of each term of the polynomial.

Example 1

Solution

Subtract: $(3n^2 + 4n + 8) - (2n^2 + n + 5)$

$(3n^2 + 4n + 8) - (2n^2 + n + 5)$
$= (3n^2 + 4n + 8) + (-2n^2 - n - 5)$
$= (3n^2 - 2n^2) + (4n - n) + (8 - 5)$
$= n^2 + 3n + 3$

The opposite of $2n^2$ is $-2n^2$.
← The opposite of n is $-n$.
The opposite of 5 is -5.

☑ **Check Your Understanding**

1. In Example 1, why was the polynomial $-2n^2 - n - 5$ added to the polynomial $3n^2 + 4n + 8$?

As with addition, you may have to subtract polynomials that have terms missing for some powers.

Example 2	Subtract: $(7a^3 + 3a^2 - 10) - (9a^3 + 4a^2 - 6a - 9)$
Solution 1	Line up like terms. Insert zero terms as needed. Add the opposite.

$$
\begin{array}{r}
7a^3 + 3a^2 + 0a - 10 \\
9a^3 + 4a^2 - 6a - 9 \\
\hline
\end{array}
\quad \longrightarrow \quad
\begin{array}{r}
7a^3 + 3a^2 + 0a - 10 \\
-9a^3 - 4a^2 + 6a + 9 \\
\hline
-2a^3 - a^2 + 6a - 1
\end{array}
$$

Solution 2

$(7a^3 + 3a^2 - 10) - (9a^3 + 4a^2 - 6a - 9)$
$= (7a^3 + 3a^2 - 10) + (-9a^3 - 4a^2 + 6a + 9)$
$= (7a^3 - 9a^3) + (3a^2 - 4a^2) + 6a + (-10 + 9)$
$= -2a^3 - a^2 + 6a - 1$

☑ **Check Your Understanding**

2. In Solution 1, why was the term $0a$ inserted in $7a^3 + 3a^2 - 10$?

Guided Practice

COMMUNICATION « *Reading*

Replace each __?__ with the correct word.

« **1.** To subtract a number, you add its __?__.

« **2.** To subtract a polynomial, you add the opposite of each __?__.

Write the opposite of each term of each polynomial.

3. $2a^2 + 5a + 6$

4. $-6x^3 - 3x^2 - x - 7$

5. $4x^4 - 6x^3 + 9x^2 + 4x - 9$

6. $-m^4 + 2m^3 - 8m^2 - m - 8$

Subtract.

7. $-4a + 7$
$2a - 3$

8. $2n^2 - 5n - 1$
$3n^2 - n + 6$

9. $7b^3 + 9b^2 - 12$
$2b^3 + 9b^2 - 2b + 10$

10. $9n^3 - 6n^2 - n - 3$
$-7n^3 + 4n^2 + n - 1$

11. $(5a^2 + 7a + 8) - (3a^2 + 4a + 2)$

12. $(v^3 + 8v^2 - 13v + 2) - (v^3 + 5v^2 + 9)$

13. $(15z^3 - 3z^2 + 6z + 13) - (-8z^3 + 7z^2 - 8z - 4)$

14. $(3w^3 - 5w^2 - 8) - (6w^3 + 2w - 18)$

Exercises

Subtract.

1. $(2x^2 + 4x + 5) - (x^2 + 3x + 1)$

2. $(5c^2 + 6c + 8) - (2c^2 + 4c + 5)$

3. $(8x^2 + 9x + 9) - (5x^2 + 2x + 3)$

4. $(5a^2 + 8a + 9) - (3a^2 + 2a + 1)$

5. $(8d^2 + 3d + 6) - (3d^2 + 4d + 8)$

6. $(9n^2 + 4n + 3) - (4n^2 + 7n + 9)$

7. $(4m^2 - 3m + 2) - (6m^2 + 7m + 6)$

8. $(6y^2 - 3y - 2) - (9y^2 + 5y + 7)$

9. $(8a^3 - 6a^2 + 2a + 4) - (5a^3 + 7a^2 - 5a - 7)$

10. $(3b^3 - 4b^2 + b - 4) - (7b^3 + 6b^2 - 4b + 8)$

11. $(-6x^2 + 5x - 9) - (-3x^2 - x + 7)$

12. $(a^2 + 2a - 3) - (-5a^2 + 2a + 4)$

13. $(8a^3 - 6a^2 - 2a + 9) - (4a^3 - 2a^2 + 6a - 8)$

14. $(3x^4 - 8x^3 - 6x^2 + 5x + 2) - (6x^4 + 2x^3 - 8x^2 + 6x + 4)$

15. $(4a^3 - 6a) - (5a^4 - 8a^2 + 2)$

16. $(9y^4 - 6y^2 + 3y - 6) - (4y^3 - 6y - 4)$

17. $(7a^3 + 9a^2 - 2) - (5a^4 - 8a^2 - 2)$

18. $(-6t^4 - 5t^3 - 8t^2 + 4) - (-2t^3 + 8t^2 + t + 1)$

19. $(z^4 - z^3 - 2z^2 - 3z) - (5z^4 + 7z^2 - z - 8)$

20. $(-2d^4 + 4d^3 + 3) - (5d^3 - 2d^2 + 3d - 7)$

21. Find the difference when $4x^3 + 7x^2 - 3x - 6$ is subtracted from $8x^3 - 7x^2 + 3x - 9$.

22. Find the difference when $-4x^3 - x^2 + 7x + 3$ is subtracted from $-6x^3 + 9x^2 - 5x - 6$.

Solve.

23. $(6a + 9) - (2a + 21) = 0$

24. $(7n + 2) + (3n - 12) = 0$

25. $(3d - 3) - (8d + 5) = -3$

26. $(7y + 1) - (-y + 7) = 3y - 26$

27. $(x^2 - 8x + 3) - (x^2 - 5x - 12) = 0$

28. $(c^2 - 5c + 12) + (-c^2 + 3c - 12) = 0$

29. $(4r + 9r^2) - (9r^2 + 2) = r + 1$

30. $(n^2 + 2n - 9) - (n^2 - 3n + 16) = n - 5$

Simplify.

31. $(7a^2 + 4a + 5) + (2a^2 - 5a + 8) - (3a^2 - 6a + 4)$

32. $(x^2 - 6x + 1) + (3x^2 + 7x + 3) - (2x^2 + 3x + 5)$

33. $(6n^2 - 7n + 9) - (3n^2 + 2n + 1) - (n^2 - 4n + 3)$

34. $(c^2 - 3c + 7) - (5c^2 + 3c - 5) - (4c^2 + 7c - 2)$

35. $(2b^2 + 5b - 8) - (3b^2 + 9b + 5) + (3b^2 - 2b + 7)$

36. $(m^2 + 6m - 2) - (4m^2 - 7m - 3) + (2m^2 + 5m - 3)$

THINKING SKILLS

37. Create two polynomials that have a difference of 2.

38. Create two polynomials that have a difference of x.

39. What can you conclude about two polynomials that have a difference of zero?

40. Subtract $5x^2 + 2x - 1$ from $3x^2 - 4x + 2$.

41. Subtract $3x^2 - 4x + 2$ from $5x^2 + 2x - 1$.

42. Compare your answers from Exercises 40 and 41. What can you conclude about reversing the order of a subtraction? Give another example to support your conclusion.

SPIRAL REVIEW

43. Make a frequency table for the given data. *(Lesson 11-1)*

Base Hits per Season

164 173 201 212 183 162 171
176 160 194 168 185 174 166

44. Find the volume of a pyramid with a base of 144 cm^2 and a height of 9 cm. *(Lesson 14-4)*

45. Find the difference: $(7d^2 + 8d - 4) - (4d^2 - 5d + 7)$ *(Lesson 15-3)*

46. Solve and graph: $2x - 7 > 9$ *(Lesson 13-7)*

Challenge

Find all whole-number values of x such that $(x^3 + 3x^2 - 8x + 1) - (x^3 + 2x^2 - 5x + 1)$ is equal to 0.

Modeling Products Using Tiles

Objective: To use algebra tiles to model products of monomials and binomials.

Materials

■ algebra tiles

You can use algebra tiles to model products of monomials and binomials or of binomials and binomials. The models use the geometric fact that the area of a rectangle equals the product of the base and the height, or $A = bh$. The base and height represent the two polynomials being multiplied, and the area represents their product.

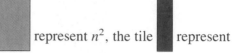

For example, let the tile ⬜ represent n^2, the tile ▮ represent n, and the tile ⬜ represent 1. Then you can represent the product $2n(n + 3)$ as shown below.

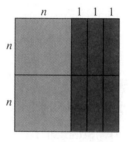

$$\text{height} = n + n = 2n$$
$$\text{base} \;\;\;= n + 1 + 1 + 1 = n + 3$$
$$\text{Area} \;\;\;= 2n(n + 3)$$

Activity I *Modeling Products of Monomials and Binomials*

1 Write the product represented by each diagram below.

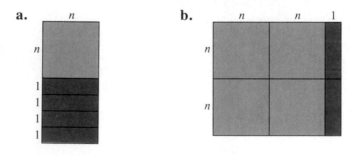

a.

b.

2 Show how to use algebra tiles to represent each product.
  a. $n(2n + 4)$ **b.** $3n(n + 1)$

3 Use your diagrams from Step 2 to complete each statement.
  a. $n(2n + 4) = \underline{\;?\;}$

  b. $3n(n + 1) = \underline{\;?\;}$

You can use algebra tiles to represent the product of a binomial and a binomial. For example, you can represent the product $(2n + 1)(n + 3)$ as shown below.

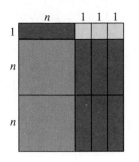

Activity II *Modeling Products of Binomials*

1 Write the product represented by each diagram below.

a.

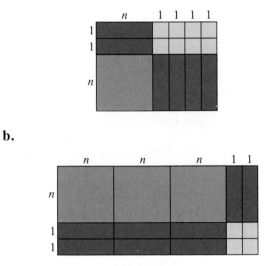

b.

2 Show how to use algebra tiles to represent each product.

    a. $(n + 2)(n + 3)$

    b. $(n + 3)(n + 3)$

    c. $(2n + 1)(n + 1)$

    d. $(2n + 1)(2n + 1)$

3 Use your diagrams from Step 2 to complete each statement.

    a. $(n + 2)(n + 3) = $ ___?___

    b. $(n + 3)(n + 3) = $ ___?___

    c. $(2n + 1)(n + 1) = $ ___?___

    d. $(2n + 1)(2n + 1) = $ ___?___

Multiplying a Polynomial by a Monomial

Objective: To multiply a polynomial by a monomial.

CONNECTION

In Chapter 2 you learned to simplify expressions such as $3(x - 7)$ by using the distributive property. Since the expression $x - 7$ is a polynomial, you have already learned how to multiply a number by a polynomial.

Example 1

Multiply: $-4(2x^2 + 5x - 3)$

Solution

$$-4(2x^2 + 5x - 3) = (-4)(2x^2) + (-4)(5x) - (-4)(3)$$
$$= -8x^2 + (-20x) - (-12)$$
$$= -8x^2 - 20x + 12$$

☑ Check Your Understanding

1. In Example 1, why can $-8x^2 + (-20x) - (-12)$ be rewritten as $-8x^2 - 20x + 12$?

Recall that to multiply powers having the same base, you add the exponents. You can use this rule to multiply monomials.

Example 2

Multiply: $(3a^3b^2)(-5a^2b^4)$

Solution

$$(3a^3b^2)(-5a^2b^4) = 3(-5)(a^3 \cdot a^2)(b^2 \cdot b^4)$$
$$= -15(a^{3 + 2})(b^{2 + 4})$$
$$= -15a^5b^6$$

← Group integers and powers having the same base.

To multiply a polynomial of two or more terms by a monomial, you use the distributive property and the rule for multiplying powers of the same base.

Example 3

Multiply: $4x^2(7x^3 + 2x^2 - 6x - 4)$

Solution

$4x^2(7x^3 + 2x^2 - 6x - 4)$
$= 4x^2(7x^3) + 4x^2(2x^2) - 4x^2(6x) - 4x^2(4)$
$= 28x^5 + 8x^4 - 24x^3 - 16x^2$

☑ Check Your Understanding

2. Explain how the distributive property was used in Example 3.

Guided Practice

COMMUNICATION « *Reading*

Refer to the text on pages 682–683.

« **1.** What is the main idea of the lesson?

« **2.** List three major points that support the main idea of the lesson.

Replace each __?__ with the expression that makes the statement true.

3. $5(2x^3 + x^2 - 4x + 6) = 10x^3 + \underline{\;?\;} - 20x + 30$

4. $-3(6x^4 - 7x^3 + 5x^2 - x + 2) = \underline{\;?\;} + 21x^3 - 15x^2 + \underline{\;?\;} - 6$

5. $7a^3b^4(5a^2b^5) = 35 \cdot \underline{\;?\;} \cdot b^9$

6. $4x^5z^2(-8x^3z^7) = \underline{\;?\;} \cdot x^8z^9$

7. $8n^3(3n^2 + 6n - 5) = 24n^5 + \underline{\;?\;} - 40n^3$

8. $-2a^4(-a^3 + 5a^2 - 7a - 8) = \underline{\;?\;} - \underline{\;?\;} + 14a^5 + 16a^4$

Multiply.

9. $5(7c^3 + 4c^2 - c - 6)$ **10.** $-7(6x^4 + 5x^3 - 3x^2 - 9x + 5)$

11. $3x^4y^5(4x^2y^6)$ **12.** $7ab^5(-3a^3b)$ **13.** $(-5m^4n^8)(-3m^3n^5)$

14. $7s^4(-5s^3 + 6s^2 - 4s - 8)$ **15.** $-2x^4(3x^3 - 4x^2 - 8x + 3)$

Exercises

Multiply.

1. $7(2x^2 + x + 2)$ **2.** $4(c^3 - 7c^2 - 6c - 5)$

3. $-6(6a^3 - 2a^2 - 4a + 2)$ **4.** $-3(-2b^4 - 7b^3 + 2b^2 + 8)$

5. $(4x^2y^3)(6x^4)$ **6.** $(7a^3c^5)(4a^8c^6)$ **7.** $(6r^3s^3)(-5r^2s^4)$

8. $(-10x^4y)(10xy^4)$ **9.** $(-6a^4x^5)(-3a^2x^6)$ **10.** $(-8c^6b^7)(-6c^2b^9)$

11. $4a(6a^2 + 4a + 5)$ **12.** $-7t(4t^2 + 2t - 9)$

13. $3d^2(5d^3 - 8d^2 + 7d - 6)$ **14.** $-8x(3x^3 - 7x^2 - 4x + 2)$

15. Find the product of $7x^3$ and $4x^3 + 6x^2 - 9x + 7$.

16. Multiply $3n^4 - n^3 + 5n^2 + 6n - 3$ by $2n^4$.

Multiply.

17. $5mn(2mn - 3mn^2 + 4n)$ **18.** $4ac(-3a^2b + 8ab^3 + 5b)$

19. $-xy(5xy + 7x^2y + 8xy^2)$ **20.** $3cd^2(7cd + 4c^2d^2 - c^3d)$

21. $m^3n(-5m^2n^2 - mn + 7m^2n)$

22. $-2a^3b^2(-2a^3 + a^3b - 7ab^4 + 9b^8)$

MENTAL MATH

In many cases multiplying a polynomial by a monomial can be done mentally. Do each of the following multiplications mentally.

23. $5(6x^3 + 7x^2 - 9x - 5)$

24. $-3(6c^4 + 8c^3 - 3c^2 - 8c + 2)$

25. $2a(4a^4 + 8a^3 - 7a^2 + a - 9)$ **26.** $-4n(-7n^3 + 2n^2 - 4n + 6)$

27. $2x^2(-9x^3 - 4x^2 + 6x + 8)$ **28.** $-5m^4(5m^3 - 2m^2 - m + 7)$

PATTERNS

Simplify.

29. $-(x^3 + 4x^2 + 7x - 5)$ **30.** $-(-x^3 - 4x^2 - 7x + 5)$

31. $-(-3x^4 + x^3 - 6x^2 - 2x + 7)$ **32.** $-(3x^4 - x^3 + 6x^2 + 2x - 7)$

33. Use the pattern in Exercises 29–32 to complete this statement: When you multiply a polynomial by -1, you __?__.

THINKING SKILLS

Combine your knowledge of exponents and fractions to complete.

34. $(a^{-2})(a^{-3}) = \left(\frac{1}{?}\right)\left(\frac{1}{?}\right)$ **35.** $\left(\frac{1}{a^2}\right)\left(\frac{1}{a^3}\right) = \frac{1}{?}$ **36.** $\frac{1}{a^5} = a^?$

37. What do the results of Exercises 34–36 seem to indicate about the product of powers rule? Create another example to support your conclusion.

Use the results of Exercise 37 to find each product. Express your answer using positive exponents.

38. $(7a^2b^{-3})(2a^{-5}b^{-1})$ **39.** $(3r^{-1}s^{-1})(6r^2s^{-4})$ **40.** $(3k^2t^{-6})(-2k^{-1}r^2t^3)$

SPIRAL REVIEW

(Lesson 12-1)

41. A cube with sides numbered 1 through 6 is rolled. Find P(2 or 5).

42. Find the volume of a rectangular aluminum container with a base area of 160 ft^2 and a height of 8 ft.   *(Lesson 14-4)*

43. Find the product: $4a(2ab + 5a - 3)$ *(Lesson 15-4)*

44. Solve the system: $x + y = 8$
$\quad\quad\quad\quad\quad\quad\quad x - y = 4$ *(Lesson 13-4)*

Multiplying Binomials

Objective: To multiply binomials.

CONNECTION

The rectangle at the right is $(2n + 3)$ units long and $(n + 2)$ units wide. You can find the area of the rectangle by counting the tiles representing n^2, n, and 1 and writing a polynomial to represent the area.

$$\text{Area} = 2n^2 + 7n + 6$$

Recall from geometry that the area of a rectangle is equal to the product of the base and the height, or $A = bh$. Applying this formula to the rectangle shown, you get the following.

$$\text{Area} = (2n + 3)(n + 2) = 2n^2 + 7n + 6$$

To multiply $(2n + 3)(n + 2)$ using algebra, you use the distributive property twice.

Example 1

Multiply: $(2n + 3)(n + 2)$

Solution

$$\begin{aligned}
(2n + 3)(n + 2) &= 2n(n + 2) + 3(n + 2) &&\longleftarrow \text{Use the distributive} \\
&= 2n^2 + 4n + 3n + 6 && \text{property.} \\
&= 2n^2 + 7n + 6
\end{aligned}$$

✓ Check Your Understanding

1. In Example 1, why can you add $4n$ and $3n$?
2. In Example 1, will the answer be the same if $(2n + 3)(n + 2)$ is written as $(2n + 3)n + (2n + 3)2$?

Some polynomials involve subtraction. You must be careful to use the correct signs when multiplying such polynomials.

Example 2

Multiply: $(3x - 2)(7x + 5)$

Solution

$$\begin{aligned}
(3x - 2)(7x + 5) &= 3x(7x + 5) - 2(7x + 5) \\
&= 21x^2 + 15x - 14x - 10 \\
&= 21x^2 + x - 10
\end{aligned}$$

✓ Check Your Understanding

3. In Example 2, why is $2(7x + 5)$ subtracted rather than added?
4. In Example 2, what number is x multiplied by in $21x^2 + x - 10$?

COMMUNICATION «Writing

Write the multiplication represented by each rectangle.

« **1.**   « **2.**

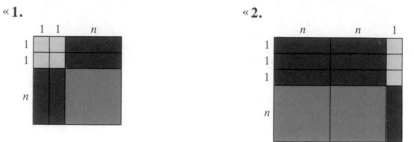

Draw a diagram similar to the ones in Exercises 1 and 2 to represent each multiplication.

« **3.** $(n + 2)(n + 3)$ « **4.** $(2x + 1)(x + 4)$

Choose the letter of the correct product.

5. $(5x - 3)(x + 2)$ **A.** $5x^2 - 7x - 6$

6. $(5x + 3)(x - 2)$ **B.** $5x^2 + 7x - 6$

7. $(5x - 3)(x - 2)$ **C.** $5x^2 + 7x + 6$

8. $(5x + 3)(x + 2)$ **D.** $5x^2 + 13x + 6$

                                     E. $5x^2 - 13x + 6$

                                     F. $5x^2 - 13x - 6$

Multiply.

9. $(x + 1)(x + 2)$ **10.** $(t + 2)(t - 3)$ **11.** $(r - 4)(r - 6)$

12. $(3x + 4)(2x + 4)$ **13.** $(2y - 1)(3y - 5)$ **14.** $(6x + 2)(x - 7)$

Exercises

Multiply.

1. $(x + 4)(x + 2)$ **2.** $(a + 6)(a + 3)$ **3.** $(2t + 7)(6t + 5)$

4. $(3x + 1)(5x + 5)$ **5.** $(4y - 2)(6y + 7)$ **6.** $(5b + 3)(b - 1)$

7. $(x + 2)(7x - 8)$ **8.** $(7k - 1)(k + 8)$ **9.** $(w - 4)(2w - 1)$

10. $(d - 2)(3d - 5)$ **11.** $(x - 4)(4x - 1)$ **12.** $(5z - 2)(2z - 5)$

13. $(4x - 6)(3x - 5)$ **14.** $(6y - 7)(3y - 5)$ **15.** $(2x + 1)(2x + 1)$

16. $(5c - 3)(5c - 3)$ **17.** $(2x - 5)(2x + 5)$ **18.** $(4d + 7)(4d - 7)$

19. Find the product of $3x + 4$ and $5x - 6$.

20. Multiply $6m - 5$ by $3m - 9$.

Multiply.

21. $(a + 3b)(a - 5b)$ **22.** $(2m - n)(m - n)$ **23.** $(x + y)(x - y)$

24. $(r^2 - 2s)(r^2 + 4s)$ **25.** $(x^2 - 3)(2x^2 - 5)$ **26.** $(x^2 + 1)(x^2 + 3)$

THINKING SKILLS

27. a. Find the products $(3x + 2)(2x - 5)$ and $(2x - 5)(3x + 2)$.
   b. Compare your answers in part (a). Create a mathematical statement relating the two products. What property is illustrated by these results?

28. a. Determine whether the product of two binomials is *always*, *sometimes*, or *never* a trinomial.
   b. Create examples to support your conclusion in part (a).

PROBLEM SOLVING/APPLICATION

A floor of an office building is being remodeled. The floor is currently divided into square work spaces. The new plans call for each work space to be 2 ft less in width and 3 ft less in length.

29. What do $x - 3$ and $x - 2$ represent?

30. What does $(x - 3)(x - 2)$ represent?

31. Find the product $(x - 3)(x - 2)$.

32. What is the original area of each square cubicle in terms of x?

33. Use your answers from Exercises 31 and 32 to find a polynomial that represents the decrease in area in terms of x.

DATA, *pages 666–667*

34. Write the sixth, seventh, and eighth rows of Pascal's triangle.

35. Use your answer to Exercise 34 to find each product.

a. $(a + b)^5$ **b.** $(a + b)^6$ **c.** $(a + b)^7$

36. **RESEARCH** Find out more about the mathematical and scientific achievements of Blaise Pascal. Then write a brief report.

COMPUTER APPLICATION

You can program a computer in BASIC to print the product $(ax + b)(cx + d)$ as a polynomial in standard form when you input values for a, b, c, and d. You can base your program on this formula.

$$(ax + b)(cx + d) = acx^2 + (ad + bc)x + bd$$

Your print statement in BASIC might be the following.

```
PRINT A*C;"X^2 + ";A*D+B*C;"X + ";B*D
```

37. What do A, B, C, and D represent?

38. Explain how the computer would find $(2x + 3)(4x + 5)$.

SPIRAL REVIEW

39. Find the product: $(x + 4)(x - 3)$ *(Lesson 15-5)*

40. Find the area of a circle with radius 8.5 cm.   *(Lesson 10-4)*

41. What percent of 60 is 15?   *(Lesson 9-7)*

Self-Test 1

Write each polynomial in standard form.

1. $4x + 5x^3 + 6x^4 - 3x^2 - 8$ **2.** $5a^2 - 7a + 4a^5 - 2 + 9a^3$ 15-1

Simplify.

3. $4x^3 + 4x^2 + 8x - x^3 + 7$ **4.** $7c^3 - 3c - 5c + 4 - 2c^2 + 1$

Find each answer.

5. $(x^2 + 6x + 2) + (x^2 + 2x + 6)$ 15-2

6. $(z^2 - 4z + 2) + (z^2 + z - 6)$

7. $(2b^3 + b^2 - 4) - (b^3 - b^2 + 2)$ 15-3

8. $(5n^2 + n - 1) - (n^2 + n + 3)$

9. $5xy^2(-2x^3y^4)$ **10.** $-3a^2(5a^3 - 3a^5)$ 15-4

11. $(2x + 3)(4x + 5)$ **12.** $(5c + 7)(4c - 8)$ 15-5

15-6

Strategy:
Working Backward

Objective: To solve problems by working backward.

In some problems, you are given an end result and are asked to find a fact needed to reach that result. One way to solve a problem of this type is to work backward.

Problem

Wendy defeated three opponents to win a chess tournament. At each stage of the tournament, the loser dropped out and the winner continued on to the next stage. How many players were in the tournament?

Solution

UNDERSTAND The problem is about players in a chess tournament.
Facts: winner defeated three players
Find: the number of players in the tournament

PLAN To solve, you work backward from the final match to the first match. Since Wendy needed three wins to complete the tournament, there must have been three stages. Start at the end of the tournament. Draw a diagram that shows the number of players at each stage of the tournament.

WORK To compete in Stage 3, each player had to win in Stage 2. To compete in Stage 2, each player had to win in Stage 1.

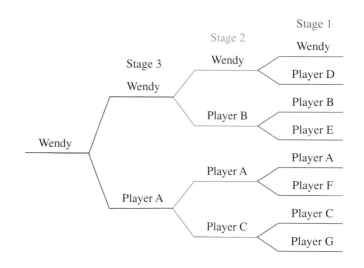

ANSWER There were 8 players in the tournament.

Look Back What if Wendy had defeated four opponents? How many players would have been in the tournament?

Guided Practice

« **1.** What is the main idea of this lesson?

« **2.** What other problem solving strategy was used to solve the problem on the previous page?

Use this problem for Exercises 3–6.

Lawanda had $1125.76 in her checking account at the end of last week. During the week she had deposited $464.55 and had written checks for $129.87, $322.51, and $47.35. She also had made an electronic withdrawal of $100. How much money was in Lawanda's checking account at the beginning of the week?

« **3.** What was the total amount of the checks?

« **4.** In the process of solving the problem, do you *add* or *subtract* the total amount of the checks?

« **5.** In the process of solving the problem, do you *add* or *subtract* the deposit?

6. Solve the problem.

Solve by working backward.

7. In a tennis tournament, the loser of each match is eliminated. Evan played five matches to win the tournament. How many players were in the tournament?

8. Dana wants to buy a turtleneck and jeans. The turtleneck costs $28 and the jeans cost $35. Dana earns $5.25 per hour. How many hours must she work to earn enough money to buy the turtleneck and the jeans?

Problem Solving Situations

Solve by working backward.

1. In each stage of a baseball tournament, a team that loses drops out. How many teams compete in the tournament if the winning team plays six games?

2. In a round-robin bridge tournament, every player plays every other player. Curtis played seven games and won the tournament. How many players were in the tournament?

3. Ming wants to buy a bicycle that costs $140 plus 5% sales tax. He works 4 h per day at a part-time job and earns $4.50 per hour. How many days will he have to work to buy the bicycle?

4. Alix has $1042.76 in her checking account at the end of the week. During the week, she wrote a check for $79.09, made a deposit of $243.87, and wrote another check for $120.97. She also transferred $225 from her savings account into her checking account and made an electronic withdrawal of $50. How much money did Alix have in her checking account at the beginning of the week?

Solve using any problem solving strategy.

5. Meredith takes a cab to a business meeting. The ride costs $1.20 for the first one-tenth of a mile, and $.90 for each additional one-tenth of a mile. At the end of the ride, Meredith gives the cab driver $15 and asks for $3 back in change, leaving the cab driver with a $1.80 tip. How many miles was the cab ride?

6. The sum of twelve and five times a number is eighty-two. Find the number.

7. Sonia needs 4 qt of oil to change the oil in her car. A gallon of oil costs $4.50, and a quart of oil costs $1.19. To get the best buy, should she buy quarts or gallons? How much can she save?

8. Kristen has a job interview at 9:45 A.M., and she wants to be there 15 min early. She needs 25 min to travel from her home to the interview, and she wants to allow an hour and a half to dress and have breakfast. At what time should Kristen get up?

> **PROBLEM SOLVING**
>
> **CHECKLIST**
>
> **Keep these in mind:**
> Using a Four-Step Plan
> Too Much or Not Enough
>   Information
> Supplying Missing Facts
> Problems with No Solution
>
> **Consider these strategies:**
> Choosing the Correct Operation
> Making a Table
> Guess and Check
> Using Equations
> Identifying a Pattern
> Drawing a Diagram
> Using Proportions
> Using Logical Reasoning
> Using a Venn Diagram
> Using a Simpler Problem
> Making a Model
> Working Backward

9. A schedule for a sales meeting showed that Pat was given half the total available time to do her presentation. Donnell was given two thirds of the remaining time, and Cody and Kim were given ten minutes each. How long was the sales meeting expected to last?

10. Anthony's company earned $40,000 in January. The earnings declined 10% each month for the next three months. Following this, the earnings increased for the next two months by 5% each month. In reporting this to the Board of Directors, Anthony decided to use a data display. Which type of display should be used? Make two data displays and compare them. Which shows the information better?

WRITING WORD PROBLEMS

Write a word problem about the given subject that you could solve by working backward. Then solve the problem.

11. The total number of ancestors Peter has within a given number of generations.

12. The price of a share of stock at the beginning of a week given its price at the end of the week.

SPIRAL REVIEW

13. A right triangle has legs of 10 cm and 24 cm. Find the length of the hypotenuse. *(Lesson 10-11)*

14. Bill wants to take $120 as spending money on his family's vacation. Bill earns $7.50 per hour from his after-school job. How many hours must Bill work to earn the $120? *(Lesson 15-6)*

15. Graph $y \geq 2x$. *(Lesson 13-8)*

16. Find the sum: $(6x^3 + 9x - 4) + (-2x^3 + 3x^2 - 4x)$ *(Lesson 15-2)*

Historical Note

Solving problems by working backward is not a new problem solving strategy. It was used at least 1500 years ago in India, most notably by the mathematician Aryabhata. Most of what is known of Aryabhata's work comes from his book on astronomy, the third chapter of which is devoted to mathematics.

Research

Indian mathematicians solved many problems by the rule of false position. Find out what this rule is and how it was used.

Factoring Polynomials

Objective: To factor polynomials.

Terms to Know

- *greatest common monomial factor*
- *factor a polynomial*

1 Find the GCF of 24 and 78.

2 Find the GCF of $24a^3b^4$ and $78a^2b^5$.

3 Complete: $24a^3b^4 + 78a^2b^5 = \underline{\ ?\ }(4a + 13b)$

4 Complete: Some polynomials can be written as the product of the $\underline{\ ?\ }$ of their terms and another polynomial.

The **greatest common monomial factor** of a polynomial is the GCF of its terms. To **factor a polynomial**, you express the polynomial as the product of other polynomials. These polynomials should contain terms involving whole numbers only, no fractions.

Example

Factor each polynomial.

a. $6t^4 + 14t^3 - 24t^2$

b. $3c^4d^2 - 24c^3d^3 - 15c^2d$

Solution

a. The GCF of $6t^4$, $14t^3$, and $24t^2$ is $2t^2$.
$$6t^4 + 14t^3 - 24t^2 = 2t^2(3t^2) + 2t^2(7t) - 2t^2(12)$$
$$= 2t^2(3t^2 + 7t - 12)$$

b. The GCF of $3c^4d^2$, $24c^3d^3$, and $15c^2d$ is $3c^2d$.
$$3c^4d^2 - 24c^3d^3 - 15c^2d = 3c^2d(c^2d) - 3c^2d(8cd^2) - 3c^2d(5)$$
$$= 3c^2d(c^2d - 8cd^2 - 5)$$

✓ Check Your Understanding

1. In part (a), why is $2t^2$ the greatest common monomial factor?

2. In part (b), why is the answer not $3(c^4d^2 - 8c^3d^3 - 5c^2d)$?

3. Describe a way to check the answers in the Example.

Guided Practice

COMMUNICATION «*Reading*

Explain the meaning of the word *factor* in the following sentences.

«**1.** The rain was a factor in his poor golf score.

«**2.** $6t$ is a factor of $18t^2$ and $30t$.

«**3.** Before Camille could state the profit her company made, she had to factor out the production costs.

Match each polynomial with its greatest common monomial factor.

4. $3a^2b + 6ab$

5. $4a^2b + 6ab^2$

6. $12a^2b + 8ab^2$

7. $4a^2b^2 + 6a^2b^3$

8. $12a^2b^3 + 8a^3b^2$

9. $3a^4b^2 + 6a^2b^3$

A. $4ab$

B. $2a^2b^2$

C. $3ab$

D. $3a^2b^2$

E. $2ab$

F. $4a^2b^2$

Factor each polynomial.

10. $4c^2 + 6c$

11. $3n^2 + 7n$

12. $16e + 2e^2$

13. $e^3 + 12e$

14. $7m^5 - 8m^3$

15. $14k^5 - 21k^2$

16. $15z^4 - 18z^2 + 12z$

17. $24p^5 - 18p^3 - 36p^2$

18. $8n^4 + 4n^2 + 5n$

19. $3t^2 - 15t + 9$

20. $6p^2t + 8pt^2 + 12pt$

21. $7rs^3t + 14r^3s^2t^2 + 21rst$

COMMUNICATION «*Discussion*

«**22.** Under what conditions does a polynomial not have a greatest common monomial factor?

«**23.** Can the greatest common monomial factor of a polynomial be equal to one of its terms? Explain.

Exercises

Factor each polynomial.

1. $7x^2 + 14x$ **2.** $18y^2 - 6y$ **3.** $4s + 5s^2$ **4.** $m^4 + 2m^2$

5. $4k^3 + 8k^2$ **6.** $9n^3 - 3n^2$ **7.** $5p^5 - 15p^2$ **8.** $9a^3 - 27a^2$

9. $9xy + 6x^2y$ **10.** $ab^2 - 3a^2b^2$ **11.** $4ac - 8bc$

12. $7rst - 5r^2t^2$ **13.** $12u^4 - 9u^3 - 24u^2$ **14.** $24t^5 - 18t^2 - 6$

15. $8s^4 - 16s^3 + 32s^2$ **16.** $18q^5 - 9q^4 - 9q^3 - 18$

17. $25qd^2 + 15q^2d + 30q^2d^2 - 5qd$ **18.** $35a^5 - 49a^3b^2 - 56a^2b^4$

19. $36y^4z^3 + 45y^3z^4 + 20y^5z^2$ **20.** $22k^6t^5 + 110k^4t^6 - 55k^7t^5$

21. $24r^2t^2u^2 - 12r^3t^3u^3 - 6r^4t^4u^4$

22. $21a^5c^2f - 14ac^2f^2 - 6a^3cf$

23. $16a^2b^2c^2 + 4ab^2c + 8a^3b^2c^4 + 12a^2b^2c$

24. $9wxy^2 + 15wx^2y^3 + 21wx^2y^2$

Factor each polynomial as the product of two binomials.

25. $r(r - 5) + 6(r - 5)$

26. $3x(x + 2) - 5(x + 2)$

27. $7d(2 - 3d) - (2 - 3d)$

28. $z(8z - 9) + (8z - 9)$

29. $(y - 3)(y - 3) + 2(y - 3)$

30. $(m + 4)(m + 4) - 3(m + 4)$

GROUP ACTIVITY

31. Create the simplest polynomial possible with four terms using the variables r, s, and t and with t as the greatest common monomial factor.

32. Create the simplest polynomial possible with three terms using the variables w, y, and z and with 1 as the greatest common monomial factor.

CONNECTING ALGEBRA AND GEOMETRY

Find the area of the shaded part of each figure in terms of π. Give your answer in factored form.

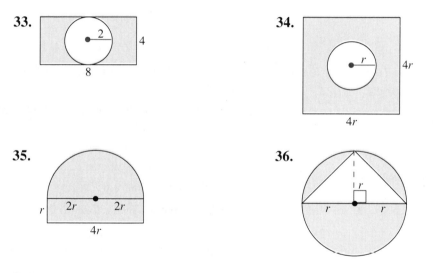

33.

34.

35.

36.

SPIRAL REVIEW

37. Find the number of permutations of 8 runners in 8 lanes. *(Lesson 12-6)*

38. Factor: $16a^2b^3 - 4a^2b^4 + 12a^3b^2$ *(Lesson 15-7)*

39. Find the difference:
$(11z^4 - 4z^3 + 2z^2 + 6z - 4) - (4z^4 - 3z^2 + 7z - 8)$
(Lesson 15-3)

40. Find the surface area of a rectangular prism with length 9 cm, height 4 cm, and width 6 cm. *(Lesson 14-2)*

Dividing a Polynomial by a Monomial

Objective: To divide a polynomial by a monomial.

CONNECTION

To add or subtract fractions with like denominators, you use the following rules.

$$\frac{a}{c} + \frac{b}{c} = \frac{a+b}{c} \quad \text{and} \quad \frac{a}{c} - \frac{b}{c} = \frac{a-b}{c}$$

By using these rules in reverse, you can divide a polynomial by a monomial.

Example 1

Divide: $\dfrac{4a^5 + 8a^4 + 6a^2}{2a}$

Solution

$$\frac{4a^5 + 8a^4 + 6a^2}{2a} = \frac{4a^5}{2a} + \frac{8a^4}{2a} + \frac{6a^2}{2a}$$

$$= \frac{4a^{5-1}}{2} + \frac{8a^{4-1}}{2} + \frac{6a^{2-1}}{2} \quad \longleftarrow \text{Use the quotient of powers rule.}$$

$$= 2a^4 + 4a^3 + 3a$$

✓ Check Your Understanding

1. In Example 1, why is 1 being subtracted from each exponent in the numerator?

2. How would the answer to Example 1 be different if the dividend were $4a^5 - 8a^4 + 6a^2$?

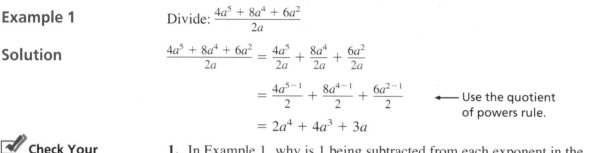

Generalization: *Dividing a Polynomial by a Monomial*

To divide a polynomial by a monomial, divide each term of the polynomial by the monomial and simplify.

Example 2

Divide: $\dfrac{5x^7y^4 - 35x^5y^5 + 20x^3y^3}{-5x^3y}$

Solution

$$\frac{5x^7y^4 - 35x^5y^5 + 20x^3y^3}{-5x^3y} = \frac{5x^7y^4}{-5x^3y} - \frac{35x^5y^5}{-5x^3y} + \frac{20x^3y^3}{-5x^3y}$$

$$= -x^4y^3 - (-7x^2y^4) + (-4y^2)$$

$$= -x^4y^3 + 7x^2y^4 - 4y^2$$

✓ Check Your Understanding

3. In Example 2, why do the addition and subtraction signs change?

Guided Practice

Replace each __?__ with the correct word or phrase.

« **1.** To divide a polynomial by a monomial, you use the rules for adding and subtracting __?__ in reverse.

« **2.** To divide a polynomial by a monomial, divide __?__ by the monomial and then __?__.

Match each division with the correct quotient.

3. $\dfrac{10ab - 15a^2}{5a}$

4. $\dfrac{10ab + 15a^2}{5a}$

5. $\dfrac{10ab - 15a^2}{-5a}$

6. $\dfrac{10ab + 15a^2}{-5a}$

A. $2b + 3a$
B. $-2b - 3a$
C. $2b - 3a$
D. $-2b + 3a$

Divide.

7. $\dfrac{14a^2 - 7a}{7}$

8. $\dfrac{16c^5 + 3c^3}{c}$

9. $\dfrac{24b^6 + 6b^2}{6b^2}$

10. $\dfrac{4b^3 + 10b^2}{-2b}$

11. $\dfrac{6d^6 - 4d^4}{d^3}$

12. $\dfrac{12x^7 - 18x^5}{-3x^4}$

Exercises

Divide.

1. $\dfrac{9x - 12y}{3}$

2. $\dfrac{16r + 8s^2}{-8}$

3. $\dfrac{7c^3 - c^2}{-c}$

4. $\dfrac{5m^7 + 4m^2}{m^2}$

5. $\dfrac{4x^2 + 8x^4 - 24x^7}{4x}$

6. $\dfrac{15k^9 - 6k^6}{-3k}$

7. $\dfrac{24t^8 + 64t^3 + 8t^2}{8t^2}$

8. $\dfrac{12s^8 - 36s^6 - 42s^4}{6s^3}$

9. $\dfrac{-15k^6 - 5k^5 + 60k^4}{-5k^2}$

10. $\dfrac{3m^6 + 4m^5 - 7m^4}{-m}$

11. $\dfrac{21de + 24de^2 + 27d^2e}{3de}$

12. $\dfrac{16k^2t^3 - 24k^3t^2 + 32k^2t^2}{8kt}$

13. $\dfrac{16r^4u^5 - 12r^7u^6}{-4r^4u^5}$

14. $\dfrac{48m^2n^2 + 42mn^4 - 54m^3n^5}{-6n^2}$

15. Divide $12x^3y^4 + 21x^4y^2 - 36x^5y^4$ by $3x^2y^2$.

16. Find the quotient when $30a^3b^7 - 15a^7b^5 + 25a^5b^5$ is divided by $5a^2b^4$.

Divide.

17. $\dfrac{3k^4 - 18k^3}{9k^3}$

18. $\dfrac{15r^3t^3 - 9r^2t^4}{21t^3}$

19. $\dfrac{24x^2y^2 + 15x^3y^2 + 4x^4y^2}{-6xy}$

20. $\dfrac{12w^4x + 15w^3x^2 + 27w^2x^3}{9w^2x}$

21. $\dfrac{3x^4 + 7x^3 + 9x^2}{2x^3}$

22. $\dfrac{50a^2b^4 + 40ab^5 + 30b^6}{-10a^2b^2}$

23. $\dfrac{25b^5 + 20b^4 + 5b^2}{15b^3}$

24. $\dfrac{8a^5 + 16a^3 + 4a}{6a^2}$

Simplify.

25. $\dfrac{8x - 6}{2} + \dfrac{12x + 9}{3}$

26. $\dfrac{5a - 10}{5} + \dfrac{6 + 9a}{3}$

27. $\dfrac{c^2 + 6c}{c} + \dfrac{4c^2 - 6c}{2c}$

28. $\dfrac{n^2 + 2n}{n} - \dfrac{2n^2 - 4n}{2n}$

29. $\dfrac{a^2b + 2a^2b^2}{ab} + \dfrac{2a^2 - 6a^2b}{2a}$

30. $\dfrac{m^3n^2 - 2m^2n^3}{m^2} + \dfrac{m^2n^4 - 4m^5n^2}{mn^2}$

CAREER APPLICATION

A *physics teacher* uses many formulas, some of which involve polynomials. For example, the height in feet at t seconds of an object thrown vertically upward with a speed of v ft/s is $h = vt - 16t^2$.

31. A ball is thrown vertically upward with a speed of 64 ft/s. Write the specific formula that gives its height at t seconds.

32. Factor the polynomial in the formula that you wrote in Exercise 31.

33. Make a table using whole-number values of t to find the maximum height of the ball.

34. Use the table you made in Exercise 33 to find the two times at which the height of the ball is 0 ft. What do these times represent?

Solve each problem. Apply the reasoning you use to solve Exercise 35 to help you solve Exercise 36.

35. A cardboard box has a height of 12 in. The volume of the box is 3192 in.^{3}. Find the area of the base of the box. Use the formula $V = Bh$, where B is the area of the base.

36. The volume of a rectangular prism is $4x^2y^2 + 8x^2y + 2xy^2 + 4xy$. The height of the prism is $2xy$. Find the area of the base. Use the formula $V = Bh$.

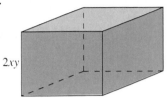

37. Make a box-and-whisker plot to display the given data. *(Lesson 11-4)*

Gasoline Mileage (mi/gal)					
26.0	21.8	26.9	26.3	30.5	19.5
23.4	20.8	22.4	24.8	22.5	22.5

38. Find the quotient: $(20b^3c^2 - 10bc^4 + 5b^2c) \div 5bc$ *(Lesson 15-8)*

39. Find the product: $3x(7x^3 - 2x)$ *(Lesson 15-4)*

40. Find the area of a parallelogram with height 16 in. and base 7 in. *(Lesson 10-3)*

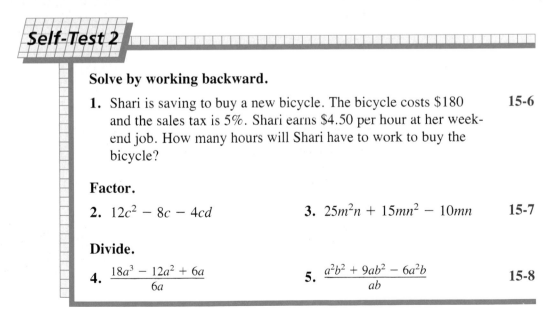

Self-Test 2

Solve by working backward.

1. Shari is saving to buy a new bicycle. The bicycle costs $180 and the sales tax is 5%. Shari earns $4.50 per hour at her weekend job. How many hours will Shari have to work to buy the bicycle?   **15-6**

Factor.

2. $12c^2 - 8c - 4cd$ **3.** $25m^2n + 15mn^2 - 10mn$ **15-7**

Divide.

4. $\dfrac{18a^3 - 12a^2 + 6a}{6a}$ **5.** $\dfrac{a^2b^2 + 9ab^2 - 6a^2b}{ab}$ **15-8**

Chapter Review

polynomial (p. 668)
monomial (p. 668)
binomial (p. 668)
trinomial (p. 668)

standard form of a polynomial
(p. 668)
greatest common monomial factor
(p. 693)
factor a polynomial (p. 693)

Choose the correct term from the list above to complete each sentence.

1. A polynomial with two terms is a __?__.

2. When you write the terms of a polynomial in order from highest to lowest, you have written the __?__.

3. The GCF of the terms of a polynomial is the __?__ of the polynomial.

4. A trinomial is a __?__ with three terms.

Write each polynomial in standard form. *(Lesson 15-1)*

5. $4x + 7x^3 + 2x^2 + 6x^4 + 7$

6. $8c^2 + 4c^5 - 5c - 9c^3 + 2 + 11c^4$

7. $8 - 4a^3 + a^6 - 2a^4 - 3a$

8. $5m^7 - 2m^3 + 4m - 9m^2 + 6m^5$

Simplify. *(Lesson 15-1)*

9. $4x^3 + 6x^2 - x + 8x^2 + 3 + 2x$

10. $b^4 + 3b^3 - 6b^2 - 3b^3 + 7b^4 + b - 5$

11. $6c^2 + 3c^3 - 5c^2 + 4 + c^3 - 5c + 9$

12. $c^4 + 4c^2 - 2c^4 + c^2 - 9c^3 + 3c + 8$

13. $-3x^3 + 6x^2 - 7x - 2x^3 + 5 + 4x$

14. $-8a^2 - 3a^3 + 7a^2 + a^3 - 5a + 11$

Add. *(Lesson 15-2)*

15. $(3a^2 + 2a + 7) + (2a^2 + 5a + 9)$

16. $(x^3 + x^2 + x + 6) + (x^3 + 3x^2 + x + 1)$

17. $(n^5 + 4n^3 + 7n) + (2n^5 + 2n + 6)$

18. $(2d^4 + 5d^2 + d + 1) + (3d^2 + 6d + 3)$

19. $(7c^3 + 5c^2 - 4c - 6) + (-2c^3 - c + 1)$

20. $(n^4 - 2n^3 + n - 1) + (n^3 + 5n^2 - n + 4)$

Subtract. *(Lesson 15-3)*

21. $(7x^2 + 5x + 9) - (3x^2 + 2x + 4)$

22. $(5a^2 + 8a + 4) - (2a^2 + 4a + 5)$

23. $(8c^2 + 5c - 2) - (3c^2 - 7c + 5)$

24. $(-b^2 - 6b + 2) - (3b^2 + 3b - 6)$

25. $(5x^4 + 2x^3 - x + 6) - (2x^4 + 3x - 4)$

26. $(5a^3 - 6a^2 + 3) - (6a^3 - 2a^2 + 5a + 9)$

Chapter 15

Multiply. *(Lessons 15-4 and 15-5)*

27. $3a(2a^3 + 6a^2 - 8a + 7)$

28. $4x^2(-3x^2 - 7x + 5)$

29. $-6c(4c^4 + 3c^2 - 5c - 8)$

30. $-4n^3(7n^3 + 5n^2 - 8)$

31. $(x + 5)(2x + 3)$

32. $(4m - 7)(2m + 4)$

33. $(3x - 8)(2x - 5)$

34. $(4d + 3)(7d - 2)$

Solve by working backward. *(Lesson 15-6)*

35. Elvin Thompson bought three compact discs and two cassettes for a total of $66.73, including sales tax. Each cassette cost $8.99 and the sales tax was $3.78. How much did each compact disc cost?

36. Lisa Goldberg earns $12.50 per hour for a 40-hour week. She earns double time for overtime. Lisa earned $575 last week. How many hours did she work?

37. Ysabel Rodriguez wants to buy a new stove. She is able to save $35 per week for that purpose. The stove costs $700, and the sales tax rate is 5%. How many weeks will it take Ysabel to save enough money to buy the stove?

38. Mike Karam has a dental appointment at 10:30 A.M., and he wants to be there 10 min early. He needs 25 min to drive to the dentist's office and wants to allow another 5 min to find a parking space. At what time should Mike leave for his appointment?

Factor. *(Lesson 15-7)*

39. $4x^3 + 8x^2 + 6x$

40. $9n^4 - 6n^3 + 18n^2$

41. $20d^5 - 10d^4 + 15d^3 - 5d^2$

42. $18m^6 - 30m^4 + 12m^3 - 24m^2$

43. $3c^3d^3 + 5c^2d^2 - 4cd$

44. $4a^3b^2 - 8a^2b^3 + 2ab^2$

45. $25x^4y^3 + 10x^3y^4 - 15x^2y^2$

46. $12mn^3 - 18m^2n^2 + 24m^2n^3$

Divide. *(Lesson 15-8)*

47. $\dfrac{14x^4 + 8x^3 - 10x^2}{2x}$

48. $\dfrac{6c^5 - 9c^3 + 18c^2 - 3c}{3c}$

49. $\dfrac{18a^3b^5 + 21a^4b^3 - 6a^2b^2}{3ab}$

50. $\dfrac{16m^4n^6 - 8m^2n^5 + 12m^7n^4}{4m^2n^3}$

Write each polynomial in standard form.

1. $c^3 - 2c^2 + 6c^4 - 9c + 7$
2. $5x - 2x^3 + 7x^5 + 3 - 8x^2$ 15-1

Simplify.

3. $6b^3 + 7b^2 - 4b^3 + 5 - 11 + b$
4. $2c - 3c^4 + 2c^3 - 5c - 7c^4 - 7$
5. $5x^3 - 8x + 12x - 9x^3 + 6x - 3$
6. $-4m^3 - 6m^3 + 8m^2 - m + 3m^2 - 1$

Add.

7. $(3x^2 + 5x + 2) + (x^2 - 2x - 1)$
8. $(z^2 - z - 7) + (3z^2 + 2z - 8)$ 15-2
9. $(c^2 + 4c - 2) + (2c^2 - 5)$
10. $(3x^4 - 5x^3 + x) + (4x^4 + 2x^3 + 4)$

Subtract.

11. $(9b^2 + 6b - 5) - (3b^2 + 5b + 7)$
12. $(3x^2 - 4x - 2) - (5x^2 + 7x - 2)$ 15-3
13. $(8c^2 + 7c + 5) - (6c^2 - 7)$
14. $(n^3 - 4n^2 + n - 1) - (3n^3 - n + 2)$

Multiply.

15. $3b(7b^2 - 4b + 3)$
16. $-4n(2n^3 - 6n^2 + n - 8)$ 15-4
17. $(4x + 3)(3x + 5)$
18. $(5a - 8)(2a + 7)$ 15-5

19. Show how to use algebra tiles to give a geometric interpretation of $(x + 2)(x + 3)$.

20. Describe a method you could use to multiply $3n(2n + 5)(4n + 7)$. Then use your method to find the product.

Solve by working backward.

21. Sara Jones has to be at work at 8:30 A.M. She has a 35 min drive to the parking lot near her office and then a 10 min walk to her office. Sara wants to arrive 5 min early. What time should she leave home?   15-6

Factor.

22. $4x^3 + 8x^2 - 16x$
23. $12a^4 - 6a^3 + 24a^2$ 15-7
24. $10z^5 - 5z^3 + 15z^2$
25. $24c^6 - 18c^4 - 12c^3$

Divide.

26. $\dfrac{7d^3 - 21d^2 + 14d}{7d}$
27. $\dfrac{-18b^5 + 12b^4 + 6b^2}{6b^2}$ 15-8
28. $\dfrac{24x^3 - 18x^2 + 12x}{-6x}$
29. $\dfrac{8m^6 - 24m^5 + 32m^3}{-8m^2}$

Multiplying Binomials Mentally

Objective: To multiply binomials mentally.

You can find the products $(a + b)(a - b)$, $(a + b)^2$, and $(a - b)^2$ mentally if you learn the following patterns.

$$(a + b)(a - b) = a^2 - b^2$$
$$(a + b)^2 = a^2 + 2ab + b^2$$
$$(a - b)^2 = a^2 - 2ab + b^2$$

Example

Find each product mentally.

a. $(x + 3)(x - 3)$ **b.** $(m + 5)^2$ **c.** $(a - 4)^2$

Solution

a. $(x + 3)(x - 3) = x^2 - 3^2$ ⟵ $(a + b)(a - b) = a^2 - b^2$
$$= x^2 - 9$$

b. $(m + 5)^2 = m^2 + 2(m)(5) + 5^2$ ⟵ $(a + b)^2 = a^2 + 2ab + b^2$
$$= m^2 + 10m + 25$$

c. $(a - 4)^2 = a^2 - 2(a)(4) + 4^2$ ⟵ $(a - b)^2 = a^2 - 2ab + b^2$
$$= a^2 - 8a + 16$$

You can also use the patterns in reverse and work backward to factor certain polynomials as the product of two binomials.

Exercises

Find each product mentally.

1. $(a + 2)(a - 2)$ **2.** $(n + 7)(n - 7)$ **3.** $(x + 8)^2$

4. $(z + 6)^2$ **5.** $(c - 5)^2$ **6.** $(m - 12)^2$

7. $(2c + 9)(2c - 9)$ **8.** $(4b + 11)(4b - 11)$ **9.** $(3x + 4)^2$

10. $(7a + 3)^2$ **11.** $(5c - 6)^2$ **12.** $(9n - 7)^2$

13. $(2x + 3y)(2x - 3y)$ **14.** $(5m + 7n)(5m - 7n)$ **15.** $(3a + 5b)^2$

16. $(7x + 4y)^2$ **17.** $(3z - 8x)^2$ **18.** $(8m - 5n)^2$

Use the patterns to factor each polynomial as the product of two binomials.

19. $x^2 - 36$ **20.** $c^2 - 81$

21. $x^2 + 2x + 1$ **22.** $100a^2 - 9b^2$

23. $c^2 - 18c + 81$ **24.** $n^2 + 14n + 49$

25. $z^2 - 16z + 64$ **26.** $4m^2 - 49n^2$

Standardized Testing Practice

Choose the letter of the correct answer.

1.

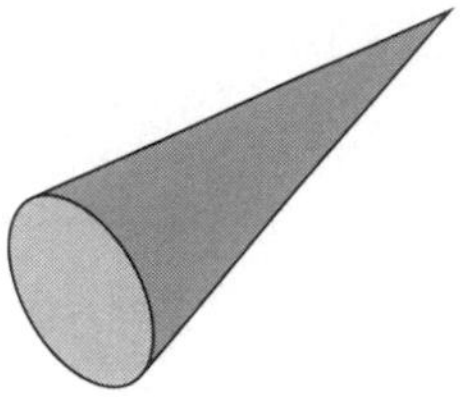

Identify the space figure.
A. cone   B. cylinder
C. sphere   D. pyramid

6.

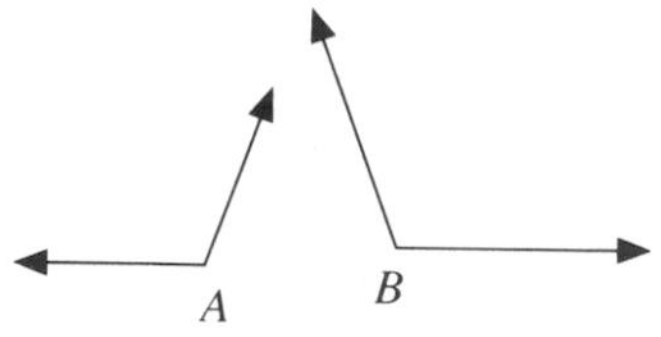

Complete: $\angle A$ and $\angle B$ are both __?__ angles.
A. adjacent   B. right
C. obtuse   D. acute

2. Write $7z^5 + 4z^9 - 6z^3 + 15$ in standard form.
A. $15 + 7z^5 - 6z^3 + 4z^9$
B. $4z^9 + 7z^5 - 6z^3 + 15$
C. $4z^9 - 6z^3 + 7z^5 + 15$
D. $15 - 6z^3 + 7z^5 + 4z^9$

7. Find the number of permutations of the letters in the word FINE.
A. 24   B. 256
C. 16   D. 120

3. 62.5% of what number is 105?
A. 62.5   B. 65.6
C. 168   D. 142.5

8. Find the sum: $6 + (-7) + 14 + (-8)$
A. 35   B. -5
C. 5   D. -35

4. Divide: $\dfrac{3x^4 + 12x^2 - 6x}{3x}$
A. $x^4 + 4x^2 - 2x$
B. $x^3 + 4x - 2$
C. $x^3 + 12x^2 - 6x$
D. $9x^5 + 36x^3 - 18x^2$

9. Subtract.
$(9y^3 - 3y - 5) - (2y^3 - y + 2)$
A. $7y^3 + 2y + 3$
B. $7y^3 - 2y - 7$
C. $7y^3 - 4y + 3$
D. $7y^3 + 2y + 7$

5. Solve for x: $y = kx$
A. $xy = k$ B. $x = \dfrac{k}{y}$
C. $x = \dfrac{y}{k}$ D. $x = ky$

10. Find the surface area of a cylinder with radius 5 mm and height 40 mm.
A. 1413 mm^2 B. 157 mm^2
C. 1256 mm^2 D. 3140 mm^2

11. **Find the product:** $(m + 7)(m - 4)$

A. $m^2 - 28$
B. $m^2 + 3m - 28$
C. $m^2 - 11m - 28$
D. $m^2 + 28$

12. **Find the mode of the data.**
18.1, 19.3, 19.7, 22.1, 21.5, 18.1

A. 19.5   B. 19.8
C. 4   D. 18.1

13. **Factor:** $18a^3b^4 + 27a^2b^2$

A. $9a^2b^2(2ab^2 + 3)$
B. $9(a^3b^4 + 4a^2b^2)$
C. $3ab(7a^2b^3 + 8ab)$
D. $18a^2b^2(ab^2 + 9)$

14. **The area of a square is 25 cm². The area of a parallelogram is 36 cm². Which of the following statements is true?**

I. One side of the square is 5 cm.
II. One side of the parallelogram is 6 cm.

A. I only   B. II only
C. I and II   D. Neither I nor II

15. **Find the volume of a sphere with a radius of 6 m.**

A. 452.16 m^3 B. 904.32 m^3
C. 75.36 m^3 D. not enough information

16. **Find the slope of the line:** $y = 3x - 7$

A. 3   B. $\frac{1}{3}$
C. 7   D. -7

17. **Find the prime factorization of 240.**

A. $2^4 \cdot 3 \cdot 5$ B. $2^3 \cdot 15$
C. $15 \cdot 16$ D. $2 \cdot 3 \cdot 5 \cdot 8$

18. **Write an equation for the sentence.**
Eight less than half a number x is five.

A. $8 - \frac{1}{2}x = 5$ B. $8 < \frac{1}{2}x + 5$
C. $\frac{1}{2}x - 8 = 5$ D. $8 < \frac{1}{2}x - 5$

19. **Daily Commuting Distances (Miles)**

What is the first quartile of the commuting distances?

A. 21   B. 32
C. 10   D. 1

20. **Find the surface area of a rectangular prism with length 5 in., height 3 in., and width 4 in.**

A. 54 in.^2 B. 45 in.^2
C. 94 in.^2 D. 74 in.^2

Looking Ahead

Transformations

Translations and Reflections

Objective: To find the image of a figure after a translation or a reflection.

Many patterns in fabric, art, tiling, and nature are the result of sliding or flipping a shape to various positions.

Movement of a figure that does not change its size and shape is called a **rigid motion.** The figure in the new location is called the **image** of the figure in the original position. *Translations* and *reflections* are rigid motions.

Terms to Know
- *rigid motion*
- *translation*
- *reflection*
- *image*

Sliding a figure from one location to another is called a **translation.**

Flipping a figure across a line is called a **reflection.**

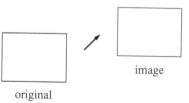

original   image

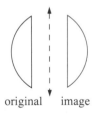

original   image

You can show rigid motions on a coordinate plane.

Example 1

The coordinates of the vertices of $\triangle ABC$ are $A(3, 4)$, $B(3, 1)$, and $C(7, 1)$. Give the coordinates of the image of each vertex after the triangle is translated 5 units to the left.

Solution

$A(3, 4) \longrightarrow A'(3 - 5, 4)$ or $A'(-2, 4)$
$B(3, 1) \longrightarrow B'(3 - 5, 1)$ or $B'(-2, 1)$
$C(7, 1) \longrightarrow C'(7 - 5, 1)$ or $C'(2, 1)$

☑ **Check Your Understanding**

1. Explain why the y-coordinates did not change.
2. Name a translation that would affect the y-coordinates.

You sometimes translate a figure both horizontally and vertically.

Example 2

A translation moves △ABC 2 units to the right and 4 units down. The image is △A′B′C′. Write the coordinates of each vertex of △A′B′C′.

Solution

$A(3, 4)$ ⟶ $A′(3 + 2, 4 − 4)$
or $A′(5, 0)$

$B(3, 1)$ ⟶ $B′(3 + 2, 1 − 4)$
or $B′(5, −3)$

$C(7, 1)$ ⟶ $C′(7 + 2, 1 − 4)$
or $C′(9, −3)$

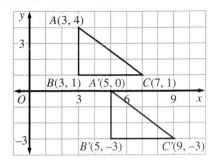

You can use a coordinate plane to show a reflection.

Example 3

Use △RST. Write the coordinates of the vertices after a reflection across the x-axis.

Solution

$R(1, 1)$ ⟶ $R′(1, −1)$
$S(3, 4)$ ⟶ $S′(3, −4)$
$T(6, 1)$ ⟶ $T′(6, −1)$

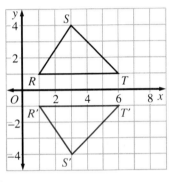

✔️ **Check Your Understanding**

3. How far is R above the x-axis? How far is R′ below the x-axis?

4. How far is S above the x-axis? How far is S′ below the x-axis?

5. If you folded the coordinate graph along the x-axis, would △RST and △R′S′T′ match exactly?

Exercises

Identify each rigid motion from A to B as a *reflection*, a *translation*, or *neither*.

1.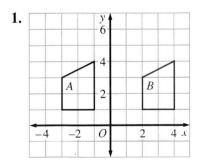

2.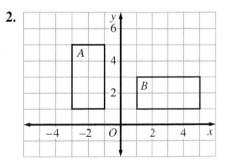

Identify each rigid motion from *A* to *B* as a *reflection*, a *translation*, or *neither*.

3.

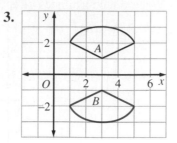

4.

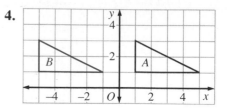

△PQR is the reflection of △ABC across the *y*-axis. Complete.

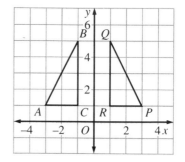

5. $A(-3, 1) \longrightarrow P(\underline{\ ?\ }, \underline{\ ?\ })$

6. $B(-1, 5) \longrightarrow Q(\underline{\ ?\ }, \underline{\ ?\ })$

7. $C(-1, 1) \longrightarrow R(\underline{\ ?\ }, \underline{\ ?\ })$

The vertices of rectangle △ABCD are A(1, 1), B(3, 1), C(3, 4), and D(1, 4).

　　a. Sketch the image of rectangle *ABCD* after each rigid motion.

　　b. State the coordinates of the images of *A*, *B*, *C*, and *D*.

8. a translation 1 unit to the right

9. a translation 3 units down

10. a translation 3 units to the left

11. a translation 2 units up

12. a reflection across the *y*-axis

13. a reflection across the *x*-axis

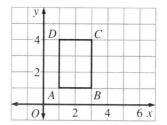

Copy each figure. Then sketch the image when the figure is reflected across the line.

14.

15.

16.

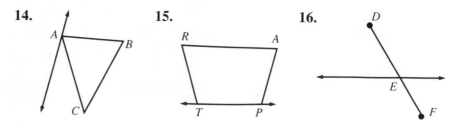

The coordinates of the vertices of $\triangle ABC$ are $A(-5, 1)$, $B(-1, 4)$, and $C(-2, 0)$. Write the coordinates of the images of the vertices after each rigid motion.

17. translation down 2 units, then to the right 4 units

18. translation to the right 3 units, then up 1 unit

19. reflection across the *y*-axis, then across the *x*-axis

20. reflection across the *x*-axis, then translation to the right 2 units

Write a description of the rigid motion that takes $\triangle DEF$ to $\triangle D'E'F'$.

21.   **22.**   **23.**

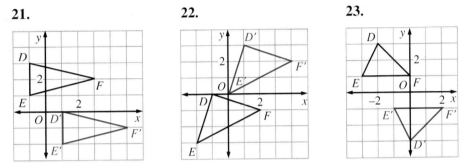

In Exercises 24–25, trace the figure and the two lines.
   a. Sketch the reflection of $\triangle ABC$ in line *m*. Call it $\triangle DEF$.
   b. Sketch the reflection of $\triangle DEF$ in line *n*. Call it $\triangle GHI$.
   c. Is the rigid motion from $\triangle ABC$ to $\triangle GHI$ a *translation* or a *reflection*?

24.   **25.**

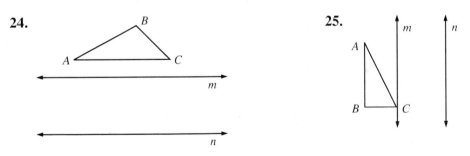

26. The coordinates of the vertices of $\triangle RST$ and the coordinates of the images of the vertices after a rigid motion are given below. Describe the rigid motion from $\triangle RST$ to $\triangle R'S'T'$.

$$R(-5, 1) \longrightarrow R'(5, 0)$$
$$S(1, 4) \longrightarrow S'(-1, 3)$$
$$T(-1, 1) \longrightarrow T'(1, 0)$$

Rotations and Symmetry

Objective: To find the image of a figure after a rotation and to find rotational symmetries.

Terms to Know
- *rotation*
- *center of rotation*
- *rotational symmetry*

Imagine that P and Q are points on a bicycle wheel. Suppose the wheel makes a quarter turn. Point P moves to P', and Q moves to Q'. P' and Q' are the images of P and Q. The quarter turn is called a **rotation** of $90°$ about O. Point O is the **center of rotation.** Notice that a point stays the same distance from the center in a rotation.

$$OP = OP' \qquad OQ = OQ'$$

Each point rotates through the same angle.

$$m\angle POP' = 90° \quad m\angle QOQ' = 90°$$

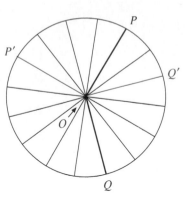

Example 1

Find the image of square $ABCD$ after a rotation of $45°$ counterclockwise about P.

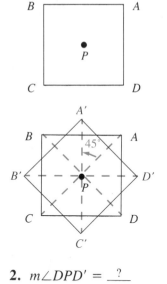

Solution

Rotate each vertex $45°$ counterclockwise to find its image. Connect the images. Square $A'B'C'D'$ is the image of square $ABCD$.

✓ Check Your Understanding

1. $m\angle APA' = \underline{\ ?\ }$
2. $m\angle DPD' = \underline{\ ?\ }$
3. If $PA = 3$, then $PA' = \underline{\ ?\ }$

In some rotations the figure rotates onto itself. We then say that the figure has **rotational symmetry.** In Example 1, if square $ABCD$ were rotated $90°$, the image would fit exactly on the original square. Square $ABCD$ has the following four rotational symmetries.

a rotation of $90°$ about P a rotation of $180°$ about P
a rotation of $270°$ about P a rotation of $360°$ about P

Of course, any figure has $360°$ rotational symmetry.

Example 2

How many rotational symmetries does the figure have?

B

● P

C   A

Solution

The image matches the original figure after rotations of 120°, 240°, and 360° about *P*. There are three rotational symmetries.

B

120° ● P 120°
120°

C   A

✓ **Check Your Understanding**

Give the image of *A* after each counterclockwise rotation about *P*.

4. 120°   **5.** 240°   **6.** 360°   **7.** 480°

A rotation is another kind of rigid motion. The image is the same size and shape as the original figure.

Exercises

Find the image of *A* after each counterclockwise rotation about *P*.

1. 120°   **2.** 240°

3. 600°   **4.** 360°

5. How many rotational symmetries does the snowflake at the right have?

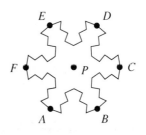

Draw the image of each figure after the given rotation about *P*.

6. 90°; counterclockwise

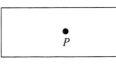

7. 270°; counterclockwise

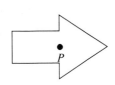

Draw the image of each figure after the given rotation about *P*.

8. 45°; counterclockwise

9. 180°; counterclockwise

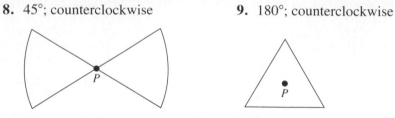

Name all rotational symmetries for each figure.

10.

11.

12.

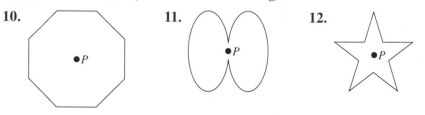

Identify the rigid motion as a *translation*, a *reflection*, or a *rotation*.

13.   14.   15.

16.   17.

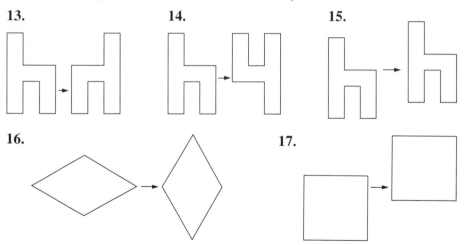

The vertices of a polygon are given below.
 a. **Plot the vertices on graph paper. Draw the polygon.**
 b. **Draw the image of the polygon after the given counterclockwise rotation.**
 c. **Give the coordinates of the image of each vertex.**

18. $A(-5, 6)$, $B(-2, 6)$, $C(-4, 1)$
  90° about the origin

19. $A(2, 2)$, $B(5, 2)$, $C(5, 7)$, $D(2, 7)$
  270° about D

20. $A(4, 4)$, $B(-4, 4)$, $C(-4, -4)$, $D(4, -4)$
  45° about the origin

Enlargements and Reductions

Objective: To find the image of a figure after a dilation.

Translations, reflections, and rotations are rigid motions; the image and the original are congruent. The image of a figure after a **dilation** is similar to the original. If the image is larger than the original, the dilation is an **enlargement.** If the image is smaller, the dilation is a **reduction.** Because the original figure and its image are similar, corresponding angles are congruent and corresponding sides are in proportion.

Suppose the image of a rectangle $ABCD$ is twice as large as the original. You can say that $A'B' = 2 \cdot AB$. This notation means *the length of segment $\overline{A'B'}$ is two times the length of segment $\overline{AB}$.*

Terms to Know

- *dilation*
- *enlargement*
- *reduction*
- *geometric transformation*

Example 1

Enlarge $\triangle ABC$ so that each side of the image is twice as long as in the original triangle.

Solution

Draw rays through A, B, and C from a point P outside the triangle.
Find A' on $\overrightarrow{PA}$ so that $PA' = 2 \cdot PA$.
Find B' on $\overrightarrow{PB}$ so that $PB' = 2 \cdot PB$.
Find C' on $\overrightarrow{PC}$ so that $PC' = 2 \cdot PC$.
Draw $\triangle A'B'C'$.
$\triangle A'B'C'$ is the image of $\triangle ABC$.

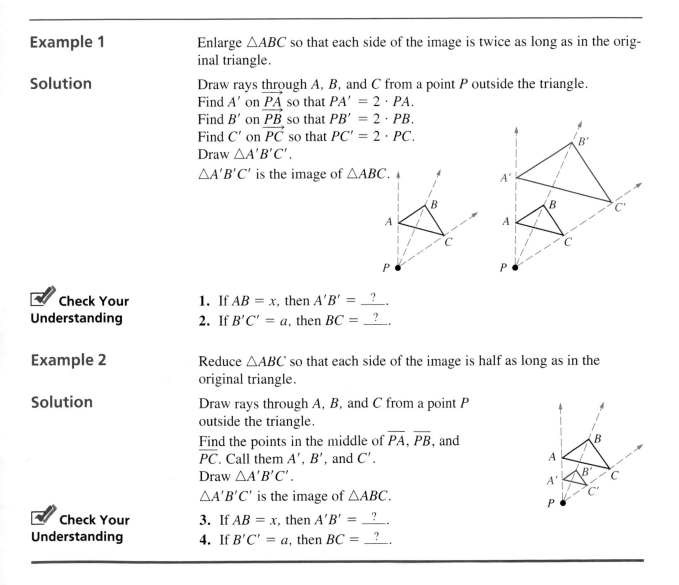

✓ Check Your Understanding

1. If $AB = x$, then $A'B' = \underline{\quad?\quad}$.
2. If $B'C' = a$, then $BC = \underline{\quad?\quad}$.

Example 2

Reduce $\triangle ABC$ so that each side of the image is half as long as in the original triangle.

Solution

Draw rays through A, B, and C from a point P outside the triangle.
Find the points in the middle of $\overline{PA}$, $\overline{PB}$, and $\overline{PC}$. Call them A', B', and C'.
Draw $\triangle A'B'C'$.
$\triangle A'B'C'$ is the image of $\triangle ABC$.

✓ Check Your Understanding

3. If $AB = x$, then $A'B' = \underline{\quad?\quad}$.
4. If $B'C' = a$, then $BC = \underline{\quad?\quad}$.

Point P in Examples 1 and 2 is called the *center* of the dilation. To enlarge a more complex figure, you may let P be a vertex of the original figure rather than a point outside the figure.

Example 3

Using vertex D as the center, enlarge quadrilateral $ABCD$ so that each side of the image is triple the length of the original.

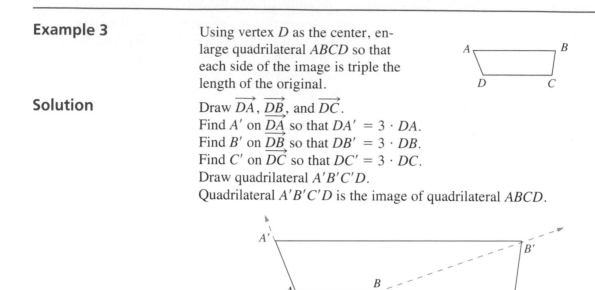

Solution

Draw $\overrightarrow{DA}$, $\overrightarrow{DB}$, and $\overrightarrow{DC}$.
Find A' on $\overrightarrow{DA}$ so that $DA' = 3 \cdot DA$.
Find B' on $\overrightarrow{DB}$ so that $DB' = 3 \cdot DB$.
Find C' on $\overrightarrow{DC}$ so that $DC' = 3 \cdot DC$.
Draw quadrilateral $A'B'C'D$.
Quadrilateral $A'B'C'D$ is the image of quadrilateral $ABCD$.

 **Check Your Understanding**

5. Compare the corresponding angles of $ABCD$ and $A'B'C'D$.

6. Compare $A'B'$ and AB. Compare $B'C'$ and BC.

When the position, size, or shape of a figure is changed according to a given rule, it is called a **geometric transformation.** Translations, reflections, rotations, and dilations are examples of transformations.

Exercises

Trace each figure. Enlarge it so that the sides of the image are twice as long.

1.

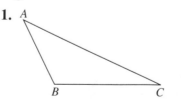

Use a center outside $\triangle ABC$.

2.

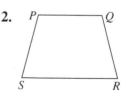

Use R as the center.

3. Reduce △*ABC* in Exercise 1 so that the sides of the image are half as long. Use a center outside △*ABC*.

4. Reduce quadrilateral *PQRS* in Exercise 2 so that the sides of the image are half as long. Use point *R* as the center.

Trace each figure. Enlarge or reduce the figure as indicated.

5.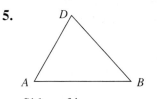

Sides of image are
3 times as long.

6.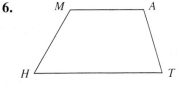

Sides of image are
3 times as long

7.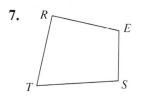

Sides of image are
$\frac{1}{2}$ times as long.

8.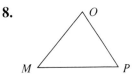

Sides of image are
$2\frac{1}{2}$ times as long.

9. Trace the figure. Enlarge the figure so that each side of the image is twice as long. Use a point *P* inside the polygon as center.

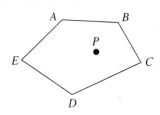

10. Trace the figure. Then reduce it so that each side of the image is one fourth as long.

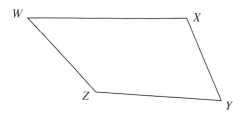

Sine, Cosine, and Tangent

Objective: To find the sine, cosine, and tangent of acute angles of a right triangle.

In the figure at the right, $\triangle ABC$ is a right triangle. In relation to $\angle A$, $\overline{BC}$ is called the **opposite side** and $\overline{AC}$ is called the **adjacent side.** $\overline{AB}$ is the hypotenuse.

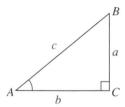

Terms to Know
- *opposite side*
- *adjacent side*
- *trigonometric ratio*
- *sine*
- *cosine*
- *tangent*

The following **trigonometric ratios** are defined for an acute angle A of a right triangle.

sine of $\angle A$: $\sin A = \dfrac{\text{length of side opposite } \angle A}{\text{length of hypotenuse}} = \dfrac{a}{c}$

cosine of $\angle A$: $\cos A = \dfrac{\text{length of side adjacent to } \angle A}{\text{length of hypotenuse}} = \dfrac{b}{c}$

tangent of $\angle A$: $\tan A = \dfrac{\text{length of side opposite } \angle A}{\text{length of side adjacent to } \angle A} = \dfrac{a}{b}$

Sometimes it is useful to use these shortened forms of the definitions.

$$\sin A = \frac{\text{opposite}}{\text{hypotenuse}} \qquad \cos A = \frac{\text{adjacent}}{\text{hypotenuse}} \qquad \tan A = \frac{\text{opposite}}{\text{adjacent}}$$

Notice that $\overline{AC}$ is the side opposite $\angle B$ and $\overline{BC}$ is the side adjacent to $\angle B$.

Example 1

Use the triangle at the right. Find each ratio in lowest terms.

a. $\tan A$ **b.** $\sin B$
c. $\cos A$ **d.** $\cos B$

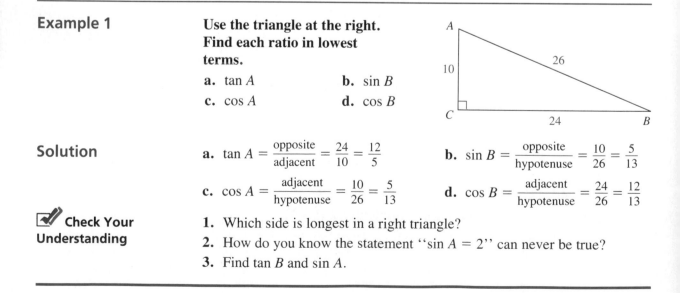

Solution

a. $\tan A = \dfrac{\text{opposite}}{\text{adjacent}} = \dfrac{24}{10} = \dfrac{12}{5}$ **b.** $\sin B = \dfrac{\text{opposite}}{\text{hypotenuse}} = \dfrac{10}{26} = \dfrac{5}{13}$

c. $\cos A = \dfrac{\text{adjacent}}{\text{hypotenuse}} = \dfrac{10}{26} = \dfrac{5}{13}$ **d.** $\cos B = \dfrac{\text{adjacent}}{\text{hypotenuse}} = \dfrac{24}{26} = \dfrac{12}{13}$

✓ **Check Your Understanding**

1. Which side is longest in a right triangle?
2. How do you know the statement "$\sin A = 2$" can never be true?
3. Find $\tan B$ and $\sin A$.

Example 2

Use the triangle at the right. Find each ratio in lowest terms.

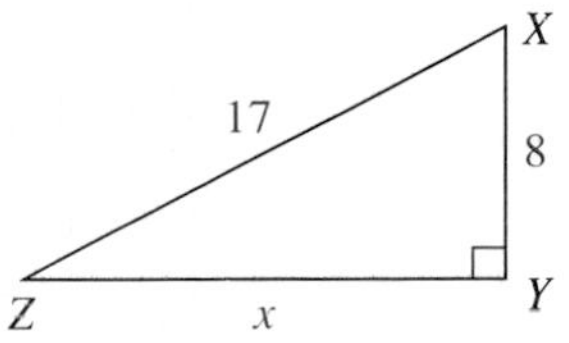

a. $\sin X$

b. $\cos Z$

c. $\tan X$

Solution

First use the Pythagorean theorem to find the length of $\overline{YZ}$.

$$x^2 + 8^2 = 17^2$$
$$x^2 + 64 = 289$$
$$x^2 = 225$$
$$x = \sqrt{225}$$
$$x = 15$$

a. $\sin X = \dfrac{\text{opposite}}{\text{hypotenuse}} = \dfrac{15}{17}$

b. $\cos Z = \dfrac{\text{adjacent}}{\text{hypotenuse}} = \dfrac{15}{17}$

c. $\tan X = \dfrac{\text{opposite}}{\text{adjacent}} = \dfrac{15}{8}$

☑ **Check Your Understanding**

4. In relation to $\angle Z$, what is the length of the opposite side? What is the length of the adjacent side? What is $\tan Z$?

5. Explain how $x^2 = 225$ is obtained from $x^2 + 64 = 289$.

6. The side opposite $\angle Z$ is the side ___?___ to $\angle X$.

Exercises

Replace each ___?___ with the correct word.

1. a. $\sin A = \dfrac{\text{opposite}}{?}$ b. $\cos A = \dfrac{?}{?}$ c. $\tan A = \dfrac{?}{?}$

2. Explain the meaning of the word *adjacent* in the following sentence.

The house adjacent to mine was recently painted.

Find the side indicated in $\triangle DEF$.

3. the hypotenuse

4. the side opposite $\angle D$

5. the side adjacent to $\angle D$

6. the side opposite $\angle E$

7. the side adjacent to $\angle E$

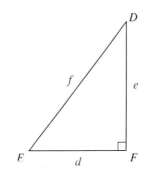

Use the triangle at the right.
Find each ratio in lowest terms.

8. $\sin A$ **9.** $\cos A$

10. $\tan A$ **11.** $\sin B$

12. $\cos B$ **13.** $\tan B$

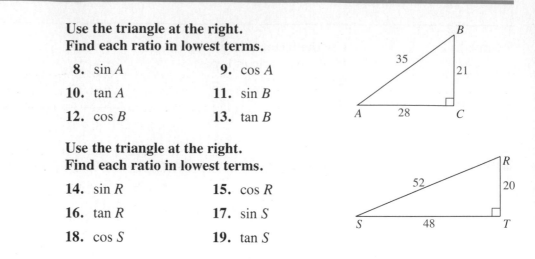

Use the triangle at the right.
Find each ratio in lowest terms.

14. $\sin R$ **15.** $\cos R$

16. $\tan R$ **17.** $\sin S$

18. $\cos S$ **19.** $\tan S$

For each right triangle, find the value of x. Then find $\sin A$, $\cos A$, and $\tan A$. Write each ratio in lowest terms.

20.   **21.**   **22.**

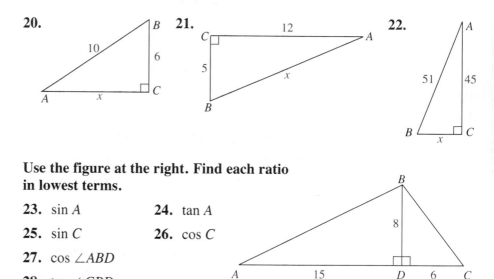

Use the figure at the right. Find each ratio in lowest terms.

23. $\sin A$ **24.** $\tan A$

25. $\sin C$ **26.** $\cos C$

27. $\cos \angle ABD$

28. $\tan \angle CBD$

29. Is triangle ABC in Exercises 23–28 a right triangle? Explain how you know.

30. K is an acute angle of right triangle JKL and $\tan K = 1$. Give the measures of the angles of $\triangle JKL$.

31. Use the triangles in Exercises 20 and 21. Find the value of $(\sin A)^2 + (\cos A)^2$ for each triangle.

32. A and B are acute angles of right triangle ABC.
   a. What is the relationship between $\sin A$ and $\cos B$? Explain your answer.
   b. What is the value of $(\tan A)(\tan B)$? Explain your answer.

Using Trigonometry

Objective: To use trigonometric ratios to find the lengths of the sides of a right triangle.

The table on page 720 lists the sines, cosines, and tangents of angles with measures from 1° to 89°. To find the value of sin 30°, locate 30° in the *Angle* column. Read the number across from 30 in the *Sine* column: .5000.

$$\sin 30° = 0.5000$$

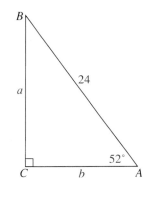

If you have a scientific calculator, you can find sin 30° by using this key sequence.

You use the TAN and COS keys in a similar manner to find the tangent and cosine of an angle.

You can use trigonometric ratios to find the lengths of the sides of a right triangle.

Example 1

Solution

Find a and b. Round to the nearest tenth.

a represents the length of the side opposite $\angle A$. The length of the hypotenuse is given. Use the sine ratio.

$$\sin A = \frac{\text{opposite}}{\text{hypotenuse}} = \frac{a}{24}$$

$$\sin 52° = \frac{a}{24}$$

$$0.7880 \approx \frac{a}{24}$$

$$24(0.7880) \approx 24\left(\frac{a}{24}\right)$$

$$18.912 \approx a$$

$$a = 18.9 \text{ to the nearest tenth}$$

b represents the length of the side adjacent to $\angle A$. Use the cosine ratio.

$$\cos A = \frac{\text{adjacent}}{\text{hypotenuse}} = \frac{b}{24}$$

$$\cos 52° = \frac{b}{24}$$

$$0.6157 \approx \frac{b}{24}$$

$$24(0.6157) \approx 24\left(\frac{b}{24}\right)$$

$$14.7768 \approx b$$

$$b = 14.8 \text{ to the nearest tenth}$$

✓ **Check Your Understanding**

1. Use the Pythagorean theorem to check the lengths.
2. How would you find a and b if you were given the measure of $\angle B$?

Table of Trigonometric Ratios

ANGLE	SINE	COSINE	TANGENT	ANGLE	SINE	COSINE	TANGENT
1°	.0175	.9998	.0175	46°	.7193	.6947	1.0355
2°	.0349	.9994	.0349	47°	.7314	.6820	1.0724
3°	.0523	.9986	.0524	48°	.7431	.6691	1.1106
4°	.0698	.9976	.0699	49°	.7547	.6561	1.1504
5°	.0872	.9962	.0875	50°	.7660	.6428	1.1918
6°	.1045	.9945	.1051	51°	.7771	.6293	1.2349
7°	.1219	.9925	.1228	52°	.7880	.6157	1.2799
8°	.1392	.9903	.1405	53°	.7986	.6018	1.3270
9°	.1564	.9877	.1584	54°	.8090	.5878	1.3764
10°	.1736	.9848	.1763	55°	.8192	.5736	1.4281
11°	.1908	.9816	.1944	56°	.8290	.5592	1.4826
12°	.2079	.9781	.2126	57°	.8387	.5446	1.5399
13°	.2250	.9744	.2309	58°	.8480	.5299	1.6003
14°	.2419	.9703	.2493	59°	.8572	.5150	1.6643
15°	.2588	.9659	.2679	60°	.8660	.5000	1.7321
16°	.2756	.9613	.2867	61°	.8746	.4848	1.8040
17°	.2924	.9563	.3057	62°	.8829	.4695	1.8807
18°	.3090	.9511	.3249	63°	.8910	.4540	1.9626
19°	.3256	.9455	.3443	64°	.8988	.4384	2.0503
20°	.3420	.9397	.3640	65°	.9063	.4226	2.1445
21°	.3584	.9336	.3839	66°	.9135	.4067	2.2460
22°	.3746	.9272	.4040	67°	.9205	.3907	2.3559
23°	.3907	.9205	.4245	68°	.9272	.3746	2.4751
24°	.4067	.9135	.4452	69°	.9336	.3584	2.6051
25°	.4226	.9063	.4663	70°	.9397	.3420	2.7475
26°	.4384	.8988	.4877	71°	.9455	.3256	2.9042
27°	.4540	.8910	.5095	72°	.9511	.3090	3.0777
28°	.4695	.8829	.5317	73°	.9563	.2924	3.2709
29°	.4848	.8746	.5543	74°	.9613	.2756	3.4874
30°	.5000	.8660	.5774	75°	.9659	.2588	3.7321
31°	.5150	.8572	.6009	76°	.9703	.2419	4.0108
32°	.5299	.8480	.6249	77°	.9744	.2250	4.3315
33°	.5446	.8387	.6494	78°	.9781	.2079	4.7046
34°	.5592	.8290	.6745	79°	.9816	.1908	5.1446
35°	.5736	.8192	.7002	80°	.9848	.1736	5.6713
36°	.5878	.8090	.7265	81°	.9877	.1564	6.3138
37°	.6018	.7986	.7536	82°	.9903	.1392	7.1154
38°	.6157	.7880	.7813	83°	.9925	.1219	8.1443
39°	.6293	.7771	.8098	84°	.9945	.1045	9.5144
40°	.6428	.7660	.8391	85°	.9962	.0872	11.4301
41°	.6561	.7547	.8693	86°	.9976	.0698	14.3007
42°	.6691	.7431	.9004	87°	.9986	.0523	19.0811
43°	.6820	.7314	.9325	88°	.9994	.0349	28.6363
44°	.6947	.7193	.9657	89°	.9998	.0175	57.2900
45°	.7071	.7071	1.0000				

Example 2

Find w to the nearest tenth.

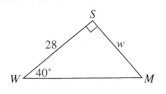

Solution

w represents the length of the side opposite $\angle W$. The length of the adjacent side is given. Use the tangent ratio.

$$\tan W = \frac{\text{opposite}}{\text{adjacent}} = \frac{w}{28}$$

$$\tan 40° = \frac{w}{28}$$

$$0.8391 \approx \frac{w}{28}$$

$$28(0.8391) \approx 28\left(\frac{w}{28}\right)$$

$$23.4948 \approx w$$

$$w = 23.5 \text{ to the nearest tenth}$$

✓ Check Your Understanding

3. Explain how to round 23.4948 to the nearest tenth.

4. Find w if $m\angle W$ had been 60°.

Exercises

Find the value of each trigonometric ratio in the table.

1. $\sin 5°$ **2.** $\tan 59°$ **3.** $\cos 75°$

4. $\cos 45°$ **5.** $\tan 80°$ **6.** $\sin 68°$

State whether you would use the *sine*, *cosine*, or *tangent* to find x. Do not solve.

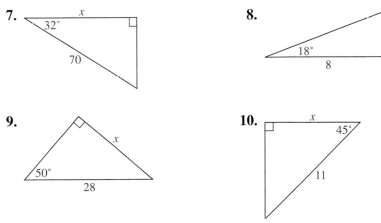

7.

8.

9.

10.

Solve for x. Round to the nearest tenth.

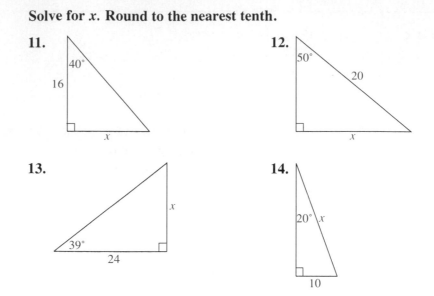

11.

40°
16
x

12.

50°
20
x

13.

x
39°
24

14.

20° x
10

Find the measures of all the sides and angles of each triangle.

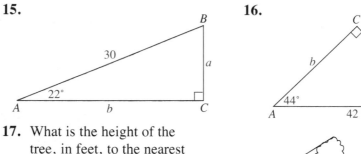

15.

B
30
a
22°
A   b   C

16.

C
b   a
44°
A   42   B

17. What is the height of the tree, in feet, to the nearest tenth?

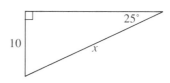

32°
40 ft

18. A cable 45ft long stretches from the top of an antenna to the ground. The cable forms an angle of 50° with the ground. What is the height of the antenna, in feet, to the nearest tenth?

19. Find x to the nearest tenth.

25°
10
x

20. The two congruent sides of an isosceles triangle measure 30 cm each. The two congruent angles measure 50° each. Find the measure of the third side. (*Hint:* Sketch the triangle and draw the height to the third side.)

Finding Angle Measures

Objective: To use trigonometric ratios to find the measure of an angle of a right triangle.

You can use trigonometry to find the measure of an angle in a right triangle. You must know the lengths of two sides.

Example 1

Find the measure of each acute angle to the nearest degree.

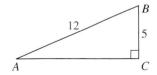

Solution

For $\angle A$, the opposite side and the hypotenuse are given.
First find the sine ratio.

$$\sin A = \frac{\text{opposite}}{\text{hypotenuse}} = \frac{5}{12} \qquad \leftarrow \text{Write } \tfrac{5}{12} \text{ as a decimal.}$$

$$\sin A \approx 0.4167$$

Then use the table on page 720. Look in the *Sine* column.
Find the closest entry to 0.4167.
$\sin 24° \approx 0.4067$, $\sin 25° \approx 0.4226$

Angle	Sine	
24°	0.4067	difference: $0.4167 - 0.4067 = 0.0100$
A	0.4167	
25°	0.4226	difference: $0.4226 - 0.4167 = 0.0059$

0.4167 is closer to 0.4226 than it is to 0.4067.
To the nearest degree, $m\angle A = 25°$.

Find the measure of $\angle B$.

$$m\angle A + m\angle B + m\angle C = 180°$$
$$25° + m\angle B + 90° = 180°$$
$$115° + m\angle B = 180°$$
$$115° + m\angle B - 115° = 180° - 115°$$
$$m\angle B = 65°$$

 **Check Your Understanding**

1. Explain why 0.4167 is closer to 0.4226 than it is to 0.4067.
2. When finding $m\angle B$ in Example 1, why does the right side of the equation equal 180°?

If you have a scientific calculator, you can find the measure of an acute angle in a triangle given two sides. To find $m\angle A$ in Example 1, you can use this key sequence if your calculator has an [INV] key.

Example 2

Find the measure of each acute angle to the nearest degree.
Find *AB* to the nearest tenth.

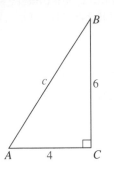

Solution

For $\angle A$, the opposite and adjacent sides are given.
First find the tangent ratio.

$$\tan A = \frac{\text{opposite}}{\text{adjacent}} = \frac{6}{4} = 1.5000$$

Then use the table on page 720.

$\tan 56° \approx 1.4826$ $\tan A = 1.5000$ $\tan 57° \approx 1.5399$

1.5000 is closer to 1.4826 than it is to 1.5399.
To the nearest degree, $m\angle A = 56°$.

Find $m\angle B$.

$$56° + m\angle B + 90° = 180°$$
$$146° + m\angle B = 180°$$
$$m\angle B = 34°$$

To find the length of the hypotenuse, $\overline{AB}$, use the Pythagorean theorem.

$$a^2 + b^2 = c^2$$
$$6^2 + 4^2 = c^2$$
$$36 + 16 = c^2$$
$$52 = c^2$$
$$c = \sqrt{52} \approx 7.211$$

$AB = 7.2$ to the nearest tenth

✓ **Check Your Understanding**

3. Explain why 1.5000 is closer to 1.4826 than it is to 1.5399.
4. If $BC = 12$ and $AC = 8$, what would be the measure of $\angle A$? $\angle B$? $\overline{AB}$?

Exercises

1. Sketch right triangle *SDW* with the right angle at *D*. To find $\tan W$, which two sides must have their lengths given?

2. How would you find the measure of an acute angle of a right triangle if the measure of the other acute angle is given?

Find $m\angle A$ to the nearest degree.

3. $\sin A = 0.3240$ **4.** $\cos A = 0.8215$ **5.** $\tan A = 1.4400$

Find the measure of each acute angle to the nearest degree.

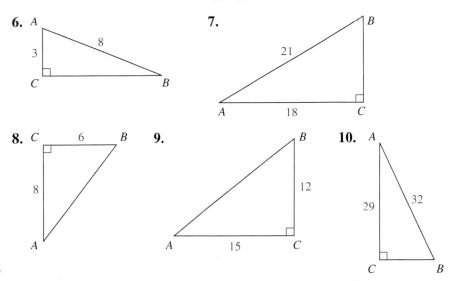

Find the measure of each acute angle to the nearest degree. Find the length of the third side to the nearest tenth.

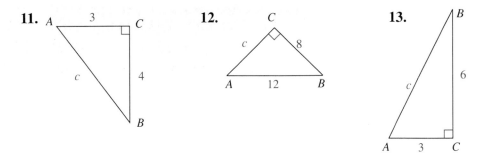

14. A 25 ft ladder is leaning against a building. The distance from the foot of the ladder to the building is 7 ft. Find the measure of the angle that the ladder makes with the ground to the nearest degree.

15. Find the measures of all three angles of isosceles triangle TRI to the nearest degree. (*Hint*: Draw the height from I to $\overline{TR}$.)

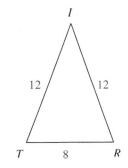

Special Right Triangles

Objective: To apply 45°-45°-90° and 30°-60°-90° right triangles.

APPLICATION

There are two special right triangles. They are often used by construction and skilled trades workers, architects, draftspersons, and engineers.

One special triangle is the *isosceles right triangle*. In an isosceles right triangle, the two legs are equal. Also, the acute angles are equal in measure and each angle measures 45°. An isosceles right triangle is often called a **45°-45°-90° right triangle.**

In the 45°-45°-90° right triangle shown, each leg is 1 unit long. Find the length of the hypotenuse by using the Pythagorean Theorem.

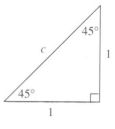

$$c^2 = 1^2 + 1^2$$
$$c^2 = 2$$
$$c = \sqrt{2}$$

The hypotenuse is $\sqrt{2}$ units long.

If each leg of a 45°-45°-90° right triangle has length 3, then the hypotenuse has length $3\sqrt{2}$. This is because all 45°-45°-90° right triangles are similar, and corresponding sides of similar triangles have the same ratio.

45°-45°-90° Right Triangle Property

If each leg of a 45°-45°-90° right triangle has length n, then the hypotenuse has length $n\sqrt{2}$.

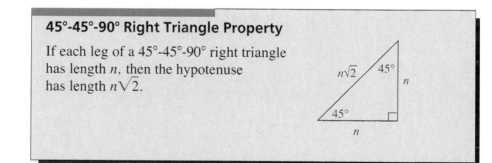

Example 1

Find the lengths of the missing sides. Round decimal answers to the nearest tenth.

Solution

$\triangle DEF$ is a 45°-45°-90° right triangle.
The legs are the same length.
Since $FE = 8$, $DE = 8$.
$\overline{FD}$ is the hypotenuse.
By the 45°-45°-90° right triangle property,
$FD = 8\sqrt{2} \approx 8(1.414) = 11.312$.
$FD = 11.3$ to the nearest tenth

1. Is $8\sqrt{2}$ the *exact length* or the *approximate length* of the hypotenuse?
2. If $FE = 14$, what are the lengths of the other two sides?
3. If $DF = 6\sqrt{2}$, what are the lengths of the other two sides?

You can use a calculator to find the length of a hypotenuse if you know the length of the shorter leg. To find FD in Example 1, use this key sequence.

$\triangle WXZ$ is an equilateral triangle. Each angle measures 60°. The lengths of the sides are equal. $\triangle WXY$ is half the size of $\triangle WXZ$. $\triangle WXY$ is a **30°-60°-90° right triangle.** If $XY = 1$, then $XW = 2$. You can find WY by using the Pythagorean theorem.

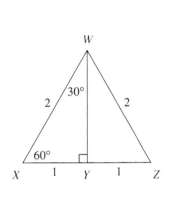

$$(XY)^2 + (WY)^2 = (XW)^2$$
$$1^2 + (WY)^2 = 2^2$$
$$1 + (WY)^2 = 4$$
$$(WY)^2 = 3$$
$$WY = \sqrt{3}$$

For this 30°-60°-90° right triangle we observe that the hypotenuse is twice as long as the shorter leg, and the longer leg is the length of the shorter leg times $\sqrt{3}$.

All 30°-60°-90° right triangles are similar, so corresponding sides have the same ratio.

30°-60°-90° Right Triangle Property

If the shorter leg of a 30°-60°-90° right triangle has length x, then the hypotenuse has length $2x$, and the longer leg has length $x\sqrt{3}$.

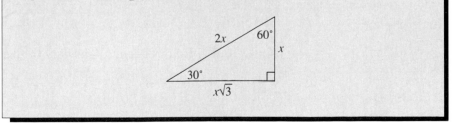

The shorter leg is always opposite the 30° angle. The longer leg is always opposite the 60° angle.

Example 2

Find a and b. Round decimal answers to the nearest tenth.

Solution

This is a 30°-60°-90° right triangle. The length of the hypotenuse is twice the length of the shorter leg.

a is the length of the side opposite the 30° angle.
$$2a = 12$$
$$a = 6$$

b is the length of the side opposite the 60° angle.
$$b = a\sqrt{3}$$
$$b = 6\sqrt{3} \approx 10.392305$$
$$b = 10.4 \text{ to the nearest tenth}$$

☑ **Check Your Understanding**

4. If the hypotenuse were 20, what would be the values of a and b?
5. If $a = 100$, what would be the value of b? What would be the length of the hypotenuse?

Exercises

Complete.

1. If each leg of a 45°-45°-90° right triangle has length x, then the hypotenuse has length ___?___.

2. In a 30°-60°-90° right triangle, the shorter leg is opposite the ___?___ angle, and the longer leg is opposite the ___?___ angle.

3. If the shorter leg of a 30°-60°-90° right triangle has length b, then the ___?___ has length $b\sqrt{3}$, and the ___?___ has length $2b$.

Find a decimal approximation to the nearest tenth.

4. $5\sqrt{2}$ 5. $6\sqrt{3}$ 6. $12\sqrt{3}$ 7. $10\sqrt{2}$

Find x and y. Round decimal answers to the nearest tenth.

8.   9.   10.

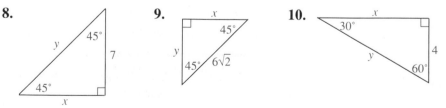

Find *x* and *y*. Round decimal answers to the nearest tenth.

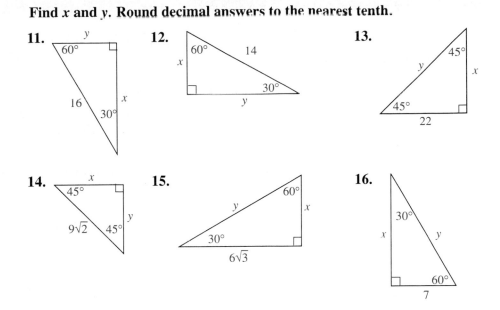

11.

12.

13.

14.

15.

16.

Find the length marked *x* or *y*. Round decimal answers to the nearest tenth.

17. *ABC* is an equilateral triangle.

18. *ABCD* is a square.

19. *ABCD* is a rectangle.

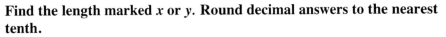

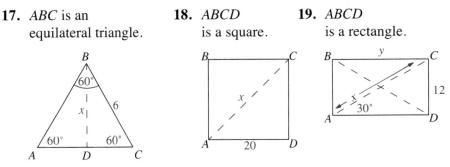

Find the lengths. Round decimal answers to the nearest tenth.

20. A leg of an isosceles right triangle has length 24. Find the lengths of the other two sides.

21. The shorter leg of a 30°-60°-90° right triangle is 40 cm long. How long are the other two sides?

22. The hypotenuse of a 45°-45°-90° right triangle has length $16\sqrt{2}$. How long are the other sides?

23. a. In △ *ABC* at the right, find *w*, *x*, *y*, and *z*.
   b. What is the area of △ *ABC*?

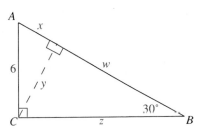

Extra Practice

Chapter 1

Evaluate each expression when $u = 6$, $v = 2$, and $w = 3$.

1. $3 + w$ **2.** $6v$ **3.** $17 - u$ **4.** $9 \div w$ 1-1

5. vw **6.** $u - w$ **7.** $u \div w$ **8.** $v + u$

Evaluate each expression when $q = 76$, $r = 47.5$, $s = 103.72$, and $t = 12.4$.

9. $114 + s$ **10.** $92 + q$ **11.** $r + 12$ **12.** $22 + t$ 1-2

13. $251 - t$ **14.** $s - r$ **15.** $q - t$ **16.** $q - r$

Evaluate each expression when $a = 24$, $b = 18.41$, $c = 7.7$, and $d = 5.2$.

17. $92b$ **18.** $10.4d$ **19.** $22a$ **20.** $61c$ 1-3

21. $20.8 \div d$ **22.** $b \div 7$ **23.** $a \div 0.12$ **24.** $b \div 21.04$

Use this paragraph for Exercises 25–28.

Chris earns $15 per hour at a chemical factory, where the standard work day is 8 h. He earns double his hourly wage for work on Sundays and holidays. Beginning Monday, July 4, he works 5 days, including July 4.

25. What is the paragraph about?   1-4

26. How many days did Chris work the week of July 4?

27. How many hours did he work during that week?

28. What facts would not be used to find the wages Chris took home that week?

Replace each ___?___ with >, <, or =.

29. 7323 ___?___ 4619 **30.** $14,135$ ___?___ $14,137$ **31.** 5.72 ___?___ 5.720 1-5

32. 110 ___?___ 11.3 **33.** 9.768 ___?___ 10.315 **34.** 0.93 ___?___ 0.94

Use the properties to find each sum mentally.

35. $37 + 21 + 313$ **36.** $45 + 9 + 1025$ **37.** $5.6 + 13 + 7.4$ 1-6

38. $10.7 + 46 + 3.3$ **39.** $97 + 0$ **40.** $83 + 2 + 17$

41. $139 + 4.8 + 91$ **42.** $2.4 + 3 + 0$ **43.** $786 + 12.7 + 94$

Use the properties to find each product mentally.

44. $5 \cdot 13 \cdot 2$ **45.** $20(7)(5)$ **46.** $98 \cdot 17 \cdot 0$ **47.** $3 \cdot 3 \cdot 1$ 1-7

48. $(96)(2)(0.5)$ **49.** $8(5)(0)$ **50.** $1 \cdot 9063 \cdot 1$ **51.** $10(2.2)(5)$

Tell whether it is most efficient to use *mental math*, *paper and pencil*, **or a** *calculator*. **Then find each answer using the method that you chose.**

52. $168 - 97$ **53.** $185 - 45$ **54.** $342 + 977$ **55.** $16.7 + 13.9 + 55.8$ 1-8

56. $83 \cdot 42$ **57.** $63.6 \div 0.2$ **58.** $(0.7)(0.3)$ **59.** $32.8 \div 100$

Find each answer.

60. 2^4 **61.** 0^{17} **62.** 1^1 **63.** 9^3 **64.** 3^2 **65.** 10^8 1-9

66. $3^3 + 3 \cdot 11$ **67.** $11^2 - 69 \div 3$ **68.** $97 - 5 \cdot 32 \div 16$ **69.** $9^2 + 8 \div 2 \cdot 7$ 1-10

70. $6^2 \div (13 - 4)$ **71.** $99 + 3(6 + 17)$ **72.** $2^5 + 9(7 - 3)$ **73.** $\dfrac{4 \cdot 12}{51 - 3}$

Chapter 2

Simplify.

1. $a^3 \cdot a^7$ **2.** $2b^4 \cdot 7b^3$ **3.** $(6x)(11x)$ **4.** $(4a)(2b)(5a)$ 2-1

5. $6(a + 7)$ **6.** $5(3b - 8)$ **7.** $2(x + 13)$ **8.** $4(6 - c)$ 2-2

9. $3(2x + 7)$ **10.** $4(5a - 6)$ **11.** $6(2 - 3a)$ **12.** $10(4n + 6)$

13. $5x + 7x$ **14.** $13z - 4z$ **15.** $5x + x + 2x$ **16.** $13y - y$ 2-3

17. $3x + 4x + x$ **18.** $3a + 4a + 6$ **19.** $2a + 5c + 8c$ **20.** $n + 4n + 3m$

21. Mikisha bought two compact discs for $13.50 each and three cassettes   2-4
for $8.98 each. What was the total cost?

22. Demiso bought three boxes of herbal tea for $1.59 each and two loaves
of bread for $1.39 each. How much more did he pay for the tea than
for the loaves of bread?

Find the next three numbers or expressions in each pattern.

23. 15, 21, 27, 33, __?__, __?__, __?__   **24.** 320, 160, 80, 40, __?__, __?__, __?__   2-5

25. $x, x + 4, x + 8,$ __?__, __?__, __?__   **26.** $x + 5, 3x + 5, 5x + 5,$ __?__, __?__, __?__

Complete each function table.   **Find the function rule.**

27.

x	$2x + 5$
0	?
2	?
4	?
6	?
8	?

28.

x	$8x$
10	?
11	?
12	?
13	?
14	?

29.   2-6

x	?
5	14
6	15
7	16
8	17
9	18

30. Sharon earns $6.50 per hour at the bike store. She also earns an extra
$12 for each bike she sells. Last week Sharon worked 30 h and sold
4 bikes. She says that she earned $207 last week. Is this correct?
Explain.   2-7

Decide whether an *estimate* or an *exact answer* is needed. Then solve.

31. Barbecue meat sells for $1.46 per pound. Alicia needs to buy 40 lb for
a school picnic. How much money should she take to the store to be
sure that she has enough to pay for the meat?   2-8

32. Charlene charges $4.50 per hour to baby-sit. Last night she baby-sat
for 6 h. How much did she earn?

Write each measure in the unit indicated.

33. 34 cm; mm	**34.** 7.5 kg; g	**35.** 4.2 L; mL 2-9
36. 120 mm; cm	**37.** 4000 mg; g	**38.** 2.6 m; mm

Write each number in scientific notation.

39. 3,500,000	**40.** 4210	**41.** 80,000 2-10

Write each number in decimal notation.

42. 3.7×10^4	**43.** 2.35×10^5	**44.** 9.9×10^7

Chapter 3

Find each absolute value.

1. $|-15|$	**2.** $|7|$	**3.** $|0|$	**4.** $|-36|$ 3-1

Replace each __?__ with >, <, or =.

5. -9 __?__ 7 **6.** 0 __?__ -3 **7.** -11 __?__ -14 **8.** 2 __?__ -5

Find each sum.

9. $-34 + (-9)$ **10.** $57 + 75$ **11.** $24 + 89$ **12.** $-45 + (-37)$ 3-2

13. Thomas withdrew $16 from his savings account last month. He with-
drew $20 and $25 this month. How much has Thomas withdrawn over
the two months?

Find each sum.

14. $37 + (-78)$ **15.** $-4 + 25$ **16.** $-6 + 32$ **17.** $8 + (-8)$ 3-3

18. Sasha recorded a temperature drop of 13°F followed by an increase of
8°F. Find the total gain or loss in temperature.

Find each answer.

19. $-17 - 6$
20. $27 - (-27)$
21. $-12 - (-43)$
22. $-18 - 31$ 3-4

23. $-56 \div (-8)$
24. $6(-10)(0)$
25. $(-7)(3)(-2)$
26. $\frac{-90}{15}$ 3-5

Evaluate each expression when $a = -4$, $b = 5$, and $c = -2$.

27. $-4c^2$
28. $c^4 - b$
29. $3a^2 - c$
30. $|a| - |c|$ 3-6

31. $6 - |a - b|$
32. $|a + b|$
33. $-ab$
34. $-bc - 7$

Solve by making a table.

35. In ice hockey, a win is scored as 2 points, a tie as 1 point, and a loss as 0 points. The Blades have played 4 games in the season. How many different point totals are possible?   3-7

Use the coordinate plane at the right. Write the coordinates of each point.

36. P
37. Q
38. R
39. S

    3-8

Graph each point on a coordinate plane.

40. $A(-5, -2)$
41. $B(3, -4)$
42. $C(0, -3)$
43. $D(3, 1)$

Graph each function.

    3-9

44.

x	$3 - x$
-2	5
-1	4
0	3
1	2

45.

x	$2x + 1$
-3	-5
-2	-3
-1	-1
0	1

46. $x \rightarrow x^2$, when $x = -3, -2, 0, 2,$ and 3

47. $x \rightarrow |x - 1|$, when $x = -5, -3, -1, 2,$ and 3

Chapter 4

Is the given number a solution of the equation? Write *Yes* or *No*.

1. $n + 7 = 9$; 2
2. $q - 4 = 18$; 21
3. $36 = \frac{k}{4}$; 4
4. $4p = 24$; 6   4-1

Use mental math to find each solution.

5. $3r + 4 = 16$
6. $8s - 1 = -25$
7. $5t - 6 = 34$
8. $\frac{y}{-7} + 9 = 5$

Solve by using the guess-and-check strategy.

9. Tickets for a game are $12 and $17. If Harriet spends $300, how many of each type of ticket does she buy?

4-2

10. In basketball, there are 3-point goals, 2-point goals, and 1-point goals. Jonathan makes 3 goals and scores 7 points. How many of each type of goal could he make to score the 7 points?

11. In hockey, one point is given for a goal and one point for an assist. Barbara has 28 points. She has four more assists than goals. How many goals and assists does she have?

12. Zack bought some pencils for $.15 each. He also bought some books for $1.99 each. He spent $8.41. How many of each item did he buy?

Solve. Check each solution.

13. $t + 19 = 31$ 14. $n + 6 = -54$ 15. $m - 4 = -11$ 16. $r - 6 = -12$

4-3

17. $-7 = q - 14$ 18. $-6 = -6 + s$ 19. $14 = h + 6$ 20. $-22 = k + 9$

21. $8t = 64$ 22. $4r = -16$ 23. $30 = 6r$ 24. $\frac{n}{4} = 16$

4-4

25. $-2 = \frac{k}{-2}$ 26. $\frac{a}{-6} = 11$ 27. $4 = -n$ 28. $-k = -6$

29. $4n + 1 = 21$ 30. $6n - 6 = 6$ 31. $2b + 7 = -9$ 32. $7d - 4 = -18$

4-5

33. $-6 = \frac{t}{-4} + 3$ 34. $7 = \frac{z}{4} - 6$ 35. $18 = \frac{v}{-2} + 8$ 36. $-4 = \frac{w}{-6} - 3$

Write a variable expression that represents each phrase. If necessary, choose a variable to represent the unknown number.

37. two less than a number x

38. seven fewer peaches than yesterday

4-6

39. three more than twice a number n

40. six less than nine times a number k

41. five times as old as Jennifer

42. double the number of points scored

43. seven less than a number t tripled

44. four more than a number r divided by two

Write an equation that represents the relationship in each sentence.

45. A hockey stick costs $25, which is $21 more than the cost of a hockey puck.

4-7

46. A laser disk costs $16, which is $9 more than the cost of an audio tape.

47. Mary's total of 20 points was 4 less than the team's points.

48. Jill's time of 47 seconds was 3 seconds less than Serena's time in the track event.

Solve using an equation.

49. Anne pays $7 each for the first two melons she buys and $5 for any others. She pays $64 in all. How many melons does she buy?

4-8

50. Sam pays $20 down for a keyboard, and $15 weekly, until he pays the total price of $290. How many weekly payments does he make?

51. A car rental agency charges $43 the first two days and $35 each day thereafter. Ted is charged $323. How many days did he rent a car?

52. A park charges $12 per person for a group of up to 100 people. After the first 100 people, the park charges $8 per person. A group spends $2000 for an outing at the park. How many people are in the group?

Solve. Check each solution.

53. $2x + 5x = 21$
54. $25 = 8t - 3t$
55. $-9n - n = 100$ 4-9

56. $3(2b + 4) = 24$
57. $5(k - 4) = 40$
58. $3q + 10 + 3q = 130$

Use the formula $D = rt$.

59. Let $r = 55$ mi/h and $t = 3$ h. Find D.
60. Let $r = 35$ mi/h and $t = 5$ h. Find D. 4-10

61. Let $t = 8$ h and $D = 376$ mi. Find r.
62. Let $D = 279$ mi and $r = 62$ mi/h. Find t.

Chapter 5

Draw a pictograph to display the data.

1. **Sales of Walking Shoes in the United States (Millions of Dollars)**

5-1

Year	1985	1986	1987	1988	1989
Sales	$263	$368	$512	$752	$888

Use the double bar graph at the right.

2. In 1983, about how many movies made $20 million or more?

3. About how many more movies earned profits of $10–$20 million in 1989 than in 1981?

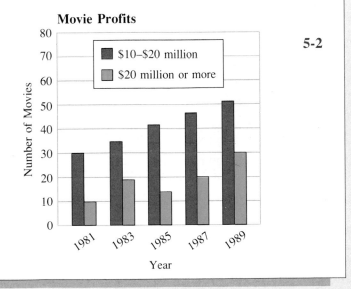

Movie Profits

5-2

Use the line graph at the right.

4. Estimate the number of house-holds with telephones in 1885.

5. Estimate the increase in house-holds with telephones from 1895 to 1900.

6. Describe the overall trend in the number of households with tele-phones.

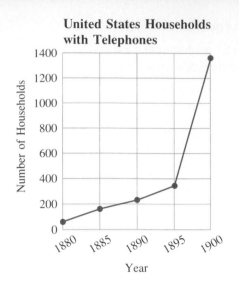

United States Households with Telephones

5-3

Draw a line graph to display the data.

7. **Average Movie Admission Prices in the United States**

5-4

Year	1981	1983	1985	1987	1989
Price	$2.78	$3.15	$3.55	$3.91	$4.45

Decide whether it would be more appropriate to draw a *bar graph* or a *line graph* to display the data. Then draw the graph.

8. **Time Spent on Volunteer Work (Hours per Week)**

5-5

Name	Jackie	Damon	Rhea	Joel	Samun
Child Care	3.5	5.0	4.2	1.7	2.3
Shopping	3.2	0.6	1.3	5.3	4.5

Use the bar graph at the right. Solve, if possible. If there is not enough information, tell what facts are needed.

9. Which store had the greatest sales of mountain bikes?

10. About how much greater were sales of touring bikes than sales of mountain bikes at Jesse's Bike Shop?

11. Does City Bikes have greater sales for touring bikes or racing bikes?

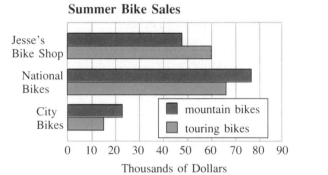

Summer Bike Sales

5-6

Use the bar graph at the right.

12. Describe the visual impression of the results of the class election.

13. The school newspaper editor wants to use a graph to accompany an article that describes the election as being very close. Explain how she might redraw the graph to support the article.

5-7

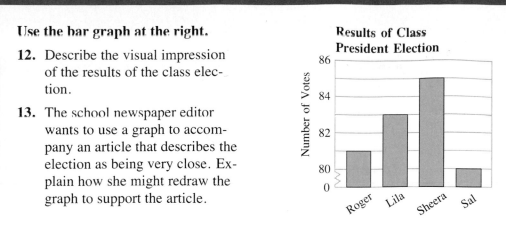

Results of Class President Election

14. The distances hiked by mountain club members (in miles) were 50, 16, 25, 23, 23, 16, 39, 43, 21, and 24. Find the mean, median, mode(s), and range of the data.

5-8

15. The Yun family's gas bills for the last seven months are $10.56, $9.22, $16.87, $9.36, $9.18, $11.13, and $9.95. Decide whether the mean or the median is the better measure of central tendency for these data. Explain.

5-9

16. Find the mean, median, mode(s), and range of the data in the frequency table below.

5-10

Bicycle Helmet Prices	Tally	Frequency
$39	ЖЖ	5
$40	ЖЖ I	6
$41	ЖЖ II	7
$44	ЖЖ	5
	Total:	23

Chapter 6

Write the name of each figure.

1.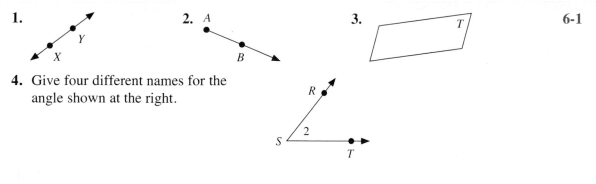

2. *A*

   B

3. *T*

6-1

4. Give four different names for the angle shown at the right.

R

S 2

T

Use a protractor to measure each angle.

5.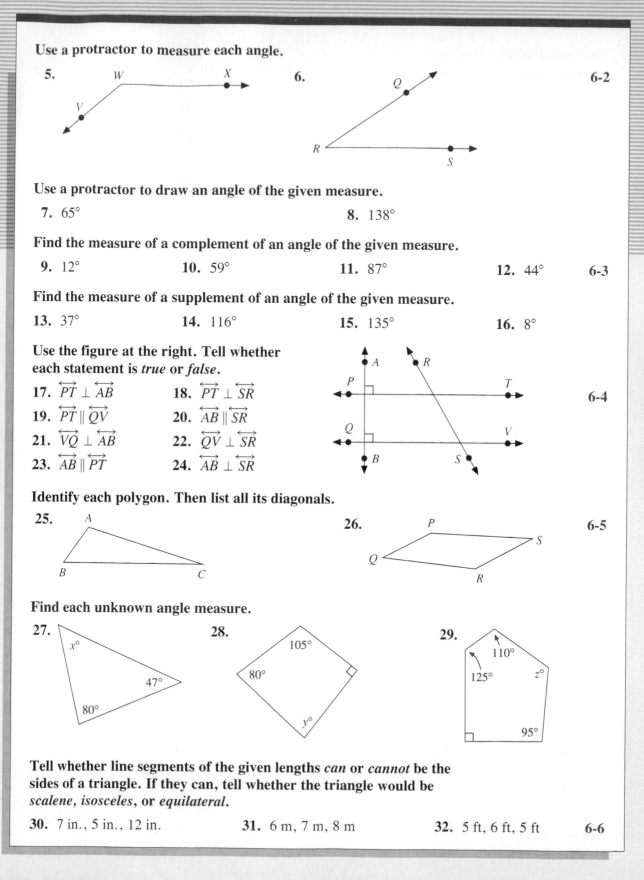

6.

6-2

Use a protractor to draw an angle of the given measure.

7. 65°

8. 138°

Find the measure of a complement of an angle of the given measure.

9. 12°

10. 59°

11. 87°

12. 44°

6-3

Find the measure of a supplement of an angle of the given measure.

13. 37°

14. 116°

15. 135°

16. 8°

Use the figure at the right. Tell whether each statement is *true* or *false*.

17. $\overleftrightarrow{PT} \perp \overleftrightarrow{AB}$

18. $\overleftrightarrow{PT} \perp \overleftrightarrow{SR}$

19. $\overleftrightarrow{PT} \parallel \overleftrightarrow{QV}$

20. $\overleftrightarrow{AB} \parallel \overleftrightarrow{SR}$

21. $\overleftrightarrow{VQ} \perp \overleftrightarrow{AB}$

22. $\overleftrightarrow{QV} \perp \overleftrightarrow{SR}$

23. $\overleftrightarrow{AB} \parallel \overleftrightarrow{PT}$

24. $\overleftrightarrow{AB} \perp \overleftrightarrow{SR}$

6-4

Identify each polygon. Then list all its diagonals.

25.

26.

6-5

Find each unknown angle measure.

27.

28.

29.

Tell whether line segments of the given lengths *can* or *cannot* be the sides of a triangle. If they can, tell whether the triangle would be *scalene*, *isosceles*, or *equilateral*.

30. 7 in., 5 in., 12 in.

31. 6 m, 7 m, 8 m

32. 5 ft, 6 ft, 5 ft

6-6

The measures of two angles of a triangle are given. Tell whether the triangle is *acute*, *right*, **or** *obtuse*.

33. 57°, 33°   **34.** 101°, 42°   **35.** 29°, 29°

In the figure at the right, *ABCD* **is a rectangle. Find each measure.**

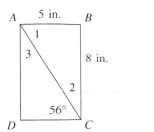

36. $m \angle 2$ **37.** $m \angle BAD$ **6-7**

38. the length of $\overline{AD}$ **39.** $m \angle 1$

40. $m \angle 3$ **41.** the length of $\overline{DC}$

Solve by identifying a pattern.

42. The figures below show the first three figures in a pattern of squares.   **6-8**
How many triangles are in the sixth figure in this pattern?

Is $\overleftrightarrow{ST}$ a line of symmetry? Write *Yes* **or** *No*.

43.   **44.**   **45.**   **6-9**

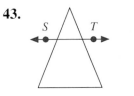   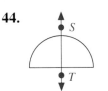

Chapter 7

Tell whether each number is *prime* **or** *composite*.

1. 7   **2.** 28   **3.** 42   **4.** 53   **7-1**

Write the prime factorization of each number.

5. 40   **6.** 99   **7.** 70   **8.** 156

Find the GCF.

9. 16 and 20   **10.** 35 and 49   **11.** 40 and 64   **12.** 75 and 100   **7-2**

13. $4a^2$ and $16a$ **14.** $8n^5$ and $32n^3$ **15.** $9np$ and $18p$ **16.** $24ab$, $48ab$, and $96ab$

Find the LCM.

17. 8 and 12   **18.** 9 and 15   **19.** 10 and 25   **20.** 6 and 14   **7-3**

21. $9a^4$ and $6a^6$ **22.** $4a^5$ and $18a^3$ **23.** $8a$ and $4a^6$ **24.** $4a$, $8a^5$, and $16a^{10}$

Replace the __?__ with the number that will make the fractions equivalent.

7-4

25. $\dfrac{6}{8} = \dfrac{18}{?}$

26. $\dfrac{9}{18} = \dfrac{?}{6}$

27. $\dfrac{?}{7} = \dfrac{8}{14}$

28. $\dfrac{8}{?} = \dfrac{2}{10}$

Write each fraction in lowest terms.

29. $\dfrac{4}{14}$

30. $\dfrac{8}{22}$

31. $\dfrac{9}{15}$

32. $\dfrac{12}{32}$

Simplify.

7-5

33. $\dfrac{9nq}{3n}$

34. $\dfrac{12ab}{4b}$

35. $\dfrac{qt^8}{t^3}$

36. $\dfrac{r^{12}}{r^7}$

37. $\dfrac{6p^4}{p^2}$

38. $\dfrac{14c^{10}}{c^5}$

39. $\dfrac{h^8}{2h^3}$

40. $\dfrac{d^{15}}{5d^{10}}$

Replace each __?__ with >, <, or =.

7-6

41. $\dfrac{2}{3} \ \underline{\ ?\ } \ \dfrac{1}{2}$

42. $\dfrac{1}{8} \ \underline{\ ?\ } \ \dfrac{1}{3}$

43. $\dfrac{3}{4} \ \underline{\ ?\ } \ \dfrac{5}{6}$

44. $\dfrac{1}{6} \ \underline{\ ?\ } \ \dfrac{1}{3}$

45. $\dfrac{3}{5} \ \underline{\ ?\ } \ \dfrac{3}{4}$

46. $\dfrac{6}{7} \ \underline{\ ?\ } \ \dfrac{6}{11}$

47. $\dfrac{3}{5} \ \underline{\ ?\ } \ \dfrac{9}{15}$

48. $\dfrac{7}{12} \ \underline{\ ?\ } \ \dfrac{9}{16}$

Write each fraction or mixed number as a decimal.

7-7

49. $\dfrac{4}{5}$

50. $\dfrac{3}{11}$

51. $2\dfrac{3}{8}$

52. $1\dfrac{9}{20}$

Write each decimal as a fraction or mixed number in lowest terms.

53. 0.32

54. 0.96

55. 3.25

56. 4.36

Solve by drawing a diagram.

7-8

57. Sonia's house is 3 blocks east and 8 blocks south of Mary's house. Joan's house is 6 blocks north and 5 blocks west of Mary's house. Where is Joan's house in relation to Sonia's house?

58. A tour bus starts at a hotel, travels 3 blocks west, 8 blocks north, 2 blocks west, 5 blocks north, and 5 blocks west. Where is the bus in relation to the hotel?

59. How many diagonals can be drawn in an octagon?

60. The lengths of three boards are 4 m, 6 m, and 9 m. How could you use these boards to mark off a length of 1 m?

Use the number line below. Name the point that represents each number.

7-9

61. $2\dfrac{1}{2}$

62. $-1\dfrac{1}{2}$

63. -0.9

64. 0.5

Express each rational number as a quotient of two integers.

65. -9 **66.** $4\frac{2}{3}$ **67.** $-6\frac{1}{5}$ **68.** $-4.\overline{3}$

Graph each open sentence.

69. $4 = 9 + x$ **70.** $-2 + x = -6$ **71.** $x - 4 = 9$ **72.** $5x + 4 = -1$ **7-10**

73. $x < 3.7$ **74.** $x > -0.5$ **75.** $x \geq -2.5$ **76.** $x \leq -0.5$

Simplify.

77. x^{-4} **78.** 2^{-4} **79.** 5^{-3} **80.** $(-6.782)^0$ **7-11**

Write each number in scientific notation.

81. 0.000082 **82.** 0.0062 **83.** 0.00007 **84.** 0.000643

Write each number in decimal notation.

85. 5×10^{-3} **86.** 6.7×10^{-4} **87.** 3.25×10^{-9} **88.** 4.223×10^{-5}

Chapter 8

Find each product. Simplify if possible.

1. $\left(-\frac{9}{10}\right)\left(-\frac{5}{12}\right)$ **2.** $\frac{5}{7} \cdot \frac{7}{5}$ **3.** $\left(2\frac{4}{5}\right)\left(-3\frac{1}{8}\right)$ **4.** $\left(-6\frac{2}{3}\right)\left(\frac{12}{13}\right)$ **8-1**

5. $(-4.3)(5.4)$ **6.** $(-2.9)(-0.4)$ **7.** $\left(\frac{5a}{7b}\right)\left(\frac{14b}{15}\right)$ **8.** $\left(\frac{x}{20}\right)\left(\frac{40y}{x}\right)$

Find each quotient. Simplify if possible.

9. $-\frac{3}{7} \div \left(-\frac{3}{14}\right)$ **10.** $\frac{3}{8} \div \frac{2}{3}$ **11.** $\frac{10}{19} \div (-2)$ **12.** $-20 \div \frac{4}{5}$ **8-2**

13. $-1.44 \div 0.45$ **14.** $-0.81 \div (-2.7)$ **15.** $\frac{4x}{7} \div \frac{8x}{9}$ **16.** $\frac{16a}{17} \div 4a$

Solve. Decide whether a *whole number*, a *fraction*, or a *decimal* is an appropriate answer.

17. Each month Junko earns $125 and saves one fourth of his earnings. How much money does he save each month? **8-3**

18. There will be 35 people at a luncheon. How many tables of 8 seats each are needed to seat everyone?

Find each answer. Simplify if possible.

19. $\frac{5}{7} + \frac{6}{7}$ **20.** $\frac{5}{8} + \frac{3}{8}$ **21.** $-\frac{4}{9} - \left(-\frac{4}{9}\right)$ **22.** $\frac{7}{10} + \left(-\frac{9}{10}\right)$ **8-4**

23. $-2\frac{1}{3} + \left(-\frac{2}{3}\right)$ **24.** $\frac{11}{12} - \left(-\frac{5}{12}\right)$ **25.** $\frac{11}{y} + \frac{9}{y}$ **26.** $\frac{12n}{5} - \frac{10n}{5}$

Find each answer. Simplify if possible.

27. $\frac{4}{9} + \frac{1}{6}$ **28.** $-4\frac{4}{7} - \frac{5}{14}$ **29.** $-5\frac{2}{3} + 3\frac{11}{12}$ **30.** $-6 + 3\frac{1}{3} + \left(-\frac{5}{6}\right)$ 8-5

31. $\frac{5a}{7} - \frac{5a}{14}$ **32.** $3x + \frac{9x}{11}$ **33.** $6.4 + (-8.7)$ **34.** $-37.5 - 0.43$

Estimate.

35. $27\frac{2}{13} - 6\frac{11}{12}$ **36.** $15\frac{1}{3} - 4\frac{5}{8}$ **37.** $6\frac{27}{28} + 3\frac{1}{5} + 2\frac{7}{12}$ 8-6

38. $11\frac{9}{10} \div \left(-\frac{8}{15}\right)$ **39.** $-17\frac{9}{10} \div \left(-6\frac{1}{5}\right)$ **40.** $-27\frac{3}{5} \div \frac{25}{99}$

Solve by supplying missing facts.

41. Elidio's health insurance costs him $85 per month. How much does his insurance cost per year?   8-7

42. It took Roseanne 3 h to complete 5 homework assignments. She spent an equal amount of time on each assignment. How many minutes did she spend on each assignment?

Solve. Check each solution.

43. $8 = -5b + 3$ **44.** $10 + 5k = 3$ **45.** $z - \frac{2}{3} = \frac{5}{6}$ 8-8

46. $w + 3\frac{1}{8} = -\frac{3}{8}$ **47.** $2a - 4.9 = -7.1$ **48.** $27.4 = 3.14 + 2x$

49. $\frac{8}{9}x = -5$ **50.** $-\frac{2}{5}y = 14$ **51.** $11 + \frac{6}{7}m = 8$ 8-9

52. $-\frac{2}{9}z + 3 = 7$ **53.** $-1 = \frac{5}{7}f + 4$ **54.** $-9 = -\frac{7}{11}d - 8$

Solve for the variable shown in color.

55. $A = lw$ **56.** $A = \frac{1}{2}bh$ **57.** $A = \frac{1}{2}h(b_1 + b_2)$ 8-10

58. $p = \frac{m}{a}$ **59.** $D = rt$ **60.** $g = a - h$

Chapter 9

Write each ratio as a fraction in lowest terms.

1. $\frac{6}{42}$ **2.** $35 : 50$ **3.** $\$.40 : \$2$ **4.** 36 in. to 2 ft   9-1

Write the unit rate.

5. $75 for 6 h   **6.** 165 mi in 3 h   **7.** 144 mi on 8 gal   **8.** 345 words in 5 min

Solve each proportion.

9. $\dfrac{25}{b} = \dfrac{5}{7}$ 10. $\dfrac{5}{8} = \dfrac{v}{24}$ 11. $\dfrac{8}{a} = \dfrac{6}{9}$ 12. $\dfrac{1.5}{1.2} = \dfrac{10}{k}$ 9-2

13. $\dfrac{b}{18} = \dfrac{20}{30}$ 14. $\dfrac{2.5}{t} = \dfrac{10}{9}$ 15. $\dfrac{0.4}{1.6} = \dfrac{n}{28}$ 16. $\dfrac{x}{7} = \dfrac{12}{3}$

17. A bedroom is 10 ft wide and 12 ft long. The scale of a floor plan of the   9-3
    bedroom is 1 in. : 12 ft. Find the width and length of the bedroom on
    the floor plan.

18. The model of a van is $1\frac{1}{2}$ ft long. The scale of the model is 1 : 10.
    What is the actual length of the van?

Solve using a proportion.

19. The ratio of juniors to seniors at a dance was 3 : 4. There were 60   9-4
    seniors at the dance. How many juniors were at the dance?

20. A computer printer can print 8 pages in 1.5 min. At this rate, how long
    will it take to print 60 pages?

Write each fraction, mixed number, or decimal as a percent.

21. $\dfrac{4}{25}$ 22. 0.09   23. $1\frac{1}{8}$ 24. 0.765   9-5

Write each percent as a fraction in lowest terms and as a decimal.

25. 45%   26. 6.5%   27. $8\frac{1}{5}\%$ 28. 160%

Find each answer.

29. What number is 20% of 48?   30. 6% of 425 is what number?   9-6

31. What percent of 60 is 4.5?   32. 270 is what percent of 600?   9-7

33. 44 is 25% of what number?   34. 150 is $66\frac{2}{3}\%$ of what number?   9-8

Find the percent of increase or percent of decrease.

35. original population: 400; new population: 520   9-9

36. original cost: $50; new cost: $40

37. original weight: 120 lb; new weight: 100 lb

Find each answer using a proportion.

38. What number is 75% of 360?   39. What percent of 120 is 45?   9-10

40. 4 is 5% of what number?   41. 12 is what percent of 18?

42. What percent of 80 is 4?   43. What number is 60% of 150?

Chapter 10

Find the perimeter of each figure.

10-1

1. a rectangle that is $5\frac{2}{3}$ ft long and $3\frac{1}{4}$ ft wide

2. a regular pentagon with one side that measures 3.5 cm

Find the radius and diameter of a circle with the given circumference.

3. 75.36 m

4. 439.6 in.

5. 40.82 m

10-2

Find the area of each figure.

6. a square with length of one side equal to 20 ft

7. a rectangle with length 9.2 m and width 6.6 m

10-3

8. a circle with radius 6.5 m

9. a circle with diameter $10\frac{1}{2}$ ft

10-4

10. Tell whether you would need to calculate *perimeter*, *circumference*, or *area* to find the amount of fencing needed to enclose a circular park.

10-5

11. Explain why this problem has no solution: Lily said she spent half of her paycheck on rent, one third on food and utilities, and one fourth on clothing. What part of her pay did she save?

10-6

Use the figures at the right to complete each statement.

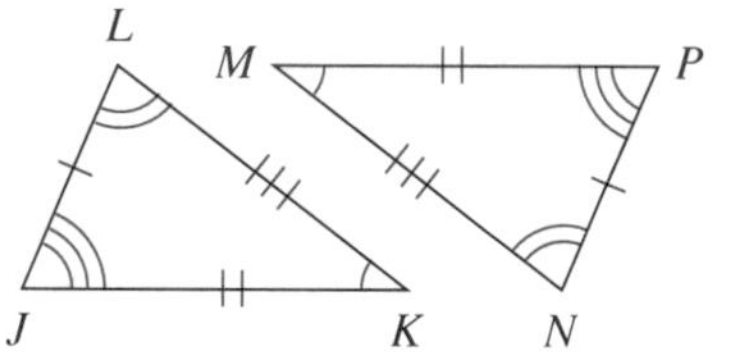

12. $\overline{LK} \cong$ ___?___

13. $\angle J \cong$ ___?___

10-7

14. In the figures at the right, quadrilateral *MNOP* ~ quadrilateral *STUV*. Find *z*.

10-8

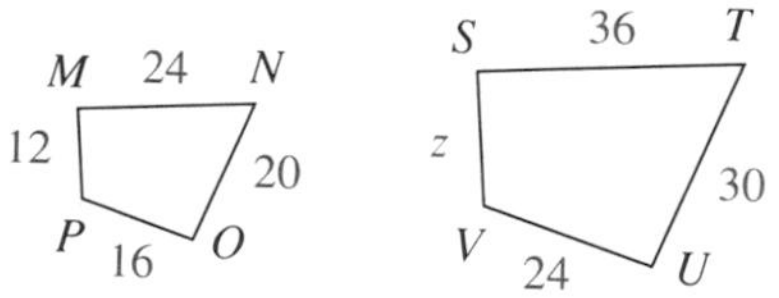

15. In the figure at the right, the triangles are similar. Find the unknown height *h*.

10-9

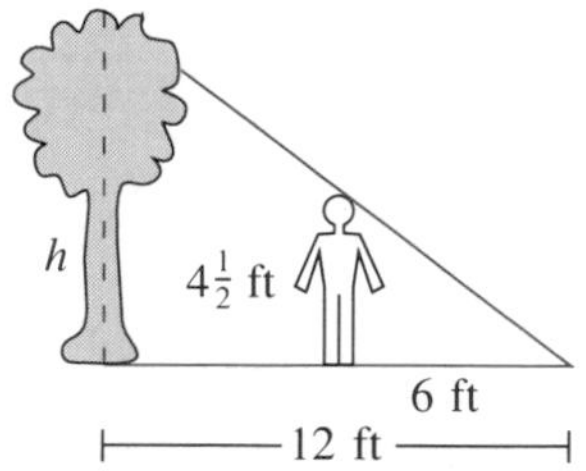

Find each square root if possible. Otherwise, approximate the square root to the nearest thousandth.

16. $\sqrt{4900}$

17. $\sqrt{\dfrac{16}{81}}$

18. $\sqrt{87}$

19. $\sqrt{139}$

10-10

20. Is a triangle with sides of 10 in., 25 in., and 27 in. a right triangle? Write *Yes* or *No*.

10-11

Chapter 11

Use the data below for Exercises 1 and 2.

Test Scores In Math
76 79 86 59 97 94 82 82 80 76 64 60 92 75 98 83 57 72 70 81

1. Make a frequency table for the data. Use intervals such as 50–59. **11-1**

2. Make a stem-and-leaf plot for the data. **11-2**

Use the stem-and-leaf plot at the right.

Ages of Teachers

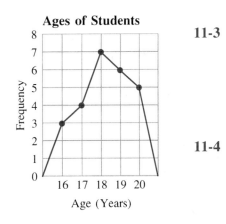

3. How many teachers are less than 45 years old?

4. How many teachers are 42 years old?

5. Find the mean, median, mode(s), and range of the data.

Use the frequency polygon at the right.

6. How many students are 19 years old?

7. Exactly 4 students are a certain age. What is this age?

8. How many students are less than 19 years old?

9. How many students are there in all?

10. Make a box-and-whisker plot for the data below. **11-4**

Prices for Compact Disc Players (Dollars)
600 400 300 350 300 500 420 550
375 310 480 500 675 380 480 590

11-3

Use the box-and-whisker plot below for Exercises 11–13.

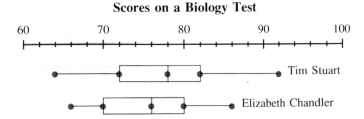

11. Find the median, the first quartile, the third quartile, and the lowest and highest scores for Tim Stuart's class.

12. What was the range of test scores for Elizabeth Chandler's class?

13. Which teacher's class had the greater median score on the test?

Use the scattergram below for Exercises 14–16.

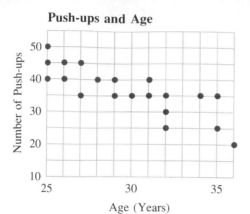

Push-ups and Age

14. How many persons did 45 push-ups?

11-5

15. How many persons who were 31 years old did push-ups?

16. Is there a *positive* or *negative* correlation between the two sets of data?

Use the circle graph at the right.

17. If the Carson family's total annual budget is $42,000, how much do they plan to spend annually on food?

18. If $16,200 is spent annually on housing, how much is the total budget?

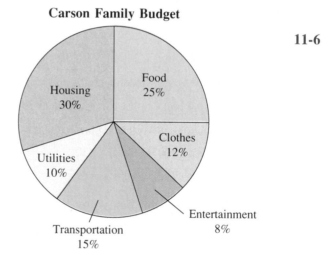

Carson Family Budget

11-6

19. Draw a circle graph to display the data below.

11-7

Videos Rented from Victor's Video

Type of Movie	Adventure	Comedy	Family
Number Rented	120	20	60

20. Decide whether it would be appropriate to draw a *circle graph* or a *stem-and-leaf plot* for the data below. Then make the display.

11-8

Scores on a History Test

75 76 89 98 57 89 73 86 90 65
67 75 80 98 72 65 83 92 76 79

21. Therese, Lucinda, and Alicia are a teacher, a lawyer, and an accountant. No one works at a job that begins with the first letter of her name. Therese and the accountant went to the same college. Who has each job?

11 9

Chapter 12

Use the spinner at the right for Exercises 1–4.

1. Find $P(3)$.   **2.** Find P(not white).

12-1

Find the odds in favor of each event.

3. an odd number   **4.** not red

5. The spinner above is spun and a coin is tossed. Make a tree diagram to show the sample space.

12-2

6. Use the tree diagram from Exercise 5 to find $P(1$, heads or tails).

7. Use the counting principle to find the number of possible outcomes of choosing an odd number from 1 through 10 and an even number from 11 through 20.

12-3

A bag of marbles contains 5 blue, 4 red, and 9 green marbles. Use this information for Exercises 8 and 9.

8. Two marbles are drawn at random from the bag with replacement. Find P(green, then blue).

12-4

9. Two marbles are drawn at random from the bag without replacement. Find P(red, then green).

10. Of 100 people surveyed, 60 own cars, 35 own trucks, and 20 own both cars and trucks. Use a Venn diagram to find how many of the 100 people do not own either a car or a truck.

12-5

11. In how many different ways can 7 pencils all with different colors be arranged in a box?

12-6

12. A committee of 4 is to be chosen from a group of 9 people. Find the number of combinations.

13. Yves tossed a thumbtack 60 times. It landed point up 27 times and point down 33 times. Find the experimental probability that the thumbtack lands point up.

12-7

14. When 40 employees at a local company were questioned, 24 said that they had worked at the company for at least 10 years. About how many of the 200 employees at the company are expected to have worked at the company for at least 10 years?

12-8

Chapter 13

Find solutions of each equation. Use $-2, -1, 0, 1,$ and 2 as values for x.

1. $y = 3x + 2$ 2. $y = \frac{1}{2}x - 1$ 3. $y = -2x + 5$ 4. $y = -5x - 2$ 13-1

5. $y - 3x = 9$ 6. $y + 4x = 3$ 7. $y - 5x = -7$ 8. $y + 3x = -4$

Graph each equation.

9. $y = \frac{1}{2}x + 4$ 10. $y = -\frac{2}{3}x + 1$ 11. $y = -\frac{3}{4}x$ 12. $y = 2x$ 13-2

13. $y - 3x = 2$ 14. $x + y = -2$ 15. $x + y = -1$ 16. $2x + y = 1$

Find the slope and the y-intercept of each line. Then use them to write an equation for each line.

17. 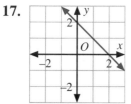   18.   13-3

19. 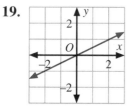   20.

Find the slope, the y-intercept, and the x-intercept of each line.

21. $y - 4 = 2x$ 22. $y - 2 = 4x$ 23. $y + 6 = -3x$ 24. $y + 1 = x$

Solve each system by graphing.

25. $y = x + 5$ 26. $y = -2x + 3$ 27. $y = -\frac{2}{3}x + 1$ 28. $y = \frac{1}{2}x + 9$ 13-4
    $y = 3x + 1$ $y = 4x - 3$ $y = \frac{1}{6}x - 4$ $y = -\frac{1}{2}x + 11$

29. $y = 3x$ 30. $y = -2x$ 31. $y = -x + 8$ 32. $y = \frac{1}{2}x + 4$
    $y = 4x + 2$ $y = 2x$ $y = 2x + 2$ $y = x + 2$

Solve using a simpler problem.

33. Find the sum of the first 25 positive odd numbers.   13-5

34. James decided to save quarters. On the first day he saved one quarter. On each of the next thirteen days he saved two more quarters than he saved the day before. How many quarters did James save in all?

Solve and graph each inequality.

35. $n + 2 \le 7$ **36.** $b - 6 \ge 9$ **37.** $p + 4 > -9$ **38.** $q - 3 < -7$ **13-6**

39. $2k < 4$ **40.** $\frac{r}{-6} \ge -2$ **41.** $-5t \le 15$ **42.** $\frac{b}{-4} > 7$

Solve each inequality.

43. $6k + 300 \ge 900$ **44.** $3p - 40 \le 80$ **45.** $-2q - 4600 > 9800$ **13-7**

46. $-5a - 800 < 7200$ **47.** $\frac{1}{3}t - 500 \ge -700$ **48.** $\frac{1}{2}q + 400 < 8000$

49. $-\frac{2}{3}n + 5 \le 707$ **50.** $-1\frac{1}{2}w + 300 > 600$

Tell whether each ordered pair is a solution of each inequality. Write _Yes_ or _No._

51. $(1, 4); y < 6x - 4$ **52.** $y \ge 3x + 2; (-2, 8)$ **13-8**

53. $(2, -2); y > -2x - 2$ **54.** $y \le \frac{1}{3}x + 2; (0, 2)$

Graph each inequality.

55. $y < 3x - 1$ **56.** $y \ge x + 1$ **57.** $y > -3x$ **58.** $y \le \frac{1}{2}x + 3$

Find values of each function. Use -1, 0, and 3 as values for x.

59. $y = 3x - 7$ **60.** $y = \frac{2}{3}x$ **61.** $y = x^2 - 4x$ **62.** $y = |x - 2|$ **13-9**

Find $f(-2), f(1),$ and $f(0)$ for each function.

63. $f(x) = -4x$ **64.** $f(x) = \frac{1}{8}x$ **65.** $f(x) = 8x^2 - 1$

66. $f(x) = |6x + 1|$ **67.** $f(x) = x^2 - 1$ **68.** $f(x) = (x - 2)^2$

Graph each function.

69. $f(x) = 3x + 1$ **70.** $f(x) = -\frac{1}{3}x + 1$ **71.** $f(x) = (x + 1)^2$ **13-10**

72. $f(x) = x^2 - 4$ **73.** $f(x) = -|x| + 3$ **74.** $f(x) = |x + 2|$

Chapter 14

Identify each space figure.

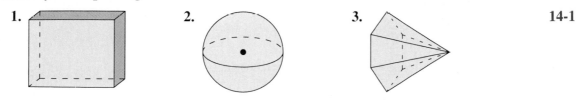

1. **2.** **3.** **14-1**

Find the surface area of each space figure.

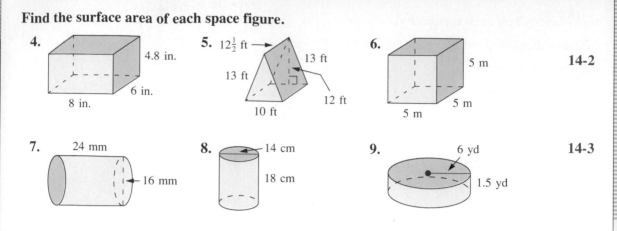

4. 4.8 in. 6 in. 8 in.

5. $12\frac{1}{2}$ ft → 13 ft 13 ft 12 ft 10 ft

6. 5 m 5 m 5 m

14-2

7. 24 mm 16 mm

8. 14 cm 18 cm

9. 6 yd 1.5 yd

14-3

10–12. Find the volume of each space figure in Exercises 4–6.

14-4

Find the volume of each space figure.

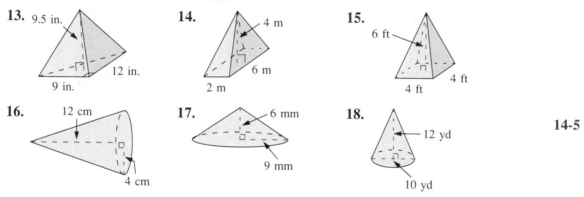

13. 9.5 in. 12 in. 9 in.

14. 4 m 6 m 2 m

15. 6 ft 4 ft 4 ft

16. 12 cm 4 cm

17. 6 mm 9 mm

18. 12 yd 10 yd

14-5

19–21. Find the volume of each space figure in Exercises 7–9.

Find the surface area of each sphere.

22. $r = 6$ in.
23. $d = 30$ ft
24. $r = 30$ cm

14-6

Find the volume of each sphere.

25. $d = 24$ m
26. $r = 3$ mm
27. $d = 36$ yd

Solve by making a model.

28. A box is designed to hold 24 cubes with edges 1 unit long. Find the dimensions of the box with the least surface area.

14-7

29. Suppose a cube is painted green on every side except the bottom. The cube is then cut into 64 smaller cubes. How many of the smaller cubes have exactly 1 green side? exactly 2 green sides? exactly 3 green sides? no green sides?

Chapter 15

Write each polynomial in standard form.

1. $3a^3 + 4a + 9 + 6a^2$

2. $8a^2 + 9a^3 + 16 + 12a^4$

15-1

3. $7a + 4 + 7a^2$

4. $6 + 3a + 9a^2 + 8a^3 + 6a^4$

Simplify.

5. $6a + 4a^3 - 2a - 2a^2 + 7$

6. $3n^2 - n^3 + 6n - 2n^2 + 4n^3 + 8 - 3n$

7. $4p^2 - 3p + 5p^2 - 6 + 8p + 8$

8. $3x + 4x^2 + 6 - x + 4x - 3 + 8x^2$

Add.

9. $(6a^2 + 3a + 9) + (4a^2 + 6a + 1)$

10. $(4b^2 + 18b + 6) + (9b^2 + 9b + 12)$

15-2

11. $(12c^2 + 19c + 9) + (6c^2 + 9c + 12)$

12. $(14y^2 + 22y + 12) + (12y^2 + 19y + 6)$

13. $(7x^4 + 3x^3 + 4x^2 - 6x + 2) + (-4x^4 - 6x^3 + 8x - 9)$

14. $(3t^4 - 4t^3 + 16t - 4) + (3t^4 + 7t^3 + 3t^2 - 7t - 3)$

15. $(8n^4 - 8n^3 - 12n^2 + 5) + (-9n^4 + 9n^3 + 14n^2 - 5n - 2)$

16. $(6q^4 + 7q^3 - 11q^2 - 8q - 12) + (-q^4 + q^3 - q - 1)$

Subtract.

17. $(5k^2 + 9k + 4) - (3k^2 + 4k + 1)$

18. $(12p^2 + 7p + 8) - (5p^2 + 6p + 5)$

15-3

19. $(3h^2 + 4h + 5) - (2h^2 + 3h + 4)$

20. $(10r^2 + 5r + 7) - (8r^2 + 3r + 5)$

21. $(9b^3 + 5b^2 - 8) - (3b^3 - 3b^2 + 5b - 6)$

22. $(6t^3 - 7t - 4) - (-8t^3 - 6t^2 - 4t + 4)$

23. $(7x^3 + 10x^2 - 4x + 6) - (10x^3 - 6x^2 + 9)$

24. $(12a^3 - 6a^2 + 9a - 4) - (5a^3 - 4a + 8)$

Multiply.

25. $3(6x^2 + 3x - 2)$

26. $-2(4q^2 - 5q - 8)$

15-4

27. $-4(-3m^3 + 6m^2 - 9m - 9)$

28. $(4x^2y^3)(-2x^5y^2)$

29. $(-6c^4d^3)(-5c^3d^7)$

30. $(-8p^4q^7)(9p^5q^6)$

31. $3a^4(4a^3 + 8a^2 - 3a - 9)$

32. $6d^2(3d^3 - 9d^2 - 6)$

33. $(n + 1)(n + 5)$

34. $(4k + 6)(5k + 2)$

15-5

35. $(5y + 6)(y + 7)$

36. $(2t + 9)(t + 1)$

37. $(3n + 1)(2n + 5)$

38. $(2g - 3)(8g + 8)$

39. $(4u - 7)(2u + 5)$

40. $(z - 6)(z - 6)$

Solve by working backward.

41. In selecting a president, the board began with a pool of people and rejected half of the applicants after each round of interviews. Sara was chosen after 5 rounds. How many people were in the initial pool?

15-6

42. At the end of January, Adrienne has $1200 in her checking account after writing checks for $350, $73, $82, $625, and making a deposit of $175. The bank charges her $3.50 as a service charge and pays her $2.75 in interest in January. How much money was in Adrienne's account at the beginning of January?

Factor.

43. $4m^4 - 16m^3 + 30m^2$

44. $8s^5 - 6s^4 + 12s^3$

15-7

45. $10v^6 - 16v^5 + 12v^4$

46. $14q^2 + 16q + 18$

47. $4a^3b^5 - 10a^4b^2 + 6a^5b^5$

48. $15c^6d^2 - 20c^2d^2 - 45c^3d^2$

49. $24p^4q^5 - 15p^3q^6 + 33p^6q^7$

50. $32r^5t^6 - 18r^4t^4 + 48r^7t^7$

Divide.

51. $\dfrac{6y^5 + 18y^4 + 9y^3}{3y^2}$

52. $\dfrac{14x^6 + 21x^4 + 42x^3}{7x^2}$

15-8

53. $\dfrac{8t^7 + 16t^5 + 24t^4}{4t^3}$

54. $\dfrac{20r^6 + 24r^4 + 6r^2}{2r}$

55. $\dfrac{7a^4b^5 - 42a^3b^7 + 14a^5b^2}{7a^2b}$

56. $\dfrac{9s^3t^8 - 24s^5t^6 - 27s^7t^3}{3s^2t^3}$

57. $\dfrac{-22m^2p^5 + 11m^7p^4 - 33m^6p^4}{11m^2p^2}$

58. $\dfrac{-63h^7k^8 - 72h^8k^9 - 81h^5k^5}{9h^3k^4}$

Estimation Skills

Skill 1 Rounding

Estimate the answer. Round each number to the place of its leading digit.

$$
\begin{array}{c}
56.9 \\
+31.4
\end{array}
\longrightarrow
\begin{array}{c}
60 \\
+30 \\
\hline
\text{about } 90
\end{array}
$$

$$
\begin{array}{c}
5439 \\
-\,782
\end{array}
\longrightarrow
\begin{array}{c}
5000 \\
-\,800 \\
\hline
\text{about } 4200
\end{array}
$$

$$
\begin{array}{c}
892 \\
\times 4.7
\end{array}
\longrightarrow
\begin{array}{c}
900 \\
\times\ \ 5 \\
\hline
\text{about } 4500
\end{array}
$$

Estimate the answer. Round each number to the place of its leading digit.

1. $\ \ 582$ -167	**2.** $\ \ 17.36$ $+42.51$	**3.** $\ \ 918$ $+\ \ 77$
4. $\ \ 4.632$ -1.299	**5.** $\ \ 635$ $+114$	**6.** $\ \ 7866$ $+\ \ 337$
7. $\ \ 769$ -278	**8.** $\ \ \ \ 2.396$ $+\ 8.535$	**9.** $\ \ 31{,}158$ $+\ 5{,}676$
10. $\ 11.784$ $-\ 3.415$	**11.** $\ \ 36$ $\times 15$	**12.** $\ \ 809$ $\times\ 74$
13. $\ 16{,}232$ $\times\ \ \ \ 198$	**14.** $\ \ 4.83$ $\times\ 3.7$	**15.** $\ \ 518.8$ $\times\ 8.63$
16. $\ \ 8808$ -2220	**17.** $\ \ 6090$ $\times\ \ \ 45$	**18.** $\ \ 3.579$ $\times\ \ \ \ 72$

Skill 2 Front-end and Adjust

Estimate the sum.

$$
\begin{array}{l}
70{,}732 \quad \text{about } 10{,}000 \\
+\ 9{,}015 \\
\hline
70{,}000 + 10{,}000 \\
\text{about } 80{,}000
\end{array}
$$

Estimate the sum.

$$
\begin{array}{l}
281.6 \quad \text{about } 100 \\
+517.1 \\
\hline
700 + 100 \quad \text{about } 800
\end{array}
$$

Estimate the answer by adjusting the sum of the front-end digits.

1. $\ \ 534$ $+607$	**2.** $\ \ 931$ $+163$	**3.** $\ \ 472$ $+901$
4. $\ \ 5310$ $+1519$	**5.** $\ \ 2093$ $+6258$	**6.** $\ \ 1580$ $+6293$
7. $\ \ 24.913$ $+\ 5.720$	**8.** $\ \ 190.30$ $+\ 86.10$	**9.** $\ \ 3750.2$ $+1223.9$
10. $\ 23{,}498$ $\ \ 10{,}976$ $+\ 8{,}231$	**11.** $\ 59.226$ $\ 14.835$ $+76.103$	**12.** $\ 62{,}018$ $\ 13{,}950$ $+71{,}004$

Skill 3 Compatible Numbers

Estimate the quotient:
$493 \div 37$

First round the divisor to its greatest place value:

$$493 \div 40$$

Then change the dividend to the closest convenient multiple of the new divisor:

$$480 \div 40$$

Now divide mentally with the compatible numbers:

$$480 \div 40 = 12$$

The quotient is about 12.

Use compatible numbers to estimate the quotient.

1. $29)\overline{587}$

2. $32)\overline{617}$

3. $26)\overline{971}$

4. $91)\overline{565}$

5. $62)\overline{1531}$

6. $57)\overline{1730}$

7. $83)\overline{4677}$

8. $86)\overline{2471}$

9. $236)\overline{7895}$

10. $421)\overline{2193}$

11. $606)\overline{38,415}$

12. $582)\overline{46,478}$

13. $156)\overline{23,199}$

14. $923)\overline{82,331}$

15. $29.6)\overline{2662.7}$

16. $463)\overline{24,129}$

17. $98.7)\overline{321.7}$

18. $27.82)\overline{208.59}$

19. $38.7)\overline{1648}$

20. $9.7)\overline{168.9}$

Skill 4 Clustering

Estimate the sum:

$62 + 59 + 55 + 67 + 61$

Observe that the numbers cluster around 60. Then the sum is about 5×60, or about 300.

Use clustering to estimate the sum.

1. $19 + 26 + 20$

2. $85 + 79 + 83$

3. $39 + 41 + 43$

4. $50 + 48 + 53 + 55$

5. $210 + 203 + 211$

6. $31 + 26 + 28 + 30$

7. $597 + 603 + 609$

8. $711 + 709 + 713$

9. $130 + 128 + 133$

10. $270 + 272 + 268$

11. $343 + 340 + 337 + 339 + 345$

12. $1110 + 1119 + 1108 + 1111 + 1102 + 1108$

13. $7.8 + 7.2 + 8.3 + 8.1 + 8 + 8.9 + 8.6$

14. $10.6 + 9.3 + 9.1 + 10 + 9.4 + 9.9 + 9.8$

15. $7.03 + 6.9 + 6.97 + 7.1 + 7.01$

16. $92 + 90.7 + 89.5 + 90.1 + 88 + 89.8$

17. $1.21 + 1.2 + 1.19 + 1.15 + 1.22 + 1.25$

Skill 5 Place Value; Writing Numbers

In 7896.513, give the place and value of each digit.

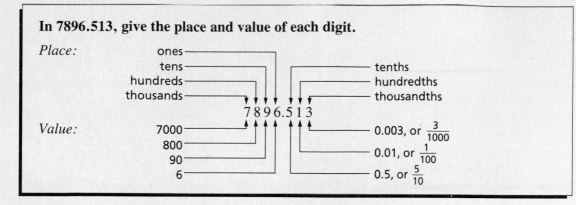

Place:

ones — tens — hundreds — thousands

tenths — hundredths — thousandths

7 8 9 6.5 1 3

Value:

7000 — 800 — 90 — 6

0.003, or $\frac{3}{1000}$

0.01, or $\frac{1}{100}$

0.5, or $\frac{5}{10}$

Give the place and value of the underlined digit.

1. 6<u>8</u>9	**2.** 11<u>3</u>	**3.** 1<u>0</u>9	**4.** <u>9</u>24
5. <u>1</u>562	**6.** 3<u>4</u>52	**7.** 537<u>1</u>	**8.** 4<u>6</u>05
9. <u>2</u>638	**10.** 0.<u>6</u>2	**11.** 10.1<u>7</u>	**12.** 9.73<u>1</u>
13. <u>9</u>5.1	**14.** 51.0<u>3</u>	**15.** 146.03<u>9</u>	**16.** 51<u>2</u>.99
17. <u>8</u>.13	**18.** 719.0<u>3</u>3	**19.** 623.0<u>9</u>	**20.** 1294.31<u>3</u>

Write the word form of each number.

208 ⟶ Two hundred eight

79 ⟶ Seventy-nine

13 ⟶ Thirteen

53.29 ⟶ Fifty-three and twenty-nine hundredths

6.418 ⟶ Six and four hundred eighteen thousandths

Write the word form of each number.

21. 15	**22.** 91	**23.** 162	**24.** 87	**25.** 493
26. 101	**27.** 258	**28.** 911	**29.** 4056	**30.** 3080
31. 32	**32.** 1473	**33.** 7006	**34.** 56	**35.** 385
36. 16.3	**37.** 4.7	**38.** 21.15	**39.** 178.61	**40.** 9.016
41. 83.097	**42.** 492	**43.** 18.315	**44.** 9.036	**45.** 6031.752
46. 24.4	**47.** 193.07	**48.** 8.949	**49.** 307.1	**50.** 12.56

Skill 6 Rounding Whole Numbers and Decimals

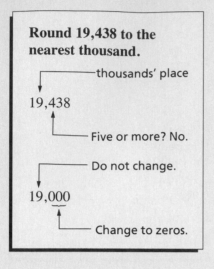

Round 19,438 to the nearest thousand.

thousands' place

19,438

Five or more? No.

Do not change.

19,000

Change to zeros.

Round to the nearest ten.

1. 18	**2.** 52	**3.** 99
4. 137	**5.** 273	**6.** 5048
7. 649	**8.** 1208	**9.** 8792

Round to the nearest hundred.

10. 652	**11.** 912	**12.** 106
13. 5329	**14.** 1493	**15.** 349
16. 1192	**17.** 2351	**18.** 9137

Round to the nearest thousand.

19. 6358	**20.** 15,491	**21.** 12,253
22. 878	**23.** 7299	**24.** 9602
25. 2549	**26.** 3197	**27.** 21,840

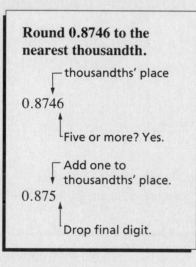

Round 0.8746 to the nearest thousandth.

thousandths' place

0.8746

Five or more? Yes.

Add one to thousandths' place.

0.875

Drop final digit.

Round to the nearest tenth.

28. 0.52	**29.** 4.01	**30.** 11.93
31. 0.312	**32.** 15.11	**33.** 4.76
34. 0.08	**35.** 12.19	**36.** 92.25
37. 107.623	**38.** 22.819	**39.** 17.853

Round to the nearest hundredth.

40. 0.739	**41.** 0.515	**42.** 127.316
43. 31.034	**44.** 0.683	**45.** 0.299
46. 2.051	**47.** 12.994	**48.** 0.625
49. 3.408	**50.** 11.003	**51.** 71.987

Round to the nearest thousandth.

52. 0.8376	**53.** 54.0178	**54.** 0.9003
55. 0.3139	**56.** 183.5069	**57.** 2.9919
58. 6.0013	**59.** 13.0297	**60.** 86.9765
61. 4.3482	**62.** 1.6105	**63.** 177.0251

Skill 7 Adding Whole Numbers

Add.

$$
\begin{array}{r} 562 \\ 103 \\ +432 \\ \end{array} \longrightarrow
\begin{array}{r} 562 \\ 103 \\ +432 \\ \hline 1097 \end{array}
$$

$$
\begin{array}{r} 5738 \\ +6425 \\ \end{array} \longrightarrow
\begin{array}{r} ^{1\ 1}5738 \\ +\ 6425 \\ \hline 12{,}163 \end{array}
$$

Add.

1. 57
  +31

2. 75
  +61

3. 19
  +80

4. 274
  +615

5. 493
  +702

6. 128
  +931

7. 546
  +323

8. 948
  + 51

9. 5463
  +7124

10. 9807
  +5161

11. 5410
  + 329

12. 8643
  + 321

13. 8263
  135
  +6401

14. 7315
  442
  +9211

15. 5471
  206
  + 112

16. 9641
  105
  + 230

17. 216
  157
  +119

Skill 8 Subtracting Whole Numbers

Subtract.

$$
\begin{array}{r} 8635 \\ -5104 \\ \end{array} \longrightarrow
\begin{array}{r} 8635 \\ -5104 \\ \hline 3531 \end{array}
$$

$$
\begin{array}{r} 6591 \\ -3185 \\ \end{array} \longrightarrow
\begin{array}{r} ^{8\ 11}6591 \\ -3185 \\ \hline 3406 \end{array}
$$

$$
\begin{array}{r} 5073 \\ -1896 \\ \end{array} \longrightarrow
\begin{array}{r} ^{4\ 9\ 16\ 13}5073 \\ -1896 \\ \hline 3177 \end{array}
$$

Subtract.

1. 971
  −531

2. 827
  −513

3. 694
  −572

4. 208
  −103

5. 1965
  − 740

6. 4835
  − 625

7. 6679
  −4238

8. 8341
  −1200

9. 1467
  −1354

10. 5438
  −4216

11. 7032
  −5011

12. 4439
  −2018

13. 39,584
  −17,160

14. 14,761
  −10,431

15. 73,899
  −21,574

16. 94,538
  −61,527

17. 141
  − 38

18. 116
  − 89

19. 147
  − 99

20. 219
  − 53

21. 289
  −193

22. 524
  −376

23. 6843
  −1951

24. 3000
  −1587

25. 61,558
  −29,709

Skill 9 Multiplying Whole Numbers

Multiply.

$$
\begin{array}{r}
332 \\
\times 132 \\
\end{array}
\longrightarrow
\begin{array}{r}
332 \\
\times 132 \\
\hline
664 \\
9960 \\
33200 \\
\hline
43,824 \\
\end{array}
$$

← 2 × 332 ← Multiply by 2 ones.
← 30 × 332 ← Multiply by 3 tens.
← 100 × 332 ← Multiply by 1 hundred.
← 132 × 332

Multiply.

1. 53
    ×31

2. 43
    ×21

3. 714
    × 22

4. 332
    × 13

5. 401
    ×32

6. 832
    × 11

7. 7034
    × 202

8. 4112
    × 331

9. 6021
    × 44

10. 641
    ×122

11. 10,021
    × 13

12. 52,022
    × 42

13. 93,122
    × 231

14. 71,301
    × 121

15. 82,113
    × 302

Multiply.

$$
\begin{array}{r}
863 \\
\times 425 \\
\end{array}
\longrightarrow
\begin{array}{r}
863 \\
\times 425 \\
\hline
4315 \\
17260 \\
345200 \\
\hline
366,775 \\
\end{array}
$$

← 5 × 863 ← Multiply by 5 ones.
← 20 × 863 ← Multiply by 2 tens.
← 400 × 863 ← Multiply by 4 hundreds.
← 425 × 863

Multiply.

16. 28
    ×76

17. 34
    ×73

18. 98
    ×43

19. 109
    × 58

20. 327
    × 66

21. 429
    × 57

22. 643
    ×157

23. 839
    ×277

24. 765
    ×415

25. 834
    ×766

26. 647
    ×519

27. 498
    ×306

28. 6318
    × 119

29. 4265
    × 539

30. 2157
    × 804

Divide. Show the remainder in fraction form.

$$\begin{array}{r} 31 \\ 24\overline{)762} \\ \underline{72} \\ 42 \\ \underline{24} \\ 18 \end{array}$$

$24\overline{)762} \longrightarrow$

The answer is $31\frac{18}{24}$, or $31\frac{3}{4}$.

Divide. Round the quotient to the nearest tenth.

$58\overline{)739} \longrightarrow$

$$\begin{array}{r} 12.74 \\ 58\overline{)739.00} \\ \underline{58} \\ 159 \\ \underline{116} \\ 43\ 0 \\ \underline{40\ 6} \\ 2\ 40 \\ \underline{2\ 32} \\ 8 \end{array}$$

To round the quotient to tenths, carry the division to the hundredths' place; then round. To the nearest tenth, the quotient is 12.7.

Divide. If there is a remainder, show it in fraction form.

1. $38\overline{)114}$ 2. $54\overline{)78}$ 3. $47\overline{)611}$ 4. $81\overline{)702}$ 5. $17\overline{)2975}$

6. $124\overline{)5932}$ 7. $34\overline{)928}$ 8. $763\overline{)28,994}$ 9. $328\overline{)63,509}$ 10. $419\overline{)92,195}$

Divide. Express the quotient as a whole number or a decimal.

11. $6\overline{)75}$ 12. $20\overline{)352}$ 13. $35\overline{)665}$ 14. $12\overline{)8694}$ 15. $25\overline{)7870}$

16. $96\overline{)4524}$ 17. $36\overline{)4122}$ 18. $90\overline{)31,644}$ 19. $64\overline{)34,192}$ 20. $85\overline{)20,111}$

Divide. Round the quotient to the nearest tenth.

21. $17\overline{)239}$ 22. $29\overline{)735}$ 23. $13\overline{)658}$ 24. $53\overline{)2947}$ 25. $61\overline{)8746}$

26. $19\overline{)564}$ 27. $44\overline{)3593}$ 28. $25\overline{)9056}$ 29. $36\overline{)10,362}$ 30. $19\overline{)24,988}$

Divide. Round the quotient to the nearest hundredth.

31. $28\overline{)563}$ 32. $78\overline{)461}$ 33. $18\overline{)4517}$ 34. $92\overline{)7639}$ 35. $61\overline{)8446}$

36. $87\overline{)6398}$ 37. $21\overline{)4283}$ 38. $23\overline{)2525}$ 39. $16\overline{)32,654}$ 40. $55\overline{)12,135}$

Skill 11 Adding Decimals

Add.

$$\begin{array}{r} 351.74 \\ 178.23 \\ +409.17 \\ \hline \end{array} \longrightarrow \begin{array}{r} \scriptstyle 111\ 1 \\ 351.74 \\ 178.23 \\ +409.17 \\ \hline 939.14 \end{array}$$

Decimal points are aligned. If necessary, use zeros as placeholders.

$$\begin{array}{r} 0.67 \\ 5.398 \\ 7.2 \\ +\ 29 \\ \hline \end{array} \longrightarrow \begin{array}{r} \scriptstyle 21\ 1 \\ 0.670 \\ 5.398 \\ 7.200 \\ +29.000 \\ \hline 42.268 \end{array}$$

Add.

1. $\begin{array}{r} 519.76 \\ +248.19 \\ \hline \end{array}$

2. $\begin{array}{r} 748.11 \\ +653.04 \\ \hline \end{array}$

3. $\begin{array}{r} 894.23 \\ +538.97 \\ \hline \end{array}$

4. $\begin{array}{r} 136.47 \\ +951.84 \\ \hline \end{array}$

5. $\begin{array}{r} 460.09 \\ +713.84 \\ \hline \end{array}$

6. $\begin{array}{r} 384.16 \\ +743.45 \\ \hline \end{array}$

7. $\begin{array}{r} 2516.3 \\ +9043.1 \\ \hline \end{array}$

8. $\begin{array}{r} 18.289 \\ +34.503 \\ \hline \end{array}$

9. $\begin{array}{r} 957.361 \\ +441.537 \\ \hline \end{array}$

10. $\begin{array}{r} 14.29 \\ 369.51 \\ +500.78 \\ \hline \end{array}$

11. $\begin{array}{r} 29.06 \\ 139.24 \\ +784.93 \\ \hline \end{array}$

12. $\begin{array}{r} 851.7 \\ 1239.3 \\ +474.6 \\ \hline \end{array}$

13. $\begin{array}{r} 14 \\ +\ 7.1 \\ \hline \end{array}$

14. $\begin{array}{r} 25.1 \\ +16.09 \\ \hline \end{array}$

15. $\begin{array}{r} 13.94 \\ +\ 8.315 \\ \hline \end{array}$

16. $\begin{array}{r} 541.006 \\ +\ 39.13 \\ \hline \end{array}$

17. $\begin{array}{r} 93.701 \\ +115.39 \\ \hline \end{array}$

18. $\begin{array}{r} 29.48 \\ 150.003 \\ +\ 4.1 \\ \hline \end{array}$

19. $\begin{array}{r} 29.504 \\ 11.06 \\ +\ 0.013 \\ \hline \end{array}$

20. $\begin{array}{r} 2.006 \\ 35.793 \\ +\ 0.16 \\ \hline \end{array}$

Skill 12 Subtracting Decimals

Subtract.

$$\begin{array}{r} 605.13 \\ -298.31 \\ \hline \end{array} \longrightarrow \begin{array}{r} \scriptstyle 5\ 9\ 14\ 11 \\ 6\ 0\ 5.1\ 3 \\ -298.31 \\ \hline 306.82 \end{array}$$

Decimal points are aligned. Use zero as a placeholder.

$$\begin{array}{r} 73.8 \\ -\ 6.93 \\ \hline \end{array} \longrightarrow \begin{array}{r} \scriptstyle 6\ 12\ 17\ 10 \\ 7\ 3.8\ 0 \\ -\ 6.93 \\ \hline 66.87 \end{array}$$

Subtract.

1. $\begin{array}{r} 63.87 \\ -54.49 \\ \hline \end{array}$

2. $\begin{array}{r} 33.17 \\ -19.64 \\ \hline \end{array}$

3. $\begin{array}{r} 72.09 \\ -48.36 \\ \hline \end{array}$

4. $\begin{array}{r} 46.139 \\ -18.096 \\ \hline \end{array}$

5. $\begin{array}{r} 100.036 \\ -\ 95.147 \\ \hline \end{array}$

6. $\begin{array}{r} 59.218 \\ -27.077 \\ \hline \end{array}$

7. $\begin{array}{r} 183.76 \\ -\ 42.09 \\ \hline \end{array}$

8. $\begin{array}{r} 193.29 \\ -\ 76.11 \\ \hline \end{array}$

9. $\begin{array}{r} 349.001 \\ -\ 60.365 \\ \hline \end{array}$

10. $\begin{array}{r} 0.6509 \\ -0.0311 \\ \hline \end{array}$

11. $\begin{array}{r} 400.31 \\ -278.74 \\ \hline \end{array}$

12. $\begin{array}{r} 112.453 \\ -103.926 \\ \hline \end{array}$

13. $\begin{array}{r} 53.2 \\ -\ 7.56 \\ \hline \end{array}$

14. $\begin{array}{r} 29.8 \\ -\ 5.17 \\ \hline \end{array}$

15. $\begin{array}{r} 96.24 \\ -\ 7.6 \\ \hline \end{array}$

16. $\begin{array}{r} 713.63 \\ -\ 5.379 \\ \hline \end{array}$

17. $\begin{array}{r} 11 \\ -\ 0.03 \\ \hline \end{array}$

Skill 13 Multiplying Decimals

Multiply.

$$\begin{array}{r} 23.78 \\ \times\ 4.6 \\ \hline 14268 \\ 95120 \\ \hline 109.388 \end{array}$$ ← 2 places
← + 1 place

← 3 places

$$\begin{array}{r} 0.07 \\ \times\ 0.3 \\ \hline 0.021 \end{array}$$ ← 2 places
← + 1 place
← 3 places

↑
Insert one zero as a placeholder.

Multiply.

1. $\begin{array}{r} 5.3 \\ \times\ 2.6 \\ \hline \end{array}$ **2.** $\begin{array}{r} 7.1 \\ \times\ 9 \\ \hline \end{array}$ **3.** $\begin{array}{r} 8.4 \\ \times\ 3.9 \\ \hline \end{array}$

4. $\begin{array}{r} 12.7 \\ \times\ 4.1 \\ \hline \end{array}$ **5.** $\begin{array}{r} 13.8 \\ \times\ 9.5 \\ \hline \end{array}$ **6.** $\begin{array}{r} 17 \\ \times\ 3.9 \\ \hline \end{array}$

7. $\begin{array}{r} 6.07 \\ \times\ 3.15 \\ \hline \end{array}$ **8.** $\begin{array}{r} 7.13 \\ \times\ 8.4 \\ \hline \end{array}$ **9.** $\begin{array}{r} 1.4 \\ \times\ 5.33 \\ \hline \end{array}$

10. $\begin{array}{r} 23.71 \\ \times\ 19.46 \\ \hline \end{array}$ **11.** $\begin{array}{r} 38.11 \\ \times\ 76.5 \\ \hline \end{array}$ **12.** $\begin{array}{r} 98 \\ \times 45.71 \\ \hline \end{array}$

13. $\begin{array}{r} 0.008 \\ \times\ 0.6 \\ \hline \end{array}$ **14.** $\begin{array}{r} 0.56 \\ \times\ 0.3 \\ \hline \end{array}$ **15.** $\begin{array}{r} 0.019 \\ \times 0.27 \\ \hline \end{array}$

Skill 14 Dividing Decimals

Divide. $3.9\overline{)22.542}$

$3.9\overline{)22.5\,42}$

Move both decimal points one place to the right.

$$\begin{array}{r} 5.78 \\ 39\overline{)225.42} \\ 195 \\ \hline 30\ 4 \\ 27\ 3 \\ \hline 3\ 12 \\ 3\ 12 \\ \hline 0 \end{array}$$

Divide. $0.076\overline{)17.48}$

Annex a zero.
↓
$0.076\overline{)17.480}$

↑
Move three places to the right.

$$\begin{array}{r} 230 \\ 76\overline{)17480} \\ 152 \\ \hline 228 \\ 228 \\ \hline 00 \\ 0 \\ \hline 0 \end{array}$$

Divide. If necessary, round the quotient to the nearest hundredth.

1. $3.5\overline{)27.3}$ **2.** $1.9\overline{)7.03}$ **3.** $1.6\overline{)10.384}$ **4.** $2.3\overline{)39.567}$

5. $7.82\overline{)148.58}$ **6.** $19.51\overline{)572.091}$ **7.** $2.04\overline{)27.336}$ **8.** $5.06\overline{)47.058}$

9. $12.53\overline{)226.793}$ **10.** $19.07\overline{)87.722}$ **11.** $0.56\overline{)21.28}$ **12.** $0.512\overline{)6.656}$

13. $0.381\overline{)24.003}$ **14.** $0.27\overline{)5.316}$ **15.** $0.63\overline{)73.081}$ **16.** $0.84\overline{)21.84}$

Skill 15 Multiplying and Dividing by 10, 100, and 1000

Multiply.

176.34×10 $= 176.34$ $= 1763.4$
Move one place
to the right.

89.6×100 $= 89.60$ $= 8960$
Move two places
to the right.

0.6043×1000 $= 0.6043$ $= 604.3$
Move three places
to the right.

Divide.

$183.24 \div 10$ $= 183.24$ $= 18.324$
Move one place
to the left.

$0.0319 \div 100$ $= 00.0319$ $= 0.000319$
Move two places
to the left.

$17.83 \div 1000$ $= 017.83$ $= 0.01783$
Move three places
to the left.

Multiply by 10.

1. 0.803	**2.** 13.8	**3.** 813.06	**4.** 924
5. 54.1	**6.** 0.519	**7.** 7.092	**8.** 0.036

Multiply by 100.

9. 28.4	**10.** 6.948	**11.** 0.3	**12.** 0.092
13. 127.51	**14.** 17.316	**15.** 8.04	**16.** 14.0052

Multiply by 1000.

17. 0.0417	**18.** 219.3	**19.** 0.012	**20.** 16.045
21. 9.31	**22.** 0.66	**23.** 82.1	**24.** 0.24054

Divide by 10.

25. 216.35	**26.** 0.31	**27.** 5.106	**28.** 23
29. 28.57	**30.** 84.19	**31.** 0.0792	**32.** 820.51

Divide by 100.

33. 29.3	**34.** 8.47	**35.** 930.1	**36.** 0.064
37. 1847.5	**38.** 506.12	**39.** 0.29	**40.** 605.8

Divide by 1000.

41. 413	**42.** 108.3	**43.** 0.6	**44.** 135.09
45. 26.9	**46.** 63,218	**47.** 0.025	**48.** 9870

Skill 16 Divisibility Tests

A number is divisible by a second number if the remainder is zero when the first number is divided by the second.

The table below shows divisibility tests you can use to determine if one number is divisible by another number.

Divisible by	Test
2	The digit in the ones' place is 0, 2, 4, 6, or 8.
5	The digit in the ones' place is 0 or 5.
10	The digit in the ones' place is 0.
3	The sum of the digits is divisible by 3.
9	The sum of the digits is divisible by 9.
4	The number formed by the last two digits is divisible by 4.
8	The number formed by the last three digits is divisible by 8.
6	The number is divisible by both 2 and 3.

Examples:

Is 5435 divisible by 5? $\longrightarrow$ Yes; the digit in the ones' place is 5.

Is 3741 divisible by 9? $\longrightarrow$ No; the sum of the digits is $3 + 7 + 4 + 1 = 15$, and 15 is not divisible by 9.

Is 3564 divisible by 8? $\longrightarrow$ No; the number formed by the last three digits is not divisible by 8.

Is 4938 divisible by 6? $\longrightarrow$ Yes, the digit in the ones' place is 8 (which is divisible by 2) and the sum of the digits is 24 (which is divisible by 3).

Test the number for divisibility. Write *Yes* or *No*.

By 2:	**1.** 138	**2.** 203	**3.** 517
By 5:	**4.** 135	**5.** 730	**6.** 219
By 10:	**7.** 90	**8.** 102	**9.** 305
By 3:	**10.** 216	**11.** 735	**12.** 889
By 9:	**13.** 258	**14.** 792	**15.** 1035
By 4:	**16.** 7248	**17.** 838	**18.** 2344
By 8:	**19.** 7688	**20.** 312	**21.** 57,680
By 6:	**22.** 8324	**23.** 6678	**24.** 504

Skill 17 Adding and Subtracting Fractions

Add or subtract. Write the answer in lowest terms.

$\dfrac{5}{7}$ $\left\{\begin{array}{l}\end{array}\right.$ The fractions have a common denominator. Add the numerators.

$+\dfrac{3}{7}$

$\dfrac{8}{7} = 1\dfrac{1}{7}$

$\dfrac{5}{9}$ $\left\{\begin{array}{l}\end{array}\right.$ The fractions have a common denominator. Subtract the numerators.

$-\dfrac{1}{9}$

$\dfrac{4}{9}$

Add or subtract. Write the answer in lowest terms.

1. $\dfrac{5}{9}$
   $+\dfrac{2}{9}$

2. $\dfrac{1}{5}$
   $+\dfrac{4}{5}$

3. $\dfrac{11}{20}$
   $+\dfrac{5}{20}$

4. $\dfrac{2}{5}$
   $\dfrac{1}{5}$
   $+\dfrac{4}{5}$

5. $\dfrac{3}{7}$
   $\dfrac{5}{7}$
   $+\dfrac{6}{7}$

6. $\dfrac{13}{100}$
   $\dfrac{3}{100}$
   $+\dfrac{9}{100}$

7. $\dfrac{11}{15}$
   $-\dfrac{8}{15}$

8. $\dfrac{11}{12}$
   $-\dfrac{5}{12}$

9. $\dfrac{23}{24}$
   $-\dfrac{11}{24}$

10. $\dfrac{7}{8}$
    $-\dfrac{3}{8}$

11. $\dfrac{9}{10}$
    $-\dfrac{3}{10}$

12. $\dfrac{29}{50}$
    $-\dfrac{9}{50}$

13. $\dfrac{91}{100}$
    $-\dfrac{31}{100}$

14. $\dfrac{5}{6}$
    $-\dfrac{1}{6}$

Add or subtract. Write the answer in lowest terms.

When the fractions have different denominators, rewrite them as equivalent fractions having the least common denominator. Then add or subtract.

$\dfrac{7}{12}$ $\longrightarrow$ $\dfrac{35}{60}$

$+\dfrac{3}{20}$ $\qquad$ $+\dfrac{9}{60}$

$\qquad\qquad\qquad \dfrac{44}{60} = \dfrac{11}{15}$

$\dfrac{5}{9}$ $\longrightarrow$ $\dfrac{10}{18}$

$-\dfrac{1}{6}$ $\qquad$ $-\dfrac{3}{18}$

$\qquad\qquad\qquad \dfrac{7}{18}$

15. $\dfrac{3}{16}$
    $+\dfrac{3}{8}$

16. $\dfrac{7}{20}$
    $+\dfrac{3}{4}$

17. $\dfrac{7}{9}$
    $+\dfrac{11}{12}$

18. $\dfrac{3}{4}$
    $\dfrac{1}{6}$
    $+\dfrac{5}{8}$

19. $\dfrac{3}{10}$
    $\dfrac{7}{45}$
    $+\dfrac{11}{90}$

20. $\dfrac{1}{2}$
    $\dfrac{5}{6}$
    $+\dfrac{4}{9}$

21. $\dfrac{6}{7}$
    $-\dfrac{1}{3}$

22. $\dfrac{4}{5}$
    $-\dfrac{3}{8}$

23. $\dfrac{9}{10}$
    $-\dfrac{7}{15}$

24. $\dfrac{7}{10}$
    $-\dfrac{1}{4}$

25. $\dfrac{11}{50}$
    $-\dfrac{1}{30}$

26. $\dfrac{5}{6}$
    $-\dfrac{5}{8}$

Skill 18 Adding Mixed Numbers

Add.
First add the fractions.
Then add the whole
numbers.

$$3\frac{4}{5}$$
$$+9\frac{3}{5}$$

$$12\frac{7}{5} = 12 + 1\frac{2}{5} = 13\frac{2}{5}$$

Add. Write each sum in lowest terms.

1. $5\frac{7}{11}$
    $+2\frac{3}{11}$

2. 8
    $+3\frac{1}{7}$

3. $6\frac{5}{9}$
    $+11$

4. $14\frac{3}{8}$
    $+ 5\frac{5}{8}$

5. $19\frac{5}{12}$
    $+ 3\frac{11}{12}$

6. $13\frac{5}{16}$
    $+ 7\frac{15}{16}$

7. $23\frac{26}{45}$
    $+15\frac{7}{45}$

8. $16\frac{21}{40}$
    $+18\frac{29}{40}$

9. $9\frac{7}{10}$
    $+2\frac{9}{10}$

If necessary, rewrite the
fractions as equivalent
fractions with a common
denominator. Then add.

$$16\frac{3}{4} = \quad 16\frac{9}{12}$$
$$+ 7\frac{2}{3} = + \; 7\frac{8}{12}$$
$$23\frac{17}{12} =$$
$$23 + 1\frac{5}{12} = 24\frac{5}{12}$$

Add. Write each sum in lowest terms.

10. $7\frac{1}{2}$
    $+5\frac{3}{4}$

11. $9\frac{7}{10}$
    $+16\frac{1}{4}$

12. $13\frac{5}{8}$
    $+ 7\frac{5}{6}$

13. $14\frac{7}{12}$
    $+ 5\frac{2}{9}$

14. $10\frac{3}{7}$
    $+ 6\frac{1}{3}$

15. $12\frac{9}{10}$
    $+ 4\frac{2}{5}$

16. $18\frac{1}{6}$
    $+ 26\frac{5}{9}$

17. $4\frac{7}{8}$
    $+25\frac{3}{4}$

18. $3\frac{5}{8}$
    $+2\frac{7}{12}$

19. $13\frac{5}{12}$
    $+17\frac{5}{6}$

20. $27\frac{9}{10}$
    $+16\frac{7}{45}$

21. $42\frac{11}{15}$
    $+40\frac{7}{12}$

22. $9\frac{91}{100}$
    $+3\frac{21}{25}$

23. $35\frac{5}{6}$
    $+15\frac{1}{8}$

24. $12\frac{11}{32}$
    $4\frac{15}{32}$
    $+ 5\frac{13}{32}$

25. $21\frac{4}{5}$
    $18\frac{3}{5}$
    $+ 6\frac{1}{5}$

26. $15\frac{13}{20}$
    $6\frac{1}{3}$
    $+ 7\frac{2}{5}$

27. $10\frac{1}{4}$
    $8\frac{1}{3}$
    $+ 7\frac{3}{8}$

28. $19\frac{3}{10}$
    $6\frac{5}{8}$
    $+11\frac{9}{16}$

Skill 19 Subtracting Mixed Numbers

Subtract.

Subtract the fractions. Then subtract the whole numbers.

$$17\frac{7}{12}$$
$$-\ 9\frac{1}{12}$$
$$\overline{\quad 8\frac{6}{12} = 8\frac{1}{2}}$$

Subtract. Write each difference in lowest terms.

1. $4\frac{6}{7}$
    $-2\frac{1}{7}$

2. $36\frac{5}{7}$
    $-13\frac{2}{7}$

3. $21\frac{19}{32}$
    $-\ 8\frac{15}{32}$

4. $8\frac{3}{4}$
    $-5\frac{1}{4}$

5. $6\frac{7}{8}$
    $-1\frac{3}{8}$

6. $14\frac{13}{20}$
    $-10\frac{7}{20}$

If necessary, first rewrite the fractions as equivalent fractions with a common denominator.

$$14\frac{9}{16} = \quad 14\frac{9}{16}$$
$$-\ 6\frac{1}{4} = -\ 6\frac{4}{16}$$
$$\overline{\qquad\qquad 8\frac{5}{16}}$$

Subtract. Write each difference in lowest terms.

7. $8\frac{3}{4}$
    $-5\frac{1}{3}$

8. $51\frac{11}{16}$
    $-36\frac{1}{8}$

9. $26\frac{7}{8}$
    $-21\frac{1}{4}$

10. $7\frac{1}{2}$
    $-5\frac{5}{12}$

11. $22\frac{19}{40}$
    $-\ 9\frac{7}{20}$

12. $13\frac{5}{18}$
    $-\ 6\frac{1}{12}$

13. $25\frac{5}{7}$
    $-10\frac{1}{6}$

14. $39\frac{8}{9}$
    $-30\frac{5}{18}$

15. $16\frac{7}{8}$
    $-12\frac{1}{5}$

If necessary, first rename a whole number or mixed number.

$$4\frac{3}{10} = \quad 4\frac{3}{10} = \quad 3\frac{13}{10}$$
$$-2\frac{4}{5} = -\ 2\frac{8}{10} = -\ 2\frac{8}{10}$$
$$\overline{\qquad\qquad\qquad\qquad\quad 1\frac{5}{10}}$$
$$= 1\frac{1}{2}$$

Subtract. Write each difference in lowest terms.

16. $9\frac{11}{20}$
    $-7\frac{3}{8}$

17. 14
    $-10\frac{3}{10}$

18. 6
    $-2\frac{5}{8}$

19. $19\frac{5}{16}$
    $-\ 6\frac{5}{8}$

20. $6\frac{2}{9}$
    $-4\frac{2}{3}$

21. $24\frac{5}{6}$
    $-10\frac{11}{12}$

22. $11\frac{3}{10}$
    $-\ 7\frac{11}{15}$

23. $6\frac{3}{14}$
    $-3\frac{5}{7}$

24. $4\frac{3}{20}$
    $-1\frac{11}{15}$

Skill 20 Multiplying Fractions and Mixed Numbers

Multiply. Write the product in lowest terms.

$$\frac{7}{8} \times \frac{4}{13} = \frac{7 \times \overset{1}{4}}{\underset{2}{8} \times 13} = \frac{7}{26} \qquad\qquad 5 \times \frac{3}{4} = \frac{5}{1} \times \frac{3}{4} = \frac{5 \times 3}{1 \times 4} = \frac{15}{4} = 3\frac{3}{4}$$

$$2\frac{1}{5} \times 3\frac{1}{3} = \frac{11}{5} \times \frac{10}{3} = \frac{11 \times \overset{2}{10}}{\underset{1}{5} \times 3} = \frac{22}{3} = 7\frac{1}{3}$$

Write each product in lowest terms.

1. $\frac{1}{6} \times \frac{2}{5}$ 2. $\frac{1}{7} \times \frac{2}{9}$ 3. $\frac{5}{11} \times \frac{3}{10}$ 4. $16 \times \frac{3}{8}$

5. $\frac{10}{13} \times 26$ 6. $\frac{1}{4} \times \frac{5}{9}$ 7. $\frac{7}{12} \times \frac{2}{21}$ 8. $\frac{5}{8} \times \frac{2}{3}$

9. $\frac{1}{2} \times \frac{10}{13}$ 10. $18 \times \frac{7}{9}$ 11. $24 \times \frac{5}{6}$ 12. $\frac{1}{3} \times \frac{9}{20}$

13. $10 \times 7\frac{3}{5}$ 14. $\frac{5}{9} \times 2\frac{7}{10}$ 15. $5\frac{1}{3} \times 1\frac{1}{8}$ 16. $5\frac{1}{2} \times \frac{8}{11}$

Skill 21 Dividing Fractions and Mixed Numbers

To divide with fractions and mixed numbers, multiply the dividend by the reciprocal of the divisor.

$$\overset{\text{reciprocal of } \frac{2}{5}}{\frac{5}{6} \div \frac{2}{5}} = \frac{5}{6} \times \frac{5}{2} = \frac{25}{12} = 2\frac{1}{12} \qquad\qquad \frac{9}{16} \div 3 = \frac{\overset{3}{9}}{16} \times \frac{1}{\underset{1}{3}} = \frac{3}{16}$$

$$15\frac{2}{3} \div 1\frac{7}{9} = \frac{47}{3} \div \frac{16}{9} = \frac{47}{\underset{1}{3}} \times \frac{\overset{3}{9}}{16} = \frac{141}{16} = 8\frac{13}{16}$$

Write each quotient in lowest terms.

1. $\frac{1}{5} \div \frac{1}{15}$ 2. $\frac{5}{6} \div \frac{1}{12}$ 3. $\frac{3}{7} \div \frac{6}{11}$ 4. $8 \div \frac{4}{7}$

5. $\frac{2}{3} \div \frac{4}{9}$ 6. $\frac{10}{13} \div \frac{25}{26}$ 7. $\frac{5}{12} \div \frac{9}{10}$ 8. $\frac{5}{9} \div 10$

9. $\frac{3}{8} \div \frac{5}{7}$ 10. $2\frac{4}{7} \div \frac{3}{4}$ 11. $1\frac{11}{19} \div 7\frac{1}{2}$ 12. $35 \div \frac{7}{10}$

Skill 22 U.S. Customary System of Measurement

Use the Table of Measures on page xiii.

To change from a larger unit to a smaller unit, you *multiply*.

7 ft 8 in. = __?__ in.

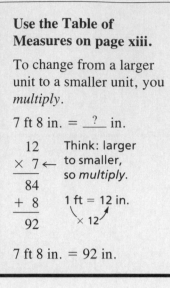

7 ft 8 in. = 92 in.

Complete by changing a larger unit to a smaller unit.

1. 12 mi = __?__ yd

2. 3 gal = __?__ qt

3. 5.5 t = __?__ lb

4. 2.5 gal = __?__ qt

5. $9\frac{1}{3}$ yd = __?__ ft

6. 3 t 350 lb = __?__ lb

7. $2\frac{1}{4}$ mi = __?__ ft

8. 2 lb 13 oz = __?__ oz

9. $4\frac{1}{2}$ t = __?__ lb

10. 39 yd = __?__ ft

11. 12 qt 1 pt = __?__ pt

12. 8.5 lb = __?__ oz

13. $\frac{1}{2}$ yd = __?__ in.

14. 7 lb 8 oz = __?__ oz

To change from a smaller unit to a larger unit, you *divide*.

36 oz = __?__ lb __?__ oz

$$16\overline{)36}$$ Think: smaller to larger, so *divide*.

32

4   16 oz = 1 lb

÷ 16

32 oz = 2 lb 4 oz

Complete by changing a smaller unit to a larger unit.

15. 37 qt = __?__ gal __?__ qt

16. 156 in. = __?__ ft

17. 110 ft = __?__ yd __?__ ft

18. 32 pt = __?__ qt

19. 72 in. = __?__ ft

20. 6000 lb = __?__ t

21. 23 c = __?__ pt __?__ c

22. 128 oz = __?__ lb

23. 10 pt = __?__ qt

24. 18 c = __?__ pt

25. 24 fl oz = __?__ pt

26. 41 ft = __?__ yd __?__ ft

27. 24 pt = __?__ gal

28. 880 ft = __?__ yd __?__ ft

Complete.

29. $3\frac{1}{2}$ c = __?__ fl oz

30. $7\frac{1}{2}$ ft = __?__ in.

31. 64 fl oz = __?__ c

32. 108 in. = __?__ yd

33. $1\frac{1}{2}$ mi = __?__ yd

34. 97 ft = __?__ yd __?__ ft

35. $1\frac{1}{2}$ t = __?__ oz

36. 19 qt = __?__ gal __?__ qt

Skill 23 Reading Pictographs

To read a pictograph, find the key to see the amount that each symbol represents. Then multiply that amount by the number of symbols on a line to get the total amount.

About how many letters did each household receive in 1940?

Each symbol represents 50 letters. The 1940 row has 2 symbols.

$$2 \times 50 = 100$$

Each household received about 100 letters in 1940.

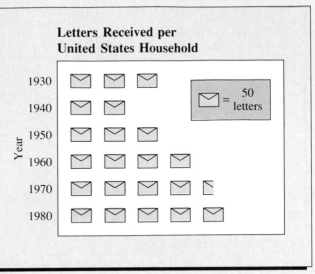

Letters Received per United States Household

Use the pictograph above.

1. In what year was the greatest number of letters received? In what year was the least number of letters received?

2. About how many letters did each household receive in 1970?

3. About how many more letters were received in 1960 than in 1940?

4. About how many fewer letters were received in 1930 than in 1960?

5. During what two years was the number of letters received the same?

6. In what year was the number of letters received twice the number received in 1940?

Use the pictograph at the right.

7. About how many pounds of garbage per citizen were collected daily in New York?

8. In which city was the least amount of garbage per citizen, per day, collected?

9. In which city was the greatest amount of garbage per citizen, per day, collected?

10. About how many more pounds of garbage per citizen were collected daily in Hartford than in Seattle?

Daily Garbage Collection per Citizen in 1967

Atlanta

Hartford

Los Angeles

New York

Seattle

Washington, D.C.

= 1 lb

To read a bar graph, find the bar that represents the information you seek. Trace an imaginary line from the end of the bar to the scale provided. Read the value represented by the bar from the scale.

About how many cars were sold in July?

The bar graph shows that about 16 cars were sold in July.

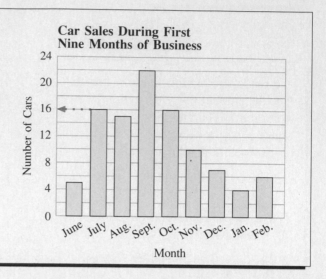

Car Sales During First Nine Months of Business

Use the bar graph above.

1. During which month were the most cars sold? the fewest?

2. During which two months was the same number of cars sold?

3. About how many more cars were sold in November than in February?

4. About how many fewer cars were sold in October than in September?

5. Were car sales increasing or decreasing from September to January?

6. About how many cars were sold altogether during September, October, and November?

Use the bar graph at the right.

7. In which city were air conditioners used for the greatest number of hours?

8. For about how many hours were air conditioners used in St. Louis?

9. For about how many fewer hours were air conditioners used in Chicago than in Dallas?

10. Were air conditioners used for more hours in New York or Atlanta?

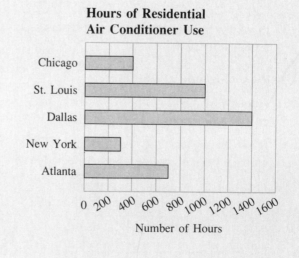

Hours of Residential Air Conditioner Use

Skill 25 Reading Line Graphs

To read a line graph, find the point that represents the information you seek. Trace an imaginary line from the point to the scale on the left. Read the value represented by the point from the scale.

About how many phone calls are placed at 8 P.M.?

The line graph shows that about 30,000 calls are placed at 8 P.M.

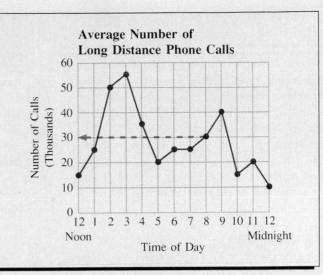

Average Number of Long Distance Phone Calls

Number of Calls (Thousands) / Time of Day (12 Noon, 1, 2, 3, 4, 5, 6, 7, 8, 9, 10, 11, 12 Midnight)

Use the line graph above.

1. At what time is the least number of long distance phone calls placed?

2. At what time is the greatest number of long distance phone calls placed?

3. About how many more long distance phone calls are placed at 9 P.M. than at 5 P.M.?

4. About how many fewer long distance phone calls are placed at 11 P.M. than at 2 P.M.?

5. From noon to 3 P.M., is the number of long distance phone calls increasing or decreasing?

6. At what time is the average number of long distance phone calls placed 50,000?

Use the line graph at the right.

7. During which month were the fewest homes for sale?

8. During which month were the most homes for sale?

9. About how many homes were for sale in August?

10. About how many homes were for sale in October?

11. About how many more homes were for sale in May than in September?

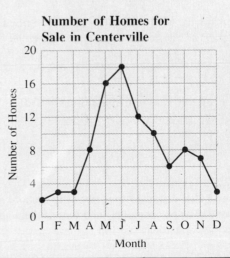

Number of Homes for Sale in Centerville

Number of Homes / Month (J F M A M J J A S O N D)

GLOSSARY

absolute value (p. 95): The distance that a number is from zero on a number line.

acute angle (p. 245): An angle whose measure is greater than 0° and less than 90°.

acute triangle (p. 262): A triangle whose angles are all acute angles.

addition property of opposites (p. 102): The sum of a number and its opposite is zero.

$$a + (-a) = 0 \text{ and } -a + a = 0$$

additive identity (p. 22): The number 0 (zero).

additive inverse (p. 102): The opposite of a number.

adjacent angles (p. 246): Two angles that share a common side but do not overlap each other.

adjacent side (p. 716): In a right triangle, the side of a given angle that is not the hypotenuse.

algebraic fraction (p. 306): A fraction that contains a variable.

alternate interior angles (p. 250): Two angles on alternate sides of a transversal that intersects two lines in the same plane, and interior to the two lines.

angle (p. 237): The figure formed by two rays (called *sides*) which share a common endpoint (called the *vertex*).

area (p. 445): The region enclosed by a plane figure.

arrow notation (p. 66): The symbol $\rightarrow$ used to show how one number is paired with another in a given function.

associative property of addition (p. 22): Changing the grouping of the terms does not change the sum.

$$(a + b) + c = a + (b + c)$$

associative property of multiplication (p. 25): Changing the grouping of the factors does not change the product.

$$(ab)c = a(bc)$$

average (p. 213): *See* mean.

axes (p. 124): The two number lines that form a coordinate plane. The horizontal number line is the *x-axis*, and the vertical number line is the *y-axis*.

bar graph (p. 192): A type of graph in which the lengths of bars are used to compare data.

base (pp. 446, 628): Either side of a pair of sides of a parallelogram; part of polyhedron used to identify a polyhedron.

base of a power (p. 30): A number that is used as a factor a given number of times. In 5^3, 5 is the base.

binomial (p. 668): A polynomial that has two terms.

box-and-whisker plot (p. 505): A display showing the median of a set of data, the median of each half of the data, and the least and greatest values of the data.

center of a circle (p. 440): The given point in a plane from which all points in a circle are a given distance.

center of rotation (p. 710): The point about which a rotation is made.

central angles (p. 517): Angles having their vertex at the center of the circle.

chord (p. 440): A line segment whose endpoints are both on the circle.

circle (p. 440): The set of all points in a plane that are a given distance from a given point.

circle graph (p. 513): A diagram used to represent data expressed as parts of a whole.

circumference (p. 441): The distance around a circle.

closure (p. 137): A given operation performed on a set of numbers results in a number also in the set.

collinear points (p. 236): Points that lie on the same line.

combination (p. 560): A group of items chosen from within a group of items without regard to order.

combining like terms (p. 56): The process of adding or subtracting like terms.

common factor (p. 293): A number that is a factor of two numbers.

common multiple (p. 296): A number that is a multiple of two numbers.

commutative property of addition (p. 22): Changing the order of the terms does not change the sum.

$$a + b = b + a$$

commutative property of multiplication (p. 25): Changing the order of the factors does not change the product.

$$ab = ba$$

comparison property (p. 18): For any two numbers a and b, exactly one of the following is true: $a > b$, $a < b$, or $a = b$.

complementary angles (p. 245): Two angles the sum of whose measures is 90°.

composite number (p. 290): A number with more than two factors.

composite space figure (p. 663): A combination of two or more space figures.

cone (p. 629): A space figure with one circular base and a vertex.

congruent figures (p. 462): Geometric figures that have the same size and shape.

coordinate plane (p. 124): A grid formed by two number lines that meet at a point.

corresponding angles (pp. 250, 463): Two angles in the same position with respect to two lines and a transversal. *Also* angles in the same position in congruent or similar polygons.

cosine (p. 716): In right triangle ABC,

$$\cos A = \frac{\text{length of side adjacent to } \angle A}{\text{length of hypotenuse}} = \frac{b}{c}.$$

cost (p. 224): The amount of money spent.

counting principle (p. 549): The number of outcomes for an event is found by multiplying the number of choices for each stage of the event.

cross products (p. 391): In a proportion, the product of the first numerator and the second denominator; also, the product of the first denominator and the second numerator.

cube (p. 628): A rectangular prism whose faces are all squares.

cylinder (p. 629): A space figure with two circular bases that are congruent and parallel.

---

d

data (p. 188): A collection of numerical facts.

decimal notation (p. 80): The writing of a number using place values that are powers of ten.

decreasing trend (p. 196): A decrease in the data over a given interval on a line graph, shown by a series of segments that slopes downward.

degree (p. 240): The unit commonly used to measure the size of an angle.

dependent events (p. 553): Events for which the occurrence of the first affects that of the second.

diagonal (p. 256): A line segment that joins two non-consecutive vertices.

diameter (p. 440): Any chord that passes through the center of a circle.

dilation (p. 713): The image of a figure similar to the original figure.

distributive property (p. 52): Each term inside a set of parentheses can be multiplied by a factor outside the parentheses. For example, $3(80 + 10) = 3(80) + 3(10)$.

domain of a function (p. 613): The set of all possible values of x in a given function.

---

e

edge (p. 628): A line segment on a space figure where two faces intersect.

endpoint (p. 237): Point included in a line segment or ray that defines that line segment or ray.

enlargement (p. 713): The image of a figure after a dilation that is larger than the original.

equally likely (p. 541): Having the same likelihood.

equation (p. 142): A statement that two numbers or two expressions are equal.

equilateral triangle (p. 261): A triangle whose sides all have the same length.

equivalent fractions (p. 302): Fractions that represent the same amount.

evaluate (p. 4): To find the value of a variable expression when a number is substituted for the variable.

event (p. 541): In probability, any group of outcomes.

expected value of an event (p. 577): The sum of the products of each outcome and its probability.

experimental probability (p. 564): Probability based on collection of actual data.

exponent (p. 30): A number used to show how many times another number is used as a factor. In 5^3, 3 is the exponent.

exponential form (p. 30): A shortened form of a multiplication expression in which all the factors are the same. 2^4 is the exponential form of $2 \cdot 2 \cdot 2 \cdot 2$.

f

face (p. 628): The flat surface of a space figure.

factor (p. 290): When a whole number is divisible by a second whole number, the second number is a *factor* of the first.

factor a polynomial (p. 693): Express a polynomial whose coefficients are whole numbers as the product of other polynomials whose coefficients are whole numbers.

factorial (p. 559): A notation for the product of a number and all nonzero whole numbers less than the given number.

Fibonacci sequence (p. 62): A pattern in nature where, beginning with the third number, 2, each number is the sum of the two numbers immediately preceding it. 1, 1, 2, 3, 5, 8, 13, 21, ...

first quartile of a set of data (p. 505): The median of the lower part of a set of data divided by its median.

formula (p. 176): An equation that states a relationship between two or more quantities.

45°-45°-90° right triangle (p. 726): An isosceles right triangle.

frequency polygon (p. 501): A type of line graph used to show frequencies.

frequency table (p. 221): A form of organizing a set of data to show how often each item in the data occurs.

function (p. 66): A relationship that pairs each number in a given set of numbers with *exactly one* number in a second set of numbers.

function notation (p. 613): A useful shorthand for showing how values are paired in a function.

function rule (p. 66): The description of a function.

function table (p. 66): A table used to show values of the variable expression for a given function.

g

geometric transformation (p. 714): Process of changing the position, size, or shape of a figure according to a given rule.

gram (p. 76): The basic unit of measure for mass in the metric system.

graph (p. 188): A picture that displays numerical facts, called data.

graph of an equation with two variables (p. 585): All points whose coordinates are solutions of an equation.

graph of a function (p. 128): The points that correspond to all the ordered pairs of a function.

graph of an open sentence (p. 324): A graph of all the solutions on a number line.

graph of a point (p. 124): The point assigned to an ordered pair on a coordinate plane.

greatest common factor (GCF) (p. 293): The greatest number in a list of common factors.

greatest common monomial factor of a polynomial (p. 693): The GCF of the terms of the polynomial.

h

height (p. 446): The perpendicular distance between bases of a parallelogram.

histogram (p. 501): A type of bar graph that is used to show frequencies.

hypotenuse (p. 481): In a right triangle, the side opposite the right angle.

i

identity property of addition (p. 23): The sum of any number and zero is the original number.

$$a + 0 = a$$

identity property of multiplication (p. 25): The product of any number and 1 is the original number.

$$a \cdot 1 = a$$

image (p. 706): A figure moved to a new location by rigid motion.

increasing trend (p. 196): An increase in the data over a given interval on a line graph, shown by a series of segments that slopes upward.

independent events (p. 552): Events for which the occurrence of the first event does not affect that of the second.

indirect measurement (p. 471): Finding an unknown measure that cannot be measured easily by direct measurement, by using similar triangles.

inequality (p. 325): A mathematical sentence that has an inequality symbol between two numbers or quantities.

inequality symbols (p. 18): The symbols $>$ (is greater than) and $<$ (is less than).

integer (p. 94): Any number in the set {. . ., -3, -2, -1, 0, 1, 2, 3, . . .}.

intersect (p. 236): To meet. Two lines intersect in a point. Two distinct planes intersect in a line.

inverse operations (p. 149): Operations that undo each other. Addition and subtraction are inverse operations; multiplication and division are inverse operations.

irrational number (p. 321): A number that cannot be written as the quotient of two integers.

isosceles triangle (p. 261): A triangle that has at least two sides of the same length.

k

key to a pictograph (p. 188): The part of the graph that tells how many items one symbol represents.

l

leaf (p. 497): The last digit on the right of a number displayed in a stem-and-leaf plot.

least common denominator (LCD) (p. 310): The least common multiple of the denominators when comparing fractions.

least common multiple (LCM) (p. 296): For two numbers, the least number in the list of their common multiples.

legend (p. 192): An identifying label on a double bar or double line graph.

legs (p. 481): The two sides of a right triangle that are shorter than the hypotenuse.

like terms (p. 56): Terms having identical variable parts.

line (p. 236): A straight arrangement of points that extends forever in opposite directions.

line graph (p. 196): A type of graph using points and line segments to show both amount and direction of change.

line segment (p. 237): A part of a line that consists of two points, called *endpoints*, and all the points between.

line symmetry (p. 275): The property of a pattern or geometric figure that "folds" along a line such that one half fits exactly over the other. The line on which it folds is called a *line of symmetry*.

liter (p. 76): The basic unit of measure for liquid capacity in the metric system.

loss (p. 224): The difference between cost and revenue when cost is greater than revenue.

lowest terms (p. 303): A fraction is in *lowest terms* if the GCF of the numerator and the denominator is 1.

m

mean (p. 213): The sum of the items in a set of data divided by the number of items; also called *average*.

measures of central tendency (p. 218): The mean, median, and mode of a set of data.

median (p. 213): The middle item in a set of data listed in numerical order; for an even number of items, the average of the two middle items.

meter (p. 76): The basic unit of measure for length in the metric system.

mode (p. 213): The item that appears most often in a set of data. A set of data can have more than one mode or no mode.

monomial (p. 668): A polynomial that has only one term.

multiple (p. 296): The product of a number and a non-zero whole number is a multiple of the given number.

multiplication property of −1 (p. 116): The product of any number and −1 is the opposite of the number.

$$-1n = -n \text{ and } -n = -1n$$

multiplication property of reciprocals (p. 340): The product of a number and its reciprocal is 1.

$$\frac{a}{b} \cdot \frac{b}{a} = 1, a \neq 0, b \neq 0$$

multiplication property of zero (p. 26): The product of any number and zero is zero.

$$a \cdot 0 = 0$$

multiplicative identity (p. 25): The number 1.

multiplicative inverse (p. 340): Reciprocal.

n

negative correlation (p. 509): The relation between sets of data in a scattergram when the trendline slopes downward.

negative integer (p. 94): Any integer less than zero.

number cube (p. 540): A cube with sides numbered 1 through 6.

o

obtuse angle (p. 245): An angle whose measure is greater than 90° and less than 180°.

obtuse triangle (p. 262): A triangle that has one obtuse angle.

odds in favor of an event (p. 542): The ratio of the number of favorable outcomes to the number of unfavorable outcomes of an event.

open sentence (p. 324): A mathematical sentence that contains a variable.

opposite side (p. 716): In a right triangle, the side opposite a given angle.

opposites (p. 94): Numbers that are the same distance from zero, but on opposite sides of zero on the number line.

order of operations (p. 36): An agreed-upon order of performing the operations in an expression that involves more than one operation.

ordered pair (p. 124): A pair of numbers assigned to any point on a coordinate plane. The first number is the *x-coordinate*, and the second number is the *y-coordinate*.

origin (p. 124): The point where the axes meet in a coordinate plane.

outcomes (p. 541): In probability, possible happenings, each of which is equally likely to happen.

p

parallel lines (p. 249): Two lines in the same plane that do not intersect.

parallelogram (p. 266): A quadrilateral with two pairs of parallel sides.

percent (p. 404): A ratio that compares a number to 100. The symbol % is read ''percent.''

percent of decrease (p. 419): The percent the amount of decrease is of the original amount. In business, often called the *discount*.

percent of increase (p. 419): The percent the amount of increase is of the original amount.

perfect square (p. 476): A number whose square root is an integer.

perimeter of a polygon (p. 436): The sum of the lengths of all the sides of the polygon.

permutation (p. 559): An arrangement of a group of things in a particular order.

perpendicular lines (p. 249): Two lines that intersect to form right angles.

pictograph (p. 188): A graph in which a symbol is used to represent a given number of items.

plane (p. 236): A flat surface that extends forever, usually represented by a four-sided figure.

point (p. 236): An exact location in space, represented by a dot.

polygon (p. 256): A closed figure formed by joining three or more line segments in a plane at their endpoints, with each line segment joining exactly two others.

polyhedron (p. 628): A space figure whose faces are polygons.

polynomial (p. 668): A variable expression that consists of one or more terms.

positive correlation (p. 509): The relation between sets of data in a scattergram when the trend line slopes upward.

positive integer (p. 94): Any integer greater than zero.

power (p. 30): The product when a number is multiplied by itself a given number of times. 64, or $4 \cdot 4 \cdot 4$, is the third power of 4.

power of a power rule (p. 89): To find the power of a power, multiply the exponents.

$$(a^m)^n = a^{mn}$$

prime factorization (p. 290): A number written as a product of prime numbers.

prime number (p. 290): A whole number greater than 1 with exactly two factors, 1 and the number itself.

prism (p. 628): A polyhedron with two parallel congruent bases.

probability of an event (p. 541): The ratio of the number of favorable outcomes to the number of possible outcomes of an event.

product of powers rule (p. 48): To multiply powers having the same base, add the exponents.

$$a^m \cdot a^n = a^{m+n}$$

profit (p. 224): The difference between revenue and cost when revenue is greater than cost.

proportion (p. 391): A statement that two ratios are equal.

protractor (p. 240): A geometric tool used to measure an angle.

pyramid (p. 628): A polyhedron with one base and triangular faces.

Pythagorean Theorem (p. 481): If the length of the hypotenuse of a right triangle is c and the lengths of the legs are a and b, then $c^2 = a^2 + b^2$.

q

quadrant (p. 124): A section of the coordinate plane.

quotient of powers rule (p. 306): To divide powers having the same base but different exponents, subtract the exponents.

$$\frac{a^m}{a^n} = a^{m-n},\ a \neq 0$$

r

radius of a circle (p. 440): A line segment whose two endpoints are the center of a circle and a point on the circle.

random sample (p. 568): A sample chosen at random from a larger group.

range (p. 213): The difference between the greatest and least values of the data in a given set of data.

range of a function (p. 613): The set of all possible values of y in a given function. Also referred to as the *values of the function*.

rate (p. 389): A ratio which compares two unlike quantities.

ratio (p. 388): A comparison of two numbers by division.

rational number (p. 321): A number that can be written as a quotient of two integers $\frac{a}{b}$, where b does not equal 0.

ray (p. 237): A part of a line that has one endpoint and extends forever in one direction.

real number (p. 321): Any number that is either rational or irrational.

reciprocals (p. 340): Two numbers whose product is 1.

rectangle (p. 266): A quadrilateral with four right angles.

reduction (p. 713): The image of a figure after a dilation that is smaller than the original.

reflection (p. 706): Flipping a figure across a line.

regular polygon (p. 257): A polygon in which all sides have the same length and all angles have the same measure.

repeating decimal (p. 313): A decimal written by dividing the numerator of a fraction by the denominator and resulting in a remainder that is not zero and a block of digits in the decimal that repeats.

revenue (p. 224): The amount of money collected.

rhombus (p. 266): A quadrilateral with four sides of equal length.

right angle (p. 245): An angle whose measure is equal to 90°.

right triangle (p. 262): A triangle that has one right angle.

rigid motion (p. 706): Movement of a figure that does not change its size and shape.

rotation (p. 710): Movement of a figure in a circular motion around a point.

rotational symmetry (p. 710): A figure that rotates onto itself has rotational symmetry.

s

sample space (p. 545): All the possible outcomes of a probability experiment.

scale (p. 395): In a scale drawing, the ratio of the size of the drawing to the actual size of the object.

scale drawing (p. 395): A drawing that represents real objects.

scale of a graph (p. 192): On a graph, numbers along an axis that show what is represented by the distances between the grid lines.

scale model (p. 396): A model of an object with dimensions in proportion to those of the actual object it represents.

scalene triangle (p. 261): A triangle with no sides of the same length.

scattergram (p. 509): A display in which the relationship between two sets of data is shown. The data are represented by unconnected points.

scientific notation (p. 80): A number in scientific notation is written as a number that is at least one but less than ten multiplied by a power of ten.

sector (p. 513): A wedge that represents part of the data in a circle graph.

side of a polygon (p. 256): A line segment that joins exactly two other line segments to form a polygon.

sides of an angle (p. 237). *See* angle.

sine (p. 716): In right triangle ABC,

$$\sin A = \frac{\text{length of side opposite } \angle A}{\text{length of hypotenuse}} = \frac{a}{c}.$$

similar figures (p. 466): Geometric figures that have the same shape but not necessarily the same size.

simplify an expression (p. 48): To perform as many of the indicated operations as possible.

simplify a fraction (p. 306): Write the fraction in lowest terms.

skew lines (p. 252): Lines that do not lie in the same plane.

slope of a line (p. 588): The ratio of *rise* to *run* of a line, describing the steepness and direction of the line.

slope-intercept form of an equation (p. 588): An equation written in the form $y = mx + b$, where m represents the slope of the line and b represents the *y-intercept*.

solution of an equation (p. 142): A value of the variable that makes an equation true.

solution of an equation with two variables (p. 582): An ordered pair of numbers that makes the equation true.

solution of an inequality with two variables (p. 609): An ordered pair of numbers that makes the inequality true.

solution of an open sentence (p.324): Any value of a variable that results in a true sentence.

solution of a system of equations (p. 594): An ordered pair that is a solution of both equations of a system of equations.

solve (p. 148): To find all values of the variable that make an equation true.

solve a proportion (p. 392): Find the value of the variable that makes the proportion true.

space figure (p. 628): A three-dimensional figure that encloses part of space.

sphere (p. 629): The set of all points in space that are the same distance from a given point called the *center*.

square (p. 266): A quadrilateral with four right angles and four sides of equal length.

square root of a number (p. 475): If $a^2 = b$, the number a is called a *square root* of b.

standard form (p. 668): The form of a polynomial whose terms are written in order from the highest to the lowest power.

statistics (p. 213): The branch of mathematics that deals with collecting, organizing, and analyzing data.

stem (p. 497): The digit or digits of the number remaining in a stem-and-leaf plot when the leaf is dropped.

stem-and-leaf plot (p. 497): A display of data where each number is represented by a *stem* and a *leaf*.

straight angle (p. 245): An angle whose measure is equal to 180°.

supplementary angles (p. 245): Two angles the sum of whose measures is 180°.

surface area of a cylinder (p. 638): The sum of the areas of the bases and the curved surface, expressed in square units.

surface area of a prism (S.A.) (p. 634): The sum of the areas of the bases and faces of the prism, expressed in square units.

system of equations (p. 594): Two equations with the same variables.

t

tangent (p. 716): In right triangle ABC,

$$\tan A = \frac{\text{length of side opposite } \angle A}{\text{length of side adjacent to } \angle A} = \frac{a}{b}.$$

terminating decimal (p. 313): A decimal written by dividing the numerator of a fraction by the denominator and resulting in a remainder of zero.

terms of a proportion (p. 391): The numbers or variables in a proportion.

tessellation (p. 285): A pattern in which identical copies of a figure cover a plane without gaps or overlaps.

third quartile of a set of data (p. 505): The median of the upper part of a set of data divided by its median.

30°-60°-90° right triangle (p. 727): A right triangle whose acute angles are 30° and 60°.

transform a formula (p. 375): To solve a formula for a particular variable by using inverse operations.

translation (p. 706): Sliding a figure from one location to another.

transversal (p. 250): A line that intersects two or more lines in the same plane at different points.

trapezoid (p. 266): A quadrilateral with exactly one pair of parallel sides.

tree diagram (p. 545): A representation of all the possible outcomes in a sample space.

trend line (p. 509): A line drawn near the points on a scattergram.

Triangle Inequality (p. 261): In any triangle, the sum of the lengths of any two sides is greater than the length of the third side.

trigonometric ratios (p. 716): Ratios of specific sides of a right triangle. *See* sine, cosine, and tangent.

trinomial (p. 668): A polynomial that has three terms.

u

unit rate (p. 389): A rate for one unit of a given quantity.

unlike terms (p. 56): Terms having different variable parts.

v

values of a function (p. 613): The set of all possible values of y for a function.

value of a variable (p. 4): Any number that is substituted for a variable.

variable (p. 4): A symbol that represents a number.

variable expression (p. 4): An expression that contains a variable.

Venn diagram (p. 556): A diagram that shows the relationships between collections of objects.

vertex of an angle (p. 237): *See* angle.

vertex of a polygon (p. 256): A point where two sides of a polygon meet.

vertex of a space figure (p. 628): A point where edges of a space figure intersect.

vertical angles (p. 246): The angles that are not adjacent to each other, equal in measure, formed by the intersection of two lines.

volume of a space figure (p. 642): The amount of space enclosed by a space figure, measured in cubic units.

w

whole numbers (p. 137): The numbers in the set $\{0, 1, 2, 3, \ldots\}$.

x

x-axis (p. 124): The horizontal number line in a coordinate plane.

x-coordinate (p. 124): The first number of an ordered pair.

x-intercept (p. 588): The x-coordinate of the point where a graph crosses the x-axis.

y

y-axis (p. 124): The vertical number line in a coordinate plane.

y-coordinate (p. 124): The second number of an ordered pair.

y-intercept (p. 588): The y coordinate of the point where a graph crosses the y-axis.

INDEX

Absolute value, 95, 97
  in adding integers, 98–104
  order of operations with, 115
Acute angle, 245
Acute triangle, 262
Addition
  associative property of, 22
  commutative property of, 22
  of decimals, 7, 357, 760
  estimation in, 7, 753, 754
  of fractions and mixed numbers,
    352, 356, 383, 764, 765
  identity property of, 23
  of integers, 98–104
  solving equations by, 148
  of whole numbers, 7, 757
Addition property of opposites, 102
Additive identity, 22
Additive inverse, 102
Alexander, Archie, 239
Algebra, 2–45, 46–91, 92–139,
    140–185, 288–337, 338–385,
    386–433, 580–625, 666–705
  history of, 55
  See also Algebraic fraction, Alge-
    bra tiles, Equations, Expo-
    nent(s), Factoring a polynomial,
    Function(s), Graphing, Inequal-
    ity(ies), Open sentence, Proper-
    ties, Systems of equations,
    Transforming formulas, *and*
    Variable(s).
Algebra tiles, 155–157
Algebraic fraction, 306, 353, 356,
    383
al-Khowarizmi, 55
Angle(s), 237
  acute, 245
  adjacent, 246
  alternate interior, 250
  central, 517
  complementary, 245
  congruent, 462
  constructing/bisecting, 243–244
  corresponding, 250, 463
  measure of, 240
  obtuse, 245
  right, 245
  sides of, 237
  straight, 245
  supplementary, 245

  vertex of, 237
  vertical, 246
Applications
  architecture and drawing, 240,
    302, 395, 436, 449, 479
  automotive, 79, 466, 551, 567,
    606
  business, 224, 418, 521, 570, 594
  carpentry, 309, 355, 362, 591
  cooking, 346, 374
  electricity, 377
  engineering, 615, 722, 725
  finances, 220, 224, 390, 411, 414,
    418, 426, 431, 513, 515,
    520–521, 690–691
  games, 689–690
  health, 221, 417
  inventory, 349
  map reading, 9, 132, 584
  optometry, 347
  overdue charges, 69
  packaging, 634, 658–659
  photography, 312
  planet dimensions, 652
  remodeling, 75, 449
  salary, 55, 59, 68, 71, 692, 730
  science, 107, 114, 159, 330–331,
    615, 649
  sewing, 352, 640–641
  space technology, 104, 654
  sports, 404, 419, 562, 690
  statistical interpretation, 21, 97,
    111, 172, 188–231, 311, 388,
    390, 494–529, 535, 572, 618,
    645
  stock, 378–379
  surveying, 474
  temperature, 110, 375
  used in lesson introduction, 4, 7,
    36, 52, 62, 98, 101, 107, 151,
    192, 200, 221, 240, 249, 293,
    296, 302, 362, 375, 395, 408,
    415, 435, 440, 466, 471, 513,
    559, 594, 606, 634, 650
  See also Career/Applications,
    Computer applications in exer-
    cises, *and* Focus on applica-
    tions.
Approximation symbol, 213
Approximation of square roots, 476
Archimedes, 653

Area
  of a circle, 452–453
  of a parallelogram, 446
  of a polygon, 445–447, 457
  of a rectangle, 445
  of a square, 445
  of a triangle, 447
  using tangrams, 450–451
Arrow notation, 66
Aryabhata, 692
Associative property
  of addition, 22
  of multiplication, 25
  to simplify expressions, 49
Average. *See* Mean.
Axis (axes), 124

Bar graph, 192–193, 204, 209, 770
  annotated, 195
  double, 192–193
  drawing, 200–201
  sliding, 195
Base
  of a parallelogram, 446
  of a triangle, 446
Base of a power, 30
BASIC, 6
  to multiply polynomials, 688
  Pythagorean triples, 485
  to simulate an experiment, 567
  variables, 6
Binomial, 668
  multiplication of, 685, 703
Box-and-whisker plot, 505

Calculator
  graphing, 586, 587, 591, 617, 618
  how to use, xviii, xix, 7, 8, 11,
    12, 31, 36–37, 108, 149, 157,
    214, 231, 263, 313, 329, 341,
    345, 392, 409, 431, 441, 453,
    476, 518, 561, 586, 617, 639,
    719
  use in exercises, 13, 28–29,
    32, 37, 63, 81, 109, 114, 150,
    153, 158, 216, 316, 330, 335,
    343, 346, 364, 370, 373, 393,
    410, 455, 519, 562, 587, 591,
    618, 640, 645, 652

Key to pictograph, 188

Lateral area, 640
Least common denominator (LCD), 310
Least common multiple (LCM), 296–297, 310
Legend, 192
Like terms, 56
combining, 56
Line(s), 236
on a geoboard, 593
graph of, 585
parallel, 249–250, 254–255
perpendicular, 249, 254–255
skew, 252
slope of, 588
y-intercept of, 588
Line segment(s), 237
congruent, 462
constructing/bisecting, 243
Line graph, 196–201, 204, 209, 771
business application, 224
double, 196
drawing, 200–201
Line of symmetry, 275
Liter, 76
Logical reasoning
exercises involving, 6, 21, 50, 97, 117, 144, 248, 264, 295, 308, 323, 465, 469, 507, 571, 699
See also Communication, Number sense, Patterns, Problem solving, Problem solving strategies, *and* Thinking skills.
Loss, 224
Lowest terms, 303, 306

Manipulatives
algebra tiles, 51, 53, 54, 56, 155, 156, 157, 671, 672, 676, 680–681
area models, 445
fraction bars, 300–301, 302, 303, 309, 310
fraction tiles, 350–351
geoboards, 592–593
integer chips, 105–106, 107, 108, 111
paper cutting/folding, 34–35, 261, 275, 445, 632–633
protractor, 240, 241–242, 253, 517

straightedge/compass, 243–244, 254–255, 256, 517
tangrams, 450–451
tessellations, 285
used to model a mean, 217
Map reading, 132
Mathematical power. *See* Applications, Career/Applications, Communication, Connections, Data analysis, Decision making, Explorations, Historical Note, Logical reasoning, Multicultural/Multiethnic contributions, Problem solving strategies, Research, Thinking skills, *and* Visualization.
Mathematical structure. *See* Integers, Irrational numbers, Number(s), Number system, Properties, *and* Rational numbers.
Mean, 213, 218
modeling, 217
Measurement(s)
converting, 84, 768
indirect, 471
metric system, 76–79
U.S. Customary system, 768
Measures, table of, xiii
Measures of central tendency, 218
Median, 213, 218, 505
Mental Math, exercises involving, 23, 26, 28–29, 52, 78, 103, 143, 190, 239, 343, 373, 411, 478, 516, 619, 637, 684, 703
Meter, 76
Metric system, 76–79
Midpoint, 470
Mixed Review, 122–123, 160–161, 360–361, 422–423
Mode, 213, 218
Modeling
area using tangrams, 450–451
equations, 155–157
expressions, 51
fractions, 300–301, 350–351
integers, 105–107, 111
the mean, 217
percents, 402
polynomials, 671, 680–681
powers of two, 34–35
ratios, 402
Monomial, 668
Multicultural/multiethnic contributions, 122–123, 160–161, 360–361, 422–423
Multiple, 296

Multiplication
of decimals, 11, 342, 761
estimation in, 11, 753
of fractions and mixed numbers, 340, 767
of integers, 111
by one, 25
by a power of 10, 762
of powers, 48
properties of, 25–26
of rational numbers, 340–342
symbols representing, 12
of whole numbers, 11, 758
by zero, 26
Multiplication property of −1, 116
Multiplication property of zero, 26
Multiplicative identity, 25
Multiplicative inverse, 340

Notation
for combination, 560
factorial, 559
for permutation, 560
Number(s)
common multiple of, 296
comparing and ordering, 18–21
compatible, 12, 754
composite, 290
in exponential form, 30
irrational, 321
least common multiple of, 296
multiple of, 296
negative, 94
oblong, 272
opposite of, 94
percent of, 408
prime, 290
rational, 321
real, 321
rounding, 8, 753, 756
square, 65
triangular, 65
whole, 137
writing, 755
Number cube, 540
Number line, 94–95
adding integers on, 98, 101
Number sense, exercises involving, 20, 78, 82, 118, 323, 355, 411, 444
Number system
Hindu-Arabic, 55
real number, 321

Credits

Cover: Concept by Martucci Studios; Photographic Special Effect Illustration by VISUAL CONSPIRACY/Martin Stein.

Technical art: Typographic Sales, Inc.; LeGwin Associates

Illustrations:

Neil Pinchin Design: xii, xiii, 122–123, 172, 188, 189, 190, 191, 192, 195, 199, 200, 201, 207, 215, 226, 228, 249, 253, 311, 312, 388, 390, 404, 419, 494, 501, 503, 513, 514, 515, 516, 517, 541, 564, 650

Mark Goldman and Charles Shields: xviii, 1, 6, 7, 9, 11, 21, 33, 61, 94, 97, 120, 121, 132, 159, 206, 215, 221, 324, 346, 352, 586, 591, 598, 618

Victor Ambrus: 10, 55, 127, 175, 208, 239, 308, 374, 414, 456, 500, 555, 608, 653, 692

PHOTOGRAPHS

xiv Comstock. **xiv** Andrew Sacks/Tony Stone Worldwide. **xiv** Texas Instruments. **xv** Texas Instruments. **xvi** The Telegraph Colour Library/F.P.G. **2–3** Michael Simpson/F.P.G. **4** Michael Newman/Photo Edit. **7** Sepp Seitz/Woodfin Camp and Associates. **13** Lawrence Migdale/Stock Boston. **14** © Paul Conklin. **16** Bill Bachman/Stock Boston. **20** Hugh Rogers/Monkmeyer Press Photos. **21** © Dave Schaefer. **24** © Susan Van Etten. **29** © Susan Van Etten. **34** Sekai Bunka. **36** © Susan Van Etten. **38** Alvis Upitis/The Image Bank. **43** Jeffry W. Myers/Stock Boston. **46–47** Tony Stone Worldwide. **50** Steve Dunwell/The Image Bank. **56** Eric Roth/The Picture Cube. **58** © Bob Daemmrich. **62** © Susan Van Etten. **64** © Bob Daemmrich. **68** © Bob Daemmrich. **69** E.R. Degginger/Animals, Animals. **72** © Bob Daemmrich. **75** Dr. E.R. Degginger. **79** © Bob Daemmrich. **82** Scott Berner/Photri. **85** Comstock. **92–93** Beryl Bidwell/Tony Stone Worldwide. **96** Herbert Lanks/Monkmeyer Press Photos. **100** © Paul Conklin. **104** N.A.S.A. **109** Chin Wai-Lan/The Image Bank. **110** Lawrence Migdale/Stock Boston. **113** © Bob Daemmrich. **119** Barrie Rokeach/The Image Bank. **122–123** L.L.T. Rhodes/earth scenes; The Granger Collection; Gerard Champlong/The Image Bank; The Granger Collection; The Granger Collection. **125** Minnesota Office of Tourism. **133** Peter Miller/The Image Bank. **140–141** Image Works. **142** © Bob Daemmrich. **146** MacDonald Photography/Photri. **151** Lawrence Migdale/Photo Researchers. **154** N.A.S.A./Peter Arnold, Inc. **159** Ron Grishaber/Photo Edit. **160–161** The Bettman Archive; UPI/Bettman; The Granger Collection; The Granger Collection; © Susan Van Etten; N.A.S.A.; Allen Green/Photo Researchers, Inc.; The Granger Collection; The Granger Collection. **162** Mike Mazzaschi/Stock Boston. **164** George Zimbel/Monkmeyer Press Photos. **166** © Bob Daemmrich. **170** © Bob Daemmrich. **178** Miro Vintoniv/The Picture Cube. **186–187** Michael Melford/The Image Bank. **193** Michal Heron/Monkmeyer Press Photos. **196** Kindra Clineff. **217** Kindra Clineff. **219** Jeffrey M. Myers/Stock Boston. **221** © Bob Daemmrich. **223** © Bob Daemmrich. **225** © Bob Daemmrich. **231** © Susan Van Etten. **234–235** Jim Larsen/West Stock. **237** Camerique. **240** Bob Daemmrich/Stock Boston. **257** Bohdan Hrynewych/Stock Boston. **261** Bill Gallery/Stock Boston. **264** Bob Crandall/Stock Boston. **272** Runk/Schoenberger from Grant Heilman. **276** Kristian Hilsen/Tony Stone Worldwide. **276** Bill Gallery/Stock Boston. **278** Odyssey/Frerck/Chicago. **285** Martin Rogers/Tony Stone Worldwide. **288–289** Tom Tracy/The Stock Market. **293** Larry Lefever from Grant Heilman. **296** Richard Wood/The Picture Cube. **302** © Bob Daemmrich. **309** © Susan Van Etten. **315** Dan Burns/Monkmeyer Press Photos. **316** Larry Lefever from Grant Heilman. **318** © Bob Daemmrich. **319** Spencer Grant/Stock Boston. **323** The University Museum, University of Pennsylvania. **331** © Bob Daemmrich. **338–339** Ted Kurihara/Ted Kurihara Studio. **343** Larry Lefever from Grant Heilman. **347** © Bob Daemmrich. **349** Martin Rogers/Stock Boston. **352** Jim Harrison/Stock Boston. **355** Kindra Clineff. **359** Stan Osolinski/F.P.G. **360–361** Bob Daemmrich/Stock Boston; Bob Daemmrich; Bob Daemmrich/Uniphoto Picture Agency; The Granger Collection; The Granger Collection. **363** Cathlyn Melloan/Tony Stone Worldwide. **365** Mark Scott/F.P.G. **367** Bob Daemmrich/The Image Works. **370** Odyssey/Frerck/Chicago. **372** Barry L. Runk from Grant Heilman. **377** Joseph Schuyler/Stock Boston. **379** Steve Leonard/Tony Stone Worldwide. **386–387** International Stock Photography. **394** © Susan Van Etten. **398** Ida Wyman/Monkmeyer Press Photos. **400** Gans/The Image Works. **411** Mavournea Hay/Daemmrich Associates. **417** Dennis Degnan/Uniphoto. **418** © Bob Daemmrich. **421** Steven E. Sutton/duomo. **422–423** John Williamson; Jack Green/Horizon; Patti Murray/Animals Animals; © Susan Van Etten; © Susan Van Etten. **426** Joseph Nettis/Stock Boston. **434–435** G.K. & Vicky HA/The Image Bank. **440** © Bob Daemmrich. **444** Michael Grecco/Stock Boston. **449** © Bob Daemmrich. **454** Don Kelly from Grant Heilman. **457** © Chris Malazorewicz/Valan Photos. **461** Martucci Studios. **463** Stephanie Dinkins/Photo Researchers, Inc. **466** Peter Menzel/Stock Boston. **471** Keith Philpott/Stockphotos, Inc. **474** © Bob Daemmrich. **479** Scala/Art Resource. **484** George Holton/Photo Researchers, Inc. **492–493** Walter Bibikow/The Image Bank. **496** Tony Freeman/Photo Edit. **499** © Bob Daemmrich. **503** Roger Dollarhide/Monkmeyer Press Photos. **508** Bob Daemmrich/Stock Boston. **512** Kennedy/TexaStock. **516** © Bob Daemmrich. **520** Richard Pasley/Stock Boston. **523** Freda Leinwand/Monkmeyer Press Photos. **526** Lee Balterman/The Picture Cube. **529** Bob Daemmrich/Stock Boston. **538–539** Stock Imagery/Tom Walker. **547** © Bob Daemmrich. **551** Bruce M. Wellman/Stock Boston. **554** Vernon Doucette/Stock Boston. **559** © Bob Daemmrich. **565** © Bob Daemmrich. **569** Richard Pasley/Stock Boston. **570** Palmer/Kane/The Stock Market. **577** Carey/The Image Works. **580–581** Robert Kristofi/The Image Bank. **584** Peter Frank/Tony Stone Worldwide. **591** © Bob Daemmrich. **599** Robert Frerck/Tony Stone Worldwide. **601** Lowell J. Georgia/Photo Researchers. **605** Stan Osolinski/F.P.G. **606** Stephen Frisch/Stock Boston. **615** © Susan Van Etten. **618** Barbara Alper/Stock Boston. **626–627** Joseph Drivas/The Image Bank. **631** © M. Greenbar/The Image Works. **634** © W. Hill/The Image Works. **640** Bob Daemmrich/Stock Boston. **644** Paul Mozell/Stock Boston. **649** E.R. Degginger/Animals, Animals. **652** Stephen Frisch/Stock Boston. **656** © Bob Daemmrich. **659** Larry Lefever from Grant Heilman. **666–667** John Martucci/Martucci Studios, Inc. **670** Scala/Art Resource. **675** © Susan Van Etten. **679** © Bob Daemmrich. **684** © Susan Van Etten. **687** E. Alan McGee/F.P.G. **689** John Running/Stock Boston. **691** Jim Pickerell/Tony Stone Worldwide. **698** DMR/The Image Works. **706** © Susan Van Etten. **706** Martin Rogers/Tony Stone Worldwide.

Answers to Check Your Understanding

CHAPTER 1

Page 4 **1.** You would substitute 20 for *n*. **2.** You would evaluate the expression 6*n*. **Page 7** **1.** You estimate before adding to see about what your actual answer should be. **2.** The sum of 87 and 16.49 is close to 100. **3.** You write 287 as 287.00 to give it the same number of decimal places as 116.49, which aids in aligning the decimal points. **Pages 11–12** **1.** The sum of the numbers of the decimal places of the factors is 1. Thus, the number of decimal places in the product is 1. **2.** You place the decimal point in the quotient directly over the decimal point in the dividend. You can also place the decimal point in the quotient by using your estimate of 3. **Pages 18–19** **1.** When you compare the numbers place-by-place from left to right, the digits 7 and 8 are the first digits that differ. **2.** Example 1(b) involves decimals. The numbers in Example 1(a) have the same number of digits, while the numbers in Example 1(b) do not. **3.** When you compare 0.47 and 0.4 place-by-place from left to right, the digits are the same in all places through the tenths' place. You need to write 0.4 as 0.40 so that you can compare the digits in the hundredths' place. **4.** $0.47 > 0.4 > 0.247$
Page 23 **1.** identity property of addition **2.** The sum of the ones' digits is 10, so the sum of $76 + 14 = 90$ is easy to find. **Page 26** **1.** multiplication property of zero; identity property of multiplication **2.** The product of 25 and 4 is 100, and it is easier to multiply 13×100 mentally. **Page 28** **1.** Paper and pencil; there is only one renaming involved. **Page 30** **1.** The expression 6^3 means that 6 is used as a factor three times. The expression $6 \cdot 3$ means that 6 and 3 are the factors. **2.** The expression 6^3 means that 6 is used as a factor three times. The expression 3^6 means that 3 is used as a factor 6 times.
3. The number 1 to any power equals 1. **Pages 36–37** **1.** You perform multiplication and division in order from left to right. In Example 1(a) the division occurs before the multiplication when you work from left to right. **2.** You would first add within parentheses $(30 + 24 = 54)$, then divide $(54 \div 6 = 9)$, and then multiply $(9 \cdot 2 = 18)$.
3. $(24 + 12) \div (13 - 4)$ **4.** You would first add within parentheses $(8 + 4 = 12)$, then square the sum $(12^2 = 144)$, then divide $(144 \div 3 = 48)$, and then add $(48 + 5 = 53)$. **5.** You would add within the first set of parentheses $(8 + 4 = 12)$, then add within the second set of parentheses $(3 + 5 = 8)$, then do the power $(12^2 = 144)$, and then divide $(144 \div 8 = 18)$.

CHAPTER 2

Page 48 Any number to the first power is equal to that number. **Pages 52–53** **1.** It is easier to use mental math to multiply $7(100 + 8)$ than $7(110 - 2)$. **2.** $7(98) = 7(100 - 2) = 7(100) - 7(2) = 700 - 14 = 686$ **3.** They show that you multiply each term inside a set of parentheses by the factor outside the parentheses when you use the

distributive property. **Page 57** **1.** They are the only like terms. **2.** identity property of multiplication **Pages 62–63** **1.** 384, 768 **2.** 8; 40 **3.** 5 **Page 66** **1.** 46 **2.** 496 **Page 74** It is easier to multiply 3×20 than 3×18. **Pages 76–77** **1.** Because a liter is three places above a milliliter in the chart, a liter is $10 \cdot 10 \cdot 10$ times as large as a milliliter. So, you multiply by $10 \cdot 10 \cdot 10$, or 1000, to change from liters to milliliters. **2.** 3 places **Pages 80–81** **1.** You multiply by 10^4 because you need to move the decimal point 4 places to the left. **2.** You move the decimal point 6 places because you are multiplying by 10 to the sixth power.
3. You would move the decimal point 2 places to the right rather than 5 places to the right. $7.16 \cdot 10^2 = 716$

CHAPTER 3

Page 95 **1.** -3 **2.** 0 **Page 98** **1.** They represent negative numbers. **2.** You would first slide 4 units left and then slide 3 more units left. The sum would be the same. **Page 102** **1.** The positive integer has the greater absolute value, so the sum would be positive.
2. $|3| = 3$ and $|-3| = 3$. Subtract: $3 - 3 = 0$. The sum is 0. **Pages 107–108** **1.** Two negative chips remain.
2. Subtracting an integer is the same as adding its opposite. **Page 112** **1.** The factors -20 and -2 have the same sign. The product of two integers with the same sign is positive. **2.** The quotient of two integers with different signs is negative. **Pages 115–116** **1.** $(-4)^3 = (-4)(-4)(-4) = 16(-4) = -64$ **2.** Absolute value signs have priority over addition in the order of operations, so you first find the absolute values of -9 and 4, and then add the absolute values. **3.** The opposite of a number is the product of the number and -1. **Pages 124–125** **1.** Point *E* is to the left of the origin. **2.** Point *F* is 0 units up or down from the *x*-axis. **3.** Point *D* is 0 units to the left or right of the origin. **4.** Point *A* would be below the *x*-axis. **Pages 128–129** **1.** The *x*-coordinates are the values of *x* in the *x*-column of the table. The *y*-coordinates are the values obtained from using the function rule.
2. You use the values given for *x*.

CHAPTER 4

Pages 142–143 **1.** The two sides of the equation are not equal. **2.** What number minus 6 equals 14? Four times what number equals 20? **3.** The first question would be "What number plus 6 equals 14?" **Pages 148–149** **1.** to keep both sides of the equation equal **2.** You would need to undo the addition. **Page 152** **1.** You need to be able to divide both sides by -1. **2.** You would not need to use the multiplication property of -1. You would just divide both sides by -3. **Pages 156–157** **1.** You would substitute 3 for *n* in the original equation. **2.** division and subtraction **3.** adding 4 to both sides, then multiplying both sides by 3 **Page 162** **1.** Yes; by the commutative property of addition.   **2.** No; subtraction is not commutative.   **Page 165** **1.** is   **2.** The cost of the financial package is given.   **Pages 172–173** **1.** $4x$ and $3x$ **2.** to simplify the left side of the equation   **Page 176** **1.** You substitute 240 for D and 40 for r in the formula $D = rt$.   **2.** The equation would be solved for r rather than t.

CHAPTER 5

Pages 188–189 **1.** The number of symbols would be multiplied by 500 instead of by 150.   **2.** Find the number of chocolate cones and the number of orange cones sold. Then subtract.   **3.** You are rounding to the nearest half-million rather than to the nearest whole million, and 5.7 is closer to 5.5 than to 6.   **Page 192** Estimate the heights of the bars for jazz and hard rock. Then subtract.   **Page 197** You would locate the point for 1982 on the *red* line.   **Page 200** Using intervals of five makes it easier to represent the data accurately.   **Page 204** Yes; average attendance changes continuously with each game throughout the baseball season.   **Page 209** The attendance in Midway appears to be about 6 times the attendance in Sunville.   **Page 213** **1.** The 6 represents the number of bowling scores.   **2.** You must average the two middle scores because there is an even number of data items.   **Page 218** Find the sum of the salaries ($252,000) and then divide by the number of salaries (7).   **Page 221** **1.** The total of the frequencies is 29.   **2.** Count the tally marks until you reach the fifteenth. Then find the rate that corresponds to that tally mark.

CHAPTER 6

Page 237 **1.** Point B is an endpoint and must come first.   **2.** No; point J is not the vertex of the angle.   **Pages 240–241** **1.** $\overrightarrow{BC}$ coincides with the 0° mark on the bottom scale.   **2.** You would use the top scale instead of the bottom scale.   **Page 246** **1.** You would subtract the given measure from 180°.   **2.** No; only acute angles have complements.   **3.** $\angle 2$ and $\angle 4$; $\angle 1$ and $\angle 3$ **4.** $\angle 1$ and $\angle 2$; $\angle 2$ and $\angle 3$; $\angle 1$ and $\angle 4$ **Page 250** **1.** Four: the right angle, the angle vertical to it, and its two supplements   **2.** $\angle 7$; 47°   **Page 258** **1.** 6; the polygon has six sides.   **2.** the sum of the known measures   **Pages 262–263** **1.** For example, change the 9 m side to 6 m.   **2.** the sum of the measures of the angles of a triangle   **3.** It has one obtuse angle, 112°.   **Page 267** **1.** Yes; $\angle DAC$ and $\angle ACB$ form a pair of alternate interior angles, so they are equal in measure.   **2.** The sum of the measures of the angles of any triangle is 180°.   **Page 275** **1.** When folded, one half does not fit exactly over the other half.   **2.** No.

CHAPTER 7

Page 290 **1.** Two   **2.** 2 and 5 are factors of both 10 and 140.   **Pages 293–294** **1.** $4 > 2$ **2.** 5 and 7 are not common factors.   **3.** a^4 is the least power of each com-

mon variable factor.   **4.** y is not a common factor.   **Pages 296–297** **1.** No; the multiples continue infinitely.   **2.** You need to use the greatest power.   **3.** You need to include the greatest power of each variable factor.   **Page 302** **1.** Each exercise involves finding equivalent fractions; in 1(a) the equivalent fraction had more parts, while in 1(b) the equivalent fraction has fewer parts.   **2.** Both calculations are used to find equivalent fractions.   **Page 307** **1.** The solution would be $5x^3$.   **2.** Writing d as d^1 makes it easier to subtract the exponents.   **Pages 309–310** **1.** The one-fourth shaded part is smaller than the one-third shaded part.   **2.** $12 = 2^2 \cdot 3$ and $18 = 2 \cdot 3^2$, so $2^2 \cdot 3^2$, or 36, is the LCM.   **3.** Because you multiply 12 by 3 to get 36.   **Pages 313–314** **1.** Both numbers repeat.   **2.** 0.6363636   **3.** You could write $\frac{3}{8}$ as 0.375 and add 6 to get 6.375.   **4.** The GCF of 111 and 200 is 1.   **Page 321** **1.** Yes; $\frac{16}{-1}$ is equal to -16.   **2.** No; $\frac{-16}{-1}$ is equal to 16.   **Pages 324–325** **1.** No.   **2.** Because -4.2 is not greater than -4.2.   **3.** You would shade in a heavy arrow to the left to graph all numbers less than -4.2.   **4.** Because $\frac{3}{4} = \frac{3}{4}$.   **5.** You would shade in a heavy arrow to the right to graph all numbers greater than $\frac{3}{4}$.   **Page 329** **1.** 6.5×10^{-1} **2.** 47,000.

CHAPTER 8

Page 341 **1.** All numbers must be written as fractions when multiplying by a fraction.   **2.** Multiplying a positive number by a negative number results in a negative number.   **3.** The answer would be positive.   **4.** $\frac{1}{3}$ is less than $\frac{1}{2}$.   **5.** -6 is the closest whole number that is compatible with $\frac{-1}{2}$.   **Page 344** **1.** Yes; the answer would be $\frac{1}{10}$.   **Page 348** **1.** Because the question is about money, $1.\overline{3}$ is rounded to the nearest cent.   **2.** You need to round up because seven tables aren't enough to seat everyone.   **Pages 352–353** **1.** The mixed number $-1\frac{1}{5}$ is written as a fraction.   **2.** Subtracting the numerators results in -40.   **3.** Instead of -40, the numerator would have become -14.   **4.** The numerator and the denominator were divided by 3.   **Pages 356–357** **1.** The fractions would be $\frac{-18}{16}$ and $\frac{44}{16}$; $\frac{-18-44}{16} = \frac{-62}{16} = \frac{-31}{8} = -3\frac{7}{8}$.   **2.** Two negatives are added.   **3.** $|-3.8| > |-1.48|$ **Page 362** If the denominator is greater than twice the numerator, then the fraction is less than $\frac{1}{2}$.   **Page 368** **1.** $\frac{(-3)(-2)}{3} = \frac{6}{3} = 2$ **2.** $\frac{2}{4}$ **3.** When subtracting rational numbers you must rewrite mixed numbers as fractions.   **Page 371** **1.** You multiply the right side, $\frac{-5}{6}b$, by the reciprocal $\frac{-6}{5}$ because you need to solve for b; you multiply the left side by $\frac{-6}{5}$ because you need to balance the equation.   **2.** When multiplying by a fraction, you should write each number as a fraction.   **Page 375**   **1.** Each side of the equation must be multiplied by $\frac{5}{9}$, and the left side of the equation is $F - 32$.   **2.** 325° is given in degrees Fahrenheit.

CHAPTER 9

Page 388   **1.** No; the ratio of wins to losses is $\frac{2}{1}$, and the ratio of losses to wins is $\frac{1}{2}$.   **2.** Yes; in lowest terms they are equal.   **Pages 391–392**   **1.** Yes; then the cross products will be equal.   **2.** The cross products would still be 60 and $4b$, and $b = 15$.   **Pages 395–396**   **1.** The length of the tray return is asked for.   **2.** Yes; the cross products and solutions would not change.   **Pages 404–405**   **1.** Per cent means "per hundred."   **2.** 0.375   **3.** The decimal point moves two places to the left.   **4.** In order to move the decimal point, you must express the fraction as a decimal.   **5.** so that $\frac{3.6}{100}$ can be expressed as a fraction which can be reduced to lowest terms   **Pages 408–409**   **1.** $\frac{30}{100}$ or $\frac{3}{10}$ **2.** 64% is a ratio that compares 64 to 100, and $\frac{64}{100} = 0.64$.   **3.** $\frac{16}{25}$ **4.** 65% is close to $66\frac{2}{3}$%, which is $\frac{2}{3}$.   **Page 412**   **1.** in order to express it as a percent   **2.** Divide 12.5 by 500; the result is 0.025.   **Page 415**   **1.** more than 126; 60% is part of the whole.   **2.** It is easier to write $66\frac{2}{3}$% as a fraction than as a decimal.   **3.** To solve $120 = \frac{2}{3}n$, multiply both sides by the reciprocal of $\frac{2}{3}$.   **Page 419**   **1.** Percent of increase $= \frac{\text{amount of increase}}{\text{original amount}}$, and 10 is the original amount.   **Page 424**   **1.** No; the cross products will be $4n = 56$, which results in $n = 14$.   **2.** The question asks for a percent.

CHAPTER 10

Pages 436–437   **1.** Another variable would be needed to represent the sixth side; $P = a + b + c + d + e + f$.
2. Both figures are pentagons; the pentagon in Example 1 is not a regular pentagon.   **3.** The formula would be $8n$.   **Pages 440–442**   **1.** It fulfills both definitions.   **2.** The value for the diameter, 21, is divisible by 7.   **3.** You would use the formula $C = 2\pi r$.   **4.** The circumference is expressed in decimal form.   **5.** You would use the formula $C = \pi d$.   **Pages 446–447**   **1.** You would use the formula $A = s^2$; $A = (3\frac{1}{2})^2 = 12\frac{1}{4}$.   **2.** The solution would be 2750 mm²; both answers are correct.   **3.** No.   **Page 453**   **1.** The radius is expressed as a decimal.
2. The radius would be 2.2 m; $A \approx 3.14\,(2.2)^2 \approx 15.2$ m².   **Page 457**   when finding the distance around something round or circular   **Page 463**   **1.** $\angle P \cong \angle C$, $\angle Q \cong \angle B$, $\angle R \cong \angle D$, $\angle S \cong \angle A$; $\overline{PQ} \cong \overline{CB}$, $\overline{QR} \cong \overline{BD}$, $\overline{RS} \cong \overline{DA}$, $\overline{SP} \cong \overline{AC}$ **2.** The vertices must be listed in the same order as the corresponding congruent angles.

Page 467   **1.** 21   **2.** The ratio would be $\frac{16}{10}$ or $\frac{8}{5}$.
Pages 471–472   **1.** $\frac{2}{3}$; they are equal.   **2.** 27 ft
3. $\frac{12}{36} = \frac{64}{h}$, $\frac{1}{3} = \frac{64}{h}$, $h = 3\,(64) = 192$ in.   **4.** You would use $5\frac{1}{2}$ instead of $5\frac{1}{3}$ and solve.
Pages 475–476   **1.** The square root of a negative number does not exist.   **2.** $\sqrt{46}$ does not exactly equal 6.782.
3. 45.995524; the result is less than 46 because 6.782 is less than $\sqrt{46}$.   **Page 482**   **1.** In part (a), the length of the hypotenuse was unknown; in part (b), the length of a leg was unknown.   **2.** Because c^2 is the sum of a^2 and b^2, c must be greater than both a and b; you should substitute the longest length for c.

CHAPTER 11

Pages 494   4   **Pages 497–498**   **1.** There would be a 6 in the 9 row.   **2.** You would count the numbers greater than 5 on the 0.73 stem and the numbers on stems 0.74 and 0.75. The total would be 5 teams.   **3.** Range equals greatest minus least value: $0.755 - 0.692 = 0.063$
Page 501   12 employees work 31–35 h and 20 employees work 36–40 h; $12 + 20 = 32$ **Page 506**   For each company find the range by subtracting the smallest value from the largest. Then compare the two ranges.
Page 509   **1.** halfway between 6 and 8   **2.** You would find the greatest and least test scores and subtract: $100 - 50$.   **Pages 513–514**   **1.** Both sides had to be divided by 0.3.   **2.** The amounts given in the graph are in thousands.   **Page 517**   The percent for each type equals $\frac{\text{type sales}}{\text{total sales}}$.   **Page 522**   Yes; you would have two sets of data to compare.

CHAPTER 12

Pages 541–542   **1.** None of the sectors is numbered 7.
2. 1, 2, 4, 5, 6   **3.** Each of the sectors is red, blue, or white, so one of these colors is certain to occur on each spin.   **4.** favorable: 3; unfavorable: 1, 2, 4, 5, 6   **5.** 1 to 1   **Page 545**   **1.** The sample space would be: HH, HT, TH, TT.   **2.** HHH, HHT, HTH, THH; $\frac{1}{2}$ **Pages 548–549**   **1.** 24   **2.** 30   **3.** $\frac{1}{20}$ **Pages 552–553**   **1.** The first marble drawn is replaced before the second marble is drawn.   **2.** There are twelve marbles in the bag, and five of them are white.   **3.** Drawing the first marble reduces the number of marbles left in the bag for the second drawing.   **4.** There are fewer marbles left in the bag after the first marble is drawn.   **Page 559**   **1.** MATH, MAHT, MTAH, MTHA, MHAT, MHTA   **2.** There are 4 letters in MATH, so find 4!   **Page 564**   **1.** the number of times the nickel landed on both red and black   **2.** A dime is smaller than a nickel, so it might land on a single color more easily.   **Page 568**   **1.** A poll taker finds the probability that a candidate will receive votes by polling a random sample of voters and multiplying this probability by the total number of voters.   **2.** about 170,500 voters

CHAPTER 13

Page 582 **1.** 9 **2.** -2 **Page 585** **1.** The values for y when x equals 3 and -3 are integers, making the solutions easy to graph. **2.** Yes.
Pages 588–589 **1.** The vertical movement is downward. **2.** (2, 1), (0, 5); yes. **3.** The slope-intercept form is $mx + b$. **4.** (0, -6) **Page 595** Yes. **Page 603** **1.** You only reverse the inequality symbol when you multiply or divide both sides by a negative number.
2. When you multiply both sides of an inequality by a negative number, you reverse the inequality symbol.
Pages 606–607 **1.** addition **2.** No; both sides were divided by a positive, not a negative number. **3.** Both sides are being divided by -16. **4.** There would be an open circle at -3. **Page 610** **1.** The region below the line would be shaded. **2.** Yes. **Pages 613–614** **1.** multiply **2.** find the power **3.** $|-3 - 2| = |-5| = 5$
Page 616 5

CHAPTER 14

Page 628 **1.** It has only one base. **2.** It has two bases and its sides are not triangular. **Page 635** **1.** The formula for the area of a triangle is $\frac{1}{2}bh$. **2.** The rectangles are not all the same size. **Page 638** **1.** The radius is a multiple of 7. **2.** the curved surface **Page 642**
1. There would be 5 layers, so the volume would be $5 \cdot 8 = 40 \text{ cm}^3$. **2.** There would be $6 \cdot 2 = 12$ cubes in one layer, so the volume would be $3 \cdot 12 = 36 \text{ cm}^3$.
Page 646 **1.** If the height were 11 in., the calculation would be the same, but the answer would be approximately 7771.5 in.^{3} **2.** The volume formula used would be $V = \frac{1}{3}Bh$, and the volume would be approximately $\frac{1}{3}$ (7771.5), or 2590.5 m^3. **3.** The radius, 14, is a multiple of 7. **Page 650** **1.** The formulas for surface area and volume use the radius, which is one half the diameter. **2.** the radius, 6, cubed

CHAPTER 15

Page 668 **1.** 1(a): 3, 2, 1; 1(b) 4, 3, 1 **2.** It has no variable factor. **3.** 3 is a higher power than 2. **4.** They are not like terms. **Page 673** **1.** $x = 1x$; therefore, $x + (-5x) = 1x + (-5x) = -4x$ **2.** Adding the x^3 terms for each equation resulted in $0x^3$ or 0.
Pages 676–677 **1.** Because $-2n^2 - n - 5$ is the opposite of $2n^2 + n + 5$ **2.** There is no a term in the first polynomial. **Page 682** **1.** Subtracting is the same as adding the opposite. **2.** Each term of the polynomial was multiplied by $4x^2$. **Page 685** **1.** They are like terms.
2. Yes. **3.** Subtracting a number is the same as adding its opposite, so $3x - 2 = 3x + (-2)$; when you multiply $3x + (-2)$ by $(7x + 5)$, you get $3x(7x + 5) + (-2)(7x + 5)$, or $3x(7x + 5) - 2(7x + 5)$. **4.** 1
Page 693 **1.** It is the GCF of the terms. **2.** The greatest common monomial factor is $3c^2d$. **3.** Multiply the polynomial by the monomial. **Page 696** **1.** When you divide powers with the same base, you subtract the exponents. **2.** The answer would be $2a^4 - 4a^3 + 3a$.
3. Adding a number is the same as subtracting its opposite.

LOOKING AHEAD

Pages 706–707 **1.** The translation is horizontal.
2. For example, move up two units. **3.** 1 unit; 1 unit
4. 4 units; 4 units **5.** Yes. **Pages 710–711** **1.** 45°
2. 45° **3.** 3 **4.** B **5.** C **6.** A **7.** B **Pages 713–714** **1.** $2x$ **2.** $\frac{1}{2}a$ **3.** $\frac{1}{2}x$ **4.** $2a$ **5.** They are congruent. **6.** $A'B' = 3AB$; $B'C' = 3BC$ **Pages 716–717**
1. the hypotenuse **2.** Because a leg of a triangle is always shorter than the hypotenuse. **3.** $\frac{5}{12}, \frac{12}{13}$ **4.** 8; 15; $\frac{8}{15}$
5. Subtract 64 from both sides **6.** adjacent
Pages 719–721 **1.** $(18.9)^2 + (14.8)^2 = 24^2$; result will not be exact due to rounding **2.** You would find sin B and cos B. **3.** The hundredths' digit is 9, and $9 \geq 5$, so you add 1 to the tenths' digit. **4.** 48.5 **Pages 723–724**
1. The difference between 0.4226 and 0.4167 is less than the difference between 0.4167 and 0.4067. **2.** The sum of the measures of the angles of any triangle is equal to 180°. **3.** The difference between 1.5000 and 1.4826 is less than the difference between 1.5399 and 1.5000.
4. $m \angle A \approx 56°$; $m \angle B \approx 34°$; $AB = \sqrt{208} \approx 14.4$
Pages 727–728 **1.** exact **2.** 14, $14\sqrt{2}$ **3.** 6, 6
4. $a = 10$, $b = 10\sqrt{3}$ **5.** $b = 100\sqrt{3}$, hypotenuse = 200

Answers to Selected Exercises

CHAPTER 1

Pages 5–6 Exercises **1.** 30 **3.** 10 **5.** 12 **7.** 2
9. 180 **11.** 20 **13.** 4 **15.** 8 **17.** 19 **19.** $b - 6$
21. $7n$ **23.** $n - 1$ **25.** $60 \div n$ **27.** 7 **29.** 20
31. $6 * M$ **33.** $J + 12 + K$ **35.** $r = 12; s = 8$
37. $r = 30; s = 15$ **39.** nine and two ten-thousandths
41. 180 **43.** $\frac{2}{5}$

Pages 9–10 Exercises **1.** 39.1 **3.** 204.21 **5.** 81.1
7. 101.64 **9.** 200.4 **11.** 85.0 **13.** $576.44
15. 1.11 mi **17.** 0.26 mi **19.** about 0.09 mi
21. $p + 43.61$ **23.** $8.7 - z$ **25.** the sum of 15.8 and a
number z **27.** a number b minus 16.4 **29.** The statement
$p + q = p - q$ is true only when $q = 0$. **31.** about 35
33. $\frac{3}{8}$ **35.** 105,153

Pages 13–14 Exercises **1.** 139.2 **3.** 1693.44
5. 0.49 **7.** 1.96 **9.** 100.8 **11.** 0.29 **13.** 187 mi
15. 42.328 **17.** 0.0532 **19.** 2021.3 **21.** unreasonable;
71.712 **23.** unreasonable; 0.61 **25.** reasonable
27. $15.4z$ **29.** $986.4 \div n$ **31.** 12.3 times a number a
33. a number x divided by 5.4 **35.** 60 **37.** 1995
39. Answers will vary. **41.** Answers will vary. Example:
840; LEN and DON **43.** about 300 **45.** hundredths

Pages 16–17 Problem Solving Situations **1.** The
paragraph is about Cathy's job as a cashier. **3.** the $6.50
per hour that Cathy earns **5.** The paragraph is about
Paul's plan to buy a video game system. **7.** No. **9.** The
paragraph is about Carol's purchase of a stereo system.
11. the $100 down payment; the 10 equal payments
13. 15.9981 **15.** 13,870

Page 17 Self-Test 1 **1.** 32 **2.** 13 **3.** 10 **4.** 64
5. 172.4 **6.** 22.71 **7.** 4.77 **8.** 66.36 **9.** 710.84
10. 379.68 **11.** 70 **12.** 6.4 **13.** The paragraph is
about Jill's movie and video game rentals last year.
14. 15 **15.** the 15 video games that Jill rented; the $1.50
cost of each video game rental **16.** Multiply 15 times
$1.50.

Pages 20–21 Exercises **1.** < **3.** = **5.** >
7. $0.2 < 0.238 < 0.26$ **9.** $12.03 < 14 < 14.36$
11. $25.04 < 25.08 < 25.60$ **13.** Mexico **15.** <
17. < **19.** < **21.** < **23.** = **25.** > **27.** > **29.** =
31. < **33.** West; West **35.** 1960: 179.4 million people;
1980: 226.6 million people; about 50 million more people
37. always **39.** sometimes **41.** always **43.** never
45. 7.2 **47.** about 900 **49.** $1\frac{3}{7}$ **51.** 327,558

Page 24 Exercises **1.** 37 **3.** 5 **5.** 29 **7.** 114
9. 9.9 **11.** 155 **13.** 115 **15.** 19 **17.** 139 **19.** 80
21. 12.2 **23.** 60 **25.** 16 **27.** b **29.** 12.35

Page 27 Exercises **1.** 26 **3.** 7 **5.** 2.1 **7.** 320
9. 89 **11.** 0 **13.** 120 **15.** 3.3 **17.** 0 **19.** 4300
21. 420 **23.** 36 **25.** 0 **27.** 48 **29.** 18 **31.** 36; 8
33. 8; 2 **35.** 5 **37.** 0 **39.** 6.6

Page 29 Exercises **1.** pp; 78 **3.** pp; 515 **5.** c; 27,405
7. pp; 468.9 **9.** c; 2.4 **11.** mm; 0.28 **13.** mm; $.90
15. mm; about $10.50 **17.** 0.002 **19.** =

Pages 32–33 Exercises **1.** 343 **3.** 9 **5.** 1000 **7.** 16
9. 128 **11.** 48 **13.** 64 **15.** 16 **17.** 512 **19.** 1024;
40,960 **21.** 25 **23.** 256 **25.** 12 **27.** 0.81 **29.** 0.125
31. 0.008 **33.** 0.001 **35.** D **37.** B **39.** 81 **41.** 11
43. < **45.** < **47.** $a = 4; b = 2$ **49.** The paragraph is
about the amount of money Emily Ling earns and the number of hours that she works. **51.** $1\frac{1}{6}$

Page 33 Challenge The sum of the first and last numbers is 50, the sum of the second and tenth numbers is 50,
and so on. There are five of these sums for a total of 250.
Add the remaining number, 25, to this sum for a total of
275.

Pages 38–39 Exercises **1.** 96 **3.** 49 **5.** 16 **7.** 82
9. 7 **11.** 5 **13.** 23 **15.** 48 **17.** 71 **19.** 4 **21.** 17
23. 4,500,000 albums **25.** 36 **27.** 63 **29.** 55
31. False; $(24 - 4) 2 = 40$ **33.** True. **35.** False;
$3 (4 - 2) 3 = 18$ **37.** False; $(12 - 2^2) \div 4 = 2$
39. 33,789 **41.** 42

Page 39 Self-Test 2 **1.** < **2.** = **3.** > **4.** 142
5. 10.9 **6.** 106 **7.** 17 **8.** 150 **9.** 0 **10-12.** Answers
may vary. Likely answers are given. **10.** c; 73.367
11. pp; 66 **12.** mm; 591 **13.** 625 **14.** 192 **15.** 196
16. 13 **17.** 8 **18.** 3

Pages 40–41 Chapter Review **1.** variable expression
2. commutative property of addition **3.** multiplicative
identity **4.** variable **5.** base **6.** multiplication property
of zero **7.** 21 **8.** 32 **9.** 31 **10.** 28 **11.** 9 **12.** 4
13. 16 **14.** 24 **15.** 37.5 **16.** 7.392 **17.** 4.4
18. 4.28 **19.** 55.2 **20.** 80.3 **21.** 90.6 **22.** 0.055
23. < **24.** = **25.** > **26.** The paragraph is about
Yvonne's job typing term papers. **27.** 6 h **28.** the $7
per hour that she earns **29.** You would add the number of
hours that Yvonne typed on Monday, Tuesday, Wednesday, and Thursday. **30.** $5642 < 12,375 < 12,456$
31. $0.078 < 0.102 < 0.62$ **32.** $0.611 < 6 < 6.10$ **33.** 0.72

34. 44 **35.** 12 **36.** 149 **37.** 0 **38.** 7 **39.** 7
40–43. Answers may vary. Likely answers are given.
40. pp; 143 **41.** c; 94.583 **42.** c; 499.28 **43.** mm; 200
44. 243 **45.** 128 **46.** 1 **47.** 1024 **48.** 10,000
49. 512 **50.** 31 **51.** 7 **52.** 64 **53.** 25 **54.** 243
55. 864 **56.** 486 **57.** 400 **58.** 216 **59.** 7 **60.** 58
61. 2

Page 43 Chapter Extension **1.** 13 **3.** 37 **5.** Yes.
Addition is commutative. **7.** 21 **9.** 70 **11.** No. Answers will vary. An example is $5 \blacksquare 4 = 21, 4 \blacksquare 5 = 29$.
13. $a \star b = a + 2b$

Pages 44–45 Cumulative Review **1.** B **3.** C **5.** A
7. D **9.** C **11.** D **13.** D **15.** B **17.** C **19.** C

CHAPTER 2
Page 50 Exercises **1.** c^{10} **3.** n^3 **5.** x^{13} **7.** $12d^5$
9. $30b^3$ **11.** $28cd$ **13.** $56w^2y$ **15.** $240wx^2$ **17.** z^5
19. c^4 **21.** $3c$ **23.** $8b^2$ **25.** 8 **27.** 2 **29.** 6 **31.** 2
33. product of powers rule **35.** Both $2^m \cdot 2^0$ and $2^m \cdot 1$
are equal to 2^m. It follows that the expressions must be
equal to each other. **37.** The paragraph is about buying
tickets for a basketball game and a hockey game.
39. $30c^3$

Pages 54–55 Exercises **1.** 560 **3.** 30 **5.** 436
7. 1773 **9.** 3184 **11.** $90 + 9a$ **13.** $5x - 45$
15. $48m + 72$ **17.** $27 - 12a$ **19.** \$20 **21.** 4 **23.** 3
25. \$240 **27.** 63 **29.** $15x + 25$ **31.** 11.6

Pages 57–58 Exercises **1.** $14x$ **3.** $6m$ **5.** $11w$
7. $5x$ **9.** $8z + 7$ **11.** $9k - 6$ **13.** $4w + r$
15. $2a + 9b$ **17.** $18n$ **19.** $10a + 13b$ **21.** $11c + 4w$
23. cannot be simplified **25.** cannot be simplified
27. $5a + 13b + 3x$ **29.** $14c + 8$ **33.** $88y + 36$
35. $8b + 23$ **37.** 625 **39.** $1\frac{5}{8}$

Page 58 Self-Test 1 **1.** x^7 **2.** c^8 **3.** a^{16} **4.** $27w^7$
5. $90xz$ **6.** $84a^2b$ **7.** 120 **8.** 535 **9.** 1568
10. $2x + 18$ **11.** $56 - 40x$ **12.** $22 + 14c$ **13.** $7n$
14. $13x$ **15.** $3x - 5y$ **16.** $4a + 11b$ **17.** $8z + 1$
18. $12w$

Pages 60–61 Problem Solving Situations **1.** 26 mi
3. \$5.43 **5.** 1688 items **7.** 163 **9.** 495 **13.** $12a - 32$
15. 119

Pages 64–65 Exercises **1.** 52, 62, 72
3. 625; 3125; 15,625 **5.** 20, 26, 33 **7.** 13, 8, 2
9. $56m, 112m, 224m$ **11.** $a + 15, a + 20, a + 25$
13. $4n + 3, 5n + 3, 6n + 3$ **15.** January **17.** 89, 144
19. **21.** Start with 1, add 2, add

3, add 4, and so on. **23.** add 3, add 5, add 7, and so on;
square 1, square 2, square 3, and so on **25.** 14, 16, 20;
20, 22, 44 **29.** 1296; 7776; 46,656 **31.** $3\frac{1}{8}$

Page 65 Challenge **a.** A, S, O **b.** E, N, T
c. , K , **d.**

Pages 68–69 Exercises **1.** 40; 60; 80 **3.** 21; 25; 29
5. $x \to 5x$ **7.** \$57.75; \$74.25; \$99; \$107.25; \$132
9. 110 times **11.** 21 times **13.** The book has been kept
out of the library past the day that its return was requested.
15. \$.45 **19.** 23.7 **21.** $5x + 5y$

Page 70 Exercises **1.** $2 * A4 + 5$ **3.** add 2 **5.** add 3
7. The pattern will be add 4; each number in the table will
be 4 greater than the number above it. **9.** The pattern will
be add 5; each number in the table will be 1 greater than
the corresponding number in the table for $x \to 5x + 2$.

Pages 72–73 Problem Solving Situations **1.** Yes.
Estimating gives $20 + 7 + 3(3) = 36$, which is close to
\$37.42. **3.** \$20.30 **5.** \$45.10 **7.** 80 points **11.** <
13. \$7.11

Page 73 Self-Test 2 **1.** \$110.01 **2.** 17, 20, 23
3. 256, 1024, 4096 **4.** 16; 24; 28 **5.** $x \to 7x$ **6.** No.
Marvin multiplied his 5 extra hours by \$7.50 instead of
\$10.75.

Page 75 Exercises **1.** estimate; \$6 **3.** exact answer;
9 flights **5.** estimate; Yes, the coach will spend about
\$210. **11.** 600

Pages 77–79 Exercises **1.** 3 cm **3.** 0.45 km
5. 25,000 mg **7.** 0.615 m **9.** 2.345 L **11.** 3400 mL
13. 74 cm **15.** 98 m **17.** 880 m **19.** 2 g **21.** =
23. < **25.** > **27.** 4.36 **29.** 0.578 **31.** 0.809
33. 0.0136 **35.** 0.0245 **37.** b **39.** c **41.** A reasonable answer is 0.25 L. **43.** 17,100 mL **45.** 2 cm
47. one millionth of a meter **49.** one billionth of a meter
51. one million liters **53.** 240 oz **55.** $10x - 20$

Pages 82–83 Exercises **1.** 1.2×10^6 **3.** 5.7×10^3
5. 4.52×10^4 **7.** 8.514×10^8 **9.** 3800 **11.** 1,520,000
13. 500,000,000 **15.** 342,500 **17.** 1.392×10^6 km
19. 36,000,000 mi **21.** 887,140,000 mi
23. 3.5×10^{11}; 350,000,000,000 **25.** 4.853×10^7;
48,530,000 **27.** D **29.** > **31.** < **33.** < **35.** 4; 5
37. Both exercises involve multiplying numbers written in
scientific notation. However, an extra step was needed in
Exercise 36 to obtain an answer that was written in scientific notation. **39.** $7\frac{7}{12}$ **41.** 4.254×10^7 **43.** $3a + 7b$

Page 83 Self-Test 3 **1.** estimate; Yes, he needs about
\$18. **2.** 48 cm **3.** 3200 mL **4.** 500 g **5.** 4.4×10^4

6. 9.87×10^6 **7.** 2.13×10^7 **8.** 5600 **9.** 378,000 **10.** 64,300,000

Pages 84–85 Exercises 1. 2.838 L **3.** 158.75 cm **5.** about 210 mL **7.** about 50 kg **9.** meter **11.** gallon **13–17.** Answers may vary. **13.** 4.0678 L; 3.2637 L **15.** 5.025 m; 4.275 m **17.** 6.35 mm; 44.45 mm

Pages 86–87 Chapter Review 1. function **2.** distributive property **3.** unlike terms **4.** gram **5.** scientific notation **6.** arrow notation **7.** a^6 **8.** c^{13} **9.** $24x + 30$ **10.** $24 + 2x$ **11.** $24n + 7m$ **12.** $17a$ **13.** $10x + 12$ **14.** $55ab$ **15.** b^{12} **16.** x^{16} **17.** $54 - 9z$ **18.** $24c - 72$ **19.** $15x^8$ **20.** $9c$ **21.** $48wx^2$ **22.** $15 + 12z$ **23.** $7d - 5c$ **24.** $42c^4$ **25.** 180 **26.** 160 **27.** 832 **28.** 282 **29.** \$93.87 **30.** 31, 38, 45 **31.** 162, 486, 1458 **32.** $a + 9, a + 10, a + 11$ **33.** $29m, 26m, 23m$ **34.** 12; 16; 20 **35.** 54; 72; 90 **36.** $x \to \frac{x}{3}$ **37.** $x \to x + 5$ **38.** Yes. **39.** exact answer; \$49.50 **40.** 1.7 kg **41.** 0.75 L **42.** 760 m **43.** 190 cm **44.** 3.5×10^5 **45.** 1.27×10^4 **46.** 6.55×10^6 **47.** 4.8×10^7 **48.** 130,000 **49.** 97,000,000 **50.** 26,400 **51.** 588,000,000

Page 89 Chapter Extension 1. b^{10} **3.** c^{28} **5.** x^{36} **7. a.** 144 **b.** 144 **9.** Power of a product rule: To find the power of a product, you find the power of each factor and then multiply. $(ab)^m = a^m b^m$

Pages 90–91 Cumulative Review 1. B **3.** A **5.** B **7.** D **9.** B **11.** C **13.** A **15.** C **17.** B **19.** A

CHAPTER 3
Pages 96–97 Exercises 1. 5 **3.** 13 **5.** 1 **7.** $>$ **9.** $=$ **11.** $<$ **13.** $>$ **15.** -754 ft **17.** $-3 < 4 < 9$ **19.** $-10 < -8 < -6$ **21.** The definition given in the lesson refers to distance from zero. This definition uses a variable to represent any integer. This definition specifically includes zero. **23.** \$10,000 **25.** Hearty Company **27.** never **29.** never **31.** always **33.** sometimes **35.** 2203 **37.** 2

Page 100 Exercises 1. 29 **3.** -30 **5.** -63 **7.** 68 **9.** -45 **11.** 112 **13.** \$38 **15.** 33°F **17.** 18 **19.** -12 **21.** -22 **23.** -17 **25.** $15y$ **27.** $-10m$ **29.** $-10r + 4s$ **31.** $8m + 8n$ **33.** about 80 **35.** $12y - 3$

Pages 103–104 Exercises 1. 0 **3.** -21 **5.** 0 **7.** 5 **9.** 10 **11.** -7 **13.** -26 **15.** 7 **17.** -14 **19.** less money; \$8 less **21.** 0 **23.** -8 **25.** 8 **27.** 0 **29.** 12 **31.** 50 s after liftoff **33.** exact times; each maneuver begins at an exact time before liftoff. **35.** T minus 45 **37.** -8 **39.** -10 **41.** -2 **43.** $-1,000,000,001$

45. infinitely many pairs; for every integer, you can add the opposite of one greater than the integer. **47.** 17, 21, 25 **49.** -28

Pages 109–110 Exercises 1. -9 **3.** 6 **5.** 25 **7.** 0 **9.** -3 **11.** -4 **13.** 0 **15.** -52 **17.** 4 **19.** 36°F **21.** 3 **23.** -16 **25.** -12 **27.** B **29.** C **31.** $-4, -9, -14$ **33.** $-7, -3, 1$ **35.** -33°F **37.** 17°F **39.** about 10°F **41.** 1070 mm **43.** about 32

Pages 113–114 Exercises 1. 4 **3.** -11 **5.** 0 **7.** -4 **9.** 154 **11.** -10 **13.** 100 **15.** -84 **17.** -16 **19.** -5°F **21.** -36 **23.** 29 **25.** 9 **27.** -41 **29.** 11 **31.** 12 **33.** $-12c + 24$ **35.** $14y + 56$ **37.** $-20 + 20b$ **39.** D **41.** C **43.** 1 aluminum, 3 chloride **47.** -12

Page 114 Self-Test 1 1. $>$ **2.** $<$ **3.** $=$ **4.** -4 **5.** -22 **6.** -18 **7.** 5 **8.** -6 **9.** 0 **10.** -31 **11.** -14 **12.** 6 **13.** -32 **14.** 70 **15.** -12

Pages 116–118 Exercises 1. 1296 **3.** -135 **5.** -166 **7.** 27 **9.** 1 **11.** 11 **13.** 14 **15.** -6 **17.** -5 **19.** 24 **21.** -1 **23.** 42 **25.** 36 **27.** -15 **29.** D **31.** A **33.** 4; 1; 16 **35.** $x \to -4x$ **37.** 1 **39.** 1 **41.** -1 **43.** If n is even, $(-1)^n$ is 1. If n is odd, $(-1)^n$ is -1. **45.** positive **47.** negative **49.** positive **51.** 150,000,000 **53.** 16 **55.** No; Julia did not include the amount that either her brother or her sister paid. **57.** 0

Page 118 Challenge Answers will vary. Examples: $-5 + 7 - 8 \div 4 \cdot 2$ and $-8 \div 2 + 5 - 1 \cdot 3$

Pages 120–121 Problem Solving Situations 1. 12 ways **3.** 15 totals **5–9.** Strategies may vary. Likely strategies are given. **5.** making a table; 12 totals **7.** choosing the correct operation; \$32 **9.** making a table; 7 ways **13.** -12 **15.** -11

Pages 126–127 Exercises 1. $(-4, 3)$ **3.** $(4, -3)$ **5.** $(0, -3)$ **7.** $(-3, 0)$ **9.** $(2, 3)$ **11.** $(-2, -4)$ **13–20.** **21.** I **23.** II

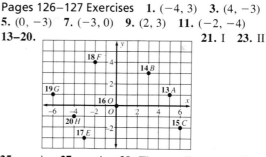

25. y-axis **27.** x-axis **29.** The coordinates are all positive. **31.** The coordinates are all negative. **33.** The x-coordinates are all negative. **35.** The y-coordinates are all zero. **37.** quadrilateral

39.

The shapes are the same. $\triangle DEF$ is 3 units above and 2 units to the right of $\triangle ABC$.   **41.** $7a - 7b$

43.

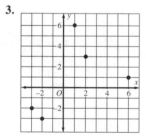

Pages 130–131 Exercises   1.

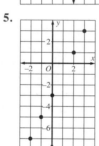

3.

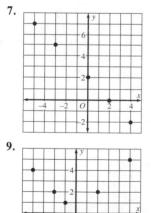

5.

7.

9.

11.

x	$2x$
-2	-4
-1	-2
0	0
1	2
2	4

$x \longrightarrow 2x$

13.

x	$-x$
-2	2
-1	1
0	0
1	-1
2	-2

$x \longrightarrow -x$

15.

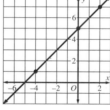

a. Answers will vary.

Examples: $(-2, 3)$, $(-1, 4)$, and $(1, 6)$.   **b.** Yes.

17.   **19.** $68

Page 131 Self-Test 2   1. 31   **2.** 3   **3.** 10   **4.** 6 ways

5–6.   **7.**

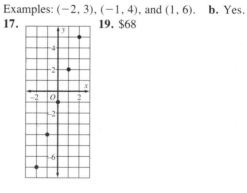

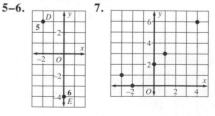

Page 133 Exercises 1. B5 **3.** Possible answers: Middletown, Paterson, Newark, and Ramsey **5.** F3 and F4 **7.** D3 and D4 **9.** Delaware and Hudson Rivers **11.** Morristown

13.

Pennsylvania Index	
Bushkill	B5
Cresco	A6
Easton	A3
Quakertown	A2
Portland	B4

15. Answers will vary.

Pages 134–135 Chapter Review 1. opposites **2.** absolute value **3.** integers **4.** addition property of opposites **5.** origin **6.** x-coordinate **7.** 11 **8.** 5 **9.** 0 **10.** 2 **11.** 4 **12.** 3 **13.** < **14.** = **15.** < **16.** < **17.a.** 7 ways **b.** 9 totals **18.** −5 **19.** 22 **20.** 76 **21.** −12 **22.** −20 **23.** 48 **24.** 11 **25.** −7 **26.** −77 **27.** 12 **28.** 108 **29.** −16 **30.** −46 **31.** 0 **32.** −45 **33.** 0 **34.** 11 **35.** −20 **36.** −8 **37.** 7 **38.** −23 **39.** −36 **40.** 0 **41.** −42 **42.** 4 **43.** 216 **44.** −24 **45.** −180 **46.** 40 **47.** 28 **48.** 13 **49.** −5 **50.** 9 **51.** 8 **52.** 4 **53.** 9 **54.** (−2, 4) **55.** (1, 0) **56.** (3, 3) **57.** (−1, −2) **58.** (4, −3) **59.** (0, −4) **60–63.**

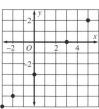

64.

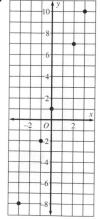

65.

66.

67.

Page 137 Exercises 1.a. closed **b.** not closed **c.** closed **d.** not closed **3.a.** closed **b.** not closed **c.** closed **d.** not closed **5.a.** not closed **b.** not closed **c.** closed **d.** closed (exclude division by zero) **7.a.** closed **b.** not closed **c.** closed **d.** not closed

Pages 138–139 Cumulative Review 1. D **3.** A **5.** B **7.** C **9.** C **11.** B **13.** A **15.** A **17.** D **19.** C

CHAPTER 4
Page 144 Exercises 1. Yes. **3.** No. **5.** No. **7.** Yes. **9.** Yes. **11.** 3 **13.** −45 **15.** −8 **17.** 5 **19.** No. **21.** B **23.** A **25.** C **27.** C **29.** A **31.** False. **33.** True. **35.** True. **37.** cannot determine **39.** Answers may vary. Examples are given. $m + 3 = 2$; $9m = −9$ **41.** Answers may vary. Examples are given. $c − 18 = 6$; $\frac{c}{8} = 3$ **43.** No; the amount $50 was not included in the total. **45.** 3

Pages 146–147 Problem Solving Situations 1. 24 **3.** 1320 **5.** 5 sweaters, 7 shirts **7.** 7 quarters, 8 dimes **9–13.** Strategies may vary. Likely strategies are given. **9.** making a table; 3 ways **11.** guess and check; 9 **13.** choosing the correct operation; $4500 **17.** 23 **19.** 4 compact discs, 3 cassette tapes

Page 150 Exercises 1. 9 **3.** −9 **5.** 4 **7.** −11 **9.** 18 **11.** 0 **13.** −11 **15.** 396 **17.** −1580 **19.** −437 **21.** 5; −5 **23.** 3; −3 **25.** 2 **27.** no solution **29.** 20; −20 **31.** no solution **33.** 43 **35.** −117

Pages 153–154 Exercises 1. 15 **3.** −8 **5.** 0 **7.** −56 **9.** −189 **11.** 62 **13.** 15 **15.** −17 **17.** −1 **19.** 0 **21.** B **23.** A **25.** b **27.** d **29.** about 1800 **31.** about 1400 **33.** 23 **35.**

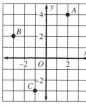

37. about 16

Page 154 Self-Test 1 1. −7 **2.** −9 **3.** 6 **4.** 5 birthday cards, 3 thank you notes **5.** −3 **6.** −2 **7.** 2 **8.** −30 **9.** −24 **10.** −7

Pages 158–159 Exercises **1.** 4 **3.** −2 **5.** −7
7. −10 **9.** 40 **11.** 39 **13.** −156 **15.** 0 **17.** 5

19.

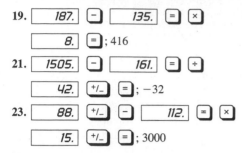

| 187. | − | 135. | = | × |

| 8. | = |; 416

21. | 1505. | − | 161. | = | ÷ |

| 42. | +/− | = |; −32

23. | 88. | +/− | − | 112. | = | × |

| 15. | +/− | = |; 3000

25. Answers may vary. Example: The first equation requires only one step to solve; $2x + 3 = 15$ requires two steps to solve. **27.** Two; a chemical reaction has a natural balance. **29.** 4; 2 **31.** 8,320,000,000 **33.** 0

Pages 163–164 Exercises **1.** $x + 5$ **3.** $2w$ **5.** $3c + 8$
7. $6d − 4$ **9.** $a − 12$ **11.** $\frac{s}{6}$ **13.** $t + 17$ **17.** $26 + 12x$
19. $26 + 12x + 7y$ **21.** 23; 31; 39 **23.** $7 + 3x$

Pages 166–167 Exercises **1.** $m + 4 = 5$ **3.** $\frac{a}{6} = 12$
5. $15k = 105$ **7.** $7 = \frac{j}{2}$ **9.** $10 = s − 15$ **11.** A **13.** D
15. C **17–19.** Answers will vary. Examples are given.
17. The number m decreased by 8 is 16. **19.** The product of 4 and a number w is 48. **21.** mental math
23.

(graph with points plotted on coordinate grid)

25. No.

Page 167 Challenge The weights 11 and 92 on one side balance the weights 47 and 56 on the other side; No.

Pages 169–171 Problem Solving Situations
1. $1000 **3.** 114 **5–11.** Strategies may vary.
5. choosing the correct operation; $91 **7.** making a table; 3 ways **9.** using equations; $159 **11.** guess and check; 4 books, 4 magazines; 1 book, 9 magazines **13.** 72 mg
19. Yes; Yes; No; No; No; No; No. **21.** 16

Page 171 Self-Test 2 **1.** −3 **2.** 24 **3.** 10 **4.** $\frac{z}{14}$
5. $3t$ **6.** $n + 7 = 35$ **7.** $t − 9 = 22$ **8.** 73 **9.** 70 km

Pages 173–175 Exercises **1.** 12 **3.** −8 **5.** −5
7. 4 **9.** 3 **11.** 3 **13.** 9 **15.** −28 **17.** 3 **19.** 4

21. 2 **23.** all values of x less than 8 **25.** 8 **27.a.** The paragraph is about Steve working at a job. **b.** 7 h **c.** the hourly wage **d.** Add 7, 8.5, 7, and 2. Multiply the sum by 9.25. **29.** about 600 **31.** 6, 3, 0

Pages 177–179 Exercises **1.** 2200 **3.** 40 **5.** 610
7. 475 mi/h **9.** 7 **11.** 36 **13.** $P = \frac{a}{2} + 110$ **15.** 80
years old **17.** 293.4 mi **19.** $E = IR$ **21.** E = force (in volts), I = current (in amperes), R = resistance (in ohms).
23. 8.4 **25.** 2 **27.** $9a + 4b$ **29.** 1024

Page 179 Self-Test 3 **1.** 12 **2.** −6 **3.** 12 **4.** 10
5. −2 **6.** 3 **7.** 180 mi **8.** 350 mi **9.** 12 h
10. 234 mi/h

Pages 180–181 Chapter Review **1.** inverse operations
2. equation **3.** solve **4.** formula **5.** solution of an
equation **6.** Yes. **7.** No. **8.** No. **9.** Yes. **10.** 5
11. −7 **12.** 50 **13.** 0 **14.** 13 haircuts, 4 perms **15.** 3
pens, 4 notebooks **16.** 3 **17.** −2 **18.** −5 **19.** 23
20. −16 **21.** −14 **22.** 45 **23.** 12 **24.** 2 **25.** 9
26. −9 **27.** −88 **28.** $8q − 6$ **29.** $\frac{h}{19}$ **30.** $7b$
31. $2t + 4$ **32.** $z − 15 = 33$ **33.** $m − 5 = 8$
34. $3x = 48$ **35.** $169 **36.** 7 fish **37.** −2 **38.** −3
39. 16 **40.** 7 **41.** 14 **42.** $47.76 **43.** $126
44. $7.50

Page 183 Chapter Extension **1.** −2 **3.** 5 **5.** −26
7. 6 **9.** −2 **11.** 2 **13.** −2 **15.** 3 **17.** 13 **19.** Answers may vary. Example: Add nine to both sides of the equation.

Pages 184–185 Cumulative Review **1.** A **3.** A
5. B **7.** A **9.** D **11.** C **13.** D **15.** C **17.** B **19.** A

CHAPTER 5
Pages 190–191 Exercises **1.** about 3,500,000
3. about 1,000,000 **5.** 1992
7.

Airport Limousine Earnings per Quarter

First	$ $ $ $ $ $ $
Second	$ $ $ $ $ $ $ $ $ $
Third	$ $ $ $ $ $ $ $ ͨ
Fourth	$ $ $ $ $	$ = $10,000

9. about 1550 **11.** about $3600 **13.** There are too few symbols, so the differences in the data are not reflected. For example, Boston and Pittsburgh seem to have the same number of passengers. **17.** 6 **19.** 136 **21.** 4
23. $40 − 24n$

Pages 194–195 Exercises **1.** Estimates may vary; about 2 million. **3.** Estimates may vary; about 3.5 million. **5.** bowling, volleyball **7.** Estimates may vary; about 2 million. **9.** Western Europe **11.** Estimates may

vary; about $1 billion.   **13.** Estimates may vary; about 35°F.   **15.** Estimates may vary; about 40°F
17. 15.4 million   **19.** about $1159.2 million   **21.** Estimates may vary; about 90.
23.

National League Standings Western Division

■ Games won   ▨ Games lost

25. $17u + 5v$

Pages 198–199 Exercises   1. Estimates may vary; about 8 million   **3.** Estimates may vary; about 9.5 million   **5.** 1977, 1981   **7.** Estimates may vary; about 2.5 million   **9.** 1973–1977   **11.** Estimates may vary; elementary: about 1.2 million; secondary: about 1.0 million.   **13.** Estimates may vary; elementary: about 1.1 million; secondary: about 0.9 million   **15.** Estimates may vary; about 0.2 million   **17.** Estimates may vary; about 2.2 million   **19.** Estimates may vary; about 0.35 million   **21.** 5 years   **23.** 1985   **25.** 50 million   **27.** 30 million   **29.** 50 million
31.

Year	Number of Sheep
1910	50
1925	40
1940	50
1955	30
1970	20
1985	10

33. $240x^2$ **35.** 2.4

Pages 202–203 Exercises
1.

Adults Participating in Leisure-Time Activities

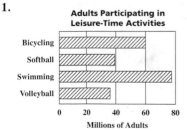

3.

Average Payment Period for Finance Company Loans on New Cars

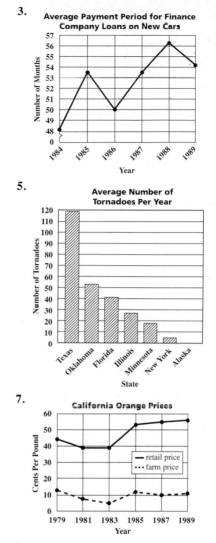

5.

Average Number of Tornadoes Per Year

7.

California Orange Prices

9. Answers may vary. Example: The retail price of oranges is always greater than the farm price. Furthermore, the direction of change in retail and farm prices is generally the same.   **11.** -22 **13.** about 8

Page 203 Self-Test 1
1.

Average Annual Snowfall in Four Locations

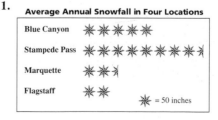

2. about 15 thousand   **3.** about 35 thousand
4. about 10 million   **5.** increasing
6.

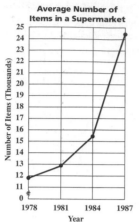

Average Number of
Items in a Supermarket

Page 205 Exercises
1. bar graph;

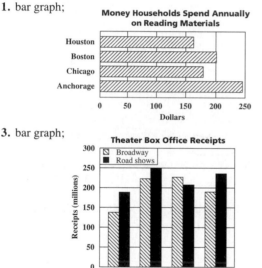

Money Households Spend Annually
on Reading Materials

3. bar graph;

Theater Box Office Receipts

5. $15n + 20$ **7.** Line graph; the population changes continuously.

Pages 207–208 Problem Solving Situations   1. 7
3. not enough information; the number of books sold in Midtown in 1980 and 1990   **5–7.** Strategies may vary. Likely strategies are given.   **5.** making a table; 3
7. choosing the correct operation; $296.52   **11.** $2\frac{5}{12}$
13. 353

Pages 211–212 Exercises   1. about the same   **3.** The gap could be eliminated and intervals of 1 from 0 to 4 could be used.   **5.** *Graph E.* It makes it appear that the advertising campaign did not significantly increase the number of subscribers.   **7.** misleading   **9.** accurate

11.

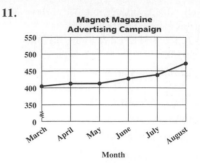

Magnet Magazine
Advertising Campaign

13. Changing the scale on a line graph can exaggerate or minimize a trend.   **15.** $9\frac{9}{10}$ **17.** It gives the false visual impression that the level of production at Plant C was about half that at Plant B.

Page 212 Self-Test 2
1. line graph;

Median Prices of New
One-Family Homes in the Northeast

2. $944.76   **3.** not enough information; the number of miles showing on the odometer before the trip   **4.** The level of production at Plant B appears to be five times that at Plant A.

Pages 214–216 Exercises   1. 18.25; 18; 16; 7
3. $43.50; $30.50; none; $467   **5.** $\approx -4.2°C$; $-4°C$; none; 7°C   **7.** 75; 60; 60; 135   **9.** 72   **11.** 64 years
13. unreasonable; ≈ 6.4 **15.** reasonable   **17.** 8.8
19. 8.7   **21.** It is the same as finding the median of the four judges' scores.   **23.** Yes; Yes; Yes; Yes; No; Yes; Yes.   **25.** 250

Page 216 Challenge   10, 10, 15, 34, and 51

Page 219 Exercises   1. Yes; the mean is 72 in. It is not distorted by an extreme value.   **3.** Mode; the data cannot be averaged or listed in numerical order.   **5.** Median; the median is 5 and the mode is 3. The mode is less than most of the family sizes.   **7.** Mean; the mean is 35 h, the median is 29 h, and the mode is 20 h. The mean is equal to the number of hours that the nurse wants to work per week.

9.

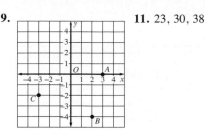

11. 23, 30, 38

Page 220 Exercises 1. Metropolitan Area, Region, Food, Housing, Transportation **3.** Atlanta, Dallas/Fort Worth, Washington, D.C. **5.** Washington, D.C. **7.** $4150.50 **9.** Answers may vary. Example: Washington, D.C.; use the database to sort records in decreasing order by the total cost for food, housing, and transportation. Washington, D.C. appears first, with a total of $23,208.

Pages 222–223 Exercises 1. 31; 31; 30; 3

3.

Number of Movies	Tally	Frequency
2	||	2
3	⊬⊦|	6
4	⊬⊦⊬⊦	10
5	||||	4
12	|	1
15	|	1
		Total 24

5. No; the mean, 4.5, is distorted by the extreme values 12 and 15. **7.** Answers may vary due to possible changes in state laws. **9.** 770 km/h **11.** 36 laps

Page 223 Self-Test 3 1. $43.75; $40; $50 and $20; $90 **2.** Yes; the mean is 14.5. It is not distorted by an extreme value. **3.** Mode; the data cannot be averaged or listed in numerical order. **4.** 7; 7; 8; 3

Page 225 Exercises 1, 5, 9. Estimates may vary. **1.** about $14,000 **3.** 0 up to 800 **5.** about $2000 **7.** 800 **9.** about $1.67 **11.** $10,000 **13.** Answers will vary. Examples: camera, film, rent for shop.

Pages 226–228 Chapter Review 1. mode **2.** line graph **3.** statistics **4.** data **5.** legend **6.** scale **7.** about 1800 million **8.** about 800 million **9.** not enough information; the amount of corn produced in 1980 **10.** about $7872 million
11.

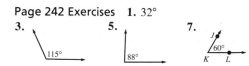

12. about $25 billion **13.** about $25 billion **14.** direct mail **15.** 1980–1985 **16.** Estimate may vary; about 5.5 visits **17.** 1985 **18.** *Graph A* gives the impression that the average number of annual visits to physicians did not change very dramatically from 1970 to 1985. *Graph B* gives the opposite impression. **19.** *Graph A*
20. bar graph;

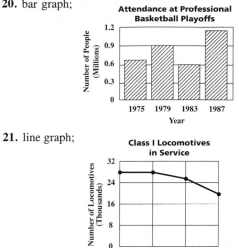

21. line graph;

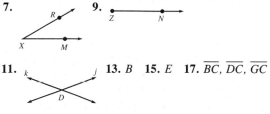

22. not enough information; the weight of the apples **23.** $.25 **24.** $43.31; $34.60; none; $53.99 **25.** 1.2; 1; 2; 3 **26.** −4°C; −4°C; none; 12°C **27.** mode; the data cannot be averaged or listed in numerical order.

Page 231 Chapter Extension 1, 3. Answers are given in the order *n*, Σ*x*, and *x̄*. **1.** 6; 36; 6 **3.** 5; 94.6; 18.92

Pages 232–233 Cumulative Review 1. B **3.** D **5.** B **7.** C **9.** B **11.** A **13.** D **15.** D **17.** A **19.** D

CHAPTER 6
Pages 238–239 Exercises 1. plane *R* **3.** ∠*TRV*, ∠*VRT*, or ∠*R* **5.** They have different endpoints.
7.   **9.**

11.   **13.** *B* **15.** *E* **17.** $\overline{BC}, \overline{DC}, \overline{GC}$

19. $\overline{BF}, \overline{GF}, \overline{EF}$ **21.** *M* **23.** $\overline{AB}, \overline{BC}, \overline{AC}$ **25.** $1\frac{7}{16}$
27. 8 adult tickets, 13 student tickets; 0 adult tickets and 36 student tickets

Page 242 Exercises 1. 32°
3.   **5.**   **7.**

9. D   **11.** C   **13.** 9   **15.**

Pages 247–248 Exercises   1. 60°   **3.** 61°   **5.** 170°
7. 78°   **9.** 28°   **11.** 28°   **13.** 45°   **15.** always
17. sometimes   **19.** A straight angle has measure 180°.
21. 18   **23.** 60°   **25.** $2n + 11 = -31$ **27.** -31

Pages 251–253 Exercises   1. False.   **3.** True.
5. False.   **7.** False.   **9.** 37°   **11.** 143°   **13.** 37°   **15.** 90
17. $m\angle 1 = 96°$; $m\angle 2 = 84°$; $m\angle 3 = 96°$; $m\angle 4 = 90°$;
$m\angle 5 = 90°$; $m\angle 6 = 90°$; intersecting   **19.** Parallel lines
lie in the same plane, and skew lines do not; both parallel
lines and skew lines do not intersect.   **21.** No, because if
two skew lines were perpendicular, they would be in the
same plane.   **23.** about 90   **25.** 200
27.

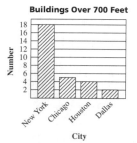

Buildings Over 700 Feet

Page 253 Self-Test 1   1. $\overrightarrow{XY}$ **2.** $\overleftrightarrow{MN}$, $\overleftrightarrow{NM}$ **3.** $\angle TSR$,
$\angle RST$, or $\angle S$ **4.**

5.   **6.**   **7.** 41°   **8.** 103°

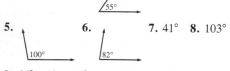

9. 65°   **10.** 115°   **11.** 65°   **12.** 115°

Pages 259–260 Exercises   1. regular triangle; no diago-
nals   **3.** quadrilateral; $\overline{LN}$, $\overline{OM}$ **5.** 48°   **7.** 33°   **9.** B
11. A   **13.** 10; 1440; 1440; 10; 144   **15.** 11   **17.** $1\frac{5}{8}$

Pages 263–265 Exercises   1. cannot   **3.** can; scalene
5. cannot   **7.** can; isosceles   **9.** can; equilateral
11. obtuse   **13.** acute   **15.** right   **17.** acute   **19.** obtuse
21. No; two sides must be equal in length.   **23.** *BAE*
25. *BAE, BED*   **27.** *BED*   **29.** 38°   **31.** 90°   **33.** 52°
35. 52°   **37.** 142°   **39.** 142°   **41.** True.   **43.** False; the
sum of the angles would be greater than 180°.   **45.** True.
47. False; an acute triangle can have three equal sides.
49. FD 40 RT 135   **51.** The three angles are each 65°.
Their sum is greater than 180°.
53.   **55.** right   **57.** 11

Page 265 Challenge   44

Pages 268–269 Exercises   1. 6 m   **3.** 12 m   **5.** 27°
7. 4 in.   **9.** 4 in.   **11.** 5 in.   **13.** 7 in.   **15.** 29°; 151°;
151°   **17.** $m\angle 1 = 102°$; $m\angle 2 = 33°$; $m\angle 3 = 57°$
21. quadrilateral, none; parallelogram, four pairs; trapezoid, two pairs; rectangle, four pairs; rhombus, four pairs;
square, four pairs   **23.** -51 **25.** False.

Page 270 Exercises   3. Answers will vary.   **5.** False.
7. parallelogram, trapezoid, rectangle, rhombus, or square;
true for all the above except trapezoid.

Pages 273–274 Problem Solving Situations   1. 21
3–5. Strategies may vary. Likely strategies are given.
3. using an equation; -21 **5.** making a table; 23   **7.** The
dots represent people; the lines represent handshakes.
9. Count the handshakes between two people, three people, four, five, and so on. Look for a pattern in the number
of handshakes. The solution is 28.   **11.** Answers will
vary. How many pieces of pizza result from four cuts
through the center of the pizza?   **13.** about 890
15.

Pages 277–279 Exercises   1. No.   **3.** No.
5.   **7.** square   **9.**

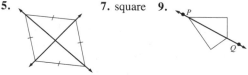

11. 3; 4; 5; 6; 7; 8   **13.** 4   **15.** 14   **17.** rotational
19. both   **23.** No.

Page 279 Self-Test 2   1. trapezoid; $\overline{AC}$, $\overline{BD}$ **2.** regular
pentagon; $\overline{VS}$, $\overline{VT}$, $\overline{RU}$, $\overline{RT}$, $\overline{SU}$ **3.** cannot   **4.** acute
5. 10 cm   **6.** 58°   **7.** 28   **8.**

Pages 280–282 Chapter Review   1. line segment
2. parallel   **3.** straight angle   **4.** square   **5.** trapezoid
6. equilateral triangle   **7.** complementary angles
8. protractor   **9.** obtuse angle   **10.** acute triangle
11. $\overline{GH}$, $\overline{HG}$ **12.** $\angle PQR$, $\angle RQP$, or $\angle Q$ **13.** Plane J
14. 37°   **15.** 115°   **16.**   **17.**

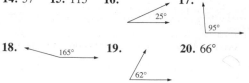

18.   **19.**   **20.** 66°

21. 45° **22.** 17° **23.** 2° **24.** 150° **25.** 105° **26.** 95°
27. 58° **28.** 40° **29.** 140° **30.** 140° **31.** 140°
32. 40° **33.** 140° **34.** False. **35.** True. **36.** True.
37. False. **38.** True. **39.** False. **40.** quadrilateral; $\overline{AC}$,
$\overline{BD}$ **41.** triangle; no diagonals **42.** pentagon; $\overline{KH}$, $\overline{KI}$,
$\overline{GJ}$, $\overline{GI}$, $\overline{HJ}$ **43.** 35° **44.** 129° **45.** 130° **46.** can;
equilateral **47.** can; scalene **48.** can; scalene **49.** ob-
tuse **50.** right **51.** acute **52.** acute **53.** 60° **54.** 60°
55. 60° **56.** 7 cm **57.** 60° **58.** 7 cm **59.** 18
60. No. **61.** Yes. **62.** Yes. **63.**

64. none **65.**

Page 285 Chapter Extension **1.** 4−4−4−4
3. Code: 3−3−3−3−3−3

5. The codes for the combinations: 6−3−6−3,
3−4−6−4, 4−4−3−3−3, 3−3−4−3−4, 6−3−3−3−3

Pages 286−287 Cumulative Review **1.** C **3.** C
5. B **7.** D **9.** A **11.** C **13.** D **15.** D **17.** D
19. B

CHAPTER 7
Pages 291−292 Exercises **1.** composite **3.** prime
5. composite **7.** prime **9.** 3^3 **11.** $2 \cdot 11$ **13.** $2^2 \cdot 5$
15. $2^2 \cdot 7$ **17.** $2^2 \cdot 3^3$ **19.** $2^2 \cdot 5^3$ **21.** $2^4 \cdot 3^2 \cdot 7$
23. $3 \cdot 5^3 \cdot 7$ **25.** $1 \cdot 60; 2 \cdot 30; 3 \cdot 20; 4 \cdot 15; 5 \cdot 12;$
$6 \cdot 10$ **27.** y^5 **29.** y^3 **31.** The factors are a^n, a^{n-1},
$a^{n-2}, \ldots, a^1, 1$ **33.** 2, 3, 5, 7, 11, 13, 17, 19, 23, 29, 31,
37, 41, 43, 47, 53, 59, 61, 67, 71, 73, 79, 83, 89, 97
35. a. fewer than 21; the number of primes decreases.
b. 211, 223, 227, 229, 233, 239, 241, 251, 257, 263, 269,
271, 277, 281, 283, 293 **c.** 16 **d.** It supports the answer to (a). **37.** 84; 84; none; 24 **39.** $2^5 \cdot 3$ **41.** 0

Pages 294−295 Exercises **1.** 2 **3.** 1 **5.** 20 **7.** 8
9. $6x$ **11.** b **13.** 2 **15.** $4r^8$ **17.** $4n^2$ **19.** 6
21. Yes. **23.** No. **25.** Answers may vary. Examples:
51, 70; 52, 55; 81, 85 **27.** True. **29.** True. **31.** 2
33. about 12

Pages 298−299 Exercises **1.** 10 **3.** 27 **5.** 40
7. 420 **9.** $6x$ **11.** $12k$ **13.** $30rst$ **15.** $42a^9$
17. $150n^4$ **19.** 24 **21.** 216 **23.** 10,080 **25.** 2304
27. A factor is any of two or more numbers multiplied to
form a product. A multiple is a product of a given number
and any nonzero whole number. **29.** 15; 90 **31.** 44; 264
33. Answers may vary. Example: 4 and 12 **35.** $14\frac{1}{3}$
37. $y + 7$

Page 299 Self-Test 1 **1.** prime **2.** prime
3. composite **4.** prime **5.** $2 \cdot 7$ **6.** $2^2 \cdot 3$ **7.** $3 \cdot 5 \cdot 7$
8. $2^4 \cdot 5^2$ **9.** 2 **10.** 9 **11.** $3c$ **12.** $8x^2$ **13.** 15
14. 80 **15.** $24awy$ **16.** $126n^7$

Pages 304−305 Exercises **1.** 2 **3.** 6 **5.** 20 **7.** 2
9. $\frac{1}{3}$ **11.** $\frac{3}{5}$ **13.** $\frac{4}{3}$ **15.** $\frac{8}{9}$ **17.** $\frac{3}{2}$ **19.** $\frac{1}{2}$ **21.** $\frac{2}{5}$ **23.** $\frac{6}{1}$
25. $\frac{4}{1}$ **27.** The denominators are 1; the fractions represent
whole numbers. **29.** always **31.** $\frac{1}{6}$
33. **35.** $\frac{1}{4}$ **37.** $\frac{1}{16}$ **39.** $\frac{4}{25}$ **41.** $\overrightarrow{MN}$

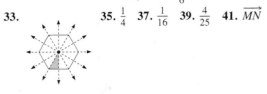

Page 305 Challenge **1.** 59 **2.** 53

Pages 307−308 Exercises **1.** $\frac{3b}{4}$ **3.** $\frac{4}{5}$ **5.** $\frac{2}{s}$ **7.** c^8
9. $6a^4$ **11.** 12 **13.** $\frac{5}{y}$ **15.** $8x^{16}$ **17.** $\frac{b^5}{6}$ **19.** $4n^2$
21. $\frac{v^3}{7}$ **23.** $\frac{4vw^3}{11}$ **25.** $7x$ **27.** $\frac{5m^5}{6n^2}$ **29.** $\frac{5s^4r^2}{2}$ **31.** $\frac{1}{a^5}$
33. a^7 **35.** 8 **37.** −8

Pages 311−312 Exercises **1.** > **3.** < **5.** < **7.** >
9. = **11.** > **13.** science book **15.** $\frac{1}{7} < \frac{1}{5} < \frac{1}{4}$
17. $\frac{3}{40} < \frac{3}{8} < \frac{7}{16}$ **19.** less than **21.** bar graph; the data
are not continuously changing. **23.** greater than; the dis-
tances are generally increasing. **25.** $\frac{1}{15}, \frac{1}{60}, \frac{1}{500}$ **27.** $\frac{1}{30}$
29. ; 1, 4, 9, 16, 25 **31.** 6
33. 192

Pages 315−316 Exercises **1.** 0.7 **3.** $0.\overline{81}$ **5.** 6.55
7. $10.\overline{6}$ **9.** $\frac{54}{125}$ **11.** $\frac{19}{100}$ **13.** $\frac{1}{6}$ **15.** $3\frac{8}{25}$ **17.** $9\frac{51}{100}$
19. $2\frac{2}{3}$ **21.** 0.875 **23.** about $\frac{1}{2}$ **25.** about $\frac{1}{3}$ **27.** about
$9\frac{3}{5}$ **29.** about $6\frac{1}{8}$ **31.** $0.\overline{1}$ **33.** $0.\overline{3}$ **35.** The numerator
is the repeating part of the decimal.

$\frac{1}{9} = 0.\overline{1}$	$\frac{5}{9} = 0.\overline{5}$
$\frac{2}{9} = 0.\overline{2}$	$\frac{6}{9} = 0.\overline{6}$
$\frac{3}{9} = 0.\overline{3}$	$\frac{7}{9} = 0.\overline{7}$
$\frac{4}{9} = 0.\overline{4}$	$\frac{8}{9} = 0.\overline{8}$

37. $0.\overline{18}$ **39.** $0.\overline{36}$ **41.** about $\frac{1}{2}$ **43.** less than

45. a. to account for the weight of the container
b. 1.01 lb **47.** $\frac{1}{45}$: repeating; $\frac{1}{50}$: terminating; $\frac{1}{45} = 0.0\overline{2}$;
$\frac{1}{50} = 0.02$ **49.** 60° **51.** about 4600

Pages 318–320 Problem Solving Situations **1.** 5
blocks due west **3.** 9 **5.** 6th floor **7–11.** Strategies
may vary. Likely strategies are given. **7.** making a table;
3 **9.** drawing a diagram; 4 **11.** drawing a diagram;
6 blocks due east **13.** choosing the correct operation;
$16.92 **17.** about 85°F **19.** 3 blocks due west

Page 320 Self-Test 2 **1.** $\frac{3}{4}$ **2.** $\frac{7}{8}$ **3.** $\frac{4}{5}$ **4.** $\frac{3}{8}$ **5.** $\frac{1}{3}$
6. $\frac{2}{3y}$ **7.** a^6 **8.** $8x^9$ **9.** < **10.** < **11.** > **12.** 0.55
13. $0.\overline{8}$ **14.** 2.68 **15.** $4.8\overline{3}$ **16.** $\frac{39}{50}$ **17.** $\frac{111}{250}$ **18.** $3\frac{14}{25}$
19. $8\frac{2}{3}$ **20.** Place the 12 m and 19 m rods end-to-end, cre-
ating a 31 m length. Place the 26 m rod parallel to these
rods, lining up the ends. The difference in length is 5 m.

Pages 322–323 Exercises **1–7.** Answers may vary.
Examples are given. **1.** $\frac{63}{5}$ **3.** $\frac{-17}{4}$ **5.** $\frac{202}{10}$ **7.** $\frac{-7}{3}$
9. N **11.** L **13.** I **15.** left **17.** < **19.** < **21.** >
23. $-\frac{1}{2}$ **25.** 0 **27.** $-\frac{1}{2}$ **29.** All **31.** Some
33. rhombus **35.** $17\frac{1}{16}$

Pages 326–327 Exercises
1.

3.

5.

7.

9.

11.

13. **15.** D **17.** B **19.** C

21. no difference **23.**

25.

27. **29.** 7; 4; 3; 0

31. **33.** about 1000

35. 1.25×10^6 **37.** 132 mi **39.** 405

Pages 330–331 Exercises **1.** $\frac{1}{y^{12}}$ **3.** 1 **5.** $\frac{1}{125}$ **7.** $\frac{1}{16}$
9. 7.2×10^{-3} **11.** 6.012×10^{-4} **13.** 2.34×10^{-6}
15. 0.00116 **17.** 0.00000001 **19.** 0.0000000061
21. -3 **23.** D **25.** 1500 times the actual size
27. 2×10^{-2} **29.** 13 mm **31.** parallel **33.** composite

Page 331 Self-Test 3 **1–4.** Answers may vary. Exam-
ples are given. **1.** $\frac{34}{10}$ **2.** $\frac{-37}{8}$ **3.** $\frac{-9}{1}$ **4.** $\frac{11}{9}$
5.

6.

7.

8.

9.

10. **11.** $-\frac{1}{8}$ **12.** 1

13. $\frac{1}{w^3}$ **14.** $\frac{1}{256}$ **15.** 3×10^{-5} **16.** 0.000000243

Pages 332–333 Chapter Review **1.** equivalent fractions **2.** rational number **3.** open sentence **4.** repeating **5.** composite **6.** greatest common factor **7.** prime
8. composite **9.** composite **10.** prime **11.** $3 \cdot 5^2$
12. $2^3 \cdot 3^3$ **13.** $2 \cdot 5^2 \cdot 7$ **14.** $2^4 \cdot 3 \cdot 5^2$ **15.** 16
16. $6a$ **17.** $3x$ **18.** $18z^2$ **19.** 30 **20.** 24 **21.** $20xy$
22. $14c^2$ **23.** $\frac{1}{2}$ **24.** $\frac{2}{3}$ **25.** $\frac{4}{5}$ **26.** $\frac{1}{7}$ **27.** $\frac{24}{5}$ **28.** $\frac{9}{2}$
29. $2a$ **30.** $\frac{4}{5}$ **31.** v^7 **32.** x^{12} **33.** $4z^3$ **34.** $\frac{x^7}{2}$ **35.** >
36. < **37.** > **38.** > **39.** 0.2 **40.** 0.3 **41.** $0.\overline{4}$
42. $0.\overline{45}$ **43.** 3.35 **44.** 5.22 **45.** $\frac{37}{100}$ **46.** $\frac{51}{200}$
47. $6\frac{7}{8}$ **48.** $4\frac{2}{3}$ **49.** Answers may vary. Example: Fill the
10 gal bucket and pour water from it into the 4 gal bucket.
This leaves 6 gal. Fill the 3 gal bucket and pour its water
into the 10 gal bucket. It contains 9 gal of water. **50.** 5
51–56. Answers may vary. Examples are given. **51.** $\frac{32}{10}$
52. $\frac{31}{5}$ **53.** $\frac{-7}{1}$ **54.** $\frac{0}{9}$ **55.** $\frac{37}{100}$ **56.** $\frac{-29}{3}$ **57.** D
58. F **59.** B **60.** A **61.** C **62.** E
63.

64.

65.

66. **67.** $\frac{1}{m^5}$ **68.** $\frac{1}{81}$

69. 1 **70.** $\frac{1}{9}$ **71.** 7.4×10^{-5} **72.** 8.21×10^{-4}

73. 6.9×10^{-7} **74.** 5.325×10^{-3} **75.** 0.0027
76. 0.0000055 **77.** 0.000000009 **78.** 0.000000324

Page 335 Chapter Extension   **1.** $\frac{5}{9}$ **3.** $\frac{35}{99}$ **5.** $\frac{7}{3}$
7. $\frac{106}{33}$ **9.** $\frac{374}{333}$ **11.** $\frac{103}{18}$ **13.** They are equivalent.

Pages 336–337 Cumulative Review   **1.** B   **3.** D
5. B   **7.** C   **9.** C   **11.** C   **13.** D   **15.** D   **17.** D
19. D

CHAPTER 8
Pages 342–343 Exercises   **1.** $\frac{32}{45}$ **3.** $-1\frac{7}{8}$ **5.** 1
7. $33\frac{1}{2}$ **9.** 56   **11.** -2.45 **13.** $\frac{1}{6}$ **15.** $\frac{5ab}{4}$
17–21. Estimates may vary.   **17.** about 10   **19.** about
-2 **21.** about 27 yd   **23.** $\frac{3}{8}$ or 0.375   **25.** 1.05 or $1\frac{1}{20}$
27. -24 **29.** $27\frac{3}{4}$ **31.** 34

33. | 4. | $a^b/_c$ | 2. | $a^b/_c$ | 7. | $\times$ |

| 2. | $a^b/_c$ | 7. | $=$ |

35. | 4. | $a^b/_c$ | 5. | $\times$ | 17. | $a^b/_c$ |

| 3. | $a^b/_c$ | 8. | $=$ |

37. about 18,000   **39.** 10   **41.** about 80

Pages 346–347 Exercises   **1.** 2   **3.** -54 **5.** 3   **7.** $\frac{2}{9y}$
9. $-\frac{16}{25}$ **11.** $8\frac{1}{3}$ **13.** $\frac{1}{5y}$ **15.** -2 **17.** 8   **19.** -0.5
21. 0.75   **23.** more than   **25.** $\frac{3}{4}$ c milk; 1 egg; $\frac{1}{4}$ c short-
ening; $2\frac{1}{4}$ c flour; 1 tsp cinnamon; $\frac{1}{3}$ c raisins; 1 tbsp baking soda; 1 c chopped apple   **27.** $\frac{1}{4}$ **29.** 5000   **31.** about
130   **33.** $\frac{z}{9} = 15$

Page 347 Self-Test 1   **1.** $-3\frac{2}{3}$ **2.** 1.08   **3.** $\frac{2c}{21}$ **4.** $\frac{2m}{5}$
5. about -48 **6.** about 33   **7.** about 4   **8.** about -7
9. $-1\frac{3}{5}$ **10.** 18   **11.** -4.5 **12.** $\frac{1}{5}$ **13.** 6
14. $\frac{2}{15}$ **15.** $\frac{z}{10}$

Page 349 Exercises   **1.** fraction; $18\frac{3}{4}$ yd   **3.** whole
number; 19   **5.** fraction; $5\frac{1}{4}$ h   **7.** decimal; $13.50
9. decimal; $10,995.54   **11.** whole number; 159   **13.** 5

Pages 354–355 Exercises   **1.** $\frac{8}{15}$ **3.** 0   **5.** $2\frac{1}{2}$ **7.** $1\frac{2}{5}$
9. -4 **11.** $-\frac{11}{12}$ **13.** $\frac{12}{r}$ **15.** $\frac{2}{m}$ **17.** $\frac{2}{x}$ **19.** $8b$ **21.** $\frac{2x}{5}$
23. $2c$ **25.** $\frac{1}{2}$ h   **27.** greater than   **29.** less than
31. $23\frac{1}{8}$ in.   **33.** $48\frac{3}{4}$ in.   **35.** $\frac{5}{9}$ **37.** $\frac{9}{16}$ **39.** $\frac{9}{11}$ **41.** m

Pages 358–359 Exercises   **1.** $1\frac{1}{2}$ **3.** $3\frac{2}{5}$ **5.** $1\frac{11}{20}$
7. $-5\frac{1}{6}$ **9.** $\frac{1}{6}$ **11.** $\frac{7}{12}$ **13.** $\frac{29m}{21}$ **15.** $\frac{17b}{4}$ **17.** $\frac{51h}{20}$
19. -3.6 **21.** 16.52   **23.** -4.37 **25.** $4\frac{5}{12}$ yd
27. $-20\frac{1}{4}$ or -20.25 **29.** $2\frac{7}{9}$ or $2.\overline{7}$ **31.** $-2\frac{3}{8}$ or -2.375
33. $9\frac{11}{12}$ **35.** $1\frac{1}{3}$ **37.** -8 **39.** 5.16   **41.** $4\frac{1}{2}$ in.
43. $-1\frac{1}{2}; -5; -10$ **45.** $x \to x + \frac{1}{3}$ **47.** Each fraction
is half the preceding fraction.   **49.** **a.** $\frac{3}{4}$ **b.** $\frac{7}{8}$ **c.** $\frac{15}{16}$
d. $\frac{31}{32}$ **51.** $\frac{63}{64}$
53.

Populations of Large Cities

New York	
London	
Tokyo	
Mexico City	
Beijing	

$=$ one million people

55. $\frac{19a}{15}$

Pages 363–364 Exercises   **1–15.** Estimates may vary.
1. about -5 **3.** about 11   **5.** about -5 **7.** about 11
9. about 4   **11.** about -48 **13.** about 32   **15.** about
33 mil   **17.** unreasonable; $-3\frac{23}{39}$ **19.** reasonable   **21.** $\frac{47}{10}$
23.

Page 364 Self-Test 2   **1.** fraction; $3\frac{1}{2}$ h   **2.** decimal;
$4.35   **3.** $\frac{1}{7}$ **4.** -17 **5.** $\frac{5a}{7}$ **6.** $-3\frac{1}{9}$ **7.** $\frac{27b}{4}$
8. -3.8 **9–11.** Estimates may vary.   **9.** about 15
10. about -19 **11.** about 5

Pages 366–367 Problem Solving Situations   **1.** $2\frac{1}{3}$ h
3. 140 years   **5–9.** Strategies may vary. Likely strategies
are given.   **5.** guess and check; 2 paperbacks, 5 children's
books   **7.** supplying missing facts; 5 lb   **9.** not enough
information; the hourly rate is needed.   **13.** $>$ **15.** -3

Page 367 Challenge   2, 3, and 6

Pages 369–370 Exercises   **1.** $-\frac{2}{3}$ **3.** $\frac{1}{9}$ **5.** $-\frac{5}{6}$
7. 4.21   **9.** -0.81 **11.** -1.1 **13.** $30\frac{1}{4}$ in.   **15.** $14.25
17. D   **19.** A   **21.** $3q - 11.3 = -8.9$ **23.** $16 = t + \frac{3}{5}$

25. 0.55; 0.8; 0.8; 3.4   **27.** $n - \left(-\frac{3}{5}\right) = 4\frac{2}{3}$; $4\frac{1}{15}$
29. 0°; 90°   **31.** $1\frac{1}{10}$

Pages 372–374 Exercises   1. 44   **3.** $-10\frac{4}{5}$ **5.** 22
7. $-17\frac{1}{2}$ **9.** $37\frac{1}{2}$ **11.** $-9\frac{1}{3}$ **13.** 7   **15.** $\frac{1}{8}$ **17.** $-\frac{1}{5}$
19. $-\frac{1}{16}$ **23.** 4   **25.** $\frac{4}{3}$ **27.** 1   **29.** $2\frac{2}{5}$ **31.** $5\frac{1}{15}$
33. $-2\frac{1}{7}$ **35.** $\frac{2}{3}t = -14$ **37.** $\frac{1}{2}g - 3 = \frac{5}{6}$ **39–43.** Answers will vary.   **39.** One third of a number g is fourteen.
41. Negative two is one third of a number x, increased by
nine.   **43.** Ten is three eighths of a number s, increased by
one half.   **45.** 20 is the LCM.   **47.** distributive property
49. 1 block due west   **51.** 138°   **53.** $-\frac{4}{5}$

Pages 376–377 Exercises   1. $h = \frac{A}{b}$ **3.** $b = P - a - c$
5. $W = Fd$ **7.** $P = \frac{I}{rt}$ **9.** $E = \frac{W}{I}$ **11.** $a = 2(P - 110)$
13. 5 lb   **15. a.** $-17\frac{7}{9}°C$ **b.** 0°C   **c.** 100°C
17. a. $I = \frac{E}{R}$ **b.** 40 amperes   **19.** $f = 22h + 30$
21. $h = \frac{1}{22}(f - 30)$ **25.** $s = \frac{2A}{p}$

Page 377 Self-Test 3   1. $31\frac{2}{3}$ yd   **2.** $\frac{1}{5}$ **3.** 4.875
4. $2\frac{5}{14}$ **5.** -108 **6.** 35   **7.** -16 **8.** $h = \frac{V}{lw}$ **9.** 2 m

Page 379 Exercises   1. BookClb   **3.** BayOil   **5.** Hi
7. $30.13   **9.** $2987.50   **11.** $27.75   **13.** $18.75
15. 800; $13.25

Pages 380–381 Chapter Review   1. multiplicative in-
verse   **2.** multiplication property of reciprocals   **3.** $\frac{15}{32}$
4. $1\frac{1}{4}$ **5.** $14\frac{2}{3}$ **6.** $1\frac{5}{6}$ **7.** $-4\frac{1}{2}$ **8.** 14   **9.** $-\frac{8}{27}$ **10.** $-\frac{4}{9}$
11. $-3\frac{3}{5}$ **12.** $-1\frac{29}{36}$ **13.** 2   **14.** $-\frac{13}{36}$ **15.** $-4\frac{1}{6}$ **16.** $3\frac{1}{3}$
17. 2   **18.** $-3\frac{7}{10}$ **19.** 0.54   **20.** -2.4 **21.** -2.1
22. 40   **23.** -6.3 **24.** 3.9   **25.** -58.75 **26.** 2.7
27. $\frac{3}{20}$ **28.** $2z$ **29.** $\frac{9}{14}$ **30.** $\frac{3}{22}$ **31.** $\frac{11}{x}$ **32.** $\frac{3a}{4}$ **33.** $\frac{r}{6}$
34. $\frac{3m}{2}$ **35.** $\frac{65n}{9}$ **36.** $\frac{29c}{7}$ **37.** $\frac{5}{2n}$ **38.** $\frac{47m}{24}$
39. fraction; $11\frac{2}{3}$ h   **40.** decimal; $52.50   **41.** whole
number; 7   **42.** fraction; $1\frac{1}{4}$ lb   **43–54.** Estimates may
vary.   **43.** about 8   **44.** about 30   **45.** about -5
46. about 3   **47.** about 6   **48.** about 11   **49.** about 26
50. about -72 **51.** about -16 **52.** about -8
53. about -5 **54.** about 3   **55.** 57 in.   **56.** $29\frac{3}{4}$ mi
57. 42 muffins   **58.** 6 min   **59.** $\frac{4}{5}$ **60.** $-\frac{1}{3}$ **61.** -3
62. 0.285   **63.** $-4\frac{2}{5}$ **64.** $3\frac{7}{8}$ **65.** $-1\frac{1}{5}$ **66.** $\frac{3}{4}$
67. -0.6 **68.** -0.2 **69.** $1\frac{1}{2}$ **70.** $-\frac{5}{8}$ **71.** 0.7

72. 1.4   **73.** $-\frac{7}{10}$ **74.** $1\frac{7}{12}$ **75.** 48   **76.** $-12\frac{1}{2}$ **77.** $2\frac{2}{5}$
78. -15 **79.** $-10\frac{2}{3}$ **80.** 21   **81.** -16 **82.** $-7\frac{1}{2}$
83. $r = \frac{d}{2}$ **84.** $m = \frac{F}{a}$ **85.** $p = \frac{2A}{s}$ **86.** $M = 2a + m$
87. 30°C   **88.** $17\frac{1}{2}$ ft

Page 383 Chapter Extension   1. $\frac{11}{5x}$ **3.** $\frac{3}{2a^2}$ **5.** $\frac{22x}{3yz}$
7. $\frac{1}{4xy}$ **9.** $\frac{3}{2n^3}$ **11.** $\frac{3r}{8s}$ **13.** $\frac{19}{12m}$ **15.** $\frac{59}{20rs}$ **17.** $\frac{52b}{15c^3}$
19. $\frac{8}{15r}$ **21.** $\frac{49r}{12s^3}$ **23.** $\frac{31u}{24vw}$ **25.** $\frac{35}{3mn}$ **27.** $\frac{15}{8x^2}$
29. No; $4y$ is not a factor of the numerator.

Pages 384–385 Cumulative Review   1. C   **3.** B
5. D   **7.** C   **9.** C   **11.** A   **13.** C   **15.** D   **17.** B
19. C

CHAPTER 9
Pages 389–390 Exercises   1. $\frac{3}{4}$ **3.** $\frac{5}{3}$ **5.** $\frac{2}{1}$ **7.** $\frac{8}{5}$
9. $\frac{2}{1}$ **11.** $\frac{3}{8}$ **13.** $\frac{3}{4}$ **15.** 50 mi/h   **17.** 25 mi/gal
19. 2.2 m/s   **21.** $\frac{9}{1}$ **23.** $\frac{x}{2y}$ **25.** $\frac{3}{1}$ **27.** 27   **29.** $\frac{5}{3}$
31. 12 pens for $4.80   **33.** Answers will vary. Example:
when the item is available only in an inconvenient size
35. 24   **37.** $2^4 \cdot 3 \cdot 5$

Pages 393–394 Exercises   1. True.   **3.** False.
5. False.   **7.** True.   **9.** 6   **11.** 11   **13.** 8   **15.** 9
17. Yes; $(4)(81) = (9)(36)$ **19.** 10.5   **21.** 10 or -10
23. 12   **25.** 4   **27.** 14   **29.** C   **31.** D   **33.** 18
35. 5.2   **37.** 14 ft   **39.** $\frac{4xy^2}{z}$ **41.** $-\frac{2}{9}$

Pages 397–398 Exercises   1. 50 ft   **3.** $18\frac{3}{4}$ ft
5. $1\frac{5}{8}$ in. wide; $1\frac{3}{4}$ in. long   **7.** 35 cm   **9.** 2 in.: 5 ft
13. Answers will vary. Example: A person is about 6 ft
tall, so a 21 ft statue indicates a scale of 6 ft : 21 ft, or 2
ft : 7 ft.   **15.** The heads on Mount Rushmore are about 60 ft
high, which makes the scale about 1 in. : 80 in.   **17.** 130 in.

Pages 400–401 Problem Solving Situations   1. 266
3. $225   **5–9.** Strategies may vary; likely strategies are
given.   **5.** drawing a diagram; 6 blocks due south
7. using equations; 34   **9.** identifying a pattern; 83

Page 401 Self-Test 1   1. $\frac{3}{5}$ **2.** $\frac{1}{4}$ **3.** $\frac{2}{1}$ **4.** $\frac{3}{2}$
5. 55 mi/h   **6.** $4/lb   **7.** $.65/comb   **8.** True.   **9.** True.
10. False.   **11.** False.   **12.** 10   **13.** 4   **14.** $33\frac{1}{3}$
15. 11.5   **16.** $7\frac{11}{12}$ ft   **17.** $99

Pages 406–407 Exercises   1. 8%   **3.** 70%   **5.** $62\frac{1}{2}$%
7. 21%   **9.** 9.9%   **11.** $15\frac{1}{5}$%   **13.** $46\frac{7}{8}$%   **15.** $228\frac{4}{7}$%
17. 0.49   **19.** 5.87   **21.** $\frac{1}{2}$ **23.** $\frac{23}{1000}$ **25.** $\frac{3}{2000}$

27. 40% **29.** 60% **31.** 2; 0.2; 0.02; 0.002; 0.0002; 0.00002 **33.** 900; 90; 9; 0.9; 0.09; 0.009 **35.** 19; 100 **37.** 100; 200 **39.** $\frac{3}{40}$; 0.075 **41.** $\frac{11}{37}$

Page 407 Challenge 125%

Pages 410–411 Exercises 1. $94.40 **3.** 240 **5.** 456 **7.** $15.75 **9–13.** Estimates may vary. **9.** about 35 **11.** about 50 **13.** about $6

15.

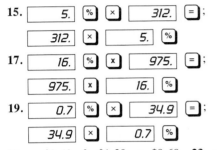

21. a. $2.40 **b.** $1.20 **c.** $9.60 **23. a.** 0.46 **b.** 0.23 **c.** 0.92 **25.** < **27.** < **29.** = **31.** = **33.** $30 **35.** $60 **37.** $32 **39.** $64 **41.** $34 **43.** $68 **45.** $187.50 **47.** 3.6

Pages 413–414 Exercises 1. 80% **3.** 70% **5.** 96% **7.** 100% **9.** $33\frac{1}{3}$% **11.** 2.5% **13.** $87\frac{1}{2}$% **15.** $62\frac{1}{2}$% **17.** 80% **19.** $59\frac{1}{11}$% **21.** $50 **23.** $487.50 **25.** 4.6% **27.** > **29.** 120

Pages 416–418 Exercises 1. 420 **3.** 34 **5.** 2000 **7.** 80 **9.** 45 **11.** 36,000 **13.** 50 **15.** $4200 **17.** 180 **19.** 100% **21. a.** 6 g **b.** 15% **c.** 40 g **23.** Since the cereal alone provides 100% of the RDA for Vitamin B_{12}, and more is provided by the milk, together they provide more than 100% of the RDA for vitamin B_{12}. **25.** $19.50 **27.** $8\frac{1}{2}$% **29.** $\frac{5}{8}$ **31.** $-1\frac{1}{3}$

Page 418 Self-Test 2 1. 65% **2.** 45% **3.** 1.3 **4.** $\frac{7}{25}$ **5.** 18.85 **6.** 6.75 **7.** $16\frac{2}{3}$% **8.** 50% **9.** 380 **10.** 160

Pages 420–421 Exercises 1. 18% **3.** 20% **5.** 100% **7.** 21% **9.** $86\frac{1}{9}$% **11.** $166\frac{2}{3}$% **13.** $62\frac{1}{2}$% **15.** increase; 125% **17.** decrease; $16\frac{2}{3}$% **19.** Answers will vary. **21.** True; half an amount x is $\frac{1}{2}x$, and a decrease of 50% is $x - 0.5x = 0.5x$. **23.** $\frac{18}{1}$ **25.** 20%

Pages 426–427 Exercises 1. 9.44 **3.** 3500 **5.** 15% **7.** 245 **9.** 1056 **11.** 7% **13.** 2800 **15.** 4500 **17.** $1.18–$1.35 **19.** $1.05 **21.** 6.5% **25.** 94.3% **27.** 5.8×10^{-5} **29.** $87\frac{1}{2}$%

Page 427 Self-Test 3 1. 250% **2.** 10% **3.** 20% **4.** $33\frac{1}{3}$% **5.** 3.2 **6.** 1200 **7.** 40 **8.** 40%

Pages 428–429 Chapter Review 1. scale **2.** unit rate **3.** ratio **4.** proportion **5.** percent **6.** terms **7.** $\frac{1}{5}$ **8.** $\frac{1}{5}$ **9.** $\frac{2}{11}$ **10.** $\frac{3}{2}$ **11.** 7 h/day **12.** 27 words/min **13.** $4.25/h **14.** False. **15.** True. **16.** False. **17.** False. **18.** 1 **19.** 4 **20.** 1 **21.** 23 **22.** 20 ft **23.** 15 in. **24.** 400 min **25.** $16.25 **26.** 55% **27.** $43\frac{3}{4}$% **28.** 275% **29.** $33\frac{1}{3}$% **30.** $62\frac{1}{2}$% **31.** 65% **32.** 3% **33.** 120% **34.** 42.8% **35.** 0.95% **36.** $\frac{3}{4}$; 0.75 **37.** $2\frac{1}{2}$; 2.5 **38.** $\frac{21}{400}$; 0.0525 **39.** $\frac{39}{200}$; 0.195 **40.** $\frac{1}{2000}$; 0.0005 **41.** about 90 **42.** about 40 **43.** about 8 **44.** about 60 **45.** 19 **46.** 14 **47.** 50% **48.** 80% **49.** 4% **50.** $2\frac{1}{2}$% **51.** 600 **52.** 42 **53.** 70 **54.** 364 **55.** 300% **56.** $6\frac{2}{3}$% **57.** 50% **58.** 20% **59.** 72 **60.** 4.5 **61.** $37\frac{1}{2}$% **62.** 2.88

Page 431 Chapter Extension 1. $3825.24 **3.** $4706.34 **5.** $6092.01

Pages 432–433 Cumulative Review 1. C **3.** B **5.** A **7.** D **9.** C **11.** D **13.** B **15.** D **17.** D **19.** B

CHAPTER 10

Pages 438–439 Exercises 1. 10 cm **3.** 11.55 m **5.** 28 cm **7.** $16\frac{1}{8}$ in. **9.** 355 ft or $118\frac{1}{3}$ yd **11.** 11.9 m **13.** $2\frac{5}{8}$ in. **15.** not enough information; the length is needed. **17.** not enough information; the perimeter or if the pentagon is regular is needed. **19.** 20 **21.** 36 **23.** 33.5 mm **25.** 20.7; 20; 18; 7 **27.** 57 in. **29.** 150°

Pages 443–444 Exercises 1. circle O **3.** 8.6 cm **5.** 6.3 m **7.** $14\frac{2}{3}$ yd **9.** 44 mm **11.** 16.3 m **13.** 9 m; 18 m **15.** 7 yd; 14 yd **17.** 50 m **19.** approximately **21.** approximately **25.** 220 in. **27.** 288 **29.** 31.5 mi **31.** 125 cm **33.** right

Pages 448–449 Exercises 1. $17\frac{1}{2}$ ft² **3.** 400 in.² or $2\frac{7}{9}$ ft² **5.** 54 cm² **7.** $6\frac{1}{4}$ ft² **9.** 63 yd² or 567 ft² **11.** 27 square units **13.** b_1 **15.** $\frac{1}{2}hb_1 + \frac{1}{2}hb_2$ **17.** 55 m² **19.** $1710 **21.** $610.45 **23.** $\frac{3x}{8}$

Pages 454–456 Exercises 1. 314 in.² **3.** 28.3 yd² **5.** $17\frac{1}{9}$ mi² **7.** 6.2 cm² **9.** 2826 mi² **11.** B **13.** C

15. Find half the area of a circle. **17.** 20 ft² **19.** 66 m²
21. 32.1 in.² **23.** 84.8 ft² **25.** 2π; 4π; 8π **27.** Each number has π as a factor; for circumference, the radius doubles; for area, the radius is squared. **29.** The circumference doubles. **31.** 0.267 **33.** about 9

Page 458 Exercises 1. circumference **3.** perimeter **5.** area; 706.5 ft² **7.** perimeter; 112 yd **9.** area; 1102.1 ft² **11.** −5 **13.** circumference

Pages 460–461 Problem Solving Situations 1. The number of books must be a whole number. **3.** There are no amounts of dimes and nickels totaling 12 that result in $.60. **5–9.** Strategies will vary. **5.** supplying missing facts; $32 **7.** choosing the correct operation; 2.5 lb **9.** using equations; 22.5 m, 22.5 m **13.** $x + 9$ **15.** $38\frac{2}{5}$

Page 461 Self-Test 1 1. 44.8 cm **2.** $31\frac{1}{2}$ in.
3. a. 14 m **b.** 44 m **4.** $30\frac{1}{4}$ in.² **5.** 36 cm²
6. 221.6 cm² **7.** circumference **8.** The sum of the lengths of the two sides given is equal to the perimeter.

Pages 464–465 Exercises 1. ∠D **3.** $\overline{BA}$ **5.** CBED **7.** ∠DEB **9.** $\overline{OM}$ **11.** False; a rectangle with sides of lengths 1 and 4 and a rectangle with sides of lengths 2 and 3. **13.** True. **15.** False; a rectangle with sides of lengths 3 and 4 and a rectangle with sides of lengths 2 and 6.
17. (1, −1) **19.** $\frac{23a}{20}$ **21.** 40

Pages 468–470 Exercises 1. 12; 12.5 **3.** 5; 12.5 **5.** 57 cm **7.** Yes; the length of $\overline{MN}$ is 10 cm because corresponding sides are congruent. **9.** 32 **11.** $\frac{8}{7}$; they are equal. **13.** always **15.** sometimes **17.** always **19.** never **21.** Answers will vary. **23.** They are similar. **25.** 6.75 **27.** 616 cm² **29.** 0.000689

Pages 473–474 Exercises 1. 15 m **3.** 10 ft **5.** 16 ft 6 in. **7.** They are vertical angles. **9.** EDC **11.** 72 ft **13.** 46 cm **15.** 16

Pages 477–479 Exercises 1. 5 **3.** 1 **5.** −30 **7.** $\frac{3}{8}$ **9.** −0.7 **11.** −9.434 **13.** 11.832 **15.** 20 **17.** 0.2 **19.** −5.745 **21.** 1.2 **23.** 14 **25.** 7 **27.** 12 **29.** 225, 256, 289 **31.** 56 m **33.** 55 **35.** 0.6 **37.** $2\frac{1}{2}$ **39.** 100 **41.** 6 and 7 **43.** −5 and −4 **45.** b **47.** b **49.** 1.618 **51.** II **53.** 16

Page 479 Challenge 20.25 cm²; 24 cm

Pages 483–485 Exercises 1. 7.2 cm **3.** 9 in. **5.** 7.1 cm **7.** Yes. **9.** Yes. **11.** Yes. **13.** 10.3 yd **15.** 6.6 ft **17.** 6.3 km **19.** 100 yd **21.** 3; 4; 5

23.

x	y	a	b	c
2	1	3	4	5
3	1	8	6	10
3	2	5	12	13
4	1	15	8	17
4	2	12	16	20
4	3	7	24	25
5	1	24	10	26
5	2	21	20	29
5	3	16	30	34
5	4	9	40	41
6	1	35	12	37
6	2	32	24	40
6	3	27	36	45
6	4	20	48	52
6	5	11	60	61

25. $-\frac{7}{8}$ **27.**

Page 485 Self-Test 2 1. $\overline{BD}$ **2.** ∠S **3.** 15.75 **4.** 20 ft **5.** 90 **6.** $\frac{5}{8}$ **7.** 11.180 **8.** No. **9.** 8.1

Pages 486–487 Chapter Review 1. circumference **2.** hypotenuse **3.** congruent **4.** radius **5.** perfect square **6.** area **7.** 37.2 cm **8.** $34\frac{1}{2}$ ft **9.** circle R **10.** radii: $\overline{RS}$, $\overline{RT}$, $\overline{RW}$; chords: $\overline{WT}$, $\overline{VU}$; diameter $\overline{WT}$ **11.** 9 cm **12.** 18 cm **13.** 56.5 cm **14.** $30\frac{1}{4}$ ft² **15.** 12.48 m² **16.** 54 in.² **17.** 803.8 cm² **18.** area **19.** perimeter **20.** There are no amounts of dimes and quarters totaling 5 which result in $1. **21.** The two lengths together equal the perimeter. **22.** $\overline{EF}$ **23.** ∠D **24.** $\overline{DE}$ **25.** ∠E **26.** 2 **27.** 9 **28.** 25 ft **29.** 30 **30.** $\frac{10}{11}$ **31.** 8.718 **32.** 11.916 **33.** Yes. **34.** 12

Page 489 Chapter Extension 1. $2\sqrt{3}$ **3.** $5\sqrt{2}$ **5.** $2\sqrt{2}$ **7.** $2\sqrt{10}$ **9.** $2\sqrt{7}$ **11.** $4\sqrt{2}$ **13.** $2\sqrt{31}$ **15.** $2\sqrt{15}$ **17. a.** 8.944 **b.** 8.944 **c.** They are equal.

Pages 490–491 Cumulative Review 1. B **3.** B **5.** C **7.** A **9.** D **11.** B **13.** A **15, 17, 19.** B

CHAPTER 11
Pages 495–496 Exercises
1. Compact Disc Prices (Dollars)

Price	Tally	Frequency
7.00-8.99	||	2
9.00-10.99	||	2
11.00-12.99	||||| ||	7
13.00-14.99	||||| ||	7
15.00-16.99	||	2

3.

Annual Tuition at Private Colleges (Dollars)

Tuition	Tally	Frequency
6000-7999	⊦⊦⊦⊦ ||	7
8000-9999	|||	3
10,000-11,999	||||	4
12,000-13,999	⊦⊦⊦⊦ |	6
14,000-15,999	|	1

5. Intervals may vary.

Cost of One Week Vacation (Dollars)

Cost	Tally	Frequency
850-1049	||||	4
1050-1249	⊦⊦⊦⊦ |	6
1250-1449	⊦⊦⊦⊦	5
1450-1649	⊦⊦⊦⊦	5
1650-1849	||||	4

7. The intervals overlap. It cannot be determined if a score of 70 is in the interval 60-70 or 70-80. **9.** $2x - 3$

11. **Number of Hours Worked Per Week**

Hours	Tally	Frequency
20-24	⊦⊦⊦⊦ ⊦⊦⊦⊦ |	11
25-29	⊦⊦⊦⊦ ||||	9
30-34		0
35-39	⊦⊦⊦⊦ ⊦⊦⊦⊦ |	11'
40-45	⊦⊦⊦⊦	5

Page 496 Challenge 16

Pages 499–500 Exercises

1. **Earned Run Averages of Leading Pitchers**

2.2	0	5		
2.3	2			
2.4				
2.5	6			
2.6	2	2	7	
2.7	7	8		
2.8	0	2	2	4
2.9	0	0	2	

3. $25 **5.** 3 **7.** 12 **9.** 13

11. **Age at Inauguration of United States Presidents**

4	2 3 6 7 8 9 9
5	0 0 1 1 1 1 2 2 4 4 4 4
	5 5 5 6 6 6 7 7 7 7 8
6	0 1 1 1 2 4 4 5 8 9

13. 19 **15.** 7 **17.** $5n + 4$; $6n + 4$; $7n + 4$ **19.** 8

Pages 503–504 Exercises **1.** 6 **3.** 14 **5.** increase **7.** 6 **9.** 7 **11.** 20 **13.**

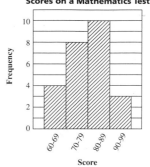

15. 8 **17.** B **19.**

21. Answers will vary. The graph in Exercise 19 shows the general trend in salaries rather than the individual salaries. **23.** 0.65 **25.** 90

Pages 507–508 Exercises
1.

Box Seat Prices at Baseball Games (Dollars)

9.00 10.00 11.00 12.00 13.00 14.00 15.00

3. 76; 70; 86; 62; 98

5. Ken Jameson's **7.** Ken Jameson's **9.** always
11. sometimes **13.**

-4 -3 -2 -1 0 1 2

15.

Number of Field Goals Scored

10 15 20 25 30

1.

Scores on an Algebra Test

Score	Tally	Frequency
60-69	|||	3
70-79	||||| ||	7
80-89	|||||		6
90-99	||||	4

2. **Scores on an Algebra Test**

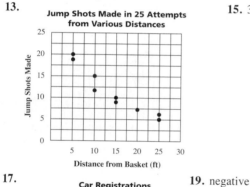

```
6 | 1   8   9
7 | 1   4   6   6   7   7   7
8 | 3   3   4   5   7   8
9 | 1   2   2   9
```

3. 18.   **4.** 22   **5.** Ken Jameson's

Pages 511–512 Exercises   1. 165 lb   **3.** 3   **5.** 64 in.
7. 145,155   **9.** negative   **11.** 14

13.

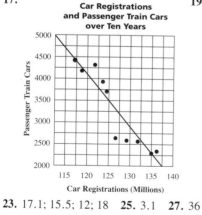

Jump Shots Made in 25 Attempts from Various Distances

15. 3

17.

Car Registrations and Passenger Train Cars over Ten Years

19. negative

23. 17.1; 15.5; 12; 18   **25.** 3.1   **27.** 36

Pages 515–516 Exercises   1. 70%   **3.** 108   **5.** 400
7. $3750   **9.** 2%   **11.** 94%   **13.** Estimates may vary.
about $\frac{1}{3}$ **15.** $\frac{3}{20}$ **17.** $\frac{33}{100}$ **19.** 9 h   **21.** 30 h
23.   **25.** 63 in.^{2}

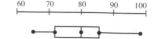

Test Scores

Pages 519–520 Exercises

1.

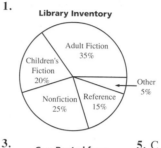

Library Inventory

Adult Fiction 35%
Children's Fiction 20%
Other 5%
Reference 15%
Nonfiction 25%

3.

Cars Rented from Yoko's Rent-A-Car

Mid Size 110
Compact 90
Mini-vans 50
Full Size 150

5. C   **7.** B   **9.** 57.4%

11. about $\frac{2}{5}$ **13.** 4.7 cm^2 **15.** 84.8 m^2 **17.** −13

Page 521 Exercises   1. 6   **3.** $\frac{1}{6}$ **5.** 1   **7.** 43%
9. The percent of total sales for each sector.

Page 523 Exercises
1. box-and-whisker plot;

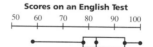

Scores on an English Test

3. circle graph;

Survey of the Number of Women in Certain Occupations

Professional 125
Technical 225
Other 60
Service 90

5. 90

Pages 525–527 Problem Solving Situations   **1.** Ann Fernandez teaches typing, Ellen Taylor teaches mathematics, Sonia Morris teaches French.   **3.** Marv is the treasurer, Shirley is the president, and Stan is the secretary.   **5–9.** Strategies may vary; likely strategies are given.   **5.** making a table; 6   **7.** using logical reasoning; Aaron   **9.** no solution; the number of pumpkins must be a whole number.   **11.** Answers will vary.   **13.** Torres is the doctor, Wong is the dentist, and Lipshaw is the journalist.

Page 527 Self-Test 2   **1.** 2   **2.** positive   **3.** $210,000
4.
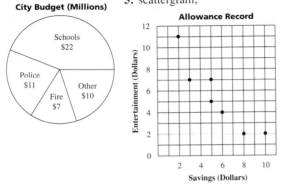

5. scattergram;

6. Harry majors in physics, Matt majors in history, and Phyllis majors in mathematics.

Page 529 Exercises   **1.** Detroit had the greatest percentage decrease, while Chicago had the greatest decrease in number of people.   **3.** Los Angeles, Houston, San Diego, Dallas, Phoenix, San Antonio; Southwest   **5.** about 1,318,919   **7.** 2,968,750   **9.** about 6,998,013

Pages 530–532 Chapter Review   **1.** histogram
2. circle graph   **3.** positive correlation   **4.** third quartile
5. leaf   **6.** frequency polygon
7.   **Points Scored Per Game**

Points	Tally	Frequency
16-20	‖‖‖ ‖	7
21-25	‖‖‖ ‖‖‖‖	9
26-30	‖	2
31-35	‖‖‖	3

8.   **Points Scored per Game**

```
1 | 6  7  8  8  8  8
2 | 0  2  2  2  3  3  4  4  4  5  7
3 | 0  1  1  3
```

9. 100   **10.** 2   **11.** 17   **12.** 84.8; 88; 88; 40

13.
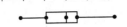
Ages of Workers in an Office (Years)

14. 4   **15.** 10

16. 3   **17.** 11   **18.** 32; 28; 38; 50; 22   **19.** Thames
20. Charles   **21.**
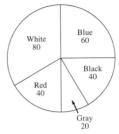
Colors of Cars Sold

22. $30,000

23. 5   **24.** $45,000   **25.** positive
26. $13,500   **27.** $35,000   **28.** $1,500,000   **29.** 15%
30. stem-and-leaf plot;
Leading Batting Averages

0.30	1	3	4	4	5	6	6	7	8
0.31	1	2	2						
0.32	2								
0.33									
0.34									
0.35	6								
0.36	6								

31. Ed is the artist, Alice is the writer, and Will is the editor.

Page 535 Chapter Extension   **1.** 96   **3.** 77   **5.** 84
7. 90   **9.** 84   **11.** 80

Pages 536–537 Cumulative Review   **1.** D   **3.** D
5. D   **7.** B   **9.** C   **11.** C   **13.** A   **15.** C   **17.** B
19. B

CHAPTER 12
Pages 543–544 Exercises   **1.** $\frac{1}{3}$ **3.** $\frac{1}{2}$ **5.** $\frac{2}{3}$ **7.** $\frac{1}{2}$
9. 0   **11.** 1 to 4   **13.** 3 to 2   **15.** 3 to 2   **17.** 3 to 2
19. 2 to 3   **21. a.** $\frac{1}{2}$ **b.** 1 to 1   **23.** $\frac{1}{6}$ **25.** $\frac{1}{2}$ **27.** $\frac{5}{6}$
29. 1   **31.** $\frac{5}{8}$ **33.** $\frac{5}{8}$ **35.** 5 to 3   **37.** 13 to 3   **39.** 7 to 9
41. about $\frac{2}{5}$ **43.** $\frac{4}{9}; \frac{5}{9}$ **45. a.** 1; P(E)   **b.** $\frac{2}{3}$ **47.** 56.5 in.

Page 544 Challenge   15

Pages 546–547 Exercises

1.

coin 1 coin 2 coin 3 coin 4 outcomes (HHHH, HHHT, HHTH, HHTT, HTHH, HTHT, HTTH, HTTT, THHH, THHT, THTH, THTT, TTHH, TTHT, TTTH, TTTT)

3. $\frac{11}{16}$ **5.** $\frac{1}{16}$
7. $\frac{1}{18}$ **9.** $\frac{1}{9}$ **11.** $\frac{5}{18}$ **13.** $\frac{1}{3}$ **15.** 0 **17.** $\frac{1}{12}$ **19.** $\frac{1}{12}$
21. $\frac{1}{24}$ **23.** 6 **25.** 0.684

Pages 550–551 Exercises
1. 1440 **3.** 216 **5.** 144
7. $\frac{1}{12}$ **9.** $\frac{2}{21}$ **11.** 30 **13.** $\frac{1}{16}$ **15. a.** $\frac{1}{6}$ **b.** $\frac{1}{2}$ **17.** $\frac{2}{45}$
19. 10 **21.** $\frac{1}{10}$ **23.** $\frac{1}{26}$ **25.** 4

Page 551 Self-Test 1
1. $\frac{1}{2}$ **2.** $\frac{5}{6}$ **3.** 1 to 1 **4.** 1 to 1
5. 5 to 1 **6.**

number cube coin outcomes (1 H, 1 T, 2 H, 2 T, 3 H, 3 T, 4 H, 4 T, 5 H, 5 T, 6 H, 6 T)

7. $\frac{1}{4}$ **8.** 1296 **9.** $\frac{1}{128}$

Pages 554–555 Exercises
1. $\frac{1}{6}$ **3.** $\frac{1}{12}$ **5.** $\frac{1}{36}$ **7.** $\frac{1}{24}$
9. $\frac{1}{72}$ **11.** $\frac{1}{22}$ **13.** $\frac{1}{66}$ **15.** $\frac{5}{33}$ **17.** $\frac{1}{216}$ **19.** $\frac{1}{8}$ **21.** $\frac{3}{8}$
23. $\frac{11}{57}$ **25.** $\frac{1}{30}$ **27.** $\frac{1}{6}$ **29.** $\frac{1}{1000}$ **31.** $\frac{1}{8}$ **33.** $\frac{63}{125}$ **35.** $\frac{1}{6}$
37. $\frac{12}{49}$

Pages 557–558 Problem Solving Situations
1. 95 **3.** 75 **5.** 15 **7.** 67 **9.** 52 **11–13.** Strategies will vary. **11.** drawing a diagram; Marvin, Kim, Julie, Lance **13.** using logical reasoning; 2 **17.** 15 **19.** 18, 23, 29

Pages 561–563 Exercises
1. 2 **3.** 720 **5.** 132
7. 20 **9.** 120 **11.** 24 **13.** 15 **15.** 120 **17.** 210
21. 120 **23.** 362,880 **25.** 28 **27.** 362,880 **29.** 21
31. $_nC_r = {_nC_{n-r}}$ **33.** 6 **35.** 15 **37.** 80 **39.** 720
41. about 31%

Page 563 Self-Test 2
1. $\frac{16}{81}$ **2.** $\frac{2}{27}$ **3.** $\frac{2}{51}$ **4.** 14
5. 120 **6.** 15

Pages 565–566 Exercises
1. $\frac{1}{5}$ **3.** $\frac{7}{40}$ **5.** $\frac{3}{5}$ **7.** $\frac{1}{4}$
9. $\frac{11}{40}$ **11.** $\frac{9}{16}$ **13.** 1 **15.** $\frac{1}{4}, \frac{13}{40}, \frac{17}{40}$ **17.** 0.9
19–25. Answers will vary. **27.** $45.50 **29.** $\frac{4}{9}$
31.

Tickets Sold by Grade Level — Sophomore 30, Freshman 60, Junior 90, Senior 120

Pages 569–571 Exercises
1. about 63,250 **3.** about 59,950 **5.** about 210 **7.** about 5400; about 3600
9. $\frac{29}{100}$ **11.** $\frac{1}{50}$ **13.** about 12,180 **15.** about 840
17. about 450 **19.** 1.25% **21.** No; the players' past results should be considered. **23.** Yes. **25.** about 156 **27.** about −45

Page 571 Self-Test 3
1. $\frac{14}{25}$ **2.** $\frac{9}{50}$ **3.** $\frac{13}{50}$ **4.** about 36,000

Pages 572–573 Exercises
1. a. 85393, 97265, 61680, 16656, 42751 **b.** They are greater than 15000.
c. 02011, 07972, 10281, 03427, 08178, 09998, 14346, 07351, 12908, 07391, 07856, 06121, 09172, 13363, 04024 **3.** Assign each card holder a number from 00001 to 85056. Randomly select a starting point and direction in the table. Discard each number greater than 85056. Continue until 25 numbers have been selected.

Pages 574–575 Chapter Review
1. dependent events **2.** event **3.** sample space **4.** permutation **5.** experimental probability **6.** factorial **7.** $\frac{1}{5}$ **8.** 0 **9.** $\frac{2}{5}$
10. 3 to 2 **11.** 2 to 3 **12.** 3 to 2
13.

spinner coin outcomes (1 (white) H, 1 (white) T, 2 (red) H, 2 (red) T, 3 (red) H, 3 (red) T, 4 (blue) H, 4 (blue) T, 5 (blue) H, 5 (blue) T)

14. $\frac{1}{10}$ **15.** $\frac{1}{5}$ **16.** $\frac{2}{5}$ **17.** 0
18. 60 **19.** $\frac{7}{30}$ **20.** $\frac{7}{45}$ **21.** $\frac{7}{75}$ **22.** $\frac{1}{5}$ **23.** $\frac{1}{14}$ **24.** 7
25. 720 **26.** 60 **27.** 126 **28.** 21 **29.** $\frac{1}{5}$ **30.** $\frac{13}{50}$
31. about 104,000

Page 577 Chapter Extension
1. $17.40 **3.** 2.5

1. D   **3.** D
5. D   **7.** D   **9.** C   **11.** B   **13.** C   **15.** C   **17.** A
19. B

CHAPTER 13
Pages 583–584 Exercises   1. $(-6, -19); (-2, -3);$
$(0, 5); (2, 13); (6, 29)$ **3.** $(-6, 7); (-2, 5\frac{2}{3}); (0, 5);$
$(2, 4\frac{1}{3}); (6, 3)$ **5.** $(-6, -39); (-2, -19); (0, -9); (2, 1);$
$(6, 21)$ **7.** $(-6, 2); (-2, -2); (0, -4); (2, -6); (6, -10)$
9. $(-6, 11); (-2, 3); (0, -1); (2, -5); (6, -13)$
11. $(-6, 42); (-2, 14); (0, 0); (2, -14); (6, -42)$
13. a. $C = 36h + 20$ **b.** $128   **c.** 4.5 h   **15.** about 9
17. about 11   **19.** Quebec   **21.** E1   **23.** about 43° north,
71° west   **25.** 120   **27.** 60

Page 587 Exercises   1.

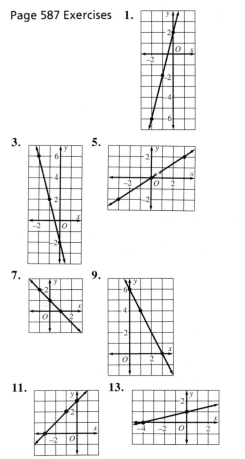

15. D   **17. a.** 2   **b.** 2   **c.** 2   **d.** 2   **19. a.** -3 **b.** -3
c. -3 **d.** -3 **21.** $y = b - mx$ **23.** 50.2 m^2

Pages 590–591 Exercises   1. $1; 1; y = x + 1$
3. $-\frac{1}{2}; 0; y = -\frac{1}{2}x$ **5.** $-3; 9; 3$ **7.** $-1; -1; -1$
9. $-2; 1; \frac{1}{2}$ **11.** $1; -4\frac{1}{2}; 4\frac{1}{2}$ **13.** $3; 0; 0$ **15.** $-3; -3; -1$
17. $4; 6; -1\frac{1}{2}$ **19.** $(3, 0); (0, 21)$ **21.** C   **23.** D   **25.** 1

27. about 17 ft   **29.** always   **31.** about 58

Pages 596–597 Exercises   1. $(1, 2)$ **3.** no solution
5. $(-1, 0)$ **7.** no solution   **9.** $(-3, -1)$ **11.** $(2, 1)$
13. $(1, 1)$ **15.** $(-3, -1)$ **17.** $(1, 6)$ **19.** triangle
21.

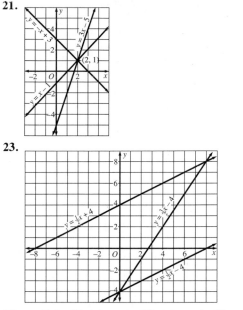

23.

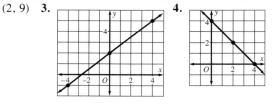

25. parallel   **27.** intersect   **29.** intersect   **31.** parallel
33. $10y^2$

Page 597 Self-Test 1   1. $(-2, -15); (-1, -9);$
$(0, -3); (1, 3); (2, 9)$ **2.** $(-2, 1); (-1, 3); (0, 5); (1, 7);$
$(2, 9)$ **3.**   **4.**

5. $5; 8; -1\frac{3}{5}$ **6.** $4; 0; 0$ **7.** $(3, 4)$ **8.** no solution

Pages 600–601 Problem Solving Situations   1. 27
3. 182   **5, 7.** Strategies may vary; likely strategies are
given.   **5.** guess and check; 14 quarters and 16 dimes
7. using a Venn diagram; 32   **11.** 34.5 in.
13.

Page 601 Challenge   a square with vertices at
$(-4, 0), (0, 4), (4, 0),$ and $(0, -4)$

Pages 604–605 Exercises
1. $n > 17$

3. $c < 6$

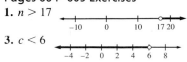

5. $b > 21$

7. $p > -7.5$

9. $g > -4$

11. $y < -18$

13. $h \le -0.8$

15. $h < 6$

17. $w < -2$

19. $n + 2 \ge 4; n \ge 2$ **21.** $\frac{n}{-6} < 12; n > -72$ **23.** b

25. b **27.** $2g < c; g < 35$ **29.** $3\frac{9}{20}$ **31.** 32 **33.** 45 mi/h

Page 608 Exercises **1.** $t > 6$

3. $a \ge -3$

5. $n \le 4$

7. $b < -1$

9. $k > -15$

11. $a \ge -3$

13. $y \ge -2\frac{2}{3}$

15. $m \le -1.7$

17. $a > -21$

19. $v > 14$ **21.** $a < 3$ **23.** $-1\frac{1}{2} > x$ **25.** $(-2, -11)$;
$(-1, -8); (0, -5); (1, -2); (2, 1)$
27. $y > -1$

Pages 611–612 Exercises **1.** Yes. **3.** No. **5.** No.
7. Yes. **9.**

11.

13.

15.

17.

19.

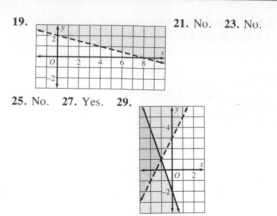

21. No. **23.** No.

25. No. **27.** Yes. **29.**

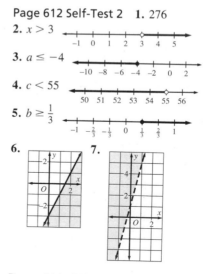

a. Answers will vary. Examples are given. $(-3, 0)$;
$(-4, 2); (-5, -3)$ **b.** $y \le 2x + 3$ **31.** $y \le 5x + 3$
33. $j > 2a$ **35, 37.** Answers will vary. Examples are
given. **35.** The value of a number y is less than three
times the number x. **37.** The value of a number k is
greater than or equal to the sum of the number m and 1.
39. 14; 18; 24 **41.**

48°

Page 612 Self-Test 2 **1.** 276
2. $x > 3$

3. $a \le -4$

4. $c < 55$

5. $b \ge \frac{1}{3}$

6.
7.

Pages 614–615 Exercises **1.** When $x = -6, y = -5$;
when $x = 2, y = 1\frac{2}{3}$; when $x = 3, y = 2\frac{1}{2}$. **3.** When
$x = -6, y = 73$; when $x = 2, y = 9$; when $x = 3, y = 19$.

5. When $x = -6, y = 8.5$; when $x = 2, y = 4.5$;
when $x = 3, y = 5.5$. **7.** $f(-4) = 8; f(-2) = 4$;
$f(5) = -10$ **9.** $f(-4) = 9; f(-2) = 1; f(5) = 36$
11. $f(-4) = 80; f(-2) = 20; f(5) = 125$ **13.** $\frac{3}{25}$ watts

15. No; the power generated by the windmill in Exercise 14 is 8 times the power generated by the windmill in Exercise 13.   **17.** No

19. 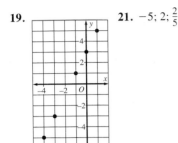   **21.** $-5; 2; \frac{2}{5}$

Pages 617–619 Exercises   1.
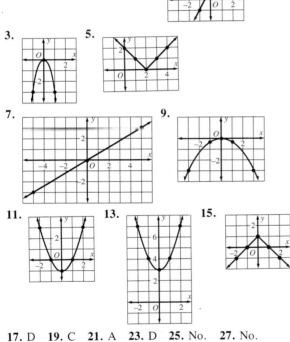

3.   **5.**

7.   **9.**

11.   **13.**   **15.**

17. D   **19.** C   **21.** A   **23.** D   **25.** No.   **27.** No.
29. No.   **31.** 137.5   **33.**
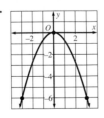

Page 619 Self-Test 3   1. When $x = -2$, $y = 0$; when $x = 0$, $y = -4$; when $x = 2$, $y = 0$.   **2.** When $x = -2$, $y = -3$; when $x = 0$, $y = -5$; when $x = 2$, $y = -3$.
3. $f(-4) = -9; f(0) = 7; f(8) = 39$ **4.** $f(-4) = 4$;

$f(0) = 0; f(8) = 16$ **5.**   **6.**

Pages 620–621 Chapter Review   1. domain of a function   **2.** solution of the system   **3.** graph of an equation with two variables   **4.** slope   **5.** slope-intercept form of an equation   **6.** y-intercept   **7.** $(-2, -10); (-1, -3);$ $(0, 4); (1, 11); (2, 18)$ **8.** $(-2, 2); (-1, 2\frac{1}{2}); (0, 3);$ $(1, 3\frac{1}{2}); (2, 4)$ **9.** $(-2, 6); (-1, 1); (0, -4); (1, -9),$ $(2, -14)$ **10.**

11.

12.   **13.** $-2; -3; y = -2x - 3$

14. $\frac{2}{3}; 1; y = \frac{2}{3}x + 1$ **15.** 8; 7; $-\frac{7}{8}$ **16.** 6; 18; -3
17. $-4; 0; 0$ **18.** $(-2, 3)$ **19.** $(1, -3)$ **20.** $(0, -2)$
21. no solution   **22.** 30   **23.** 900
24. $x \le 6$
25. $c > -1$
26. $z \ge 4$
27. $m > -4$
28. $a > -2$
29. $x \le 12$
30. $a < -1$
31. $m \le -5$
32. $y > 0.7$
33. No.   **34.** No.

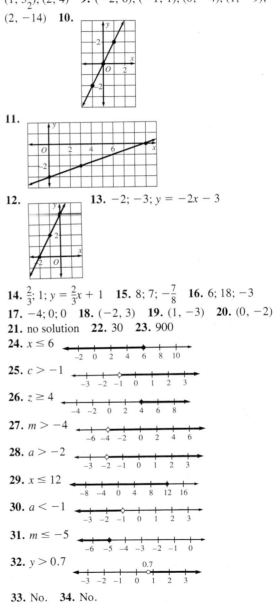

35. Yes. **36.** 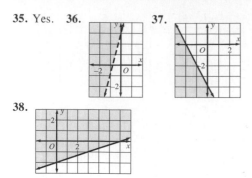 **37.**

38.

39. When $x = -3$, $y = 32$; when $x = 0$; $y = 11$; when $x = 4$, $y = -17$. **40.** When $x = -3$, $y = -3$; when $x = 0$, $y = -6$; when $x = 4$, $y = -2$. **41.** When $x = -3$, $y = 16$; when $x = 0$, $y = 1$; when $x = 4$, $y = 9$. **42.** $f(-2) = -18; f(0) = -8; f(3) = 7$ **43.** $f(-2) = 9; f(0) = 5; f(3) = 14$ **44.** $f(-2) = 2; f(0) = 4; f(3) = 7$ **45.**

46. **47.**

Page 623 Chapter Extension **1.** right **3.** left **5.** right **7.** above **9.** below **11.** below **13.** **15.**

17. **19.**

Pages 624–625 Cumulative Review **1.** D **3.** B **5.** C **7.** A **9.** A **11.** D **13.** C **15.** B **17.** A **19.** C

CHAPTER 14
Pages 630–631 Exercises **1.** triangular prism **3.** pentagonal pyramid **5.** cone **7.** hexagonal prism **9.** cylinder **11.** **13.**

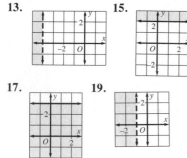

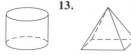

15.

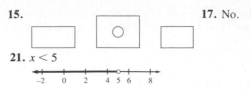

17. No.

21. $x < 5$

Pages 636–637 Exercises **1.** 384 ft^2 **3.** 75 m^2 **5.** 1062 m^2 **7.** 1288 in.^{2} **9.** 842 ft^2 **11.** 411.1 cm^2 **13.** 106 cm^2 **15.** 24 ft^2 **17.** 1 : 9 **19.** c **21.** 336 ft^2 **23.** 1.7 **25.**

54°

Page 637 Challenge
Answers may vary.

Pages 639–641 Exercises **1.** 828.96 yd^2 **3.** 1780.38 m^2 **5.** 528 in.^{2} **7.** 1610.82 mm^2 **9.** 151.62 cm^2 **11.** B **13.** 113 in.^{2} **15.** 19 in. **17.** a, b, c **19.** 45 cm^2 **21.** $\frac{1}{3}$

Page 641 Self Test 1 **1.** square pyramid **2.** cylinder **3.** sphere **4.** 184.5 mm^2 **5.** 264 yd^2 **6.** 320.28 in.^{2} **7.** 1848 cm^2

Pages 644–645 Exercises **1. a.** 132 in.^{3} **b.** 44 in.^{3} **3. a.** 336 ft^3 **b.** 112 ft^3 **5.** 7644 in.^{3} **7.** 405 mm^3 **9.** 132 cm^3 **11.** 512 m^3 **13.** 600 ft^3 **15.** 1575 cm^3 **17.** 8.5 yd^2 **19.** 49 m^2 **21.** 58,969,350 ft^3 **23.** the volume of the Great Pyramid **25.** about 30,702 gal **27.** 224 **29.** 98.5 cm^2

Pages 648–649 Exercises **1.** 4019.2 m^3 **3.** 565.2 yd^3 **5.** 847.8 yd^3 **7.** 6867.18 cm^3 **9.** 792 ft^3 **11.** 264,000 in.^{3} **13.** 502.4 in.^{3} **15.** 5 cm **19. a.** 141.3 cm^2 **b.** 94.2 cm^2 **c.** short legs **21.** using a π key: 150.79645 cm^3; using 3.14 for π: 150.72 cm^3 **23.** 5652 m^3 **25.** (1, 2)

Pages 651–653 Exercises **1.** 1256 mm^2 **3.** 803.84 m^2 **5.** 38,808 yd^3 **7.** 1808.64 cm^2 **9.** 1519.76 cm^2 **11.** 9156.24 m^2; 82,406.16 m^3 **13.** 22,176 cm^2; 310,464 cm^3 **15.** 7 units **17.** $\frac{1017.38}{3052.14} = \frac{3}{r}$; 9 cm **19.** 452.16 cm^3 **21. a.** 1.9741×10^8 mi^2 **b.** 2.6081×10^{11} mi^3 **23.** S.A. of sphere is $4(3.14)(3^2) = 113.04$ cm^2. S.A. of curved surface of cylinder is $2(3.14)(3)(6) = 113.04$ cm^2.

25. 56.52 cm^3 **27.** 51.48 in.^3
29. 8.1×10^{-6} **31.** 120 ft   **33.** 210

Pages 655–657 Problem Solving Situations   **1. a.** 8
b. 12   **c.** 6   **d.** 1   **3.** 5; $2 \times 3 \times 5$; $30 \times 1 \times 1$
5–9. Strategies may vary.   **5.** using proportions;
$4.90   **7.** no solution   **9.** supplying missing facts; $3\frac{1}{2}$ h
13. 2 units $\times$ 3 units $\times$ 2 units   **15.** $\frac{1}{6}$

Page 657 Self Test 2   **1.** 180 in.^3 **2.** 72 m^3
3. $69{,}300 \text{ cm}^3$ **4.** 669.9 ft^3 **5.** 3052.08 in.^3
6. 1017.36 in.^2 **7.** 10

Page 659 Exercises   **1.** A   **3.** C   **5.** $30 \text{ cm} \times 22.5 \text{ cm}$
$\times 11.3 \text{ cm}$ **7.** $15 \text{ cm} \times 45 \text{ cm} \times 11.3 \text{ cm}$ **9–11.** Answers will vary.

Pages 660–661 Chapter Review   **1.** pyramid
2. polyhedron   **3.** vertex   **4.** volume   **5.** sphere
6. cube   **7.** triangular prism   **8.** cylinder   **9.** octagonal
pyramid   **10.** 726 m^2 **11.** 765 in.^2 **12.** 920 yd^2
13. 471 m^2 **14.** 1012 ft^2 **15.** 621.72 m^2 **16.** 729 in.^3
17. 300 cm^3 **18.** 420 ft^3 **19.** 1540 m^3 **20.** 4019.2 in.^3
21. 1526.04 yd^3 **22.** $11{,}304 \text{ ft}^2$; $113{,}040 \text{ ft}^3$
23. 1017.36 mm^2; 3052.08 mm^3 **24.** 2826 cm^2;
$14{,}130 \text{ cm}^3$ **25.** 8   **26.** 3; 8 units $\times$ 1 unit $\times$ 1 unit

Page 663 Chapter Extension   **1.** 39 ft^3 **3.** 226 cm^3
5. 5595 m^3

Pages 664–665 Cumulative Review   **1.** D   **3.** D
5. C   **7.** B   **9.** A   **11.** C   **13.** A   **15.** A   **17.** A
19. D

CHAPTER 15
Pages 669–670 Exercises   **1.** $x^2 + x - 2$
3. $m^3 + 3m^2 + m - 3$ **5.** $3z^4 + 8z^3 + 4z^2 - 6z - 7$
7. $c^4 + c^3 - 5c - 2$ **9.** $-8k^4 + 3k^3 + k^2 - 6$
11. $x^2 + 4x + 7$ **13.** $e^2 + 11e - 20$
15. $-4x^2 + 9x + 12$ **17.** $-7x^2 - 11x + 1$
19. $8w^3 + 4w^2 - 12w + 1$ **21.** $x^5 + 4x^2 + 2x + 1$
23. $2a^4 - 10a^3 - 7a^2$ **25.** $-5x^3$
27. $ab^3 + a^2b^2 + a^2b + 4$ **29.** $x^3y^3 - xy^2 + x^2y - 4$
31. $a^3b^2 - a^2b + ab^3 - 4$ **33.** $a^3c^4 + ac^3 + a^4c^2 - a^2c$
35. many; polyester   **37.** two; bicycle
39. $x^3yz^2 + x^2y^2z + xy^3z^3$; $xy^3z^3 + x^2y^2z + x^3yz^2$;
$xy^3z^3 + x^3yz^2 + x^2y^2z$ **41.** 3052.08 cm^3
43. $x \le -6$

Pages 674–675 Exercises   **1.** $5a^2 + 4a + 10$
3. $7x^2 + 11x + 11$ **5.** $9v^2 + 6v + 7$
7. $14y^3 + 12y^2 + 9y + 8$ **9.** $8b^4 + 5b^3 + 12b^2 + 7b + 7$
11. $y^2 - 4y + 9$ **13.** $6x^2 + 3x + 5$
15. $-k^4 - 7k^3 - 3k^2 + k - 2$ **17.** $5k^5 - k^4$

19. $5x^5 + 3x^4 + 3x^3 + x^2 - 5$
21. $6a^4 + 5a^3 + 9a^2 + 11a + 7$
23. $-3d^3 + 8d^2 - 4d - 4$
25. $2b^4 - 2b^3 + 3b^2 - b - 1$ **29.** 0
31.

12	-12	0
5	-1	4
0	4	4
-3	9	6
-4	20	16

33.

35. $7a^4 - 3a^3 + 6a^2 - 3$

Pages 678–679 Exercises   **1.** $x^2 + x + 4$
3. $3x^2 + 7x + 6$ **5.** $5d^2 - d - 2$ **7.** $-2m^2 - 10m - 4$
9. $3a^3 - 13a^2 + 7a + 11$ **11.** $-3x^2 + 6x - 16$
13. $4a^3 - 4a^2 - 8a + 17$
15. $-5a^4 + 4a^3 + 8a^2 - 6a - 2$
17. $-5a^4 + 7a^3 + 17a^2$
19. $-4z^4 - z^3 - 9z^2 - 2z + 8$ **21.** $4x^3 - 14x^2 + 6x - 3$
23. 3   **25.** -1 **27.** 5   **29.** 1   **31.** $6a^2 + 5a + 9$
33. $2n^2 - 5n + 5$ **35.** $2b^2 - 6b - 6$ **37.** Answers will
vary. Example: $x^2 + 3x + 7$ and $x^2 + 3x + 5$ **39.** They
are the same.   **41.** $2x^2 + 6x - 3$
43.   Base Hits Per Season

Hits	Tally	Frequency
160–169	⊬⊬⊬	5
170–179	‖‖	4
180–189	‖	2
190–199	‖	1
200–209	‖	1
210–219	‖	1

45. $3d^2 + 13d - 11$

Page 679 Challenge   0, 3

Pages 683–684 Exercises   **1.** $14x^2 + 7x + 14$
3. $-36a^3 + 12a^2 + 24a - 12$ **5.** $24x^6y^3$ **7.** $-30r^5s^7$
9. $18a^6x^{11}$ **11.** $24a^3 + 16a^2 + 20a$
13. $15d^5 - 24d^4 + 21d^3 - 18d^2$
15. $28x^6 + 42x^5 - 63x^4 + 49x^3$
17. $10m^2n^2 - 15m^2n^3 + 20mn^2$

19. $-5x^2y^2 - 7x^3y^2 - 8x^2y^3$
21. $-5m^5n^3 - m^4n^2 + 7m^5n^2$
23. $30x^3 + 35x^2 - 45x - 25$
25. $8a^5 + 16a^4 - 14a^3 + 2a^2 - 18a$
27. $-18x^5 - 8x^4 + 12x^3 + 16x^2$
29. $-x^3 - 4x^2 - 7x + 5$ 31. $3x^4 - x^3 + 6x^2 + 2x - 7$
33. change each sign 35. a^5 37. The product of powers rule applies to negative exponents;
$(a^{-4})(a^{-5}) = \frac{1}{a^4} \cdot \frac{1}{a^5} = \frac{1}{a^9} = a^{-9}$ 39. $\frac{18r}{s^5}$ 41. $\frac{1}{3}$
43. $8a^2b + 20a^2 - 12a$

Pages 686–688 Exercises 1. $x^2 + 6x + 8$
3. $12t^2 + 52t + 35$ 5. $24y^2 + 16y - 14$
7. $7x^2 + 6x - 16$ 9. $2w^2 - 9w + 4$ 11. $4x^2 - 17x + 4$
13. $12x^2 - 38x + 30$ 15. $4x^2 + 4x + 1$ 17. $4x^2 - 25$
19. $15x^2 + 2x - 24$ 21. $a^2 - 2ab - 15b^2$ 23. $x^2 - y^2$
25. $2x^4 - 11x^2 + 15$ 27. a. $6x^2 - 11x - 10$;
$6x^2 - 11x - 10$ b. They are equal; $(3x + 2)(2x - 5) = (2x - 5)(3x + 2)$; commutative property of multiplication
29. the new length and width 31. $x^2 - 5x + 6$
33. $5x - 6$ 35. a. $a^5 + 5a^4b + 10a^3b^2 + 10a^2b^3 + 5ab^4 + b^5$ b. $a^6 + 6a^5b + 15a^4b^2 + 20a^3b^3 + 15a^2b^4 + 6ab^5 + b^6$ c. $a^7 + 7a^6b + 21a^5b^2 + 35a^4b^3 + 35a^3b^4 + 21a^2b^5 + 7ab^6 + b^7$ 37. A and C represent the numbers the variable is multiplied by. B and D represent the numbers added to the variable. 39. $x^2 + x - 12$ 41. 25%

Page 688 Self-Test 1 1. $6x^4 + 5x^3 - 3x^2 + 4x - 8$
2. $4a^5 + 9a^3 + 5a^2 - 7a - 2$ 3. $3x^3 + 4x^2 + 8x + 7$
4. $7c^3 - 2c^2 - 8c + 5$ 5. $2x^2 + 8x + 8$
6. $2z^2 - 3z - 4$ 7. $b^3 + 2b^2 - 6$ 8. $4n^2 - 4$
9. $-10x^4y^6$ 10. $9a^7 - 15a^5$ 11. $8x^2 + 22x + 15$
12. $20c^2 - 12c - 56$

Pages 690–692 Problem Solving Situations 1. 64
3. 9 5–9. Strategies will vary. 5. working backward;
1.1 mi 7. supplying missing facts; gallons; $.26
9. using equations; 2 h 13. 26 cm 15.

Pages 694–695 Exercises 1. $7x(x + 2)$ 3. $s(4 + 5s)$
5. $4k^2(k + 2)$ 7. $5p^2(p^3 - 3)$ 9. $3xy(3 + 2x)$
11. $4c(a - 2b)$ 13. $3u^2(4u^2 - 3u - 8)$
15. $8s^2(s^2 - 2s + 4)$ 17. $5qd(5d + 3q + 6qd - 1)$
19. $y^3z^2(36yz + 45z^2 + 20y^2)$
21. $6r^2t^2u^2(4 - 2rtu - r^2t^2u^2)$
23. $4ab^2c(4ac + 1 + 2a^2c^3 + 3a)$ 25. $(r - 5)(r + 6)$
27. $(2 - 3d)(7d - 1)$ 29. $(y - 3)(y - 1)$
31. $rt + st + t^2 + t$ 33. $4(8 - \pi)$ 35. $2r^2(2 + \pi)$
37. 8! or 40,320 39. $7z^4 - 4z^3 + 5z^2 - z + 4$

Pages 697–699 Exercises 1. $3x - 4y$ 3. $-7c^2 + c$
5. $x + 2x^3 - 6x^6$ 7. $3t^6 + 8t + 1$ 9. $3k^4 + k^3 - 12k^2$
11. $7 + 8e + 9d$ 13. $3r^3u - 4$
15. $4xy^2 + 7x^2 - 12x^3y^2$ 17. $\frac{1}{3}k - 2$
19. $-4xy - \frac{5}{2}x^2y - \frac{2}{3}x^3y$ 21. $\frac{3}{2}x + \frac{7}{2} + \frac{9}{2x}$
23. $\frac{5}{3}b^2 + \frac{4}{3}b + \frac{1}{3b}$ 25. $8x$ 27. $3c + 3$ 29. $2a - ab$
31. $h = 64t - 16t^2$
33.

t (seconds)	$h = 16t(4 - t)$ (feet)
0	0
1	48
2	64
3	48
4	0

35. 266 in.^{2}

37. Gasoline Mileage (mi/gal)

18 19 20 21 22 23 24 25 26 27 28 29 30 31 32

39. $21x^4 - 6x^2$

Page 699 Self-Test 2 1. 42 h 2. $4c(3c - 2 - d)$
3. $5mn(5m + 3n - 2)$ 4. $3a^2 - 2a + 1$
5. $ab + 9b - 6a$

Pages 700–701 Chapter Review 1. binomial
2. standard form of a polynomial 3. greatest common monomial factor 4. polynomial 5. $6x^4 + 7x^3 + 2x^2 + 4x + 7$ 6. $4c^5 + 11c^4 - 9c^3 + 8c^2 - 5c + 2$
7. $a^6 - 2a^4 - 4a^3 - 3a + 8$
8. $5m^7 + 6m^5 - 2m^3 - 9m^2 + 4m$ 9. $4x^3 + 14x^2 + x + 3$
10. $8b^4 - 6b^2 + b - 5$ 11. $4c^3 + c^2 - 5c + 13$
12. $-c^4 - 9c^3 + 5c^2 + 3c + 8$ 13. $-5x^3 + 6x^2 - 3x + 5$
14. $-2a^3 - a^2 - 5a + 11$ 15. $5a^2 + 7a + 16$
16. $2x^3 + 4x^2 + 2x + 7$ 17. $3n^5 + 4n^3 + 9n + 6$
18. $2d^4 + 8d^2 + 7d + 4$ 19. $5c^3 + 5c^2 - 5c - 5$
20. $n^4 - n^3 + 5n^2 + 3$ 21. $4x^2 + 3x + 5$
22. $3a^2 + 4a - 1$ 23. $5c^2 + 12c - 7$
24. $-4b^2 - 9b + 8$ 25. $3x^4 + 2x^3 - 4x + 10$
26. $-a^3 - 4a^2 - 5a - 6$ 27. $6a^4 + 18a^3 - 24a^2 + 21a$
28. $-12x^4 - 28x^3 + 20x^2$
29. $-24c^5 - 18c^3 + 30c^2 + 48c$
30. $-28n^6 - 20n^5 + 32n^3$ 31. $2x^2 + 13x + 15$
32. $8m^2 + 2m - 28$ 33. $6x^2 - 31x + 40$
34. $28d^2 + 13d - 6$ 35. $14.99 36. 43 h
37. 21 weeks 38. 9:50 A.M. 39. $2x(2x^2 + 4x + 3)$
40. $3n^2(3n^2 - 2n + 6)$ 41. $5d^2(4d^3 - 2d^2 + 3d - 1)$
42. $6m^2(3m^4 - 5m^2 + 2m - 4)$ 43. $cd(3c^2d^2 + 5cd - 4)$
44. $2ab^2(2a^2 - 4ab + 1)$ 45. $5x^2y^2(5x^2y + 2xy^2 - 3)$
46. $6mn^2(2n - 3m + 4mn)$ 47. $7x^3 + 4x^2 - 5x$
48. $2c^4 - 3c^2 + 6c - 1$ 49. $6a^2b^4 + 7a^3b^2 - 2ab$
50. $4m^2n^3 - 2n^2 + 3m^5n$

Page 703 Chapter Extension
1. $a^2 - 4$ **3.** $x^2 + 16x + 64$ **5.** $c^2 - 10c + 25$
7. $4c^2 - 81$ **9.** $9x^2 + 24x + 16$ **11.** $25c^2 - 60c + 36$
13. $4x^2 - 9y^2$ **15.** $9a^2 + 30ab + 25b^2$
17. $9z^2 - 48xz + 64x^2$ **19.** $(x - 6)(x + 6)$
21. $(x + 1)^2$ **23.** $(c - 9)^2$ **25.** $(z - 8)^2$

Pages 704–705 Cumulative Review **1.** A **3.** C
5. C **7.** A **9.** B **11.** B **13.** A **15.** B **17.** A
19. C

LOOKING AHEAD
Pages 707–709 Exercises **1.** translation **3.** reflection
5. (3, 1) **7.** (1, 1) **9. a.**

b. A' (1, −2); B' (3, −2); C' (3, 1); D' (1,1)
11. a.

b. A' (1, 3); B' (3, 3), C' (3, 6), D' (1, 6)
13. a.

b. A' (1, −1); B' (3, −1); C' (3, −4); D' (1, −4)
15.

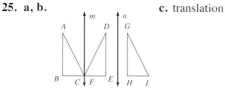

17. A' (−1, −1); B' (3, 2); C' (2, −2) **19.** A' (5, −1);
B' (1, −4); C' (2, 0) **21–23.** Answers may vary.
21. translation down 3 units and right 2 units **23.** reflection across the x-axis and a translation right 2 units
25. a, b.   **c.** translation

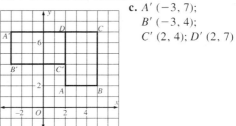

Pages 711–712 Exercises **1.** C **3.** E **5.** 6
7.   **9.**

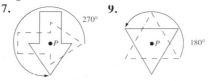

11. 180° and 360° about P **13.** reflection or
reflection and translation **15.** translation **17.** translation, rotation, or rotation and translation
19. a, b.   **c.** A' (−3, 7);
B' (−3, 4);
C' (2, 4); D' (2, 7)

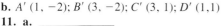

Pages 717–718 Exercises **1. a.** hypotenuse
b. adjacent; hypotenuse **c.** opposite; adjacent **3.** f
5. e **7.** d **9.** $\frac{4}{5}$ **11.** $\frac{4}{5}$ **13.** $\frac{4}{3}$ **15.** $\frac{5}{13}$ **17.** $\frac{5}{13}$ **19.** $\frac{5}{12}$
21. $x = 13$; $\frac{5}{13}, \frac{12}{13}, \frac{5}{12}$ **23.** $\frac{8}{17}$ **25.** $\frac{4}{5}$ **27.** $\frac{8}{17}$
29. No; the Pythagorean theorem does not hold true.
31. 1; 1

Pages 721–722 Exercises **1.** 0.0872 **3.** 0.2588
5. 5.6713 **7.** cosine **9.** sine **11.** 13.4 **13.** 19.4
15. $a = 11.2$; $b = 27.8$; $m\angle B = 68°$ **17.** 25.0 ft
19. 23.7

Pages 724–725 Exercises **1.** $\overline{DW}$ and $\overline{SD}$ **3.** 19°
5. 55° **7.** $m\angle A = 31°$; $m\angle B = 59°$ **9.** $m\angle A = 39°$;
$m\angle B = 51°$ **11.** $m\angle A = 53°$; $m\angle B = 37°$; $c = 5$
13. $m\angle A = 63°$; $m\angle B = 27°$; $c = 6.7$
15. $m\angle T = 71°$; $m\angle R = 71°$; $m\angle I = 38°$

Pages 728–729 Exercises **1.** $x\sqrt{2}$ **3.** longer leg;
hypotenuse **5.** 10.4 **7.** 14.1 **9.** $x = 6$; $y = 6$
11. 13.9; 8 **13.** 22; 31.1 **15.** 6; 12 **17.** 5.2
19. 24; 20.8 **21.** 69.3 cm; 80 cm
23. a. 9; 3; 5.2; 10.4 **b.** 31.2 square units

EXTRA PRACTICE
Pages 730–731 Chapter 1 **1.** 6 **3.** 11 **5.** 6 **7.** 2
9. 217.72 **11.** 59.5 **13.** 238.6 **15.** 63.6 **17.** 1693.72
19. 528 **21.** 4 **23.** 200 **25.** Chris's wages **27.** 40
29. > **31.** = **33.** < **35.** 371 **37.** 26 **39.** 97
41. 234.8 **43.** 892.7 **45.** 700 **47.** 9 **49.** 0 **51.** 110
53–59. Answers may vary. Likely answers are given.
53. mm; 140 **55.** c; 86.4 **57.** pp; 318 **59.** mm; 0.328

61. 0 **63.** 729 **65.** 100,000,000 **67.** 98 **69.** 109
71. 168 **73.** 1

Pages 731–732 Chapter 2 **1.** a^{10} **3.** $66x^2$
5. $6a + 42$ **7.** $2x + 26$ **9.** $6x + 21$ **11.** $12 - 18a$
13. $12x$ **15.** $8x$ **17.** $8x$ **19.** $2a + 13c$ **21.** $53.94
23. 39, 45, 51 **25.** $x + 12, x + 16, x + 20$
27. 5; 9; 13; 17; 21 **29.** $x \rightarrow x + 9$ **31.** estimate; about
$60 **33.** 340 mm **35.** 4200 mL **37.** 4 g
39. 3.5×10^6 **41.** 8×10^4 **43.** 235,000

Pages 732–733 Chapter 3 **1.** 15 **3.** 0 **5.** $<$ **7.** $>$
9. -43 **11.** 113 **13.** $61 **15.** 21 **17.** 0 **19.** -23
21. 31 **23.** 7 **25.** 42 **27.** -16 **29.** 50 **31.** -3
33. 20 **35.** 9 **37.** $(-2, -1)$ **39.** $(2, 0)$
40–43. **45.**

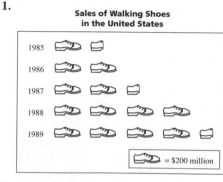

47.

Pages 733–735 Chapter 4 **1.** Yes. **3.** No. **5.** 4
7. 8 **9.** 8 tickets for $12 and 12 tickets for $17
11. 12 goals and 16 assists **13.** 12 **15.** -7 **17.** 7
19. 8 **21.** 8 **23.** 5 **25.** 4 **27.** -4 **29.** 5 **31.** -8
33. 36 **35.** -20 **37.** $x - 2$ **39.** $2n + 3$ **41.** $5j$
43. $3t - 7$ **45.** $25 = u + 21$ **47.** $20 = t - 4$ **49.** 12
51. 10 **53.** 3 **55.** -10 **57.** 12 **59.** 165 mi
61. 47 mi/h

Pages 735–737 Chapter 5
1.

**Sales of Walking Shoes
in the United States**

1985

1986

1987

1988

1989

= $200 million

3. about 20 **5.** about 100

8.

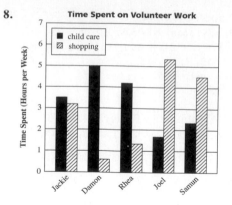

9. National Bikes **11.** not enough information; sales figures for racing bikes **13.** Remove the gap in the scale; change the intervals on the scale. **15.** median; the mean is greater than 5 of the 7 bills.

Pages 737–739 Chapter 6 **1.** $\overleftrightarrow{XY}$ or $\overleftrightarrow{YX}$ **3.** Plane T
5. 140° **7.** **9.** 78° **11.** 3° **13.** 143°

65°

15. 45° **17.** True. **19.** True. **21.** True. **23.** False.
25. triangle; no diagonals **27.** 53° **29.** 120° **31.** can;
scalene **33.** right **35.** obtuse **37.** 90° **39.** 56°
41. 5 in. **43.** No. **45.** Yes.

Pages 739–741 Chapter 7 **1.** prime **3.** composite
5. $2^3 \cdot 5$ **7.** $2 \cdot 5 \cdot 7$ **9.** 4 **11.** 8 **13.** $4a$ **15.** $9p$
17. 24 **19.** 50 **21.** $18a^6$ **23.** $8a^6$ **25.** 24 **27.** 4
29. $\frac{2}{7}$ **31.** $\frac{3}{5}$ **33.** $3q$ **35.** qt^5 **37.** $6p^2$ **39.** $\frac{h^5}{2}$ **41.** $>$
43. $<$ **45.** $<$ **47.** $=$ **49.** 0.8 **51.** 2.375 **53.** $\frac{8}{25}$
55. $3\frac{1}{4}$ **57.** 14 blocks north and 8 blocks west **59.** 20
61. G **63.** C **65.** $\frac{-9}{1}$ **67.** $\frac{-31}{5}$
69.

$-6 \quad -5 \quad -4 \quad -3 \quad -2 \quad -1 \quad 0$

71.

$10 \quad 11 \quad 12 \quad 13 \quad 14 \quad 15 \quad 16$

73.

3.7
$2 \quad 3 \quad 4 \quad 5$

75.

$-3 \quad -2.5 \quad -2 \quad -1.5 \quad -1 \quad -0.5 \quad 0$ **77.** $\frac{1}{x^4}$ **79.** $\frac{1}{125}$
81. 8.2×10^{-5} **83.** 7×10^{-5} **85.** 0.005
87. 0.00000000325

Pages 741–742 Chapter 8 **1.** $\frac{3}{8}$ **3.** $-8\frac{3}{4}$
5. -23.22 **7.** $\frac{2a}{3}$ **9.** 2 **11.** $-\frac{5}{19}$ **13.** -3.2 **15.** $\frac{9}{14}$
17. decimal; $31.25 **19.** $1\frac{4}{7}$ **21.** 0 **23.** -3 **25.** $\frac{20}{y}$

27. $\frac{11}{18}$ **29.** $-1\frac{3}{4}$ **31.** $\frac{5a}{14}$ **33.** -2.3 **35.** about 20

37. about $12\frac{1}{2}$ **39.** about 3 **41.** \$1020 **43.** -1 **45.** $1\frac{1}{2}$

47. -1.1 **49.** $-5\frac{5}{8}$ **51.** $-3\frac{1}{2}$ **53.** -7 **55.** $w = \frac{A}{l}$

57. $h = \frac{2A}{b_1 + b_2}$ **59.** $t = \frac{D}{r}$

Pages 742–743 Chapter 9 **1.** $\frac{1}{7}$ **3.** $\frac{1}{5}$ **5.** \$12.50/h

7. 18 mi/gal **9.** 35 **11.** 12 **13.** 12 **15.** 7 **17.** $\frac{5}{6}$ in.;

1 in. **19.** 45 **21.** 16% **23.** 112.5% **25.** $\frac{9}{20}$; 0.45

27. $\frac{41}{500}$; 0.082 **29.** 9.6 **31.** 7.5% **33.** 176 **35.** 30%

increase **37.** $16\frac{2}{3}$% decrease **39.** 37.5% **41.** $66\frac{2}{3}$%

43. 90

Page 744 Chapter 10 **1.** $17\frac{5}{6}$ ft **3.** 12 m; 24 m

5. 6.5 m; 13 m **7.** 60.72 m^2 **9.** 86.5 ft^2

11. $\frac{1}{2} + \frac{1}{3} + \frac{1}{4} = 1\frac{1}{12}$; this is more than one whole

paycheck. **13.** $\angle P$ **15.** 9 ft **17.** $\frac{4}{9}$ **19.** 11.790

Pages 745–747 Chapter 11
1. Test Scores in Math

Score	Tally	Frequency
50–59	||	2
60–69	||	2
70–79	$\cancel{||||}$ |	6
80–89	$\cancel{||||}$ |	6
90–99	||||	4

3. 7 **5.** 43.6; 44; 33, 42, 50; 27 **7.** 17 **9.** 25 **11.** 78;
72; 82; 64; 92 **13.** Tim Stuart's **15.** 2 **17.** \$10,500

19.

Videos Rented from Victor's Videos
Family 60
Comedy 20
Adventure 120

21. Therese is the lawyer, Lucinda is the accountant, and Alicia is the teacher.

Page 747 Chapter 12 **1.** $\frac{1}{5}$ **3.** 3 to 2

5.

spinner	coin	outcomes
1	H	1 (red) H
	T	1 (red) T
2	H	2 (white) H
	T	2 (white) T
3	H	3 (blue) H
	T	3 (blue) T
4	H	4 (blue) H
	T	4 (blue) T
5	H	5 (white) H
	T	5 (white) T

7. 25 **9.** $\frac{2}{17}$ **11.** 5040 **13.** $\frac{9}{20}$

Pages 748–749 Chapter 13 **1.** $(-2, -4)$; $(-1, -1)$;
$(0, 2)$; $(1, 5)$; $(2, 8)$ **3.** $(-2, 9)$; $(-1, 7)$; $(0, 5)$; $(1, 3)$;
$(2, 1)$ **5.** $(-2, 3)$; $(-1, 6)$; $(0, 9)$; $(1, 12)$; $(2, 15)$
7. $(-2, -17)$; $(-1, -12)$; $(0, -7)$; $(1, -2)$; $(2, 3)$
9. 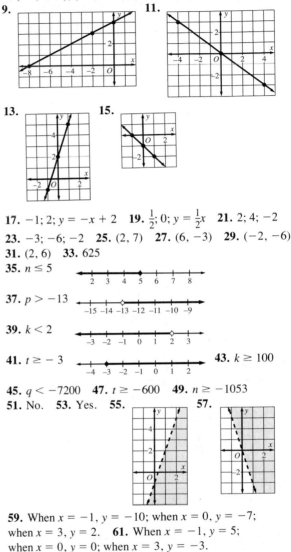 **11.**

13. **15.**

17. -1; 2; $y = -x + 2$ **19.** $\frac{1}{2}$; 0; $y = \frac{1}{2}x$ **21.** 2; 4; -2
23. -3; -6; -2 **25.** $(2, 7)$ **27.** $(6, -3)$ **29.** $(-2, -6)$
31. $(2, 6)$ **33.** 625
35. $n \le 5$

37. $p > -13$

39. $k < 2$

41. $t \ge -3$ **43.** $k \ge 100$

45. $q < -7200$ **47.** $t \ge -600$ **49.** $n \ge -1053$
51. No. **53.** Yes. **55.** **57.**

59. When $x = -1$, $y = -10$; when $x = 0$, $y = -7$;
when $x = 3$, $y = 2$. **61.** When $x = -1$, $y = 5$;
when $x = 0$, $y = 0$; when $x = 3$, $y = -3$.

63. $f(-2) = 8$; $f(1) = -4$; $f(0) = 0$

65. $f(-2) - 31$; $f(1) - 7$; $f(0) = -1$

67. $f(-2) = 3$; $f(1) = 0$; $f(0) = -1$ **69.**

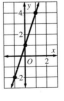

71. **73.**

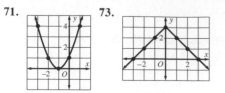

Pages 749–750 Chapter 14 **1.** rectangular prism
3. hexagonal pyramid **5.** 570 ft^2 **7.** 1607.68 mm^2
9. 282.6 yd^2 **11.** 750 ft^3 **13.** 171 in.^{3} **15.** 32 ft^3
17. 508.68 mm^3 **19.** 4823.04 mm^3 **21.** 169.56 yd^3
23. 2826 ft^2 **25.** 7234.56 m^3 **27.** 24,416.64 yd^3
29. 28; 20; 4; 12

Pages 751–752 Chapter 15 **1.** $3a^3 + 6a^2 + 4a + 9$
3. $7a^2 + 7a + 4$ **5.** $4a^3 - 2a^2 + 4a + 7$
7. $9p^2 + 5p + 2$ **9.** $10a^2 + 9a + 10$
11. $18c^2 + 28c + 21$ **13.** $3x^4 - 3x^3 + 4x^2 + 2x - 7$
15. $-n^4 + n^3 + 2n^2 - 5n + 3$ **17.** $2k^2 + 5k + 3$
19. $h^2 + h + 1$ **21.** $6b^3 + 8b^2 - 5b - 2$
23. $-3x^3 + 16x^2 - 4x - 3$ **25.** $18x^2 + 9x - 6$
27. $12m^3 - 24m^2 + 36m + 36$ **29.** $30c^7d^{10}$
31. $12a^7 + 24a^6 - 9a^5 - 27a^4$ **33.** $n^2 + 6n + 5$
35. $5y^2 + 41y + 42$ **37.** $6n^2 + 17n + 5$
39. $8u^2 + 6u - 35$ **41.** 32 **43.** $2m^2(2m^2 - 8m + 15)$
45. $2v^4(5v^2 - 8v + 6)$ **47.** $2a^3b^2(2b^3 - 5a + 3a^2b^3)$
49. $3p^3q^5(8p - 5q + 11p^3q^2)$ **51.** $2y^3 + 6y^2 + 3y$
53. $2t^4 + 4t^2 + 6t$ **55.** $a^2b^4 - 6ab^6 + 2a^3b$
57. $-2p^3 + m^5p^2 - 3m^4p^2$

TOOLBOX SKILLS PRACTICE

Page 753 Skill 1 **1.** about 400 **3.** about 980
5. about 700 **7.** about 500 **9.** about 36,000
11. about 800 **13.** about 4,000,000 **15.** about 4500
17. about 300,000 **Skill 2** **1.** about 1100
3. about 1400 **5.** about 8000 **7.** about 30
9. about 5000 **11.** about 150

Page 754 Skill 3 **1.** about 20 **3.** about 30
5. about 30 **7.** about 60 **9.** about 40 **11.** about 60
13. about 120 **15.** about 90 **17.** about 3 **19.** about 40
Skill 4 **1.** about 60 **3.** about 120 **5.** about 630
7. about 1800 **9.** about 390 **11.** about 1700
13. about 56 **15.** about 35 **17.** about 7.2

Page 755 Skill 5 **1.** tens, 80 **3.** tens, 0 **5.** thousands,
1000 **7.** ones, 1 **9.** thousands, 2000 **11.** hundredths,
0.07 **13.** tens, 90 **15.** thousandths, 0.009 **17.** ones,
8 **19.** hundredths, 0.09 **21.** fifteen **23.** one hundred
sixty-two **25.** four hundred ninety-three **27.** two

hundred fifty-eight **29.** four thousand, fifty-six
31. thirty-two **33.** seven thousand, six **35.** three
hundred eighty-five **37.** four and seven tenths
39. one hundred seventy-eight and sixty-one hundredths
41. eighty-three and ninety-seven thousandths
43. eighteen and three hundred fifteen thousandths
45. six thousand thirty-one and seven hundred fifty-two
thousandths **47.** one hundred ninety-three and seven
hundredths **49.** three hundred seven and one tenth

Page 756 Skill 6 **1.** 20 **3.** 100 **5.** 270 **7.** 650
9. 8790 **11.** 900 **13.** 5300 **15.** 300 **17.** 2400
19. 6000 **21.** 12,000 **23.** 7000 **25.** 3000 **27.** 22,000
29. 4.0 **31.** 0.3 **33.** 4.8 **35.** 12.2 **37.** 107.6
39. 17.9 **41.** 0.52 **43.** 31.03 **45.** 0.30 **47.** 12.99
49. 3.41 **51.** 71.99 **53.** 54.018 **55.** 0.314 **57.** 2.992
59. 13.030 **61.** 4.348 **63.** 177.025

Page 757 Skill 7 **1.** 88 **3.** 99 **5.** 1195 **7.** 869
9. 12,587 **11.** 5739 **13.** 14,799 **15.** 5789
17. 492 **Skill 8** **1.** 440 **3.** 122 **5.** 1225 **7.** 2441
9. 113 **11.** 2021 **13.** 22,424 **15.** 52,325 **17.** 103
19. 48 **21.** 96 **23.** 4892 **25.** 31,849

Page 758 Skill 9 **1.** 1643 **3.** 15,708 **5.** 12,832
7. 1,420,868 **9.** 264,924 **11.** 130,273
13. 21,511,182 **15.** 24,798,126 **17.** 2482 **19.** 6322
21. 24,453 **23.** 232,403 **25.** 638,844 **27.** 152,388
29. 2,298,835

Page 759 Skill 10 **1.** 3 **3.** 13 **5.** 175 **7.** $27\frac{5}{17}$
9. $193\frac{5}{8}$ **11.** 12.5 **13.** 19 **15.** 314.8 **17.** 114.5
19. 534.25 **21.** 14.1 **23.** 50.6 **25.** 143.4 **27.** 81.7
29. 287.8 **31.** 20.11 **33.** 250.94 **35.** 138.46
37. 203.95 **39.** 2040.88

Page 760 Skill 11 **1.** 767.95 **3.** 1433.20 **5.** 1173.93
7. 11,559.4 **9.** 1398.898 **11.** 953.23 **13.** 21.1
15. 22.255 **17.** 209.091 **19.** 40.577 **Skill 12** **1.** 9.38
3. 23.73 **5.** 4.889 **7.** 141.67 **9.** 288.636 **11.** 121.57
13. 45.64 **15.** 88.64 **17.** 10.97

Page 761 Skill 13 **1.** 13.78 **3.** 32.76 **5.** 131.1
7. 19.1205 **9.** 7.462 **11.** 2915.415 **13.** 0.0048
15. 0.00513 **Skill 14** **1.** 7.8 **3.** 6.49 **5.** 19 **7.** 13.4
9. 18.1 **11.** 38 **13.** 63 **15.** 116.00

Page 762 Skill 15 **1.** 8.03 **3.** 8130.6 **5.** 541
7. 70.92 **9.** 2840 **11.** 30 **13.** 12,751 **15.** 804
17. 41.7 **19.** 12 **21.** 9310 **23.** 82,100 **25.** 21.635
27. 0.5106 **29.** 2.857 **31.** 0.00792 **33.** 0.293
35. 9.301 **37.** 18.475 **39.** 0.0029 **41.** 0.413
43. 0.0006 **45.** 0.0269 **47.** 0.000025

Page 763 Skill 16   **1.** Yes.   **3.** No.   **5.** Yes.   **7.** Yes.
9. No.   **11.** Yes.   **13.** No.   **15.** Yes.   **17.** No.
19. Yes.   **21.** Yes.   **23.** Yes.

Page 764 Skill 17   **1.** $\frac{7}{9}$ **3.** $\frac{4}{5}$ **5.** 2   **7.** $\frac{1}{5}$ **9.** $\frac{1}{2}$ **11.** $\frac{3}{5}$
13. $\frac{3}{5}$ **15.** $\frac{9}{16}$ **17.** $1\frac{25}{36}$ **19.** $\frac{26}{45}$ **21.** $\frac{11}{21}$ **23.** $\frac{13}{30}$ **25.** $\frac{14}{75}$

Page 765 Skill 18   **1.** $7\frac{10}{11}$ **3.** $17\frac{5}{9}$ **5.** $23\frac{1}{3}$ **7.** $38\frac{11}{15}$
9. $12\frac{3}{5}$ **11.** $25\frac{19}{20}$ **13.** $19\frac{29}{36}$ **15.** $17\frac{3}{10}$ **17.** $30\frac{5}{8}$
19. $31\frac{1}{4}$ **21.** $83\frac{19}{60}$ **23.** $50\frac{23}{24}$ **25.** $46\frac{3}{5}$ **27.** $25\frac{23}{24}$

Page 766 Skill 19   **1.** $2\frac{5}{7}$ **3.** $13\frac{1}{8}$ **5.** $5\frac{1}{2}$ **7.** $3\frac{5}{12}$
9. $5\frac{5}{8}$ **11.** $13\frac{1}{8}$ **13.** $15\frac{23}{42}$ **15.** $4\frac{27}{40}$ **17.** $3\frac{7}{10}$ **19.** $12\frac{11}{16}$
21. $13\frac{11}{12}$ **23.** $2\frac{1}{2}$

Page 767 Skill 20   **1.** $\frac{1}{13}$ **3.** $\frac{3}{22}$ **5.** 20   **7.** $\frac{1}{18}$ **9.** $\frac{5}{13}$
11. 20   **13.** 76   **15.** 6   **Skill 21**   **1.** 3   **3.** $\frac{11}{14}$ **5.** $1\frac{1}{2}$
7. $\frac{25}{54}$ **9.** $\frac{21}{40}$ **11.** $\frac{4}{19}$

Page 768 Skill 22   **1.** 21,120   **3.** 11,000   **5.** 28
7. 11,880   **9.** 9000   **11.** 25   **13.** 18   **15.** 9; 1
17. 36; 2   **19.** 6   **21.** 11; 1   **23.** 5   **25.** $1\frac{1}{2}$ **27.** 3
29. 28   **31.** 8   **33.** 2640   **35.** 48,000

Page 769 Skill 23   **1.** 1980; 1940   **3.** about 100
5. 1930 and 1950   **7.** about 4 lb   **9.** Los Angeles

Page 770 Skill 24   **1.** September; January   **3.** about 4
5. decreasing   **7.** Dallas   **9.** about 1000 h

Page 771 Skill 25   **1.** midnight   **3.** about 20,000
5. increasing   **7.** January   **9.** about 10   **11.** about 10